# INTRODUCTION TO ENGINEERING THERMODYNAMICS

# INTRODUCTION TO ENGINEERING THERMODYNAMICS

## SECOND EDITION

### RICHARD E. SONNTAG
*University of Michigan*
*Mechanical Engineering*

### CLAUS BORGNAKKE
*University of Michigan*
*Mechanical Engineering*

JOHN WILEY & SONS, INC

ASSOCIATE PUBLISHER    Daniel Sayre
ACQUISITIONS EDITOR    Joseph Hayton
SENIOR PRODUCTION EDITOR    Valerie A. Vargas
MARKETING MANAGER    Phyllis Cerys
SENIOR DESIGNER    Kevin Murphy
COVER DESIGN    David Levy
COVER AND CHAPTER OPENING PHOTOS    courtesy of NASA: Left cover image and opener photo (engine test fire) Space Shuttle Main Engine Test Firing, Stennis Space Center, NASA; Right cover image (thermal image) Thermal Image Test of Space Shuttle Main Engine, Stennis Space Center, NASA
EDITORIAL ASSISTANT    Mary Moran-McGee
MEDIA EDITOR    Stefanie Liebman
PRODUCTION SERVICES    mb editorial services

This book was set in "TimesNewRomanPS 10/12" by TechBooks/GTS Companies, York, PA and printed and bound by R.R. Donnelley. The cover was printed by Phoenix Color, Inc.

This book is printed on acid free paper. ∞

To order books or for customer service please, call 1-800-CALL WILEY (225-5945).

ISBN-13    978-0-471-73759-9
ISBN-10    0-471-73759-3

Printed in the United States of America

10 9 8 7 6 5 4 3

# Preface

This book was originally written as a text for two different courses: a one-semester introductory course in engineering thermodynamics for both majors and non-majors; and also, the first course in an introductory sequence in engineering thermal-fluid sciences. This book has also been written to cover a broad range of background, interests, and applications as well as to provide a great deal of flexibility in terms of topics covered and units. In the second edition, we have broadened the coverage somewhat to accommodate a second course for mechanical engineering majors, primarily by inclusion of a chapter on chemical reactions. This chapter focuses on combustion processes, with an introduction to chemical equilibrium.

As in our earlier thermodynamics books, we have deliberately directed our presentation to students. New concepts and definitions are presented in the context where they are first relevant. The first thermodynamic properties to be defined (Chapter 2) are those that can be readily measured: pressure, specific volume, and temperature. In Chapter 3, tables of thermodynamic properties are introduced, but only in regard to these measurable properties. Internal energy and enthalpy are introduced in connection with the first law, and entropy with the second law. Many more examples have been included in the second edition to assist the student in gaining an understanding of thermodynamics, and the problems at the end of each chapter have been carefully sequenced to correlate with the subject, matter, and are grouped and identified as such.

This book is based on our most recent textbook, *Fundaments of Thermodynamics, Sixth Edition,* Richard E. Sonntag, Claus Borgnakke, and Gordon J. Van Wylen, John Wiley & Sons, Inc. (2003). The present book, however, is written at a somewhat lower level, including the chapter-end problems, is directed at a broader audience, and is specifically intended for the multipurpose use mentioned above.

## NEW FEATURES IN THE SECOND EDITION

### New Chapter on Chemical Reactions

A new chapter on chemical reactions, focused on fuels and combustion processes, and including an introduction to chemical equilibrium, has been included especially for a second-semester course.

### Revised Coverage of Heat Transfer

The chapter on heat transfer has been rewritten to decrease coverage on conduction, and to emphasize applications involving the cooling of electronic equipment. This material is intended for use in a single course offered to non-majors, or as part of a first course in engineering thermal-fluid sciences.

### Inclusion of End-of-Chapter Material from
### *Fundamentals, Sixth Edition*

End-of-chapter material from the authors' *Fundamentals, Sixth Edition* has been revised and included in this edition: Chapter Summaries; Key Concepts and formulas; and Concept Problems. All this material has been found to be beneficial to and appreciated by the students in their understanding of the subject.

### Chapter In-Text Concept Questions and How-To Sections

Groups of In-Text Concept Questions have been included in all chapters having introductory material, to direct basic points and procedures to the students. Also, How-To Sections are included at the end of the chapters to help answer commonly asked questions.

### Chapter Examples, Illustrations, and Homework Problems

The number of chapter examples, illustrations, and homework problems have all been expanded and revised; a large number of new problems are included, especially introductory-level problems.

### Reorganization of Entropy and Second Law Presentation

A significant change in this edition is the reorganization of material in Chapter 8, Entropy. Following the introduction of the property entropy, its relevance in reversible processes, and the thermodynamic property relation, we now discuss its calculation for various thermodynamic models and its use in certain thermodynamic processes. Then the significance and behavior of entropy in irreversible processes is considered, including a revised development of entropy generation and the general statement of the second law of thermodynamics.

### Other New or Revised Material and Developments

There are a number of new or revised developments: separate section on compressibility factor and equations of state in Chapter 3; new section on thermodynamic temperature scale and comparison with ideal-gas scale in Chapter 7; new development of steady-state single flow process in Chapter 9; new development of reversible work and irreversibility in Chapter 9; new section on entropy and chaos in Chapter 9; streamlined discussion of Rankine cycle variables in Chapter 11; new section of reciprocating engine power cycles in Chapter 11.

### Expanded Software Included

In this edition we have included the expanded software *Tables of Thermodynamic and Transport Properties* (expanded CATT2) that includes many additional substances besides those included in the printed tables in Appendices B and F. A number of hydrocarbon fuels, refrigerants, and cryogenic fluids are included. The software is available as a download from the student companion site (www.wiley.com/college/sonntag) by using the registration card enclosed in the text. A printed version is available in the book *Thermodynamic and Transport Properties,* Claus Borgnakke and Richard E. Sonntag, John Wiley & Sons, Inc. (©1997)

## FLEXIBILITY IN COVERAGE AND SCOPE

We have attempted to cover fairly comprehensively the basic introductory subject matter of engineering thermodynamics and believe that the book provides adequate preparation for study of the application of thermodynamics to the various professional fields as well as for study of more advanced topics in thermodynamics. We recognize that a number of schools offer a single introductory course in thermodynamics or engineering thermal-fluid sciences for a number of programs, and have tried to cover those topics that the different programs might wish to have included. Recognizing that specific courses vary considerably in background of the students, prerequisites, and level of course introduction, we have arranged the text material, particularly in the later chapters, so that there is considerable flexibility in the amount of material that may be included and also in the order of coverage.

### Units

This edition has been organized so that the course or sequence can be taught entirely in SI units (Le Systeme International d'Unites). Thus, all the text examples, complete problem sets, and thermodynamics tables are given in SI units. In recognition of the continuing need for engineering graduates to be familiar with the corresponding Engineering English units, we have included an Appendix that includes an introduction to this system, examples, problems, and several thermodynamics tables, so that this material can be introduced at the appropriate times in the course(s) as desired. When dealing with English units, the force-mass conversion between pound force and pound mass is treated simply as a units conversion, without using an explicit conversion constant. Throughout the text, symbols, units, and sign conversions are all treated as in our previous books.

### Supplements and Additional Support

Additional support is available through a companion website at www.wiley.com/college/sonntag. Instructors may want to visit the Web site for information and suggestions on possible course structure and schedules, additional study problem material, and current errata for the book.

## ACKNOWLEDGMENTS

We acknowledge with appreciation the suggestions, counsel, and encouragement of many colleagues, both at the University of Michigan and elsewhere. In particular, we acknowledge the background and roots of this book lying in our earlier efforts with our colleague and co-author Gordon J. Van Wylen and his important contributions to those projects over the years. We also appreciate the contributions and suggestions of many of our former students, who have also been of particular assistance in a number of areas. The comments and suggestions of the several reviewers of an early version of the manuscript were important to us in rethinking the organization and content of the book. We also appreciate the efforts of our editor at John Wiley, Joseph Hayton in the production of this edition. Finally, for each of us, the encouragement and patience of our wives and families have been indispensable, and have made this an enjoyable undertaking.

We hope that this book will contribute to the effective teaching of thermodynamics to students who face significant challenges and opportunities during their professional careers. Your comments, criticism, and suggestions will also be appreciated.

RICHARD E. SONNTAG   rsonntag@umich.edu

CLAUS BORGNAKKE   claus@umich.edu

*Ann Arbor, Michigan*

*August, 2005*

# Brief Contents

# Contents

# Symbols

| | |
|---|---|
| $a$ | acceleration |
| A | area |
| $AF$ | air-fuel ratio |
| c | mass fraction |
| $C_p$ | constant-pressure specific heat |
| $C_v$ | constant-volume specific heat |
| $C_{po}$ | zero-pressure constant-pressure specific heat |
| $C_{vo}$ | zero-pressure constant-volume specific heat |
| $e, E$ | specific energy and total energy |
| $F$ | force |
| $FA$ | fuel-air ratio |
| $g$ | acceleration due to gravity |
| $g, G$ | specific Gibbs function and total Gibbs function |
| $h, H$ | specific enthalpy and total enthalpy |
| $h$ | convection heat transfer coefficient |
| $i$ | electrical current |
| $I$ | irreversibility |
| $k$ | conductivity |
| $k$ | specific heat ratio: $C_p/C_v$ |
| $K$ | equilibrium constant |
| KE | kinetic energy |
| $L$ | length |
| $m$ | mass |
| $\dot{m}$ | mass flow rate |
| $M$ | molecular weight |
| $n$ | number of moles |
| $n$ | polytropic exponent |
| $P$ | pressure |
| $P_i$ | partial pressure of component $i$ in a mixture |
| PE | potential energy |
| $P_r$ | relative pressure as used in gas tables |
| Pr | Prandtl number |
| $q, Q$ | heat transfer per unit mass and total heat transfer |
| $\dot{Q}$ | rate of heat transfer |
| $Q_H, Q_L$ | heat transfer with high-temperature body and heat transfer with low-temperature body; sign determined from context |
| $R$ | gas constant |
| $\overline{R}$ | universal gas constant |
| $R$ | thermal resistance |
| Re | Reynolds number |

| | |
|---|---|
| $s, S$ | specific entropy and total entropy |
| $S_{gen}$ | entropy generation |
| $\dot{S}_{gen}$ | rate of entropy generation |
| $t$ | time |
| $T$ | temperature |
| $u, U$ | specific internal energy and total internal energy |
| $v, V$ | specific volume and total volume |
| $v_r$ | relative specific volume as used in gas tables |
| $\mathbf{V}$ | velocity |
| $w, W$ | work per unit mass and total work |
| $\dot{W}$ | rate of work, or power |
| $w^{rev}$ | reversible work between two states |
| $x$ | quality |
| $y$ | gas-phase mole fraction |
| $Z$ | elevation |
| $Z$ | compressibility factor |
| $Z$ | electrical charge |

## SCRIPT LETTERS

| | |
|---|---|
| $\mathcal{E}$ | electrical potential |
| $\mathcal{S}$ | surface tension |
| $\mathcal{T}$ | tension |

## GREEK LETTERS

| | |
|---|---|
| $\alpha$ | temperature coefficient |
| $\beta$ | coefficient of performance for a refrigerator |
| $\beta'$ | coefficient of performance for a heat pump |
| $\eta$ | efficiency |
| $\theta$ | temperature variable |
| $\mu$ | absolute viscosity |
| $\nu$ | stoichiometric coefficient |
| $\nu$ | kinematic viscosity |
| $\rho$ | density |
| $\Phi$ | equivalence ratio |
| $\phi$ | relative humidity |
| $\psi$ | flow availability or exergy |
| $\omega$ | humidity ratio or specific humidity |
| $\omega$ | acentric factor |

## SUBSCRIPTS

| | |
|---|---|
| $c$ | property at the critical point |
| $c$ | cross-sectional |
| $cond$ | conduction |
| $conv$ | convection |

| | |
|---|---|
| c.v. | control volume |
| *e* | state of a substance leaving a control volume |
| *f* | formation |
| *f* | property of saturated liquid |
| *fg* | difference in property for saturated vapor and saturated liquid |
| *g* | property of saturated vapor |
| *i* | state of a substance entering a control volume |
| *i* | property of saturated solid |
| *if* | difference in property for saturated liquid and saturated solid |
| *ig* | difference in property for saturated vapor and saturated solid |
| *r* | reduced property |
| *s* | isentropic process |
| *s* | surface value |
| 0 | property of the surroundings free stream value |

## SUPERSCRIPTS

| | |
|---|---|
| ‾ | bar over symbol denotes property on a molal basis |
| ° | property at standard-state condition |
| rev | reversible |
| ′ | property per unit length |
| ″ | property per unit area |
| ‴ | property per unit volume |

# Chapter 1

# Some Introductory Comments

In the course of our study of thermodynamics, a number of the examples and problems presented refer to processes that occur in equipment such as a steam power plant, a fuel cell, a vapor-compression refrigerator, a thermoelectric cooler, a turbine or rocket engine, and an air separation plant. In this introductory chapter, a brief description of this equipment is given. There are at least two reasons for including such a chapter. First, many students have had limited contact with such equipment, and the solution of problems will be more meaningful when they have some familiarity with the actual processes and the equipment. Second, this chapter will provide an introduction to thermodynamics, including the use of certain terms (which will be more formally defined in later chapters), some of the problems to which thermodynamics can be applied, and some of the things that have been accomplished, at least in part, from the application of thermodynamics.

Thermodynamics is relevant to many other processes than those cited in this chapter. It is basic to the study of materials, chemical reactions, and plasmas. The student should bear in mind that this chapter is only a brief and necessarily very incomplete introduction to the subject of thermodynamics.

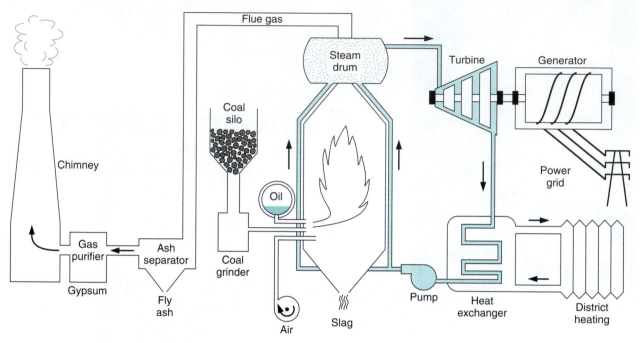

**FIGURE 1.1** Schematic diagram of a steam power plant.

## 1.1 THE SIMPLE STEAM POWER PLANT

A schematic diagram of a recently installed steam power plant is shown in Fig. 1.1. High-pressure superheated steam leaves the steam drum at the top of the boiler, also referred to as a steam generator, and enters the turbine. The steam expands in the turbine and in doing so does work, which enables the turbine to drive the electric generator. The steam, now at low pressure, exits the turbine and enters the heat exchanger, where heat is transferred from the steam (causing it to condense) to the cooling water. Since large quantities of cooling water are required, power plants have traditionally been located near rivers or lakes, leading to thermal pollution of those water supplies. More recently, condenser cooling water is recycled by evaporating a fraction of the water in large cooling towers, thereby cooling the remainder of the water that remains as a liquid. In the power plant shown in Fig. 1.1, the plant is designed to recycle the condenser cooling water by using the heated water for district space heating.

The pressure of the condensate leaving the condenser is increased in the pump, enabling it to return to the steam generator for reuse. In many cases, an economizer or water preheater is used in the steam cycle, and in many power plants, the air that is used for combustion of the fuel may be preheated by the exhaust combustion-product gases. These exhaust gases must also be purified before being discharged to the atmosphere, such that there are many complications to the simple cycle.

Fig. 1.2 is a photograph of the power plant depicted in Fig. 1.1. The tall building shown at the left is the boiler house, next to which are buildings housing the turbine and other components. Also noted are the tall chimney or stack, and the coal supply ship at dock. This particular power plant is located in Denmark, and at the time of its installation, set a world record of efficiency, converting 45% of the 850 MW of coal combustion energy into electricity. Another 47% is reusable for district space heating, an amount that in older plants was simply thrown away to the environment with no benefit.

**FIGURE 1.2** The Esbjerg, Denmark, power station. (Courtesy Vestkraft 1996.)

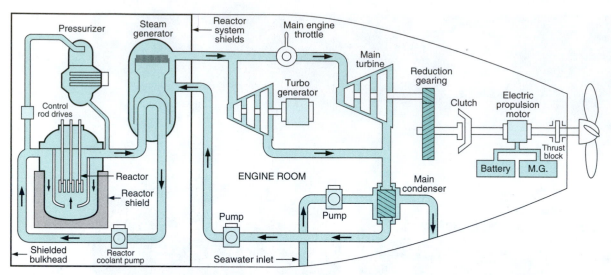

**FIGURE 1.3** Schematic diagram of a shipboard nuclear propulsion system. (Courtesy Babcock & Wilcox Co.)

The steam power plant described utilizes coal as the combustion fuel. Other plants use natural gas, fuel oil, or biomass as the fuel. A number of power plants around the world operate on the heat released from nuclear reactions instead of fuel combustion. Fig. 1.3 is a schematic diagram of a nuclear marine propulsion power plant. A secondary fluid circulates through the reactor, picking up heat generated by the nuclear reaction inside. This heat is then transferred to the water in the steam generator. The steam cycle processes are the same as in the previous example, but in this application the condenser cooling water is seawater, that is then returned at higher temperature to the sea.

## 1.2 FUEL CELLS

When a conventional power plant is viewed as a whole, as shown in Fig. 1.4, fuel and air enter the power plant and products of combustion leave the unit. There is also a transfer of heat to the cooling water, and work is done in the form of the electrical energy leaving the power plant. The overall objective of a power plant is to convert the availability

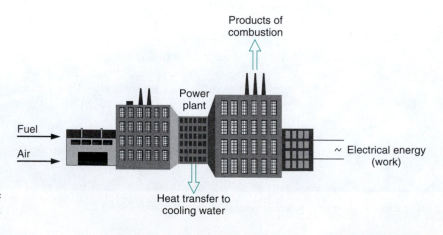

**FIGURE 1.4** Schematic diagram of a power plant.

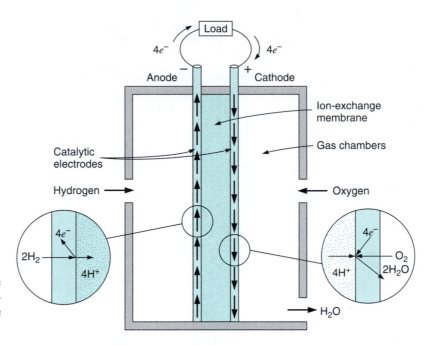

**FIGURE 1.5** Schematic arrangement of an ion-exchange membrane type of fuel cell.

(to do work) of the fuel into work (in the form of electrical energy) in the most efficient manner, taking into consideration cost, space, safety, and environmental concerns.

We might well ask whether all the equipment in the power plant, such as the steam generator, the turbine, the condenser, and the pump, is necessary. Is it possible to produce electrical energy from the fuel in a more direct manner?

The fuel cell accomplishes this objective. Fig. 1.5 shows a schematic arrangement of a fuel cell of the ion-exchange-membrane type. In this fuel cell, hydrogen and oxygen react to form water. Hydrogen gas enters at the anode side and is ionized at the surface of the ion-exchange membrane, as indicated in Fig. 1.5. The electrons flow through the external circuit to the cathode while the positive hydrogen ions migrate through the membrane to the cathode, where both react with oxygen to form water.

There is a potential difference between the anode and cathode, and thus there is a flow of electricity through a potential difference; this, in thermodynamic terms, is called work. There may also be a transfer of heat between the fuel cell and the surroundings.

At the present time the fuel used in fuel cells is usually either hydrogen or a mixture of gaseous hydrocarbons and hydrogen. The oxidizer is usually oxygen. However, current development is directed toward the production of fuel cells that use hydrogen or hydrocarbon fuels and air. Although the conventional (or nuclear) steam power plant is still used in large-scale power-generating systems and conventional piston engines and gas turbines are still used in most transportation power systems, the fuel cell may eventually become a serious competitor. The fuel cell is already being used to produce power for space and other special applications.

Thermodynamics plays a vital role in the analysis, development, and design of all power-producing systems, including reciprocating internal-combustion engines and gas turbines. Considerations such as the increase of efficiency, improved design, optimum operating conditions, environmental pollution, and alternate methods of power generation involve, among other factors, the careful application of the fundamentals of thermodynamics.

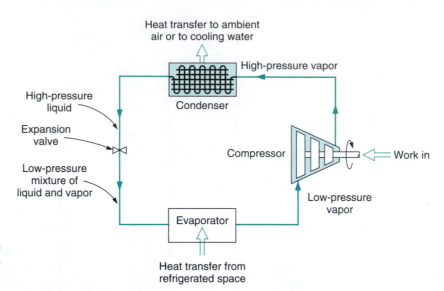

**FIGURE 1.6** Schematic diagram of a simple refrigeration cycle.

## 1.3 THE VAPOR-COMPRESSION REFRIGERATION CYCLE

A simple vapor-compression refrigeration cycle is shown schematically in Fig. 1.6. The refrigerant enters the compressor as a slightly superheated vapor at a low pressure. It then leaves the compressor and enters the condenser as a vapor at some elevated pressure, where the refrigerant is condensed as heat is transferred to cooling water or to the surroundings. The refrigerant then leaves the condenser as a high-pressure liquid. The pressure of the liquid is decreased as it flows through the expansion valve, and as a result, some of the liquid flashes into cold vapor. The remaining liquid, now at a low pressure and temperature, is vaporized in the evaporator as heat is transferred from the refrigerated space. This vapor then reenters the compressor.

In a typical home refrigerator the compressor is located in the rear near the bottom of the unit. The compressors are usually hermetically sealed; that is, the motor and compressor are mounted in a sealed housing, and the electric leads for the motor pass through this housing. This seal prevents leakage of the refrigerant. The condenser is also located at the back of the refrigerator and is arranged so that the air in the room flows past the condenser by natural convection. The expansion valve takes the form of a long capillary tube, and the evaporator is located around the outside of the freezing compartment inside the refrigerator.

Fig. 1.7 shows a large centrifugal unit that is used to provide refrigeration for an air-conditioning unit. In this unit, water is cooled and then circulated to provide cooling where needed.

## 1.4 THE THERMOELECTRIC REFRIGERATOR

We may well ask the same question about the vapor-compression refrigerator that we asked about the steam power plant—is it possible to accomplish our objective in a more direct manner? Is it possible, in the case of a refrigerator, to use the electrical energy (which goes to the electric motor that drives the compressor) to produce cooling in a

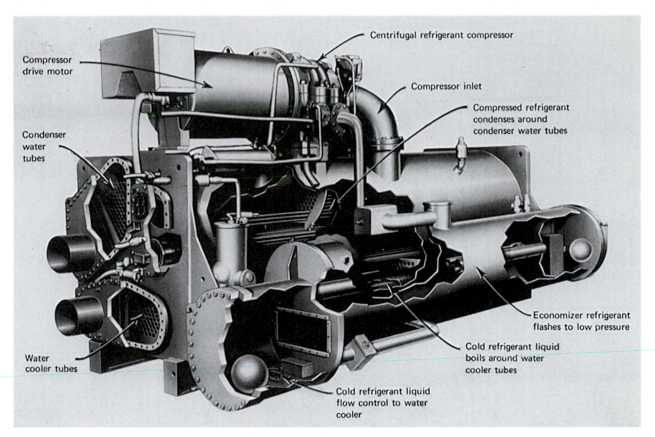

Compressor drive motor

Centrifugal refrigerant compressor

Compressor inlet

Compressed refrigerant condenses around condenser water tubes

Condenser water tubes

Economizer refrigerant flashes to low pressure

Cold refrigerant liquid boils around water cooler tubes

Water cooler tubes

Cold refrigerant liquid flow control to water cooler

**FIGURE 1.7**   A refrigeration unit for an air-conditioning system. (Courtesy Carrier Air Conditioning Co.)

more direct manner and thereby to avoid the cost of the compressor, condenser, evaporator, and all the related piping?

The thermoelectric refrigerator is such a device. This is shown schematically in Fig. 1.8a. The thermoelectric device, like the conventional thermocouple, uses two dissimilar materials. There are two junctions between these two materials in a thermoelectric refrigerator. One is located in the refrigerated space and the other in ambient surroundings. When a potential difference is applied, as indicated, the temperature of the junction located in the refrigerated space will decrease and the temperature of the other junction will increase. Under steady-state operating conditions, heat will be transferred from the refrigerated space to the cold junction. The other junction will be at a temperature above the ambient, and heat will be transferred from the junction to the surroundings.

A thermoelectric device can also be used to generate power by replacing the refrigerated space with a body that is at a temperature above the ambient. Such a system is shown in Fig. 1.8b.

The thermoelectric refrigerator cannot yet compete economically with conventional vapor-compression units. However, in certain special applications, the thermoelectric refrigerator is already is use and, in view of research and development efforts under way in this field, it is quite possible that thermoelectric refrigerators will be much more extensively used in the future.

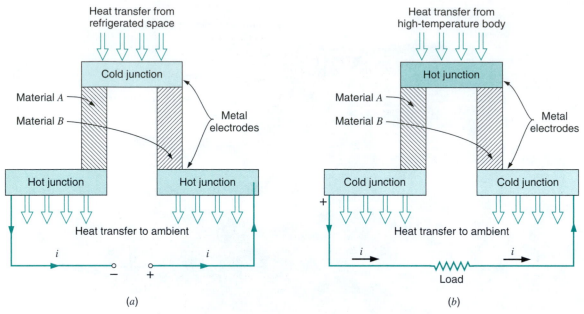

**FIGURE 1.8**    (a) A thermoelectric refrigerator. (b) A thermoelectric power generation device.

## 1.5 THE AIR SEPARATION PLANT

One process of great industrial significance is the air separation. In an air separation plant, air is separated into its various components. The oxygen, nitrogen, argon, and rare gases so produced are used extensively in various industrial, research, space, and consumer-goods applications. The air separation plant can be considered an example from two major fields, the chemical process industry and cryogenics. Cryogenics is a term applied to technology, processes, and research at very low temperatures (in general, below 150 K). In both chemical processing and cryogenics, thermodynamics is basic to an understanding of many phenomena that occur and to the design and development of processes and equipment.

A number of different designs of air separation plants have been developed. Consider Fig. 1.9, which shows a somewhat simplified sketch of a type of plant that is frequently used. Air from the atmosphere is compressed to a pressure of 2 to 3 MPa. It is then purified, particularly to remove carbon dioxide (which would plug the flow passages as it solidifies when the air is cooled to its liquefaction temperature). The air is then compressed to a pressure of 15 to 20 MPa, cooled to the ambient temperature in the after-cooler, and dried to remove the water vapor (which would also plug the flow passages as it freezes).

The basic refrigeration in the liquefaction process is provided by two different processes. In one process the air in the expansion engine expands. During this process the air does work and as a result the temperature of the air is reduced. In the other refrigeration process air passes through a throttle valve that is so designed and so located that there is a substantial drop in the pressure of the air and, associated with this, a substantial drop in the temperature of the air.

As shown in Fig. 1.9, the dry, high-pressure air enters a heat exchanger. The air temperature drops as it flows through the heat exchanger. At some intermediate point in the heat exchanger, part of the air is bled off and flows through the expansion engine. The remaining air flows through the rest of the heat exchanger and through the throttle valve. The two streams join (both are at the pressure of 0.5 to 1 MPa) and enter the bottom of the

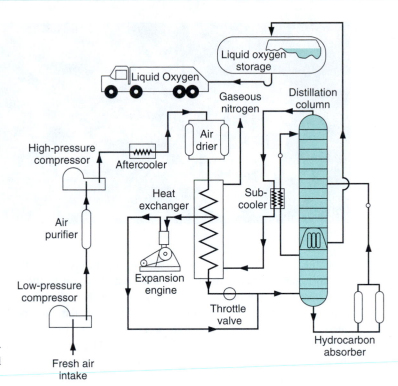

**FIGURE 1.9** A simplified diagram of a liquid oxygen plant.

distillation column, which is referred to as the high-pressure column. The function of the distillation column is to separate the air into its various components, principally oxygen and nitrogen. Two streams of different composition flow from the high-pressure column through throttle valves to the upper column (also called the low-pressure column). One of these streams is an oxygen-rich liquid that flows from the bottom of the lower column, and the other is a nitrogen-rich stream that flows through the subcooler. The separation is completed in the upper column. Liquid oxygen leaves from the bottom of the upper column, and gaseous nitrogen leaves from the top of the column. The nitrogen gas flows through the subcooler and the main heat exchanger. It is the transfer of heat to this cold nitrogen gas that causes the high-pressure air entering the heat exchanger to become cooler.

Not only is a thermodynamic analysis essential to the design of the system as a whole, but essentially every component of such a system, including the compressors, the expansion engine, the purifiers and driers, and the distillation column, operates according to the principles of thermodynamics. In this separation process we are also concerned with the thermodynamic properties of mixtures and the principles and procedures by which these mixtures can be separated. This is the type of problem encountered in petroleum refining and many other chemical processes. It should also be noted that cryogenics is particularly relevant to many aspects of the space program, and a thorough knowledge of thermodynamics is essential for creative and effective work in cryogenics.

## 1.6 THE GAS TURBINE

The basic operation of a gas turbine is similar to that of the steam power plant, except that air is used instead of water. Fresh atmospheric air flows through a compressor that brings it to a high pressure. Energy is then added by spraying fuel into the air and igniting

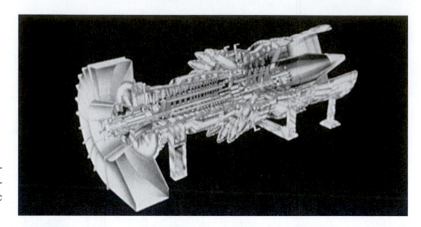

**FIGURE 1.10** A 150-MW gas turbine. (Courtesy Westinghouse Electric Corporation.)

it so the combustion generates a high-temperature flow. This high-temperature, high-pressure gas enters a turbine, where it expands down to the exhaust pressure, producing a shaft work output in the process. The turbine shaft work is used to drive the compressor and other devices, such as an electric generator that may be coupled to the shaft. The energy that is not used for shaft work comes out in the exhaust gases, so these have either a high temperature or a high velocity. The purpose of the gas turbine determines the design so that the most desirable energy form is maximized. An example of a large gas turbine for stationary power generation is shown in Fig. 1.10. The unit has sixteen stages of compression and four stages in the turbine and is rated at 150 MW. Notice that since the combustion of fuel uses the oxygen in the air, the exhaust gases cannot be recirculated as the water is in the steam power plant.

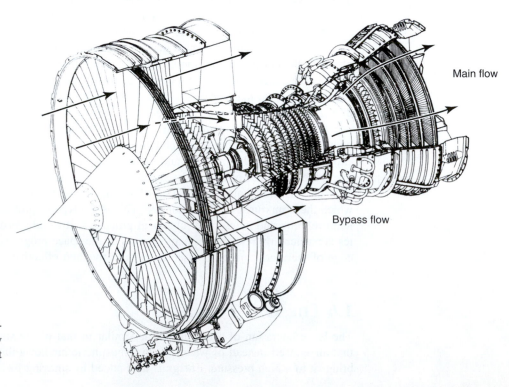

Main flow

Bypass flow

**FIGURE 1.11** A turbo-fan jet engine. (Courtesy General Electric Aircraft Engines.)

A gas turbine is often the preferred power-generating device where a large amount of power is needed, but only a small physical size is possible. Examples are jet engines, turbofan jet engines, offshore oilrig power plants, ship engines, helicopter engines, smaller local power plants, or peak-load power generators in larger power plants. Since the gas turbine has relatively high exhaust temperatures, it can also be arranged so the exhaust gases are used to heat water that runs in a steam power plant before it exhausts to the atmosphere.

In the examples mentioned previously, the jet engine and turboprop applications utilize part of the power to discharge the gases at high velocity. This is what generates the thrust of the engine that moves the airplane forward. The gas turbines in these applications are therefore designed differently than for the stationary power plant, where the energy is taken out as shaft work to an electric generator. An example of a turbofan jet engine used in a commercial airplane is shown in Fig. 1.11. The large front-end fan also blows air past the engine, providing cooling and giving additional thrust.

## 1.7 THE CHEMICAL ROCKET ENGINE

The advent of missiles and satellites brought to prominence the use of the rocket engine as a propulsion power plant. Chemical rocket engines may be classified as either liquid propellant or solid propellant, according to the fuel used.

Fig. 1.12 shows a simplified schematic diagram of a liquid-propellant rocket. The oxidizer and fuel are pumped through the injector plate into the combustion chamber where combustion takes place at high pressure. The high-pressure, high-temperature products of combustion expand as they flow through the nozzle, and as a result they leave the nozzle with a high velocity. The momentum change associated with this increase in velocity gives rise to the forward thrust on the vehicle.

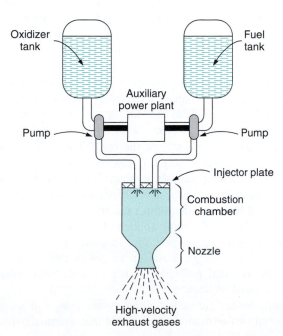

**FIGURE 1.12** Simplified schematic diagram of a liquid-propellant rocket engine.

The oxidizer and fuel must be pumped into the combustion chamber, and some auxiliary power plant is necessary to drive the pumps. In a large rocket this auxiliary power plant must be very reliable and have a relatively high power output, yet it must be light in weight. The oxidizer and fuel tanks occupy the largest part of the volume of an actual rocket, and the range and payload of a rocket are determined largely by the amount of oxidizer and fuel that can be carried. Many different fuels and oxidizers have been considered and tested, and much effort has gone into the development of fuels and oxidizers that will give a higher thrust per unit mass rate of flow of reactants. Liquid oxygen is frequently used as the oxidizer in liquid-propellant rockets, and liquid hydrogen is frequently used as the fuel.

Much work has also been done on solid-propellant rockets. They have been very successfully used for jet-assisted takeoffs of airplanes, military missiles, and space vehicles. They are much simpler in both the basic equipment required for operation and the logistic problems involved in their use, but they are more difficult to control.

## 1.8 OTHER APPLICATIONS AND ENVIRONMENTAL ISSUES

There are many other applications in which thermodynamics is relevant. Many municipal landfill operations are now utilizing the heat produced by the decomposition of biomass waste to produce power, and they also capture the methane gas produced by these chemical reactions for use as a fuel. Geothermal sources of heat are also being utilized, as are solar- and windmill-produced electricity. Sources of fuel are being converted from one form to another, more usable or convenient form, such as in the gasification of coal or the conversion of biomass to liquid fuels. Hydroelectric plants have been in use for many years, as have other applications involving water power. Thermodynamics is also relevant to such processes as the curing of a poured concrete slab, which produces heat, the cooling of electronic equipment, in various applications in cryogenics (cryo-surgery, food fast-freezing), and many other diverse applications.

We must also be concerned with environmental issues related to these many devices and applications of thermodynamics. For example, the construction and operation of the steam power plant creates electricity, which is so deeply entrenched in our society that we take its ready availability for granted. In recent years, however, it has become increasingly apparent that we need to consider seriously the effects of such an operation on our environment. Combustion of hydrocarbon fuels releases carbon dioxide into the atmosphere, where its concentration is increasing. Carbon dioxide, as well as other gases, absorbs infrared radiation from the surface of the earth, holding it close to the planet and creating the "greenhouse effect," which in turn is believed to cause global warming and critical climatic changes around the earth. Power plant combustion, particularly of coal, releases sulfur dioxide, which is absorbed in clouds and later falls as acid rain in many areas. Combustion processes in power plants and gasoline and diesel engines also generate pollutants other than these two. Species such as carbon monoxide, nitric oxides, and partly burned fuels together with particulates all contribute to atmospheric pollution and are regulated by law for many applications. Catalytic converters on automobiles help to minimize the air pollution problem. In power plants, Fig. 1.1 indicates the fly ash cleanup and also the flue gas clean up processes that are now incorporated to address these problems. Thermal pollution associated with power plant cooling water requirements was discussed in Section 1.1.

Refrigeration and air-conditioning systems, as well as other industrial processes, have used certain chlorofluorocarbon fluids that eventually find their way to the upper atmosphere and destroy the protective ozone layer. Many countries have already banned the production of some of these compounds, and the search for improved replacement fluid continues.

These are only some of the many environmental problems caused by our efforts to produce goods and effects intended to improve our way of life. During our study of thermodynamics, which is the science of the conversion of energy from one form to another, we must continue to reflect on these issues. We must consider how we can eliminate or at least minimize damaging effects, as well as use our natural resources efficiently and responsibly.

# Chapter **2**

# Some Concepts and Definitions

One excellent definition of thermodynamics is that it is the science of energy and entropy. Since we have not yet defined these terms, an alternate definition in already familiar terms is: Thermodynamics is the science that deals with heat and work and those properties of substances that bear a relation to heat and work. Like all sciences, the basis of thermodynamics is experimental observation. In thermodynamics these findings have been formalized into certain basic laws, which are known as the first, second, and third laws of thermodynamics. In addition to these laws, the zeroth law of thermodynamics, which in the logical development of thermodynamics precedes the first law, has been set forth.

In the chapters that follow, we will present these laws and the thermodynamic properties related to these laws and apply them to a number of representative examples. The objective of the student should be to gain both a thorough understanding of the fundamentals and an ability to apply these fundamentals to thermodynamic problems. The examples and problems further this twofold objective. It is not necessary for the student

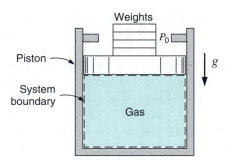

**FIGURE 2.1** Example of a control mass.

to memorize numerous equations, for problems are best solved by the application of the definitions and laws of thermodynamics. In this chapter some concepts and definitions basic to thermodynamics are presented.

## 2.1 A Thermodynamic System and the Control Volume

A thermodynamic system comprises a device or combination of devices containing a quantity of matter that is being studied. To define this more precisely, a control volume is chosen so that it contains the matter and devices inside a control surface. Everything external to the control volume is the surroundings, with the separation given by the control surface. The surface may be open or closed to mass flows, and it may have flows of energy in terms of heat transfer and work across it. The boundaries may be movable or stationary. In the case of a control surface that is closed to mass flow, so that no mass can escape or enter the control volume, it is called a control mass containing the same amount of matter at all times.

Selecting the gas in the cylinder of Fig. 2.1 as a control volume by placing a control surface around it, we recognize this as a control mass. If a Bunsen burner is placed under the cylinder, the temperature of the gas will increase and the piston will rise. As the piston rises, the boundary of the control mass moves. As we will see later, heat and work cross the boundary of the control mass during this process, but the matter that composes the control mass can always be identified and remains the same.

An isolated system is one that is not influenced in any way by the surroundings. This means that no mass, heat, or work cross the boundary of the system. In many cases a thermodynamic analysis must be made of a device, such as an air compressor, which has a flow of mass into it, out of it, or both, as shown schematically in Fig. 2.2. The

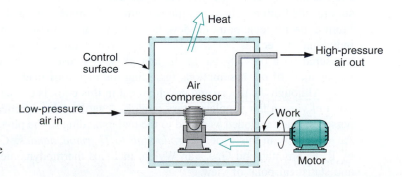

**FIGURE 2.2** Example of a control volume.

procedure followed in such an analysis is to specify a control volume that surrounds the device under consideration. The surface of this control volume is the control surface, which may have mass, momentum, and also heat and work, cross it.

Thus the more general control surface defines a control volume, where mass may flow in or out, with a control mass as the special case of no mass flow in or out. Hence the control mass contains a fixed mass at all times, which explains its name. The difference in the formulation of the analysis is considered in detail in Chapter 6. The terms closed system (fixed mass) and open system (involving a flow of mass) are sometimes used to make this distinction. Here, we use the term system as a more general and loose description for a mass, device, or combination of devices that then is more precisely defined, when a control volume is selected. The procedure that will be followed in the presentation of the first and the second laws of thermodynamics is first to present each law for a control mass and then to extend the analysis to the more general control volume.

## 2.2 MACROSCOPIC VERSUS MICROSCOPIC POINT OF VIEW

An investigation into the behavior of a system may be undertaken from either a microscopic or a macroscopic point of view. Let us briefly describe a system from a microscopic point of view. Consider a system consisting of a cube 25 mm on a side and containing a monatomic gas at atmospheric pressure and temperature. This volume contains approximately $10^{20}$ atoms. To describe the position of each atom, we need to specify three coordinates; to describe the velocity of each atom, we specify three velocity components.

Thus, to describe completely the behavior of this system from a microscopic point of view we must deal with at least $6 \times 10^{20}$ equations. Even with the capacity of modern computers, this is a quite hopeless computational task. However, there are two approaches to this problem that reduce the number of equations and variables to a few that can be computed relatively easily. One approach is the statistical approach, in which, on the basis of statistical considerations and probability theory, we deal with "average" values for all particles under consideration. This is usually done in connection with a model of the atom under consideration. This is the approach used in the disciplines known as kinetic theory and statistical mechanics.

The other approach to reducing the number of variables to a few that can be handled is the macroscopic point of view of classical thermodynamics. As the word macroscopic implies, we are concerned with the gross or average effects of many molecules. These effects can be perceived by our senses and measured by instruments. However, what we really perceive and measure is the time-averaged influence of many molecules. For example, consider the pressure a gas exerts on the walls of its container. This pressure results from the change in momentum of the molecules as they collide with the wall. From a macroscopic point of view, however, we are not concerned with the action of the individual molecules but with the time-averaged force on a given area, which can be measured by a pressure gauge. In fact, these macroscopic observations are completely independent of our assumptions regarding the nature of matter.

Although the theory and development in this book is presented from a macroscopic point of view, a few supplementary remarks regarding the significance of the microscopic perspective are included as an aid to the understanding of the physical processes involved. Another book in this series, *Introduction to Thermodynamics: Classical and Statistical,* by R. E. Sonntag and G. J. Van Wylen, includes thermodynamics from the microscopic and statistical point of view.

A few remarks should be made regarding the continuum. From the macroscopic view, we are always concerned with volumes that are very large compared to molecular dimensions and, therefore, with systems that contain many molecules. Because we are not concerned with the behavior of individual molecules, we can treat the substance as being continuous, disregarding the action of individual molecules. This continuum concept, of course, is only a convenient assumption that loses validity when the mean free path of the molecules approaches the order of magnitude of the dimensions of the vessel, as, for example, in high-vacuum technology. In much engineering work the assumption of a continuum is valid and convenient, going hand in hand with the macroscopic view.

## 2.3 PROPERTIES AND STATE OF A SUBSTANCE

If we consider a given mass of water, we recognize that this water can exist in various forms. If it is a liquid initially, it may become a vapor when it is heated or a solid when it is cooled. Thus, we speak of the different phases of a substance. A phase is defined as a quantity of matter that is homogeneous throughout. When more than one phase is present, the phases are separated from each other by the phase boundaries. In each phase the substance may exist at various pressures and temperatures or, to use the thermodynamic term, in various states. The state may be identified or described by certain observable, macroscopic properties; some familiar ones are temperature, pressure, and density. In later chapters other properties will be introduced. Each of the properties of a substance in a given state has only one definite value, and these properties always have the same value for a given state, regardless of how the substance arrived at the state. In fact, a property can be defined as any quantity that depends on the state of the system and is independent of the path (that is, the prior history) by which the system arrived at the given state. Conversely, the state is specified or described by the properties. Later we will consider the number of independent properties a substance can have, that is, the minimum number of properties that must be specified to fix the state of the substance.

Thermodynamic properties can be divided into two general classes, intensive and extensive properties. An intensive property is independent of the mass; the value of an extensive property varies directly with the mass. Thus, if a quantity of matter in a given state is divided into two equal parts, each part will have the same value of intensive properties as the original and half the value of the extensive properties. Pressure, temperature, and density are examples of intensive properties. Mass and total volume are examples of extensive properties. Extensive properties per unit mass, such as specific volume, are intensive properties.

Frequently we will refer not only to the properties of a substance but to the properties of a system. When we do so we necessarily imply that the value of the property has significance for the entire system, and this implies equilibrium. For example, if the gas that composes the system (control mass) in Fig. 2.1 is in thermal equilibrium, the temperature will be the same throughout the entire system, and we may speak of the temperature as a property of the system. We may also consider mechanical equilibrium, which is related to pressure. If a system is in mechanical equilibrium, there is no tendency for the pressure at any point to change with time as long as the system is isolated from the surroundings. There will be a variation in pressure with elevation because of the influence of gravitational forces, although under equilibrium conditions there will be no tendency for the pressure at any location to change. However, in many thermodynamic problems, this variation in pressure with elevation is so small that is can be neglected. Chemical equilibrium is also important and will be considered in Chapter 12. When a system is in equilibrium regarding all possible changes of state, we say that the system is in thermodynamic equilibrium.

## 2.4 PROCESSES AND CYCLES

Whenever one or more of the properties of a system change, we say that a change in state has occurred. For example, when one of the weights on the piston in Fig. 2.3 is removed, the piston rises and a change in state occurs, for the pressure decreases and the specific volume increases. The path of the succession of states through which the system passes is called the process.

Let us consider the equilibrium of a system as it undergoes a change in state. The moment the weight is removed from the piston in Fig. 2.3, mechanical equilibrium does not exist, and as a result the piston is moved upward until mechanical equilibrium is again restored. The question is this: Since the properties describe the state of a system only when it is in equilibrium, how can we describe the states of a system during a process if the actual process occurs only when equilibrium does not exist? One step in the answer to this question concerns the definition of an ideal process, which we call a quasi-equilibrium process. A quasi-equilibrium process is one in which the deviation from thermodynamic equilibrium is infinitesimal, and all the states the system passes through during a quasi-equilibrium process may be considered equilibrium states. Many actual processes closely approach a quasi-equilibrium process and may be so treated with essentially no error. If the weights on the piston in Fig. 2.3 are small and are taken off one by one, the process could be considered quasi-equilibrium. However, if all the weights were removed at once, the piston would rise rapidly until it hit the stops. This would be a nonequilibrium process, and the system would not be in equilibrium at any time during this change of state.

For nonequilibrium processes, we are limited to a description of the system before the process occurs and after the process is completed and equilibrium is restored. We are unable to specify each state through which the system passes or the rate at which the process occurs. However, as we will see later, we are able to describe certain overall effects that occur during the process.

Several processes are described by the fact that one property remains constant. The prefix iso- is used to describe such a process. An isothermal process is a constant-temperature process, an isobaric (sometimes called isopiestic) process is a constant-pressure process, and an isochoric process is a constant-volume process.

When a system in a given initial state goes through a number of different changes of state or processes and finally returns to its initial state, the system has undergone a

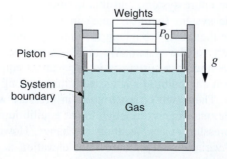

**FIGURE 2.3** Example of a system that may undergo a quasi-equilibrium process.

cycle. Therefore, at the conclusion of a cycle, all the properties have the same value they had at the beginning. Steam (water) that circulates through a steam power plant undergoes a cycle.

A distinction should be made between a thermodynamic cycle, which has just been described, and a mechanical cycle. A four-stroke-cycle internal-combustion engine goes through a mechanical cycle once every two revolutions. However, the working fluid does not go through a thermodynamic cycle in the engine, since air and fuel are burned and changed to products of combustion that are exhausted to the atmosphere. In this text the term cycle will refer to a thermodynamic "cycle" unless otherwise designated.

## 2.5 UNITS FOR MASS, LENGTH, TIME, AND FORCE

Since we are considering thermodynamic properties from a macroscopic perspective, we are dealing with quantities that can, either directly or indirectly, be measured and counted. Therefore, the matter of units becomes an important consideration. In the remaining sections of this chapter we will define certain thermodynamic properties and the basic units. Because the relation between force and mass is often a difficult matter for students, it is considered in this section in some detail.

Force, mass, length, and time are related by Newton's second law of motion, which states that the force acting on a body is proportional to the product of the mass and the acceleration in the direction of the force:

$$F \propto ma$$

The concept of time is well established. The basic unit of time is the second (s), which in the past was defined in terms of the solar day, the time interval for one complete revolution of the earth relative to the sun. Since this period varies with the season of the year, an average value over a one-year period is called the mean solar day, and the mean solar second is 1/86 400 of the mean solar day. (The measurement of the earth's rotation is sometimes made relative to a fixed star, in which case the period is called a sidereal day.) In 1967, the General Conference of Weights and Measures (CGPM) adopted a definition of the second as the time required for a beam of cesium-133 atoms to resonate 9 192 631 770 cycles in a cesium resonator.

For periods of time less than a second, the prefixes milli, micro, nano, or pico, as listed in Table 2.1, are commonly used. For longer periods of time, the units minute (min), hour (h), or day (day) are frequently used. It should be pointed out that the prefixes in Table 2.1 are used with many other units as well.

The concept of length is also well established. The basic unit of length is the meter (m). For many years the accepted standard was the International Prototype Meter, the

**TABLE 2.1**
*Unit Prefixes*

| Factor | Prefix | Symbol | Factor | Prefix | Symbol |
|--------|--------|--------|--------|--------|--------|
| $10^{12}$ | tera | T | $10^{-3}$ | milli | m |
| $10^{9}$ | giga | G | $10^{-6}$ | micro | $\mu$ |
| $10^{6}$ | mega | M | $10^{-9}$ | nano | n |
| $10^{3}$ | kilo | k | $10^{-12}$ | pico | p |

distance between two marks on a platinum–iridium bar under certain prescribed conditions. This bar is maintained at the International Bureau of Weights and Measures, in Sevres, France. In 1960, the CGPM adopted a definition of the meter as a length equal to 1 650 763.73 wavelengths in a vacuum of the orange-red line of krypton-86. Then in 1983, the CGPM adopted a more precise definition of the meter in terms of the speed of light (which is now a fixed constant): The meter is the length of the path traveled by light in a vacuum during a time interval of 1/299 792 458 of a second.

The fundamental unit of mass is the kilogram (kg). As adopted by the first CGPM in 1889 and restated in 1901, it is the mass of a certain platinum–iridium cylinder maintained under prescribed conditions at the International Bureau of Weights and Measures. A related unit that is used frequently in thermodynamics is the mole (mol), defined as an amount of substance containing as many elementary entities as there are atoms in 0.012 kg of carbon-12. These elementary entities must be specified; they may be atoms, molecules, electrons, ions, or other particles or specific groups. For example, one mole of diatomic oxygen, having a molecular weight of 32 (compared to 12 for carbon), has a mass of 0.032 kg. The mole is often termed a gram mole, since it is an amount of substance in grams numerically equal to the molecular weight. In this text, when using the metric SI system we will find it preferable to use the kilomole (kmol), the amount of substance in kilograms numerically equal to the molecular weight, rather than the mole.

The system of units in use presently throughout most of the world is the metric International System, commonly referred to as SI units (from Le Système International d'Unités). In this system, the second, meter, and kilogram are the basic units for time, length, and mass, respectively, as just defined, and the unit of force is defined directly from Newton's second law.

Therefore, a proportionality constant is unnecessary, and we may write that law as an equality:

$$F = ma \tag{2.1}$$

The unit of force is the newton (N), which by definition is the force required to accelerate a mass of one kilogram at the rate of one meter per second per second:

$$1 \text{ N} = 1 \text{ kg m/s}^2$$

It is worth noting that SI units derived from proper nouns use capital letters for symbols; others use the lowercase letters. The liter, with the symbol L, is an exception.

The term weight is often used with respect to a body and is sometimes confused with mass. Weight is really correctly used only as a force. When we say a body weighs so much, we mean that this is the force with which it is attracted to the earth (or some other body), that is, the product of its mass and the local gravitational acceleration. The mass of a substance remains constant with elevation, but its weight varies with elevation.

EXAMPLE 2.1

What is the weight of a one kg mass at an altitude where the local acceleration of gravity is 9.75 m/s²?

**Solution**

Weight is the force acting on the mass, which from Newton's second law is

$$F = mg = 1 \text{ kg} \times 9.75 \text{ m/s}^2 \times [1 \text{ N s}^2/\text{kg m}] = 9.75 \text{ N}$$

The English Engineering System, the system of units traditionally used in the United States, is not presented here but is instead included in Appendixes F and G. A set of examples and chapter-end problems appear in Appendix F and thermodynamic data and tables appear in Appendix G, for the use of those who wish to incorporate English unit applications as part of their study of thermodynamics.

## 2.6 ENERGY

One of the very important concepts in a study of thermodynamics is that of energy. Energy is a fundamental concept, such as mass or force and, as is often the case with such concepts, is very difficult to define. Energy has been defined as the capability to produce an effect. Fortunately the word energy and the basic concept that this word represents are familiar to us in everyday usage, and a precise definition is not essential at this point.

Energy can be stored within a system and can be transferred (as heat, for example) from one system to another. In a study of statistical thermodynamics we would examine, from a molecular view, the ways in which energy can be stored. Because it is helpful in a study of classical thermodynamics to have some notion of how this energy is stored, a brief introduction is presented here.

Consider as a system a certain gas at a given pressure and temperature contained within a tank or pressure vessel. When considered from the molecular view, we identify three general forms of energy:

1. Intermolecular potential energy, which is associated with the forces between molecules.
2. Molecular kinetic energy, which is associated with the translational velocity of individual molecules.
3. Intramolecular energy (that within the individual molecules), which is associated with the molecular and atomic structure and related forces.

The first of these forms of energy, the intermolecular potential energy, depends on the magnitude of the intermolecular forces and the position the molecules have relative to each other at any instant of time. It is impossible to determine accurately the magnitude of this energy because we do not know either the exact configuration and orientation of the molecules at any time or the exact intermolecular potential function. However, there are two situations for which we can make good approximations. The first situation is at low or moderate densities. In this case the molecules are relatively widely spaced, so that only two-molecule or two- and three-molecule interactions contribute to the potential energy. At these low and moderate densities, techniques are available for determining, with reasonable accuracy, the potential energy of a system composed of reasonably simple molecules. The second situation is at very low densities; under these conditions the average intermolecular distance between molecules is so large that the potential energy may be assumed to be zero. Consequently, we have in this case a system of independent particles (an ideal gas) and, therefore, from a statistical point of view, we are able to concentrate our efforts on evaluating the molecular translational and internal energies.

The translational energy, which depends only on the mass and velocities of the molecules, is determined by using the equations of mechanics—either quantum or classical.

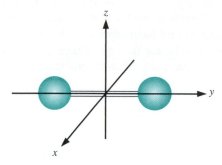

**FIGURE 2.4**  The coordinate system for a diatomic molecule.

The intramolecular internal energy is more difficult to evaluate because, in general, it may result from a number of contributions. Consider a simple monatomic gas such as helium. Each molecule consists of a helium atom. Such an atom possesses electronic energy as a result of both orbital angular momentum of the electrons about the nucleus and angular momentum of the electrons spinning on their axes. The electronic energy is commonly very small compared with the translational energies. (Atoms also possess nuclear energy, which, except in the case of nuclear reactions, is constant. We are not concerned with nuclear energy at this time.) When we consider more complex molecules, such as those composed of two or three atoms, additional factors must be considered. In addition to having electronic energy, a molecule can rotate about its center of gravity and thus have rotational energy. Furthermore, the atoms may vibrate with respect to each other and have vibrational energy. In some situations there may be an interaction between the rotational and vibrational modes of energy.

Consider a diatomic molecule, such as oxygen, as shown in Fig. 2.4. In addition to translation of the molecule as a solid body, the molecule can rotate about its center of mass in two normal directions, about the $x$ axis and about the $z$ axis (rotation about the $y$ axis is negligible), and the two atoms can also vibrate, that is, stretch the bond joining the atoms along the $y$ axis. A more rapid rotation increases the rotational energy, and a stronger vibration results in an increase of vibrational energy of the molecule.

More complex molecules, such as typical polyatomic molecules, are usually three-dimensional in structure and have multiple vibrational modes, each of which contributes to the energy storage of the molecule. The more complicated the molecule is, the larger the number of degrees of freedom that exist for energy storage. This subject of the modes of energy storage and their evaluation is discussed in some detail in Appendix C, for those interested in a further development of the quantitative effects from a molecular viewpoint.

This general discussion can be summarized by referring to Fig. 2.5. Let heat be transferred to the water. During this process the temperature of the liquid and vapor (steam) will increase, and eventually all the liquid will become vapor. From the macroscopic view we are concerned only with the energy that is transferred as heat, the change in properties, such as temperature and pressure, and the total amount of energy (relative to some base) that the $H_2O$ contains at any instant. Thus, questions about how energy is stored in the $H_2O$ do not concern us. From a microscopic viewpoint we are concerned about the way in which energy is stored in the molecules. We might be interested in developing a model of the molecule so that we could predict the amount of energy required to change the temperature a given amount. Although the focus in this book is on the macroscopic or classical viewpoint, it is helpful to keep in mind the microscopic or statistical perspective as well, as the relationship between the two helps us in understanding basic concepts such as energy.

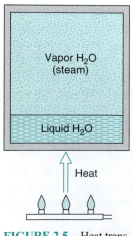

**FIGURE 2.5**  Heat transfer to water.

**a.** Make a control volume around the turbine in the steam power plant in Fig. 1.1 and list the flows of mass and energy that are there.

**b.** Take a control volume around your kitchen refrigerator and indicate where the components shown in Fig. 1.6 are located and show all flows of energy transfer.

## 2.7 SPECIFIC VOLUME AND DENSITY

The specific volume of a substance is defined as the volume per unit mass and is given the symbol $v$. The density of a substance is defined as the mass per unit volume, and it is therefore the reciprocal of the specific volume. Density is designated by the symbol $\rho$. Specific volume and density are intensive properties.

The specific volume of a system in a gravitational field may vary from point to point. For example, if the atmosphere is considered a system, the specific volume increases as the elevation increases. Therefore, the definition of specific volume involves the specific volume of a substance at a point in a system.

Consider a small volume $\delta V$ of a system, and let the mass be designated $\delta m$. The specific volume is defined by the relation

$$v = \lim_{\delta V \to \delta V'} \frac{\delta V}{\delta m}$$

where $\delta V'$ is the smallest volume for which the mass can be considered a continuum. Volumes smaller than this will lead to the recognition that mass is not evenly distributed in space but is concentrated in particles as molecules, atoms, electrons, etc. This is tentatively indicated in Fig. 2.6, where in the limit of a zero volume the specific volume may be infinite (the volume does not contain any mass) or very small (the volume is part of a nucleus).

Thus, in a given system, we should speak of the specific volume or density at a point in the system, and recognize that this may vary with elevation. However, most of the systems that we consider are relatively small, and the change in specific volume with elevation is not significant. Therefore, we can speak of one value of specific volume or density for the entire system.

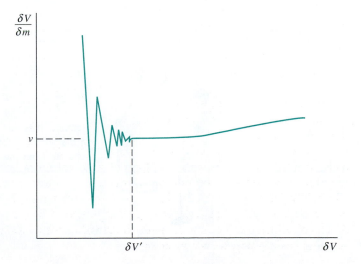

**FIGURE 2.6** The continuum limit for the specific volume.

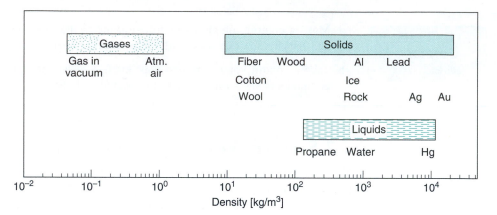

**FIGURE 2.7** Density of common substances.

In this text, the specific volume and density will be given either on a mass or on a mole basis. A bar over the symbol (lowercase) will be used to designate the property on a mole basis. Thus, $\bar{v}$ will designate molal specific volume and $\bar{\rho}$ will designate the molal density. In SI units, those for specific volume are $m^3/kg$ and $m^3/mol$ (or $m^3/kmol$); for density the corresponding units are $kg/m^3$ and $mol/m^3$ (or $kmol/m^3$).

Although the SI unit for volume is the cubic meter, a commonly used volume unit is the liter (L), which is a special name given to a volume of 0.001 cubic meters, that is, $1\ L = 10^{-3}\ m^3$. The general ranges of density for some common solids, liquids, and gases are shown in Fig. 2.7. Specific values for various solids, liquids and gases in SI units are listed in Tables A.3, A.4, and A.5, respectively, and in English units in Tables F.2, F.3, and F.4.

---

**EXAMPLE 2.2**

A 1-$m^3$ container, Fig. 2.8, is filled with 0.12 $m^3$ of granite, 0.15 $m^3$ of sand, 0.2 $m^3$ of liquid 25°C water, and the rest of the volume, 0.53 $m^3$, is air with a density of 1.15 $kg/m^3$. Find the overall (average) specific volume and density.

**Solution**

From the definition of specific volume and density we have:

$$v = V/m \quad \text{and} \quad \rho = m/V = 1/v$$

Remark: It is misleading to include air in the numbers for $\rho$ and $V$, as the air is separate from the rest of the mass.

**FIGURE 2.8** Sketch for Example 2.2.

We need to find the total mass taking density from Tables A.3 and A.4

$$m_{\text{granite}} = \rho V_{\text{granite}} = 2750 \text{ kg/m}^3 \times 0.12 \text{ m}^3 = 330 \text{ kg}$$

$$m_{\text{sand}} = \rho_{\text{sand}} V_{\text{sand}} = 1500 \text{ kg/m}^3 \times 0.15 \text{ m}^3 = 225 \text{ kg}$$

$$m_{\text{water}} = \rho_{\text{water}} V_{\text{water}} = 997 \text{ kg/m}^3 \times 0.2 \text{m}^3 = 199.4 \text{ kg}$$

$$m_{\text{air}} = \rho_{\text{air}} V_{\text{air}} = 1.15 \text{ kg/m}^3 \times 0.53 \text{ m}^3 = 0.61 \text{ kg}$$

Now the total mass becomes

$$m_{\text{tot}} = m_{\text{granite}} + m_{\text{sand}} + m_{\text{water}} + m_{\text{air}} = 755 \text{ kg}$$

and the specific volume and density can be calculated

$$v = V_{\text{tot}}/m_{\text{tot}} = 1 \text{ m}^3/755 \text{ kg} = 0.001\ 325 \text{ m}^3/\text{kg}$$

$$\rho = m_{\text{tot}}/V_{\text{tot}} = 755 \text{ kg}/1 \text{ m}^3 = 755 \text{ kg/m}^3$$

**In-Text Concept Questions**

**c.** Density of fibers, rock wool insulation, foams, and cotton is fairly low. Why is that?

**d.** Why do people float "high" in the water in the Dead Sea when they swim as compared with swimming in a fresh water lake?

**e.** Density of liquid water is $\rho = 1008 - T/2 \ [\text{kg/m}^3]$ with T in °C. If the temperature increases, what happens to the density and specific volume?

## 2.8 PRESSURE

When dealing with liquids and gases, we ordinarily speak of pressure; for solids we speak of stresses. The pressure in a fluid at rest at a given point is the same in all directions, and we define pressure as the normal component of force per unit area. More specifically, if $\delta A$ is a small area, $\delta A'$ the smallest area over which we can consider the fluid a continuum, and $\delta F_n$ the component of force normal to $\delta A$, we define pressure, $P$, as

$$P = \lim_{\delta A \to \delta A'} \frac{\delta F_n}{\delta A}$$

where the lower limit corresponds to sizes as mentioned for the specific volume, shown in Fig. 2.6. The pressure $P$ at a point in a fluid in equilibrium is the same in all directions. In a viscous fluid in motion, the variation in the state of stress with orientation becomes an important consideration. These considerations are beyond the scope of this book, and we will consider pressure only in terms of a fluid in equilibrium.

The unit for pressure in the International System is the force of one newton acting on a square meter area, which is called the pascal (Pa). That is,

$$1 \text{ Pa} = 1 \text{ N/m}^2$$

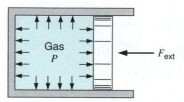

**FIGURE 2.9** The balance of forces on a movable boundary relates to inside gas pressure.

Two other units, not part of the International System, continue to be widely used. These are the bar, where

$$1 \text{ bar} = 10^5 \text{ Pa} = 0.1 \text{ MPa}$$

and the standard atmosphere, where

$$1 \text{ atm} = 101\ 325 \text{ Pa}$$

which is slightly larger than the bar. In this text, we will normally use the SI unit, the pascal, and especially the multiples of kilopascal and megapascal. The bar will be utilized often in the examples and problems, but the atmosphere will not be used, except in specifying certain reference points.

Consider a gas contained in a cylinder fitted with a movable piston, as shown in Fig. 2.9. The pressure exerted by the gas on all its boundaries is the same, assuming that the gas is in an equilibrium state. This pressure is fixed by the external force acting on the piston, since there must be a balance of forces for the piston to remain stationary. Thus, the product of the pressure and the movable piston area must be equal to the external force. If the external force is now changed, in either direction, the gas pressure inside must accordingly adjust, with appropriate movement of the piston, to establish a force balance at a new equilibrium state. As another example, if the gas in the cylinder is heated by an outside body, which tends to increase the gas pressure, the piston will move instead, such that the pressure remains equal to whatever value is required by the external force.

**EXAMPLE 2.3**

The hydraulic piston/cylinder system shown in Fig. 2.10 has a cylinder diameter of $D = 0.1$ m with a piston and rod mass of 25 kg. The rod has a diameter of 0.01 m with an outside atmospheric pressure of 101 kPa. The inside hydraulic fluid pressure is 250 kPa. How large a force can the rod push within the upward direction?

**Solution**

We will assume a static balance of forces on the piston (positive upward) so

$$F_{\text{net}} = ma = 0$$

$$= P_{\text{cyl}}A_{\text{cyl}} - P_0(A_{\text{cyl}} - A_{\text{rod}}) - F - m_p g$$

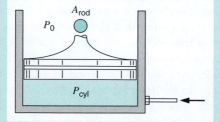

**FIGURE 2.10** Sketch for Example 2.3.

solve for $F$

$$F = P_{cyl}A_{cyl} - P_0(A_{cyl} - A_{rod}) - m_p g$$

The areas are:

$$A_{cyl} = \pi r^2 = \pi D^2/4 = \frac{\pi}{4} 0.1^2 \, m^2 = 0.007\ 854 \, m^2$$

$$A_{rod} = \pi r^2 = \pi D^2/4 = \frac{\pi}{4} 0.01^2 \, m^2 = 0.000\ 078\ 54 \, m^2$$

So the force becomes

$$F = [250 \times 0.007\ 854 - 101(0.007\ 854 - 0.000\ 078\ 54)] 1000 - 25 \times 9.81$$

$$= 1963.5 - 785.32 - 245.25$$

$$= 932.9 \, N$$

Note that we must convert kPa to Pa to get units of N.

In most thermodynamic investigations we are concerned with absolute pressure. Most pressure and vacuum gauges, however, read the difference between the absolute pressure and the atmospheric pressure existing at the gauge. This is referred to as gauge pressure. This is shown graphically in Fig. 2.11, and the following examples illustrate the principles. Pressures below atmospheric and slightly above atmospheric, and pressure differences (for example, across an orifice in a pipe), are frequently measured with a manometer, which contains water, mercury, alcohol, oil, or other fluids.

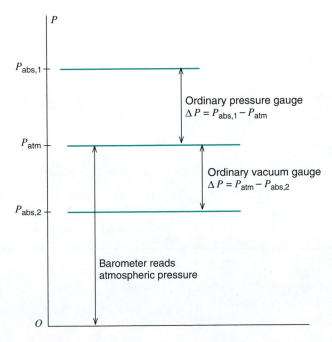

**FIGURE 2.11** Illustration of terms used in pressure measurement.

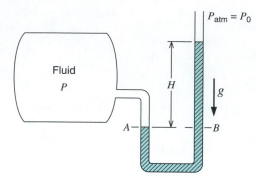

**FIGURE 2.12** Example of pressure measurement using a column of fluid.

Consider the column of fluid of height $H$ standing above point $B$ in the manometer shown in Fig. 2.12. The force acting downward at the bottom of the column is

$$P_0 A + mg = P_0 A + \rho A g H$$

where $m$ is the mass of the fluid column, $A$ is its cross-sectional area, and $\rho$ is its density. This force must be balanced by the upward force at the bottom of the column, which is $P_B A$. Therefore,

$$P_B - P_0 = \rho g H$$

Since points $A$ and $B$ are at the same elevation in columns of the same fluid, their pressures must be equal (the fluid being measured in the vessel has a much lower density, such that its pressure $P$ is equal to $P_A$). Overall,

$$\Delta P = P - P_0 = \rho g H \tag{2.2}$$

For distinguishing between absolute and gauge pressure in this text, the term pascal will always refer to absolute pressure. Any gauge pressure will be indicated as such.

---

**EXAMPLE 2.4**

A mercury (Hg) manometer is used to measure the pressure in a vessel as shown in Fig. 2.12. The mercury has a density of 13 590 kg/m³, and the height difference between the two columns is measured to be 24 cm. We want to determine the pressure inside the vessel.

**Solution**

The manometer measures the gauge pressure as a pressure difference. From Eq. 2.2,

$$\Delta P = P_{\text{gauge}} = \rho g H = 13\,590 \times 9.806\,65 \times 0.24$$

$$= 31\,985 \, \frac{\text{kg}}{\text{m}^3} \frac{\text{m}}{\text{s}^2} \text{m} = 31\,985 \text{ Pa} = 31.985 \text{ kPa}$$

$$= 0.316 \text{ atm}$$

To get the absolute pressure inside the vessel we have

$$P_A = P_{\text{vessel}} = P_B = \Delta P + P_{\text{atm}}$$

We need to know the atmospheric pressure measured by a barometer (absolute pressure). Assume this pressure is known as 750 mm Hg, being measured with a setup similar to the one in Fig. 2.12 with one side open to the atmosphere and the other side closed so there is mercury vapor with a very small pressure on top of the liquid column. The absolute pressure in the vessel becomes

$$P_{vessel} = \Delta P + P_{atm} = 31\ 985 + 13\ 590 \times 0.750 \times 9.806\ 65$$

$$= 31\ 985 + 99\ 954 = 131\ 940\ \text{Pa} = 1.302\ \text{atm}$$

**EXAMPLE 2.5**

What is the pressure at the bottom of the 7.5-m-tall storage tank of fluid at 25°C shown in Fig. 2.13? Assume the fluid is gasoline with atmospheric pressure 101 kPa on the top surface. Repeat the question for liquid refrigerant R-134a when the top surface pressure is 1 MPa.

**Solution**

The densities of the liquids are listed in Table A.4:

$$\rho_{gasoline} = 750\ \text{kg/m}^3; \qquad \rho_{R\text{-}134a} = 1206\ \text{kg/m}^3$$

The pressure difference due to the gravity is, from Eq. 2.2,

$$\Delta P = \rho g H$$

The total pressure is

$$P = P_{top} + \Delta P$$

For the gasoline we get

$$\Delta P = \rho g H = 750\ \text{kg/m}^3 \times 9.807\ \text{m/s}^2 \times 7.5\ \text{m} = 55\ 164\ \text{Pa}$$

Now convert all pressures to kPa

$$P = 101 + 55.164 = 156.2\ \text{kPa}$$

For the R-134a we get

$$\Delta P = \rho g H = 1206\ \text{kg/m}^3 \times 9.807\ \text{m/s}^2 \times 7.5\ \text{m} = 88\ 704\ \text{Pa}$$

Now convert all pressures to kPa

$$P = 1000 + 88.704 = 1089\ \text{kPa}$$

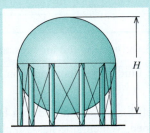

**FIGURE 2.13** Sketch for Example 2.5.

**FIGURE 2.14** Sketch for Example 2.6.

**EXAMPLE 2.6**

A piston/cylinder with cross-sectional area of 0.01 m$^2$ is connected with a hydraulic line to another piston cylinder of cross-sectional area of 0.05 m$^2$. Assume both chambers and the line are filled with hydraulic fluid of density 900 kg/m$^3$ and the larger second piston/cylinder is 6 m higher up in elevation. The telescope arm and the buckets have hydraulic piston/cylinders moving them, as seen in Fig. 2.14. With an outside atmospheric pressure of 100 kPa and a net force of 25 kN on the smallest piston, what is the balancing force on the second larger piston?

**Solution**

When the fluid is stagnant and at the same elevation we have the same pressure throughout the fluid. The force balance on the smaller piston is then related to the pressure (we neglect the rod area) as

$$F_1 + P_0 A_1 = P_1 A_1$$

from which the fluid pressure is

$$P_1 = P_0 + F_1/A_1 = 100 \text{ kPa} + 25 \text{ kN}/0.01 \text{ m}^2 = 2600 \text{ kPa}$$

The pressure at the higher elevation in piston/cylinder 2 is, from Eq. 2.2,

$$P_2 = P_1 - \rho g H = 2600 \text{ kPa} - 900 \text{ kg/m}^3 \times 9.81 \text{ m/s}^2 \times 6 \text{ m}/(1000 \text{ Pa/kPa})$$

$$= 2547 \text{ kPa}$$

where the second term is divided by 1000 to convert from Pa to kPa. Then the force balance on the second piston gives

$$F_2 + P_0 A_2 = P_2 A_2$$

$$F_2 = (P_2 - P_0)A_2 = (2547 - 100) \text{ kPa} \times 0.05 \text{ m}^2 = 122.4 \text{ kN}$$

**In-Text Concept Questions**

**f.** A car tire gauge indicates 195 kPa; what is the air pressure?

**g.** Can I always neglect $\Delta P$ in the fluid above location $A$ in Fig. 2.12? What circumstances does that depend on?

**h.** A U tube manometer has the left branch connected to a box with a pressure of 110 kPa and the right branch open. Which side has a higher column of fluid?

## 2.9 Equality of Temperature

Although temperature is a familiar property, defining it exactly is difficult. We are aware of "temperature" first of all as a sense of hotness or coldness when we touch an object. We also learn early that when a hot body and a cold body are brought into contact, the hot body becomes cooler and the cold body becomes warmer. If these bodies remain in contact for some time, they usually appear to have the same hotness or coldness. However, we also realize that our sense of hotness or coldness is very unreliable. Sometimes very cold bodies may seem hot, and bodies of different materials that are at the same temperature appear to be at different temperatures.

Because of these difficulties in defining temperature, we define equality of temperature. Consider two blocks of copper, one hot and the other cold, each of which is in contact with a mercury-in-glass thermometer. If these two blocks of copper are brought into thermal communication, we observe that the electrical resistance of the hot block decreases with time and that of the cold block increases with time. After a period of time has elapsed, however, no further changes in resistance are observed. Similarly, when the blocks are first brought in thermal communication, the length of a side of the hot block decreases with time, but the length of a side of the cold block increases with time. After a period of time, no further change in length of either of the blocks is perceived. In addition, the mercury column of the thermometer in the hot block drops at first and that in the cold block rises, but after a period of time no further changes in height are observed. We may say, therefore, that two bodies have equality of temperature if, when they are in thermal communication, no change in any observable property occurs.

## 2.10 The Zeroth Law of Thermodynamics

Now consider the same two blocks of copper and another thermometer. Let one block of copper be brought into contact with the thermometer until equality of temperature is established, and then remove it. Then let the second block of copper be brought into contact with the thermometer. Suppose that no change in the mercury level of the thermometer occurs during this operation with the second block. We then can say that both blocks are in thermal equilibrium with the given thermometer.

The zeroth law of thermodynamics states that when two bodies have equality of temperature with a third body, they in turn have equality of temperature with each other. This seems obvious to us because we are so familiar with this experiment. Because the principle is not derivable from other laws, and because it precedes the first and second laws of thermodynamics in the logical presentation of thermodynamics, it is called the zeroth law of thermodynamics. This law is really the basis of temperature measurement. Every time a body has equality of temperature with the thermometer, we can say that the body has the temperature we read on the thermometer. The problem remains of how to relate temperatures that we might read on different mercury thermometers or obtain from different temperature-measuring devices, such as thermocouples and resistance thermometers. This observation suggests the need for a standard scale for temperature measurements.

## 2.11 Temperature Scales

The scale used for measuring temperature in SI units is the Celsius scale, the symbol for which is °C. This was formerly called the centigrade scale, but it is now designated after Anders Celsius (1701–1744), the Swedish astronomer who devised this scale.

Until 1954, the Celsius scale was based on two fixed, easily duplicated points, the ice point and the steam point. The temperature of the ice point is defined as the temperature of a mixture of ice and liquid water that is in equilibrium with saturated air at a pressure of 1 atm (0.101 325 MPa). The temperature of the steam point is the temperature of liquid water and steam that are in equilibrium at a pressure of 1 atm. These two points are designated 0 and 100 on the Celsius scale.

At the tenth CGPM in 1954, the Celsius scale was redefined in terms of a single fixed point and the ideal-gas temperature scale. The single fixed point is the triple point of water (the state in which the solid, liquid, and vapor phases of water exist together in equilibrium). The magnitude of the degree is defined in terms of the ideal-gas temperature scale, which is discussed in Chapter 7. The essential features of this new scale are a single fixed point and a definition of the magnitude of the degree. The triple point of water is assigned the value of 0.01°C. On this scale the steam point is experimentally found to be 100.00°C. Thus, there is essential agreement between the old and new temperature scales.

We have not yet considered an absolute scale of temperature. The possibility of such a scale comes from the second law of thermodynamics and is discussed in Chapter 7. On the basis of the second law of thermodynamics, a temperature scale that is independent of any thermometric substance can be defined. This absolute scale is usually referred to as the thermodynamic scale of temperature.

The absolute scale related to the Celsius scale is the Kelvin scale (after William Thomson, 1824–1907, who was also known as Lord Kelvin), and is designated K (without the degree symbol). The relation between these scales is

$$K = °C + 273.15$$

In 1967, the CGPM defined the kelvin as 1/273.16 of the temperature at the triple point of water. The Celsius scale is now defined by this equation instead of by its earlier definition.

A number of empirically based temperature scales, to standardize temperature measurement and calibration, have been in use over the past seventy years. The most recent of these is the International Temperature Scale of 1990, or ITS-90. It is based on a number of fixed and easily reproducible points that are assigned definite numerical values of temperature, and on specified formulas relating temperature to the readings on certain temperature-measuring instruments for the purpose of interpolation between the defining fixed points. Details of the ITS-90 are not considered further in this text. It is noted that this scale is a practical means for establishing measurements that conform closely to the absolute thermodynamic temperature scale.

## SUMMARY

We introduce a thermodynamic system as a control volume, which for a fixed mass is a control mass. Such a system can be isolated, exchanging neither mass, momentum, or energy with its surroundings. A closed system versus an open system refers to the ability of mass exchange with the surroundings. If properties for a substance change, the state changes and a process occurs. When a substance has gone through several processes returning to the same initial state it has completed a cycle.

Basic units for thermodynamic and physical properties are mentioned and most are covered in Table A. 1. Thermodynamic properties such as density $\rho$, specific volume $v$, pressure $P$, and temperature $T$ are introduced together with units for these. Properties are classified as intensive, independent of mass (like $v$), or extensive, proportional to mass (like $V$). Students should already be familiar with other concepts from physics such as force $F$, velocity $\mathbf{V}$, and acceleration $a$. Application of Newton's law of motion leads to the variation of the static pressure in a column of fluid and the measurements of pressure (absolute and gauge) by barometers and manometers. The normal temperature scale and the absolute temperature scale are introduced.

You should have learned a number of skills and acquired abilities from studying this chapter that will allow you to

- Define (choose) a control volume (C.V.) around some matter; sketch the content and identify storage locations for mass; and identify mass and energy flows crossing the C.V. surface.
- Know properties $P$, $T$, $v$, and $\rho$ and their units.
- Know how to look up conversion of units in Table A.1.
- Know that energy is stored as kinetic, potential, or internal (in molecules).
- Know that energy can be transferred.
- Know the difference between $(v, \rho)$ and $(V, m)$ intensive versus extensive.
- Apply a force balance to a given system and relate it to pressure $P$.
- Know the difference between a relative (gauge) and absolute pressure $P$.
- Understand the working of a manometer or a barometer and get $\Delta P$ or $P$ from height $H$.
- Know the difference between a relative and absolute temperature $T$.
- Have an idea about magnitudes $(v, \rho, P, T)$.

Most of these concepts will be repeated and reinforced in the following chapters such as properties in Chapter 3, energy transfer as heat and work in Chapter 4, and internal energy in Chapter 5 together with their applications.

## KEY CONCEPTS AND FORMULAS

| | |
|---|---|
| Control volume | everything inside a control surface |
| Pressure definition | $P = \dfrac{F}{A}$ (mathematical limit for small $A$) |
| Specific volume | $v = \dfrac{V}{m}$ |
| Density | $\rho = \dfrac{m}{V}$ (Tables A.3, A.4, A.5, F.2, F.3, and F.4) |
| Static pressure variation | $\Delta P = \rho g H$ (depth $H$ in fluid of density $\rho$) |
| Absolute temperature | $T[\text{K}] = T[°\text{C}] + 273.15$ |
| Units | Table A. 1 |

### Concepts from Physics

| | |
|---|---|
| Newton's law of motion | $F = ma$ |
| Acceleration | $a = \dfrac{d^2x}{dt^2} = \dfrac{d\mathbf{V}}{dt}$ |
| Velocity | $\mathbf{V} = \dfrac{dx}{dt}$ |

# HOMEWORK PROBLEMS

## Concept-study guide problems

**2.1** Make a control volume around the whole power plant in Fig. 1.2 and with the help of Fig. 1.1 list what flows of mass and energy are in or out and any storage of energy. Make sure you know what is inside and what is outside your chosen C.V.

**2.2** Make a control volume that includes the steam flow around in the main turbine loop in the nuclear propulsion system in Fig. 1.3. Identify mass flows (hot or cold) and energy transfers that enter or leave the C.V.

**2.3** Separate the list $P$, $F$, $V$, $v$, $\rho$, $T$, $a$, $m$, $L$, $t$, and $\mathbf{V}$ into intensive, extensive, and non-properties.

**2.4** Water in nature exists in different phases such as solid, liquid, and vapor (gas). Indicate the relative magnitude of density and specific volume for the three phases.

**2.5** An electric dip heater is put into a cup of water and heats it from 20°C to 80°C. Show the energy flow(s) and storage and explain what changes.

**2.6** An escalator brings four people of total 300 kg, 25 m up in a building. Explain what happens with respect to energy transfer and stored energy.

**2.7** Is density a unique measure of mass distribution in a volume? Does it vary? If so, on what magnitude of scale (distance)?

**2.8** Can you carry 1 m³ of liquid water?

**2.9** The pressure at the bottom of a swimming pool is evenly distributed. Suppose we look at a cast iron plate of 7272 kg lying on the ground with an area of 100 m². What is the average pressure below that? Is it just as evenly distributed as the pressure at the bottom of the pool?

**2.10** I put four adjustable feet on a heavy cabinet. What feature of the feet will ensure that the cabinet does not make marks in the floor?

**2.11** A manometer with water shows a $\Delta P$ of $P_0/10$; what is the column height difference?

**2.12** Two divers swim at 20 m depth. One of them swims right in under a supertanker; the other stays away from the tanker. Who feels a greater pressure?

**2.13** A water skier does not sink too far down in the water if his speed is high enough. What makes that situation different from our static pressure calculations?

**2.14** What is the smallest temperature in degrees Celsius you can have? Rankine?

**2.15** Convert the formula for water density in In-Text Concept Question "e" (p. 25) to be for T in degrees Kelvin.

## Properties and units

**2.16** A steel cylinder of mass 2 kg contains 4 L of liquid water at 25°C at 200 kPa. Find the total mass and volume of the system. List two extensive and three intensive properties of the water.

**2.17** An apple "weighs" 80 g and has a volume of 100 cm³ in a refrigerator at 8°C. What is the apple density? List three intensive and two extensive properties of the apple.

**2.18** A pressurized steel bottle is charged with 5 kg of oxygen gas and 7 kg of nitrogen gas. How many kmoles are in the bottle?

## Force and energy

**2.19** The "standard" acceleration (at sea level and 45° latitude) due to gravity is 9.806 65 m/s². What is the force needed to hold a mass of 2 kg at rest in this gravitational field? How much mass can a force of 1 N support?

**2.20** When you move up from the surface of the earth, the gravitation is reduced as $g = 9.807 - 3.32 \times 10^{-6}\, z$, with $z$ as the elevation in meters. By how many percent is the weight of an airplane reduced when it cruises at 11 000 m?

**2.21** A car is driven at 60 km/h and is brought to a full stop with constant deceleration in 5 s. If the total car and driver mass is 1075 kg, find the necessary force.

**2.22** A car of mass 1775 kg travels with a velocity of 100 km/h. Find the kinetic energy. How high should the car be lifted in the standard gravitational field to have a potential energy that equals the kinetic energy?

**2.23** A steel plate of 950 kg accelerates from rest at 3 m/s² for a period of 10 s. What force is needed and what is the final velocity?

**2.24** A 15-kg steel container has 1.75 kmole of liquid propane inside. A force of 2 kN now accelerates this system. What is the acceleration?

## Specific volume

**2.25** A tank has two rooms separated by a membrane. Room A has 1 kg of air and a volume of 0.5 m³; room B has 0.75 m³ of air with density 0.8 kg/m³. The membrane is broken and the air comes to a uniform state. Find the final density of the air.

**2.26** A 1-m³ container is filled with 400 kg of granite stone, 200 kg of dry sand, and 0.2 m³ of liquid 25°C water. Use properties from Tables A.3 and A.4. Find the average specific volume and density of the masses when you exclude air mass and volume.

**2.27** A 15-kg steel gas tank holds 300 L of liquid gasoline, having a density of 800 kg/m³. If the system is decelerated at 6 m/s² what is the needed force?

**2.28** How much mass is there approximately in 1 L of mercury (Hg)? Atmospheric air?

## Pressure

**2.29** A hydraulic lift has a maximum fluid pressure of 500 kPa. What should the piston/cylinder diameter be so it can lift a mass of 850 kg?

**2.30** A piston/cylinder with a cross-sectional area of 0.01 m² has a piston mass of 100 kg resting on the stops, as shown in Fig. P2.30. With an outside atmospheric pressure of 100 kPa, what should the water pressure be to lift the piston?

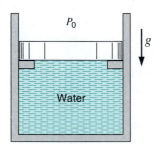

**FIGURE P2.30**

**2.31** A cannnonball of 5 kg acts as a piston in a cylinder of 0.15 m diameter. As the gunpowder is burned, a pressure of 7 MPa is created in the gas behind the ball. What is the acceleration of the ball if the cylinder (cannon) is pointing horizontally?

**2.32** A vertical hydraulic cylinder has a 125-mm-diameter piston with hydraulic fluid inside the cylinder and an ambient pressure of 1 bar. Assuming standard gravity, find the piston mass that will create a pressure inside of 1500 kPa.

**2.33** A valve in the cylinder shown in Fig. P2.33 has a cross-sectional area of 11 cm² with a pressure of 735 kPa inside the cylinder and 99 kPa outside. How large a force is needed to open the valve?

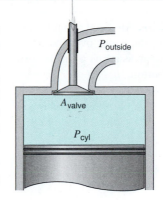

**FIGURE P2.33**

**2.34** You dive 5 m down in the ocean. What is the absolute pressure there?

**2.35** A large exhaust fan in a laboratory room keeps the pressure inside at 10 cm of water relative vacuum to the hallway. What is the net force on the door measuring 1.9 m by 1.1 m?

**2.36** The hydraulic lift in an auto-repair shop has a cylinder diameter of 0.2 m. To what pressure should the hydraulic fluid be pumped to lift 40 kg of piston/arms and 700 kg of a car?

**2.37** A 2.5-m-tall steel cylinder has a cross-sectional area of 1.5 m². At the bottom with a height of 0.5 m is liquid water on top of which is a 1-m-high layer of gasoline. This is shown in Fig. P2.37. The gasoline surface is exposed to atmospheric air at 101 kPa. What is the highest pressure in the water?

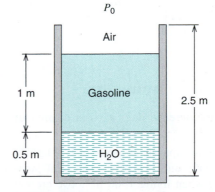

**FIGURE P2.37**

**2.38** At the beach, atmospheric pressure is 1025 mbar. You dive 15 m down in the ocean and you later climb a hill up to 250 m in elevation. Assume the density of water is about 1000 kg/m³ and the density of air is 1.18 kg/m³. What pressure do you feel at each place?

**2.39** Liquid water with density ρ is filled on top of a thin piston in a cylinder with cross-sectional area A and total height $H$, as shown in Fig. P2.39. Air is let in under the piston so it pushes up, spilling the water over the edge. Deduce the formula for the air pressure as a function of the piston elevation from the bottom, $h$.

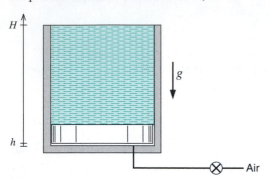

**FIGURE P2.39**

**2.40** A tornado rips off a 100-m² roof with a mass of 1000 kg. What is the minimum vacuum pressure needed to do that if we neglect the anchoring forces?

**2.41** A steel tank of cross-sectional area 3 m² and 16 m tall weighs 10 000 kg and is open at the top as shown in Fig. P2.41. We want to float it in the ocean so it sticks 10 m straight down by pouring concrete into the bottom of it. How much concrete should we put in?

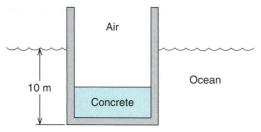

**FIGURE P2.41**

# Manometers and barometers

**2.42** What pressure difference does a 10-m column of atmospheric air show?

**2.43** The density of atmospheric air is about 1.15 kg/m³, which we assume is constant. How large an absolute pressure will a pilot see when flying 1500 m above ground level where the pressure is 101 kPa?

**2.44** A manometer shows a pressure difference of 1 m of liquid mercury. Find $\Delta P$ in kPa.

**2.45** Blue manometer fluid of density 925 kg/m³ shows a column height difference of 3 cm vacuum with one end attached to a pipe and the other open to $P_0 = 101$ kPa. What is the absolute pressure in the pipe?

**2.46** The pressure gauge on an air tank shows 75 kPa when the diver is 10 m down in the ocean. At what depth will the gauge pressure be zero? What does that mean?

**2.47** The absolute pressure in a tank is 85 kPa and the local ambient absolute pressure is 97 kPa. If a U-tube with mercury (density = 13 550 kg/m³) is attached to the tank to measure the vacuum, what column height difference would it show?

**2.48** The difference in height between the columns of a manometer is 200 mm with a fluid of density 900 kg/m³. What is the pressure difference? What is the height difference if the same pressure difference is measured using mercury (density = 13 600 kg/m³) as manometer fluid?

**2.49** A barometer to measure absolute pressure shows a mercury column height of 725 mm. The temperature is such that the density of the mercury is 13 550 kg/m³. Find the ambient pressure.

**2.50** An absolute pressure gauge attached to a steel cylinder shows 135 kPa. We want to attach a manometer using liquid water a day that $P_{atm} = 101$ kPa. How high a fluid level difference must we plan for?

**2.51** A differential pressure gauge mounted on a vessel shows 1.25 MPa and a local barometer gives atmospheric pressure as 0.96 bar. Find the absolute pressure inside the vessel.

**2.52** A submarine maintains 101 kPa inside it and dives 240 m down in the ocean having an average density of 1030 kg/m³. What is the pressure difference between the inside and the outside of the submarine hull?

**2.53** A barometer measures 760 mm Hg at street level and 735 mm Hg on top of a building. How tall is the building if we assume air density of 1.15 kg/m³?

**2.54** Assume we use a pressure gauge to measure the air pressure at street level and at the roof of a tall building. If the pressure difference can be determined with an accuracy of 1 mbar (0.001 bar), what uncertainty in the height estimate does that correspond to?

**2.55** A pipe flowing light oil has a manometer attached as shown in Fig. P2.55. What is the absolute pressure in the pipe flow?

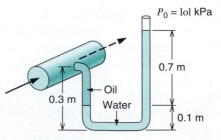

**FIGURE P2.55**

**2.56** A U-tube manometer filled with water (density = 1000 kg/m³) shows a height difference of 25 cm. What is the gauge pressure? If the right branch is tilted to make an angle of 30° with the horizontal, as shown in Fig. P2.56, what should the length of the column in the tilted tube be relative to the U-tube?

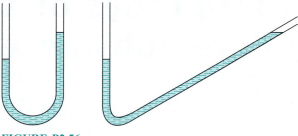

**FIGURE P2.56**

**2.57** In the city water tower, water is pumped up to a level 25 m above ground in a pressurized tank with air at 125 kPa over the water surface. This is illustrated in Fig. P2.57. Assuming the water density is 1000 kg/m³ and standard gravity, find the pressure required to pump more water in at ground level.

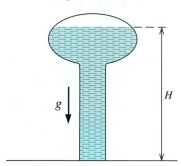

**FIGURE P2.57**

# Temperature

**2.58** What is a temperature of −5°C in degrees Kelvin?

**2.59** Density of liquid water is $\rho = 1008 - T/2 \ [\text{kg/m}^3]$ with $T$ in °C. If the temperature increases 10°C, how much deeper does a 1-m layer of water become?

**2.60** The density of mercury changes approximately linearly with temperature as

$$\rho_{\text{Hg}} = 13\ 595 - 2.5\ T \, \text{kg/m}^3 \quad (T \text{ in Celsius})$$

so the same pressure difference will result in a manometer reading that is influenced by temperature. If a pressure difference of 100 kPa is measured in the summer at 35°C and in the winter at −15°C, what is the difference in column height between the two measurements?

**2.61** A mercury thermometer measures temperature by measuring the volume expansion of a fixed mass of liquid Hg due to a change in the density (see Problem 2.60). Find the relative change (%) in volume for a change in temperature from 10°C to 20°C.

**2.62** The atmosphere becomes colder at higher elevation. As an average, the standard atmospheric absolute temperature can be expressed as $T_{\text{atm}} = 288 - 6.5 \times 10^{-3} z$, where $z$ is the elevation in meters. How cold is it outside an airplane cruising at 12 000 m expressed in Kelvin and in Celsius?

# Chapter 3

# Properties of a Pure Substance

In the previous chapter we considered three familiar properties of a substance—specific volume, pressure, and temperature. We now turn our attention to pure substances and consider some of the phases in which a pure substance may exist, the number of independent properties a pure substance may have, and methods of presenting thermodynamic properties.

Properties and the behavior of substances are very important for our studies of devices and thermodynamic systems. The steam power plant in Fig. 1.1 and the nuclear propulsion system in Fig. 1.3 have very similar processes, using water as the working substance. Water vapor (steam) is made by boiling at high pressure in the steam generator followed by an expansion in the turbine to a lower pressure, a cooling in the condenser, and a return to the boiler by a pump that raises the pressure. We must know the water properties to properly size the equipment such as the burners or heat exchangers, turbine, and pump for the desired transfer of energy and the flow of water. As the water is brought from liquid to vapor we need to know the temperature for the given pressure, and we must know the density or specific volume so that the piping can be properly dimensioned for the flow. If the pipes are too small, the expansion creates excessive velocities, leading to pressure losses and increased friction, and thus demanding a larger pump and reducing the turbine work output.

Another example is a refrigerator, shown in Fig. 1.6, where we need a substance that will boil from liquid to vapor at a low temperature, say −20°C. This absorbs energy from the cold space, keeping it cold. Inside the black grille in the back or at the bottom, the now hot substance is cooled by air flowing around the grille, so it condenses from vapor to liquid at a temperature slightly higher than room temperature. When such a system is designed, we need to know the pressures at which these processes take place and the amount of energy, covered in Chapter 5, that is involved. We also need to know how much volume the substance occupies, the specific volume, so that the piping diameters can be selected as mentioned for the steam power plant. The substance is selected so that the pressure is reasonable during these processes; it should not be too high, due to leakage and safety concerns, and not too low either, as air might leak into the system.

A final example of a situation where we need to know the substance properties is the gas turbine and a variation thereof, namely the jet engine shown in Fig. 1.11. In these systems, the working substance is a gas (very similar to air) and no phase change takes place. A combustion process burns fuel and air, freeing a large amount of energy, which heats the gas so that it expands. We need to know how hot the gas gets and how much the expansion is so that we can analyze the expansion process in the turbine and the exit nozzle of the jet engine. In this device, we do need large velocities inside the turbine section and for the exit of the jet engine. This high-velocity flow pushes on the blades in the turbine to create shaft work or pushes on the jet engine (something called thrust) to move the aircraft forward.

These are just a few examples of complete thermodynamic systems where a substance goes through several processes involving changes of its thermodynamic state and therefore its properties. As your studies progress, many other examples will be used to illustrate the general subjects.

## 3.1 THE PURE SUBSTANCE

A pure substance is one that has a homogeneous and invariable chemical composition. It may exist in more than one phase, but the chemical composition is the same in all phases. Thus, liquid water, a mixture of liquid water and water vapor (steam), and a mixture of ice and liquid water are all pure substances; every phase has the same chemical composition. In contrast, a mixture of liquid air and gaseous air is not a pure substance because the composition of the liquid phase is different from that of the vapor phase.

Sometimes a mixture of gases, such as air, is considered a pure substance as long as there is no change of phase. Strictly speaking, this is not true. As we will see later, we should say that a mixture of gases such as air exhibits some of the characteristics of a pure substance as long as there is no change of phase.

In this text the emphasis will be on simple compressible substances. This term designates substances whose surface effects, magnetic effects, and electrical effects are insignificant when dealing with the substances. But changes in volume, such as those associated with the expansion of a gas in a cylinder, are very important. Reference will be made, however, to other substances for which surface, magnetic, and electrical effects are important. We will refer to a system consisting of a simple compressible substance as a simple compressible system.

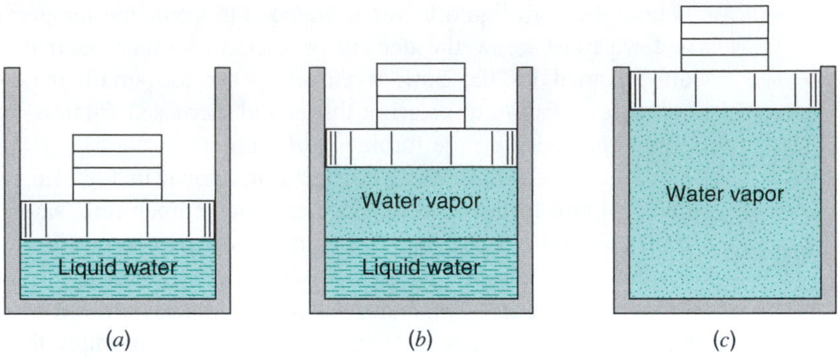

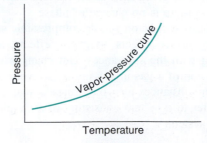

**FIGURE 3.1** Constant-pressure change from liquid to vapor phase for a pure substance.

Water vapor

Liquid water

Water vapor

Liquid water

Water vapor

(a)          (b)          (c)

## 3.2 Vapor–Liquid–Solid-Phase Equilibrium in a Pure Substance

Consider as a system 1 kg of water contained in the piston/cylinder arrangement shown in Fig. 3.1a. Suppose that the piston and weight maintain a pressure of 0.1 MPa in the cylinder and that the initial temperature is 20°C. As heat is transferred to the water, the temperature increases appreciably, the specific volume increases slightly, and the pressure remains constant. When the temperature reaches 99.6°C, additional heat transfer results in a change of phase, as indicated in Fig. 3.1b. That is, some of the liquid becomes vapor, and during this process both the temperature and pressure remain constant, but the specific volume increases considerably. When the last drop of liquid has vaporized, further transfer of heat results in an increase in both temperature and specific volume of the vapor, as shown in Fig. 3.1c.

The term saturation temperature designates the temperature at which vaporization takes place at a given pressure. This pressure is called the saturation pressure for the given temperature. Thus, for water at 99.6°C the saturation pressure is 0.1 MPa, and for water at 0.1 MPa the saturation temperature is 99.6°C. For a pure substance there is a definite relation between saturation pressure and saturation temperature. A typical curve, called the vapor-pressure curve, is shown in Fig. 3.2.

If a substance exists as liquid at the saturation temperature and pressure, it is called saturated liquid. If the temperature of the liquid is lower than the saturation temperature for the existing pressure, it is called either a subcooled liquid (implying that the temperature is lower than the saturation temperature for the given pressure) or a compressed liquid (implying that the pressure is greater than the saturation pressure for the given temperature). Either term may be used, but the latter term will be used in this text.

When a substance exists as part liquid and part vapor at the saturation temperature, its quality is defined as the ratio of the mass of vapor to the total mass. Thus, in Fig. 3.1b, if the mass of the vapor is 0.2 kg and the mass of the liquid is 0.8 kg, the

Pressure

Vapor-pressure curve

Temperature

**FIGURE 3.2** Vapor-pressure curve of a pure substance.

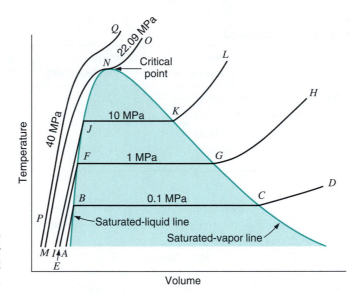

**FIGURE 3.3** Temperature–volume diagram for water showing liquid and vapor phases (not to scale).

quality is 0.2 or 20%. The quality may be considered an intensive property and has the symbol $x$. Quality has meaning only when the substance is in a saturated state, that is, at saturation pressure and temperature.

If a substance exists as vapor at the saturation temperature, it is called saturated vapor. (Sometimes the term dry saturated vapor is used to emphasize that the quality is 100%.) When the vapor is at a temperature greater than the saturation temperature, it is said to exist as superheated vapor. The pressure and temperature of superheated vapor are independent properties, since the temperature may increase while the pressure remains constant. Actually, the substances we call gases are highly superheated vapors.

Consider Fig. 3.1 again. Let us plot on the temperature-volume diagram of Fig. 3.3 the constant-pressure line that represents the states through which the water passes as it is heated from the initial state of 0.1 MPa and 20°C. Let state $A$ represent the initial state, $B$ the saturated-liquid state (99.6°C), and line $AB$ the process in which the liquid is heated from the initial temperature to the saturation temperature. Point $C$ is the saturated-vapor state, and line $BC$ is the constant-temperature process in which the change of phase from liquid to vapor occurs. Line $CD$ represents the process in which the steam is superheated at constant pressure. Temperature and volume both increase during this process.

Now let the process take place at a constant pressure of 1 MPa, starting from an initial temperature of 20°C. Point $E$ represents the initial state, in which the specific volume is slightly less than that at 0.1 MPa and 20°C. Vaporization begins at point $F$, where the temperature is 179.9°C. Point $G$ is the saturated-vapor state, and line $GH$ is the constant-pressure process in which the steam in superheated.

In a similar manner, a constant pressure of 10 MPa is represented by line $IJKL$, for which the saturation temperature is 311.1°C.

At a pressure of 22.09 MPa, represented by line $MNO$, we find, however, that there is no constant-temperature vaporization process. Instead, point $N$ is a point of inflection with a zero slope. This point is called the critical point. At the critical point the saturated-liquid and saturated-vapor states are identical. The temperature, pressure, and specific volume at the critical point are called the critical temperature, critical pressure, and critical volume. The critical-point data for some substances are given in Table 3.1. More extensive data are given in Table A.2 in the appendix.

TABLE 3.1
*Some Critical-Point Data*

|  | Critical Temperature, °C | Critical Pressure, MPa | Critical Volume, m³/kg |
|---|---|---|---|
| Water | 374.14 | 22.09 | 0.003 155 |
| Carbon dioxide | 31.05 | 7.39 | 0.002 143 |
| Oxygen | −118.35 | 5.08 | 0.002 438 |
| Hydrogen | −239.85 | 1.30 | 0.032 192 |

A constant-pressure process at a pressure greater than the critical pressure is represented by line *PQ*. If water at 40 MPa and 20°C is heated in a constant-pressure process in a cylinder as shown in Fig. 3.1, there will never be two phases present and the state shown in Fig. 3.1*b* will never exist. Instead, there will be a continuous change in density and at all times there will be only one phase present. The question then is when do we have a liquid and when do we have a vapor? The answer is that this is not a valid question at supercritical pressures. We simply term the substance a fluid. However, rather arbitrarily, at temperatures below the critical temperature we usually refer to it as a compressed liquid and at temperatures above the critical temperature as a superheated vapor. It should be emphasized, however, that at pressures above the critical pressure we never have a liquid and vapor phase of a pure substance existing in equilibrium.

In Fig. 3.3, line *NJFB* represents the saturated-liquid line and line *NKGC* represents the saturated-vapor line.

By convention, the subscript *f* is used to designate a property of a saturated liquid and the subscript *g* a property of a saturated vapor (subscript *g* being used to denote saturation temperature and pressure). Thus, a saturation condition involving part liquid and part vapor such as in Fig. 3.1*b* can be shown on *T–v* coordinates as in Fig. 3.4. All of the liquid present is at state *f* with specific volume $v_f$ and all of the vapor present is at state *g* with $v_g$. The total volume is the sum of the liquid volume and the vapor volume, or

$$V = V_{\text{liq}} + V_{\text{vap}} = m_{\text{liq}}v_f + m_{\text{vap}}v_g$$

The average specific volume of the system *v* is then

$$v = \frac{V}{m} = \frac{m_{\text{liq}}}{m}v_f + \frac{m_{\text{vap}}}{m}v_g = (1 - x)v_f + xv_g \qquad (3.1)$$

in terms of the definition of quality $x = m_{\text{vap}}/m$.

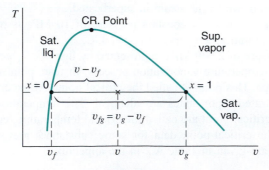

FIGURE 3.4  *T–v* diagram for the two-phase, liquid–vapor region to show the quality specific volume relation.

Using the definition

$$v_{fg} = v_g - v_f$$

Equation 3.1 can also be written as

$$v = v_f + xv_{fg} \tag{3.2}$$

Now, the quality $x$ can be viewed as the fraction $(v - v_f)/v_{fg}$ of the distance between saturated liquid and saturated vapor, as indicated in Fig. 3.4.

Let us now consider another experiment with the piston/cylinder arrangement. Suppose that the cylinder contains 1 kg of ice at $-20°C$, 100 kPa. When heat is transferred to the ice, the pressure remains constant, the specific volume increases slightly, and the temperature increases until it reaches $0°C$, at which point the ice melts and the temperature remains constant. In this state the ice is called a saturated solid. For most substances the specific volume increases during this melting process, but for water the specific volume of the liquid is less than the specific volume of the solid. When all the ice has melted, a further heat transfer causes an increase in temperature of the liquid.

If the initial pressure of the ice at $-20°C$ is 0.260 kPa, heat transfer to the ice results in an increase in temperature to $-10°C$. At this point, however, the ice passes directly from the solid phase to the vapor phase in the process known as sublimation. Further heat transfer results in superheating of the vapor.

Finally, consider an initial pressure of the ice of 0.6113 kPa and a temperature of $-20°C$. Through heat transfer let the temperature increase until it reaches $0.01°C$. At this point, however, further heat transfer may cause some of the ice to become vapor and some to become liquid, for at this point it is possible to have the three phases in equilibrium. This point is called the triple point, which is defined as the state in which all three phases may be present in equilibrium. The pressure and temperature at the triple point for a number of substances are given in Table 3.2.

This whole matter is best summarized by the diagram of Fig. 3.5, which shows how the solid, liquid, and vapor phases may exist together in equilibrium. Along the sublimation line the solid and vapor phases are in equilibrium, along the fusion line the solid and liquid phases are in equilibrium, and along the vaporization line the liquid and vapor phases are in equilibrium. The only point at which all three phases may exist in equilibrium is the triple point. The vaporization line ends at the critical point because there is no distinct change from the liquid phase to the vapor phase above the critical point.

**TABLE 3.2**

*Some Solid–Liquid–Vapor Triple-Point Data*

|  | Temperature, °C | Pressure, kPa |
|---|---|---|
| Hydrogen (normal) | −259 | 7.194 |
| Oxygen | −219 | 0.15 |
| Nitrogen | −210 | 12.53 |
| Carbon dioxide | −56.4 | 520.8 |
| Mercury | −39 | 0.000 000 13 |
| Water | 0.01 | 0.6113 |
| Zinc | 419 | 5.066 |
| Silver | 961 | 0.01 |
| Copper | 1083 | 0.000 079 |

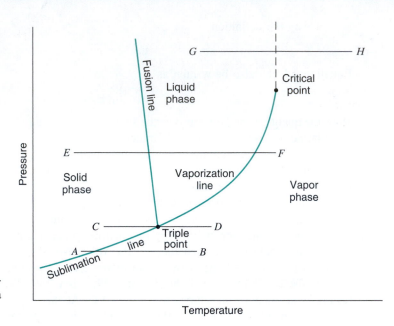

**FIGURE 3.5** Pressure-temperature diagram for a substance such as water.

Consider a solid in state *A*, as shown in Fig. 3.5. When the temperature increases but the pressure (which is less than the triple-point pressure) is constant, the substance passes directly from the solid to the vapor phase. Along the constant-pressure line *EF*, the substance passes from the solid to the liquid phase at one temperature, and then from the liquid to the vapor phase at a higher temperature. Constant-pressure line *CD* passes through the triple point, and it is only at the triple point that the three phases may exist together in equilibrium. At a pressure above the critical pressure, such as *GH*, there is no sharp distinction between the liquid and vapor phases.

Although we have made these comments with rather specific reference to water (only because of our familiarity with water), all pure substances exhibit the same general behavior. However, the triple-point temperature and critical temperature vary greatly from one substance to another. For example, the critical temperature of helium, as given in Table A.2, is 5.3 K. Therefore, the absolute temperature of helium at ambient conditions is over 50 times greater than the critical temperature. In contrast, water has a critical temperature of 374.14°C (647.29 K), and at ambient conditions the temperature of water is less than half the critical temperature. Most metals have a much higher critical temperature than water. When we consider the behavior of a substance in a given state, it is often helpful to think of this state in relation to the critical state or triple point. For example, if the pressure is greater than the critical pressure, it is impossible to have a liquid phase and a vapor phase in equilibrium. Or, to consider another example, the states at which vacuum-melting a given metal is possible can be ascertained by a consideration of the properties at the triple point. Iron at a pressure just above 5 Pa (the triple-point pressure) would melt at a temperature of about 1535°C (the triple-point temperature).

Figure 3.6 shows the three-phase diagram for carbon dioxide, in which it is seen (see also Table 3.2) that the triple-point pressure is greater than normal atmospheric pressure, which is very unusual. Therefore, the commonly observed phase transition under conditions of atmospheric pressure of about 100 kPa is a sublimation from solid directly to vapor, without passing through a liquid phase, which is why solid carbon dioxide is commonly referred to as dry ice. We note from Fig. 3.6 that this phase transformation at 100 kPa occurs at a temperature below 200 K.

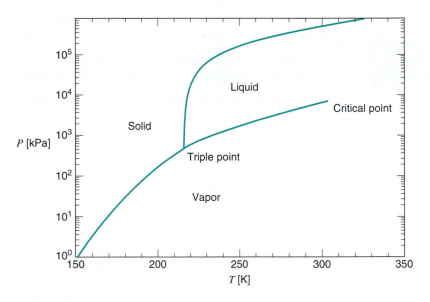

**FIGURE 3.6** Carbon dioxide phase diagram.

Finally, it should be pointed out that a pure substance can exist in a number of different solid phases. A transition from one solid phase to another is called an allotropic transformation. Figure 3.7 shows a number of solid phases for water. A pure substance can have a number of triple points, but only one triple point has a solid, liquid, and vapor equilibrium. Other triple points for a pure substance can have two solid phases and a liquid phase, two solid phases and a vapor phase, or three solid phases.

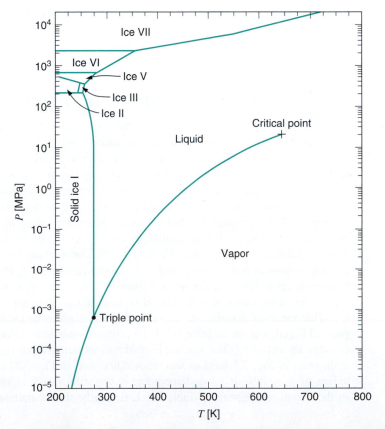

**FIGURE 3.7** Water phase diagram.

**a.** If the pressure is smaller than $P_{sat}$ at a given $T$, what phase do I have?

**b.** An external water tap has the valve activated by a long spindle so the closing mechanism is located well inside the wall. Why is that?

**c.** What is the lowest temperature (approximately) at which water can be liquid?

## 3.3 Independent Properties of a Pure Substance

One important reason for introducing the concept of a pure substance is that the state of a simple compressible pure substance (that is, a pure substance in the absence of motion, gravity, and surface, magnetic, or electrical effects) is defined by two independent properties. For example, if the specific volume and temperature of superheated steam are specified, the state of the steam is determined.

To understand the significance of the term independent property, consider the saturated-liquid and saturated-vapor states of a pure substance. These two states have the same pressure and the same temperature, but they are definitely not the same state. In a saturation state, therefore, pressure and temperature are not independent properties. Two independent properties such as pressure and specific volume or pressure and quality are required to specify a saturation state of a pure substance.

The reason for mentioning previously that a mixture of gases, such as air, has the same characteristics as a pure substance as long as only one phase is present, concerns precisely this point. The state of air, which is a mixture of gases of definite composition, is determined by specifying two properties as long as it remains in the gaseous phase. Air then can be treated as a pure substance.

## 3.4 Tables of Thermodynamic Properties

Tables of thermodynamic properties of many substances are available, and in general, all these tables have the same form. In this section we will refer to the steam tables. The steam tables are selected both because they are a vehicle for presenting thermodynamic tables and because steam is used extensively in power plants and industrial processes. Once the steam tables are understood, other thermodynamic tables can be readily used.

Several different versions of steam tables have been published over the years. The set included in Appendix B, Table B.1, is a summary based on a complicated fit to the behavior of water. It is very similar to the *Steam Tables* by Keenan, Keyes, Hill, and Moore, published in 1969 and 1978. We will concentrate here on the three properties already discussed in Chapter 2 and in Section 3.2, namely $T$, $P$, and $v$, and note that the other properties listed in the set of Tables B.1, $u$, $h$, and $s$, will be introduced later.

The steam tables in Appendix B consist of five separate tables, as indicated in Fig. 3.8. The region of superheated vapor in Fig. 3.5 is given in Table B.1.3, and that of compressed liquid is given in Table B.1.4. The compressed-solid region shown in Fig. 3.5 is not listed in the appendix. The saturated-liquid and saturated-vapor region, as seen in the $T$ and $v$ diagram of Fig. 3.3 (and as the vaporization line in Fig. 3.5), is listed according to the values of $T$ in Table B.1.1 and according to the values of $P$ ($T$ and $P$ are not independent in the two-phase regions) in Table B.1.2. Similarly, the saturated-solid and saturated-vapor

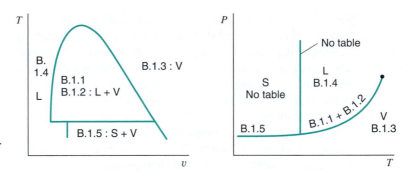

**FIGURE 3.8** Listing of the steam tables.

region is listed according to $T$ in Table B.1.5, but the saturated-solid and saturated-liquid region, the third phase boundary line shown in Fig. 3.5 is not listed in the appendix.

In Table B.1.1, the first column after the temperature gives the corresponding saturation pressure in kilopascals. The next three columns give specific volume in cubic meters per kilogram. The first of these columns gives the specific volume of the saturated liquid, $v_f$; the third column gives the specific volume of the saturated vapor $v_g$; and the second column gives the difference between the two, $v_{fg}$, as defined in Section 3.2. Table B.1.2 lists the same information as Table B.1.1, but the data are listed according to pressure, as mentioned earlier.

As an example, let us calculate the specific volume of saturated steam at 200°C having a quality of 70%. Using Eq. 3.1 gives

$$v = 0.3(0.001\ 156) + 0.7(0.127\ 36)$$

$$= 0.0895\ \text{m}^3/\text{kg}$$

Table B.1.3 gives the properties of superheated vapor. In the superheated region, pressure and temperature are independent properties; therefore, for each pressure a large number of temperatures are given, and for each temperature four thermodynamic properties are listed, the first one being specific volume. Thus, the specific volume of steam at a pressure of 0.5 MPa and 200°C is 0.4249 m³/kg.

Table B.1.4 gives the properties of the compressed liquid. To demonstrate the use of this table, consider a piston and a cylinder (as shown in Fig. 3.9) that contains 1 kg of saturated-liquid water at 100°C. Its properties are given in Table B.1.1, and we note that the pressure is 0.1013 MPa and the specific volume is 0.001 044 m³/kg. Suppose the pressure is increased to 10 MPa while the temperature is held constant at 100°C by the necessary transfer of heat, $Q$. Since water is slightly compressible, we would expect a slight decrease in specific volume during this process. Table B.1.4 gives this specific

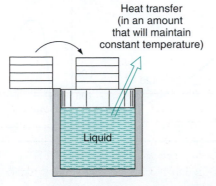

**FIGURE 3.9** Illustration of compressed-liquid state.

volume as 0.001 039 m³/kg. This is only a slight decrease, and only a small error would be made if one assumed that the volume of a compressed liquid is equal to the specific volume of the saturated liquid at the same temperature. In many situations this is the most convenient procedure, particularly when compressed-liquid data are not available. It is very important to note, however, that the specific volume of saturated liquid at the given pressure, 10 MPa, does not give a good approximation. This value, from Table B.1.2, at a temperature of 311.1°C, is 0.001 452 m³/kg, which is in error by almost 40%.

Table B.1.5 of the steam tables gives the properties of saturated solid and saturated vapor that are in equilibrium. The first column gives the temperature, and the second column gives the corresponding saturation pressure. As would be expected, all these pressures are less than the triple-point pressure. The next two columns give the specific volume of the saturated solid and saturated vapor.

Appendix B also includes thermodynamic tables for several other substances; refrigerant fluids ammonia, R-12, R-22, and R-134a and the cryogenic fluids nitrogen and methane. In each case, only two tables are given—saturated liquid–vapor listed by temperature (equivalent to Table B.1.1 for water), and superheated vapor (equivalent to Table B.1.3).

Let us now consider a number of examples to illustrate the use of thermodynamic tables for water and also the other substances listed in Appendix B.

**Example 3.1**

Determine the phase for each of the following water states using the Appendix B tables and indicate the relative position in the $P–v$, $T–v$, and $P–T$ diagrams.

**a.** 120°C, 500 kPa

**b.** 120°C, 0.5 m³/kg

**Solution**

**a.** Enter Table B.1.1 with 120°C. The saturation pressure is 198.5 kPa, so we have a compressed liquid, point $a$ in Fig. 3.10. That is above the saturation line for 120°C. We could also have entered Table B.1.2 with 500 kPa and found the saturation temperature as 151.86°C, so we would say it is subcooled liquid. That is to the left of the saturation line for 500 kPa as seen in the $P–T$ diagram.

**b.** Enter Table B.1.1 with 120°C and notice

$$v_f = 0.00106 < v < v_g = 0.89186 \text{ m}^3/\text{kg}$$

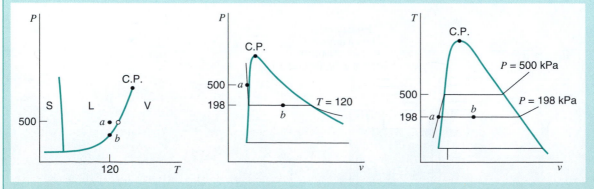

**FIGURE 3.10**  Diagram for Example 3.1.

so the state is a two phase mixture of liquid and vapor, point *b* in Fig. 3.10. The state is to the left of the saturated vapor state and to the right of the saturated liquid state both seen in the *T–v* diagram.

**EXAMPLE 3.2**   Determine the phase for each of the following states using the Appendix B tables and indicate the relative position in the *P–v*, *T–v* and *P–T* diagrams, as in Figs. 3.11 and 3.12.

  **a.** Ammonia 30°C, 1000 kPa
  **b.** R-22 200 kPa, 0.15 m³/kg

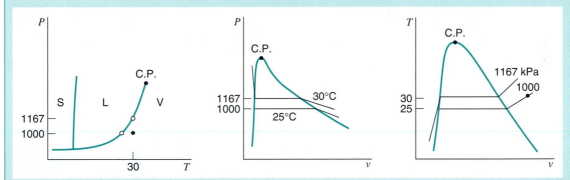

**FIGURE 3.11**   Diagram for Example 3.2*a*.

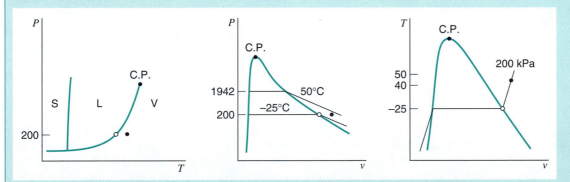

**FIGURE 3.12**   Diagram for Example 3.2*b*.

**Solution**

  **a.** Enter Table B.2.1 with 30°C. The saturation pressure is 1167 kPa. As we have a lower *P*, it is a superheated vapor state. We could also have entered with 1000 kPa and found a saturation temperature of slightly less than 25°C, so we have a state that is superheated about 5°C.

  **b.** Enter Table B.4.1 with 200 kPa and notice

$$v > v_g \approx 0.1119 \text{ m}^3/\text{kg}$$

so from the *P–v* diagram the state is superheated vapor. We can find the state in Table B.4.2 between 40 and 50°C.

**EXAMPLE 3.3**

Determine the temperature and quality (if defined) for water at a pressure of 300 kPa and at each of these specific volumes:

**a.** 0.5 m³/kg

**b.** 1.0 m³/kg

**Solution**

For each state, it is necessary to determine what phase or phases are present, in order to know which table is the appropriate one to find the desired state information. That is, we must compare the given information with the appropriate phase boundary values. Consider a $T$–$v$ diagram (or a $P$–$v$ diagram) such as in Fig. 3.8. For the constant-pressure line of 300 kPa shown in Fig. 3.13, the values for $v_f$ and $v_g$ shown there are found from the saturation table, Table B.1.2.

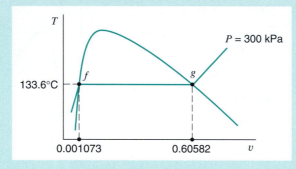

**FIGURE 3.13** A $T$–$v$ diagram for water at 300 kPa.

**a.** By comparison with the values in Fig. 3.13, the state at which $v$ is 0.5 m³/kg is seen to be in the liquid–vapor two-phase region, at which $T = 133.6°$C, and the quality $x$ is found from Eq. 3.2 as

$$0.5 = 0.001\ 073 + x\ 0.604\ 75, \qquad x = 0.825$$

Note that if we did not have Table B.1.2 (as would be the case with the other substances listed in Appendix B), we could have interpolated in Table B.1.1 between the 130°C and 135°C entries to get the $v_f$ and $v_g$ values for 300 kPa.

**b.** By comparison with the values in Fig. 3.13, the state at which $v$ is 1.0 m³/kg is seen to be in the superheated vapor region, in which quality is undefined, and the temperature for which is found from Table B.1.3. In this case, $T$ is found by linear interpolation between the 300 kPa specific-volume values at 300°C and 400°C, as shown in Fig. 3.14. This is an approximation for $T$, since the actual relation along the 300 kPa constant-pressure line is not exactly linear.

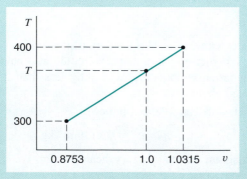

**FIGURE 3.14** $T$ and $v$ values for superheated vapor water at 300 kPa.

From the figure we have

$$\text{slope} = \frac{T - 300}{1.0 - 0.8753} = \frac{400 - 300}{1.0315 - 0.8753}$$

solving this gives $T = 379.8°C$.

---

**EXAMPLE 3.4**

A closed vessel contains 0.1 m$^3$ of saturated liquid and 0.9 m$^3$ of saturated vapor R-134a in equilibrium at 30°C. Determine the percent vapor on a mass basis.

**Solution**

Values of the saturation properties for R-134a are found from Table B.5.1. The mass–volume relations then give

$$V_{\text{liq}} = m_{\text{liq}}v_f, \qquad m_{\text{liq}} = \frac{0.1}{0.000\ 843} = 118.6 \text{ kg}$$

$$V_{\text{vap}} = m_{\text{vap}}v_g, \qquad m_{\text{vap}} = \frac{0.9}{0.026\ 71} = 33.7 \text{ kg}$$

$$m = 152.3 \text{ kg}$$

$$x = \frac{m_{\text{vap}}}{m} = \frac{33.7}{152.3} = 0.221$$

That is, the vessel contains 90% vapor by volume but only 22.1% vapor by mass.

---

**EXAMPLE 3.5**

A rigid vessel contains saturated ammonia vapor at 20°C. Heat is transferred to the system until the temperature reaches 40°C. What is the final pressure?

**Solution**

Since the volume does not change during this process, the specific volume also remains constant. From the ammonia tables, Table B.2.1, we have

$$v_1 = v_2 = 0.149\ 22 \text{ m}^3/\text{kg}$$

Since $v_g$ at 40°C is less than 0.149 22 m$^3$/kg, it is evident that in the final state the ammonia is superheated vapor. By interpolating between the 800- and 1000-kPa columns of Table B.2.2, we find that

$$P_2 = 945 \text{ kPa}$$

---

**EXAMPLE 3.6**

Determine the missing property of $P$–$v$–$T$ and $x$ if applicable for the following states.

**a.** Nitrogen: $-53.2°C$, 600 kPa
**b.** Nitrogen: 100 K, 0.008 m$^3$/kg

**Solution**

For nitrogen the properties are listed in Table B.6 with temperature in Kelvin.

**a.** Enter in Table B.6.1 with $T = 273.2 - 53.2 = 220$ K, which is higher than the critical $T$ in the last entry. Then proceed to the superheated vapor tables. We would also have realized this by looking at the critical properties in Table A.2. From Table B.6.2 in the subsection for 600 kPa ($T_{sat} = 96.37$ K)

$$v = 0.10788 \text{ m}^3/\text{kg}$$

shown as point $a$ in Fig. 3.15.

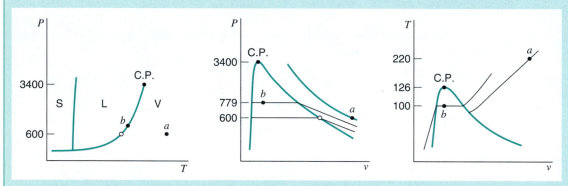

**FIGURE 3.15** Diagram for Example 3.6.

**b.** Enter in Table B.6.1 with $T = 100$ K, and we see

$$v_f = 0.001\ 452 < v < v_g = 0.0312 \text{ m}^3/\text{kg}$$

so we have a two-phase state with a pressure as the saturation pressure, shown as $b$ in Fig. 3.15

$$P_{sat} = 779.2 \text{ kPa}$$

and the quality from Eq. 3.2 becomes

$$x = (v - v_f)/v_{fg} = (0.008 - 0.001\ 452)/0.029\ 75 = 0.2201$$

---

**EXAMPLE 3.7**

Determine the pressure for water at 200°C with $v = 0.4$ m³/kg.

**Solution**

Start in Table B.1.1 with 200°C and note that $v > v_g = 0.127\ 36$ m³/kg so we have superheated vapor. Proceed to Table B.1.3 at any subsection with 200°C; say we start at 200 kPa. There the $v = 1.080\ 34$, which is too large so the pressure must be higher. For 500 kPa, $v = 0.424\ 92$, and for 600 kPa, $v = 0.352\ 02$, so it is bracketed. This is shown in Fig. 3.16.

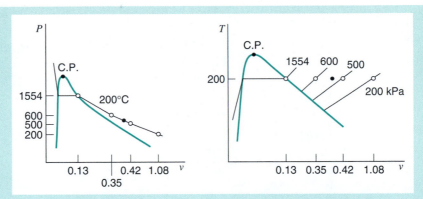

**FIGURE 3.16** Diagram for Example 3.7.

A linear interpolation, Fig. 3.17, between the two pressures is done to get $P$ at the desired $v$.

$$P = 500 + (600 - 500)\frac{0.4 - 0.424\,92}{0.352\,02 - 0.424\,92} = 534.2 \text{ kPa}$$

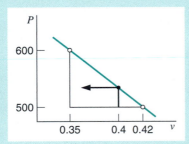

The real constant-$T$ curve is slightly curved and not linear, but for manual interpolation we assume a linear variation.

**FIGURE 3.17** Linear interpolation for Example 3.7.

---

**In-Text Concept Questions**

**d.** Some tools should be cleaned in liquid water at 150°C. How high a $P$ is needed?

**e.** Water at 200 kPa has a quality of 50%. Is the volume fraction $V_g/V_{tot} < 50\%$ or $> 50\%$?

**f.** Why are most of the compressed liquid or solid regions not included in the printed tables?

**g.** Why is it not typical to find tables for Ar, He, Ne or air like a B-section table?

**h.** What is the percent change in volume as liquid water freezes? Mention some effects the volume change can have in nature and for our households.

## 3.5 THERMODYNAMIC SURFACES

The matter discussed to this point can be well summarized by a consideration of a pressure-specific volume–temperature surface. Two such surfaces are shown in Figs. 3.18 and 3.19. Fig. 3.18 shows a substance such as water in which the specific volume increases during freezing. Fig. 3.19 shows a substance in which the specific volume decreases during freezing.

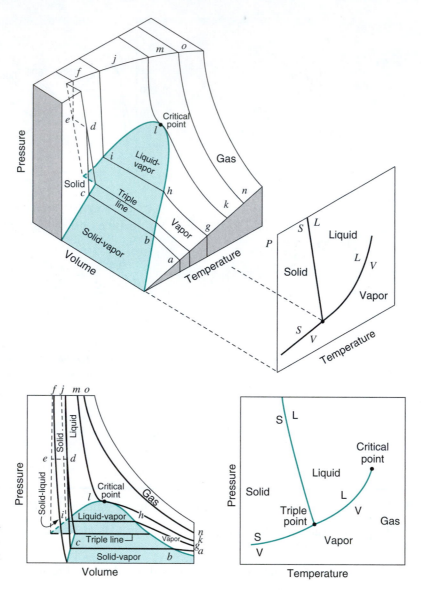

**FIGURE 3.18** Pressure–volume–temperature surface for a substance that expands on freezing.

In these diagrams the pressure, specific volume, and temperature are plotted on mutually perpendicular coordinates, and each possible equilibrium state is thus represented by a point on the surface. This follows directly from the fact that a pure substance has only two independent intensive properties. All points along a quasi-equilibrium process lie on the $P–v–T$ surface, since such a process always passes through equilibrium states.

The regions of the surface that represent a single phase—the solid, liquid, and vapor phases—are indicated. These surfaces are curved. The two-phase regions—the solid–liquid, solid–vapor, and liquid–vapor regions—are ruled surfaces. By this we understand that they are made up of straight lines parallel to the specific-volume axis. This, of course, follows from the fact that in the two-phase region, lines of constant pressure are also lines of constant temperature, although the specific volume may change. The triple point actually appears as the triple line on the $P–v–T$ surface, since the pressure and temperature

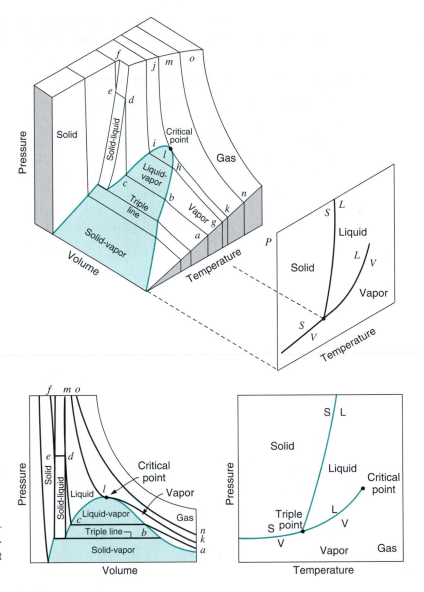

**FIGURE 3.19** Pressure–volume–temperature surface for a substance that contracts on freezing.

of the triple point are fixed, but the specific volume may vary, depending on the proportion of each phase.

It is also of interest to note the pressure–temperature and pressure–volume projections of these surfaces. We have already considered the pressure–temperature diagram for a substance such as water. It is on this diagram that we observe the triple point. Various lines of constant temperature are shown on the pressure–volume diagram, and the corresponding constant-temperature sections are lettered identically on the $P$–$v$–$T$ surface. The critical isotherm has a point of inflection at the critical point.

One notices that for a substance such as water, which expands on freezing, the freezing temperature decreases with an increase in pressure. For a substance that contracts on freezing, the freezing temperature increases as the pressure increases. Thus, as the pressure of vapor is increased along the constant-temperature line *abcdef* in Fig. 3.18, a substance that expands on freezing first becomes solid and then liquid. For the substance that contracts on freezing, the corresponding constant-temperature line (Fig. 3.19) indicates that as the pressure on the vapor is increased, it first becomes liquid and then solid.

## 3.6 THE $P$–$V$–$T$ BEHAVIOR OF LOW- AND MODERATE-DENSITY GASES

One form of energy possession by a system discussed in Section 2.6 was intermolecular (IM) potential energy, that associated with the forces between molecules. It was stated there that at very low densities the average distances between molecules is so large that the IM potential energy may effectively be neglected. In such a case, the particles would be independent of one another, a situation referred to as an ideal gas. Under this approximation, it has been observed experimentally that, to a close degree, a very low density gas behaves according to the ideal gas equation of state

$$PV = n\bar{R}T, \qquad P\bar{v} = \bar{R}T \tag{3.3}$$

in which $n$ is the number of kmol of gas, or

$$n = \frac{m}{M} = \frac{\text{kg}}{\text{kg/kmol}} \tag{3.4}$$

In Eq. 3.3, $\bar{R}$ is the universal gas constant, the value of which is, for any gas,

$$\bar{R} = 8.3145 \frac{\text{kN m}}{\text{kmol K}} = 8.3145 \frac{\text{kJ}}{\text{kmol K}}$$

and $T$ is the absolute (ideal gas scale) temperature in kelvins (i.e., $T(\text{K}) = T(°\text{C}) + 273.15$). It is important to note that $T$ must always be the absolute temperature whenever it is being used to multiply or divide in an equation. The ideal gas absolute temperature scale will be discussed in more detail in Chapter 7.

Substituting Eq. 3.4 into Eq. 3.3 and rearranging, we find that the ideal gas equation of state can be written conveniently in the form

$$PV = mRT, \qquad Pv = RT \tag{3.5}$$

where

$$R = \frac{\bar{R}}{M} \tag{3.6}$$

in which $R$ is a different constant for each particular gas. The value of $R$ for a number of substances is given in Table A.5 of Appendix A.

---

**EXAMPLE 3.8**

What is the mass of air contained in a room 6 m × 10 m × 4 m if the pressure is 100 kPa and the temperature is 25°C?

**Solution**

Assume air to be an ideal gas. By using Eq. 3.5 and the value of $R$ from Table A.5, we have

$$m = \frac{PV}{RT} = \frac{100 \text{ kN/m}^2 \times 240 \text{ m}^3}{0.287 \text{ kN m/kg K} \times 298.2 \text{ K}} = 280.5 \text{ kg}$$

**EXAMPLE 3.9**

A tank has a volume of 0.5 m$^3$ and contains 10 kg of an ideal gas having a molecular weight of 24. The temperature is 25°C. What is the pressure?

**Solution**

The gas constant is determined first:

$$R = \frac{\overline{R}}{M} = \frac{8.3145 \text{ kN m/kmol K}}{24 \text{ kg/kmol}}$$

$$= 0.346\,44 \text{ kN m/kg K}$$

We now solve for *P*:

$$P = \frac{mRT}{V} = \frac{10 \text{ kg} \times 0.346\,44 \text{ kN m/kg K} \times 298.2 \text{ K}}{0.5 \text{ m}^3}$$

$$= 2066 \text{ kPa}$$

**EXAMPLE 3.10**

A gas-bell is submerged in liquid water with its mass counterbalanced with rope and pulleys as shown in Fig. 3.20. The pressure inside is measured carefully to be 105 kPa, and the temperature is 21°C. A volume increase is measured to be 0.75 m$^3$ over a period of 185 s. What is the volume flow rate and the mass flow rate of the flow into the bell assuming it is carbon dioxide gas?

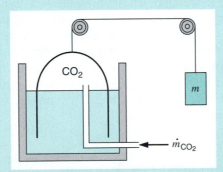

$CO_2$

$m$

$\dot{m}_{CO_2}$

**FIGURE 3.20**  Sketch for Example 3.10.

**Solution**

The volume flow rate is

$$\dot{V} = \frac{dV}{dt} = \frac{\Delta V}{\Delta t} = \frac{0.75}{185} = 0.040\,54 \text{ m}^3/\text{s}$$

and the mass flow rate is  $\dot{m} = p\dot{V} = \dot{V}/v$. At close to room conditions the carbon dioxide is an ideal gas, so $PV = mRT$ or $v = RT/P$, and from Table A.5 we have the ideal gas constant $R = 0.1889$ kJ/kg K. The mass flow rate becomes

$$\dot{m} = \frac{P\dot{V}}{RT} = \frac{105 \times 0.040\,54}{0.1889\,(273.15 + 21)} \frac{\text{kPa m}^3/\text{s}}{\text{kJ/kg}} = 0.0766 \text{ kg/s}$$

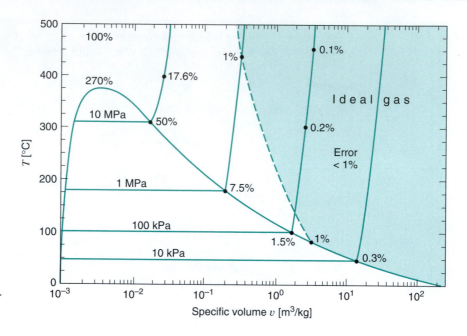

**FIGURE 3.21**
Temperature-specific volume diagram for water.

Because of its simplicity, the ideal-gas equation of state is very convenient to use in thermodynamic calculations. However, two questions are now appropriate. The ideal-gas equation of state is a good approximation at low density. But what constitutes low density? Or, expressed in other words, over what range of density will the ideal-gas equation of state hold with accuracy? The second question is, how much does an actual gas at a given pressure and temperature deviate from ideal-gas behavior?

One specific example in response to these questions is shown in Fig. 3.21, a $T$–$v$ diagram for water that indicates the error in assuming ideal gas for saturated vapor and for superheated vapor. As would be expected, at very low pressure or high temperature the error is small, but this becomes severe as the density increases. The same general trend would be the case in referring to Fig. 3.18 or 3.19. As the state becomes further removed from the saturation region (i.e., high $T$ or low $P$), the gas behavior becomes closer to the ideal-gas model.

## 3.7 THE COMPRESSIBILITY FACTOR

A more quantitative study of the question of the ideal-gas approximation can be conducted by introducing the compressibility factor $Z$, defined as

$$Z = \frac{Pv}{RT}$$

or

$$Pv = ZRT \tag{3.7}$$

Note that for an ideal gas $Z = 1$, and the deviation of $Z$ from unity is a measure of the deviation of the actual relation from the ideal-gas equation of state.

Fig. 3.22 shows a skeleton compressibility chart for nitrogen. From this chart we make three observations. The first is that at all temperatures $Z \to 1$ as $P \to 0$. That is,

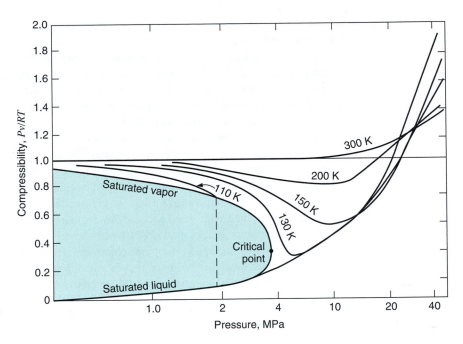

**FIGURE 3.22** Compressibility of nitrogen.

as the pressure approaches zero, the $P–v–T$ behavior closely approaches that predicted by the ideal-gas equation of state. Note also that at temperatures of 300 K and above (that is, room temperature and above) the compressibility factor is near unity up to pressure of about 10 MPa. This means that the ideal-gas equation of state can be used for nitrogen (and, as it happens, air) over this range with considerable accuracy.

We further note that at lower temperatures or at very high pressures, the compressibility factor deviates significantly from the ideal-gas value. Moderate-density forces of attraction tend to pull molecules together, resulting in a value of $Z < 1$, whereas very high density forces of repulsion tend to have the opposite effect.

If we examine compressibility diagrams for other pure substances, we find that the diagrams are all similar in the characteristics described above for nitrogen, at least in a qualitative sense. Quantitatively the diagrams are all different, since the critical temperatures and pressures of different substances vary over wide ranges, as evidenced from the values listed in Table A.2. Is there a way in which we can put all of these substances on a common basis? To do so, we "reduce" the properties with respect to the values at the critical point. The reduced properties are defined as

$$\text{reduced pressure} = P_r = \frac{P}{P_c}, \qquad P_c = \text{critical pressure}$$

$$\text{reduced temperature} = T_r = \frac{T}{T_c}, \qquad T_c = \text{critical temperature} \qquad (3.8)$$

These equations state that the reduced property for a given state is the value of this property in this state divided by the value of this same property at the critical point.

If lines of constant $T_r$ are plotted on a $Z$ versus $P_r$ diagram, a plot such as that in Fig. D.1 is obtained. The striking fact is that when such $Z$ versus $P_r$ diagrams are prepared for a number of different substances, all of them very nearly coincide, especially when the substances have simple, essentially spherical molecules. Correlations for substances with more complicated molecules are reasonably close, except near or at saturation or at high density. Thus, Fig. D.1 is actually a generalized diagram for simple molecules, which means that it represents the average behavior for a number of different simple substances. When such a diagram is used for a particular substance, the results will generally be somewhat in error. However, if $P–v–T$ information is required for a substance in a region where no experimental measurements have been made, this generalized compressibility diagram will give reasonably accurate results. We need know only the critical pressure and critical temperature to use this basic generalized chart.

In our study of thermodynamics, we will use Fig. D.1 primarily to help us decide whether, in a given circumstance, it is reasonable to assume ideal-gas behavior as a model. For example, we note from the chart that if the pressure is very low (that is, $\ll P_c$), the ideal-gas model can be assumed with good accuracy, regardless of the temperature. Furthermore, at high temperatures (that is, greater than about twice $T_c$), the ideal-gas model can be assumed with good accuracy to pressures as high as four or five times $P_c$. When the temperature is less than about twice the critical temperature and the pressure is not extremely low, we are in a region, commonly termed superheated vapor, in which the deviation from ideal-gas behavior may be considerable. In this region it is preferable to use tables of thermodynamic properties or charts for a particular substance, as discussed in Section 3.4.

---

**EXAMPLE 3.11**

Is it reasonable to assume ideal-gas behavior at each of the given states?

**a.** Nitrogen at 20°C, 1.0 MPa

**b.** Carbon dioxide at 20°C, 1.0 MPa

**c.** Ammonia at 20°C, 1.0 MPa

**Solution**

In each case it is first necessary to check phase boundary and critical state data.

**a.** For nitrogen, the critical properties are, from Table A.2, 126.2 K, 3.39 MPa. Since the given temperature, 293.2 K is more than twice $T_c$ and the reduced pressure is less than 0.3, ideal gas behavior is a very good assumption.

**b.** For carbon dioxide, the critical properties are 304.1 K, 7.38 MPa. Therefore, the reduced properties are 0.96 and 0.136. From Appendix Fig. D.1, $CO_2$ is a gas (although $T < T_c$) with a $Z$ of about 0.95, so the ideal-gas model is accurate to within about 5% in this case.

**c.** The ammonia tables, Appendix B.2, give the most accurate information. From Table B.2.1 at 20°C, $P_g = 858$ kPa. Since the given pressure of 1 MPa is greater than $P_g$, this state is a compressed liquid, and not a gas.

---

**EXAMPLE 3.12**

Determine the specific volume for R-134a at 100°C, 3.0 MPa, for the following models:

**a.** The R-134a tables, Table B.5

**b.** Ideal gas

**c.** The generalized chart, Fig. D.1

**Solution**

**a.** From Table B.5.2 at 100°C, 3 MPa,

$$v = 0.006\ 65\ \text{m}^3/\text{kg} \quad \text{(most accurate value)}$$

**b.** Assuming ideal gas, we have

$$R = \frac{\bar{R}}{M} = \frac{8.3145}{102.03} = 0.081\ 49\ \frac{\text{kJ}}{\text{kg K}}$$

$$v = \frac{RT}{P} = \frac{0.081\ 49 \times 373.2}{3000} = 0.010\ 14\ \text{m}^3/\text{kg}$$

which is more than 50% too large.

**c.** Using the generalized chart, Fig. D.1, we obtain

$$T_r = \frac{373.2}{374.2} = 1.0, \qquad P_r = \frac{3}{4.06} = 0.74, \qquad Z = 0.67$$

$$v = Z \times \frac{RT}{P} = 0.67 \times 0.010\ 14 = 0.006\ 79\ \text{m}^3/\text{kg}$$

which is only 2% too large.

---

**EXAMPLE 3.13**

Propane in a steel bottle of volume 0.1 m³ has a quality of 10% at a temperature of 15°C. Use the generalized compressibility chart to estimate the total propane mass and to find the pressure.

**Solution**

To use the generalized chart we need the reduced pressure and temperature. From Table A.2 for propane, $P_c = 4250$ kPa and $T_c = 369.8$ K. The reduced temperature is, from Eq. 3.8,

$$T_r = \frac{T}{T_c} = \frac{273.15 + 15}{369.8} = 0.779\ 2 = 0.78$$

From Figure D.1, shown in Fig. 3.23, we can read for the saturated states

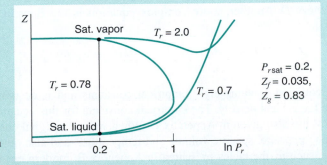

**FIGURE 3.23** Diagram for Example 3.13.

For the two-phase state the pressure is the saturated pressure

$$P = P_{r\,sat} \times P_c = 0.2 \times 4250 \text{ kPa} = 850 \text{ kPa}$$

The overall compressibility factor becomes, as Eq. 3.1 for $v$

$$Z = (1 - x)Z_f + xZ_g = 0.9 \times 0.035 + 0.1 \times 0.83 = 0.1145$$

The gas constant from Table A.5 is $R = 0.1886$ kJ/kg K, so the gas law is Eq. 3.7.

$$PV = mZRT$$

$$m = \frac{PV}{ZRT} = \frac{850 \times 0.1}{0.1145 \times 0.1886 \times 288.15} \frac{\text{kPa m}^3}{\text{kJ/kg}} = 13.66 \text{ kg}$$

---

**In-Text Concept Questions**

  **i.** How accurate is it to assume methane is an ideal gas at room conditions?

  **j.** I want to determine a state of some substance and I know $P = 200$ kPa; does it help to write $PV = mRT$ to find the second property?

  **k.** A bottle at 298 K should have liquid propane; how high a pressure is needed? (Use Fig. D.1.)

---

## 3.8 EQUATIONS OF STATE

Instead of the ideal-gas model to represent gas behavior, or even the generalized compressibility chart, which is approximate, it is desirable to have an equation of state that accurately represents the $P$–$v$–$T$ behavior for a particular gas over the entire superheated vapor region. Such an equation is necessarily more complicated and consequently more difficult to use. Many such equations have been proposed and used to correlate the observed behavior of gases. As an example, consider the class of relatively simple equations known as cubic equations of state:

$$P = \frac{RT}{v - b} - \frac{a}{v^2 + cbv + db^2} \tag{3.9}$$

in terms of the four parameters $a$, $b$, $c$ and $d$. (Note that if all four are zero, this reduces to the ideal gas model.) Several other different models in this class are given in Appendix D. Another, more complex equation is the Benedict-Webb-Rubin equation of state:

$$P = \frac{RT}{v} + \frac{RTB_0 - A_0 - C_0/T^2}{v^2} + \frac{RTb - a}{v^3} + \frac{a\alpha}{v^6} + \frac{c}{v^3 T^2}\left(1 + \frac{\gamma}{v^2}\right)e^{-\gamma/v^2} \tag{3.10}$$

This equation contains eight empirical constants and is accurate to densities of about twice the critical density. The empirical constants for this equation for a number of substances are also given in Appendix D. When we use a digital computer to determine and tabulate pressure, specific volume and temperature, and also other thermodynamic properties, such as the tables presented in Appendix B, modern equations of state are much more complicated, often containing 40 or more empirical constants.

## 3.9 COMPUTERIZED TABLES

Most of the tables in the appendix are supplied in a computer program on the CD accompanying this book. The main program operates with a visual interface in the Windows environment on a PC-type computer and is generally self-explanatory.

The main program covers the full set of tables for water, refrigerants, and cryogenic fluids, as in Tables B.1 to B.7 including the compressed liquid region, which is only printed for water. For these substances a small graph with the $P-v$ diagram shows the region around the critical point down toward the triple line covering the compressed liquid, two-phase liquid–vapor, dense fluid, and superheated vapor regions. As a state is selected and the properties computed, a thin crosshair set of lines indicates the state in the diagram so this can be seen with a visual impression of the state's location.

Ideal gases are covered corresponding to the Tables A.7 for air and A.8 or A.9 for other ideal gases. You are able to select the substance and the units to work in for all the various table sections giving a wider choice than the printed tables. Metric units (SI) or standard English units for the properties can be used as well as a mass basis (kg or lbm) or a mole basis, satisfying the need for the most common applications.

The generalized chart, Fig. D.1, with the compressibility factor, is included to allow a more accurate value of $Z$ to be obtained than can be read from the graph. This is particularly useful for the case of a two-phase mixture where the saturated liquid and saturated vapor values are needed. Besides the compressibility factor, this part of the program includes correction terms beyond ideal-gas approximations for changes in the other thermodynamic properties.

The only mixture application that is included with the program is moist air.

---

**EXAMPLE 3.14**

Find the states in Examples 3.1 and 3.2 with the computer-aided thermodynamics tables, CATT, and list the missing property of $P-v-T$ and $x$ if applicable.

**Solution**

Water states from Example 3.1: Click Water, click Calculator and then select Case 1 $(T, P)$. Input $(T, P) = (120, 0.5)$. The result is as shown in Fig. 3.24.

$\Rightarrow$ Compressed liquid $\qquad v = 0.0106$ m$^3$/kg (same as in Table B.1.4)

Click Calculator and then select Case 2 $(T, v)$. Input $(T, v) = (120, 0.5)$

$\Rightarrow$ Two-phase $\qquad x = 0.5601$, $P = 198.5$ kPa

Ammonia state from Example 3.2: Click Cryogenics; check that it is ammonia. Otherwise select Ammonia, click Calculator, and then select Case 1 $(T, P)$. Input $(T, P) = (30, 1)$

$\Rightarrow$ Superheated vapor $\qquad v = 0.1321$ m$^3$/kg (same as in Table B.2.2)

R-22 state from Example 3.2: Click Refrigerants; check that it is R-22. Otherwise select R-22 (Alt-R) click Calculator and then select Case 5 $(P, v)$. Input $(P, v) = (0.2, 0.15)$

$\Rightarrow$ Superheated vapor $\qquad T = 46.26°C$

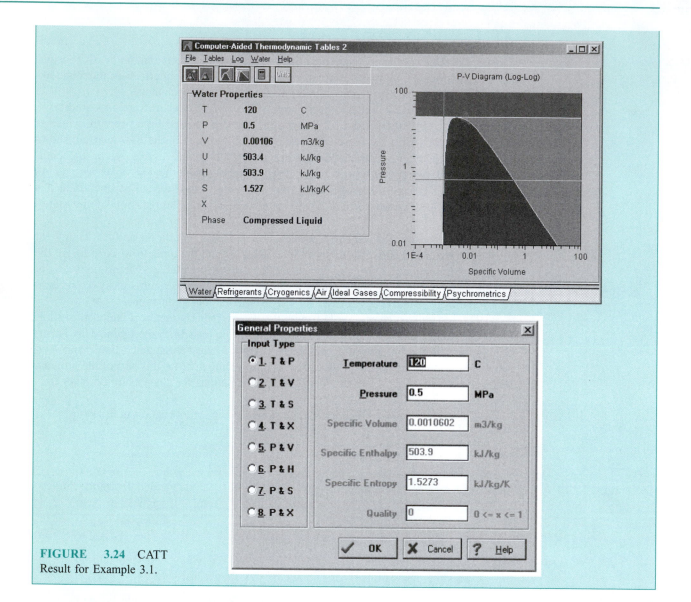

**FIGURE 3.24** CATT Result for Example 3.1.

---

**In-Text Concept Question**

**I.** A bottle at 298 K should have liquid propane; how high a pressure is needed? (Use the software.)

---

## 3.10 HOW-TO SECTION

### How do I determine the phase for a given state?

You must have a phase diagram like Figs. 3.6–3.7 or a B-section table for the given substance. The figures can be used if the state is specified with a pair (*T*, *P*); otherwise, you will need the printed tables. Sometimes the critical point Table A.2 or the triple point Table 3.2 may give you enough information.

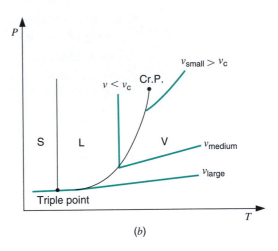

**FIGURE 3.25**

On these diagrams or the printed table you have ($T_3 = T$ at triple point)

$$P \geq P_c \text{ dense fluid}$$

$$T > T_c: P < P_c \text{ superheated vapor}$$

$$P \ll P_c \text{ ideal gas}$$

$$P > P_{\text{sat } T}: \text{ compressed liquid } (T > T_3 \text{ or solid } T < T_3)$$

$$T < T_c: P = P_{\text{sat } T}: \text{ mixture of liquid } (T > T_3) \text{ or solid } (T < T_3) \text{ and vapor}$$

$$P < P_{\text{sat } T}: \text{ superheated vapor}$$

### How do I determine a volume fraction of vapor in a two-phase mixture (L–V)?

The volume fraction is different from the mass fraction, so we set $y = V_g/V_{\text{tot}}$ and get the relation

$$y = \frac{V_g}{V_{\text{tot}}} = \frac{V_g}{V_f + V_g} = \frac{m_g v_g}{m_f v_f + m_g v_g} = \frac{x v_g}{(1 - x)v_f + x v_g}$$

where the last expression divided both numerator and denominator with total mass and used the mass fraction $x = m_g/m_{\text{tot}}$.

### How do I find a state in the B-section tables?

Assume we have a state given by two properties out of $(T, P, v, x)$.

**a. (T, P)** Given. Enter B.?.1 with given $T$ and find $P_{\text{sat}}$.

$P > P_{\text{sat}}$    then we have a liquid; proceed to compressed liquid table if available. If no table, then assume the properties of saturated liquid same $T$.

$P < P_{\text{sat}}$    then we have a superheated vapor; proceed to that table. If $P$ is lower than the lowest $P$, then you have an ideal gas.

**b. (T, v)** Given. Enter B.?.1 with given $T$ and find $v_f, v_g$.

   **(P, v)** Given. Enter B.?.1 and interpolate to the given $P$ and find $v_f, v_g$.

$v < v_f$  then we have a compressed liquid; go to that table if printed.

   **(*T*, *v*):** If table is not printed, we cannot find *P*.

   **(*P*, *v*):** Find *T* so $v = v_f$; it is compressed liquid at that *T* and the given *P*.

$v_f < v < v_g$  this is a two-phase state between the saturated liquid line and the saturated vapor line in a *T–v* diagram. Find the quality *x* as

$$x = \frac{v - v_f}{v_{fg}}$$

$v > v_g$  then we have a superheated vapor; go to that table.

**c. (*T*, *x*) or (*P*, *x*)** Given. Enter B.?.1 You know it is two-phase, so find or interpolate to get to the given *T* or *P* and with the same interpolation get the $v_f$ and $v_g$. Then we have for the overall specific volume

$$v = v_f + x\,(v_g - v_f)$$

**d. (*v*, *x*)** Given. Enter B.?.1 You know it is two-phase but you do not have a *P* or *T* to enter the table. This would be trial and error—guess a *T* then read $v_f$, $v_g$ and calculate

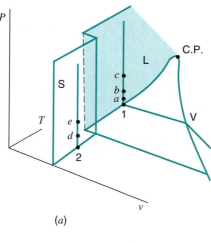

The liquid surface of water where *v* for ice is larger than *v* for liquid.

1: Saturated liquid
*a*, *b*, *c*: Compressed liquid
2: Saturated solid
*d*, *e*: Compressed solid

(a)

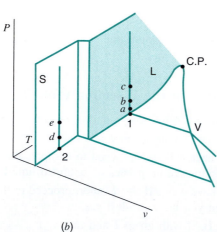

The liquid surface for most substances has *v* for solid less than *v* for liquid

1: Saturated liquid
*a*, *b*, *c*: Compressed liquid
2: Saturated solid
*d*, *e*: Compressed solid

**FIGURE 3.26**

(b)

$v$ for the given quality $x$ and compare with the given $v$. For first guess, assume $v = x\, v_g$ (i.e., neglect $v_f$) and find the $T$ for which you get this $v_g$.

At lower $T$ the surfaces are very steep and flat so $v$ is nearly constant (independent on $P$).

$$v_a \approx v_b \approx v_c \approx v_1 = v_f(T) \qquad \text{and} \qquad v_d \approx v_e \approx v_2 = v_i(T)$$

Other properties that will be covered later ($u$ and $h$ in Chapter 5, $s$ in Chapter 8) are

$$u_a \approx u_b \approx u_c \approx u_1 = u_f(T) \qquad \text{and} \qquad u_d \approx u_e \approx u_2 = u_i(T)$$

$$h_a \approx h_f + (P_a - P_{\text{sat}})v_f = u_f + P_a v_f \qquad \text{similarly for states } b \text{ and } c$$

$$h_d \approx h_i + (P_d - P_{\text{sat}})v_i = u_i + P_d v_i \qquad \text{similarly for state } e$$

$$s_a \approx s_b \approx s_c \approx s_1 = s_f(T) \qquad \text{and} \qquad s_d \approx s_e \approx s_2 = s_i(T)$$

For higher $T$, the liquid surface starts to curve and you must know how compressible it becomes; then a table, an equation of state, or a generalized chart is needed.

### How do I make a linear interpolation between two table values?

Assume we have two table entries of a property $f$ at $x_1$ and $x_2$. We want a value for a given $x$, so we interpolate as

$$f(x) = f_1 + (f_2 - f_1)\frac{x - x_1}{x_2 - x_1}$$

which are the first two terms in a Taylor series expansion of the function. The derivative (slope) is $(f_2 - f_1)/(x_2 - x_1)$ assumed constant between the two states.

The first case in Fig. 3.27 shows a negative slope and the second case a positive slope. The fraction of the difference $(f_2 - f_1)$ that you need is equal to the size ratio of the two triangles in the figures. The small triangle is shaded; the large triangle is 1-2-a.

### How do I determine if a state is an ideal gas state?

To get an idea, look at Fig. 3.21 where the ideal gas (error less than 1%) region for water is shaded. Water vapor is an ideal gas for any $T$ within the vapor region and all $P$s less than about 40 kPa. The smaller $P$ is, the more accurate the ideal gas model is (density

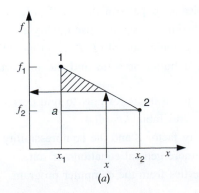

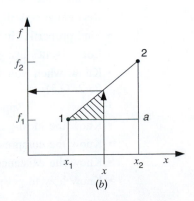

**FIGURE 3.27**  $\qquad\qquad\qquad\qquad (a)$ $\qquad\qquad\qquad\qquad\qquad\qquad\qquad\qquad (b)$

is smaller). For larger pressures, the temperature must be correspondingly higher to stay within the ideal gas region.

Relative to the critical point, we notice that if $T > T_c$ and $P < 0.1\,P_c$ we have an ideal gas and if $T > 2T_c$ then $P$ can be up to $P_c$ and we still have an ideal gas.

Calculate the reduced properties, $P_r = P/P_c$, $T_r = T/T_c$, and $v_r = v/v_c$ with the critical constants from Table A.2. The compressibility factor in Eq. 3.7 reduces to

$$P_r v_r = Z\,T_r/Z_c \qquad \text{with} \qquad Z_c = \frac{P_c V_c}{RT_c}$$

Use the compressibility chart in Fig. D.1 and find $Z$. If you do not have $(P, T)$, iterations may be needed to find the state. The compressibility factor $Z$ directly expresses how close to an ideal gas the state is.

## SUMMARY

Thermodynamic properties of a pure substance and the phase boundaries for solid, liquid, and vapor states are discussed. Phase equilibrium for vaporization (boiling liquid to vapor), with the opposite direction being condensation (vapor to liquid); sublimation (solid to vapor) or the opposite solidification (vapor to solid); and melting (solid to liquid) or the opposite solidifying (liquid to solid) should be recognized. The three-dimensional $P$–$v$–$T$ surface and the two-dimensional representations in the $(P, T)$, $(T, v)$ and $(P, v)$ diagrams, and the vaporization, sublimation, and fusion lines are related to the printed tables in Appendix B. Properties from printed and computer tables covering a number of substances are introduced, including two-phase mixtures, for which we use the mass fraction of vapor (quality). The ideal-gas law approximates the limiting behavior for low density. An extension of the ideal-gas law is shown with the compressibility factor $Z$, and other more complicated equations of state are mentioned.

You should have learned a number of skills and acquired abilities from studying this chapter that will allow you to

- Know phases and the nomenclature used for states and interphases.
- Identify a phase given a state $(T, P)$.
- Locate states relative to the critical point and know Tables A.2 (G.1) and 3.2.
- Recognize phase diagrams and interphase locations.
- Locate states in the Appendix B tables with any entry: $(T, P)$, $(T, v)$ or $(P, v)$
- Recognize how the tables show parts of the $(T, P)$, $(T, v)$ or $(P, v)$ diagrams.
- Find properties in the two-phase regions; use quality $x$.
- Locate states using any combination of $(T, P, v, x)$ including linear interpolation.
- Know when you have a liquid or solid and the properties in Tables A.3, A.4 (G.2, G.3).
- Know when a vapor is an ideal gas (or how to find out).
- Know the ideal-gas law and Table A.5 (G.4).
- Know the compressibility factor $Z$ and the compressibility chart Fig. D.1.
- Know the existence of more general equations of state.
- Know how to get properties from the computer program.

## KEY CONCEPTS AND FORMULAS

| | |
|---|---|
| Phases | Solid, liquid, and vapor (gas) |
| Phase equilibrium | $T_{sat}$, $P_{sat}$, $v_f$, $v_g$, $v_i$ |
| Multiphase boundaries | Vaporization, sublimation, and fusion lines: Figs. 3.5 (general), 3.6 ($CO_2$) and 3.7 (water) |
| | Critical point: Table 3.1, Table A.2; Triple point: Table 3.2 |
| Equilibrium state | Two independent properties (#1, #2) |
| Quality | $x = m_{vap}/m$ (vapor mass fraction) |
| | $1 - x = m_{liq}/m$  (liquid mass fraction) |
| Average specific volume | $v = (1 - x)v_f + xv_g$  (only two-phase mixture) |
| Equilibrium surface | $P$–$v$–$T$   Tables or equation of state |
| Ideal-gas law | $Pv = RT$   $PV = mRT = n\bar{R}T$ |
| Universal gas constant | $\bar{R} = 8.3145$ kJ/kmol K |
| Gas constant | $R = \bar{R}/M$   kJ/kg K, Table A.5 or $M$ from Table A.2 |
| Compressibility factor $Z$ | $Pv = ZRT$   Chart for $Z$ in Fig. D.1 |
| Reduced properties | $P_r = \dfrac{P}{P_c}$   $T_r = \dfrac{T}{T_c}$   Entry to compressibility chart |
| Equations of state | Cubic, pressure explicit: Appendix D, Table D.1 |
| | $B$–$W$–$R$: Eq. 3.7 and Table D.2 for various substances |

# HOMEWORK PROBLEMS

## Concept problems

**3.1** Are the pressures in the tables absolute or gauge pressures?

**3.2** When you skate on ice, a thin liquid film forms under the skate. How can that be?

**3.3** If I have 1 L ammonia at room pressure and temperature (100 kPa, 20°C), how much mass is that?

**3.4** Sketch two constant-pressure curves (500 kPa and 30 000 kPa) in a $T$–$v$ diagram and indicate on the curves where in the water tables you see the properties.

**3.5** Locate the state of ammonia at 200 kPa, −10°C. Indicate in both the $P$–$v$ and the $T$–$v$ diagrams the location of the nearest states listed in the printed Table B.2.

**3.6** Calculate the ideal gas constant for argon and hydrogen based on Table A.2 and verify the value with Table A.5.

**3.7** Water at room temperature and room pressure has $v \approx 1 \times 10^n$ m³/kg. What is n?

**3.8** How accurate is it to assume propane is an ideal gas at room conditions?

**3.9** With $T_r = 0.85$ and a quality of 0.6, find the compressibility factor using Fig. D.1.

## Phase diagrams, triple and critical points

**3.10** Modern extraction techniques can be based on dissolving material in supercritical fluids such as carbon dioxide. How high are the pressure and density of carbon dioxide when the pressure and temperature are around the critical point? Repeat for ethyl alcohol.

**3.11** Find the lowest temperature at which it is possible to have water in the liquid phase. At what pressure must the liquid exist?

**3.12** What is the lowest temperature in Kelvins for which you can see metal as a liquid if the metal is a. silver or b. copper?

**3.13** Water at 27°C can exist in different phases dependent on the pressure. Give the approximate pressure range in kPa for water being in each one of the three phases, vapor, liquid, or solid.

**3.14** If density of ice is 920 kg/m³, find the pressure at the bottom of a 1000-m-thick ice cap on the North Pole. What is the melting temperature at that pressure?

**3.15** Dry ice is the name of solid carbon dioxide. How cold must it be at atmospheric (100 kPa) pressure? If it is heated at 100 kPa what eventually happens?

**3.16** A substance is at 2 MPa and 17°C in a rigid tank. Using only the critical properties, can the phase of the mass be determined if the substance is nitrogen, water, or propane?

**3.17** Give the phase for the following states:
a. $CO_2$ at $T = 267$°C and $P = 0.5$ MPa
b. Air at $T = 20$°C and $P = 200$ kPa
c. $NH_3$ at $T = 170$°C and $P = 600$ kPa

## General tables

**3.18** Give the phase for the following states:
a. $H_2O$ at $T = 275$°C and $P = 5$ MPa
b. $H_2O$ at $T = -2$°C and $P = 100$ kPa

**3.19** Determine whether water at each of the following states is a compressed liquid, a superheated vapor, or a mixture of saturated liquid and vapor.
a. 10 MPa, 0.003 m³/kg    c. 200°C, 0.1 m³/kg
b. 1 MPa, 190°C    d. 10 kPa, 10°C

**3.20** Determine the phase of the substance at the given state using Appendix B tables.
a. Water: 100°C, 500 kPa
b. Ammonia: $-10$°C, 150 kPa
c. R-12: 0°C, 350 kPa

**3.21** Fill out the following table for substance water:

| | P[kPa] | T[°C] | v[m³/kg] | x |
|---|---|---|---|---|
| a. | 500 | 20 | | |
| b. | 500 | | 0.20 | |
| c. | 1400 | 200 | | |
| d. | | 300 | | 0.8 |

**3.22** Place the four states *a–d* listed in Problem 3.21 as labeled dots in a sketch of the P–v and T–v diagrams.

**3.23** Refer to Fig. 3.18. How much is the change in liquid specific volume for water at 20°C as you move up from state *i* toward state *j*, reaching 15 000 kPa?

**3.24** Determine whether refrigerant R-22 in each of the following states is a compressed liquid, a superheated vapor, or a mixture of saturated liquid and vapor.
a. 50°C, 0.05 m³/kg    c. 0.1 MPa, 0.1 m³/kg
b. 1.0 MPa, 20°C    d. $-20$°C, 200 kPa

**3.25** Place the states in Problem 3.24 in a sketch of the P–v diagram.

**3.26** For water at 100 kPa with a quality of 10%, find the volume fraction of vapor.

**3.27** Fill out the following table for substance ammonia:

| | P[kPa] | T[°C] | v[m³/kg] | x |
|---|---|---|---|---|
| a. | | 50 | 0.1185 | |
| b. | | 50 | | 0.5 |

**3.28** Place the two states *a–b* listed in Problem 3.27 as labeled dots in a sketch of the P–v and T–v diagrams.

**3.29** Give the phase and the missing property of $P, T, v$, and $x$.
a. R-134a, $T = -20$°C, $P = 150$ kPa
b. R-134a, $P = 300$ kPa, $v = 0.072$ m³/kg

**3.30** Give the phase and $P, x$ for nitrogen at
a. $T = 120$ K, $v = 0.006$ m³/kg
b. $T = 140$ K, $v = 0.002$ m³/kg

**3.31** Determine the phase and the specific volume for ammonia at these states using the Appendix B table.
a. $-10$°C, 150 kPa
b. 20°C, 100 kPa
c. 60°C, quality 25%

**3.32** Water at 120°C with a quality of 25% has its temperature raised 20°C in a constant volume process. What is the new quality and pressure?

**3.33** A sealed rigid vessel has volume of 1 m³ and contains 2 kg of water at 100°C. The vessel is now heated. If a safety pressure valve is installed, at what pressure should the valve be set to have a maximum temperature of 200°C?

**3.34** Saturated liquid water at 60°C is put under pressure to decrease the volume by 1% while keeping the temperature constant. To what pressure should it be compressed?

**3.35** Water at 200 kPa with a quality of 25% has its temperature raised 20°C in a constant pressure process. What is the new quality and volume?

**3.36** You want a pot of water to boil at 105°C. How heavy a lid should you put on the 15-cm-diameter pot when $P_{atm} = 101$ kPa?

**3.37** In your refrigerator the working substance evaporates from liquid to vapor at $-20$°C inside a pipe around the cold section. Outside (on the back or below) is a black grille, inside of which the working substance condenses from vapor to liquid at $+40$°C. For each location find the pressure and the change in specific volume ($v$) if
a. the substance is R-12
b. the substance is ammonia

**3.38** A boiler feed pump delivers 0.05 m$^3$/s of water at 240°C, 20 MPa. What is the mass flow rate (kg/s)? What would be the percent error if the properties of saturated liquid at 240°C were used in the calculation? What if the properties of saturated liquid at 20 MPa were used?

**3.39** Saturated vapor R-134a at 50°C changes volume at constant temperature. Find the new pressure, and quality if saturated, if the volume doubles. Repeat the question for the case where the volume is reduced to half the original volume.

**3.40** A glass jar is filled with saturated water at 500 kPa of quality 25%, and a tight lid is put on. Now it is cooled to −10°C. What is the mass fraction of solid at this temperature?

**3.41** A sealed rigid vessel of 2 m$^3$ contains a saturated mixture of liquid and vapor R-134a at 10°C. If it is heated to 50°C, the liquid phase disappears. Find the pressure at 50°C and the initial mass of the liquid.

**3.42** A pressure cooker (closed tank) contains water at 100°C with the liquid volume being 1/10 of the vapor volume. It is heated until the pressure reaches 2.0 MPa. Find the final temperature. Has the final state more or less vapor than the initial state?

**3.43** Saturated water vapor at 60°C has its pressure decreased to increase the volume by 10% while keeping the temperature constant. To what pressure should it be expanded?

**3.44** Ammonia at 20°C with a quality of 50% and total mass 2 kg is in a rigid tank with an outlet valve at the bottom. How much liquid (mass) can you take out through the valve assuming the temperature stays constant?

**3.45** A steel tank contains 6 kg of propane (liquid + vapor) at 20°C with a volume of 0.015 m$^3$. The tank is now slowly heated. Will the liquid level inside eventually rise to the top or drop to the bottom of the tank? What if the initial mass is 1 kg instead of 6 kg?

**3.46** Two tanks are connected as shown in Fig. P3.46, both containing water. Tank $A$ is at 200 kPa, $v = 0.5$ m$^3$/kg, $V_A = 1$ m$^3$, and tank $B$ contains 3.5 kg at 0.5 MPa and 400°C. The valve is now opened and the two come to a uniform state. Find the final specific volume.

**3.47** Ammonia at 10°C with a mass of 10 kg is in a piston/cylinder assembly with an initial volume of 1 m$^3$. The piston initially resting on the stops has a mass such that a pressure of 900 kPa will float it. Now the ammonia is slowly heated to 50°C. Find the final pressure and volume.

## Ideal gas

**3.48** What is the relative (%) change in $P$ if we double the absolute temperature of an ideal gas, keeping mass and volume constant? What will it be if we double $V$ having $m$, $T$ constant?

**3.49** A spherical helium balloon 10 m in diameter is at ambient $T$ and $P$, 15°C and 100 kPa. How much helium does it contain? It can lift a total mass that equals the mass of displaced atmospheric air. How much mass of the balloon fabric and cage can then be lifted?

**3.50** Is it reasonable to assume that at the given states the substance behaves as an ideal gas?

a. Oxygen at 30°C, 3 MPa
b. Methane at 30°C, 3 MPa
c. Water at 30°C, 3 MPa
d. R-134a at 30°C, 3 MPa
e. R-134a at 30°C, 100 kPa

**3.51** A 1-m$^3$ tank is filled with a gas at room temperature (20°C) and pressure (100 kPa). How much mass is there if the gas is a. air, b. neon, or c. propane?

**3.52** Helium in a steel tank is at 250 kPa, 300 K with a volume of 0.1 m$^3$. It is used to fill a balloon. When the pressure drops to 150 kPa, the flow of helium stops by itself. If all the helium still is at 300 K, how big a balloon did I get?

**3.53** A 1-m$^3$ rigid tank as shown in Fig. P3.53 has propane at 100 kPa, 300 K and connected by a valve to another tank of 0.5 m$^3$ with propane at 250 kPa, 400 K. The valve is opened and the two tanks come to a uniform state at 325 K. What is the final pressure?

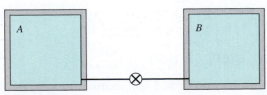

**FIGURE P3.46**

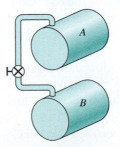

**FIGURE P3.53**

**3.54** A glass is cleaned in 45°C hot water and placed on the table bottom up. The room air at 20°C that was trapped in the glass gets heated up to 40°C and some of it leaks out so that the net resulting pressure inside is 2 kPa above the ambient pressure of 101 kPa. Now the glass and the air inside cool down to room temperature. What is the pressure inside the glass?

**3.55** A cylindrical gas tank 1 m long, with inside diameter of 20 cm, is evacuated and then filled with carbon dioxide gas at 25°C. To what pressure should it be charged if there should be 1.2 kg of carbon dioxide?

**3.56** A pneumatic cylinder (a piston cylinder with air) must close a door with a force of 500 N. The cylinder cross-sectional area is 5 cm$^2$ and its volume is 50 cm$^3$. What is the air pressure and its mass?

**3.57** Air in an internal combustion engine has 227°C, 1000 kPa with a volume of 0.1 m$^3$. Now combustion heats it to 1500 K in a constant volume process. What is the mass of air and how high does the pressure become?

**3.58** Air in an automobile tire is initially at −10°C and 190 kPa. After the automobile is driven awhile, the temperature gets up to 10°C. Find the new pressure. You must make one assumption on your own.

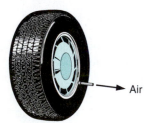

Air

**FIGURE P3.58**

**3.59** Verify the accuracy of the ideal-gas model when it is used to calculate specific volume for saturated water vapor as shown in Fig. 3.21. Do the calculation for 10 kPa and 1 MPa.

**3.60** Assume we have three states of saturated vapor R-134a at +40°C, 0°C, and −40°C. Calculate the specific volume at the set of temperatures and corresponding saturated pressure assuming ideal-gas behavior. Find the percent relative error $= 100(v - v_g)/v_g$ with $v_g$ from the saturated R-134a table.

**3.61** A rigid tank of 1 m$^3$ contains nitrogen gas at 600 kPa, 400 K. By mistake someone lets 0.5 kg flow out. If the final temperature is 375 K, what is then the final pressure?

**3.62** Do Problem 3.60, but for the substance ammonia.

**3.63** Do Problem 3.60, but for the substance R-12.

**3.64** A vacuum pump is used to evacuate a chamber where some specimens are dried at 50°C. The pump rate of volume displacement is 0.5 m$^3$/s with an inlet pressure of 0.1 kPa and temperature 50°C. How much water vapor has been removed over a 30-min period?

## Compressibility factor

**3.65** How close to ideal gas behavior (find $Z$) is ammonia at saturated vapor, 100 kPa? How about saturated vapor at 2000 kPa?

**3.66** A cylinder fitted with a frictionless piston contains butane at 25°C, 500 kPa. Can the butane reasonably be assumed to behave as an ideal gas at this state?

**3.67** Find the volume of 2 kg of ethylene at 270 K, 2500 kPa using $Z$ from Fig. D.1.

**3.68** Argon is kept in a rigid 5-m$^3$ tank at −30°C and 3 MPa. Determine the mass using the compressibility factor. What is the error (%) if the ideal-gas model is used?

**3.69** What is the percent error in specific volume if the ideal-gas model is used to represent the behavior of superheated ammonia at 40°C and 500 kPa? What if the generalized compressibility chart, Fig. D.1, is used instead?

**3.70** A new refrigerant R-125 is stored as a liquid at −20°C with a small amount of vapor. For a total of 1.5 kg R-125 find the pressure and the volume.

**3.71** Estimate the saturation pressure of chlorine at 300 K.

**3.72** Many substances that normally do not mix well do so easily under supercritical pressures. A mass of 125 kg ethylene at 7.5 MPa and 296.5 K is stored for such a process. How much volume does it occupy?

**3.73** Carbon dioxide at 330 K is pumped at a very high pressure, 10 MPa, into an oil well. As it penetrates the rock/oil, the oil viscosity is lowered so it flows out easily. For this process we need to know the density of the carbon dioxide being pumped.

**3.74** A bottle with a volume of 0.1 m$^3$ contains butane with a quality of 75% and a temperature of 300 K. Estimate the total butane mass in the bottle using the generalized compressibility chart.

**3.75** Refrigerant R-32 is at −10°C with a quality of 15%. Find the pressure and specific volume.

## Linear interpolation

**3.76** Find the pressure and temperature for saturated vapor R-12 with $v = 0.1$ m$^3$/kg.

**3.77** Use a linear interpolation to estimate properties of ammonia to fill out the table below.

|      | P[kPa] | T[°C] | v[m³/kg] | x    |
|------|--------|-------|----------|------|
| a.   | 550    |       |          | 0.75 |
| b.   | 80     | 20    |          |      |
| c.   |        | 10    | 0.4      |      |

**3.78** Use a linear interpolation to estimate $T_{sat}$ at 900 kPa for nitrogen. Sketch by hand the curve $P_{sat}(T)$ by using a few table entries around 900 kPa from Table B.6.1. Is your linear interpolation above or below the actual curve?

**3.79** Find the specific volume of ammonia at 140 kPa and 0°C.

**3.80** Find the pressure of water at 200°C and specific volume of 1.5 m³/kg.

## Computer tables

**3.81** Use the computer software to find the properties for water at the four states in Problem 3.21.

**3.82** Use the computer software to find the properties for ammonia at the two states listed in Problem 3.27.

**3.83** Use the computer software to find the properties for ammonia at the three states listed in Problem 3.77.

**3.84** Find the value of the saturated temperature for nitrogen by linear interpolation in Table B.6.1 for a pressure of 900 kPa. Compare this to the value given by the computer software.

# Chapter 4

# Work and Heat

In this chapter we consider work and heat. It is essential for the student of thermodynamics to understand clearly the definitions of both work and heat, because the correct analysis of many thermodynamic problems depends on distinguishing between them.

Work and heat are energy in transfer from one system to another and thus play a crucial role in most thermodynamic systems or devices. As we want to analyze such systems, we need to model the heat and work as functions of properties and parameters characteristic of the system or how it functions. An understanding of the physics involved allows us to construct a model for the heat and work and use the result in our analysis of the energy transfers and changes, which we will do with the first law of thermodynamics in Chapter 5.

To facilitate an understanding of the basic concepts we present a number of physical arrangements that will enable us to express the work done from changes in the system during a process. We will also examine work that is the result of a given process without going into details about how the process physically can be made to occur. This is done because such a description will be too complex and involve concepts that are not covered so far, but at least we can examine the result of the process.

A general description of heat transfer in different situations is a subject that usually is studied separately. However, a very simple introduction is beneficial so that the heat transfer does not become too abstract and we can relate it to the processes we examine. Heat transfer by conduction, convection (flow), and radiation is presented in terms of very simple models, emphasizing that it is driven by a temperature difference.

## 4.1 DEFINITION OF WORK

Work is usually defined as a force $F$ acting through a displacement $x$, where the displacement is in the direction of the force. That is,

$$W = \int_1^2 F\,dx \tag{4.1}$$

This is a very useful relationship because it enables us to find the work required to raise a weight, to stretch a wire, or to move a charged particle through a magnetic field.

However, when treating thermodynamics from a macroscopic point of view, it is advantageous to tie in the definition of work with the concepts of systems, properties, and processes. We therefore define work as follows: Work is done by a system if the sole effect on the surroundings (everything external to the system) could be the raising of a weight. Notice that the raising of a weight is in effect a force acting through a distance. Notice also that our definition does not state that a weight was actually raised or that a force actually acted through a given distance, but that the sole effect external to the system could be the raising of a weight. Work done *by* a system is considered positive and work done *on* a system is considered negative. The symbol $W$ designates the work done by a system.

In general, work is a form of energy in transit, that is, energy being transferred across a system boundary. The concept of energy and energy storage or possession has been discussed in some detail in Section 2.6. Work is the form of energy that fulfills the definition given in the preceding paragraph.

Let us illustrate this definition of work with a few examples. Consider as a system the battery and motor of Fig. 4.1a and let the motor drive a fan. Does work cross the boundary of the system? To answer this question using the definition of work given earlier, replace the fan with the pulley and weight arrangement shown in Fig. 4.1b. As the motor turns, the weight is raised, and the sole effect external to the system is the raising

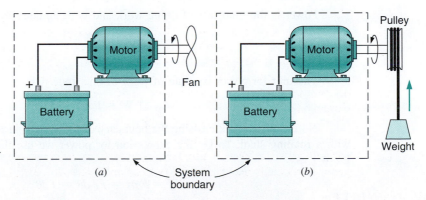

**FIGURE 4.1** Example of work crossing the boundary of a system.

(a)    System boundary    (b)

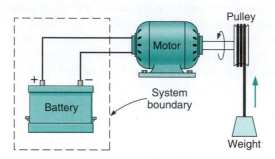

**FIGURE 4.2** Example of work crossing the boundary of a system because of a flow of an electric current across the system boundary.

of a weight. Thus, for our original system of Fig. 4.1*a*, we conclude that work is crossing the boundary of the system, since the sole effect external to the system could be the raising of a weight.

Let the boundaries of the system be changed now to include only the battery shown in Fig. 4.2. Again we ask the question, does work cross the boundary of the system? To answer this question, we need to ask a more general question: Does the flow of electrical energy across the boundary of a system constitute work?

The only limiting factor in having the sole external effect be the raising of a weight is the inefficiency of the motor. However, as we design a more efficient motor, with lower bearing and electrical losses, we recognize that we can approach a certain limit that meets the requirement of having the only external effect be the raising of a weight. Therefore, we can conclude that when there is a flow of electricity across the boundary of a system, as in Fig. 4.2, it is work.

## 4.2 UNITS FOR WORK

As already noted, work done *by* a system, such as that done by a gas expanding against a piston, is positive, and work done *on* a system, such as that done by a piston compressing a gas, is negative. Thus, positive work means that energy leaves the system, and negative work means that energy is added to the system.

Our definition of work involves raising of a weight, that is, the product of a unit force (one newton) acting through a unit distance (one meter). This unit for work in SI units is called the joule (J).

$$1\,J = 1\,N\,m$$

Power is the time rate of doing work and is designated by the symbol $\dot{W}$:

$$\dot{W} \equiv \frac{\delta W}{dt}$$

The unit for power is a rate of work of one joule per second, which is a watt (W):

$$1\,W = 1\,J/s$$

Note that the work crossing the boundary of the system in Fig. 4.1 is that associated with a rotating shaft. To get the expression for power we use the differential work from Eq. 4.1 as

$$\delta W = F\,dx = Fr\,d\theta = T\,d\theta$$

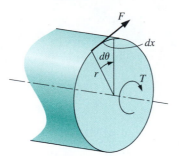

**FIGURE 4.3** Force acting at radius $r$ gives a torque $T = Fr$.

that is, force acting through a distance $dx$ or a torque ($T = Fr$) acting through an angle of rotation as shown in Fig. 4.3. Now the power becomes

$$\dot{W} = \frac{\delta W}{dt} = F\frac{dx}{dt} = F\,\mathbf{V} = Fr\frac{d\theta}{dt} = T\omega \qquad (4.2)$$

that is, a force times rate of displacement (velocity) or a torque times angular velocity.

It is often convenient to speak of the work per unit mass of the system, often termed "specific work." This quantity is designated $w$ and is defined

$$w \equiv \frac{W}{m}$$

---

**In-Text Concept Questions**

**a.** The electric company charges the customers per kW-hour. What is that in SI units?

**b.** Torque, energy, and work have the same units (Nm). Explain the difference.

---

## 4.3 WORK DONE AT THE MOVING BOUNDARY OF A SIMPLE COMPRESSIBLE SYSTEM

We have already noted that there are a variety of ways in which work can be done on or by a system. These include work done by a rotating shaft, electrical work, and the work done by the movement of the system boundary, such as the work done in moving the piston in a cylinder. In this section we will consider in some detail the work done at the moving boundary of a simple compressible system during a quasi-equilibrium process.

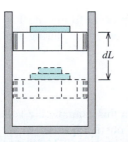

**FIGURE 4.4** Example of work done at the moving boundary of a system in a quasi-equilibrium process.

Consider as a system the gas contained in a cylinder and piston, as in Fig. 4.4. Let one of the small weights be removed from the piston, which will cause the piston to move upward a distance $dL$. We can consider this quasi-equilibrium process and calculate the amount of work $W$ done by the system during this process. The total force on the piston is $PA$, where $P$ is the pressure of the gas and $A$ is the area of the piston. Therefore, the work $\delta W$ is

$$\delta W = PA\,dL$$

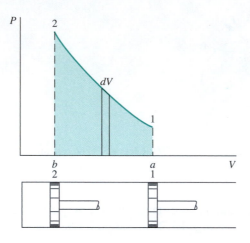

**FIGURE 4.5** Use of pressure–volume diagram to show work done at the moving boundary of a system in a quasi-equilibrium process.

But $A\,dL = dV$, the change in volume of the gas. Therefore,

$$\delta W = P\,dV \tag{4.3}$$

The work done at the moving boundary during a given quasi-equilibrium process can be found by integrating Eq. 4.3. However, this integration can be performed only if we know the relationship between $P$ and $V$ during this process. This relationship may be expressed in the form of an equation, or it may be shown in the form of a graph.

Let us consider a graphical solution first. We use as an example a compression process such as occurs during the compression of air in a cylinder, Fig. 4.5. At the beginning of the process the piston is at position 1, and the pressure is relatively low. This state is represented on a pressure–volume diagram (usually referred to as a $P$–$V$ diagram). At the conclusion of the process the piston is in position 2, and the corresponding state of the gas is shown at point 2 on the $P$–$V$ diagram. Let us assume that this compression was a quasi-equilibrium process and that during the process the system passed through the states shown by the line connecting states 1 and 2 on the $P$–$V$ diagram. The assumption of a quasi-equilibrium process is essential here because each point on line 1–2 represents a definite state, and these states will correspond to the actual state of the system only if the deviation from equilibrium is infinitesimal. The work done on the air during this compression process can be found by integrating Eq. 4.3:

$$_1W_2 = \int_1^2 \delta W = \int_1^2 P\,dV \tag{4.4}$$

The symbol $_1W_2$ is to be interpreted as the work done during the process from state 1 to state 2. It is clear from examining the $P$–$V$ diagram that the work done during this process,

$$\int_1^2 P\,dV$$

is represented by the area under the curve 1–2, area $a$–1–2–$b$–$a$. In this example the volume decreased, and the area $a$–1–2–$b$–$a$ represents work done on the system. If the process had proceeded from state 2 to state 1 along the same path, the same area would represent work done by the system.

Further consideration of a $P$–$V$ diagram, such as Fig. 4.6, leads to another important conclusion. It is possible to go from state 1 to state 2 along many different quasi-equilibrium

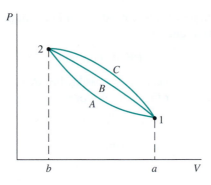

**FIGURE 4.6** Various quasi-equilibrium processes between two given states, indicating that work is a path function.

paths, such as $A$, $B$, or $C$. Since the area underneath each curve represents the work for each process, the amount of work done during each process not only is a function of the end states of the process but depends on the path that is followed in going from one state to another. For this reason work is called a path function or, in mathematical parlance, $\delta W$ is an inexact differential.

This concept leads to a brief consideration of point and path functions or, to use another term, exact and inexact differentials. Thermodynamic properties are point functions, a name that comes from the fact that for a given point on a diagram (such as Fig. 4.6) or surface (such as Fig. 3.18), the state is fixed, and thus there is a definite value of each property corresponding to this point. The differentials of point functions are exact differentials, and the integration is simply

$$\int_1^2 dV = V_2 - V_1$$

Thus, we can speak of the volume in state 2 and the volume in state 1, and the change in volume depends only on the initial and final states.

Work, however, is a path function, for, as has been indicated, the work done in a quasi-equilibrium process between two given states depends on the path followed. The differentials of path functions are inexact differentials, and the symbol $\delta$ will be used in this text to designate inexact differentials (in contrast to $d$ for exact differentials). Thus, for work, we write

$$\int_1^2 \delta W = {}_1W_2$$

It would be more precise to use the notation ${}_1W_{2A}$, which would indicate the work done during the change from state 1 to state 2 along path $A$. However, it is implied in the notation ${}_1W_2$ that the process between states 1 and 2 has been specified. It should be noted that we never speak about the work in the system in state 1 or state 2, and thus we would never write $W_2 - W_1$.

In evaluating the integral of Eq. 4.4, we should always keep in mind that we wish to determine the area under the curve in Fig. 4.6. In connection with this point, we identify the following two classes of problems.

1. The relationship between $P$ and $V$ is given in terms of experimental data or in graphical form (as, for example, the trace on an oscilloscope). Therefore, we may evaluate the integral, Eq. 4.4, by graphical or numerical integration.

2. The relationship between $P$ and $V$ makes it possible to fit an analytical relationship between them. We may then integrate directly.

One common example of this second type of functional relationship is a process called a polytropic process, one in which

$$PV^n = \text{constant}$$

throughout the process. The exponent $n$ may possibly be any value from $-\infty$ to $+\infty$, depending on the particular process. For this type of process, we can integrate Eq. 4.4 as follows:

$$PV^n = \text{constant} = P_1 V_1^n = P_2 V_2^n$$

$$P = \frac{\text{constant}}{V^n} = \frac{P_1 V_1^n}{V^n} = \frac{P_2 V_2^n}{V^n}$$

$$\int_1^2 P\, dV = \text{constant} \int_1^2 \frac{dV}{V^n} = \text{constant} \left( \frac{V^{-n+1}}{-n+1} \right) \bigg|_1^2$$

$$\int_1^2 P\, dV = \frac{\text{constant}}{1-n} (V_2^{1-n} - V_1^{1-n}) = \frac{P_2 V_2^n V_2^{1-n} - P_1 V_1^n V_1^{1-n}}{1-n}$$

$$= \frac{P_2 V_2 - P_1 V_1}{1-n} \tag{4.5}$$

Note that the resulting Eq. 4.5 is valid for any exponent $n$, except $n = 1$. Where $n = 1$,

$$PV = \text{constant} = P_1 V_1 = P_2 V_2$$

and

$$\int_1^2 P\, dV = P_1 V_1 \int_1^2 \frac{dV}{V} = P_1 V_1 \ln \frac{V_2}{V_1} \tag{4.6}$$

Note that in Eqs. 4.5 and 4.6 we did not say that the work is equal to the expressions given in these equations. These expressions give us the value of a certain integral, that is, a mathematical result. Whether or not that integral equals the work in a particular process depends on the result of a thermodynamic analysis of that process. It is important to keep the mathematical result separate from the thermodynamic analysis, for there are many situations in which work is not given by Eq. 4.4.

The polytropic process as described demonstrates one special functional relationship between $P$ and $V$ during a process. There are many other possible relations, some of which will be examined in the problems at the end of this chapter.

---

**EXAMPLE 4.1**

Consider as a system the gas in the cylinder shown in Fig. 4.7; the cylinder is fitted with a piston on which a number of small weights are placed. The initial pressure is 200 kPa, and the initial volume of the gas is 0.04 m³.

**a.** Let a Bunsen burner be placed under the cylinder, and let the volume of the gas increase to 0.1 m³ while the pressure remains constant. Calculate the work done by the system during this process.

$$_1W_2 = \int_1^2 P\, dV$$

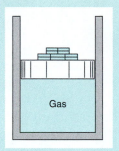

Gas

**FIGURE 4.7** Sketch for Example 4.1.

Since the pressure is constant, we conclude from Eq. 4.4 that

$$_1W_2 = P \int_1^2 dV = P(V_2 - V_1)$$

$$_1W_2 = 200 \text{ kPa} \times (0.1 - 0.04)\text{m}^3 = 12.0 \text{ kJ}$$

**b.** Consider the same system and initial conditions, but at the same time as the Bunsen burner is under the cylinder and the piston is rising, let weights be removed from the piston at such a rate that, during the process, the temperature of the gas remains constant.

If we assume that the ideal-gas model is valid, then, from Eq. 3.5,

$$PV = mRT$$

We note that this is a polytropic process with exponent $n = 1$. From our analysis, we conclude that the work is given by Eq. 4.4 and that the integral in this equation is given by Eq. 4.6. Therefore,

$$_1W_2 = \int_1^2 P\, dV = P_1 V_1 \ln \frac{V_2}{V_1}$$

$$= 200 \text{ kPa} \times 0.04 \text{ m}^3 \times \ln \frac{0.10}{0.04} = 7.33 \text{ kJ}$$

**c.** Consider the same system, but during the heat transfer let the weights be removed at such a rate that the expression $PV^{1.3} = $ constant describes the relation between pressure and volume during the process. Again the final volume is 0.1 m³. Calculate the work.

This is a polytropic process in which $n = 1.3$. Analyzing the process, we conclude again that the work is given by Eq. 4.4 and that the integral is given by Eq. 4.5. Therefore,

$$P_2 = 200 \left(\frac{0.04}{0.10}\right)^{1.3} = 60.77 \text{ kPa}$$

$$_1W_2 = \int_1^2 P\, dV = \frac{P_2 V_2 - P_1 V_1}{1 - 1.3} = \frac{60.77 \times 0.1 - 200 \times 0.04}{1 - 1.3} \text{ kPa m}^3$$

$$= 6.41 \text{ kJ}$$

**d.** Consider the system and initial state given in the first three examples, but let the piston be held by a pin so that the volume remains constant. In addition, let heat be transferred from the system until the pressure drops to 100 kPa. Calculate the work.

Since $\delta W = P\, dV$ for a quasi-equilibrium process, the work is zero, because there is no change in volume.

The process for each of the four examples is shown on the $P$–$V$ diagram of Fig. 4.8. Process 1–2$a$ is a constant-pressure process, and area 1–2$a$–$f$–$e$–1 represents the work. Similarly, line 1–2$b$ represents the process in which $PV = $ constant, line 1–2$c$ the process in which $PV^{1.3} = $ constant, and line 1–2$d$ the constant-volume process. The student should compare the relative areas under each curve with the numerical results obtained for the amounts of work done.

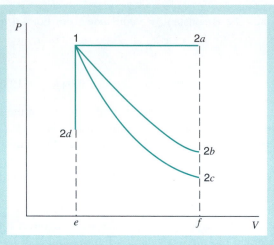

**FIGURE 4.8** Pressure–volume diagram showing work done in the various processes of Example 4.1.

EXAMPLE 4.2

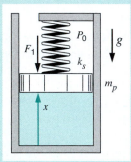

**FIGURE 4.9** Sketch of physical system for Example 4.2.

Consider a slightly different piston/cylinder arrangement as shown in Fig. 4.9. In this example the piston is loaded with a mass, $m_p$, the outside atmosphere $P_0$, a linear spring, and a single point force $F_1$. The piston traps the gas inside with a pressure $P$. A force balance on the piston in the direction of motion yields

$$m_p a \cong 0 = \sum F_\uparrow - \sum F_\downarrow$$

with a zero acceleration in a quasi-equilibrium process. The forces, when the spring is in contact with the piston, are

$$\sum F_\uparrow = PA, \qquad \sum F_\downarrow = m_p g + P_0 A + k_s(x - x_0) + F_1$$

with the linear spring constant, $k_s$. The piston position for a relaxed spring is $x_0$, which depends on how the spring is installed. The force balance then gives the gas pressure by division with the area $A$ as

$$P = P_0 + [m_p g + F_1 + k_s(x - x_0)]/A$$

To illustrate the process in a $P$–$V$ diagram, the distance $x$ is converted to volume by division and multiplication with $A$:

$$P = P_0 + \frac{m_p g}{A} + \frac{F_1}{A} + \frac{k_s}{A^2}(V - V_0) = C_1 + C_2 V$$

This relation gives the pressure as a linear function of the volume, with the line having a slope of $C_2 = k_s/A^2$. Possible values of $P$ and $V$ are as shown in Fig. 4.10 for an expansion. Regardless of what substance is inside, any process must proceed along the line in the $P$–$V$ diagram. The work term in a quasi-equilibrium process then follows as

$$_1 W_2 = \int_1^2 P \, dV = \text{area under the process curve}$$

$$_1 W_2 = \frac{1}{2}(P_1 + P_2)(V_2 - V_1)$$

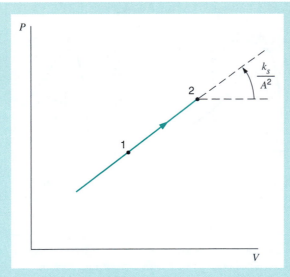

**FIGURE 4.10** The process curve showing possible $P$–$V$ combinations for Example 4.2.

For a contraction instead of expansion, the process would proceed in the opposite direction from the initial point 1 along a line of the same slope shown in Fig. 4.10.

---

**EXAMPLE 4.3**

The cylinder/piston setup of Example 4.2 contains 0.5 kg ammonia at $-20°C$ with a quality of 25%. The ammonia is now heated to $+20°C$, at which state the volume is observed to be 1.41 times larger. Find the final pressure and the work the ammonia produced.

**Solution**

The forces acting on the piston, gravitation constant, external atmosphere at constant pressure and the linear spring give a linear relation between $P$ and $v(V)$.

> *State 1*: $(T_1, x_1)$ from Table B.2.1
> $P_1 = P_{sat} = 190.2 \text{ kPa}$
> $v_1 = v_f + x_1 v_{fg} = 0.001\,504 + 0.25 \times 0.621\,84 = 0.156\,96 \text{ m}^3/\text{kg}$
> *State 2*: $(T_2, v_2 = 1.41\,v_1 = 1.41 \times 0.156\,96 = 0.2213 \text{ m}^3/\text{kg})$
> Table B.2.2 state very close to $P_2 = 600 \text{ kPa}$.
> *Process*: $P = C_1 + C_2 v$

The work term can now be integrated knowing $P$ versus $v$ and can be seen as the area in the $P$–$v$ diagram, shown in Fig. 4.11.

$$_1W_2 = \int_1^2 P\,dV = \int_1^2 Pm\,dv = \text{area} = m\frac{1}{2}(P_1 + P_2)(v_2 - v_1)$$

$$= 0.5 \text{ kg} \frac{1}{2}(190.2 + 600) \text{ kPa} (0.2213 - 0.156\,96) \text{ m}^3/\text{kg}$$

$$= 12.71 \text{ kJ}$$

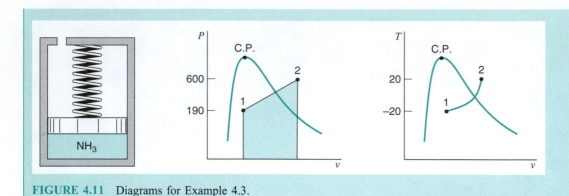

**FIGURE 4.11**   Diagrams for Example 4.3.

**EXAMPLE 4.4**

The piston/cylinder setup shown in Figure 4.12 contains 0.1 kg of water at 1000 kPa, 500°C. The water is now cooled with a constant force on the piston until it reaches half the initial volume. After this it cools to 25°C while the piston is against the stops. Find the final water pressure and the work in the overall process, and show the process in a *P–v* diagram.

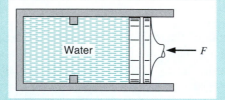

**FIGURE 4.12**   Sketch for Example 4.4.

**Solution**

We recognize this is a two-step process, one of constant *P* and one of constant *V*. This behavior is dictated by the construction of the device.

$$\text{State 1:} \quad (P, T) \quad \text{From Table B.1.3;} \ v_1 = 0.354\,11 \ \text{m}^3/\text{kg}.$$
$$\text{Process 1-1a:} \quad P = \text{constant} = F/A$$
$$1a\text{-}2: \quad v = \text{constant} = v_{1a} = v_2 = v_1/2$$
$$\text{State 2:} \quad (T, v_2 = v_1/2 = 0.177\,06 \ \text{m}^3/\text{kg})$$

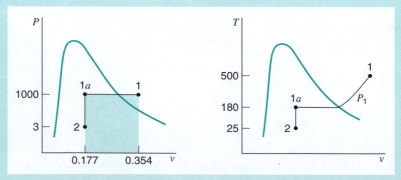

**FIGURE 4.13**
Diagrams for Example 4.4.

From Table B.1.1, $v_2 < v_g$, so the state is two phase and $P_2 = P_{sat} = 3.169$ kPa.

$$_1W_2 = \int_1^2 P\, dV = m \int_1^2 P\, dv = mP_1(v_{1a} - v_1) + 0$$

$$= 0.1 \text{ kg} \times 1000 \text{ kPa}\, (0.177\,06 - 0.354\,11)\, \text{m}^3/\text{kg} = -17.7 \text{ kJ}$$

Note that the work done from $1a$ to 2 is zero (no change in volume) as shown in Fig. 4.13.

In this section we have discussed boundary movement work in a quasi-equilibrium process. We should also realize that there may very well be boundary movement work in a nonequilibrium process. Then the total force exerted on the piston by the gas inside the cylinder, $PA$, does not equal the external force, $F_{ext}$, and the work is not given by Eq. 4.3. The work can, however, be evaluated in terms of $F_{ext}$ or, dividing by area, an equivalent external pressure, $P_{ext}$. The work done at the moving boundary in this case is

$$\delta W = F_{ext}\, dL = P_{ext}\, dV \qquad (4.7)$$

Evaluation of Eq. 4.7 in any particular instance requires a knowledge of how the external force or pressure changes during the process.

**EXAMPLE 4.5**

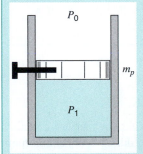

**FIGURE 4.14** Example of a nonequilibrium process.

Consider the system shown in Fig. 4.14 in which the piston of mass $m_p$ is initially held in place by a pin. The gas inside the cylinder is initially at pressure $P_1$ and volume $V_1$. When the pin is now released, the external force per unit area acting on the system (gas) boundary is comprised of two parts:

$$P_{ext} = F_{ext}/A = P_0 + m_p g/A$$

Calculate the work done by the system when the piston has come to rest.

After the piston is released, the system is exposed to the boundary pressure equal to $P_{ext}$, which dictates the pressure inside the system, as discussed in Section 2.8 in connection with Fig. 2.9. We further note that neither of the two components of this external force will change with a boundary movement, since the cylinder is vertical (gravitational force) and the top is open to the ambient surroundings (movement upward merely pushes the air out of the way). If the initial pressure $P_1$ is greater than that resisting the boundary, the piston will move upward at a finite rate, that is, in a nonequilibrium process, with the cylinder pressure eventually coming to equilibrium at the value $P_{ext}$. If we were able to trace the average cylinder pressure as a function of time, it would typically behave as shown in Fig. 4.15. However, the work done by the system during this process is done against the force resisting the boundary movement and is therefore given by Eq. 4.7. Also, since the external force is constant during this process, the result is

$$_1W_2 = \int_1^2 P_{ext}\, dV = P_{ext}(V_2 - V_1)$$

where $V_2$ is greater than $V_1$, and the work by the system is positive. If the initial pressure had been less than the boundary pressure, the piston would have moved downward,

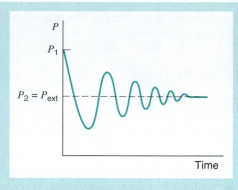

**FIGURE 4.15** Cylinder pressure as a function of time.

compressing the gas, with the system eventually coming to equilibrium at $P_{\text{ext}}$, at a volume less than the initial volume, and the work would be negative, that is, done on the system by its surroundings.

---

**In-Text Concept Questions**

**c.** What is roughly the relative magnitude of the work in the process l-2c versus the process l-2a shown in Fig. 4.8?

**d.** Helium gas expands from 125 kPa, 350 K, and 0.25 m$^3$ to 100 kPa in a polytropic process with $n = 1.667$. Is the work positive, negative, or zero?

**e.** An ideal gas goes through an expansion process where the volume doubles. Which process will lead to the larger work output: an isothermal process or a polytropic process with $n = 1.25$?

## 4.4 OTHER SYSTEMS THAT INVOLVE WORK

In the preceding section we considered the work done at the moving boundary of a simple compressible system during a quasi-equilibrium process and also during a nonequilibrium process. There are other types of systems in which work is done at a moving boundary. In this section we briefly consider three such systems, a stretched wire, a surface film, and electrical work.

Consider as a system a stretched wire that is under a given tension $\mathcal{T}$. When the length of the wire changes by the amount $dL$, the work done by the system is

$$\delta W = -\mathcal{T}\, dL \tag{4.8}$$

The minus sign is necessary because work is done by the system when $dL$ is negative. This equation can be integrated to have

$$_1W_2 = -\int_1^2 \mathcal{T}\, dL \tag{4.9}$$

The integration can be performed either graphically or analytically if the relation between $\mathcal{T}$ and $L$ is known. The stretched wire is a simple example of the type of problem in solid-body mechanics that involves the calculation of work.

**EXAMPLE 4.6**

A metallic wire of initial length $L_0$ is stretched. Assuming elastic behavior, determine the work done in terms of the modulus of elasticity and the strain.

Let $\sigma$ = stress, $e$ = strain, and $E$ = the modulus of elasticity.

$$\sigma = \frac{\mathcal{T}}{A} = Ee$$

Therefore,

$$\mathcal{T} = AEe$$

From the definition of strain,

$$de = \frac{dL}{L_0}$$

Therefore,

$$\delta W = -\mathcal{T}\, dL = -AEeL_0\, de$$

$$W = -AEL_0 \int_{e=0}^{e} e\, de = -\frac{AEL_0}{2}(e)^2$$

Now consider a system that consists of a liquid film having a surface tension $\mathcal{S}$. A schematic arrangement of such a film, maintained on a wire frame, one side of which can be moved, is shown in Fig. 4.16. When the area of the film is changed, for example, by sliding the movable wire along the frame, work is done on or by the film. When the area changes by an amount $dA$, the work done by the system is

$$\delta W = -\mathcal{S}\, dA \tag{4.10}$$

For finite changes,

$$_1W_2 = -\int_1^2 \mathcal{S}\, dA \tag{4.11}$$

We have already noted that electrical energy flowing across the boundary of a system is work. We can gain further insight into such a process by considering a system in which the only work mode is electrical. As examples of such a system, we can think of a charged condenser, an electrolytic cell, and the type of fuel cell described in Chapter 1. Consider a quasi-equilibrium process for such a system, and during this process let the

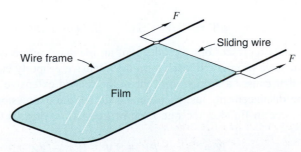

**FIGURE 4.16** Schematic arrangement showing work done on a surface film.

potential difference be $\mathscr{E}$ and the amount of electrical charge that flows into the system be $dZ$. For this quasi-equilibrium process the work is given by the relation

$$\delta W = -\mathscr{E} \, dZ \tag{4.12}$$

Since the current, $i$, equals $dZ/dt$ (where $t$ = time), we can also write

$$\delta W = -\mathscr{E}i \, dt$$

$$_1W_2 = -\int_1^2 \mathscr{E}i \, dt \tag{4.13}$$

Equation 4.13 may also be written as a rate equation for work (the power).

$$\dot{W} = \frac{\delta W}{dt} = -\mathscr{E}i \tag{4.14}$$

Since the ampere (electric current) is one of the fundamental units in the International System, and the watt has been defined previously, this relation serves as the definition of the unit for electric potential, the volt (V), which is one watt divided by one ampere.

## 4.5 Concluding Remarks Regarding Work

The similarity of the expressions for work in the three processes discussed in Section 4.4 and in the processes in which work is done at a moving boundary should be noted. In each of these quasi-equilibrium processes, the work is given by the integral of the product of an intensive property and the change of an extensive property. The following is a summary list of these processes and their work expressions

Simple compressible system $\qquad\qquad _1W_2 = \int_1^2 P \, dV$

Stretched wire $\qquad\qquad\qquad\qquad _1W_2 = -\int_1^2 \mathscr{T} \, dL$

Surface film $\qquad\qquad\qquad\qquad\quad _1W_2 = -\int_1^2 \mathscr{S} \, dA$

System in which the work is completely electrical $\quad _1W_2 = -\int_1^2 \mathscr{E} \, dZ \quad$ (4.15)

Although we will deal primarily with systems in which there is only one mode of work, it is quite possible to have more than one work mode in a given process. Thus, we could write

$$\delta W = P \, dV - \mathscr{T} \, dL - \mathscr{S} \, dA - \mathscr{E} \, dZ + \cdots \tag{4.16}$$

where the dots represent other products of an intensive property and the derivative of a related extensive property. In each term the intensive property can be viewed as the driving force that causes a change to occur in the related extensive property, which is often termed the displacement. Just as we could derive the expression for power for the single point force in Eq. 4.2, the rate form of Eq. 4.16 expresses the power as

$$\dot{W} = \frac{dW}{dt} = P\dot{V} - \mathscr{T}\mathbf{V} - \mathscr{S}\dot{A} - \mathscr{E}\dot{Z} + \cdots \tag{4.17}$$

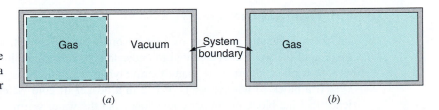

**FIGURE 4.17** Example of process involving a change of volume for which the work is zero.

It should also be noted that many other forms of work can be identified in processes that are not quasi-equilibrium processes. For example, there is the work done by shearing forces in the friction in a viscous fluid or the work done by a rotating shaft that crosses the system boundary.

The identification of work is an important aspect of many thermodynamic problems. We have already noted that work can be identified only at the boundaries of the system. For example, consider Fig. 4.17, which shows a gas separated from the vacuum by a membrane. Let the membrane rupture and the gas fill the entire volume. Neglecting any work associated with the rupturing of the membrane, we can ask whether work is done in the process. If we take as our system the gas and the vacuum space, we readily conclude that no work is done because no work can be identified at the system boundary. If we take the gas as a system, we do have a change of volume, and we might be tempted to calculate the work from the integral

$$\int_1^2 P \, dV$$

However, this is not a quasi-equilibrium process, and therefore the work cannot be calculated from this relation. Because there is no resistance at the system boundary as the volume increases, we conclude that for this system no work is done in this process of filling the vacuum.

Another example can be cited with the aid of Fig. 4.18. In Fig. 4.18*a* the system consists of the container plus the gas. Work crosses the boundary of the system at the point where the system boundary intersects the shaft, and this work can be associated with the shearing forces in the rotating shaft. In Fig. 4.18*b* the system includes the shaft and weight as well as the gas and the container. Therefore, no work crosses the system boundary as the weight moves downward. As we will see in the next chapter, we can identify a change of potential energy within the system, but this should not be confused with work crossing the system boundary.

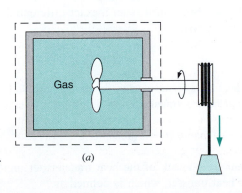

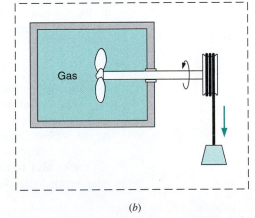

**FIGURE 4.18** Example showing how selection of the system determines whether work is involved in a process.

## 4.6 DEFINITION OF HEAT

The thermodynamic definition of heat is somewhat different from the everyday understanding of the word. It is essential to understand clearly the definition of heat given here, because it plays a part in so many thermodynamic problems.

If a block of hot copper is placed in a beaker of cold water, we know from experience that the block of copper cools down and the water warms up until the copper and water reach the same temperature. What causes this decrease in the temperature of the copper and the increase in the temperature of the water? We say that it is the result of the transfer of energy from the copper block to the water. It is out of such a transfer of energy that we arrive at a definition of heat.

Heat is defined as the form of energy that is transferred across the boundary of a system at a given temperature to another system (or the surroundings) at a lower temperature by virtue of the temperature difference between the two systems. That is, heat is transferred from the system at the higher temperature to the system at the lower temperature, and the heat transfer occurs solely because of the temperature difference between the two systems. Another aspect of this definition of heat is that a body never contains heat. Rather, heat can be identified only as it crosses the boundary. Thus, heat is a transient phenomenon. If we consider the hot block of copper as one system and the cold water in the beaker as another system, we recognize that originally neither system contains any heat (they do contain energy, of course). When the copper block is placed in the water and the two are in thermal communication, heat is transferred from the copper to the water until equilibrium of temperature is established. At this point we no longer have heat transfer, because there is no temperature difference. Neither system contains heat at the conclusion of the process. It also follows that heat is identified at the boundary of the system, for heat is defined as energy being transferred across the system boundary.

Heat, like work, is a form of energy transfer to or from a system. Therefore, the units for heat, and to be more general, for any other form of energy as well, are the same as the units for work, or are at least directly proportional to them. In the International System the unit for heat (energy) is the joule.

Heat transferred *to* a system is considered positive, and heat transferred *from* a system is negative. Thus, positive heat represents energy transferred to a system, and negative heat represents energy transferred from a system. The symbol $Q$ represents heat. A process in which there is no heat transfer ($Q = 0$) is called an adiabatic process.

From a mathematical perspective, heat, like work, is a path function and is recognized as an inexact differential. That is, the amount of heat transferred when a system undergoes a change from state 1 to state 2 depends on the path that the system follows during the change of state. Since heat is an inexact differential, the differential is written $\delta Q$. On integrating, we write

$$\int_1^2 \delta Q = {}_1Q_2$$

In words, ${}_1Q_2$ is the heat transferred during the given process between states 1 and 2.

The rate at which heat is transferred to a system is designated by symbol $\dot{Q}$

$$\dot{Q} \equiv \frac{\delta Q}{dt}$$

It is also convenient to speak of the heat transfer per unit mass of the system, $q$, often termed specific heat transfer, which is defined as

$$q \equiv \frac{Q}{m}$$

## 4.7 HEAT TRANSFER MODES

Heat transfer is the transport of energy due to a temperature difference between different amounts of matter. We know that an ice cube taken out of the freezer will melt as it is placed in a warmer environment such as a glass of liquid water or on a plate with room air around it. From the discussion about energy in Section 2.6 we realize that molecules of matter have translational (kinetic), rotational, and vibrational energy. Energy in these modes can be transmitted to the nearby molecules by interactions (collisions) or by exchange of molecules such that energy is given out by molecules that have more in the average (higher temperature) to those that have less in the average (lower temperature). This energy exchange between molecules is heat transfer by conduction, and it increases with the temperature difference and the ability of the substance to make the transfer. This is expressed in Fourier's law of conduction

$$\dot{Q} = -kA \frac{dT}{dx} \qquad (4.18)$$

giving the rate of heat transfer as proportional to the conductivity, $k$, the total area, $A$, and the temperature gradient. The minus sign gives a direction of the heat transfer from a higher temperature to a lower temperature region. Often the gradient is evaluated as a temperature difference divided by a distance when an estimate has to be done if a mathematical or numerical solution is not available.

Values of the conductivity, $k$, range from the order of 100 W/m K for metals, 1 to 10 for nonmetallic solids as glass, ice and rock, from 0.1 to 10 for liquids, around 0.1 for insulation materials, and from 0.1 down to less than 0.01 for gases.

A different mode of heat transfer takes place when a medium is flowing, called convective heat transfer. In this mode the bulk motion of a substance moves matter with a certain energy level over or near a surface with a different temperature. Now the heat transfer by conduction is dominated by the manner in which the bulk motion brings the two substances in contact or close proximity. Examples of this are the wind blowing over a building or flow through heat exchangers, which can be air flowing over/through a radiator with water flowing inside the radiator piping. The overall heat transfer is typically correlated with Newton's law of cooling as

$$\dot{Q} = Ah \, \Delta T \qquad (4.19)$$

where the transfer properties are lumped into the heat transfer coefficient, $h$, which then becomes a function of the media properties, the flow and geometry. A more detailed study of fluid mechanics and heat transfer aspects of the overall process is necessary to evaluate the heat transfer coefficient for a given situation.

Typical values for the convection coefficient (all in W/m$^2$ K) are

| | | |
|---|---|---|
| Natural convection | $h$ = 5–25, gas | $h$ = 50–1000, liquid |
| Forced convection | $h$ = 25–250, gas | $h$ = 50–20 000, liquid |
| Boiling phase change | $h$ = 2500–100 000 | |

The final mode of heat transfer is radiation, which transmits energy as electromagnetic waves in space. The transfer can happen in empty space and does not require any matter, but the emission (generation) of the radiation and the absorption does require a substance to be present. Surface emission is usually written as a fraction, emissivity $\varepsilon$, of a perfect black body emission as

$$\dot{Q} = \varepsilon \sigma A T_s^4 \qquad (4.20)$$

with the surface temperature, $T_s$, and the Stefan-Boltzmann constant, $\sigma$. Typical values of the emissivity range from 0.92 for nonmetallic surfaces to 0.6 to 0.9 for nonpolished metallic surfaces, to less than 0.1 for highly polished metal surfaces. Radiation is distributed over a range of wavelengths and it is emitted and absorbed differently for different surfaces, but such a description is beyond the scope of the present text.

**EXAMPLE 4.7**

Consider the constant transfer of energy from a warm room at 20°C inside a house to the colder ambient at −10°C through a single-pane window as shown in Fig. 4.19. The temperature variation with distance from the outside glass surface is shown with an outside convection heat transfer layer, but no such layer is inside the room (as a simplification). The glass pane has a thickness of 5 mm (0.005 m) with a conductivity of 1.4 W/m K and a total surface area of 0.5 m². The outside wind is blowing so that the convective heat transfer coefficient is 100 W/m² K. With an outer glass surface temperature of 12.1°C we would like to know the rate of heat transfer in the glass and the convective layer.

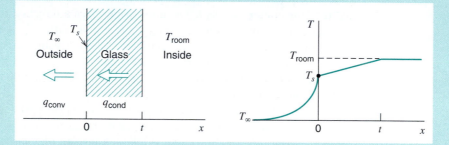

**FIGURE 4.19** Conduction and convection heat transfer through a window pane.

For the conduction through the glass we have

$$\dot{Q} = -kA\,\frac{dT}{dx} = -kA\,\frac{\Delta T}{\Delta x} = -1.4\,\frac{\text{W}}{\text{m K}} \times 0.5\,\text{m}^2\,\frac{20-12.1}{0.005}\,\frac{\text{K}}{\text{m}} = -1106\,\text{W}$$

and the negative sign shows that energy is leaving the room. For the outside convection layer we have

$$\dot{Q} = hA\,\Delta T = 100\,\frac{\text{W}}{\text{m}^2\,\text{K}} \times 0.5\,\text{m}^2\,[12.1-(-10)]\,\text{K} = 1105\,\text{W}$$

with a direction from the higher to the lower temperature, i.e., toward the outside.

## 4.8 COMPARISON OF HEAT AND WORK

At this point it is evident that there are many similarities between heat and work.

1. Heat and work are both transient phenomena. Systems never possess heat or work, but either or both cross the system boundary when a system undergoes a change of state.
2. Both heat and work are boundary phenomena. Both are observed only at the boundaries of the system, and both represent energy crossing the boundary of the system.
3. Both heat and work are path functions and inexact differentials.

It should also be noted that in our sign convention, $+Q$ represents heat transferred *to* the system and thus is energy added to the system, and $+W$ represents work done *by* the system and thus represents energy leaving the system.

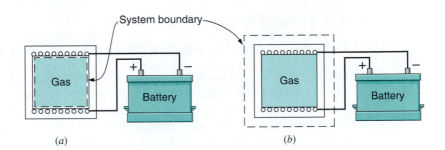

**FIGURE 4.20** An example showing the difference between heat and work.

A final illustration may help explain the difference between heat and work. Fig. 4.20 shows a gas contained in a rigid vessel. Resistance coils are wound around the outside of the vessel. When current flows through the resistance coils, the temperature of the gas increases. Which crosses the boundary of the system, heat or work?

In Fig. 4.20*a* we consider only the gas as the system. The energy crosses the boundary of the system because the temperature of the walls is higher than the temperature of the gas. Therefore, we recognize that heat crosses the boundary of the system.

In Fig. 4.20*b* the system includes the vessel and the resistance heater. Electricity crosses the boundary of the system and, as indicated earlier, this is work.

Consider a gas in a cylinder fitted with a movable piston, as shown in Fig. 4.21. There is a positive heat transfer to the gas, which tends to make the temperature increase. It also tends to increase the gas pressure. However, the pressure is dictated by the external force acting on its movable boundary, as discussed in Section 2.8. If this remains constant, then the volume increases instead. There are also the opposite tendencies for a negative heat transfer, that is, one out of the gas. Consider again the positive heat transfer, except that in this case the external force simultaneously decreases. This causes the gas pressure to decrease, such that the temperature tends to go down. In this case, there are simultaneous tendencies for temperature change in the opposite direction, which effectively decouples directions of heat transfer and temperature change.

Often when we want to evaluate a finite amount of energy transferred as either work or heat we must integrate the instantaneous rate over time.

$$_1W_2 = \int_1^2 \dot{W}\,dt, \qquad _1Q_2 = \int_1^2 \dot{Q}\,dt$$

In order to perform the integration we must know how the rate varies with time. For time periods where the rate does not change significantly, a simple average may be of sufficient accuracy to allow us to write

$$_1W_2 = \int_1^2 \dot{W}\,dt = \dot{W}_{\text{avg}}\,\Delta t \tag{4.21}$$

which is similar to the information given on your electric utility bill as kilowatt-hours.

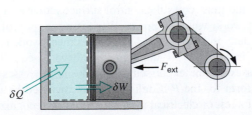

**FIGURE 4.21** The effects of heat addition to a control volume that also can give out work.

## 4.9 HOW-TO SECTION

### How do I determine which kind of process takes place?

Work and heat transfer are both functions of the process path, that is, they depend on how the state changes from beginning to end of the process. In other words, we need to know the process path to find those quantities. This is a device-dependent equation that expresses something about the device behavior. If we have a rigid tank, the volume does not change and there cannot be any boundary work. If the tank is insulated, the heat transfer is small and for a reasonably fast process we then assume that it is adiabatic. The opposite would be a metal tank in an oil or water bath, where you control the temperature, and have a relatively slow process so enough heat transfer can take place to keep the temperature constant.

Boundary work: $\quad {}_1W_2 = \int F\,dx \;$ or $\; \int \mathscr{S}\,dA \;$ or $\; \int P\,dV \quad$ or $\quad \dot{W} = P\dot{V} = F\mathbf{V}$

You must move the C.V. surface. If it does not move, the boundary work is zero.

Heat Transfer: $\quad {}_1Q_2 = C_1\,\Delta T \quad$ or $\quad \dot{Q} = C_2\,\Delta T$

You must have a temperature difference to the local ambient and the $C$'s $(C_1 = C_2\,\Delta t)$.

$\dot{Q} \to 0 \quad$ as $\quad \Delta T \to 0 \quad$ **or** $\quad C_2 \to 0$ (more insulation).

${}_1Q_2 \approx 0 \quad$ as $\quad \Delta T \approx 0 \quad$ **or** $\quad C_1 \approx 0$ (lots of insulation **or** $\Delta t \approx 0$ fast process)

${}_1Q_2 \neq 0 \quad$ as $\quad \Delta T \neq 0 \quad$ **and** $\quad C_1 \neq 0$ (no insulation **or** $\Delta t \to \infty$ slow process)

### How do I relate a rate and a finite amount?

The rates we are dealing with are rates of work (power), rates of heat transfer, mass flow rate, volume flow rate, and storage rates such as rate of stored mass, energy, and volume to mention most of them. These rates in general describe the instantaneous process as it progresses, leading to final transfers (mass, work, and heat) and changes in the stored amount over a certain period of time.

The summation over time is done for the heat transfer as

$${}_1Q_2 = \int \dot{Q}\,dt = \dot{Q}_{\mathrm{avg}}\,\Delta t = \dot{Q}_{\mathrm{avg}}\,(t_2 - t_1)$$

with similar forms for all the other terms. This requires that we can estimate a suitable average and thus depends on how the rate changes with time.

## SUMMARY

Work and heat are energy transfers between a control volume and its surroundings. Work is energy that can be transferred mechanically (or electrically, or chemically) from one system to another and must cross the control surface either as a transient phenomenon or as a steady rate of work, which is power. Work is a function of the process path as well as the beginning state and end state. The displacement work is equal to the area below the process curve drawn in a $P$–$V$ diagram if we have an equilibrium process. A number of ordinary processes can be expressed as polytropic processes having a particular simple mathematical form for the $P$–$V$ relation. Work involved by the action of surface tension, single-point forces, or electrical systems should be recognized and treated separately.

Any nonequilibrium processes (say, dynamic forces, which are important due to accelerations) should be identified so that only equilibrium force or pressure is used to evaluate the work term.

Heat transfer is energy transferred due to a temperature difference, and the conduction, convection, and radiation modes are discussed.

You should have learned a number of skills and acquired abilities from studying this chapter that will allow you to

- Recognize force and displacement in a system.
- Know power as rate of work (force × velocity, torque × angular velocity)
- Know work is a function of the end states and the path followed in process
- Calculate the work term knowing the *P–V* or the *F–x* relationship
- Evaluate the work involved in a polytropic process between two states
- Know work is the area under the process curve in a *P–V* diagram
- Apply a force balance on a mass and determine work in a process from it
- Distinguish between an equilibrium process and a nonequilibrium process
- Recognize the three modes of heat transfer: conduction, convection, and radiation
- Be familiar with Fourier's law of conduction and its use in simple applications
- Know the simple models for convection and radiation heat transfer
- Understand the difference between the rates $(\dot{W}, \dot{Q})$ and the amounts $({}_1W_2, {}_1Q_2)$.

## KEY CONCEPTS AND FORMULAS

| | |
|---|---|
| Work | Energy in transfer—mechanical, electrical, and chemical |
| Heat | Energy in transfer, caused by a $\Delta T$ |
| Displacement work | $W = \int_1^2 F\, dx = \int_1^2 P\, dV = \int_1^2 \mathscr{S}\, dA = \int_1^2 T\, d\theta$ |
| Specific work | $w = W/m$ (work per unit mass) |
| Power, rate of work | $\dot{W} = F\mathbf{V} = P\dot{V} = T\omega$ ($\dot{V}$ displacement rate) |
| | Velocity $\mathbf{V} = r\omega$, torque $T = Fr$, angular velocity $= \omega$ |
| Polytropic process | $PV^n = \text{constant}$ or $Pv^n = \text{constant}$ |
| Polytropic process work | ${}_1W_2 = \dfrac{1}{1-n}(P_2V_2 - P_1V_1)$ (if $n \neq 1$) |
| | ${}_1W_2 = P_1V_1 \ln\dfrac{V_2}{V_1}$ (if $n = 1$) |
| Conduction heat transfer | $\dot{Q} = -kA\,\dfrac{dT}{dx}$ |
| Conductivity | $k$ (W/m K) |
| Convection heat transfer | $\dot{Q} = hA\,\Delta T$ |
| Convection coefficient | $h$ (W/m² K) |
| Radiation heat transfer (net to ambient) | $\dot{Q} = \epsilon\sigma A(T_s^4 - T_{\text{amb}}^4)$ ($\sigma = 5.67 \times 10^{-8}$ W/m² K⁴) |
| Rate integration | ${}_1Q_2 = \int \dot{Q}\, dt \approx \dot{Q}_{\text{avg}}\Delta t$ |

# Homework Problems

## Concept problems

**4.1** A car engine is rated at 160 hp. What is the power in SI units?

**4.2** Normally pistons have a flat head, but in diesel engines pistons can have bowls in them and protruding ridges. Does this geometry influence the work term?

**4.3** CV $A$ is the mass inside a piston-cylinder, CV $B$ is that plus the piston outside, which is the standard atmosphere. See Fig. P4.3. Write the process equation and the work term for the two CVs assuming we have a non-zero $Q$ between a state 1 and a state 2.

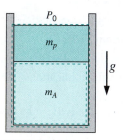

**FIGURE P4.3**

**4.4** Air at 290 K, 100 kPa in a rigid box is heated to 325 $K$. How does each of the properties $P$ and $v$ change (increase, about the same, or decrease) and what transfers do we have for $Q$ and $W$ (pos., neg., or zero)?

**4.5** The sketch in Fig. P4.5 shows a physical setup. We now heat the cylinder. What happens to $P$, $T$ and $v$ (up, down, or constant)? What transfers do we have for $Q$ and $W$ (pos., neg., or zero)?

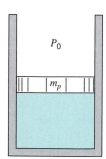

**FIGURE P4.5**

**4.6** Two hydraulic piston/cylinders are connected through a hydraulic line so they have roughly the same pressure. If they have diameters of $D_1$ and $D_2 = 2D_1$ respectively, what can you say about the piston forces $F_1$ and $F_2$?

**4.7** For a buffer storage of natural gas ($CH_4$) a large bell in a container can move up and down keeping a pressure of 105 kPa inside. The sun then heats the container and the gas from 280 K to 300 K during 4 hours. What happens to the volume and what is the sign of the work term?

**4.8** Three physical situations are in Fig. P4.9. Show the possible process in a $P-v$ diagram for a, b, and c.

**4.9** For the indicated physical set-up in $a–b$ and $c$ shown in Fig. P4.9 write a process equation and the expression for work.

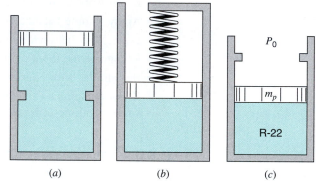

(a)          (b)          (c)

**FIGURE P4.9**

**4.10** The sketch in Fig. P4.10 shows a physical situation; what is the work term $a$, $b$, $c$ or $d$?

a: $_1w_2 = P_1(v_2 - v_1)$

b: $_1w_2 = v_1(P_2 - P_1)$

c: $_1w_2 = \dfrac{1}{2}(P_1 + P_2)(v_2 - v_1)$

d: $_1w_2 = \dfrac{1}{2}(P_1 - P_2)(v_2 + v_1)$

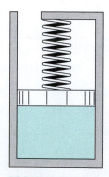

**FIGURE P4.10**

**4.11** The sketch in Fig. P4.11 shows a physical situation; show the possible process in a *P–v* diagram.

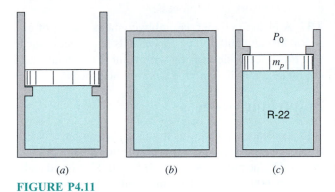

| (a) | (b) | (c) |

**FIGURE P4.11**

**4.12** Show how the polytropic exponent *n* can be evaluated if you know the end state properties, $(P_1, V_1)$ and $(P_2, V_2)$.

**4.13** A drag force on an object moving through a medium (like a car through air or a submarine through water) is $F_d = 0.225 A \rho V^2$. Verify the unit becomes Newton.

## Force displacement work

**4.14** A piston of mass 2 kg is lowered 0.5 m in the standard gravitational field. Find the required force and the work involved in the process.

**4.15** An escalator raises a 100-kg bucket of sand 10 m in 1 min. Determine the total amount of work done during the process.

**4.16** A hydraulic cylinder of area 0.01 m² must push a 1000-kg arm and shovel 0.5 m straight up. What pressure is needed and how much work is done?

**4.17** A hydraulic cylinder has a piston of cross-sectional area 25 cm² and a fluid pressure of 2 MPa. If the piston is moved 0.25 m, how much work is done?

**4.18** Two hydraulic cylinders maintain a pressure of 1200 kPa. One has a cross-sectional area of 0.01 m², the other one of 0.03 m². To deliver 1 kJ of work to the piston, how large a displacement (*V*) and piston motion *H* are needed for each cylinder? Neglect $P_{atm}$.

**4.19** A linear spring, $F = k_s(x - x_0)$ with spring constant $k_s = 500$ N/m is stretched until it is 100 mm longer. Find the required force and the work input.

**4.20** A work of 2.5 kJ must be delivered on a rod from a pneumatic piston/cylinder where the air pressure is limited to 500 kPa. What diameter cylinder should I have to restrict the rod motion to maximum 0.5 m?

**4.21** The rolling resistance of a car depends on its weight as $F = 0.006$ mg. How long will a car of 1400 kg drive for a work input of 25 kJ?

**4.22** The air drag force on a car is $0.225 A\rho V^2$. Assume air at 290 K, 100 kPa and a car frontal area of 4 m² driving at 90 km/h. How much energy is used to overcome the air drag driving for 30 min?

## Boundary work: simple one-step process

**4.23** A constant-pressure piston/cylinder assembly contains 0.2 kg of water as saturated vapor at 400 kPa. It is now cooled so that the water occupies half the original volume. Find the work done in the process.

**4.24** A steam radiator in a room at 25°C has saturated water vapor at 110 kPa flowing through it when the inlet and exit valves are closed. What are the pressure and the quality of the water when it has cooled to 25°C? How much work is done?

**4.25** Find the specific work in Problem 3.35.

**4.26** A 400-L tank, *A* (see Fig. P4.26), contains argon gas at 250 kPa and 30°C. Cylinder *B*, having a frictionless piston of such mass that a pressure of 150 kPa will float it, is initially empty. The valve is opened and argon flows into *B* and eventually reaches a uniform state of 150 kPa and 30°C throughout. What is the work done by the argon?

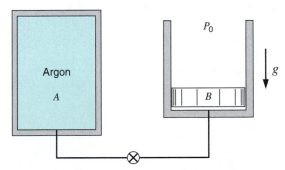

**FIGURE P4.26**

**4.27** A piston cylinder contains air at 600 kPa, 290 K and a volume of 0.01 m³. A constant pressure process gives 18 kJ of work out. Find the final volume and temperature of the air.

**4.28** Saturated water vapor at 200 kPa is in a constant-pressure piston/cylinder. In this state, the piston is 0.1 m from the cylinder bottom and the cylinder area is 0.25 m². The temperature is then changed to 200°C. Find the work in the process.

**4.29** A cylinder fitted with a frictionless piston contains 5 kg of superheated refrigerant R-134a vapor at 1000 kPa and 140°C. The setup is cooled at constant pressure until the R-134a reaches a quality of 25%. Calculate the work done in the process.

**4.30** A piston cylinder contains 1.5 kg water at 200 kPa, 150°C. It is now heated in a process where pressure is linearly related to volume to a state of 600 kPa, 350°C. Find the final volume and the work in the process.

**4.31** Find the specific work in Problem 3.39 for the case where the volume is reduced.

**4.32** A piston/cylinder has 5 m of liquid 20°C water on top of the piston ($m = 0$) with cross-sectional area of 0.1 m$^2$, see Fig. P2.39. Air is let in under the piston that rises and pushes the water out over the top edge. Find the necessary work to push all the water out and plot the process in a P–V diagram.

**4.33** A piston/cylinder contains 1 kg of water at 20°C with volume 0.1 m$^3$. By mistake someone locks the piston, preventing it from moving while we heat the water to saturated vapor. Find the final temperature and volume, and the process work.

**4.34** Ammonia (0.5 kg) is in a piston cylinder at 200 kPa, −10°C is heated in a process where the pressure varies linear with the volume to a state of 120°C, 300 kPa. Find the work the ammonia gives out in the process.

**4.35** A piston/cylinder assembly contains 1 kg of liquid water at 20°C and 300 kPa, as shown in Fig. P4.35. There is a linear spring mounted on the piston such that when the water is heated the pressure reaches 3 MPa with a volume of 0.1 m$^3$.

a. Find the final temperature.

b. Plot the process in a P–$v$ diagram.

c. Find the work in the process.

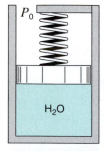

**FIGURE P4.35**

**4.36** A piston/cylinder assembly contains 3 kg of air at 20°C and 300 kPa. It is now heated in a constant pressure process to 600 K.

a. Find the final volume.

b. Plot the process path in a P–$v$ diagram.

c. Find the work in the process.

**4.37** A piston/cylinder assembly contains 0.5 kg of air at 500 kPa and 500 K. The air expands in a process such that P is linearly decreasing with volume to a final state of 100 kPa, 300 K. Find the work in the process.

## Polytropic process

**4.38** Consider a mass going through a polytropic process where pressure is directly proportional to volume ($n = −1$). The process starts with $P = 0$, $V = 0$ and ends with $P = 600$ kPa, $V = 0.01$ m$^3$. Find the boundary work done by the mass.

**4.39** Air at 1500 K, 1000 kPa expands in a polytropic process, $n = 1.5$, to a pressure of 200 kPa. How cold does the air become and what is the specific work out?

**4.40** The piston/cylinder arrangement shown in Fig. P4.40 contains carbon dioxide at 300 KPa and 100°C with a volume of 0.2 m$^3$. Weights are added to the piston such that the gas compresses according to the relation $PV^{1.2} =$ constant to a final temperature of 200°C. Determine the work done during the process.

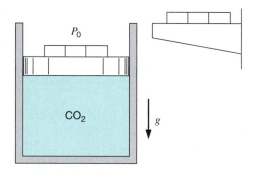

**FIGURE P4.40**

**4.41** A gas initially at 1 MPa and 500°C is contained in a piston and cylinder arrangement with an initial volume of 0.1 m$^3$. The gas is then slowly expanded according to the relation $PV =$ constant until a final pressure of 100 kPa is reached. Determine the work for this process.

**4.42** A spring loaded piston/cylinder assembly contains 1 kg water at 500°C, 3 MPa. The setup is such that the pressure is proportional to volume, $P = CV$. It is now cooled until the water becomes saturated vapor. Sketch the P–$v$ diagram and find the work in the process.

**4.43** Helium gas expands from 125 kPa, 350 K, and 0.25 m$^3$ to 100 kPa in a polytropic process with $n = 1.667$. How much work does it give out?

**4.44** A balloon behaves so the pressure is $P = C_2 V^{1/3}$, $C_2 = 100$ kPa/m. The balloon is blown up with air from a starting volume of 1 m³ to a volume of 3 m³. Find the final mass of the air, assuming it is at 25°C, and the work done by the air.

**4.45** A piston cylinder contains 0.1 kg nitrogen at 100 kPa, 27°C and it is now compressed in a polytropic process with $n = 1.25$ to a pressure of 250 kPa. What is the work involved?

**4.46** A balloon behaves such that the pressure inside is proportional to the diameter squared. It contains 2 kg of ammonia at 0°C, with 60% quality. The balloon and ammonia are now heated so that a final pressure of 600 kPa is reached. Considering the ammonia as a control mass, find the amount of work done in the process.

**4.47** Consider a piston/cylinder setup with 0.5 kg of R-134a as saturated vapor at −10°C. It is now compressed to a pressure of 500 kPa in a polytropic process with $n = 1.5$. Find the final volume and temperature, and determine the work done during the process.

**4.48** Air goes through a polytropic process from 125 kPa and 325 K to 300 kPa and 500 K. Find the polytropic exponent $n$ and the specific work in the process.

**4.49** A piston/cylinder contains water at 500°C, 3 MPa. It is cooled in a polytropic process to 200°C, 1 MPa. Find the polytropic exponent and the specific work in the process.

## Boundary work: multi-step process

**4.50** Consider a two-part process with an expansion from 0.1 to 0.2 m³ at a constant pressure of 150 kPa followed by an expansion from 0.2 to 0.4 m³ with a linearly rising pressure from 150 kPa ending at 300 kPa. Show the process in a P–V diagram and find the boundary work.

**4.51** A cylinder containing 1 kg of ammonia has an externally loaded piston. Initially the ammonia is at 2 MPa and 180°C. It is now cooled to saturated vapor at 40°C and then further cooled to 20°C, at which point the quality is 50%. Find the total work for the process, assuming a piecewise linear variation of $P$ versus $V$.

**4.52** A helium gas is heated at constant volume from a state of 100 kPa, 300 K to 500 K. A following process expands the gas at constant pressure to three times the initial volume. What is the specific work in the combined process?

**4.53** A piston/cylinder arrangement shown in Fig. P4.53 initially contains air at 150 kPa and 400°C. The setup is allowed to cool to the ambient temperature of 20°C.

a. Is the piston resting on the stops in the final state? What is the final pressure in the cylinder?

b. What is the specific work done by the air during the process?

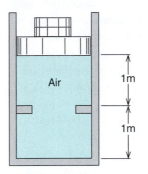

**FIGURE P4.53**

**4.54** A piston/cylinder has 1.5 kg of air at 300 K and 150 kPa. It is now heated up in a two step process. First constant volume to 1000 K (state 2) then followed by a constant pressure process to 1500 K, state 3. Find the final volume and the work in the process.

**4.55** A piston cylinder contains air at 1000 kPa, 800 K with a volume of 0.05 m³. The piston is pressed against the upper stops and it will float at a pressure of 750 kPa. Then the air is cooled to 400 K. What is the process work?

**4.56** The refrigerant R-22 is contained in a piston/cylinder as shown in Fig. P4.56, where the volume is 11 L when the piston hits the stops. The initial state is −30°C, 150 kPa, with a volume of 10 L. This system is brought indoors and warms up to 15°C.

a. Is the piston at the stops in the final state?

b. Find the work done by the R-22 during this process.

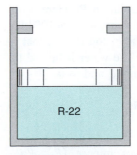

**FIGURE P4.56**

**4.57** A piston/cylinder assembly contains 50 kg of water at 200 kPa with a volume of 0.1 m³. Stops in the cylinder restrict the enclosed volume to 0.5 m³, similar to the setup in Problem 4.56. The water is now heated to 200°C. Find the final pressure, volume, and work done by the water.

**4.58** A piston/cylinder setup (Fig. P4.60) contains 1 kg of water at 20°C with a volume of 0.1 m³. Initially, the piston rests on some stops with the top surface open to the atmosphere, $P_0$, and a mass such that a water pressure of 400 kPa will lift it. To what temperature should the water be heated to lift the piston? If it is heated to saturated vapor, find the final temperature, volume, and work, $_1W_2$.

**4.59** A piston/cylinder assembly contains 1 kg of liquid water at 20°C and 300 kPa. Initially the piston floats, similar to the setup in Problem 4.56, with a maximum enclosed volume of 0.002 m³ if the piston touches the stops. Now heat is added so that a final pressure of 600 kPa is reached. Find the final volume and the work in the process.

**4.60** Ten kilograms of water in a piston/cylinder arrangement exist as saturated liquid/vapor at 100 kPa, with a quality of 50%. It is now heated so the volume triples. The mass of the piston is such that a cylinder pressure of 200 kPa will float it (see Fig. P4.60).

a. Find the final temperature and volume of the water.

b. Find the work given out by the water.

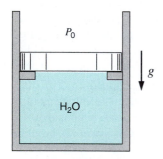

**FIGURE P4.60**

**4.61** Find the work in Problem 3.47

**4.62** A piston/cylinder setup similar to Problem 4.60 contains 0.1 kg saturated liquid and vapor water at 100 kPa with quality 25%. The mass of the piston is such that a pressure of 500 kPa will float it. The water is heated to 300°C. Find the final pressure, volume, and work, $_1W_2$.

## Other types of work and general concepts

**4.63** Electric power is volts times ampere $(P = V\,i)$. When a car battery at 12 V is charged with 6 amp for 3 hours how much energy is delivered?

**4.64** A 0.5-m-long steel rod with a 1-cm diameter is stretched in a tensile test. What is the work required to obtain a relative strain of 0.1%? The modulus of elasticity of steel is $2 \times 10^8$ kPa.

**4.65** A film of ethanol at 20°C has a surface tension of 22.3 mN/m and is maintained on a wire frame as shown in Fig. P4.65. Consider the film with two surfaces as a control mass and find the work done when the wire is moved 10 mm to make the film 20 × 40 mm.

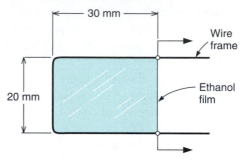

**FIGURE P4.65**

**4.66** Assume a balloon material with a constant surface tension of $\mathcal{S} = 2$ N/m. What is the work required to stretch a special balloon up to a radius of $r = 0.5$ m? Neglect any effect from atmospheric pressure.

**4.67** A soap bubble has a surface tension of $\mathcal{S} = 3 \times 10^{-4}$ N/cm as it sits flat on a rigid ring of diameter 5 cm. You now blow on the film to create a half-sphere surface of diameter 5 cm. How much work was done?

**4.68** A sheet of rubber is stretched out over a ring of radius 0.25 m. I pour liquid water at 20°C on it, as in Fig. P4.68, so that the rubber forms a half sphere (cup). Neglect the rubber mass and find the surface tension near the ring.

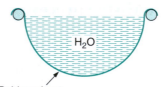

**FIGURE P4.68**

**4.69** Consider a light bulb that is on. Explain where we have rates of work and heat transfer (include modes) that moves energy.

**4.70** Consider a window-mounted air-conditioning unit used in the summer to cool incoming air. Examine the system boundaries for rates of work and heat transfer, including signs.

**4.71** A room is heated with an electric space heater on a winter day. Examine the following control volumes, regarding heat transfer and work, including sign:

a. The space heater

b. Room

c. The space heater and the room together

# Rates of work

**4.72** A force of 1.2 kN moves a truck with 60 km/h up a hill. What is the power?

**4.73** An escalator raises a 100-kg bucket 10 m in 1 min. Determine the rate of work in the process.

**4.74** A car uses 25 hp to drive at a horizontal level at a constant speed of 100 km/h. What is the traction force between the tires and the road?

**4.75** A piston/cylinder of cross-sectional area 0.01 m$^2$ maintains constant pressure. It contains 1 kg of water with a quality of 5% at 150°C. If we heat so that 1 g/s of liquid turns into vapor, what is the rate of work out?

**4.76** Consider the car with the rolling resistance as in Problem 4.21. How fast can it drive using 30 hp?

**4.77** Consider the car with the air drag force as in Problem 4.22. How fast can it drive using 30 hp?

**4.78** A battery is well insulated while being charged by 12.3 V at a current of 6 A. Take the battery as a control mass and find the instantaneous rate of work and the total work done over 4 h.

**4.79** A current of 10 A runs through a resistor with a resistance of 15 Ω. Find the rate of work that heats the resistor up.

**4.80** A pressure of 650 kPa pushes a piston of diameter 0.25 m with **V** = 5 m/s. What is the volume displacement rate, the force, and the transmitted power?

**4.81** Assume the process in Problem 4.50 takes place with a constant rate of change in volume over 2 minutes. Show the power (rate of work) as a function of time.

**4.82** Air at a constant pressure in a piston/cylinder is at 300 kPa, 300 K and has a volume of 0.1 m$^3$. It is heated to 600 K over 30 s in a process with constant piston velocity. Find the power delivered to the piston.

# Heat transfer rates

**4.83** Find the rate of conduction heat transfer through a 1.5 cm thick hardwood board, $k = 0.16$ W/m K, with a temperature difference between the two sides of 20°C.

**4.84** The sun shines on a 150-m$^2$ road surface so that it is at 45°C. Below the 5-cm-thick asphalt, with average conductivity of 0.06 W/m K, is a layer of compacted rubble at a temperature of 15°C. Find the rate of heat transfer to the rubble.

**4.85** A water heater is covered up with insulation boards over a total surface area of 3 m$^2$. The inside board surface is at 75°C, the outside surface is at 20°C, and the board material has a conductivity of 0.08 W/m K. How thick should the board be to limit the heat transfer loss to 200 W?

**4.86** A large condenser (heat exchanger) in a power plant must transfer a total of 100 MW from steam running in a pipe to seawater being pumped through the heat exchanger. Assume the wall separating the steam and seawater is 4 mm of steel, with conductivity of 15 W/m K, and that a maximum of 5°C difference between the two fluids is allowed in the design. Find the required minimum area for the heat transfer, neglecting any convective heat transfer in the flows.

**4.87** The black grille on the back of a refrigerator has a surface temperature of 35°C with a total surface area of 1 m$^2$. Heat transfer to the room air at 20°C takes place with an average convective heat transfer coefficient of 15 W/m$^2$ K. How much energy can be removed during 15 minutes of operation?

**4.88** A 2 m$^2$ window has a surface temperature of 15°C and the outside wind is blowing air at 2°C across it with a convection heat transfer coefficient of $h = 125$ W/m$^2$ K. What is the total heat transfer loss?

**4.89** A wall surface on a house is 30°C with an emissivity of $\varepsilon = 0.7$. The surrounding ambient air is at 15°C with an average emissivity of 0.9. Find the rate of radiation energy from each of those surfaces per unit area.

**4.90** A log of burning wood in the fireplace has a surface temperature of 450°C. Assume the emissivity is 1 (perfect black body) and find the radiant emission of energy per unit surface area.

**4.91** A radiant heating lamp has a surface temperature of 1000 K with $\varepsilon = 0.8$. How large a surface area is needed to provide 250 W of radiation heat transfer?

# Chapter 5

# The First Law of Thermodynamics

Having completed our consideration of basic definitions and concepts, we are ready to proceed to a discussion of the first law of thermodynamics. This law is often called the conservation of energy law and, as we will see later, this is essentially true. Our procedure will be to state this law for a system (control mass) undergoing a cycle and then for a change of state of a system.

After the energy equation is formulated we will use it to relate change of state inside a control volume to the amount of energy that is transferred in a process as work or heat transfer. When a car engine has transferred some work to the car, the car's speed is increased, so we can relate the kinetic energy increase to the work, or if a stove provides a certain amount of heat transfer to a pot with water we can relate the water temperature increase to the heat transfer. More complicated processes can also occur, such as the expansion of very hot gases in a piston cylinder, as in a car engine, in which work is given out and at the same time heat is transferred to the colder walls. In other applications we can also see a change in the state without any work or heat transfer, such as a falling object that changes kinetic energy at the same time it is changing elevation. The energy equation then relates the two forms of energy of the object.

## 5.1 THE FIRST LAW OF THERMODYNAMICS FOR A CONTROL MASS UNDERGOING A CYCLE

The first law of thermodynamics states that during any cycle a system (control mass) undergoes, the cyclic integral of the heat is proportional to the cyclic integral of the work.

To illustrate this law, consider as a control mass the gas in the container shown in Fig. 5.1. Let this system go through a cycle that is made up of two processes. In the first process work is done on the system by the paddle that turns as the weight is lowered. Let the system then return to its initial state by transferring heat from the system until the cycle has been completed.

Historically, work was measured in mechanical units of force times distance, such as foot pounds force or joules, and heat was measured in thermal units, such as the British thermal unit or the calorie. Measurements of work and heat were made during a cycle for a wide variety of systems and for various amounts of work and heat. When the amounts of work and heat were compared, it was found that they were always proportional. Such observations led to the formulation of the first law of thermodynamics, which in equation form is written

$$J \oint \delta Q = \oint \delta W \tag{5.1}$$

The symbol $\oint \delta Q$, which is called the cyclic integral of the heat transfer, represents the net heat transfer during the cycle, and $\oint \delta W$, the cyclic integral of the work, represents the net work during the cycle. Here, $J$ is a proportionality factor that depends on the units used for work and heat.

The basis of every law of nature is experimental evidence, and this is true also of the first law of thermodynamics. Many different experiments have been conducted on the first law, and every one thus far has verified it either directly or indirectly. The first law has never been disproved.

As was discussed in Chapter 4, the unit for work and heat, and for any other unit of energy as well, in the International System of units is the joule. Thus in this system

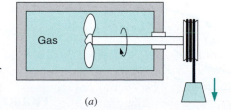
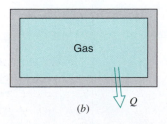

**FIGURE 5.1** Example of a control mass undergoing a cycle.

*(a)*          *(b)*

we do not need the proportionality factor $J$, and we may write Eq. 5.1 as

$$\oint \delta Q = \oint \delta W \qquad (5.2)$$

which can be considered the basic statement of the first law of thermodynamics.

## 5.2 The First Law of Thermodynamics For a Change in State of a Control Mass

Equation 5.2 states the first law of thermodynamics for a control mass during a cycle. Many times, however, we are concerned with a process rather than a cycle. We now consider the first law of thermodynamics for a control mass that undergoes a change of state. We begin by introducing a new property, the energy, which is given the symbol $E$. Consider a system that undergoes a cycle in which it changes from state 1 to state 2 by process $A$ and returns from state 2 to state 1 by process $B$. This cycle is shown in Fig. 5.2 on a pressure (or other intensive property)–volume (or other extensive property) diagram. From the first law of thermodynamics, Eq. 5.2, we have

$$\oint \delta Q = \oint \delta W$$

Considering the two separate processes, we have

$$\int_1^2 \delta Q_A + \int_2^1 \delta Q_B = \int_1^2 \delta W_A + \int_2^1 \delta W_B$$

Now consider another cycle in which the control mass changes from state 1 to state 2 by process $C$ and returns to state 1 by process $B$, as before. For this cycle we can write

$$\int_1^2 \delta Q_C + \int_2^1 \delta Q_B = \int_1^2 \delta W_C + \int_2^1 \delta W_B$$

Subtracting the second of these equations from the first, we obtain

$$\int_1^2 \delta Q_A - \int_1^2 \delta Q_C = \int_1^2 \delta W_A - \int_1^2 \delta W_C$$

or, by rearranging,

$$\int_1^2 (\delta Q - \delta W)_A = \int_1^2 (\delta Q - \delta W)_C \qquad (5.3)$$

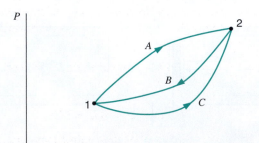

**FIGURE 5.2** Demonstration of the existence of thermodynamic property $E$.

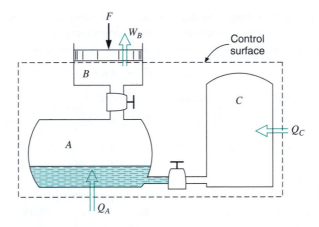

**FIGURE 5.3** A control mass with several different subsystems.

Since $A$ and $C$ represent arbitrary processes between states 1 and 2, the quantity $\delta Q - \delta W$ is the same for all processes between states 1 and 2. Therefore, $\delta Q - \delta W$ depends only on the initial and final states and not on the path followed between the two states. We conclude that this is a point function, and therefore it is the differential of a property of the mass. This property is the energy of the mass and is given the symbol $E$. Thus we can write

$$dE = \delta Q - \delta W \tag{5.4}$$

Because $E$ is a property, its derivative is written $dE$. When Eq. 5.4 is integrated from an initial state 1 to a final state 2, we have

$$E_2 - E_1 = {}_1Q_2 - {}_1W_2 \tag{5.5}$$

where $E_1$ and $E_2$ are the initial and final values of the energy $E$ of the control mass, ${}_1Q_2$ is the heat transferred to the control mass during the process from state 1 to state 2, and ${}_1W_2$ is the work done by the control mass during the process.

Note that a control mass may be made up of several different subsystems, as shown in Fig. 5.3. In this case, each part must be analyzed and included separately in applying the first law, Eq. 5.5. We further note that Eq. 5.5 is an expression of the general form

$$\Delta \text{ Energy} = +\text{in} - \text{out}$$

in terms of the standard sign conventions for heat and work.

The physical significance of the property $E$ is that it represents all the energy of the system in the given state. This energy might be present in a variety of forms, such as the kinetic or potential energy of the system as a whole with respect to the chosen coordinate frame, energy associated with the motion and position of the molecules, energy associated with the structure of the atom, chemical energy present in a storage battery, energy present in a charged condenser, or any of a number of other forms.

In the study of thermodynamics, it is convenient to consider the bulk kinetic and potential energy separately and then to consider all the other energy of the control mass in a single property that we call the internal energy and to which we give the symbol $U$. Thus, we would write

$$E = \text{Internal energy} + \text{kinetic energy} + \text{potential energy}$$

or

$$E = U + \text{KE} + \text{PE}$$

The kinetic and potential energy of the control mass are associated with the coordinate frame that we select and can be specified by the macroscopic parameters of mass, velocity, and elevation. The internal energy $U$ includes all other forms of energy of the control mass and is associated with the thermodynamic state of the system.

Since the terms comprising $E$ are point functions, we can write

$$dE = dU + d(\text{KE}) + d(\text{PE}) \tag{5.6}$$

The first law of thermodynamics for a change of state may therefore be written

$$dE = dU + d(\text{KE}) + d(\text{PE}) = \delta Q - \delta W \tag{5.7}$$

In words this equation states that as a control mass undergoes a change of state, energy may cross the boundary as either heat or work, and each may be positive or negative. The net change in the energy of the system will be exactly equal to the net energy that crosses the boundary of the system. The energy of the system may change in any of three ways—by a change in internal energy, in kinetic energy, or in potential energy.

This section concludes by deriving an expression for the kinetic and potential energy of a control mass. Consider a mass that is initially at rest relative to the earth, which is taken as the coordinate frame. Let this system be acted on by an external horizontal force $F$ that moves the mass a distance $dx$ in the direction of the force. Thus, there is no change in potential energy. Let there be no heat transfer and no change in internal energy. Then from the first law, Eq. 5.7, we have

$$\delta W = -F\,dx = -d\text{KE}$$

But

$$F = ma = m\frac{d\mathbf{V}}{dt} = m\frac{dx}{dt}\frac{d\mathbf{V}}{dx} = m\mathbf{V}\frac{d\mathbf{V}}{dx}$$

Then

$$d\text{KE} = F\,dx = m\mathbf{V}\,d\mathbf{V}$$

Integrating, we obtain

$$\int_{\text{KE}=0}^{\text{KE}} d\text{KE} = \int_{\mathbf{V}=0}^{\mathbf{V}} m\mathbf{V}\,d\mathbf{V}$$

$$\text{KE} = \frac{1}{2}m\mathbf{V}^2 \tag{5.8}$$

A similar expression for potential energy can be found. Consider a control mass that is initially at rest and at the elevation of some reference level. Let this mass be acted on by a vertical force $F$ of such magnitude that it raises (in elevation) the mass with constant velocity an amount $dZ$. Let the acceleration due to gravity at this point be $g$. From the first law, Eq. 5.7, we have

$$\delta W = -F\,dZ = -d\,\text{PE}$$

$$F = ma = mg$$

Then

$$d\,\text{PE} = F\,dZ = mg\,dZ$$

Integrating gives

$$\int_{\text{PE}_1}^{\text{PE}_2} d\,\text{PE} = m\int_{Z_1}^{Z_2} g\,dZ$$

Assuming that $g$ does not vary with $Z$ (which is a very reasonable assumption for moderate changes in elevation), we obtain

$$PE_2 - PE_1 = mg(Z_2 - Z_1) \qquad (5.9)$$

**EXAMPLE 5.1**

A car of mass 1100 kg drives with a velocity such that it has a kinetic energy of 400 kJ (see Fig. 5.4). Find the velocity. If the car is raised with a crane how high should it be lifted in the standard gravitational field to have a potential energy that equals the kinetic energy?

**FIGURE 5.4**  Sketch for Example 5.1.

**Solution**

The standard kinetic energy of the mass is

$$KE = \frac{1}{2} m\mathbf{V}^2 = 400 \text{ kJ}$$

From this we can solve for the velocity

$$\mathbf{V} = \sqrt{\frac{2 \text{ KE}}{m}} = \sqrt{\frac{2 \times 400 \text{ kJ}}{1100 \text{ kg}}}$$

$$= \sqrt{\frac{800 \times 1000 \text{ N m}}{1100 \text{ kg}}} = \sqrt{\frac{8000 \text{ kg m s}^{-2}\text{ m}}{11 \text{ kg}}} = 27 \text{ m/s}$$

Standard potential energy is

$$PE = mgH$$

so when this is equal to the kinetic energy we get

$$H = \frac{KE}{mg} = \frac{400\,000 \text{ N m}}{1100 \text{ kg} \times 9.807 \text{ m s}^{-2}} = 37.1 \text{ m}$$

Notice the necessity of converting the kJ to J in both calculations.

Now, substituting the expressions for kinetic and potential energy into Eq. 5.6, we have

$$dE = dU + m\mathbf{V}\,d\mathbf{V} + mg\,dZ$$

Integrating for a change of state from state 1 to state 2 with constant $g$, we get

$$E_2 - E_1 = U_2 - U_1 + \frac{m\mathbf{V}_2^2}{2} - \frac{m\mathbf{V}_1^2}{2} + mg\,Z_2 - mgZ_1$$

Similarly, substituting these expressions for kinetic and potential energy into Eq. 5.7, we have

$$dE = dU + \frac{d(m\mathbf{V}^2)}{2} + d(mgZ) = \delta Q - \delta W \qquad (5.10)$$

Assuming $g$ is a constant, in the integrated form of this equation,

$$U_2 - U_1 + \frac{m(\mathbf{V}_2^2 - \mathbf{V}_1^2)}{2} + mg(Z_2 - Z_1) = {}_1Q_2 - {}_1W_2 \tag{5.11}$$

Three observations should be made regarding this equation. The first observation is that the property $E$, the energy of the control mass, was found to exist, and we were able to write the first law for a change of state using Eq. 5.5. However, rather than deal with this property $E$, we find it more convenient to consider the internal energy and the kinetic and potential energies of the mass. In general, this procedure will be followed in the rest of this book.

The second observation is that Eqs. 5.10 and 5.11 are in effect a statement of the con-servation of energy. The net change of the energy of the control mass is always equal to the net transfer of energy across the boundary as heat and work. This is somewhat analogous to a joint checking account shared by a husband and wife. There are two ways in which deposits and withdrawals can be made—either by the husband or by the wife—and the balance will always reflect the net amount of the transaction. Similarly, there are two ways in which energy can cross the boundary of a control mass—either as heat or as work—and the energy of the mass will change by the exact amount of the net energy crossing the boundary. The concept of energy and the law of the conservation of energy are basic to thermodynamics.

The third observation is that Eqs. 5.10 and 5.11 can give only changes in internal energy, kinetic energy, and potential energy. We can learn nothing about absolute values of these quantities from these equations. If we wish to assign values to internal energy, kinetic energy, and potential energy, we must assume reference states and assign a value to the quantity in this reference state. The kinetic energy of a body with zero velocity relative to the earth is assumed to be zero. Similarly, the value of the potential energy is assumed to be zero when the body is at some reference elevation. With internal energy, therefore, we must also have a reference state if we wish to assign values of this property. This matter is considered in the following section.

---

**EXAMPLE 5.2**

A tank containing a fluid is stirred by a paddle wheel. The work input to the paddle wheel is 5090 kJ. The heat transfer from the tank is 1500 kJ. Consider the tank and the fluid inside a control surface and determine the change in internal energy of this control mass.

The first law of thermodynamics is (Eq. 5.11)

$$U_2 - U_1 + \frac{1}{2}m(\mathbf{V}_2^2 - \mathbf{V}_1^2) + mg(Z_2 - Z_1) = {}_1Q_2 - {}_1W_2$$

Since there is no change in kinetic and potential energy, this reduces to

$$U_2 - U_1 = {}_1Q_2 - {}_1W_2$$
$$U_2 - U_1 = -1500 - (-5090) = 3590 \text{ kJ}$$

---

**EXAMPLE 5.3**

Consider a stone having a mass of 10 kg and a bucket containing 100 kg of liquid water. Initially the stone is 10.2 m above the water, and the stone and the water are at the same temperature, state 1. The stone then falls into the water.

Determine $\Delta U$, $\Delta KE$, $\Delta PE$, $Q$, and $W$ for the following changes of state, assum-ing standard gravitational acceleration of 9.806 65 m/s$^2$.

**a.** The stone is about to enter the water, state 2.

**b.** The stone has just come to rest in the bucket, state 3.

**c.** Heat has been transferred to the surroundings in such an amount that the stone and water are at the same temperature, $T_1$, state 4.

**Analysis and Solution**

The first law for any of the steps is

$$Q = \Delta U + \Delta \text{KE} + \Delta \text{PE} + W$$

and each term can be identified for each of the changes of state.

**a.** The stone has fallen from $Z_1$ to $Z_2$, and we assume no heat transfer as it falls. The water has not changed state; thus

$$\Delta U = 0, \qquad {}_1Q_2 = 0, \qquad {}_1W_2 = 0$$

and the first law reduces to

$$\Delta \text{KE} + \Delta \text{PE} = 0$$

$$\Delta \text{KE} = -\Delta \text{PE} = -mg(Z_2 - Z_1)$$

$$= -10 \text{ kg} \times 9.806\,65 \text{ m/s}^2 \times (-10.2 \text{ m})$$

$$= 1000 \text{ J} = 1 \text{ kJ}$$

That is, for the process from state 1 to state 2,

$$\Delta \text{KE} = 1 \text{ kJ} \qquad \text{and} \qquad \Delta \text{PE} = -1 \text{ kJ}$$

**b.** For the process from state 2 to state 3 with zero kinetic energy, we have

$$\Delta \text{PE} = 0, \qquad {}_2Q_3 = 0, \qquad {}_2W_3 = 0$$

Then

$$\Delta U + \Delta \text{KE} = 0$$

$$\Delta U = -\Delta \text{KE} = 1 \text{ kJ}$$

**c.** In the final state, there is no kinetic, nor potential energy, and the internal energy is the same as in state 1.

$$\Delta U = -1 \text{ kJ}, \qquad \Delta \text{KE} = 0, \qquad \Delta \text{PE} = 0, \qquad {}_3W_4 = 0$$

$$\,_3Q_4 = \Delta U = -1 \text{ kJ}$$

---

**In-Text Concept Questions**

**a.** In a complete cycle, what is the net change in energy and in volume?

**b.** Explain in words what happens with the energy terms for the stone in Example 5.3. What would happen if it were a bouncing ball falling to a hard surface?

**c.** Make a list of at least 5 systems that store energy, explaining which form of energy.

**d.** A constant mass goes through a process where 100 J of heat transfer comes in and 100 J of work leaves. Does the mass change state?

## 5.3 INTERNAL ENERGY—A THERMODYNAMIC PROPERTY

Internal energy is an extensive property because it depends on the mass of the system. Similarly, kinetic and potential energies are extensive properties.

The symbol $U$ designates the internal energy of a given mass of a substance. Following the convention used with other extensive properties, the symbol $u$ designates the internal energy per unit mass. We could speak of $u$ as the specific internal energy, as we do with specific volume. However, because the context will usually make it clear whether $u$ or $U$ is referred to, we will simply use the term internal energy to refer to both internal energy per unit mass and the total internal energy.

In Chapter 3 we noted that in the absence of motion, gravity, surface effects, electricity, or other effects, the state of a pure substance is specified by two independent properties. It is very significant that, with these restrictions, the internal energy may be one of the independent properties of a pure substance. This means, for example, that if we specify the pressure and internal energy (with reference to an arbitrary base) of super-heated steam, the temperature is also specified.

Thus, in a table of thermodynamic properties such as the steam tables, the value of internal energy can be tabulated along with other thermodynamic properties. Tables 1 and 2 of the steam tables (Tables B.1.1 and B.1.2) list the internal energy for saturated states. Included are the internal energy of saturated liquid $u_f$, the internal energy of saturated vapor $u_g$, and the difference between the internal energy of saturated liquid and saturated vapor $u_{fg}$. The values are given in relation to an arbitrarily assumed reference state, which, for water in the steam tables, is taken as zero for saturated liquid at the triple-point temperature, 0.01°C. All values of internal energy in the steam tables are then calculated relative to this reference (note that the reference state cancels out when finding a difference in $u$ between any two states). Values for internal energy are found in the steam tables in the same manner as for specific volume. In the liquid–vapor saturation region,

$$U = U_{\text{liq}} + U_{\text{vap}}$$

or

$$mu = m_{\text{liq}}u_f + m_{\text{vap}}u_g$$

Dividing by $m$ and introducing the quality $x$ gives

$$u = (1 - x)u_f + xu_g$$

$$u = u_f + xu_{fg}$$

As an example, the specific internal energy of saturated steam having a pressure of 0.6 MPa and a quality of 95% can be calculated as

$$u = u_f + xu_{fg} = 669.9 + 0.95(1897.5) = 2472.5 \text{ kJ/kg}$$

Values for $u$ in the superheated vapor region are tabulated in Table B.1.3, for compressed liquid in Table B.1.4, and for solid–vapor in Table B.1.5.

| EXAMPLE 5.4 | Determine the missing property ($P$, $T$ or $x$) and also $v$ for water at each of the following states: |
|---|---|

**a.** $T = 300°C$, $u = 2780$ kJ/kg

**b.** $P = 2000$ kPa, $u = 2000$ kJ/kg

For each case, the two properties given are independent properties and therefore fix the state. For each, we must first determine the phase by comparison of the given information with phase boundary values.

**a.** At 300°C, from Table B.1.1, $u_g$ = 2563.0 kJ/kg. The given $u > u_g$, so the state is in the superheated vapor region at some $P$ less than $P_g$, which is 8581 kPa. Searching through Table B.1.3 at 300°C, we find that the value $u$ = 2780 is between given values of $u$ at 1600 kPa (2781.0) and 1800 kPa (2776.8). Interpolating linearly, we obtain

$$P = 1648 \text{ kPa}.$$

Note that quality is undefined in the superheated vapor region. At this pressure, by linear interpolation, we have $v$ = 0.1542 m³/kg.

**b.** At $P$ = 2000 kPa, from Table B.1.2, the given $u$ of 2000 kJ/kg is greater than $u_f$ (906.4) but less than $u_g$ (2600.3). Therefore, this state is in the two-phase region with $T = T_g$ = 212.4°C, and

$$u = 2000 = 906.4 + x1693.8, \qquad x = 0.6456$$

Then,

$$v = 0.001\ 177 + 0.6456 \times 0.098\ 45 = 0.064\ 74 \text{ m}^3/\text{kg}.$$

---

**In-Text Concept Questions**

**e.** Water is heated from 100 kPa, 20°C to 1000 kPa, 200°C. In one case pressure is raised at $T = C$, then $T$ is raised at $P = C$. In a second case the opposite order is done. Does that make a difference for $_1Q_2$ and $_1W_2$?

**f.** A rigid insulated tank $A$ contains water at 400 kPa, 800°C. A pipe and valve connects this to another rigid insulated tank $B$ of equal volume having saturated water vapor at 100 kPa. The valve is opened and stays open while the water in the two tanks comes to a uniform final state. Which two properties determine the final state?

## 5.4 PROBLEM ANALYSIS AND SOLUTION TECHNIQUE

At this point in our study of thermodynamics, we have progressed sufficiently far (that is, we have accumulated sufficient tools with which to work) that it is worthwhile to develop a somewhat formal technique or procedure for analyzing and solving thermodynamic problems. For the time being it may not seem entirely necessary to use such a rigorous procedure for many of our problems, but we should keep in mind that as we acquire more analytical tools the problems that we are capable of dealing with will become much more complicated. Thus, it is appropriate that we begin to practice this technique now in anticipation of these future problems.

Our problem analysis and solution technique is contained within the framework of the following set of questions that must be answered in the process of an orderly solution of a thermodynamic problem.

**1.** What is the control mass or control volume? Is it useful, or necessary, to choose more than one? It may be helpful to draw a sketch of the system at this point, illustrating all heat and work flows, and indicating forces such as external pressures and gravitation.

2. What do we know about the initial state (i.e., which properties are known)?

3. What do we know about the final state?

4. What do we know about the process that takes place? Is anything constant or zero? Is there some known functional relation between two properties?

5. Is it helpful to draw a diagram of the information in steps 2 to 4 (for example, a $T$–$v$ or $P$–$v$ diagram)?

6. What is our thermodynamic model for the behavior of the substance (for example, steam tables, ideal gas, and so on)?

7. What is our analysis of the problem (i.e., do we examine control surfaces for various work modes or use the first law or conservation of mass)?

8. What is our solution technique? In other words, from what we have done so far in steps 1–7, how do we proceed to find whatever it is that is desired? Is a trial-and-error solution necessary?

It is not always necessary to write out all these steps, and in the majority of the examples throughout this text we will not do so. However, when faced with a new and unfamiliar problem, the student should always at least think through this set of questions to develop the ability to solve more challenging problems. In solving the following example, we will use this technique in detail.

---

**Example 5.5**

A vessel having a volume of 5 m³ contains 0.05 m³ of saturated liquid water and 4.95 m³ of saturated water vapor at 0.1 MPa. Heat is transferred until the vessel is filled with saturated vapor. Determine the heat transfer for this process.

| | |
|---:|:---|
| *Control mass*: | All the water inside the vessel. |
| *Sketch*: | Fig. 5.5. |
| *Initial state*: | Pressure, volume of liquid, volume of vapor; therefore, state 1 is fixed. |
| *Final state*: | Somewhere along the saturated-vapor curve; the water was heated, so $P_2 > P_1$. |
| *Process*: | Constant volume and mass; therefore, constant specific volume. |
| *Diagram*: | Fig. 5.6. |
| *Model*: | Steam tables. |

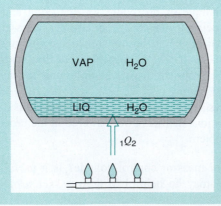

**FIGURE 5.5** Sketch for Example 5.5.

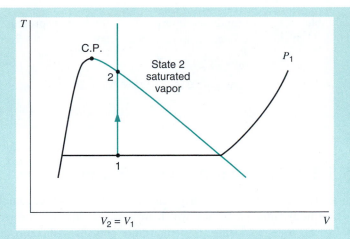

**FIGURE 5.6** Diagram for Example 5.5.

**Analysis**

From the first law we have

$$_1Q_2 = U_2 - U_1 + m\frac{\mathbf{V}_2^2 - \mathbf{V}_1^2}{2} + mg(Z_2 - Z_1) + {_1W_2}$$

From examining the control surface for various work modes, we conclude that the work for this process is zero. Furthermore, the system is not moving, so there is no change in kinetic energy. There is a small change in the center of mass of the system but we will assume that the corresponding change in potential energy is negligible (in kilojoules). Therefore,

$$_1Q_2 = U_2 - U_1$$

**Solution**

The heat transfer will be found from the first law. State 1 is known, so $U_1$ can be calculated. The specific volume at state 2 is also known (from state 1 and the process). Since state 2 is saturated vapor, state 2 is fixed, as is seen from Fig. 5.6. Therefore, $U_2$ can also be found.

The solution proceeds as follows:

$$m_{1\,liq} = \frac{V_{liq}}{v_f} = \frac{0.05}{0.001\,043} = 47.94 \text{ kg}$$

$$m_{1\,vap} = \frac{V_{vap}}{v_g} = \frac{4.95}{1.6940} = 2.92 \text{ kg}$$

Then

$$U_1 = m_{1\,liq}u_{1\,liq} + m_{1\,vap}u_{1\,vap}$$

$$= 47.94(417.36) + 2.92(2506.1) = 27\,326 \text{ kJ}$$

To determine $u_2$ we need to know two thermodynamic properties, since this determines the final state. The properties we know are the quality, $x = 100\%$, and $v_2$, the final specific volume, which can readily be determined.

$$m = m_{1\,liq} + m_{1\,vap} = 47.94 + 2.92 = 50.86 \text{ kg}$$

$$v_2 = \frac{V}{m} = \frac{5.0}{50.86} = 0.098\,31 \text{ m}^3/\text{kg}$$

In Table B.1.2 we find, by interpolation, that at a pressure of 2.03 MPa, $v_g =$ 0.098 31 m$^3$/kg. The final pressure of the steam is therefore 2.03 MPa. Then

$$u_2 = 2600.5 \text{ kJ/kg}$$

$$U_2 = mu_2 = 50.86(2600.5) = 132\ 261 \text{ kJ}$$

$$_1Q_2 = U_2 - U_1 = 132\ 261 - 27\ 326 = 104\ 935 \text{ kJ}$$

## 5.5 THE THERMODYNAMIC PROPERTY ENTHALPY

In analyzing specific types of processes, we frequently encounter certain combinations of thermodynamic properties, which are therefore also properties of the substance undergoing the change of state. To demonstrate one such situation, let us consider a control mass undergoing a quasi-equilibrium constant-pressure process, as shown in Fig. 5.7. Assume that there are no changes in kinetic or potential energy and that the only work done during the process is that associated with the boundary movement. Taking the gas as our control mass and applying the first law, Eq. 5.11, we have, in terms of $Q$,

$$_1Q_2 = U_2 - U_1 + _1W_2$$

The work done can be calculated from the relation

$$_1W_2 = \int_1^2 P\,dV$$

Since the pressure is constant,

$$_1W_2 = P \int_1^2 dV = P(V_2 - V_1)$$

Therefore,

$$_1Q_2 = U_2 - U_1 + P_2V_2 - P_1V_1$$
$$= (U_2 + P_2V_2) - (U_1 + P_1V_1)$$

We find that, in this very restricted case, the heat transfer during the process is given in terms of the change in the quantity $U + PV$ between the initial and final states. Because all these quantities are thermodynamic properties, that is, functions only of the state of the system, their combination must also have these same characteristics. Therefore, we find it convenient to define a new extensive property, the enthalpy,

$$H \equiv U + PV \tag{5.12}$$

or, per unit mass,

$$h \equiv u + Pv \tag{5.13}$$

As for internal energy, we could speak of specific enthalpy, $h$, and total enthalpy, $H$. However, we will refer to both as enthalpy, since the context will make it clear which is being discussed.

The heat transfer in a constant-pressure quasi-equilibrium process is equal to the change in enthalpy, which includes both the change in internal energy and the work for this particular process. This is by no means a general result. It is valid for this special case only because the work done during the process is equal to the difference in the $PV$

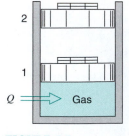

**FIGURE 5.7**
The constant-pressure quasi-equilibrium process.

product for the final and initial states. This would not be true if the pressure had not remained constant during the process.

The significance and use of enthalpy is not restricted to the special process just described. Other cases in which this same combination of properties $u + Pv$ appear will be developed later, notably in Chapter 6 in which we discuss control volume analyses. Our reason for introducing enthalpy at this time is that although the tables in Appendix B list values for internal energy, many other tables and charts of thermodynamic properties give values for enthalpy but not for the internal energy. Therefore, it is necessary to calculate the internal energy at a state using the tabulated values and Eq. 5.13:

$$u = h - Pv$$

Students often become confused about the validity of this calculation when analyzing system processes that do not occur at constant pressure, for which enthalpy has no physical significance. We must keep in mind that enthalpy, being a property, is a state or point function, and its use in calculating internal energy at the same state is not related to, or dependent on, any process that may be taking place.

Tabular values of internal energy and enthalpy, such as those included in Tables B.1 through B.6, are all relative to some arbitrarily selected base. In the steam tables, the internal energy of saturated liquid at 0.01°C is the reference state and is given a value of zero. For refrigerants, such as ammonia and chlorofluorocarbons R-12 and R-22, the reference state is arbitrarily taken as saturated liquid at −40°C. The enthalpy in this reference state is assigned the value of zero. Cryogenic fluids, such as nitrogen, have other arbitrary reference states chosen for enthalpy values listed in their tables. Because each of these reference states is arbitrarily selected, it is always possible to have negative values for enthalpy, as for saturated-solid water in Table B.1.5. When enthalpy and internal energy are given values relative to the same reference state, as they are in essentially all thermodynamic tables, the difference between internal energy and enthalpy at the reference state is equal to $Pv$. Since the specific volume of the liquid is very small, this product is negligible as far as the significant figures of the tables are concerned, but the principle should be kept in mind, for in certain cases it is significant.

In many thermodynamic tables, values of the specific internal energy $u$ are not given. As mentioned earlier, these values can be readily calculated from the relation $u = h - Pv$, though it is important to keep the units in mind. As an example, let us calculate the internal energy $u$ of superheated R-134a at 0.4 MPa, 70°C.

$$u = h - Pv$$

$$= 460.55 - 400 \times 0.066\,48$$

$$= 433.96 \text{ kJ/kg}$$

The enthalpy of a substance in a saturation state and with a given quality is found in the same way as the specific volume and internal energy. The enthalpy of saturated liquid has the symbol $h_f$, saturated vapor $h_g$, and the increase in enthalpy during vaporization $h_{fg}$. For a saturation state, the enthalpy can be calculated by one of the following relations:

$$h = (1 - x)h_f + xh_g$$

$$h = h_f + xh_{fg}$$

The enthalpy of compressed liquid water may be found from Table B.1.4. For substances for which compressed-liquid tables are not available, the enthalpy is taken as that of saturated liquid at the same temperature.

**EXAMPLE 5.6**

A cylinder fitted with a piston has a volume of 0.1 m³ and contains 0.5 kg of steam at 0.4 MPa. Heat is transferred to the steam until the temperature is 300°C, while the pressure remains constant.

Determine the heat transfer and the work for this process.

*Control mass*: Water inside cylinder.

*Initial state*: $P_1$, $V_1$, $m$; therefore $v_1$, is known, state 1 is fixed (at $P_1$, $v_1$, check steam tables—two-phase region).

*Final state*: $P_2$, $T_2$; therefore state 2 is fixed (superheated).

*Process*: Constant pressure.

*Diagram*: Fig. 5.8.

*Model*: Steam tables.

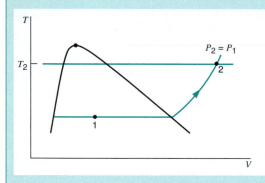 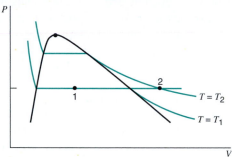

**FIGURE 5.8**   The constant-pressure quasi-equilibrium process.

### Analysis

There is no change in kinetic energy or change in potential energy. Work is done by movement at the boundary. Assume the process to be quasi-equilibrium. Since the pressure is constant, we have

$$_1W_2 = \int_1^2 P\,dV = P\int_1^2 dV = P(V_2 - V_1) = m(P_2 v_2 - P_1 v_1)$$

Therefore, the first law is, in terms of $Q$,

$$_1Q_2 = m(u_2 - u_1) + {}_1W_2$$
$$= m(u_2 - u_1) + m(P_2 v_2 - P_1 v_1) = m(h_2 - h_1)$$

### Solution

There is a choice of procedures to follow. State 1 is known, so $v_1$ and $h_1$ (or $u_1$) can be found. State 2 is also known, so $v_2$ and $h_2$ (or $u_2$) can be found. Using the first law and the work equation, we can calculate the heat transfer and work. Using the enthalpies, we have

$$v_1 = \frac{V_1}{m} = \frac{0.1}{0.5} = 0.2 = 0.001\,084 + x_1 0.4614$$

$$x_1 = \frac{0.1989}{0.4614} = 0.4311$$

$$h_1 = h_f + x_1 h_{fg}$$

$$= 604.74 + 0.4311 \times 2133.8 = 1524.7 \text{ kJ/kg}$$

$$h_2 = 3066.8 \text{ kJ/kg}$$

$$_1Q_2 = 0.5(3066.8 - 1524.7) = 771.1 \text{ kJ}$$

$$_1W_2 = mP(v_2 - v_1) = 0.5 \times 400(0.6548 - 0.2) = 91.0 \text{ kJ}$$

Therefore,

$$U_2 - U_1 = {_1Q_2} - {_1W_2} = 771.1 - 91.0 = 680.1 \text{ kJ}$$

The heat transfer could also have been found from $u_1$ and $u_2$:

$$u_1 = u_f + x_1 u_{fg}$$

$$= 604.31 + 0.4311 \times 1949.3 = 1444.7 \text{ kJ/kg}$$

$$u_2 = 2804.8 \text{ kJ/kg}$$

and

$$_1Q_2 = U_2 - U_1 + {_1W_2}$$

$$= 0.5(2804.8 - 1444.7) + 91.0 = 771.1 \text{ kJ}$$

## 5.6 THE CONSTANT-VOLUME AND CONSTANT-PRESSURE SPECIFIC HEATS

In this section we will consider a homogeneous phase of a substance of constant composition. This phase may be a solid, a liquid, or a gas, but no change of phase will occur. We will then define a variable termed the specific heat, the amount of heat required per unit mass to raise the temperature by one degree. Since it would be of interest to examine the relation between the specific heat and other thermodynamic variables, we note first that the heat transfer is given by Eq. 5.10. Neglecting changes in kinetic and potential energies, and assuming a simple compressible substance and quasi-equilibrium process, for which the work in Eq. 5.10 is given by Eq. 4.2, we have

$$\delta Q = dU + \delta W = dU + P\,dV$$

We find that this expression can be evaluated for two separate special cases.

1. Constant volume, for which the work term ($P\,dV$) is zero, so that the specific heat (at constant volume) is

$$C_v = \frac{1}{m}\left(\frac{\delta Q}{\delta T}\right)_v = \frac{1}{m}\left(\frac{\partial U}{\partial T}\right)_v = \left(\frac{\partial u}{\partial T}\right)_v \tag{5.14}$$

2. Constant pressure, for which the work term can be integrated and the resulting $PV$ terms at the initial and final states can be associated with the internal energy terms, as in Section 5.5, thereby leading to the conclusion that the heat transfer can be expressed in terms of the enthalpy change. The corresponding specific heat (at constant pressure) is

$$C_p = \frac{1}{m}\left(\frac{\delta Q}{\delta T}\right)_p = \frac{1}{m}\left(\frac{\partial H}{\partial T}\right)_p = \left(\frac{\partial h}{\partial T}\right)_p \tag{5.15}$$

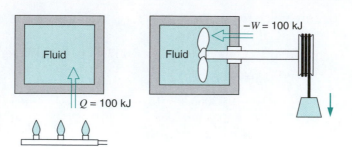

**FIGURE 5.9** Sketch showing two ways in which a given $\Delta U$ may be achieved.

Note that in each of these special cases, the resulting expression, Eq. 5.14 or 5.15, contains only thermodynamic properties, from which we conclude that the constant-volume and constant-pressure specific heats must themselves be thermodynamic properties. This means that, although we began this discussion by considering the amount of heat transfer required to cause a unit temperature change and then proceeded through a very specific development leading to Eq. 5.14 (or 5.15), the result ultimately expresses a relation among a set of thermodynamic properties and therefore constitutes a definition that is independent of the particular process leading to it (in the same sense that the definition of enthalpy in the previous section is independent of the process used to illustrate one situation in which the property is useful in a thermodynamic analysis). As an example, consider the two identical fluid masses shown in Fig. 5.9. In the first system 100 kJ of heat is transferred to it, and in the second system 100 kJ of work is done on it. Thus, the change of internal energy is the same for each, and therefore the final state and the final temperature are the same in each. In accordance with Eq. 5.14, therefore, exactly the same value for the average constant-volume specific heat would be found for this substance for the two processes, even though the two processes are very different as far as heat transfer is concerned.

## Solids and Liquids

As a special case, consider either a solid or a liquid. Since both of these phases are nearly incompressible,

$$dh = du + d(Pv) \approx du + v \, dP \qquad (5.16)$$

Also, for both of these phases, the specific volume is very small, such that in many cases

$$dh \approx du \approx C \, dT \qquad (5.17)$$

where $C$ is either the constant-volume or the constant-pressure specific heat, as the two would be nearly the same. In many processes involving a solid or a liquid, we might further assume that the specific heat in Eq. 5.17 is constant (unless the process occurs at low temperature or over a wide range of temperatures). Equation 5.17 can then be integrated to

$$h_2 - h_1 \simeq u_2 - u_1 \simeq C(T_2 - T_1) \qquad (5.18)$$

Specific heats for various solids and liquids are listed in Tables A.3 and A.4.

In other processes for which it is not possible to assume constant specific heat, there may be a known relation for $C$ as a function of temperature. Equation 5.17 could then also be integrated.

## 5.7 THE INTERNAL ENERGY, ENTHALPY, AND SPECIFIC HEAT OF IDEAL GASES

In general, for any substance the internal energy $u$ depends on the two independent properties specifying the state. For a low-density gas, however, $u$ depends primarily on $T$ and much less on the second property, $P$ or $v$. For example, consider several values for superheated vapor steam from Table B.1.3, shown in Table 5.1. From these values, it is evident that $u$ depends strongly on $T$, but not much on $P$. Also, we note that the dependence of $u$ on $P$ is less at low pressure and is much less at high temperature; that is, as the density decreases, so does dependence of $u$ on $P$ (or $v$). It is therefore reasonable to extrapolate this behavior to very low density and to assume that as gas density becomes so low that the ideal-gas model is appropriate, internal energy does not depend on pressure at all but is a function only of temperature. That is, for an ideal gas,

$$Pv = RT \qquad \text{and} \qquad u = f(T) \text{ only} \tag{5.19}$$

The relation between the internal energy $u$ and the temperature can be established by using the definition of constant-volume specific heat given by Eq. 5.14:

$$C_v = \left(\frac{\partial u}{\partial T}\right)_v$$

Because the internal energy of an ideal gas is not a function of specific volume, for an ideal gas we can write

$$C_{v0} = \frac{du}{dT}$$

$$du = C_{v0} \, dT \tag{5.20}$$

where the subscript 0 denotes the specific heat of an ideal gas. For a given mass $m$,

$$dU = mC_{v0} \, dT \tag{5.21}$$

From the definition of enthalpy and the equation of state of an ideal gas, it follows that

$$h = u + Pv = u + RT \tag{5.22}$$

Since $R$ is a constant and $u$ is a function of temperature only, it follows that the enthalpy, $h$, of an ideal gas is also a function of temperature only. That is,

$$h = f(T) \tag{5.23}$$

### TABLE 5.1
*Internal Energy for Superheated Vapor Steam*

| T, °C | P, kPa | | | |
|---|---|---|---|---|
| | 10 | 100 | 500 | 1000 |
| 200 | 2661.3 | 2658.1 | 2642.9 | 2621.9 |
| 700 | 3479.6 | 3479.2 | 3477.5 | 3475.4 |
| 1200 | 4467.9 | 4467.7 | 4466.8 | 4465.6 |

The relation between enthalpy and temperature is found from the constant-pressure specific heat as defined by Eq. 5.15:

$$C_p = \left(\frac{\partial h}{\partial T}\right)_p$$

Since the enthalpy of an ideal gas is a function of the temperature only and is independent of the pressure, it follows that

$$C_{p0} = \frac{dh}{dT}$$

$$dh = C_{p0}dT \qquad (5.24)$$

For a given mass $m$,

$$dH = mC_{p0}dT \qquad (5.25)$$

The consequences of Eqs. 5.20 and 5.24 are demonstrated in Fig. 5.10, which shows two lines of constant temperature. Since internal energy and enthalpy are functions of temperature only, these lines of constant temperature are also lines of constant internal energy and constant enthalpy. From state 1 the high temperature can be reached by a variety of paths, and in each case the final state is different. However, regardless of the path, the change in internal energy is the same, as is the change in enthalpy, for lines of constant temperature are also lines of constant $u$ and constant $h$.

Because the internal energy and enthalpy of an ideal gas are functions of temperature only, it also follows that the constant-volume and constant-pressure specific heats are also functions of temperature only. That is,

$$C_{v0} = f(T), \qquad C_{p0} = f(T) \qquad (5.26)$$

Because all gases approach ideal-gas behavior as the pressure approaches zero, the ideal-gas specific heat for a given substance is often called the zero-pressure specific heat, and the zero-pressure, constant-pressure specific heat is given the symbol $C_{p0}$. The zero-pressure, constant-volume specific heat is given the symbol $C_{v0}$. Figure 5.11 shows $C_{p0}$ as a function of temperature for a number of different substances. These values are determined by the techniques of statistical thermodynamics and will not be discussed here. A brief summary presentation of this subject is given in Appendix C. It is noted there that the principal factor causing specific heat to vary with temperature is molecular vibration. More complex molecules have multiple vibrational modes and therefore show a greater temperature dependency, as is seen in Fig. 5.11. This is an important consideration when deciding whether or not to account for specific heat variation with temperature in any particular application.

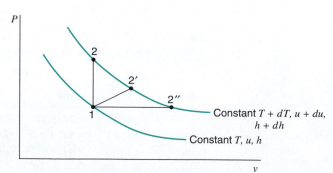

**FIGURE 5.10** Pressure-volume diagram for an ideal gas.

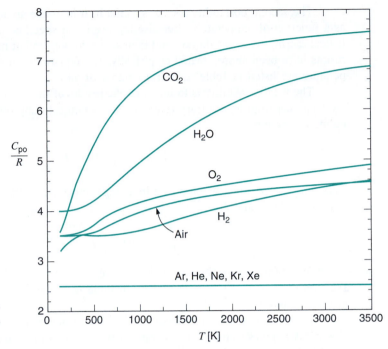

**FIGURE 5.11** Heat capacity for some gases as function of temperature.

A very important relation between the constant-pressure and constant-volume specific heats of an ideal gas may be developed from the definition of enthalpy:

$$h = u + Pv = u + RT$$

Differentiating and substituting Eqs. 5.20 and 5.24, we have

$$dh = du + R\,dT$$

$$C_{p0}\,dT = C_{v0}\,dT + R\,dT$$

Therefore,

$$C_{p0} - C_{v0} = R \tag{5.27}$$

On a mole basis this equation is written

$$\overline{C}_{p0} - \overline{C}_{v0} = \overline{R} \tag{5.28}$$

This tells us that the difference between the constant-pressure and constant-volume specific heats of an ideal gas is always constant, though both are functions of temperature. Thus, we need examine only the temperature dependency of one, and the other is given by Eq. 5.27.

Let us consider the specific heat $C_{p0}$. There are three possibilities to examine. The situation is simplest if we assume constant specific heat, that is, no temperature dependence. Then it is possible to integrate Eq. 5.24 directly to

$$h_2 - h_1 = C_{p0}(T_2 - T_1) \tag{5.29}$$

We note from Fig. 5.11 the circumstances under which this will be an accurate model. It should be added, however, that it may be a reasonable approximation under other conditions, especially if an average specific heat in the particular temperature range is used in Eq. 5.29. Values of specific heat at room temperature and gas constants for various gases are given in Table A.5.

The second possibility for the specific heat is to use an analytical equation for $C_{p0}$ as a function of temperature. Because the results of specific-heat calculations from statistical thermodynamics do not lend themselves to convenient mathematical forms, these results have been approximated empirically. The equations for $C_{p0}$ as a function of temperature are listed in Table A.6 for a number of gases.

The third possibility is to integrate the results of the calculations of statistical thermodynamics from an arbitrary reference temperature to any other temperature $T$ and to define a function

$$h_T = \int_{T_0}^{T} C_{p0}\, dT$$

This function can then be tabulated in a single-entry (temperature) table. Then, between any two states 1 and 2,

$$h_2 - h_1 = \int_{T_0}^{T_2} C_{p0}\, dT - \int_{T_0}^{T_1} C_{p0}\, dT = h_{T_2} - h_{T_1} \tag{5.30}$$

and it is seen that the reference temperature cancels out. This function $h_T$ (and a similar function $u_T = h_T - RT$) is listed for air in Table A.7. These functions are listed for other gases in Table A.8.

To summarize the three possibilities, we note that using the ideal-gas tables, Tables A.7 and A.8, gives us the most accurate answer, but that the equations in Table A.6 would give a close empirical approximation. Constant specific heat would be less accurate, except for monatomic gases and gases below room temperature. It should be remembered that all these results are a part of the ideal-gas model, which in many of our problems is not a valid assumption for the behavior of the substance.

---

**EXAMPLE 5.7**

Calculate the change of enthalpy as 1 kg of oxygen is heated from 300 to 1500 K. Assume ideal-gas behavior.

**Solution**

For an ideal gas, the enthalpy change is given by Eq. 5.24. However, we also need to make an assumption about the dependence of specific heat on temperature. Let us solve this problem in several ways and compare the answers.

Our most accurate answer for the ideal-gas enthalpy change for oxygen between 300 and 1500 K would be from the ideal-gas tables, Table A.8. This result is, using Eq. 5.30,

$$h_2 - h_1 = 1540.2 - 273.2 = 1267.0 \text{ kJ/kg}$$

The empirical equation from Table A.6 should give a good approximation to this result. Integrating Eq. 5.24, we have

$$h_2 - h_1 = \int_{T_1}^{T_2} C_{p0}\, dT = \int_{\theta_1}^{\theta_2} C_{p0}(\theta) \times 1000\, d\theta$$

$$= 1000 \left[ 0.88\theta - \frac{0.0001}{2}\theta^2 + \frac{0.54}{3}\theta^3 - \frac{0.33}{4}\theta^4 \right]_{\theta_1 = 0.3}^{\theta_2 = 1.5}$$

$$= 1241.5 \text{ kJ/kg}$$

which is lower than the first result by 2.0%.

If we assume constant specific heat, we must be concerned about what value we are going to use. If we use the value at 300 K from Table A.5, we find, from Eq. 5.29, that

$$h_2 - h_1 = C_{p0}(T_2 - T_1) = 0.922 \times 1200 = 1106.4 \text{ kJ/kg}$$

which is low by 12.7%. However, suppose we assume that the specific heat is constant at its value at 900 K, the average temperature. Substituting 900 K into the equation for specific heat from Table A.6, we have

$$C_{p0} = 0.88 - 0.0001(0.9) + 0.54(0.9)^2 - 0.33(0.9)^3$$

$$= 1.0767 \text{ kJ/kg K}$$

Substituting this value into Eq. 5.29 gives the result

$$h_2 - h_1 = 1.0767 \times 1200 = 1292.1 \text{ kJ/kg}$$

which is high by about 2.0%, a much closer result than the one using the room temperature specific heat. It should be kept in mind that part of the model involving ideal gas with constant specific heat also involves a choice of what value is to be used.

---

**EXAMPLE 5.8**

A cylinder fitted with a piston has an initial volume of 0.1 m³ and contains nitrogen at 150 kPa, 25°C. The piston is moved, compressing the nitrogen until the pressure is 1 MPa and the temperature is 150°C. During this compression process heat is transferred from the nitrogen, and the work done on the nitrogen is 20 kJ. Determine the amount of this heat transfer.

> *Control mass*: Nitrogen.
> *Initial state*: $P_1$, $T_1$, $V_1$; state 1 fixed.
> *Final state*: $P_2$, $T_2$; state 2 fixed.
> *Process*: Work input known.
> *Model*: Ideal gas, constant specific heat with value at 300 K, Table A.5.

**Analysis**

From the first law we have

$$_1Q_2 = m(u_2 - u_1) + {_1}W_2$$

**Solution**

The mass of nitrogen is found from the equation of state with the value of $R$ from Table A.5:

$$m = \frac{PV}{RT} = \frac{150 \text{ kPa} \times 0.1 \text{ m}^3}{0.2968 \dfrac{\text{kJ}}{\text{kg K}} \times 298.15 \text{ K}} = 0.1695 \text{ kg}$$

Assuming constant specific heat as given in Table A.5, we have

$$_1Q_2 = mC_{v0}(T_2 - T_1) + {_1}W_2$$

$$= 0.1695 \text{ kg} \times 0.745 \frac{\text{kJ}}{\text{kg K}} \times (150 - 25) \text{ K} - 20.0 \text{ kJ}$$

$$= 15.8 - 20.0 = -4.2 \text{ kJ}$$

It would, of course, be somewhat more accurate to use Table A.8 than to assume constant specific heat (room temperature value), but often the slight increase in accuracy does not warrant the added difficulties of manually interpolating the tables.

**In-Text Concept Questions**

**g.** To determine $v$ or $u$ for some liquid or solid is it more important that I know $P$ or $T$?

**h.** To determine $v$ or $u$ for an ideal gas is it more important that I know $P$ or $T$?

**i.** I heat 1 kg substance at constant pressure (200 kPa) 1 degree. How much heat is needed if the substance is: water at 10°C, steel at 25°C, air at 325 K, or ice at −10°C.

## 5.8 THE FIRST LAW AS A RATE EQUATION

We frequently find it desirable to use the first law as a rate equation that expresses either the instantaneous or average rate at which energy crosses the control surface as heat and work and the rate at which the energy of the control mass changes. In so doing we are departing from a strictly classical point of view, because basically classical thermodynamics deals with systems that are in equilibrium, and time is not a relevant parameter for systems that are in equilibrium. However, since these rate equations are developed from the concepts of classical thermodynamics and are used in many applications of thermodynamics, they are included in this book. This rate form of the first law will be used in the development of the first law for the control volume in Section 6.2, and in this form the first law finds extensive applications in thermodynamics, fluid mechanics, and heat transfer.

Consider a time interval $\delta t$ during which an amount of heat $\delta Q$ crosses the control surface, an amount of work $\delta W$ is done by the control mass, the internal energy change is $\Delta U$, the kinetic energy change is $\Delta \text{KE}$, and the potential energy change is $\Delta \text{PE}$. From the first law we can write

$$\Delta U + \Delta \text{KE} + \Delta \text{PE} = \delta Q - \delta W$$

Dividing by $\delta t$ we have the average rate of energy transfer as heat work and increase of the energy of the control mass:

$$\frac{\Delta U}{\delta t} + \frac{\Delta \text{KE}}{\delta t} + \frac{\Delta \text{PE}}{\delta t} = \frac{\delta Q}{\delta t} - \frac{\delta W}{\delta t}$$

Taking the limit for each of these quantities as $\delta t$ approaches zero, we have

$$\lim_{\delta t \to 0} \frac{\Delta U}{\delta t} = \frac{dU}{dt}, \qquad \lim_{\delta t \to 0} \frac{\Delta(\text{KE})}{\delta t} = \frac{d(\text{KE})}{dt}, \qquad \lim_{\delta t \to 0} \frac{\Delta(\text{PE})}{\delta t} = \frac{d(\text{PE})}{dt}$$

$$\lim_{\delta t \to 0} \frac{\delta Q}{\delta t} = \dot{Q} \qquad \text{(the heat transfer rate)}$$

$$\lim_{\delta t \to 0} \frac{\delta W}{\delta t} = \dot{W} \qquad \text{(the power)}$$

Therefore, the rate equation form of the first law is

$$\frac{dU}{dt} + \frac{d(\text{KE})}{dt} + \frac{d(\text{PE})}{dt} = \dot{Q} - \dot{W} \tag{5.31}$$

We could also write this in the form

$$\frac{dE}{dt} = \dot{Q} - \dot{W} \tag{5.32}$$

**EXAMPLE 5.9**

During the charging of a storage battery, the current $i$ is 20 A and the voltage $\mathscr{E}$ is 12.8 V. The rate of heat transfer from the battery is 10 W. At what rate is the internal energy increasing?

**Solution**

Since changes in kinetic and potential energy are insignificant, the first law can be written as a rate equation in the form of Eq. 5.31:

$$\frac{dU}{dt} = \dot{Q} - \dot{W}$$

$$\dot{W} = \mathscr{E}i = -12.8 \times 20 = -256 \text{ W} = -256 \text{ J/s}$$

Therefore,

$$\frac{dU}{dt} = \dot{Q} - \dot{W} = -10 - (-256) = 246 \text{ J/s}$$

**EXAMPLE 5.10**

**FIGURE 5.12** Sketch for Example 5.10.

A 25-kg cast-iron wood-burning stove, shown in Fig. 5.12, contains 5 kg of soft pine wood and 1 kg of air. All the masses are at room temperature, 20°C, and pressure, 101 kPa. The wood now burns and heats all the mass uniformly, releasing 1500 watts. Neglect any air flow and changes in mass and heat losses. Find the rate of change of the temperature ($dT/dt$) and estimate the time it will take to reach a temperature of 75°C.

**Solution**

C.V.: The iron, wood and air.
This is a control mass.

Energy equation rate form: $\dot{E} = \dot{Q} - \dot{W}$

We have no changes in kinetic or potential energy and no change in mass, so

$$U = m_{\text{air}} u_{\text{air}} + m_{\text{wood}} u_{\text{wood}} + m_{\text{iron}} u_{\text{iron}}$$

$$\dot{E} = \dot{U} = m_{\text{air}} \dot{u}_{\text{air}} + m_{\text{wood}} \dot{u}_{\text{wood}} + m_{\text{iron}} \dot{u}_{\text{iron}}$$

$$= (m_{\text{air}} C_{V\text{air}} + m_{\text{wood}} C_{\text{wood}} + m_{\text{iron}} C_{\text{iron}}) \frac{dT}{dt}$$

Now the energy equation has zero work, an energy release of $\dot{Q}$, and becomes

$$(m_{air}C_{Vair} + m_{wood}C_{wood} + m_{iron}C_{iron})\frac{dT}{dt} = \dot{Q} - 0$$

$$\frac{dT}{dt} = \frac{\dot{Q}}{(m_{air}C_{Vair} + m_{wood}C_{wood} + m_{iron}C_{iron})}$$

$$= \frac{1500}{1 \times 0.717 + 5 \times 1.38 + 25 \times 0.42} \frac{W \, K}{kg \, (kJ/kg)} = 0.0828 \text{ K/s}$$

Assuming the rate of temperature rise is constant, we can find the elapsed time as

$$\Delta T = \int \frac{dT}{dt} \, dt = \frac{dT}{dt}\Delta t$$

$$\Rightarrow \Delta t = \frac{\Delta T}{\dfrac{dT}{dt}} = \frac{75 - 20}{0.0828} = 664 \text{ s} = 11 \text{ min}$$

## 5.9 Conservation of Mass

In the previous sections we considered the first law of thermodynamics for a control mass undergoing a change of state. A control mass is defined as a fixed quantity of mass. The question now is whether the mass of such a system changes when its energy changes. If it does, our definition of a control mass as a fixed quantity of mass is no longer valid when the energy changes.

We know from relativistic considerations that mass and energy are related by the well-known equation

$$E = mc^2 \tag{5.33}$$

where $c$ = velocity of light and $E$ = energy. We conclude from this equation that the mass of a control mass does change when its energy changes. Let us calculate the magnitude of this change of mass for a typical problem and determine whether this change in mass is significant.

Consider a rigid vessel that contains a 1-kg stoichiometric mixture of a hydrocarbon fuel (such as gasoline) and air. From our knowledge of combustion, we know that after combustion takes place it will be necessary to transfer about 2900 kJ from the system to restore it to its initial temperature. From the first law

$$_1Q_2 = U_2 - U_1 + {_1W_2}$$

we conclude that since $_1W_2 = 0$ and $_1Q_2 = -2900$ kJ, the internal energy of this system decreases by 2900 kJ during the heat transfer process. Let us now calculate the decrease in mass during this process using Eq. 5.33.

The velocity of light, $c$, is $2.9979 \times 10^8$ m/s. Therefore,

$$2900 \text{ kJ} = 2\,900\,000 \text{ J} = m \text{ (kg)} \times (2.9979 \times 10^8 \text{ m/s})^2$$

and so

$$m = 3.23 \times 10^{-11} \text{ kg}$$

Thus, when the energy of the control mass decreases by 2900 kJ, the decrease in mass is $3.23 \times 10^{-11}$ kg.

A change in mass of this magnitude cannot be detected by even our most accurate chemical balance. Certainly, a fractional change in mass of this magnitude is beyond the accuracy required in essentially all engineering calculations. Therefore, if we use the laws of conservation of mass and conservation of energy as separate laws, we will not introduce significant error into most thermodynamic problems and our definition of a control mass as having a fixed mass can be used even though the energy changes.

## 5.10 HOW-TO SECTION

### How is $U$ (or $E$) evaluated when there are several different masses?

When a control volume is a combination of domains with different states and substances the internal energy (and $E = U + \frac{1}{2} mV^2 + mgZ$) is a summation over those as

$$U = \sum mu = m_a u_a + m_b u_b + m_c u_c$$

We then have assumed a uniform state for all mass $m_a$ and so forth. If the state varies through the mass, then we understand "$u_a$" to be the mass averaged quantity. This is the basis for the lumped (integral or averaged) analysis we normally do. The same summation must be done for all other stored quantities like mass ($m$).

### When can I use specific heats?

The specific heats $C_p$ and $C_v$ are derivatives (slopes) of the $h(T, P)$ and $u(T, v)$ functions with respect to $T$ in the direction of constant $P$ and $v$, respectively. They are really only useful for parts of the single phase regions where they do not depend on $P$ or $v$. This will be the case for a solid or a liquid (colder than $T_c$). For a gas it is true for low densities (that is, for a low enough pressure at a given $T$). To get an idea, look at Fig. 3.21 where the ideal gas (error less than 1%) region for water is shaded. Water vapor is an ideal gas for any $T$ and all $P$s less than about 40 kPa. The smaller $P$ is, the more accurate the ideal gas model is (density is smaller).

If you have a phase transformation like melting (from solid to liquid) or evaporation (from liquid to vapor) you cannot use the specific heats. In that case, you need to know the change in "$u$" as $u_{fg}$, for instance, or in "$h$" as $h_{fg}$ in case of the evaporation process.

### When can I use constant specific heats?

For a limited temperature range and a single phase you can use a specific heat evaluated at an average temperature. If $T$ is not too high, we use the reference $T_0$ with values from Table A.3, A.4 or A.5. The values of specific heats do not change much for the liquid and solid phases so we can easily evaluate changes in $u$ and $h$ using constant specific heat over even larger temperature ranges. See formulas on pages 121 and 128–129.

## SUMMARY

Conservation of energy is expressed for a cycle, and changes of total energy are then written for a control mass. Kinetic and potential energy can be changed through the work of a force acting on the control mass, and they are part of the total energy.

The internal energy and the enthalpy are introduced as substance properties with the specific heats (heat capacity) as derivatives of these with temperature. Property variations for limited cases are presented for incompressible states of a substance such as liquids and solids, and for a highly compressible state as an ideal gas. The specific heat for solids and liquids changes little with temperature, whereas the specific heat for a gas can change substantially with temperature.

The energy equation is also shown in a rate form to cover transient processes.

You should have learned a number of skills and acquired abilities from studying this chapter that will allow you to

- Recognize the components of total energy stored in a control mass
- Write the energy equation for a single uniform control mass
- Find the properties $u$ and $h$ for a given state in the Appendix B tables
- Locate a state in the tables with an entry such as $(P, h)$
- Find changes in $u$ and $h$ for liquid or solid states using Tables A.3 and A.4
- Find changes in $u$ and $h$ for ideal-gas states using Table A.5
- Find changes in $u$ and $h$ for ideal-gas states using Tables A.7 and A.8
- Recognize that forms for $C_p$ in Table A.6 are approximations to what is shown in Fig. 5.11 and the more accurate tabulations in Tables A.7 and A.8
- Formulate the conservation of mass and energy for a control mass that goes through a process involving work and heat transfers and different states
- Formulate the conservation of mass and energy for a more complex control mass where there are different masses with different states
- Use the energy equation in a rate form
- Know the difference between the general laws as the conservation of mass (continuity equation), conservation of energy (first law) and the specific laws that describes a device behavior or process

## KEY CONCEPTS AND FORMULAS

Total energy
$$E = U + \text{KE} + \text{PE} = mu + \frac{1}{2}m\mathbf{V}^2 + mgZ$$

Kinetic energy
$$\text{KE} = \frac{1}{2}m\mathbf{V}^2$$

Potential energy
$$\text{PE} = mgZ$$

Specific energy
$$e = u + \frac{1}{2}\mathbf{V}^2 + gZ$$

Enthalpy
$$h \equiv u + Pv$$

Two-phase mass average
$$u = u_f + xu_{fg} = (1 - x)u_f + xu_g$$
$$h = h_f + xh_{fg} = (1 - x)h_f + xh_g$$

Specific heat, heat capacity
$$C_v = \left(\frac{\partial u}{\partial T}\right)_v; \quad C_p = \left(\frac{\partial h}{\partial T}\right)_p$$

Solids and liquids
Incompressible, so $v = $ constant $\cong v_f$ and $v$ very small
$$C = C_v = C_p \quad \text{(Tables A.3 and A.4)}$$
$$u_2 - u_1 = C(T_2 - T_1)$$

Ideal gas

$$h_2 - h_1 = u_2 - u_1 + v(P_2 - P_1) \quad \text{(Often the second term is small.)}$$

$$h = h_f + v_f(P - P_{sat}); u \cong u_f \quad \text{(saturated at same } T\text{)}$$

$$h = u + Pv = u + RT \quad \text{(only functions of } T\text{)}$$

$$C_v = \frac{du}{dT}; C_p = \frac{dh}{dT} = C_v + R$$

$$u_2 - u_1 = \int C_v \, dT \cong C_v(T_2 - T_1)$$

$$h_2 - h_1 = \int C_p \, dT \cong C_p(T_2 - T_1)$$

Left-hand side from Table A.7 or A.8, middle from Table A.6 and right-hand side from Table A.6 at a $T_{avg}$ or from Table A.5 at 25°C

Energy equation rate form    $\dot{E} = \dot{Q} - \dot{W}$ (rate = +in − out)

Energy equation integrated    $E_2 - E_1 = {}_1Q_2 - {}_1W_2$ (change = +in − out)

$$m(e_2 - e_1) = m(u_2 - u_1) + \frac{1}{2}m(\mathbf{V}_2^2 - \mathbf{V}_1^2) + mg(Z_2 - Z_1)$$

Multiple masses, states    $E = m_A e_A + m_B e_B + m_C e_C + \cdots$

# HOMEWORK PROBLEMS

## Concept problems

**5.1** What is 1 cal in SI units and what is the name given to 1 N–m?

**5.2** Why do we write $\Delta E$ or $E_2 - E_1$ whereas we write ${}_1Q_2$ and ${}_1W_2$?

**5.3** When you wind up a spring in a toy or stretch a rubber band, what happens in terms of work, energy, and heat transfer? Later, when they are released, what happens then?

**5.4** CV $A$ is the mass inside a piston-cylinder, CV $B$ is that plus the piston, outside of which is the standard atmosphere. See Fig. P5.4. Write the energy equation and work term for the two CVs assuming we have a non-zero $Q$ between state 1 and state 2.

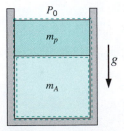

**FIGURE P5.4**

**5.5** Saturated water vapor has a maximum for $u$ and $h$ at around 235°C. Is it similar for other substances?

**5.6** A rigid tank with pressurized air is used to a) increase the volume of a linear spring loaded piston cylinder (cylindrical geometry) arrangement and b) to blow up a spherical balloon. Assume that in both cases $P = A + BV$ with the same $A$ and $B$. What is the expression for the work term in each situation?

**5.7** You heat a gas 10 K at $P = C$. Which one in Table A.5 requires most energy? Why?

**5.8** A 500 W electric space heater with a small fan inside heats air by blowing it over a hot electrical wire. For each control volume: a) wire only; b) all the room air; and c) total room plus the heater, specify the storage, work, and heat transfer terms as +500 W or −500 W or 0. Neglect any $\dot{Q}$ through the room walls or windows.

## Kinetic and potential energy

**5.9** A hydraulic hoist raises a 1750-kg car 1.8 m in an auto repair shop. The hydraulic pump has a constant pressure of 800 kPa on its piston. What is the increase in potential energy of the car and how much volume should the pump displace to deliver that amount of work?

**5.10**  A 1200 kg car is accelerated from 30 to 50 km/h in 5 s. How much work is that? If you continue from 50 to 70 km/h in 5 s, is that the same?

**5.11**  Airplane takeoff from an aircraft carrier is assisted by a steam-driven piston/cylinder with an average pressure of 1250 kPa. A 17 500-kg airplane should be accelerated from zero to a speed of 30 m/s with 30% of the energy coming from the steam piston. Find the needed piston displacement volume.

**5.12**  Solve Problem 5.11, but assume the steam pressure in the cylinder starts at 1000 kPa, dropping linearly with volume to reach 100 kPa at the end of the process.

**5.13**  A 1200-kg car accelerates from zero to 100 km/h over a distance of 400 m. The road at the end of the 400 m is at 10 m higher elevation. What is the total increase in the car kinetic and potential energy?

**5.14**  A 25 kg piston is above a gas in a long vertical cylinder. Now the piston is released from rest and accelerates up in the cylinder reaching the end 5 m higher at a velocity of 25 m/s. The gas pressure drops during the process, so the average is 600 kPa with an outside atmosphere at 100 kPa. Neglect the change in gas kinetic and potential energy, and find the needed change in the gas volume.

**5.15**  A mass of 5 kg is tied to an elastic cord 5 m long and dropped from a tall bridge. Assume the cord, once straight, acts as a spring with $k = 100$ N/m. Find the velocity of the mass when the cord is straight (5 m down). At what level does the mass come to rest after bouncing up and down?

## Properties (*u, h*) from general tables

**5.16**  Find the missing properties of $T$, $P$, $v$, $u$, $h$, and $x$ if applicable and plot the location of the three states as points in the $T$–$v$ and the $P$–$v$ diagrams.

a. Water at 5000 kPa, $u = 800$ kJ/kg
b. Water at 5000 kPa, $v = 0.06$ m³/kg
c. R-134a at 35°C, $v = 0.01$ m³/kg

**5.17**  Find the phase and the missing properties of $P$, $T$, $v$, $u$ and $x$.

a. Water at 5000 kPa, $u = 3000$ kJ/kg
b. Ammonia at 50°C, $v = 0.08506$ m³/kg
c. Ammonia at 28°C, 1200 kPa
d. R-134a at 20°C, $u = 350$ kJ/kg

**5.18**  Find the missing properties.

a. H₂O,     $T = 250°C$,      $P = ?\ u = ?$
              $v = 0.02$ m³/kg,

b. N₂       $T = 120$ K,       $x = ?\ h = ?$
              $P = 0.8$ MPa,

c. H₂O,     $T = -2°C$,        $u = ?\ v = ?$
              $P = 100$ kPa,

d. R-134a,  $P = 200$ kPa,     $u = ?\ T = ?$
              $v = 0.12$ m³/kg,

**5.19**  Two kg water at 120°C with a quality of 25% has its temperature raised 20°C in a constant volume process. What are the new quality and specific internal energy?

**5.20**  Two kg water at 200 kPa with a quality of 25% has its temperature raised 20°C in a constant pressure process. What is the change in enthalpy?

**5.21**  Saturated liquid water at 20°C is compressed to a higher pressure with constant temperature. Find the changes in $u$ and $h$ from the initial state when the final pressure is

a. 500 kPa
b. 2000 kPa
c. 20 000 kPa

**5.22**  Find the missing properties and give the phase of the ammonia, NH₃.

a. $T = 65°C$, $P = 600$ kPa      $u = ?$   $v = ?$
b. $T = 20°C$, $P = 100$ kPa      $u = ?$   $v = ?$   $x = ?$
c. $T = 50°C$, $v = 0.1185$ m³/kg   $u = ?$   $v = ?$   $x = ?$

**5.23**  i.  Find the phase and missing properties of $P$, $T$, $v$, $u$, and $x$.

a. H₂O at $P = 5000$ kPa,   $u = 1000$ kJ/kg   (steam table reference)
b. R-134a at $T = 20°C$, $u = 300$ kJ/kg
c. N₂ at 250 K, $P = 200$ kPa

ii.  Show the three states as labeled dots in a $T$–$v$ diagram with correct position relative to the two-phase region.

**5.24**  Find the missing properties and give the phase of the substance.

a. H₂O,     $T = 120°C$,       $u = ?\ P = ?\ x = ?$
              $v = 0.5$ m³/kg,

b. H₂O,     $T = 100°C$,       $u = ?\ x = ?\ v = ?$
              $P = 10$ MPa,

c. N₂,      $T = 200$ K,        $v = ?\ u = ?$
              $P = 200$ kPa,

d. NH₃,     $T = 100°C$,        $P = ?\ x = ?$
              $v = 0.1$ m³/kg,

e. N₂,      $T = 100$ K,        $v = ?\ u = ?$
              $x = 0.75$,

**5.25** Find the missing properties among $T, P, v, u, h$, and $x$ (if applicable), give the phase of the substance, and indicate the states relative to the two-phase region in both a $T–v$ and a $P–v$ diagram.

a. R-12,     $P = 500$ kPa, $h = 230$ kJ/kg
b. R-22,     $T = 10°C, u = 200$ kJ/kg
c. R-134a,    $T = 40°C, h = 400$ kJ/kg

## Energy equation: simple process

**5.26** A 100-L rigid tank contains nitrogen ($N_2$) at 900 K and 3 MPa. The tank is now cooled to 100 K. What are the work and heat transfer for the process?

**5.27** A rigid container has 0.75 kg of water at 300°C, 1200 kPa. The water is now cooled to a final pressure of 300 kPa. Find the final temperature, the work, and the heat transfer in the process.

**5.28** I have 2 kg of liquid water at 20°C, 100 kPa. I now add 20 kJ of energy at a constant pressure. How hot does it get if it is heated? How fast does it move if it is pushed by a constant horizontal force? How high does it go if it is raised straight up?

**5.29** A cylinder fitted with a frictionless piston contains 2 kg of superheated refrigerant R-134a vapor at 350 kPa, 100°C. The cylinder is now cooled so that the R-134a remains at constant pressure until it reaches a quality of 75%. Calculate the heat transfer in the process.

**5.30** Ammonia at 0°C with a quality of 60% is contained in a rigid 200-L tank. The tank and ammonia are now heated to a final pressure of 1 MPa. Determine the heat transfer for the process.

**5.31** A piston/cylinder contains 1 kg of water at 20°C with volume 0.1 m³. By mistake someone locks the piston, preventing it from moving while we heat the water to saturated vapor. Find the final temperature and the amount of heat transfer in the process.

**5.32** A test cylinder with constant volume of 0.1 L contains water at the critical point. It now cools down to room temperature of 20°C. Calculate the heat transfer from the water.

**5.33** Find the heat transfer for the process in Problem 4.30.

**5.34** A 10-L rigid tank contains R-22 at −10°C with a quality of 80%. A 10-A electric current (from a 6-V battery) is passed through a resistor inside the tank for 10 min, after which the R-22 temperature is 40°C. What was the heat transfer to or from the tank during this process?

**5.35** A constant-pressure piston/cylinder assembly contains 0.2 kg of water as saturated vapor at 400 kPa. It is now cooled so that the water occupies half the original volume. Find the heat transfer in the process.

**5.36** Two kilograms of water at 120°C with a quality of 25% has its temperature raised 20°C in a constant-volume process as in Fig. P5.36. What are the heat transfer and work in the process?

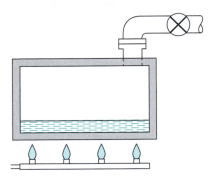

**FIGURE P5.36**

**5.37** A 25-kg mass moves at 25 m/s. Now a brake system brings the mass to a complete stop with a constant deceleration over a period of 5 s. The brake energy is absorbed by 0.5 kg of water initially at 20°C and 100 kPa. Assume the mass is at constant $P$ and $T$. Find the energy the brake removes from the mass and the temperature increase of the water, assuming its pressure is constant.

**5.38** An insulated cylinder fitted with a piston contains R-12 at 25°C with a quality of 90% and a volume of 45 L. The piston is allowed to move, and the R-12 expands until it exists as saturated vapor. During this process the R-12 does 7.0 kJ of work against the piston. Determine the final temperature, assuming the process is adiabatic.

**5.39** Find the heat transfer for the process in Problem 4.34.

**5.40** A water-filled reactor with volume of 1 m³ is at 20 MPa and 360°C and placed inside a containment room, as shown in Fig. P5.40. The room is well insulated and initially evacuated. Due to a failure, the reactor ruptures and the water fills the containment room. Find the minimum room volume so that the final pressure does not exceed 200 kPa.

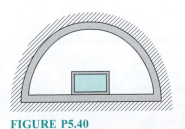

**FIGURE P5.40**

**5.41** A piston/cylinder arrangement has the piston loaded with outside atmospheric pressure and the piston mass to a pressure of 150 kPa, as shown in Fig. P5.41. It contains water at $-2°C$, which is then heated until the water becomes saturated vapor. Find the final temperature and specific work and heat transfer for the process.

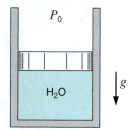

**FIGURE P5.41**

**5.42** A piston/cylinder assembly contains 1 kg of liquid water at 20°C and 300 kPa. There is a linear spring mounted on the piston such that when the water is heated the pressure reaches 1 MPa with a volume of 0.1 $m^3$. Find the final temperature and the heat transfer in the process.

**5.43** Find the heat transfer for the process in Problem 4.35.

**5.44** A closed steel bottle shown in Fig. P5.44 contains ammonia at $-20°C$, $x = 20\%$ and the volume is 0.05 $m^3$. It has a safety valve that opens at a pressure of 1.4 MPa. By accident, the bottle is heated until the safety valve opens. Find the temperature and heat transfer when the valve first opens.

**FIGURE P5.44**

**5.45** Two kilograms of water at 200 kPa with a quality of 25% has its temperature raised 20°C in a constant-pressure process. What are the heat transfer and work in the process?

**5.46** Two kilograms of nitrogen at 100 K, $x = 0.5$ are heated in a constant pressure process to 300 K in a piston/cylinder arrangement. Find the initial and final volumes and the total heat transfer required.

**5.47** Two tanks, each with a volume of 1 $m^3$, are connected by a valve and line, as shown in Fig. P5.47. Tank $A$ is filled with R-134a at 20°C with a quality of 15%. Tank $B$ is evacuated. The valve is opened and saturated vapor flows from $A$ into $B$ until the pressures become equal. The process occurs slowly enough that all temperatures stay at 20°C during the process. Find the total heat transfer to the R-134a during the process.

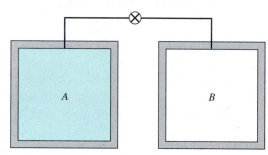

**FIGURE P5.47**

**5.48** Consider the same system as in the previous problem. Let the valve be opened and transfer enough heat to both tanks so that all the liquid disappears. Find the necessary heat transfer.

**5.49** Superheated refrigerant R-134a at 20°C and 0.5 MPa is cooled in a piston/cylinder arrangement at constant temperature to a final two-phase state with quality of 50%. The refrigerant mass is 5 kg, and during this process 500 kJ of heat is removed. Find the initial and final volumes and the necessary work.

**5.50** A 1-L capsule of water at 700 kPa and 150°C is placed in a larger insulated and otherwise evacuated vessel. The capsule breaks and its contents fill the entire volume. If the final pressure should not exceed 125 kPa, what should the vessel volume be?

**5.51** A rigid tank is divided into two rooms, both containing water, by a membrane, as shown in Fig. P5.51. Room $A$ is at 200 kPa, $v = 0.5$ $m^3/kg$, $V_A = 1$ $m^3$, and room $B$ contains 3.5 kg at 0.5 MPa, 400°C. The membrane now ruptures and heat transfer takes place so the water comes to a uniform state at 100°C. Find the heat transfer during the process.

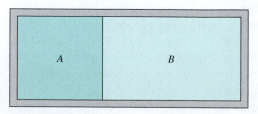

**FIGURE P5.51**

**5.52** A 10-m-high open cylinder, with $A_{cyl} = 0.1$ m$^2$, contains 20°C water above and 2 kg of 20°C water below a 198.5-kg thin insulated floating piston, as shown in Fig. P5.52. Assume standard $g$, $P_0$. Now heat is added to the water below the piston so that it expands, pushing the piston up, causing the water on top to spill over the edge. This process continues until the piston reaches the top of the cylinder. Find the final state of the water below the piston $(T, P, v)$ and the heat added during the process.

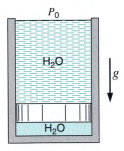

**FIGURE P5.52**

## Energy equation: multi-step solution

**5.53** Ten kilograms of water in a piston/cylinder arrangement exist as saturated liquid/vapor at 100 kPa, with a quality of 50%. The system is now heated so that the volume triples. The mass of the piston is such that a cylinder pressure of 200 kPa will float it, as in Fig. P5.59. Find the final temperature and the heat transfer in the process.

**5.54** A vertical cylinder fitted with a piston contains 5 kg of R-22 at 10°C, as shown in Fig. P5.54. Heat is transferred to the system, causing the piston to rise until it reaches a set of stops, at which point the volume has doubled. Additional heat is transferred until the temperature inside reaches 50°C, at which point the pressure inside the cylinder is 1.3 MPa.

a. What is the quality at the initial state?
b. Calculate the heat transfer for the overall process.

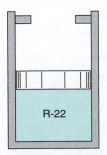

**FIGURE P5.54**

**5.55** Find the heat transfer in Problem 4.59.

**5.56** Refrigerant-12 is contained in a piston/cylinder arrangement at 2 MPa and 150°C with a massless piston against the stops, at which point $V = 0.5$ m$^3$. The side above the piston is connected by an open valve to an air line at 10°C and 450 kPa, as shown in Fig. P5.56. The whole setup now cools to the surrounding temperature of 10°C. Find the heat transfer and show the process in a $P$–$v$ diagram.

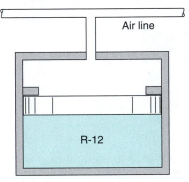

**FIGURE P5.56**

**5.57** Find the heat transfer in Problem 4.58.

**5.58** Calculate the heat transfer for the process described in Problem 4.62.

**5.59** A cylinder/piston arrangement contains 5 kg of water at 100°C with $x = 20\%$ and the piston, of $m_p = 75$ kg, resting on some stops, similar to Fig. P5.59. The outside pressure is 100 kPa, and the cylinder area is $A = 24.5$ cm$^2$. Heat is now added until the water reaches a saturated vapor state. Find the initial volume, final pressure, work, and heat transfer terms and show the $P$–$v$ diagram.

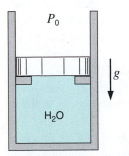

**FIGURE P5.59**

## Energy equation: solids and liquids

**5.60** In a sink, 5 liters of water at 70°C is combined with 1 kg aluminum pots, 1 kg of steel flatware, and 1 kg of glass—all put in at 20°C. What is the final uniform temperature neglecting any heat loss and work?

**5.61.** Because a hot water supply must also heat some pipe mass as it is turned on, the water does not come out hot right away. Assume 80°C liquid water at 100 kPa is cooled to 45°C as it heats 15 kg of copper pipe from 20 to 45°C. How much mass (kg) of water is needed?

**5.62** A copper block of volume 1 L is heat treated at 500°C and now cooled in a 200-L oil bath initially at 20°C, as shown in Fig. P5.62. Assuming no heat transfer with the surroundings, what is the final temperature?

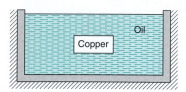

**FIGURE P5.62**

**5.63** A house is being designed to use a thick concrete floor mass as thermal storage material for solar energy heating. The concrete is 30 cm thick and the area exposed to the sun during the daytime is 4 m × 6 m. It is expected that this mass will undergo an average temperature rise of about 3°C during the day. How much energy will be available for heating during the nighttime hours?

**5.64** A 1-kg steel pot contains 1 kg of liquid water, both at 15°C. It is now put on the stove, where it is heated to the boiling point of the water. Neglect any air being heated and find the total amount of energy needed.

**5.65** A car with mass 1275 kg is driven at 60 km/h when the brakes are applied quickly to decrease its speed to 20 km/h. Assume that the brake pads have a 0.5-kg mass with a heat capacity of 1.1 kJ/kg K and that the brake disks/drums are 4.0 kg of steel. Further assume that both masses are heated uniformly. Find the temperature increase in the brake assembly.

**5.66** A computer cpu chip consists of 50 g silicon, 20 g copper, 50 g polyvinyl chloride (plastic). It heats from 15°C to 70°C as the computer is turned on. How much energy does the heating require?

**5.67** A 25-kg steel tank initially at −10°C is filled up with 100 kg of milk (assumed to have the same properties as water) at 30°C. The milk and the steel come to a uniform temperature of +5°C in a storage room. How much heat transfer is needed for this process?

**5.68** An engine, shown in Fig. P5.68, consists of a 100-kg cast iron block with a 20-kg aluminum head, 20 kg of steel parts, 5 kg of engine oil, and 6 kg of glycerine (antifreeze). Everything begins at 5°C, and as the engine starts we want to know how hot it becomes if it absorbs a net of 7000 kJ before it reaches a steady uniform temperature.

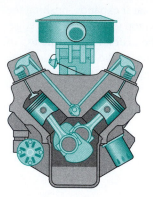

Automobile engine
**FIGURE P5.68**

# Properties ($u$, $h$, $C_v$ and $C_p$), ideal gas

**5.69** Use the ideal-gas air Table A.7 to evaluate the heat capacity $C_p$ at 300 K as a slope of the curve $h(T)$ by $\Delta h/\Delta T$. How much larger is it at 1000 K and 1500 K?

**5.70** We want to find the change in $u$ for carbon dioxide between 600 K and 1200 K.

a. Find it from a constant $C_{v0}$ from Table A.5.
b. Find it from a $C_{v0}$ evaluated from the equation in Table A.6 at the average $T$.
c. Find it from the values of $u$ listed in Table A.8.

**5.71** Water at 400 kPa is brought from 150°C to 1200°C in a constant pressure process. Evaluate the change in specific internal energy using a) the steam tables, b) the ideal gas Tables A.8, and c) the specific heat A.5.

**5.72** We want to find the increase in temperature of nitrogen gas at 1200 K when the specific internal energy is increased with 40 kJ/kg.

a. Find it from a constant $C_{v0}$ from Table A.5.
b. Find it from the values of $u$ listed in Table A.8.

**5.73** For a particular application the change in enthalpy of carbon dioxide from 30 to 1500°C at 100 kPa is needed. Consider the following methods and indicate the most accurate one.

a. Using a constant specific heat and reading the value from Table A.5.
b. Reading the enthalpy from ideal gas tables in Table A.8.

**5.74** An ideal gas is heated from 500 to 1500 K. Find the change in enthalpy using constant specific heat from Table A.5 (room temperature value) and discuss the accuracy of the result if the gas is

a. Argon
b. Oxygen
c. Carbon dioxide

## Energy equation: ideal gas

**5.75** Air inside a rigid tank is heated from 300 to 350 K. Find $_1q_2$. What if from 1300 to 1350 K?

**5.76** A rigid insulated tank is separated into two rooms by a stiff plate. Room $A$ of 0.5 m³ contains air at 250 kPa and 300 K and room $B$ of 1 m³ has air at 150 kPa and 1000 K. The plate is removed and the air comes to a uniform state without any heat transfer. Find the final pressure and temperature.

**5.77** A 250 L rigid tank contains methane at 500 K, 1500 kPa. It is now cooled down to 300 K. Find the mass of methane and the heat transfer using ideal gas.

**5.78** A rigid container has 2 kg of carbon dioxide gas at 100 kPa and 1200 K that is heated to 1400 K. Solve for the heat transfer using (a) the heat capacity from Table A.5 and (b) properties from Table A.8.

**5.79** Do the previous problem for nitrogen, $N_2$, gas.

**5.80** Find the heat transfer in Problem 4.36.

**5.81** An insulated cylinder is divided into two parts of 1 m³ each by an initially locked piston, as shown in Fig. P5.81. Side $A$ has air at 200 kPa, 300 K, and side $B$ has air at 1.0 MPa, 1000 K. The piston is now unlocked so that it is free to move, and it conducts heat so that the air comes to a uniform temperature $T_A = T_B$. Find the mass in both $A$ and $B$ and the final $T$ and $P$.

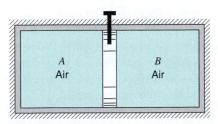

**FIGURE P5.81**

**5.82** A piston cylinder contains air at 600 kPa, 290 K, and a volume of 0.01 m³. A constant pressure process gives 18 kJ of work out. Find the final temperature of the air and the heat transfer input.

**5.83** A cylinder with a piston restrained by a linear spring contains 2 kg of carbon dioxide at 500 kPa and 400°C. It is cooled to 40°C, at which point the pressure is 300 kPa. Calculate the heat transfer for the process.

**5.84** Water at 100 kPa and 400 K is heated electrically adding 700 kJ/kg in a constant pressure process. Find the final temperature using

a. The water Table B. 1
b. The ideal-gas Table A.8
c. Constant specific heat from Table A.5

**5.85** Find the specific heat transfer for the helium in Problem 4.52.

**5.86** A piston/cylinder contains 1.5 kg of air at 300 K and 150 kPa. It is now heated up in a two-step process. First constant volume to 1000 K (state 2) and then followed by a constant pressure process to 1500 K, state 3. Find the heat transfer for the process.

**5.87** A piston/cylinder has 0.5 kg of air at 2000 kPa, 1000 K as shown in Fig. P5.87. The cylinder has stops, so $V_{min} = 0.03$ m³. The air now cools to 400 K by heat transfer to the ambient. Find the final volume and pressure of the air (does it hit the stops?) and the work and heat transfer in the process.

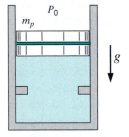

**FIGURE P5.87**

**5.88** A spring-loaded piston/cylinder contains 1.5 kg of air at 27°C and 160 kPa. It is now heated to 900 K in a process where the pressure is linear in volume to a final volume of twice the initial volume. Plot the process in a $P$–$v$ diagram and find the work and heat transfer.

## Energy equation: polytropic process

**5.89** A piston/cylinder device contains 0.1 kg of air at 300 K and 100 kPa. The air is now slowly compressed in an isothermal ($T$ = constant) process to a final pressure of 250 kPa. Show the process in a $P$–$V$ diagram and find both the work and heat transfer in the process.

**5.90** Find the heat transfer for Problem 4.45.

**5.91** Oxygen at 300 kPa and 100°C is in a piston/cylinder arrangement with a volume of 0.1 m³. It is now compressed in a polytropic process with exponent $n = 1.2$ to a final temperature of 200°C. Calculate the heat transfer for the process.

**5.92** Helium gas expands from 125 kPa, 350 K and 0.25 m³ to 100 kPa in a polytropic process with $n = 1.667$. How much heat transfer is involved?

**5.93** Find the heat transfer for Problem 4.48.

**5.94** A piston/cylinder has nitrogen gas at 750 K and 1500 kPa shown in Fig. P5.94. Now it is expanded in a polytropic process with $n = 1.2$ to $P = 750$ kPa. Find the final temperature, the specific work and specific heat transfer in the process.

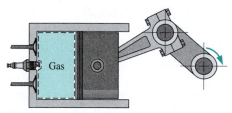

**FIGURE P5.94**

**5.95** A piston/cylinder setup contains 0.001 m³ air at 300 K and 150 kPa. The air is now compressed in a process in which $PV^{1.25} = C$ to a final pressure of 600 kPa. Find the work performed by the air and the heat transfer.

**5.96** A piston/cylinder assembly in a car contains 0.2 L of air at 90 kPa and 20°C, as shown in Fig. P5.96. The air is compressed in a quasi-equilibrium polytropic process with polytropic exponent $n = 1.25$ to a final volume six times smaller. Determine the final pressure and temperature, and the heat transfer for the process.

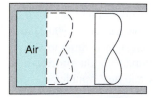

**FIGURE P5.96**

**5.97** A piston/cylinder arrangement of initial volume 0.025 m³ contains saturated water vapor at 180°C. The steam now expands in a polytropic process with exponent $n = 1$ to a final pressure of 200 kPa while it does work against the piston. Determine the heat transfer for this process.

**5.98** Air is expanded from 400-kPa and 600 K in a polytropic process to 150 kPa and 400 K in a piston/cylinder arrangement. Find the polytropic exponent $n$ and the work and heat transfer per kg of air using constant heat capacity from Table A.5.

**5.99** A piston/cylinder assembly has 1 kg of propane gas at 700 kPa and 40°C. The piston cross-sectional area is 0.5 m², and the total external force restraining the piston is directly proportional to the cylinder volume squared. Heat is transferred to the propane until its temperature reaches 700°C. Determine the final pressure inside the cylinder, the work done by the propane, and the heat transfer during the process.

**5.100** A piston/cylinder setup contains argon gas at 140 kPa and 10°C, and the volume is 100 L. The gas is compressed in a polytropic process to 700 kPa and 280°C. Calculate the polytropic exponent and the heat transfer during the process.

## Energy equation in rate form

**5.101** A crane uses 2 kW to raise a 100 kg box 20 m. How much time does it take?

**5.102** A crane lifts a load of 450 kg vertically upward with a power input of 1 kW. How fast can the crane lift the load?

**5.103** A pot of 1.2 kg water at 20°C is put on a stove supplying 1250 W to the water. What is the rate of temperature increase (K/s)?

**5.104** A pot of 1.2 kg water at 20°C is put on a stove supplying 1250 W to the water. After how long time can I expect it to come to a boil (100°C)?

**5.105** A computer in a closed room of volume 200 m³ dissipates energy at a rate of 10 kW. The room has 50 kg of wood, 25 kg of steel, and air, with all material at 300 K and 100 kPa. Assuming all the mass heats up uniformly, how long will it take to increase the temperature 10°C?

**5.106** The rate of heat transfer to the surroundings from a person at rest is about 400 kJ/h. Suppose that the ventilation system fails in an auditorium containing 100 people. Assume the energy goes into the air of volume 1500 m³ initially at 300 K and 101 kPa. Find the rate (degrees per minute) of the air temperature change.

**5.107** A mass of 3 kg nitrogen gas at 2000 K, $V = C$, cools with 500 W. What is $dT/dt$?

**5.108** The heaters in a spacecraft suddenly fail. Heat is lost by radiation at the rate of 100 kJ/h, and the electric instruments generate 75 kJ/h. Initially, the air is at 100 kPa and 25°C with a volume of 10 m³. How long will it take to reach an air temperature of −20°C?

**5.109** A steam-generating unit heats saturated liquid water at constant pressure of 200 kPa in a piston/cylinder device. If 1.5 kW of power is added by heat transfer, find the rate (kg/s) at which saturated vapor is made.

**5.110** A pot of water is boiling on a stove supplying 325 W to the water. What is the rate of mass (kg/s) vaporizing assuming a constant pressure process?

**5.111** A small elevator is being designed for a construction site. It is expected to carry four 75-kg workers to the top of a 100-m-tall building in less than 2 min. The elevator cage will have a counterweight to balance its mass. What is the smallest size (power) electric motor that can drive this unit?

**5.112** As fresh poured concrete hardens, the chemical transformation releases energy at a rate of 2 W/kg. Assume the center of a poured layer does not have any heat loss and that it has an average heat capacity of 0.9 kJ/kg K. Find the temperature rise during 1 h of the hardening (curing) process.

**5.113** Water is in a piston/cylinder maintaining constant $P$ at 700 kPa, quality 90% with a volume of 0.1 $m^3$. A heater is turned on, heating the water with 2.5 kW. How long does it take to vaporize all the liquid?

**5.114** A drag force on a car, with frontal area $A = 2\ m^2$, driving at 80 km/h in air at 20°C is $F_d = 0.225\ A\ \rho_{air} \mathbf{V}^2$. How much power is needed and what is the traction force?

# Chapter 6

# First-Law Analysis for a Control Volume

## TOPICS DISCUSSED

CONSERVATION OF MASS AND THE CONTROL VOLUME

THE FIRST LAW OF THERMODYNAMICS FOR A
CONTROL VOLUME

THE STEADY-STATE PROCESS

EXAMPLES OF STEADY-STATE PROCESSES

THE TRANSIENT PROCESS

HOW-TO SECTION

In the preceding chapter we developed the first-law analysis (energy balance) for a control mass going through a process. Many applications in thermodynamics do not readily lend themselves to a control mass approach but are conveniently handled by the more general control volume technique, as discussed in Chapter 2. The present chapter is concerned with development of the control volume forms of the conservation of mass and energy in situations where there are flows of substance present.

## 6.1 CONSERVATION OF MASS AND THE CONTROL VOLUME

A control volume is a volume in space in which one has interest for a particular study or analysis. The surface of this control volume is referred to as a control surface and always consists of a closed surface. The size and shape of the control volume are completely arbitrary and are so defined as to best suit the analysis to be made. The surface may be fixed, or it may move so that it expands or contracts. However, the surface must be defined relative to some coordinate system. In some analyses it may be desirable to consider a rotating or moving coordinate system and to describe the position of the control surface relative to such a coordinate system.

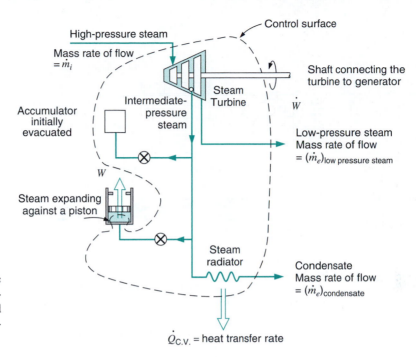

**FIGURE 6.1** Schematic diagram of a control volume showing mass and energy transfers and accumulation.

$\dot{Q}_{C.V.}$ = heat transfer rate

Mass as well as heat and work can cross the control surface, and the mass in the control volume, as well as the properties of this mass, can change with time. Fig. 6.1 shows a schematic diagram of a control volume that includes heat transfer, shaft work, moving boundary work, accumulation of mass within the control volume, and several mass flows. It is important to identify and label each flow of mass and energy and the parts of the control volume that can store (accumulate) mass.

Let us consider the conservation of mass law as it relates to the control volume. The physical law concerning mass, recalling Section 5.9, says that we cannot create or destroy mass. We will express this law in a mathematical statement about the mass in the control volume. To do this we must consider all the mass flows into and out of the control volume and the net increase of mass within the control volume. As a somewhat simpler control volume we consider a tank with a cylinder and piston and two pipes attached as shown in Fig. 6.2. The rate of change of mass inside the control volume can be different from zero if we add or take a flow of mass out as

$$\text{Rate of change} = +\text{in} - \text{out}$$

With several possible flows this is written as

$$\frac{dm_{C.V.}}{dt} = \sum \dot{m}_i - \sum \dot{m}_e \tag{6.1}$$

stating that if the mass inside the control volume changes with time it is because we add some mass or take some mass out. There are no other means by which the mass inside the control volume could change. Equation 6.1 expressing the conservation of mass is commonly termed the continuity equation. While this form of the equation is sufficient for the majority of applications in thermodynamics, it is frequently rewritten in terms of the local fluid properties in the study of fluid mechanics and heat transfer. In this text we are mainly concerned with the overall mass balance and thus consider Eq. 6.1 as the general expression for the continuity equation.

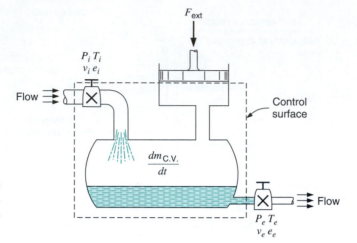

**FIGURE 6.2** Schematic diagram of a control volume for the analysis of the continuity equation.

Since Eq. 6.1 is written for the total mass (lumped form) inside the control volume we may have to consider several contributions to the mass as

$$m_{\text{C.V.}} = \int \rho \, dV = \int (1/v) dV = m_A + m_B + m_C + \cdots$$

Such a summation is needed when the control volume has several accumulation units with different states of the mass.

Let us now consider the mass flow rates across the control volume surface in a little more detail. For simplicity we assume the fluid is flowing in a pipe or duct as illustrated in Fig. 6.3. We wish to relate the total flow rate that appears in Eq. 6.1 to the local properties of the fluid state. The flow across the control volume surface can be indicated with an average velocity shown to the left of the valve or with a distributed velocity over the cross section as shown to the right of the valve.

The volume flow rate is

$$\dot{V} = \mathbf{V}A = \int \mathbf{V}_{\text{local}} \, dA \tag{6.2}$$

so the mass flow rate becomes

$$\dot{m} = \rho_{\text{avg}} \dot{V} = \dot{V}/v = \int (\mathbf{V}_{\text{local}}/v) dA = \mathbf{V}A/v \tag{6.3}$$

where often the average velocity is used. It should be noted that this result, Eq. 6.3, has been developed for a stationary control surface and we tacitly assumed the flow was normal to the surface. This expression for the mass flow rate applies to any of the various flow streams entering or leaving the control volume, subject to the assumptions mentioned.

**FIGURE 6.3**    The flow across a control volume surface with a flow cross-sectional area of $A$. Left of valve shown as an average velocity and to the right of valve shown as a distributed flow across area.

**EXAMPLE 6.1**

Air is flowing in a 0.2-m-diameter pipe at a uniform velocity of 0.1 m/s. The temperature is 25°C and the pressure 150 kPa. Determine the mass flow rate.

**Solution**

From Eq. 6.3 the mass flow rate is

$$\dot{m} = \mathbf{V}A/v$$

For air, using $R$ from Table A.5, we have

$$v = \frac{RT}{P} = \frac{0.287 \text{ kJ/kg K} \times 298.2 \text{ K}}{150 \text{ kPa}} = 0.5705 \text{ m}^3/\text{kg}$$

The cross-sectional area is

$$A = \frac{\pi}{4}(0.2)^2 = 0.0314 \text{ m}^2$$

Therefore,

$$\dot{m} = \mathbf{V}A/v = 0.1 \text{ m/s} \times 0.0314 \text{ m}^2/0.5705 \text{ m}^3/\text{kg} = 0.0055 \text{ kg/s}$$

| In-Text Concept Question |

**a.** A mass flow rate into a control volume requires a normal velocity component. Why?

## 6.2 THE FIRST LAW OF THERMODYNAMICS FOR A CONTROL VOLUME

We have already considered the first law of thermodynamics for a control mass, which consists of a fixed quantity of mass, and noted, Eq. 5.5, that it may be written

$$E_2 - E_1 = {}_1Q_2 - {}_1W_2$$

We have also noted that this may be written as an instantaneous rate equation as

$$\frac{dE_{\text{C.M.}}}{dt} = \dot{Q} - \dot{W} \tag{6.4}$$

To write the first law as a rate equation for a control volume, we proceed in a manner analogous to that used in developing a rate equation for the law of conservation of mass. For this purpose a control volume is shown in Fig. 6.4 that involves rate of heat transfer, rates of work, and mass flows. The fundamental physical law states that we cannot create or destroy energy such that any rate of change of energy must be caused by rates of energy into or out of the control volume. We have already included rates of heat transfer and work in Eq. 6.4, so the additional explanations we need are associated with the mass flow rates.

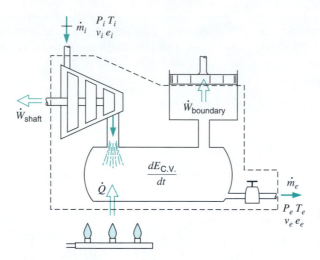

**FIGURE 6.4** Schematic diagram to illustrate terms in the energy equation for a general control volume.

The fluid flowing across the control surface enters or leaves with an amount of energy per unit mass as

$$e = u + \frac{1}{2}\mathbf{V}^2 + gZ$$

relating to the state and position of the fluid. Whenever a fluid mass enters a control volume at state *i*, or exits at state *e*, there is a boundary movement work associated with that process.

To explain this in more detail consider an amount of mass flowing into the control volume. As this mass flows in there is a pressure at *its* back surface, so as this mass moves into the control volume *it* is being pushed by the mass behind *it*, which is the surroundings. The net effect is that after the mass has entered the control volume the surroundings have pushed it in against the local pressure with a velocity giving it a rate of work in the process. Similarly a fluid exiting the control volume at state *e* must push the surrounding fluid ahead of it, doing work on it, which is work leaving the control volume. The velocity and the area correspond to a certain volume per unit time entering the control volume, enabling us to relate that to the mass flow rate and the specific volume at the state of the mass going in. Now we are able to express the rate of flow work as

$$\dot{W}_{\text{flow}} = F\mathbf{V} = \int P\mathbf{V}\,dA = P\dot{V} = Pv\dot{m} \qquad (6.5)$$

For the flow that leaves the control volume work is being done by the control volume, $P_e v_e \dot{m}_e$, and for the mass that enters, the surroundings do the rate of work, $P_i v_i \dot{m}_i$. The flow work per unit mass is then $Pv$, and the total energy associated with the flow of mass is

$$e + Pv = u + Pv + \frac{1}{2}\mathbf{V}^2 + gZ = h + \frac{1}{2}\mathbf{V}^2 + gZ \qquad (6.6)$$

In this equation we have used the definition of the thermodynamic property enthalpy, and it is the appearance of the combination $(u + Pv)$ for the energy in connection with a mass flow that is the primary reason for the definition of the property enthalpy. Its introduction earlier in conjunction with the constant-pressure process was to facilitate use of the tables of thermodynamic properties at that time.

| EXAMPLE 6.2 | Assume we are standing next to the local city's main water line. The liquid water inside flows at a pressure of say 600 kPa (6 atm) with a temperature of about 10°C. We want to add a smaller amount, 1 kg, of liquid to the line through a side pipe and valve mounted on the main line. How much work will be involved in this process? |

If the 1 kg of liquid water is in a bucket and we open the valve to the water main trying to pour it down into the pipe opening, we will realize that the water flows the other way. The water will flow from a higher to a lower pressure, that is, from inside the main line to the atmosphere (from 600 kPa to 101 kPa).

We must take the 1 kg of liquid water and put it into a piston cylinder (like a handheld pump) and attach the cylinder to the water pipe. Now we can press on the piston until the water pressure inside is 600 kPa and then open the valve to the main line and slowly squeeze the 1 kg of water in. The work done at the piston surface to the water is

$$W = \int P \, dV = P_{\text{water}} \, mv = 600 \, \text{kPa} \times 1 \, \text{kg} \times 0.001 \, \text{m}^3/\text{kg} = 0.6 \, \text{kJ}$$

and this is the necessary flow work for adding the 1 kg of liquid.

The extension of the first law of thermodynamics from Eq. 6.4 becomes

$$\frac{dE_{\text{C.V.}}}{dt} = \dot{Q}_{\text{C.V.}} - \dot{W}_{\text{C.V.}} + \dot{m}_i e_i - \dot{m}_e e_e + \dot{W}_{\text{flow in}} - \dot{W}_{\text{flow out}}$$

and the substitution of Eq. 6.5 gives

$$\frac{dE_{\text{C.V.}}}{dt} = \dot{Q}_{\text{C.V.}} - \dot{W}_{\text{C.V.}} + \dot{m}_i (e_i + P_i v_i) - \dot{m}_e (e_e + P_e v_e)$$

$$= \dot{Q}_{\text{C.V.}} - \dot{W}_{\text{C.V.}} + \dot{m}_i \left( h_i + \frac{1}{2} \mathbf{V}_i^2 + g Z_i \right) - \dot{m}_e \left( h_e + \frac{1}{2} \mathbf{V}_e^2 + g Z_e \right)$$

In this form of the energy equation the rate of work term is the sum of all shaft work terms and boundary work terms and any other types of work given out by the control volume; however, the flow work is now listed separately and included with the mass flow rate terms.

For the general control volume we may have several entering or leaving mass flow rates, so a summation over those terms is often needed. The final form of the first law of thermodynamics then becomes

$$\frac{dE_{\text{C.V.}}}{dt} = \dot{Q}_{\text{C.V.}} - \dot{W}_{\text{C.V.}} + \sum \dot{m}_i \left( h_i + \frac{1}{2} \mathbf{V}_i^2 + g Z_i \right) - \sum \dot{m}_e \left( h_e + \frac{1}{2} \mathbf{V}_e^2 + g Z_e \right)$$

(6.7)

expressing that the rate of change of energy inside the control volume is due to a net rate of heat transfer, a net rate of work (measured positive out), and the summation of energy fluxes due to mass flows into and out of the control volume. As with the conservation of mass, this equation can be written for the total control volume and can therefore be put in the lumped or integral form where

$$E_{\text{C.V.}} = \int \rho e \, dV = me = m_A e_A + m_B e_B + m_C e_C + \cdots$$

As the kinetic and potential energy terms per unit mass appear together with the enthalpy in all the flow terms, a shorter notation is often used

$$h_{\text{tot}} = h + \frac{1}{2}\mathbf{V}^2 + gZ$$

$$h_{\text{stag}} = h + \frac{1}{2}\mathbf{V}^2$$

defining the total enthalpy and the stagnation enthalpy (used in fluid mechanics). The shorter equation then becomes

$$\frac{dE_{\text{C.V.}}}{dt} = \dot{Q}_{\text{C.V.}} - \dot{W}_{\text{C.V.}} + \sum \dot{m}_i h_{\text{tot.}i} - \sum \dot{m}_e h_{\text{tot.}e} \qquad (6.8)$$

giving the general energy equation on a rate form. All applications of the energy equation start with the form in Eq. 6.8, and for special cases this will result in a slightly simpler form as shown in the subsequent sections.

## 6.3 THE STEADY-STATE PROCESS

Our first application of the control volume equations will be to develop a suitable analytical model for the long-term steady operation of devices such as turbines, compressors, nozzles, boilers, condensers—a very large class of problems of interest in thermodynamic analysis. This model will not include the short-term transient start-up or shutdown of such devices, but only the steady operating period of time.

Let us consider a certain set of assumptions (beyond those leading to Eqs. 6.1 and 6.7) that lead to a reasonable model for this type of process, which we refer to as the steady-state process.

1. The control volume does not move relative to the coordinate frame.
2. The state of the mass at each point in the control volume does not vary with time.
3. As for the mass that flows across the control surface, the mass flux and the state of this mass at each discrete area of flow on the control surface do not vary with time. The rates at which heat and work cross the control surface remain constant.

As an example of a steady-state process consider a centrifugal air compressor that operates with constant mass rate of flow into and out of the compressor, constant properties at each point across the inlet and exit ducts, a constant rate of heat transfer to the surroundings, and a constant power input. At each point in the compressor the properties are constant with time, even though the properties of a given elemental mass of air vary as it flows through the compressor. Often, such a process is referred to as a steady-flow process, since we are concerned primarily with the properties of the fluid entering and leaving the control volume. However, in the analysis of certain heat transfer problems in which the same assumptions apply, we are primarily interested in the spatial distribution of properties, particularly temperature, and such a process is referred to as a steady-state process. Since this is an introductory book we will use the term steady-state process for both. The student should realize that the terms steady-state process and steady-flow process are both used extensively in the literature.

Let us now consider the significance of each of these assumptions for the steady-state process.

1. The assumption that the control volume does not move relative to the coordinate frame means that all velocities measured relative to the coordinate frame are also velocities relative to the control surface, and there is no work associated with the acceleration of the control volume.

2. The assumption that the state of the mass at each point in the control volume does not vary with time requires that

$$\frac{dm_{\text{C.V.}}}{dt} = 0$$

and also

$$\frac{dE_{\text{C.V.}}}{dt} = 0$$

Therefore, we conclude that for the steady-state process we can write, from Eqs. 6.1 and 6.7,

$$\text{Continuity equation: } \sum \dot{m}_i = \sum \dot{m}_e \tag{6.9}$$

$$\text{First law: } \dot{Q}_{\text{C.V.}} + \sum \dot{m}_i \left( h_i + \frac{\mathbf{V}_i^2}{2} + gZ_i \right) = \sum \dot{m}_e \left( h_e + \frac{\mathbf{V}_e^2}{2} + gZ_e \right) + \dot{W}_{\text{C.V.}} \tag{6.10}$$

3. The assumption that the various mass flows, states, and rates at which heat and work cross the control surface remain constant requires that every quantity in Eqs. 6.9 and 6.10 be steady with time. This means that application of Eqs. 6.9 and 6.10 to the operation of some device is independent of time.

Many of the applications of the steady-state model are such that there is only one flow stream entering and one leaving the control volume. For this type of process, we can write

$$\text{Continuity equation: } \dot{m}_i = \dot{m}_e = \dot{m} \tag{6.11}$$

$$\text{First law: } \dot{Q}_{\text{C.V.}} + \dot{m} \left( h_i + \frac{\mathbf{V}_i^2}{2} + gZ_i \right) = \dot{m} \left( h_e + \frac{\mathbf{V}_e^2}{2} + gZ_e \right) + \dot{W}_{\text{C.V.}} \tag{6.12}$$

Rearranging this equation, we have

$$q + h_i + \frac{\mathbf{V}_i^2}{2} + gZ_i = h_e + \frac{\mathbf{V}_e^2}{2} + gZ_e + w \tag{6.13}$$

where, by definition,

$$q = \frac{\dot{Q}_{\text{C.V.}}}{\dot{m}} \quad \text{and} \quad w = \frac{\dot{W}_{\text{C.V.}}}{\dot{m}} \tag{6.14}$$

Note that the units for $q$ and $w$ are kJ/kg. From their definition, $q$ and $w$ can be thought of as the heat transfer and work (other than flow work) per unit mass flowing into and out of the control volume for this particular steady-state process.

The symbols $q$ and $w$ are also used for the heat transfer and work per unit mass of a control mass. However, since it is always evident from the context whether it is a control mass (fixed mass) or control volume (involving a flow of mass) with which we are concerned, the significance of the symbols $q$ and $w$ will also be readily evident in each situation.

The steady-state process is often used in the analysis of reciprocating machines, such as reciprocating compressors or engines. In this case the rate of flow, which may actually be pulsating, is considered to be the average rate of flow for an integral number of cycles. A similar assumption is made regarding the properties of the fluid flowing across the control surface and the heat transfer and work crossing the control surface. It is also assumed that for an integral number of cycles the reciprocating device undergoes, the energy and mass within the control volume do not change.

A number of examples are now given to illustrate the analysis of steady-state processes.

---

**In-Text Concept Questions**

**b.** Can a steady-state device have boundary work?

**c.** Can you say something about changes in $\dot{m}$ and $\dot{V}$ through a steady flow device?

**d.** In a multiple device flow system, I want to determine a state property. Where should I be looking for information—upstream or downstream?

---

## 6.4 Examples of Steady-State Processes

In this section, we consider a number of examples of steady-state processes in which there is one fluid stream entering and one leaving the control volume, such that the first law can be written in the form of Eq. 6.13. Some may instead utilize control volumes that include more than one fluid stream, such that it is necessary to write the first law in the more general form of Eq. 6.10.

### Heat Exchanger

A steady-state heat exchanger is a simple fluid flow through a pipe or system of pipes, where heat is transferred to or from the fluid. The fluid may be heated or cooled, and may or may not boil, liquid to vapor, or condense, vapor to liquid. One such example is the condenser in an R-134a refrigeration system, as shown in Fig. 6.5. Superheated vapor enters the condenser, and liquid exits. The process tends to occur at constant pressure, since a fluid flowing in a pipe usually undergoes only a small pressure drop, because of fluid friction at the walls. The pressure drop may or may not be taken into account in a particular analysis. There is no means for doing any work (shaft work, electrical work, etc.), and changes in kinetic and potential energies are commonly negligibly small. (One

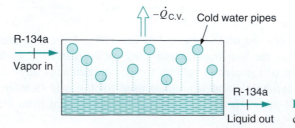

**FIGURE 6.5** A refrigeration system condenser.

exception may be a boiler tube in which liquid enters and vapor exits at a much larger specific volume. In such a case, it may be necessary to check the exit velocity using Eq. 6.3.) The heat transfer in most heat exchangers is then found from Eq. 6.13 as the change in enthalpy of the fluid. In the condenser shown in Fig. 6.5, the heat transfer out of the condenser then goes to whatever is receiving it, perhaps a stream of air or of cooling water. It is often simpler to write the first law around the entire heat exchanger, including both flow streams, in which case there is little or no heat transfer with the surroundings. Such a situation is the subject of the following example.

**EXAMPLE 6.3**

Consider a water-cooled condenser in a large refrigeration system in which R-134a is the refrigerant fluid. The refrigerant enters the condenser at 1.0 MPa and 60°C, at the rate of 0.2 kg/s, and exits as a liquid at 0.95 MPa and 35°C. Cooling water enters the condenser at 10°C and exits at 20°C. Determine the rate at which cooling water flows through the condenser.

$$
\begin{aligned}
&\textit{Control volume}: && \text{Condenser.} \\
&\textit{Sketch}: && \text{Fig. 6.6} \\
&\textit{Inlet states}: && \text{R-134a—fixed; water—fixed.} \\
&\textit{Exit states}: && \text{R134a—fixed; water—fixed.} \\
&\textit{Process}: && \text{Steady-state.} \\
&\textit{Model}: && \text{R-134a tables; steam tables.}
\end{aligned}
$$

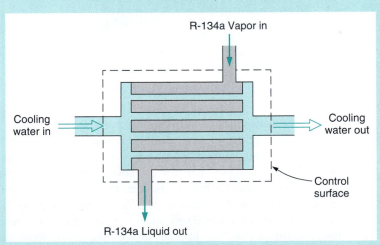

**FIGURE 6.6**
Schematic diagram of an R-134a condenser.

**Analysis**

With this control volume we have two fluid streams, the R-134a and the water, entering and leaving the control volume. It is reasonable to assume that both kinetic and potential energy changes are negligible. We note that the work is zero, and we make the other reasonable assumption that there is no heat transfer across the control surface. Therefore, the first law, Eq. 6.10, reduces to

$$\sum \dot{m}_i h_i = \sum \dot{m}_e h_e$$

Using the subscript $r$ for refrigerant and $w$ for water, we write

$$\dot{m}_r(h_i)_r + \dot{m}_w(h_i)_w = \dot{m}_r(h_e)_r + \dot{m}_w(h_e)_w$$

**Solution**

From the R-134a and steam tables, we have

$$(h_i)_r = 441.89 \text{ kJ/kg}, \qquad (h_i)_w = 42.00 \text{ kJ/kg}$$

$$(h_e)_r = 249.10 \text{ kJ/kg}, \qquad (h_e)_w = 83.95 \text{ kJ/kg}$$

Solving the above equation for $\dot{m}_w$, the rate of flow of water, we obtain

$$\dot{m}_w = \dot{m}_r \frac{(h_i - h_e)_r}{(h_e - h_i)_w} = 0.2 \text{ kg/s} \frac{(441.89 - 249.10) \text{ kJ/kg}}{(83.95 - 42.00) \text{ kJ/kg}} = 0.919 \text{ kg/s}$$

This problem can also be solved by considering two separate control volumes, one having the flow of R-134a across its control surface and the other having the flow of water across its control surface. Further, there is heat transfer from one control volume to the other.

The heat transfer for the control volume involving R-134a is calculated first. In this case the steady-state energy equation, Eq. 6.10, reduces to

$$\dot{Q}_{\text{C.V.}} = \dot{m}_r (h_e - h_i)_r$$

$$= 0.2 \text{ kg/s} \times (249.10 - 441.89) \text{ kJ/kg} = -38.558 \text{ kW}$$

This is also the heat transfer to the other control volume, for which $\dot{Q}_{\text{C.V.}} = +38.558$ kW.

$$\dot{Q}_{\text{C.V.}} = \dot{m}_w (h_e - h_i)_w$$

$$\dot{m}_w = \frac{38.558 \text{ kW}}{(83.95 - 42.00) \text{ kJ/kg}} = 0.919 \text{ kg/s}$$

## Nozzle

A nozzle is a steady-state device whose purpose is to create a high-velocity fluid stream at the expense of the fluid's pressure. It is contoured in an appropriate manner to expand a flowing fluid smoothly to a lower pressure, thereby increasing its velocity. There is no means to do any work—there are no moving parts. There is little or no change in potential energy and usually little or no heat transfer. An exception is the large nozzle on a liquid-propellant rocket, such as was described in Section 1.7, in which the cold propellant is commonly circulated around the outside of the nozzle walls before going to the combustion chamber, in order to keep the nozzle from melting. This case, a nozzle with significant heat transfer, is the exception and would be noted in such an application. In addition, the kinetic energy of the fluid at the nozzle inlet is usually small and would be neglected if its value is not known.

| | |
|---|---|
| **EXAMPLE 6.4** | Steam at 0.6 MPa and 200°C enters an insulated nozzle with a velocity of 50 m/s. It leaves at a pressure of 0.15 MPa and a velocity of 600 m/s. Determine the final temperature if the steam is superheated in the final state and the quality if it is saturated. |

*Control volume*:   Nozzle.

*Inlet state*:   Fixed (see Fig. 6.7).

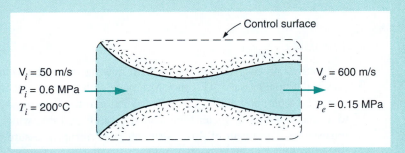

*Exit state*:  $P_e$ known.

*Process*:  Steady-state.

*Model*:  Steam tables.

**FIGURE 6.7** Illustration for Example 6.4.

### Analysis

We have

$$\dot{Q}_{\text{C.V.}} = 0 \qquad \text{(nozzle insulated)}$$

$$\dot{W}_{\text{C.V.}} = 0$$

$$PE_i \approx PE_e$$

The first law (Eq. 6.13) yields

$$h_i + \frac{\mathbf{V}_i^2}{2} = h_e + \frac{\mathbf{V}_e^2}{2}$$

### Solution

Solving for $h_e$ we obtain

$$h_e = 2850.1 + \left[ \frac{(50)^2}{2 \times 1000} - \frac{(600)^2}{2 \times 1000} \right] \frac{\text{m}^2/\text{s}^2}{\text{J/kJ}} = 2671.4 \text{ kJ/kg}$$

The two properties of the fluid leaving that we now know are pressure and enthalpy, and therefore the state of this fluid is determined. Since $h_e$ is less than $h_g$ at 0.15 MPa, the quality is calculated.

$$h = h_f + x h_{fg}$$

$$2671.4 = 467.1 + x_e 2226.5$$

$$x_e = 0.99$$

## Diffuser

A steady-state diffuser is a device constructed to decelerate a high-velocity fluid in a manner that results in an increase in pressure of the fluid. In essence, it is the exact opposite of a nozzle, and it may be thought of as a fluid flowing in the opposite direction through a nozzle, with the opposite effects. The assumptions are similar to those for a nozzle, with a large kinetic energy at the diffuser inlet and a small, but usually not negligible, kinetic energy at the exit being the only terms besides the enthalpies remaining in the first law, Eq. 6.13.

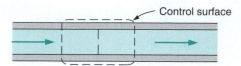

**FIGURE 6.8** The throttling process.

## Throttle

A throttling process occurs when a fluid flowing in a line suddenly encounters a restriction in the flow passage. This may be a plate with a small hole in it, as shown in Fig. 6.8, it may be a partially closed valve protruding into the flow passage, or it may be a change to a much smaller diameter tube, called a capillary tube, which is normally found on a refrigerator. The result of this restriction is an abrupt pressure drop in the fluid, as it is forced to find its way through a suddenly smaller passageway. This process is drastically unlike the smoothly contoured nozzle expansion and area change, which results in a significant velocity increase. There is typically some increase in velocity in a throttle, but both inlet and exit kinetic energies are usually small enough to be neglected. There is no means for doing work and little or no change in potential energy. Usually, there is neither time nor opportunity for an appreciable heat transfer, such that the only terms left in the first law, Eq. 6.13, are the inlet and exit enthalpies. We conclude that a steady-state throttling process is approximately a pressure drop at constant enthalpy, and we will assume this to be the case unless otherwise noted.

Frequently, a throttling process involves a change in the phase of the fluid. A typical example is the flow through the expansion valve of a vapor-compression refrigeration system. The following example deals with this problem.

---

**EXAMPLE 6.5**

Consider the throttling process across the expansion valve or through the capillary tube in a vapor-compression refrigeration cycle. In this process the pressure of the refrigerant drops from the high pressure in the condenser to the low pressure in the evaporator, and during this process some of the liquid flashes into vapor. If we consider this process to be adiabatic, the quality of the refrigerant entering the evaporator can be calculated.

Consider the following process, in which ammonia is the refrigerant. The ammonia enters the expansion valve at a pressure of 1.50 MPa and a temperature of 35°C. Its pressure on leaving the expansion valve is 291 kPa. Calculate the quality of the ammonia leaving the expansion valve.

| | |
|---:|:---|
| *Control volume*: | Expansion valve or capillary tube. |
| *Inlet state*: | $P_i$, $T_i$ known; state fixed. |
| *Exit state*: | $P_e$ known. |
| *Process*: | Steady-state. |
| *Model*: | Ammonia tables. |

**Analysis**

We can use standard throttling process analysis and assumptions. The first law reduces to

$$h_i = h_e$$

**Solution**

From the ammonia tables we get

$$h_i = 346.8 \text{ kJ/kg}$$

(The enthalpy of a slightly compressed liquid is essentially equal to the enthalpy of saturated liquid at the same temperature.)

$$h_e = h_i = 346.8 = 134.4 + x_e(1296.4)$$

$$x_e = 0.1638 = 16.38\%$$

## Turbine

A turbine is a rotary steady-state machine whose purpose is to produce shaft work (power, on a rate basis) at the expense of the pressure of the working fluid. Two general classes of turbines are steam (or other working fluid) turbines, in which the steam exiting the turbine passes to a condenser, where it is condensed to liquid, and gas turbines, in which the gas usually exhausts to the atmosphere from the turbine. In either type, the turbine exit pressure is fixed by the environment into which the working fluid exhausts, and the turbine inlet pressure has been reached by previously pumping or compressing the working fluid in another process. Inside the turbine, there are two distinct processes. In the first, the working fluid passes through a set of nozzles, or the equivalent—fixed blade passages contoured to expand the fluid to a lower pressure and to a high velocity. In the second process inside the turbine, this high-velocity fluid stream is directed onto a set of moving (rotating) blades, in which the velocity is reduced before being discharged from the passage. This directed velocity decrease produces a torque on the rotating shaft, resulting in a shaft work output. The low-velocity, low-pressure fluid then exhausts from the turbine.

The first law for this process is either Eq. 6.10 or 6.13. Usually, changes in potential energy are negligible, as is the inlet kinetic energy. Often, the exit kinetic energy is neglected, and any heat rejection from the turbine is undesirable and is commonly small. We therefore normally assume that a turbine process is adiabatic, and the work output in this case reduces to the decrease in enthalpy form the inlet to exit states. In the following example, however, we include all the terms in the first law and study their relative importance.

**EXAMPLE 6.6**

The mass rate of flow into a steam turbine is 1.5 kg/s, and the heat transfer from the turbine is 8.5 kW. The following data are known for the steam entering and leaving the turbine.

|  | Inlet Conditions | Exit Conditions |
|---|---|---|
| Pressure | 2.0 MPa | 0.1 MPa |
| Temperature | 350°C | |
| Quality | | 100% |
| Velocity | 50 m/s | 100 m/s |
| Elevation above reference plane | 6 m | 3 m |
| $g = 9.8066 \text{ m/s}^2$ | | |

Determine the power output of the turbine.

| | |
|---:|:---|
| *Control volume*: | Turbine (Fig. 6.9). |
| *Inlet state*: | Fixed (above). |
| *Exit state*: | Fixed (above). |
| *Process*: | Steady-state. |
| *Model*: | Steam tables. |

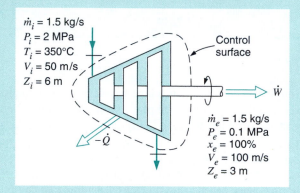

**FIGURE 6.9** Illustration for Example 6.6.

### Analysis

From the first law (Eq. 6.12) we have

$$\dot{Q}_{C.V.} + \dot{m}\left(h_i + \frac{\mathbf{V}_i^2}{2} + gZ_i\right) = \dot{m}\left(h_e + \frac{\mathbf{V}_e^2}{2} + gZ_e\right) + \dot{W}_{C.V.}$$

with

$$\dot{Q}_{C.V.} = -8.5 \text{ kW}$$

### Solution

From the steam tables, $h_i = 3137.0$ kJ/kg. Substituting inlet conditions gives

$$\frac{\mathbf{V}_i^2}{2} = \frac{50 \times 50}{2 \times 1000} = 1.25 \text{ kJ/kg}$$

$$gZ_i = \frac{6 \times 9.8066}{1000} = 0.059 \text{ kJ/kg}$$

Similarly, for the exit $h_e = 2675.5$ kJ/kg and

$$\frac{\mathbf{V}_e^2}{2} = \frac{100 \times 100}{2 \times 1000} = 5.0 \text{ kJ/kg}$$

$$gZ_e = \frac{3 \times 9.8066}{1000} = 0.029 \text{ kJ/kg}$$

Therefore, substituting into Eq. 6.12, we obtain

$$-8.5 + 1.5(3137 + 1.25 + 0.059) = 1.5(2675.5 + 5.0 + 0.029) + \dot{W}_{C.V.}$$

$$\dot{W}_{C.V.} = -8.5 + 4707.5 - 4020.8 = 678.2 \text{ kW}$$

If Eq. 6.13 is used, the work per kilogram of fluid flowing is found first.

$$q + h_i + \frac{\mathbf{V}_i^2}{2} + gZ_i = h_e + \frac{\mathbf{V}_e^2}{2} + gZ_e + w$$

$$q = \frac{-8.5}{1.5} = -5.667 \text{ kJ/kg}$$

Therefore, substituting into Eq. 6.13, we get

$$-5.667 + 3137 + 1.25 + 0.059 = 2675.5 + 5.0 + 0.029 + w$$

$$w = 452.11 \text{ kJ/kg}$$

$$\dot{W}_{\text{C.V.}} = 1.5 \text{ kg/s} \times 452.11 \text{ kJ/kg} = 678.2 \text{ kW}$$

Two further observations can be made by referring to this example. First, in many engineering problems, potential energy changes are insignificant when compared with the other energy quantities. In the above example the potential energy change did not affect any of the significant figures. In most problems where the change in elevation is small the potential energy terms may be neglected.

Second, if velocities are small—say, under 20 m/s—in many cases the kinetic energy is insignificant compared with other energy quantities. Furthermore, when the velocities entering and leaving the system are essentially the same, the change in kinetic energy is small. Since it is the change in kinetic energy that is important in the steady-state energy equation, the kinetic energy terms can usually be neglected when there is no significant difference between the velocity of the fluid entering and that leaving the control volume. Thus, in many thermodynamic problems, one must make judgments as to which quantities may be negligible for a given analysis.

The preceding discussion and example concerned the turbine, which is a rotary work-producing device. There are other nonrotary devices that produce work, which can be called expanders as a general name. In such devices, the first-law analysis and assumptions are generally the same as for turbines, except that in a piston/cylinder type expander, there would in most cases be a larger heat loss or rejection during the process.

## Compressor and Pump

The purpose of a steady-state compressor (gas) or pump (liquid) is the same: to increase the pressure of a fluid by putting in shaft work (power, on a rate basis). There are two fundamentally different classes of compressors. The most common is a rotary-type compressor (either axial flow or radial/centrifugal flow), in which the internal processes are essentially the opposite of the two processes occurring inside a turbine. The working fluid enters the compressor at low pressure, moving into a set of rotation blades, from which it exits at high velocity, a result of the shaft work input to the fluid. The fluid then passes through a diffuser section, in which it is decelerated in a manner that results in a pressure increase. The fluid then exits the compressor at high pressure.

The first law for the compressor is either Eq. 6.10 or 6.13. Usually, changes in potential energy are negligible, as is the inlet kinetic energy. Often the exit kinetic energy is neglected, as well. Heat rejection from the working fluid during compression would be desirable, but it is usually small in a rotary compressor, which is a high-volume flow-rate machine, and there is not sufficient time to transfer much heat from the working fluid. We therefore normally assume that a rotary compressor process is adiabatic, and the work input in this case reduces to the change in enthalpy from the inlet to exit states.

In a piston/cylinder-type compressor, the cylinder usually contains fins to promote heat rejection during compression (or the cylinder may be water-jacketed in a large compressor for even greater cooling rates). In this type of compressor, the heat transfer from the working fluid is significant and is not neglected in the first law. As a general rule, in any example or problem in this text, we will assume that a compressor is adiabatic unless otherwise noted.

**EXAMPLE 6.7**

The compressor in a plant (see Fig. 6.10) receives carbon dioxide at 100 kPa, 280 K, with a low velocity. At the compressor discharge, the carbon dioxide exits at 1100 kPa, 500 K, with velocity of 25 m/s and then flows into a constant-pressure aftercooler (heat exchanger) where it is cooled down to 350 K. The power input to the compressor is 50 kW. Determine the heat transfer rate in the aftercooler.

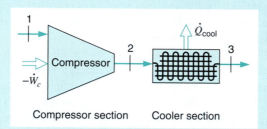

**FIGURE 6.10**   Sketch for Example 6.7.

**Solution**

C.V. compressor, steady state, single inlet and exit flow.

Energy Eq. 6.13:       $q + h_1 + \frac{1}{2}\mathbf{V}_1^2 = h_2 + \frac{1}{2}\mathbf{V}_2^2 + w$

Here we assume $q \cong 0$ and $\mathbf{V}_1 \cong 0$, so, getting $h$ from Table A.8,

$$-w = h_2 - h_1 + \frac{1}{2}\mathbf{V}_2^2 = 401.52 - 198 + \frac{(25)^2}{2 \times 1000} = 203.5 + 0.3 = 203.8 \text{ kJ/kg}$$

Remember here to convert kinetic energy J/kg to kJ/kg by division by 1000.

$$\dot{m} = \frac{\dot{W}_c}{w} = \frac{-50}{-203.8} = 0.245 \text{ kg/s}$$

C.V. aftercooler, steady state, single inlet and exit flow, and no work.

Energy Eq. 6.13:       $q + h_2 + \frac{1}{2}\mathbf{V}_2^2 = h_3 + \frac{1}{2}\mathbf{V}_3^2$

Here we assume no significant change in kinetic energy (notice how unimportant it was) and again we look for $h$ in Table A.8

$$q = h_3 - h_2 = 257.9 - 401.5 = -143.6 \text{ kJ/kg}$$

$$\dot{Q}_{cool} = -\dot{Q}_{C.V.} = -\dot{m}q = 0.245 \text{ kg/s} \times 143.6 \text{ kJ/kg} = 35.2 \text{ kW}$$

**EXAMPLE 6.8**

A small liquid water pump is located 15 m down in a well (see Fig. 6.11), taking water in at 10°C, 90 kPa at a rate of 1.5 kg/s. The exit line is a pipe of diameter 0.04 m that goes up to a receiver tank maintaining a gauge pressure of 400 kPa. Assume the process is adiabatic with the same inlet and exit velocities and that the water stays at 10°C. Find the required pump work.

C. V. pump + pipe. Steady state, 1 inlet, 1 exit flow. Assume same velocity in and out and no heat transfer.

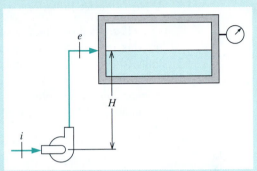

**FIGURE 6.11** Sketch for Example 6.8.

**Solution**

Continuity equation: $\dot{m}_{in} = \dot{m}_{ex} = \dot{m}$

Energy Eq. 6.12: $\dot{m}\left(h_{in} + \frac{1}{2}\mathbf{V}_{in}^2 + gZ_{in}\right) = \dot{m}\left(h_{ex} + \frac{1}{2}\mathbf{V}_{ex}^2 + gZ_{ex}\right) + \dot{W}$

States: $h_{ex} = h_{in} + (P_{ex} - P_{in})v$    ($v$ is constant and $u$ is constant.)

From the energy equation

$$\dot{W} = \dot{m}(h_{in} + gZ_{in} - h_{ex} - gZ_{ex}) = \dot{m}[g(Z_{in} - Z_{ex}) - (P_{ex} - P_{in})v]$$

$$= 1.5\,\frac{kg}{s} \times \left[9.807\,\frac{m}{s^2} \times \frac{-15 - 0}{1000}\,m - (400 + 101.3 - 90)\,kPa\,0.001\,001\,\frac{m^3}{kg}\right]$$

$$= 1.5 \times (-0.147 - 0.412) = -0.84\,kW)$$

That is, the pump requires a power input of 840 W.

## Power Plant and Refrigerator

The following examples illustrate the incorporation of several of the devices and machines already discussed in this section into a complete thermodynamic system, which is built for a specific purpose.

**EXAMPLE 6.9**

Consider the simple steam power plant, as shown in Fig. 6.12. The following data are for such a power plant.

| Location | Pressure | Temperature or Quality |
|---|---|---|
| Leaving boiler | 2.0 MPa | 300°C |
| Entering turbine | 1.9 MPa | 290°C |
| Leaving turbine, entering condenser | 15 kPa | 90% |
| Leaving condenser, entering pump | 14 kPa | 45°C |
| Pump work = 4 kJ/kg | | |

Determine the following quantities per kilogram flowing through the unit:

**a.** Heat transfer in line between boiler and turbine.

**b.** Turbine work.

**c.** Heat transfer in condenser.

**d.** Heat transfer in boiler.

There is a certain advantage in assigning a number to various points in the cycle. For this reason the subscripts $i$ and $e$ in the steady-state energy equation are often replaced by appropriate numbers.

Since there are several control volumes to be considered in the solution to this problem, let us consolidate our solution procedure somewhat in this example. Using the notation of Fig. 6.12, we have:

> *All processes*:   Steady-state.
> *Model*:   Steam tables.

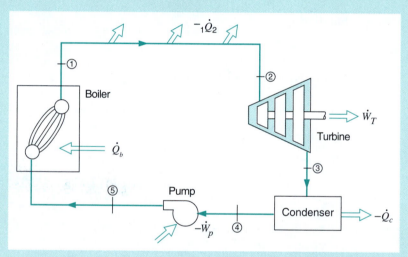

**FIGURE 6.12** Simple steam power plant.

From the steam tables:

$$h_1 = 3023.5 \text{ kJ/kg}$$

$$h_2 = 3002.5 \text{ kJ/kg}$$

$$h_3 = 226.0 + 0.9(2373.1) = 2361.8 \text{ kJ/kg}$$

$$h_4 = 188.5 \text{ kJ/kg}$$

> *All analyses*:   No changes in kinetic or potential energy will be considered in the solution. In each case, the first law is given by Eq. 6.13.

Now, we proceed to answer the specific questions raised in the problem statement.

**a.** For the control volume for the pipe line between the boiler and the turbine, the first law and solution are

$$_1q_2 + h_1 = h_2$$

$$_1q_2 = h_2 - h_1 = 3002.5 - 3023.5 = -21.0 \text{ kJ/kg}$$

**b.** A turbine is essentially an adiabatic machine. Therefore, it is reasonable to neglect heat transfer in the first law, so that

$$h_2 = h_3 + {_2}w_3$$

$${_2}w_3 = 3002.5 - 2361.8 = 640.7 \text{ kJ/kg}$$

**c.** There is no work for the control volume enclosing the condenser. Therefore, the first law and solution are

$${_3}q_4 + h_3 = h_4$$

$${_3}q_4 = 188.5 - 2361.8 = -2173.3 \text{ kJ/kg}$$

**d.** If we consider a control volume enclosing the boiler, the work is equal to zero, so that the first law becomes

$${_5}q_1 + h_5 = h_1$$

A solution requires a value for $h_5$, which can be found by taking a control volume around the pump:

$$h_4 = h_5 + {_4}w_5$$

$$h_5 = 188.5 - (-4) = 192.5 \text{ kJ/kg}$$

Therefore, for the boiler,

$${_5}q_1 + h_5 = h_1$$

$${_5}q_1 = 3023.5 - 192.5 = 2831 \text{ kJ/kg}$$

---

**EXAMPLE 6.10**

The refrigerator shown in Fig. 6.13 uses R-134a as the working fluid. The mass flow rate through each component is 0.1 kg/s, and the power input to the compressor is 5.0 kW. The following state data are known, using the state notation of Fig. 6.13.

$$P_1 = 100 \text{ kPa}, \qquad T_1 = -20°C$$

$$P_2 = 800 \text{ kPa}, \qquad T_2 = 50°C$$

$$T_3 = 30°C, \qquad x_3 = 0.0$$

$$T_4 = -25°C$$

Determine the following:

**a.** The quality at the evaporator inlet.
**b.** The rate of heat transfer to the evaporator.
**c.** The rate of heat transfer from the compressor.

*All processes*: Steady-state.
*Model*: R-134a tables.
*All analyses*: No changes in kinetic or potential energy. The first law in each case is given by Eq. 6.10.

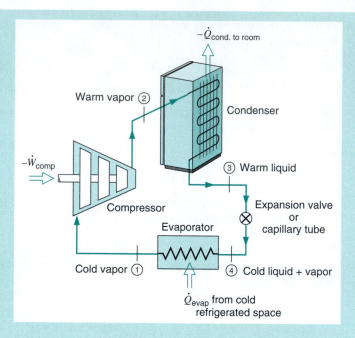

$-\dot{Q}_{\text{cond. to room}}$

Warm vapor ②

Condenser

$-\dot{W}_{\text{comp}}$

③ Warm liquid

Compressor

Expansion valve
or
capillary tube

Evaporator

Cold vapor ①

④ Cold liquid + vapor

$\dot{Q}_{\text{evap}}$ from cold
refrigerated space

**FIGURE 6.13** Refrigerator.

**Solution**

**a.** For a control volume enclosing the throttle, the first law gives

$$h_4 = h_3 = 241.8 \text{ kJ/kg}$$

$$h_4 = 241.8 = h_{f4} + x_4 h_{fg4} = 167.4 + x_4 \times 215.6$$

$$x_4 = 0.345$$

**b.** For a control volume enclosing the evaporator, the first law gives

$$\dot{Q}_{\text{evap}} = \dot{m}(h_1 - h_4)$$

$$= 0.1(387.2 - 241.8) = 14.54 \text{ kW}$$

**c.** And for the compressor, the first law gives

$$\dot{Q}_{\text{comp}} = \dot{m}(h_2 - h_1) + \dot{W}_{\text{comp}}$$

$$= 0.1(435.1 - 387.2) - 5.0 = -0.21 \text{ kW}$$

---

**In-Text Concept Questions**

**e.** How does a nozzle or sprayhead generate kinetic energy?

**f.** What is the difference between a nozzle flow and a throttle process?

**g.** If you throttle a saturated liquid, what happens to the fluid state? What if this is done to an ideal gas?

**h.** A turbine at the bottom of a dam has a flow of liquid water through it. How does that produce power? Which terms in the energy equation are important if the *CV* is the turbine only? If the *CV* is the turbine plus the upstream flow up to the top of the lake, which terms in the energy equation are then important?

**i.** If you compress air, the temperature goes up. Why? When the hot air, high *P* flows in long pipes, it eventually cools to ambient *T*. How does that change the flow?

**j.** A mixing chamber has all flows at the same *P*, neglecting losses. A heat exchanger has separate flows exchanging energy, but they do not mix. Why have both kinds?

## 6.5 THE TRANSIENT PROCESS

In Sections 6.3 and 6.4 we considered the steady-state process and several examples of its application. Many processes of interest in thermodynamics involve unsteady flow and do not fit into this category. A certain group of these—for example, filling closed tanks with a gas or liquid, or discharge from closed vessels—can be reasonably represented to a first approximation by another simplified model. We call this process the transient process, for convenience, recognizing that our model includes specific assumptions that are not always valid. Our transient model assumptions are as follows:

1. The control volume remains constant relative to the coordinate frame.
2. The state of the mass within the control volume may change with time, but at any instant of time the state is uniform throughout the entire control volume (or over several identifiable regions that make up the entire control volume).
3. The state of the mass crossing each of the areas of flow on the control surface is constant with time although the mass flow rates may be time varying.

Let us examine the consequence of these assumptions and derive an expression for the first law that applies to this process. The assumption that the control volume remains stationary relative to the coordinate frame has already been discussed in Section 6.3. The remaining assumptions lead to the following simplifications for the continuity equation and the first law.

The overall process occurs during time *t*. At any instant of time during the process, the continuity equation is

$$\frac{dm_{\text{C.V.}}}{dt} + \sum \dot{m}_e - \sum \dot{m}_i = 0$$

where the summation is over all areas on the control surface through which flow occurs. Integrating over time *t* gives the change of mass in the control volume during the overall process:

$$\int_0^t \left(\frac{dm_{\text{C.V.}}}{dt}\right) dt = (m_2 - m_1)_{\text{C.V.}}$$

The total mass leaving the control volume during time *t* is

$$\int_0^t \left(\sum \dot{m}_e\right) dt = \sum m_e$$

and the total mass entering the control volume during time $t$ is

$$\int_0^t \left( \sum \dot{m}_i \right) dt = \sum m_i$$

Therefore, for this period of time $t$, we can write the continuity equation for the transient process as

$$(m_2 - m_1)_{\text{C.V.}} + \sum m_e - \sum m_i = 0 \qquad (6.15)$$

In writing the first law of the transient process we consider Eq. 6.7, which applies at any instant of time during the process:

$$\dot{Q}_{\text{C.V.}} + \sum \dot{m}_i \left( h_i + \frac{\mathbf{V}_i^2}{2} + gZ_i \right) = \frac{dE_{\text{C.V.}}}{dt} + \sum \dot{m}_e \left( h_e + \frac{\mathbf{V}_e^2}{2} + gZ_e \right) + \dot{W}_{\text{C.V.}}$$

Since at any instant of time the state within the control volume is uniform, the first law for the transient process becomes

$$\dot{Q}_{\text{C.V.}} + \sum \dot{m}_i \left( h_i + \frac{\mathbf{V}_i^2}{2} + gZ_i \right) = \sum \dot{m}_e \left( h_e + \frac{\mathbf{V}_e^2}{2} + gZ_e \right)$$

$$+ \frac{d}{dt} \left[ m \left( u + \frac{\mathbf{V}^2}{2} + gZ \right) \right]_{\text{C.V.}} + \dot{W}_{\text{C.V.}}$$

Let us now integrate this equation over time $t$, during which time we have

$$\int_0^t \dot{Q}_{\text{C.V.}} \, dt = Q_{\text{C.V.}}$$

$$\int_0^t \left[ \sum \dot{m}_i \left( h_i + \frac{\mathbf{V}_i^2}{2} + gZ_i \right) \right] dt = \sum m_i \left( h_i + \frac{\mathbf{V}_i^2}{2} + gZ_i \right)$$

$$\int_0^t \left[ \sum \dot{m}_e \left( h_e + \frac{\mathbf{V}_e^2}{2} + gZ_e \right) \right] dt = \sum m_e \left( h_e + \frac{\mathbf{V}_e^2}{2} + gZ_e \right)$$

$$\int_0^t \dot{W}_{\text{C.V.}} \, dt = W_{\text{C.V.}}$$

$$\int_0^t \frac{d}{dt} \left[ m \left( u + \frac{\mathbf{V}^2}{2} + gZ \right) \right]_{\text{C.V.}} dt = \left[ m_2 \left( u_2 + \frac{\mathbf{V}_2^2}{2} + gZ_2 \right) - m_1 \left( u_1 + \frac{\mathbf{V}_1^2}{2} + gZ_1 \right) \right]_{\text{C.V.}}$$

Therefore, for this period of time $t$, we can write the first law for the transient process as

$$Q_{\text{C.V.}} + \sum m_i \left( h_i + \frac{\mathbf{V}_i^2}{2} + gZ_i \right)$$

$$= \sum m_e \left( h_e + \frac{\mathbf{V}_e^2}{2} + gZ_e \right)$$

$$+ \left[ m_2 \left( u_2 + \frac{\mathbf{V}_2^2}{2} + gZ_2 \right) - m_1 \left( u_1 + \frac{\mathbf{V}_1^2}{2} + gZ_1 \right) \right]_{\text{C.V.}} + W_{\text{C.V.}} \qquad (6.16)$$

As an example of the type of problem for which these assumptions are valid and Eq. 6.16 is appropriate, let us consider the classic problem of flow into an evacuated vessel. This is the subject of Example 6.11.

**EXAMPLE 6.11**

Steam at a pressure of 1.4 MPa and temperature of 300°C is flowing in a pipe (Fig. 6.14). Connected to this pipe through a valve is an evacuated tank. The valve is opened and the tank fills with steam until the pressure is 1.4 MPa, and then the valve is closed. The process takes place adiabatically and kinetic energies and potential energies are negligible. Determine the final temperature of the steam.

*Control volume*: Tank, as shown in Fig. 6.14.

*Initial state* (in tank): Evacuated, mass $m_1 = 0$.

*Final state*: $P_2$ known.

*Inlet state*: $P_i$, $T_i$ (in line) known.

*Process*: Transient.

*Model*: Steam tables.

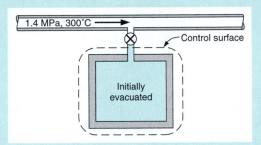

**FIGURE 6.14** Flow into an evacuated vessel—control volume analysis.

**Analysis**

From the first law, Eq. 6.16, we have

$$Q_{C.V.} + \sum m_i \left( h_i + \frac{\mathbf{V}_i^2}{2} + gZ_i \right)$$

$$= \sum m_e \left( h_e + \frac{\mathbf{V}_e^2}{2} + gZ_e \right)$$

$$+ \left[ m_2 \left( u_2 + \frac{\mathbf{V}_2^2}{2} + gZ_2 \right) - m_1 \left( u_1 + \frac{\mathbf{V}_1^2}{2} + gZ_1 \right) \right]_{C.V.} + W_{C.V.}$$

We note that $Q_{C.V.} = 0$, $W_{C.V.} = 0$, $m_e = 0$, and $(m_1)_{C.V.} = 0$. We further assume that changes in kinetic and potential energy are negligible. Therefore, the statement of the first law for this process reduces to

$$m_i h_i = m_2 u_2$$

From the continuity equation for this process, Eq. 6.15, we conclude that

$$m_2 = m_i$$

Therefore, combining the continuity equation with the first law, we have

$$h_i = u_2$$

That is, the final internal energy of the steam in the tank is equal to the enthalpy of the steam entering the tank.

**Solution**

From the steam tables we obtain

$$h_i = u_2 = 3040.4 \text{ kJ/kg}$$

Since the final pressure is given as 1.4 MPa, we know two properties at the final state and therefore the final state is determined. The temperature corresponding to a pressure of 1.4 MPa and an internal energy of 3040.4 kJ/kg is found to be 452°C.

This problem can also be solved by considering the steam that enters the tank and the evacuated space as a control mass, as indicated in Fig. 6.15.

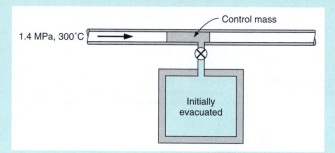

**FIGURE 6.15** Flow into an evacuated vessel—control mass.

The process is adiabatic, but we must examine the boundaries for work. If we visualize a piston between the steam that is included in the control mass and the steam that flows behind, we readily recognize that the boundaries move and that the steam in the pipe does work on the steam that comprises the control mass. The amount of this work is

$$-W = P_1 V_1 = m P_1 v_1$$

Writing the first law for the control mass, Eq. 5.11, and noting that kinetic and potential energies can be neglected, we have

$$_1Q_2 = U_2 - U_1 + {_1}W_2$$
$$0 = U_2 - U_1 - P_1 V_1$$
$$0 = m u_2 - m u_1 - m P_1 v_1 = m u_2 - m h_1$$

Therefore,

$$u_2 = h_1$$

which is the same conclusion that was reached using a control volume analysis.

The two other examples that follow illustrate further the transient process.

**EXAMPLE 6.12**

Let the tank of the previous example have a volume of 0.4 m³ and initially contain saturated vapor at 350 kPa. The valve is then opened and steam from the line at 1.4 MPa and 300°C flows into the tank until the pressure is 1.4 MPa.

Calculate the mass of steam that flows into the tank.

*Control volume*:  Tank, as in Fig. 6.14.

*Initial state*:  $P_1$, saturated vapor; state fixed.

*Final state*: $P_2$.

*Inlet state*: $P_i$, $T_i$; state fixed.

*Process*: Transient.

*Model*: Steam tables.

**Analysis**

The situation is the same as in Example 6.11, except that the tank is not evacuated initially. Again we note that $Q_{C.V.} = 0$, $W_{C.V.} = 0$, and $m_e = 0$, and we assume that changes in kinetic and potential energy are zero. The statement of the first law for this process, Eq. 6.16, reduces to

$$m_i h_i = m_2 u_2 - m_1 u_1$$

The continuity equation, Eq. 6.15, reduces to

$$m_2 - m_1 = m_i$$

Therefore, combining the continuity equation with the first law, we have

$$(m_2 - m_1)h_i = m_2 u_2 - m_1 u_1$$

$$m_2(h_i - u_2) = m_1(h_i - u_1) \tag{a}$$

There are two unknowns in this equation—$m_2$ and $u_2$. However, we have one additional equation:

$$m_2 v_2 = V = 0.4 \ m^3 \tag{b}$$

Substituting (b) into (a) and rearranging, we have

$$\frac{V}{v_2}(h_i - u_2) - m_1(h_i - u_1) = 0 \tag{c}$$

in which the only unknowns are $v_2$ and $u_2$, both functions of $T_2$ and $P_2$. Since $T_2$ is unknown, it means that there is only one value of $T_2$ for which Eq. (c) will be satisfied, and we must find it by trial and error.

**Solution**

We have

$$v_1 = 0.5243 \ \text{m}^3/\text{kg}, \qquad m_1 = \frac{0.4}{0.5243} = 0.763 \ \text{kg}$$

$$h_i = 3040.4 \ \text{kJ/kg}, \qquad u_1 = 2548.9 \ \text{kJ/kg}$$

Assume that

$$T_2 = 300°C$$

For this temperature and the known value of $P_2$, we get

$$v_2 = 0.1823 \ \text{m}^3/\text{kg}, \qquad u_2 = 2785.2 \ \text{kJ/kg}$$

Substituting into (c), we obtain

$$\frac{0.4}{0.1823}(3040.4 - 2785.2) - 0.763(3040.4 - 2548.9) = +185.0 \ \text{kJ}$$

Now assume instead that

$$T_2 = 350°C$$

For this temperature and the known $P_2$, we get

$$v_2 = 0.2003 \text{ m}^3/\text{kg}, \qquad u_2 = 2869.1 \text{ kJ/kg}$$

Substituting these values into (c), we obtain

$$\frac{0.4}{0.2003}(3040.4 - 2869.1) - 0.763(3040.4 - 2548.9) = -32.9 \text{ kJ}$$

and we find that the actual $T_2$ must be between these two assumed values, in order that (c) be equal to zero. By interpolation,

$$T_2 = 342°C \qquad \text{and} \qquad v_2 = 0.1974 \text{ m}^3/\text{kg}$$

The final mass inside the tank is

$$m_2 = \frac{0.4}{0.1974} = 2.026 \text{ kg}$$

and the mass of steam that flows into the tank is

$$m_i = m_2 - m_1 = 2.026 - 0.763 = 1.263 \text{ kg}$$

---

**EXAMPLE 6.13**

A tank of 2 m³ volume contains saturated ammonia at a temperature of 40°C. Initially the tank contains 50% liquid and 50% vapor by volume. Vapor is withdrawn from the top of the tank until the temperature is 10°C. Assuming that only vapor (i.e., no liquid) leaves and that the process is adiabatic, calculate the mass of ammonia that is withdrawn.

Control volume: Tank.
Initial state: $T_1$, $V_{liq}$, $V_{vap}$; state fixed.
Final state: $T_2$.
Exit state: Saturated vapor (temperature changing).
Process: Transient.
Model: Ammonia tables.

**Analysis**

In the first law, Eq. 6.16, we note that $Q_{C.V.} = 0$, $W_{C.V.} = 0$, and $m_i = 0$, and we assume that changes in kinetic and potential energy are negligible. However, the enthalpy of saturated vapor varies with temperature, and therefore we cannot simply assume that the enthalpy of the vapor leaving the tank remains constant. However, we note that at 40°C, $h_g = 1470.2$ kJ/kg and at 10°C, $h_g = 1452.0$ kJ/kg. Since the change in $h_g$ during this process is small, we may accurately assume that $h_e$ is the average of the two values given above. Therefore,

$$(h_e)_{av} = 1461.1 \text{ kJ/kg}$$

and the first law reduces to

$$m_e h_e + m_2 u_2 - m_1 u_1 = 0$$

and the continuity equation (from Eq. 6.15) becomes

$$(m_2 - m_1)_{\text{C.V.}} + m_e = 0$$

Combining these two equations we have

$$m_2(h_e - u_2) = m_1 h_e - m_1 u_1$$

**Solution**

The following values are from the ammonia tables:

$$v_{f1} = 0.001\ 725\ \text{m}^3/\text{kg}, \qquad v_{g1} = 0.083\ 13\ \text{m}^3/\text{kg}$$

$$v_{f2} = 0.001\ 60, \qquad v_{fg2} = 0.203\ 81$$

$$u_{f1} = 368.7\ \text{kJ/kg}, \qquad u_{g1} = 1341.0$$

$$u_{f2} = 226.0, \qquad u_{fg2} = 1099.7$$

Calculating first the initial mass, $m_1$, in the tank, we find that the mass of the liquid initially present, $m_{f1}$, is

$$m_{f1} = \frac{V_f}{v_{f1}} = \frac{1.0}{0.001\ 725} = 579.7\ \text{kg}$$

Similarly, the initial mass of vapor, $m_{g1}$, is

$$m_{g1} = \frac{V_g}{v_{g1}} = \frac{1.0}{0.083\ 13} = 12.0\ \text{kg}$$

$$m_1 = m_{f1} + m_{g1} = 579.7 + 12.0 = 591.7\ \text{kg}$$

$$m_1 h_e = 591.7 \times 1461.1 = 864\ 533\ \text{kJ}$$

$$m_1 u_1 = (mu)_{f1} + (mu)_{g1} = 579.7 \times 368.7 + 12.0 \times 1341.0$$

$$= 229\ 827\ \text{kJ}$$

Substituting these into the first law, we obtain

$$m_2(h_e - u_2) = m_1 h_e - m_1 u_1 = 864\ 533 - 229\ 827 = 634\ 706$$

There are two unknowns, $m_2$ and $u_2$, in this equation. However,

$$m_2 = \frac{V}{v_2} = \frac{2.0}{0.001\ 60 + x_2(0.203\ 81)}$$

and

$$u_2 = 226.0 + x_2(1099.7)$$

and thus both are functions only of $x_2$, the quality at the final state. Consequently,

$$\frac{2.0(1461.1 - 226.0 - 1099.7x_2)}{0.001\ 60 + 0.203\ 81x_2} = 634\ 706$$

Solving for $x_2$ we get

$$x_2 = 0.011\ 057$$

Therefore,

$$v_2 = 0.001\ 60 + 0.011\ 057 \times 0.203\ 81 = 0.003\ 853\ 5\ \text{m}^3/\text{kg}$$

$$m_2 = \frac{V}{v_2} = \frac{2}{0.003\ 853\ 5} = 519\ \text{kg}$$

and the mass of ammonia withdrawn, $m_e$, is

$$m_e = m_1 - m_2 = 591.7 - 519 = 72.7\ \text{kg}$$

---

**In-Text Concept Question**

k. An initially empty cylinder is filled with air from 20°C, 100 kPa until it is full. Assuming no heat transfer, is the final temperature larger than, equal to, or smaller than 20°C? Does the final $T$ depend on the size of the cylinder?

## 6.6 HOW-TO SECTION

### Which form of the energy equation should I use?

There really is only one form of the energy equation (the same is true for the other balance equations for mass, momentum, and entropy). The general rate equation is

$$\text{Rate of storage} = +\text{in} - \text{out} + \text{gen}$$

and only entropy (Chapter 8) can have a generation term. For energy, the equation becomes

Energy Rate Eq.: $\dot{E}_{\text{C.V.}} = \dot{Q}_{\text{C.V.}} - \dot{W}_{\text{C.V.}} + \sum \dot{m}_i h_{\text{tot}\,i} - \sum \dot{m}_e h_{\text{tot}\,e}$

and integrated in time it is

Energy Eq.: $E_2 - E_1 = {}_1Q_2 - {}_1W_2 + \sum m_i\,h_{\text{tot}\,i} - \sum m_e\,h_{\text{tot}\,e}$

The left-hand side expresses the storage of energy, like the change in a bank account balance, inside the chosen control volume. The right-hand side gives all the transfers of energy to/from the surroundings crossing the control surface, like deposits to and withdrawals from the bank account. Any particular case is a simplified version of the above equation(s) removing terms not present.

For instance if there are no flows, then the summation terms disappear and it reduces to Eq. 5.11. For a steady-state flow situation, the left-hand side disappears and it reduces to Eq. 6.10 (Eq. 6.12 for single flow). Finally, the transient process has both a storage effect and the flow transfers so it is written as in Eq. 6.16.

### When do I use "$u$" and when do I use "$h$"?

The appearance of $u$ or $h$ in the energy equation is not arbitrary. The left-hand side of the general energy equation, Eq. 6.8, has the storage of energy $E_{CV}$ that includes a $(mu)$ term, whereas the flow terms on the right have $(\dot{m}h)$ terms. For the steady-flow problems, the left-hand side becomes zero, leaving the $h$ terms in the energy equation as in

Eqs. 6.12–13. When there are no flow terms, then the time integration gives the "$u$" terms from the left-hand side as in Eq. 5.11 and, finally, if you have both set of terms as in the transient process, the "$u$" and "$h$" appears as in Eq. 6.16. So remember, "$u$" is associated with storage of energy and "$h$" is associated with a flow of energy due to a mass flow.

### With a flow, do I always account for the kinetic and potential energy?

Not necessarily—it depends on the magnitude of the energy relative to the other energy terms. Assume a flow of air at 10 m/s at a 10 m elevation and that we heat it 10 degrees. What are the kinetic and potential energies relative to the increase in $h$?

$$KE = \tfrac{1}{2}\mathbf{V}^2 = \tfrac{1}{2}10^2 \text{ (m/s)}^2 = 50 \text{ J/kg} = 0.05 \text{ kJ/kg}$$

$$PE = gZ = 9.807 \times 10 \text{ (m/s)}^2 = 98.1 \text{ J/kg} = 0.098 \text{ kJ/kg}$$

$$\Delta h = C_P \Delta T = 1.0 \times 10 \text{ kJ/kg} = 10 \text{ kJ/kg}$$

We now realize that a velocity of more than 50 m/s or an elevation change of more than 100 m are necessary to have kinetic or potential energy terms of 1 kJ/kg, usually very small in comparison to other terms in the first law.

## SUMMARY

Conservation of mass is expressed as a rate of change of total mass due to mass flows in or out of the control volume. The control mass energy equation is extended to include mass flows that also carry energy (internal, kinetic, and potential) and the flow work needed to push the flow in or out of the control volume against the prevailing pressure. The conservation of mass (continuity equation) and the conservation of energy (first law) are applied to a number of standard devices.

A steady-state device has no storage effects, with all properties constant with time, and constitutes the majority of all flow-type devices. A combination of several devices forms a complete system built for a specific purpose, such as a power plant, jet engine, or refrigerator.

A transient process with a change in mass (storage) such as filling or emptying of a container is considered based on an average description. It is also realized that the start-up or shutdown of a steady-state device leads to a transient process.

You should have learned a number of skills and acquired abilities from studying this chapter that will allow you to

- Understand the physical meaning of the conservation equations. Rate $= +$in $-$ out
- Understand the concepts of mass flow rate, volume flow rate, and local velocity
- Recognize the flow and nonflow terms in the energy equation
- Know how the most typical devices work and if they have heat or work transfers
- Have a sense about devices where kinetic and potential energies are important
- Analyze steady-state single-flow devices such as nozzles, throttles, turbines, or pumps
- Extend the application to a multiple-flow device such as a heat exchanger, mixing chamber, or turbine, given the specific setup
- Apply the conservation equations to complete systems as a whole or to the individual devices and recognize their connections and interactions

- Recognize and use the proper form of the equations for transient problems
- Be able to assume a proper average value for any flow term in a transient
- Recognize the difference between storage of energy ($dE/dt$) and flow ($\dot{m}h$)

A number of steady-flow devices are listed in Table 6.1 with a very short statement of the device's purpose, known facts about work and heat transfer, and a common assumption if appropriate. This list is not complete with respect to the number of devices nor with respect to the facts listed but is meant as a short list of typical devices, some of which may be unfamiliar to many readers.

## KEY CONCEPTS AND FORMULAS

Volume flow rate $\qquad \dot{V} = \int \mathbf{V}\, dA = A\mathbf{V}$  (using average velocity)

Mass flow rate $\qquad \dot{m} = \int \rho \mathbf{V}\, dA = \rho A \mathbf{V} = A\mathbf{V}/v$  (using average values)

Flow work rate $\qquad \dot{W}_{\text{flow}} = P\dot{V} = \dot{m}Pv$

Flow direction $\qquad$ From higher $P$ to lower $P$ unless significant KE or PE.

### Instantaneous Process

Continuity equation $\qquad \dot{m}_{\text{C.V.}} = \sum \dot{m}_i - \sum \dot{m}_e$

Energy equation $\qquad \dot{E}_{\text{C.V.}} = \dot{Q}_{\text{C.V.}} - \dot{W}_{\text{C.V.}} + \sum \dot{m}_i h_{\text{tot}\,i} - \sum \dot{m}_e h_{\text{tot}\,e}$

Total enthalpy $\qquad h_{\text{tot}} = h + \dfrac{1}{2}\mathbf{V}^2 + gZ = h_{\text{stagnation}} + gZ$

### Steady State

$\qquad$ No storage: $\quad \dot{m}_{\text{C.V.}} = 0; \quad \dot{E}_{\text{C.V.}} = 0$

Continuity equation $\qquad \sum \dot{m}_i = \sum \dot{m}_e$  (in = out)

Energy equation $\qquad \dot{Q}_{\text{C.V.}} + \sum \dot{m}_i h_{\text{tot}\,i} = \dot{W}_{\text{C.V.}} + \sum \dot{m}_e h_{\text{tot}\,e}$  (in = out)

Specific heat transfer $\qquad q = \dot{Q}_{\text{C.V.}}/\dot{m}$  (steady state only)

Specific work $\qquad w = \dot{W}_{\text{C.V.}}/\dot{m}$  (steady state only)

Steady-state single flow energy equation $\qquad q + h_{\text{tot}\,i} = w + h_{\text{tot}\,e}$  (in = out)

### Transient Process

Continuity equation $\qquad m_2 - m_1 = \sum m_i - \sum m_e$

Energy equation $\qquad E_2 - E_1 = {}_1Q_2 - {}_1W_2 + \sum m_i h_{\text{tot}\,i} - \sum m_e h_{\text{tot}\,e}$

$$E_2 - E_1 = m_2\left(u_2 + \frac{1}{2}\mathbf{V}_2^2 + gZ_2\right) - m_1\left(u_1 + \frac{1}{2}\mathbf{V}_1^2 + gZ_1\right)$$

$$h_{\text{tot}\,e} = h_{\text{tot exit average}} \approx \frac{1}{2}\left(h_{\text{hot}\,e1} + h_{\text{tot}\,e2}\right)$$

**TABLE 6.1**

*Typical Steady-Flow Devices*

| Device | Purpose | Given | Assumption |
|---|---|---|---|
| Aftercooler | Cool a flow after a compressor | $w = 0$ | $P$ = constant |
| Boiler | Bring substance to a vapor state | $w = 0$ | $P$ = constant |
| Condenser | Take $q$ out to bring substance to liquid state | $w = 0$ | $P$ = constant |
| Combustor | Burn fuel; acts like heat transfer in | $w = 0$ | $P$ = constant |
| Compressor | Bring a substance to higher pressure | $w$ in | $q = 0$ |
| Deaerator | Remove gases dissolved in liquids | $w = 0$ | $P$ = constant |
| Dehumidifier | Remove water from air | | $P$ = constant |
| Desuperheater | Add liquid water to superheated vapor steam to make it saturated vapor | $w = 0$ | $P$ = constant |
| Diffuser | Convert KE energy to higher $P$ | $w = 0$ | $q = 0$ |
| Economizer | Low-$T$, low-$P$ heat exchanger | $w = 0$ | $P$ = constant |
| Evaporator | Bring a substance to vapor state | $w = 0$ | $P$ = constant |
| Expander | Similar to a turbine, but may have a $q$ | | |
| Fan/blower | Move a substance, typically air | $w$ in, KE up | $P = C, q = 0$ |
| Feedwater heater | Heat liquid water with another flow | $w = 0$ | $P$ = constant |
| Flash evaporator | Generate vapor by expansion (throttling) | $w = 0$ | $q = 0$ |
| Heat engine | A device that converts part of heat into work | $q$ in, $w$ out | |
| Heat exchanger | Transfer heat from one medium to another | $w = 0$ | $P$ = constant |
| Heat pump | A device moving a $Q$ from $T_{\text{low}}$ to $T_{\text{high}}$, requires a work input, refrigerator | $w$ in | |
| Heater | Heat a substance | $w = 0$ | $P$ = constant |
| Humidifier | Add water to air–water mixture | $w = 0$ | $P$ = constant |
| Intercooler | Heat exchanger between compressor stages | $w = 0$ | $P$ = constant |
| Nozzle | Create KE; $P$ drops | $w = 0$ | $q = 0$ |
| | Measure flow rate | | |
| Mixing chamber | Mix two or more flows | $w = 0$ | $q = 0$ |
| Pump | Same as compressor, but handles liquid | $w$ in, $P$ up | $q = 0$ |
| Reactor | Allow reaction between two or more substances | $w = 0$ | $q = 0, P = C$ |
| Regenerator | Usually a heat exchanger to recover energy | $w = 0$ | $P$ = constant |
| Steam generator | Same as boiler, heat liquid water to superheat vapor | $w = 0$ | $P$ = constant |
| Supercharger | A compressor driven by engine shaft work to drive air into an automotive engine | $w$ in | |
| Superheater | A heat exchanger that brings $T$ up over $T_{\text{sat}}$ | $w = 0$ | $P$ = constant |
| Turbine | Create shaft work from high $P$ flow | $w$ out | $q = 0$ |
| Turbocharger | A compressor driven by an exhaust flow turbine to charge air into an engine | $\dot{W}_{\text{turbine}} = \dot{W}_{\text{C.V.}}$ | |
| Throttle | Same as valve | | |
| Valve | Control flow by restriction; $P$ drops | $w = 0$ | $q = 0$ |

# HOMEWORK PROBLEMS

## Concept problems

**6.1** A temperature difference drives a heat transfer. Does a similar concept apply to $\dot{m}$?

**6.2** What kind of effect can be felt upstream in a flow?

**6.3** Which of the properties $(P, v, T)$ can be controlled in a flow? How?

**6.4** Air at 500 K, 500 kPa is expanded to 100 kPa in two steady-flow cases. Case 1 is a throttle and case 2 is a turbine. Which has the highest exit $T$? Why?

**6.5** A windmill takes a fraction of the wind kinetic energy out as power on a shaft. In what manner does the temperature and wind velocity influence the power? *Hint:* Write the power as mass flow rate times specific work.

**6.6** You blow a balloon up with air. What kind of work terms including flow work do you see in that case? Where is energy stored?

## Continuity equation and flow rates

**6.7** An empty bath tub has its drain closed and is being filled with water from the faucet at a rate of 10 kg/min. After 10 min the drain is opened and 4 kg/min flows out, and at the same time the inlet flow is reduced to 2 kg/min. Plot the mass of the water in the bathtub versus time and determine the time from the very beginning when the tub will be empty.

**6.8** Saturated vapor R-134a leaves the evaporator in a heat pump system at 10°C, with a steady mass flow rate of 0.1 kg/s. What is the smallest diameter tubing that can be used at this location if the velocity of the refrigerant is not to exceed 7 m/s?

**6.9** In a boiler, you vaporize some liquid water at 100 kPa flowing at 1 m/s. What is the velocity of the saturated vapor at 100 kPa if the pipe size is the same? Can the flow then be constant $P$?

**6.10** A boiler receives a constant flow of 5000 kg/h liquid water at 5 MPa and 20°C, and it heats the flow such that the exit state is 450°C with a pressure of 4.5 MPa. Determine the necessary minimum pipe flow area in both the inlet and exit pipe(s) if there should be no velocities larger than 20 m/s.

**6.11** Nitrogen gas flowing in a 50-mm-diameter pipe at 15°C and 200 kPa, at the rate of 0.05 kg/s, encounters a partially closed valve. If there is a pressure drop of 30 kPa across the valve and essentially no temperature change, what are the velocities upstream and downstream of the valve?

**6.12** A hot-air home heating system takes 0.25 m³/s air at 100 kPa, 17°C into a furnace, heats it to 52°C, and delivers the flow to a square duct 0.2 m by 0.2 m at 110 kPa (see Fig. P6.12). What is the velocity in the duct?

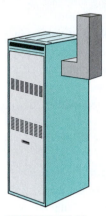

**FIGURE P6.12**

**6.13** Steam at 3 MPa and 400°C enters a turbine with a volume flow rate of 5 m³/s. An extraction of 15% of the inlet mass flow rate exits at 600 kPa and 200°C. The rest exits the turbine at 20 kPa with a quality of 90% and a velocity of 20 m/s. Determine the volume flow rate of the extraction flow and the diameter of the final exit pipe.

**6.14** A household fan of diameter 0.75 m takes air in at 98 kPa, 22°C and delivers it at 105 kPa, 23°C with a velocity of 1.5 m/s (see Fig. P6.14). What are the mass flow rate (kg/s), the inlet velocity, and the outgoing volume flow rate in m³/s?

**FIGURE P6.14**

## Single flow, single device processes

### Nozzles, diffusers

**6.15** Liquid water at 15°C flows out of a nozzle straight up 15 m. What is nozzle $\mathbf{V}_{exit}$?

**6.16** Nitrogen gas flows into a convergent nozzle at 200 kPa and 400 K and very low velocity. It flows out of the nozzle at 100 kPa and 330 K. If the nozzle is insulated, find the exit velocity.

**6.17** A nozzle receives 0.1 kg/s of steam at 1 MPa and 400°C with negligible kinetic energy. The exit is at 500 kPa and 350°C, and the flow is adiabatic. Find the nozzle exit velocity and the exit area.

**6.18** In a jet engine a flow of air at 1000 K, 200 kPa, and 30 m/s enters a nozzle, as shown in Fig. P6.18, where the air exits at 850 K, 90 kPa. What is the exit velocity assuming no heat loss?

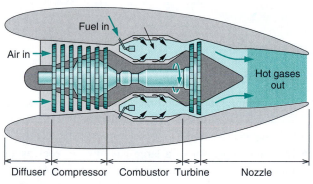

FIGURE P6.18

**6.19** In a jet engine a flow of air at 1000 K, 200 kPa, and 40 m/s enters a nozzle, where the air exits at 500 m/s, 90 kPa. What is the exit temperature, assuming no heat loss?

**6.20** A sluice gate dams water up 5 m. A 1-cm-diameter hole at the bottom of the gate allows liquid water at 20°C to come out. Neglect any changes in internal energy and find the exit velocity and mass flow rate.

**6.21** A diffuser, shown in Fig. P6.21, has air entering at 100 kPa and 300 K with a velocity of 200 m/s. The inlet cross-sectional area of the diffuser is 100 mm². At the exit, the area is 860 mm², and the exit velocity is 20 m/s. Determine the exit pressure and temperature of the air.

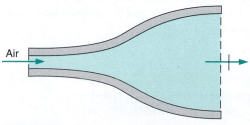

FIGURE P6.21

**6.22** A diffuser receives an ideal-gas flow at 100 kPa and 300 K with a velocity of 250 m/s, and the exit velocity is

25 m/s. Determine the exit temperature if the gas is argon, helium, or nitrogen.

**6.23** Superheated vapor ammonia enters an insulated nozzle at 20°C and 800 kPa, as shown in Fig. P6.23, with a low velocity and at a steady rate of 0.01 kg/s. The ammonia exits at 300 kPa with a velocity of 450 m/s. Determine the temperature (or quality, if saturated) and the exit area of the nozzle.

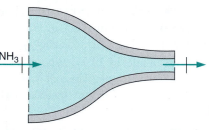

FIGURE P6.23

**6.24** Air flows into a diffuser at 300 m/s, 300 K, and 100 kPa. At the exit the velocity is very small but the pressure is high. Find the exit temperature assuming zero heat transfer.

**Throttle flow**

**6.25** R-134a at 30°C, 800 kPa is throttled so it becomes cold at −10°C. What is exit $P$?

**6.26** Helium is throttled from 1.2 MPa and 20°C to a pressure of 100 kPa. The diameter of the exit pipe is so much larger than that of the inlet pipe that the inlet and exit velocities are equal. Find the exit temperature of the helium and the ratio of the pipe diameters.

**6.27** Saturated vapor R-134a at 500 kPa is throttled to 200 kPa in a steady flow through a valve. The kinetic energy in the inlet and exit flow is the same. What is the exit temperature?

**6.28** Water is flowing in a line at 400 kPa, and saturated vapor is taken out through a valve to 100 kPa. What is the temperature as it leaves the valve, assuming no changes in kinetic energy and no heat transfer?

**6.29** Liquid water at 180°C and 2000 kPa is throttled into a flash evaporator chamber having a pressure of 500 kPa. Neglect any change in the kinetic energy. What is the fraction of liquid and vapor in the chamber?

**6.30** Saturated liquid R-12 at 25°C is throttled to 150.9 kPa in your refrigerator. What is the exit temperature? Find the percent increase in the volume flow rate.

**6.31** Water at 1.5 MPa and 150°C is throttled adiabatically through a valve to 200 kPa. The inlet velocity is

5 m/s, and the inlet and exit pipe diameters are the same. Determine the state (neglecting kinetic energy in the energy equation) and the velocity of the water at the exit.

**6.32** Methane at 3 MPa, 300 K is throttled through a valve to 100 kPa. Calculate the exit temperature assuming no changes in the kinetic energy and ideal gas behavior.

## Turbines, expanders

**6.33** Air at 20 m/s, 260 K, 75 kPa with 5 kg/s flows into a jet engine and it flows out at 500 m/s, 800 K, 75 kPa. What is the change (power) in flow of kinetic energy?

**6.34** A steam turbine has an inlet of 2 kg/s water at 1000 kPa and 350°C with velocity of 15 m/s. The exit is at 100 kPa, 150°C and very low velocity. Find the specific work and the power produced.

**6.35** A small, high-speed turbine operating on compressed air produces a power output of 100 W. The inlet state is 400 kPa, 50°C, and the exit state is 150 kPa, −30°C. Assuming the velocities to be low and the process to be adiabatic, find the required mass flow rate of air through the turbine.

**6.36** A small expander (a turbine with heat transfer) has 0.05 kg/s helium entering at 1000 kPa, 550 K and it leaves at 250 kPa, 300 K. The power output on the shaft is measured to 55 kW. Find the rate of heat transfer neglecting kinetic energies.

**6.37** A liquid water turbine receives 2 kg/s water at 2000 kPa and 20°C with a velocity of 15 m/s. The exit is at 100 kPa, 20°C, and very low velocity. Find the specific work and the power produced.

**6.38** Hoover Dam across the Colorado River dams up Lake Mead 200 m higher than the river downstream (see Fig. P6.38). The electric generators driven by water-powered turbines deliver 1300 MW of power. If the water is 17.5°C, find the minimum amount of water running through the turbines.

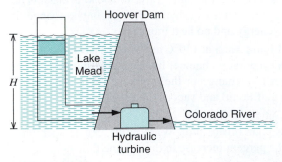

**FIGURE P6.38**

**6.39** A windmill with rotor diameter of 30 m takes 40% of the kinetic energy out as shaft work on a day with 20°C and wind speed of 30 km/h. What power is produced?

**6.40** A small turbine, shown in Fig. P6.40, is operated at part load by throttling a 0.25-kg/s steam supply at 1.4 MPa and 250°C down to 1.1 MPa before it enters the turbine, and the exhaust is at 10 kPa. If the turbine produces 110 kW, find the exhaust temperature (and quality if saturated).

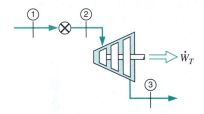

**FIGURE P6.40**

## Compressors, fans

**6.41** A compressor in a commercial refrigerator receives R-22 at −25°C and $x = 1$. The exit is at 1000 kPa and 60°C. Neglect kinetic energies and find the specific work.

**6.42** The compressor of a large gas turbine receives air from the ambient surroundings at 95 kPa and 20°C with a low velocity. At the compressor discharge, air exits at 1.52 MPa and 430°C with velocity of 90 m/s. The power input to the compressor is 5000 kW. Determine the mass flow rate of air through the unit.

**6.43** A compressor brings R-134a from 150 kPa, −10°C to 1200 kPa, 50°C. It is water cooled with a heat loss estimated as 40 kW, and the shaft work input is measured to be 150 kW. How much is the mass flow rate through the compressor?

**6.44** An ordinary portable fan blows 0.2 kg/s of room air with a velocity of 18 m/s (see Fig. P6.14). What is the minimum power electric motor that can drive it? Hint: Are there any changes in $P$ or $T$?

**6.45** An air compressor takes in air at 100 kPa and 17°C, and delivers it at 1 MPa and 600 K to a constant-pressure cooler, which the air exits at 300 K (see Fig. P6.45). Find the specific compressor work and the specific heat transfer in the cooler.

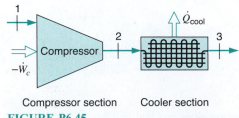

**FIGURE P6.45**

**6.46** An exhaust fan in a building should be able to move 2.5 kg/s of air at 98 kPa and 20°C through a 0.4-m-diameter vent hole. How high a velocity must it generate and how much power is required to do that?

**6.47** How much power is needed to run the fan in Problem 6.14?

## Heaters, coolers

**6.48** Carbon dioxide enters a steady-state, steady-flow heater at 300 kPa, 300 K, and exits at 275 kPa, 1500 K, as shown in Fig. P6.48. Changes in kinetic and potential energies are negligible. Calculate the required heat transfer per kilogram of carbon dioxide flowing through the heater.

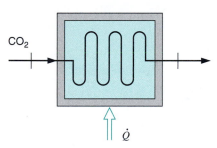

**FIGURE P6.48**

**6.49** A condenser (cooler) receives 0.05 kg/s of R-22 at 800 kPa and 40°C, and cools it to 15°C. There is a small pressure drop so that the exit state is saturated liquid. What cooling capacity (kW) must the condenser have?

**6.50** A chiller cools liquid water for air-conditioning purposes. Assume 2.5 kg/s water at 20°C and 100 kPa is cooled to 5°C in a chiller. How much heat transfer (kW) is needed?

**6.51** Saturated liquid nitrogen at 600 kPa enters a boiler at a rate of 0.005 kg/s and exits as saturated vapor (see Fig. P6.51). It then flows into a super heater also at 600 kPa where it exits at 600 kPa, 280 K. Find the rate of heat transfer in the boiler and the super heater.

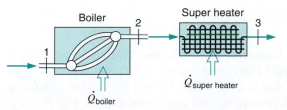

**FIGURE P6.51**

**6.52** In a steam generator, compressed liquid water at 10 MPa and 30°C enters a 30-mm-diameter tube at a rate of 3 L/s. Steam at 9 MPa and 400°C exits the tube. Find the rate of heat transfer to the water.

**6.53** The air conditioner in a house or a car has a cooler that brings atmospheric air from 30°C to 10°C with both states at 101 kPa. If the flow rate is 0.5 kg/s, find the rate of heat transfer.

**6.54** A flow of liquid glycerine flows around an engine, cooling it as it absorbs energy. The glycerine enters the engine at 60°C and receives 19 kW of heat transfer. What is the required mass flow rate if the glycerine should come out at maximum 95°C?

**6.55** Liquid nitrogen at 90 K and 400 kPa flows into a probe used in cryogenic survey. In the return line the nitrogen is then at 160 K and 400 kPa. Find the specific heat transfer to the nitrogen. If the return line has a cross-sectional area 100 times larger than that of the inlet line, what is the ratio of the return velocity to the inlet velocity?

## Pumps, pipe, and channel flows

**6.56** A steam pipe for a 300-m-tall building receives superheated steam at 200 kPa at ground level. At the top floor the pressure is 125 kPa, and the heat loss in the pipe is 110 kJ/kg. What should the inlet temperature be so that no water will condense inside the pipe?

**6.57** A small stream with 20°C water runs out over a cliff, creating a 100-m-tall waterfall. Estimate the downstream temperature when you neglect the horizontal flow velocities upstream and downstream from the waterfall. How fast was the water dropping just before it splashed into the pool at the bottom of the waterfall?

**6.58** A small water pump is used in an irrigation system. The pump takes water in from a river at 10°C and 100 kPa at a rate of 5 kg/s. The exit line enters a pipe that goes up to an elevation 20 m above the pump and river, where the water runs into an open channel. Assume the process is adiabatic and that the water stays at 10°C. Find the required pump work.

**6.59** A pipe flows water at 15°C from one building to another. In winter, the pipe loses an estimated 500 W of heat transfer. What is the minimum required mass flow rate that will ensure that the water does not freeze (i.e. reach 0°C)?

**6.60** The main water line into a tall building has a pressure of 600 kPa at 5 m below ground level, as shown in Fig. P6.60. A pump brings the pressure up so the water can be delivered at 200 kPa at the top floor 150 m above ground level. Assume a flow rate of 10 kg/s liquid water at 10°C and neglect any difference in kinetic energy and internal energy $u$. Find the pump work.

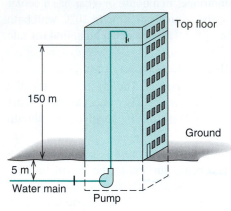

**FIGURE P6.60**

**6.61** Consider a water pump that receives liquid water at 15°C and 100 kPa, and delivers it to a same diameter short pipe having a nozzle with exit diameter of 1 cm (0.01 m) to the atmosphere at 100 kPa (see Fig. P6.61). Neglect the kinetic energy in the pipes and assume constant $u$ for the water. Find the exit velocity and the mass flow rate if the pump draws 1 kW of power.

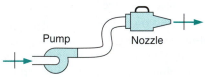

**FIGURE P6.61**

**6.62** A cutting tool uses a nozzle that generates a high speed jet of liquid water. Assume an exit velocity of 500 m/s of 20°C liquid water with a jet diameter of 2 mm (0.002 m). How much mass flow rate is this? What size (power) pump is needed to generate this from a steady supply of 20°C liquid water at 200 kPa?

## Multiple flow, single device processes

### Turbines, compressors, expanders

**6.63** A steam turbine receives steam from two boilers (see Fig. P6.63). One flow is 5 kg/s at 3 MPa and 700°C, and the other flow is 15 kg/s at 800 kPa and 500°C. The exit state is 10 kPa, with a quality of 96%. Find the total power out of the adiabatic turbine.

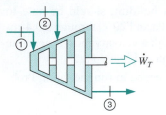

**FIGURE P6.63**

**6.64** A steam turbine receives water at 15 MPa and 600°C at a rate of 100 kg/s, as shown in Fig. P6.64. In the middle section 20 kg/s is withdrawn at 2 MPa and 350°C, and the rest exits the turbine at 75 kPa, with 95% quality. Assuming no heat transfer and no changes in kinetic energy, find the total turbine power output.

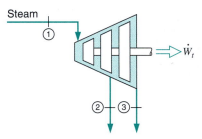

**FIGURE P6.64**

**6.65** Two steady flows of air enter a control volume, shown in Fig. P6.65. One is 0.025 kg/s flow at 350 kPa, 150°C, state 1, and the other enters at 450 kPa, 15°C, state 2. A single flow exits at 100 kPa, −40°C, state 3. The control volume rejects 1 kW heat to the surroundings and produces 4 kW of power output. Neglect kinetic energies and determine the mass flow rate at state 2.

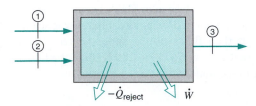

**FIGURE P6.65**

**6.66** Cogeneration is often used where a steam supply is needed for industrial process energy. Assume a supply of 5 kg/s steam at 0.5 MPa is needed. Rather than generating this from a pump and boiler, the setup in Fig. P6.66 is used to extract the supply from the high-pressure turbine.

Find the power the turbine now cogenerates in this process.

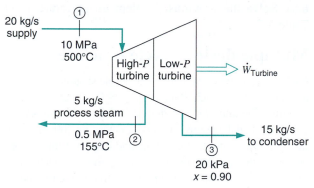

**FIGURE P6.66**

**6.67** A compressor receives 0.1 kg/s R-134a at 150 kPa, −10°C and delivers it at 1000 kPa, 40°C. The power input is measured to be 3 kW. The compressor has heat transfer to air at 100 kPa coming in at 20°C and leaving at 25°C. How much is the mass flow rate of air?

## Heat exchangers

**6.68** In a co-flowing (same direction) heat exchanger, 1 kg/s air at 500 K flows into one channel and 2 kg/s air flows into the neighboring channel at 300 K. If it is infinitely long, what is the exit temperature? Sketch the variation of $T$ in the two flows.

**6.69** Air at 600 K flows with 3 kg/s into a heat exchanger. How much (kg/s) water coming in at 100 kPa, 20°C can the air heat to the boiling point?

**6.70** A condenser (heat exchanger) brings 1 kg/s water flow at 10 kPa from 300°C to saturated liquid at 10 kPa, as shown in Fig. P6.70. The cooling is done by lake water at 20°C that returns to the lake at 30°C. For an insulated condenser, find the flow rate of cooling water.

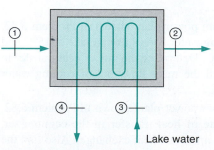

**FIGURE P6.70**

**6.71** Steam at 500 kPa, 300°C is used to heat cold water at 15°C to 75°C for domestic hot water supply. How much steam per kg liquid water is needed if the steam should not condense?

**6.72** An automotive radiator has glycerine at 95°C enter and return at 55°C as shown in Fig. P6.72. Air flows in at 20°C and leaves at 25°C. If the radiator should transfer 25 kW, what is the mass flow rate of the glycerine and what is the volume flow rate of air in at 100 kPa?

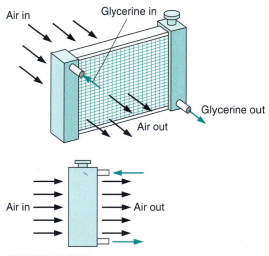

**FIGURE P6.72**

**6.73** A copper wire has been heat treated to 1000 K and is now pulled into a cooling chamber that has 1.5 kg/s air coming in at 20°C; the air leaves the other end at 60°C. If the wire moves 0.25 kg/s copper, how hot is the copper as it comes out?

## Mixing processes

**6.74** An open feedwater heater in a power plant heats 4 kg/s water at 45°C and 100 kPa by mixing it with steam from the turbine at 100 kPa and 250°C as in Fig. P6.74. Assume the exit flow is saturated liquid at the given pressure and find the mass flow rate from the turbine.

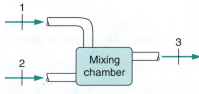

**FIGURE P6.74**

**6.75** A desuperheater mixes superheated water vapor with liquid water in a ratio that produces saturated water vapor as output without any external heat transfer. A flow of 0.5 kg/s superheated vapor at 5 MPa and 400°C, and a flow of liquid water at 5 MPa and 40°C enters a desuperheater. If saturated water vapor at 4.5 MPa is produced, determine the flow rate of the liquid water.

**6.76** Two air flows are combined into a single flow. Flow 1 is 1 m³/s at 20°C and flow 2 is 2 m³/s at 200°C, both at 100 kPa. They mix without any heat transfer to produce an exit flow at 100 kPa. Neglect kinetic energies and find the exit temperature and volume flow rate.

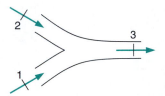

**FIGURE P6.76**

**6.77** A mixing chamber with heat transfer receives 2 kg/s of R-22 at 1 MPa and 40°C in one line and 1 kg/s of R-22 at 30°C with a quality of 50% in a line with a valve. The outgoing flow is at 1 MPa and 60°C. Find the rate of heat transfer to the mixing chamber.

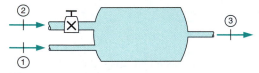

**FIGURE P6.77**

**6.78** A flow of water at 2000 kPa, 20°C is mixed with a flow of 2 kg/s water at 2000 kPa, 180°C. What should the flow rate of the first flow be to produce an exit state of 200 kPa and 100°C?

**6.79** An insulated mixing chamber receives 2 kg/s of R-134a at 1 MPa and 100°C in a line with low velocity. Another line with R-134a as saturated liquid at 60°C flows through a valve to the mixing chamber at 1 MPa after the valve, as shown in Fig. P6.77. The exit flow is saturated vapor at 1 MPa flowing at 20 m/s. Find the flow rate for the second line.

**6.80** To keep a jet engine cool, some intake air bypasses the combustion chamber. Assume 2 kg/s hot air at 2000 K

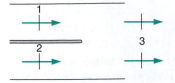

**FIGURE P6.80**

and 500 kPa is mixed with 1.5 kg/s air at 500 K, 500 kPa without any external heat transfer. Find the exit temperature using constant heat capacity from Table A.5.

**6.81** Solve the previous problem using values from Table A.7.

# Multiple devices, cycle processes

**6.82** The following data are for a simple steam power plant as shown in Fig. P6.82. State 6 has $x_6 = 0.92$ and velocity of 200 m/s. The rate of steam flow is 25 kg/s, with 300 kW of power input to the pump. Piping diameters are 200 mm from the steam generator to the turbine and 75 mm from the condenser to the economizer and steam generator. Determine the velocity at state 5 and the power output of the turbine.

| State | 1 | 2 | 3 | 4 | 5 | 6 | 7 |
|-------|------|------|------|------|------|----|-----|
| $P$, kPa | 6200 | 6100 | 5900 | 5700 | 5500 | 10 | 9 |
| $T$, °C | | 45 | 175 | 500 | 490 | | 40 |
| $h$, kJ/kg | | 194 | 744 | 3426 | 3404 | | 168 |

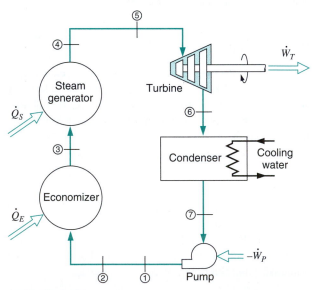

**FIGURE P6.82**

**6.83** For the steam power plant shown in Problem 6.82, assume the cooling water comes from a lake at 15°C and is returned at 25°C. Determine the rate of heat transfer in the condenser and the mass flow rate of cooling water from the lake.

**6.84** For the steam power plant shown in Problem 6.82, determine the rate of heat transfer in the economizer, which is a low-temperature heat exchanger. Also find the rate of heat transfer needed in the steam generator.

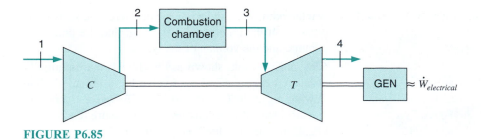

**FIGURE P6.85**

**6.85** A gas turbine set-up to produce power during peak demand is shown in Fig. P6.85. The turbine provides power to the air compressor and the electric generator. If the electric generator should provide 5 MW what is the needed air flow at state 1 and the combustion heat transfer between states 2 and 3? The states are

1. 90 kPa, 290 K
2. 900 kPa, 560 K
3. 900 kPa, 1400 K
4. 100 kPa, 850 K

**6.86** A proposal is made to use a geothermal supply of hot water to operate a steam turbine, as shown in Fig. P6.86. The high-pressure water at 1.5 MPa and 180°C is throttled into a flash evaporator chamber, which forms liquid and vapor at a lower pressure of 400 kPa. The liquid is discarded while the saturated vapor feeds the turbine and exits at 10 kPa with a 90% quality. If the turbine should produce 1 MW, find the required mass flow rate of hot geothermal water in kilograms per hour.

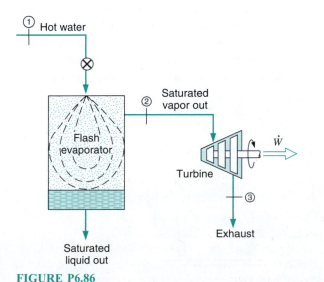

**FIGURE P6.86**

**6.87** An R-12 heat pump cycle shown in Fig. P6.87 has an R-12 flow rate of 0.05 kg/s with 4 kW into the compressor. The following data are given:

| State | 1 | 2 | 3 | 4 | 5 | 6 |
|---|---|---|---|---|---|---|
| $P$, kPa | 1250 | 1230 | 1200 | 320 | 300 | 290 |
| $T$, °C | 120 | 110 | 45 | | 0 | 5 |
| $h$, kJ/kg | 260 | 253 | 79.7 | | 188 | 191 |

Calculate the heat transfer from the compressor, the heat transfer from the R-12 in the condenser, and the heat transfer to the R-12 in the evaporator.

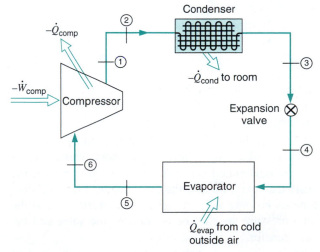

**FIGURE P6.87**

**6.88** A modern jet engine has a temperature after combustion of about 1500 K at 3200 kPa as it enters the turbine section (see state 3, Fig. P6.88). The compressor inlet is at 80 kPa, 260 K (state 1) and the outlet (state 2) is at 3300 kPa, 780 K; the turbine outlet (state 4) into the nozzle is at 400 kPa, 900 K and the nozzle exit (state 5) is at 80 kPa, 640 K. Neglect any heat transfer and neglect kinetic energy except out of the nozzle. Find the compressor and turbine specific work terms and the nozzle exit velocity.

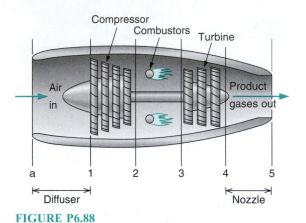

**FIGURE P6.88**

## Transient processes

**6.89** A cylinder has 0.1 kg air at 25°C, 200 kPa with a 5 kg piston on top. A valve at the bottom is opened to let the air out and the piston drops 0.25 m to the bottom. What is the work involved in this process? What happens to the energy?

**6.90** A 1-m³, 40-kg rigid steel tank contains air at 500 kPa, and both tank and air are at 20°C. The tank is connected to a line flowing air at 2 MPa and 20°C. The valve is opened, allowing air to flow into the tank until the pressure reaches 1.5 MPa, and is then closed. Assume the air and tank are always at the same temperature and the final temperature is 35°C. Find the final air mass and the heat transfer.

**6.91** A 2.5 liter tank is initially empty and we want to have 10 g of ammonia in it. The ammonia comes from a line with saturated vapor at 25°C. To end up with the desired amount, we cool the can while we fill it in a slow process keeping the can and content at 30°C. Find the final pressure to reach before closing the valve and the heat transfer.

**6.92** An evacuated 150-L tank is connected to a line flowing air at room temperature, 25°C, and 8 MPa pressure. The valve is opened, allowing air to flow into the tank until the pressure inside is 6 MPa. At this point the valve is closed. This filling process occurs rapidly and is essentially adiabatic. The tank is then placed in storage where it eventually returns to room temperature. What is the final pressure?

**6.93** An initially empty bottle is filled with water from a line at 0.8 MPa and 350°C. Assume no heat transfer and

that the bottle is closed when the pressure reaches the line pressure. If the final mass is 0.75 kg, find the final temperature and the volume of the bottle.

**6.94** A 25-L tank, shown in Fig. P6.94, that is initially evacuated is connected by a valve to an air supply line flowing air at 20°C and 800 kPa. The valve is opened, and air flows into the tank until the pressure reaches 600 kPa. Determine the final temperature and mass inside the tank, assuming the process is adiabatic. Develop an expression for the relation between the line temperature and the final temperature using constant specific heats.

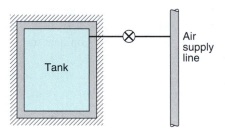

**FIGURE P6.94**

**6.95** A rigid 100-L tank contains air at 1 MPa and 200°C. A valve on the tank is now opened, and air flows out until the pressure drops to 100 kPa. During this process, heat is transferred from a heat source at 200°C, such that when the valve is closed, the temperature inside the tank is 50°C. What is the heat transfer?

**6.96** An empty canister of volume 1 L is filled with R-134a from a line flowing saturated liquid R-134a at 0°C. The filling is done quickly, so it is adiabatic. How much mass of R-134a is there after filling? The canister is placed on a storage shelf, where it slowly heats up to room temperature 20°C. What is the final pressure?

**6.97** A 200-L tank (see Fig. P6.97) initially contains water at 100 kPa and a quality of 1%. Heat is transferred to the water, thereby raising its pressure and temperature.

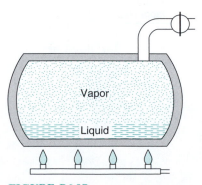

**FIGURE P6.97**

At a pressure of 2 MPa a safety valve opens and saturated vapor at 2 MPa flows out. The process continues, maintaining 2 MPa inside until the quality in the tank is 90%, then stops. Determine the total mass of water that flowed out and the total heat transfer.

**6.98** A 100-L rigid tank contains carbon dioxide gas at 1 MPa and 300 K. A valve is cracked open, and carbon dioxide escapes slowly until the tank pressure has dropped to 500 kPa. At this point the valve is closed. The gas remaining inside the tank may be assumed to have undergone a polytropic expansion, with polytropic exponent $n = 1.15$. Find the final mass inside and the heat transferred to the tank during the process.

**6.99** A nitrogen line, at 300 K and 0.5 MPa, shown in Fig. P6.99, is connected to a turbine that exhausts to a closed initially empty tank of 50 m$^3$. The turbine operates to a tank pressure of 0.5 MPa, at which point the temperature is 250 K. Assuming the entire process is adiabatic, determine the turbine work.

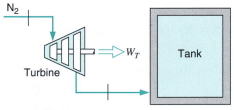

**FIGURE P6.99**

**6.100** A 2-m$^3$ insulated vessel, shown in Fig. P6.100, contains saturated vapor steam at 4 MPa. A valve on the top of the tank is opened, and steam is allowed to escape. During the process any liquid formed collects at the bottom of the vessel, so only saturated vapor exits. Calculate the total mass that has escaped when the pressure inside reaches 1 MPa.

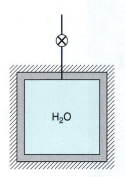

**FIGURE P6.100**

**6.101** A 750-L rigid tank, shown in Fig. P6.101, initially contains water at 250°C, which is 50% liquid and 50% vapor, by volume. A valve at the bottom of the tank is opened, and liquid is slowly withdrawn. Heat transfer takes place such that the temperature remains constant. Find the amount of heat transfer required to reach the state where half the initial mass is withdrawn.

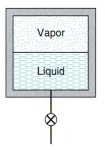

**FIGURE P6.101**

**6.102** Consider the previous problem but let the line and valve be located in the top of the tank. Now saturated vapor is slowly withdrawn while heat transfer keeps temperature inside constant. Find the heat transfer required to reach a state where half the original mass is withdrawn.

# Chapter 7

# The Second Law of Thermodynamics

The first law of thermodynamics states that during any cycle that a system undergoes, the cyclic integral of the heat is equal to the cyclic integral of the work. The first law, however, places no restrictions on the direction of flow of heat and work. A cycle in which a given amount of heat is transferred from the system and an equal amount of work is done on the system satisfies the first law just as well as a cycle in which the flows of heat and work are reversed. However, we know from our experience that because a proposed cycle does not violate the first law does not ensure that the cycle will actually occur. It is this kind of experimental evidence that led to the formulation of the second law of thermodynamics. Thus, a cycle will occur only if both the first and second laws of thermodynamics are satisfied.

In its broader significance, the second law acknowledges that processes proceed in a certain direction but not in the opposite direction. A hot cup of coffee cools by virtue of heat transfer to the surroundings, but heat will not flow from the cooler surroundings to the hotter cup of coffee. Gasoline is used as a car drives up a hill, but the fuel level

180

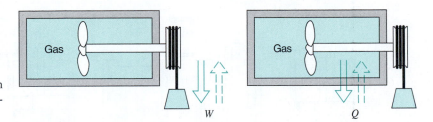

**FIGURE 7.1** A system that undergoes a cycle involving work and heat.

in the gasoline tank cannot be restored to its original level when the car coasts down the hill. Such familiar observations as these, and a host of others, are evidence of the validity of the second law of thermodynamics.

In this chapter, we consider the second law for a system undergoing a cycle, and in the next two chapters we extend the principles to a system undergoing a change of state and then to a control volume.

## 7.1 HEAT ENGINES AND REFRIGERATORS

Consider the system and the surroundings previously cited in the development of the first law, as shown in Fig. 7.1. Let the gas constitute the system and, as in our discussion of the first law, let this system undergo a cycle in which work is first done on the system by the paddle wheel as the weight is lowered. Then let the cycle be completed by transferring heat to the surroundings.

We know from our experience that we cannot reverse this cycle. That is, if we transfer heat to the gas, as shown by the dotted arrow, the temperature of the gas will increase, but the paddle wheel will not turn and raise the weight. With the given surroundings (the container, the paddle wheel, and the weight), this system can operate in a cycle in which the heat transfer and work are both negative, but it cannot operate in a cycle in which both the heat transfer and work are positive, even though this would not violate the first law.

Consider another cycle, known from our experience to be impossible actually to complete. Let two systems, one at a high temperature and the other at a low temperature, undergo a process in which a quantity of heat is transferred from the high-temperature system to the low-temperature system. We know that this process can take place. We also know that the reverse process, in which heat is transferred from the low-temperature system to the high-temperature system, does not occur, and that it is impossible to complete the cycle by heat transfer only. This impossibility is illustrated in Fig. 7.2.

These two examples lead us to a consideration of the heat engine and the refrigerator, which is also referred to as a heat pump. With the heat engine we can have a system that operates in a cycle and performs a net positive work and a net positive heat transfer.

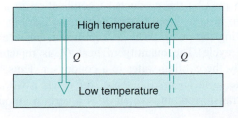

**FIGURE 7.2** An example showing the impossibility of completing a cycle by transferring heat from a low-temperature body to a high-temperature body.

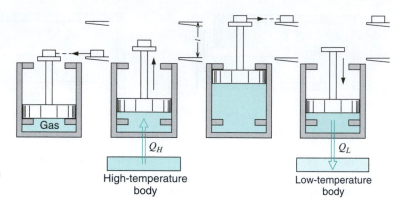

**FIGURE 7.3** A simple heat engine.

With the heat pump we can have a system that operates in a cycle and has heat transferred to it from a low-temperature body and heat transferred from it to a high-temperature body, though work is required to do this. Three simple heat engines and two simple refrigerators will be considered.

The first heat engine is shown in Fig. 7.3. It consists of a cylinder fitted with appropriate stops and a piston. Let the gas in the cylinder constitute the system. Initially the piston rests on the lower stops, with a weight on the platform. Let the system now undergo a process in which heat is transferred from some high-temperature body to the gas, causing it to expand and raise the piston to the upper stops. At this point the weight is removed. Now let the system be restored to its initial state by transferring heat from the gas to a low-temperature body, thus completing the cycle. Since the weight was raised during the cycle, it is evident that work was done by the gas during the cycle. From the first law we conclude that the net heat transfer was positive and equal to the work done during the cycle.

Such a device is called a heat engine, and the substance to which and from which heat is transferred is called the working substance or working fluid. A heat engine may be defined as a device that operates in a thermodynamic cycle and does a certain amount of net positive work through the transfer of heat from a high-temperature body to a low-temperature body. Often the term heat engine is used in a broader sense to include all devices that produce work, either through heat transfer or through combustion, even though the device does not operate in a thermodynamic cycle. The internal combustion engine and the gas turbine are examples of such devices, and calling them heat engines is an acceptable use of the term. In this chapter, however, we are concerned with the more restricted form of heat engine, as just defined, one that operates on a thermodynamic cycle.

A simple steam power plant is an example of a heat engine in this restricted sense. Each component in this plant may be analyzed individually as a steady-state, steady-flow process, but as a whole it may be considered a heat engine (Fig. 7.4) in which water (steam) is the working fluid. An amount of heat, $Q_H$, is transferred from a high-temperature body, which may be the products of combustion in a furnace, a reactor, or a secondary fluid that in turn has been heated in a reactor. In Fig. 7.4 the turbine is shown schematically as driving the pump. What is significant, however, is the net work that is delivered during the cycle. The quantity of heat $Q_L$ is rejected to a low-temperature body, which is usually the cooling water in a condenser. Thus, the simple steam power plant is a heat engine in the restricted sense, for it has a working fluid, to which and from which heat is transferred, and which does a certain amount of work as it undergoes a cycle.

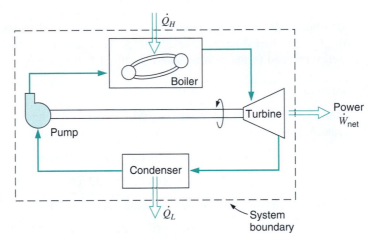

**FIGURE 7.4** A heat engine involving steady-state processes.

Another example of a heat engine is the thermoelectric power generation device that was discussed in Chapter 1 and shown schematically in Fig. 1.8b. Heat is transferred from a high-temperature body to the hot junction ($Q_H$), and heat is transferred from the cold junction to the surroundings ($Q_L$). Work is done in the form of electrical energy. Since there is no working fluid, we do not usually think of this as a device that operates in a cycle. However, if we adopt a microscopic point of view, we could regard a cycle as the flow of electrons. Furthermore, as with the steam power plant, the state at each point in the thermoelectric power generator does not change with time under steady-state conditions.

Thus, by means of a heat engine, we are able to have a system operate in a cycle and have both the net work and the net heat transfer positive, which we were not able to do with the system and surroundings of Fig. 7.1.

We note that in using the symbols $Q_H$ and $Q_L$, we have departed from our sign connotation for heat, because for a heat engine $Q_L$ is negative when the working fluid is considered as the system. In this chapter it will be advantageous to use the symbol $Q_H$ to represent the heat transfer to or from the high-temperature body, and $Q_L$ to represent the heat transfer to or from the low-temperature body. The direction of the heat transfer will be evident from the context.

At this point, it is appropriate to introduce the concept of thermal efficiency of a heat engine. In general, we say that efficiency is the ratio of output, the energy sought, to input, the energy that costs, but the output and input must be clearly defined. At the risk of oversimplification, we may say that in a heat engine the energy sought is the work, and the energy that costs money is the heat from the high-temperature source (indirectly, the cost of the fuel). Thermal efficiency is defined as

$$\eta_{\text{thermal}} = \frac{W(\text{energy sought})}{Q_H(\text{energy that costs})} = \frac{Q_H - Q_L}{Q_H} = 1 - \frac{Q_L}{Q_H} \qquad (7.1)$$

Heat engines vary greatly in size and shape, from large steam engines, gas turbines, or jet engines, to gasoline engines for cars and diesel engines for trucks or cars, to much smaller engines for lawn mowers or hand-held devices such as chain saws or trimmers. Typical values for the thermal efficiency of real engines are about 35–50% for large power plants, 30–35% for gasoline engines, and 35–40% for diesel engines. Smaller utility-type engines may have only about 20% efficiency, owing to their simple carburetion and controls and to the fact that some losses scale differently with size and therefore represent a larger fraction for smaller machines.

**EXAMPLE 7.1**

An automobile engine produces 136 hp on the output shaft with a thermal efficiency of 30%. The fuel it burns gives 35 000 kJ/kg as energy release. Find the total rate of energy rejected to the ambient and the rate of fuel consumption in kg/s.

**Solution**

From the definition of a heat engine efficiency, Eq. 7.1, and the conversion of hp from Table A.1 we have:

$$\dot{W} = \eta_{eng}\dot{Q}_H = 136 \text{ hp} \times 0.7355 \text{ kW/hp} = 100 \text{ kW}$$

$$\dot{Q}_H = \dot{W}/\eta_{eng} = 100/0.3 = 333 \text{ kW}$$

The energy equation for the overall engine gives:

$$\dot{Q}_L = \dot{Q}_H - \dot{W} = (1 - 0.3)\dot{Q}_H = 233 \text{ kW}$$

From the energy release in the burning we have: $\dot{Q}_H = \dot{m}q_H$, so

$$\dot{m} = \dot{Q}_H/q_H = \frac{333 \text{ kW}}{35 \ 000 \text{ kJ/kg}} = 0.0095 \text{ kg/s}$$

An actual engine shown in Fig. 7.5 rejects energy to the ambient through the radiator cooled by atmospheric air, as heat transfer from the exhaust system and the exhaust flow of hot gases.

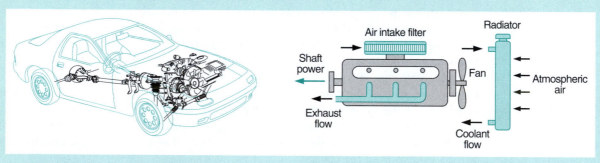

**FIGURE 7.5** Sketch for Example 7.1.

The second cycle that we were not able to complete was the one indicating the impossibility of transferring heat directly from a low-temperature body to a high-temperature body. This can of course be done with a refrigerator or heat pump. A vapor-compression refrigerator cycle, which was introduced in Chapter 1 and shown in Fig. 1.7, is shown again in Fig. 7.6. The working fluid is the refrigerant, such as R-134a or ammonia, which goes through a thermodynamic cycle. Heat is transferred to the refrigerant in the evaporator, where its pressure and temperature are low. Work is done on the refrigerant in the compressor, and heat is transferred from it in the condenser, where its pressure and temperature are high. The pressure drops as the refrigerant flows through the throttle valve or capillary tube.

Thus, in a refrigerator or heat pump, we have a device that operates in a cycle, that requires work, and that accomplishes the objective of transferring heat from a low-temperature body to a high-temperature body.

The thermoelectric refrigerator, which was discussed in Chapter 1 and shown schematically in Fig. 1.8a, is another example of a device that meets our definition of a

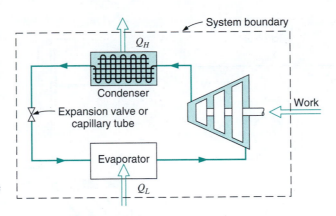

**FIGURE 7.6** A simple refrigeration cycle.

refrigerator. The work input to the thermoelectric refrigerator is in the form of electrical energy, and heat is transferred from the refrigerated space to the cold junction ($Q_L$) and from the hot junction to the surroundings ($Q_H$).

The "efficiency" of a refrigerator is expressed in terms of the coefficient of performance, which we designate with the symbol $\beta$. For a refrigerator the objective, that is, the energy sought, is $Q_L$, the heat transferred from the refrigerated space. The energy that costs is the work $W$. Thus, the coefficient of performance, $\beta$,[1] is

$$\beta = \frac{Q_L(\text{energy sought})}{W(\text{energy that costs})} = \frac{Q_L}{Q_H - Q_L} = \frac{1}{Q_H/Q_L - 1} \tag{7.2}$$

A household refrigerator may have a coefficient of performance (often referred to as COP) of about 2.5, whereas a deep freeze unit will be closer to 1.0. Lower cold temperature space or higher warm temperature space will result in lower values of COP, as will be found in Section 7.6. For a heat pump operating over a moderate temperature range, a value of its COP can be around 4, with this value decreasing sharply as the heat pump's operating temperature range is broadened.

[1]It should be noted that a refrigeration or heat pump cycle can be used with either of two objectives. It can be used as a refrigerator, in which case the primary objective is $Q_L$, the heat transferred to the refrigerant from the refrigerated space. It can also be used as a heating system (in which case it is usually referred to as a heat pump), the objective being $Q_H$, the heat transferred from the refrigerant to the high-temperature body, which is the space to be heated. $Q_L$ is transferred to the refrigerant from the ground, the atmospheric air, or well water. The coefficient of performance for this case, $\beta'$, is

$$\beta' = \frac{Q_H(\text{energy sought})}{W(\text{energy that costs})} = \frac{Q_H}{Q_H - Q_L} = \frac{1}{1 - Q_L/Q_H}$$

It also follows that for a given cycle.

$$\beta' - \beta = 1$$

Unless otherwise specified, the term coefficient of performance will always refer to a refrigerator as defined by Eq. 7.2.

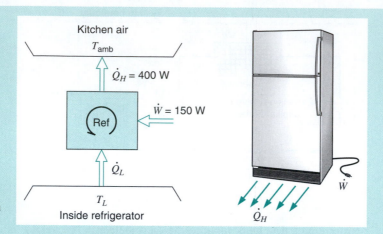

**FIGURE 7.7** Sketch for Example 7.2.

**Solution**

C.V. refrigerator. Assume steady state so there is no storage of energy. The information provided is $\dot{W} = 150$ W, and the heat rejected is $\dot{Q}_H = 400$ W.

The energy equation gives:

$$\dot{Q}_L = \dot{Q}_H - \dot{W} = 400 - 150 = 250 \text{ W}$$

This is also the rate of energy transfer into the cold space from the warmer kitchen due to heat transfer and exchange of cold air inside with warm air when you open the door. From the definition of the coefficient of performance, Eq. 7.2

$$\beta_{\text{REFRIG}} = \frac{\dot{Q}_L}{\dot{W}} = \frac{250}{150} = 1.67$$

Before we state the second law, the concept of a thermal reservoir should be introduced. A thermal reservoir is a body to which and from which heat can be transferred indefinitely without change in the temperature of the reservoir. Thus, a thermal reservoir always remains at constant temperature. The ocean and the atmosphere approach this definition very closely. Frequently it will be useful to designate a high-temperature reservoir and a low-temperature reservoir. Sometimes a reservoir from which heat is transferred is called a source, and a reservoir to which heat is transferred is called a sink.

## 7.2 THE SECOND LAW OF THERMODYNAMICS

On the basis of the matter considered in the previous section, we are now ready to state the second law of thermodynamics. There are two classical statements of the second law, known as the Kelvin–Planck statement and the Clausius statement.

*The Kelvin–Planck statement*: It is impossible to construct a device that will operate in a cycle and produce no effect other than the raising of a weight and the exchange of heat with a single reservoir. See Fig. 7.8.

This statement ties in with our discussion of the heat engine. In effect, it states that it is impossible to construct a heat engine that operates in a cycle, receives a given

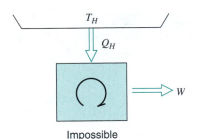

**FIGURE 7.8**   The Kelvin–Planck statement.

amount of heat from a high-temperature body, and does an equal amount of work. The only alternative is that some heat must be transferred from the working fluid at a lower temperature to a low-temperature body. Thus, work can be done by the transfer of heat only if there are two temperature levels, and heat is transferred from the high-temperature body to the heat engine and also from the heat engine to the low-temperature body. This implies that it is impossible to build a heat engine that has a thermal efficiency of 100%.

> *The Clausius statement*: It is impossible to construct a device that operates in a cycle and produces no effect other than the transfer of heat from a cooler body to a hotter body. See Fig. 7.9.

This statement is related to the refrigerator or heat pump. In effect, it states that it is impossible to construct a refrigerator that operates without an input of work. This also implies that the coefficient of performance is always less than infinity.

Three observations should be made about these two statements. The first observation is that both are negative statements. It is of course impossible to "prove" these negative statements. However, we can say that the second law of thermodynamics (like every other law of nature) rests on experimental evidence. Every relevant experiment that has been conducted either directly or indirectly verifies the second law, and no experiment has ever been conducted that contradicts the second law. The basis of the second law is therefore experimental evidence.

A second observation is that these two statements of the second law are equivalent. Two statements are equivalent if the truth of each statement implies the truth of the other, or if the violation of each statement implies the violation of the other. That a violation of the Clausius statement implies a violation of the Kelvin–Planck statement may be shown. The device at the left in Fig. 7.10 is a refrigerator that requires no work and

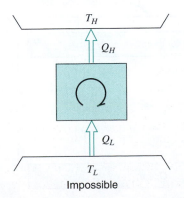

**FIGURE 7.9**   The Clausius statement.

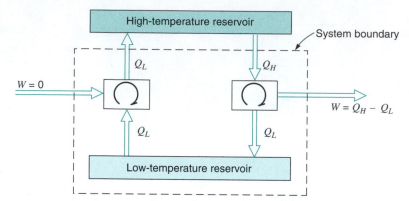

**FIGURE 7.10** Demonstration of the equivalence of the two statements of the second law.

thus violates the Clausius statement. Let an amount of heat $Q_L$ be transferred from the low-temperature reservoir to this refrigerator, and let the same amount of heat $Q_L$ be transferred to the high-temperature reservoir. Let an amount of heat $Q_H$ that is greater than $Q_L$ be transferred from the high-temperature reservoir to the heat engine, and let the engine reject the amount of heat $Q_L$ as it does an amount of work $W$, which equals $Q_H - Q_L$. Because there is no net heat transfer to the low-temperature reservoir, the low-temperature reservoir, along with the heat engine and the refrigerator, can be considered together as a device that operates in a cycle and produces no effect other than the raising of a weight (work) and the exchange of heat with a single reservoir. Thus, a violation of the Clausius statement implies a violation of the Kelvin–Planck statement. The complete equivalence of these two statements is established when it is also shown that a violation of the Kelvin–Planck statement implies a violation of the Clausius statement. This is left as an exercise for the student.

The third observation is that frequently the second law of thermodynamics has been stated as the impossibility of constructing a perpetual-motion machine of the second kind. A perpetual-motion machine of the first kind would create work from nothing or create mass or energy, thus violating the first law. A perpetual-motion machine of the second kind would extract heat from a source and then convert this heat completely into other forms of energy, thus violating the second law. A perpetual-motion machine of the third kind would have no friction, and thus would run indefinitely but produce no work.

A heat engine that violated the second law could be made into a perpetual-motion machine of the second kind by taking the following steps. Consider Fig. 7.11, which

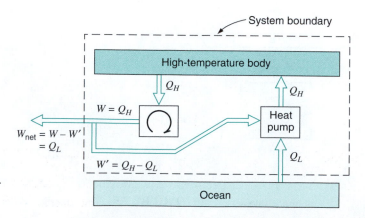

**FIGURE 7.11** A perpetual-motion machine of the second kind.

might be the power plant of a ship. An amount of heat $Q_L$ is transferred from the ocean to a high-temperature body by means of a heat pump. The work required is $W'$, and the heat transferred to the high-temperature body is $Q_H$. Let the same amount of heat be transferred to a heat engine that violates the Kelvin–Planck statement of the second law and does an amount of work $W = Q_H$. Of this work an amount $Q_H - Q_L$ is required to drive the heat pump, leaving the net work ($W_{net} = Q_L$) available for driving the ship. Thus, we have a perpetual-motion machine in the sense that work is done by utilizing freely available sources of energy such as the ocean or atmosphere.

| | |
|---|---|
| **In-Text Concept Questions** | **a.** Electrical appliances (TV, stereo) use electric power as input. What happens to the power? Are those heat engines? What does the second law say about those devices? |

**b.** Geothermal underground hot water or steam can be used to generate electric power. Does that violate the second law?

**c.** A windmill produces power on a shaft taking kinetic energy out of the wind. Is it a heat engine? Is it a perpetual-motion machine? Explain.

**d.** Heat engines and heat pumps (refrigerators) are energy conversion devices altering amounts of energy transfer between $Q$ and $W$. Which conversion direction ($Q \rightarrow W$ or $W \rightarrow Q$) is limited and which is unlimited according to the second law?

## 7.3 THE REVERSIBLE PROCESS

The question that can now logically be posed is this: If it is impossible to have a heat engine of 100% efficiency, what is the maximum efficiency one can have? The first step in the answer to this question is to define an ideal process, which is called a reversible process.

A reversible process for a system is defined as a process that once having taken place can be reversed and in so doing leave no change in either system or surroundings.

Let us illustrate the significance of this definition for a gas contained in a cylinder that is fitted with a piston. Consider first Fig. 7.12, in which a gas, which we define as the system, is restrained at high pressure by a piston that is secured by a pin. When

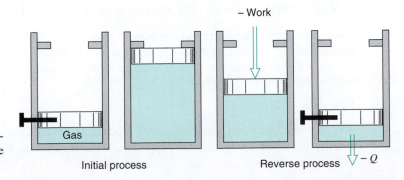

**FIGURE 7.12** An example of an irreversible process.

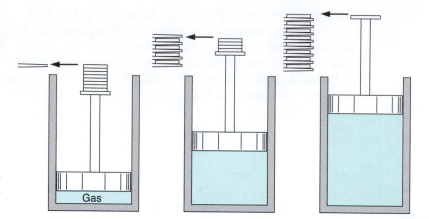

**FIGURE 7.13** An example of a process that approaches being reversible.

the pin is removed, the piston is raised and forced abruptly against the stops. Some work is done by the system, since the piston has been raised a certain amount. Suppose we wish to restore the system to its initial state. One way of doing this would be to exert a force on the piston and thus compress the gas until the pin can be reinserted in the piston. Since the pressure on the face of the piston is greater on the return stroke than on the initial stroke, the work done on the gas in this reverse process is greater than the work done by the gas in the initial process. An amount of heat must be transferred from the gas during the reverse stroke so that the system has the same internal energy as it had originally. Thus, the system is restored to its initial state, but the surroundings have changed by virtue of the fact that work was required to force the piston down and heat was transferred to the surroundings. The initial process therefore is an irreversible one because it could not be reversed without leaving a change in the surroundings.

In Fig. 7.13 let the gas in the cylinder comprise the system, and let the piston be loaded with a number of weights. Let the weights be slid off horizontally one at a time, allowing the gas to expand and do work in raising the weights that remain on the piston. As the size of the weights is made smaller and their number is increased, we approach a process that can be reversed, for at each level of the piston during the reverse process there will be a small weight that is exactly at the level of the platform and thus can be placed on the platform without requiring work. In the limit, therefore, as the weights become very small, the reverse process can be accomplished in such a manner that both the system and surroundings are in exactly the same state they were initially. Such a process is a reversible process.

## 7.4 FACTORS THAT RENDER PROCESSES IRREVERSIBLE

There are many factors that make processes irreversible. Four of those factors—friction, unrestrained expansion, heat transfer through a finite temperature difference, and mixing of two different substances—are considered in this section.

### Friction

It is readily evident that friction makes a process irreversible, but a brief illustration may amplify the point. Let a block and an inclined plane make up a system, as in Fig. 7.14,

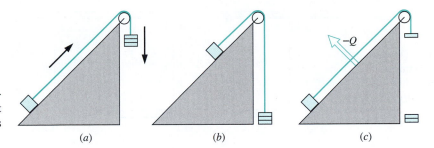

**FIGURE 7.14** Demonstration of the fact that friction makes processes irreversible.

and let the block be pulled up the inclined plane by weights that are lowered. A certain amount of work is needed to do this. Some of this work is required to overcome the friction between the block and the plane, and some is required to increase the potential energy of the block. The block can be restored to its initial position by removing some of the weights and thus allowing the block to slide back down the plane. Some heat transfer from the system to the surroundings will no doubt be required to restore the block to its initial temperature. Since the surroundings are not restored to their initial state at the conclusion of the reverse process, we conclude that friction has rendered the process irreversible. Another type of frictional effect is that associated with the flow of viscous fluids in pipes and passages and in the movement of bodies through viscous fluids.

## Unrestrained Expansion

The classic example of an unrestrained expansion, as shown in Fig. 7.15, is a gas separated from a vacuum by a membrane. Consider what happens when the membrane breaks and the gas fills the entire vessel. It can be shown that this is an irreversible process by considering what would be necessary to restore the system to its original state. The gas would have to be compressed and heat transferred from the gas until its initial state is reached. Since the work and heat transfer involve a change in the surroundings, the surroundings are not restored to their initial state, indicating that the unrestrained expansion was an irreversible process. The process described in Fig. 7.12 is also an example of an unrestrained expansion.

In the reversible expansion of a gas, there must be only an infinitesimal difference between the force exerted by the gas and the restraining force, so that the rate at which the boundary moves will be infinitesimal. In accordance with our previous definition, this is a quasi-equilibrium process. However, actual systems have a finite difference in forces, which causes a finite rate of movement of the boundary, and thus the processes are irreversible in some degree.

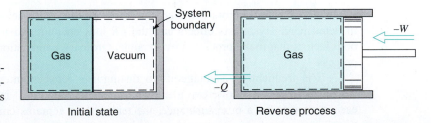

**FIGURE 7.15** Demonstration of the fact that unrestrained expansion makes processes irreversible.

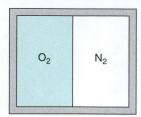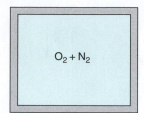

FIGURE 7.16 Demonstration of the fact that the mixing of two different substances is an irreversible process.

## Heat Transfer through a Finite Temperature Difference

Consider as a system a high-temperature body and a low-temperature body, and let heat be transferred from the high-temperature body to the low-temperature body. The only way in which the system can be restored to its initial state is to provide refrigeration, which requires work from the surroundings, and some heat transfer to the surroundings will also be necessary. Because of the heat transfer and the work, the surroundings are not restored to their original state, indicating that the process was irreversible.

An interesting question is now posed. Heat is defined as energy that is transferred through a temperature difference. We have just shown that heat transfer through a temperature difference is an irreversible process. Therefore, how can we have a reversible heat-transfer process? A heat-transfer process approaches a reversible process as the temperature difference between the two bodies approaches zero. Therefore, we define a reversible heat-transfer process as one in which the heat is transferred through an infinitesimal temperature difference. We realize of course that to transfer a finite amount of heat through an infinitesimal temperature difference would require an infinite amount of time or infinite area. Therefore, all actual heat transfers are through a finite temperature difference and hence are irreversible, with the greater the temperature difference, the greater the irreversibility. We will find, however, that the concept of reversible heat transfer is very useful in describing ideal processes.

## Mixing of Two Different Substances

Fig. 7.16 illustrates the process of mixing two different gases separated by a membrane. When the membrane is broken, a homogeneous mixture of oxygen and nitrogen fills the entire volume. This process will be considered in some detail in Chapter 10. We can say here that this may be considered a special case of an unrestrained expansion, for each gas undergoes an unrestrained expansion as it fills the entire volume. A certain amount of work is necessary to separate these gases. Thus, an air separation plant such as described in Chapter 1 requires an input of work to accomplish the separation.

## Other Factors

A number of other factors make processes irreversible, but they will not be considered in detail here. Hysteresis effects and the $i^2R$ loss encountered in electrical circuits are both factors that make processes irreversible. Ordinary combustion is also an irreversible process.

It is frequently advantageous to distinguish between internal and external irreversibility. Fig. 7.17 shows two identical systems to which heat is transferred. Assuming each system to be a pure substance, the temperature remains constant during the heat-transfer process. In one system the heat is transferred from a reservoir at a temperature

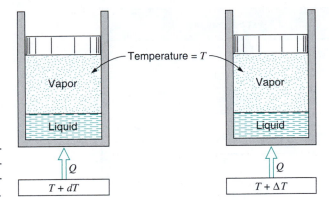

**FIGURE 7.17** Illustration of the difference between an internally and externally reversible process.

$T + dT$, and in the other the reservoir is at a much higher temperature, $T + \Delta T$, than the system. The first is a reversible heat-transfer process, and the second is an irreversible heat-transfer process. However, as far as the system itself is concerned, it passes through exactly the same states in both processes, which we assume are reversible. Thus, we can say for the second system that the process is internally reversible but externally irreversible because the irreversibility occurs outside the system.

We should also note the general interrelation of reversibility, equilibrium, and time. In a reversible process, the deviation from equilibrium is infinitesimal, and therefore it occurs at an infinitesimal rate. Since it is desirable that actual processes proceed at a finite rate, the deviation from equilibrium must be finite, and therefore the actual process is irreversible in some degree. The greater the deviation from equilibrium, the greater the irreversibility, and the more rapidly the process will occur. It should also be noted that the quasi-equilibrium process, which was described in Chapter 2, is a reversible process, and hereafter the term reversible process will be used.

**In-Text Concept Questions**

e. Ice cubes in a glass of liquid water will eventually melt and all the water will approach room temperature. Is this a reversible process? Why?

f. Does a process become more or less reversible with respect to the heat transfer if it is fast rather than slow? *Hint:* Recall from Chapter 4 that $\dot{Q} = CA\,\Delta T$.

g. If you generated hydrogen from, say, solar power, which of these would be more efficient: (1) transport it and then burn it in an engine or (2) convert the solar power to electricity and transport that? What else would you need to know in order to give a definite answer?

## 7.5 THE CARNOT CYCLE

Having defined the reversible process and considered some factors that make processes irreversible, let us again pose the question raised in Section 7.3. If the efficiency of all heat engines is less than 100%, what is the most efficient cycle we can have? Let us answer this question for a heat engine that receives heat from a high-temperature reservoir and rejects heat to a low-temperature reservoir. Since we are dealing with reservoirs,

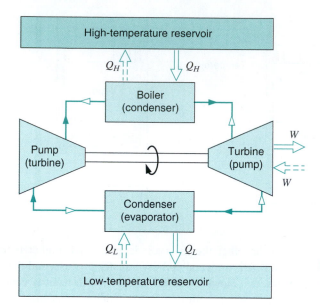

**FIGURE 7.18** Example of a heat engine that operates on a Carnot cycle.

we recognize that both the high temperature and the low temperature of the reservoirs are constant and remain constant regardless of the amount of heat transferred.

Let us assume that this heat engine, which operates between the given high-temperature and low-temperature reservoirs, does so in a cycle in which every process is reversible. If every process is reversible, the cycle is also reversible; and if the cycle is reversed, the heat engine becomes a refrigerator. In the next section we will show that this is the most efficient cycle that can operate between two constant-temperature reservoirs. It is called the Carnot cycle and is named after a French engineer, Nicolas Leonard Sadi Carnot (1796–1832), who expressed the foundations of the second law of thermodynamics in 1824.

We now turn our attention to the Carnot cycle. Fig. 7.18 shows a power plant that is similar in many respects to a simple steam power plant and, we assume, operates on the Carnot cycle. Consider the working fluid to be a pure substance, such as steam. Heat is transferred from the high-temperature reservoir to the water (steam) in the boiler. For this process to be a reversible heat transfer, the temperature of the water (steam) must be only infinitesimally lower than the temperature of the reservoir. This result also implies, since the temperature of the reservoir remains constant, that the temperature of the water must remain constant. Therefore, the first process in the Carnot cycle is a reversible isothermal process in which heat is transferred from the high-temperature reservoir to the working fluid. A change of phase from liquid to vapor at constant pressure is of course an isothermal process for a pure substance.

The next process occurs in the turbine without heat transfer and is therefore adiabatic. Since all processes in the Carnot cycle are reversible, this must be a reversible adiabatic process, during which the temperature of the working fluid decreases from the temperature of the high-temperature reservoir to the temperature of the low-temperature reservoir.

In the next process heat is rejected from the working fluid to the low-temperature reservoir. This must be a reversible isothermal process in which the temperature of the working fluid is infinitesimally higher than that of the low-temperature reservoir. During this isothermal process some of the steam is condensed.

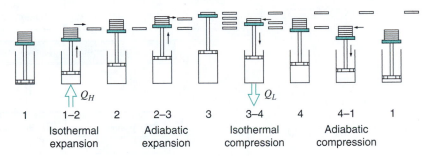

**FIGURE 7.19** Example of a gaseous system operating on a Carnot cycle.

The final process, which completes the cycle, is a reversible adiabatic process in which the temperature of the working fluid increases from the low temperature to the high temperature. If this were to be done with water (steam) as the working fluid, a mixture of liquid and vapor would have to be taken from the condenser and compressed. (This would be very inconvenient in practice, and therefore in all power plants the working fluid is completely condensed in the condenser. The pump handles only the liquid phase.)

Sine the Carnot heat engine cycle is reversible, every process could be reversed, in which case it would become a refrigerator. The refrigerator is shown by the dotted lines and parentheses in Fig. 7.18. The temperature of the working fluid in the evaporator would be infinitesimally lower than the temperature of the low-temperature reservoir, and in the condenser it is infinitesimally higher than that of the high-temperature reservoir.

It should be emphasized that the Carnot cycle can, in principle, be executed in many different ways. Many different working substances can be used, such as a gas or a thermoelectric device such as described in Chapter 1. There are also various possible arrangements of machinery. For example, a Carnot cycle can be devised that takes place entirely within a cylinder, using a gas as a working substance, as shown in Fig. 7.19.

The important point to be made here is that the Carnot cycle, regardless of what the working substance may be, always has the same four basic processes. These processes are:

1. A reversible isothermal process in which heat is transferred to or from the high-temperature reservoir.
2. A reversible adiabatic process in which the temperature of the working fluid decreases from the high temperature to the low temperature.
3. A reversible isothermal process in which heat is transferred to or from the low-temperature reservoir.
4. A reversible adiabatic process in which the temperature of the working fluid increases from the low temperature to the high temperature.

## 7.6 TWO PROPOSITIONS REGARDING THE EFFICIENCY OF A CARNOT CYCLE

There are two important propositions regarding the efficiency of a Carnot cycle.

### First Proposition

It is impossible to construct an engine that operates between two given reservoirs and is more efficient than a reversible engine operating between the same two reservoirs.

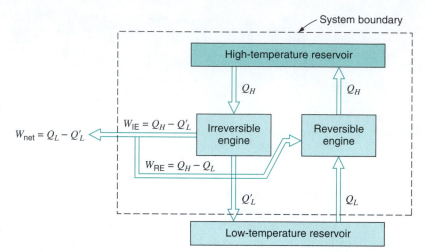

**FIGURE 7.20** Demonstration of the fact that the Carnot cycle is the most efficient cycle operating between two fixed-temperature reservoirs.

The proof of this statement is accomplished through a "thought experiment." An initial assumption is made, and it is then shown that this assumption leads to impossible conclusions. The only possible conclusion is that the initial assumption was incorrect.

Let us assume that there is an irreversible engine operating between two given reservoirs that has a greater efficiency than a reversible engine operating between the same two reservoirs. Let the heat transfer to the irreversible engine be $Q_H$, the heat rejected be $Q'_L$, and the work be $W_{IE}$ (which equals $Q_H - Q'_L$) as shown in Fig. 7.20. Let the reversible engine operate as a refrigerator (this is possible since it is reversible). Finally, let the heat transfer with the low-temperature reservoir be $Q_L$, the heat transfer with the high-temperature reservoir be $Q_H$, and the work required be $W_{RE}$ (which equals $Q_H - Q_L$).

Since the initial assumption was that the irreversible engine is more efficient, it follows (because $Q_H$ is the same for both engines) that $Q'_L < Q_L$ and $W_{IE} > W_{RE}$. Now the irreversible engine can drive the reversible engine and still deliver the net work $W_{net}$, which equals $W_{IE} - W_{RE} = Q_L - Q'_L$. If we consider the two engines and the high-temperature reservoir as a system, as indicated in Fig. 7.20, we have a system that operates in a cycle, exchanges heat with a single reservoir, and does a certain amount of work. However, this would constitute a violation of the second law, and we conclude that our initial assumption (that the irreversible engine is more efficient than a reversible engine) is incorrect. Therefore, we cannot have an irreversible engine that is more efficient than a reversible engine operating between the same two reservoirs.

## Second Proposition

All engines that operate on the Carnot cycle between two given constant-temperature reservoirs have the same efficiency. The proof of this proposition is similar to the proof just outlined, which assumes that there is one Carnot cycle that is more efficient than another Carnot cycle operating between the same temperature reservoirs. Let the Carnot cycle with the higher efficiency replace the irreversible cycle of the previous argument, and let the Carnot cycle with the lower efficiency operate as the refrigerator. The proof proceeds with the same line of reasoning as in the first proposition. The details are left as an exercise for the student.

## 7.7 THE THERMODYNAMIC TEMPERATURE SCALE

In discussing the matter of temperature in Chapter 2, we pointed out that the zeroth law of thermodynamics provides a basis for temperature measurement, but that a temperature scale must be defined in terms of a particular thermometer substance and device. A temperature scale that is independent of any particular substance, which might be called an absolute temperature scale, would be most desirable. In the last paragraph we noted that the efficiency of a Carnot cycle is independent of the working substance and depends only on the temperature. This fact provides the basis for an absolute temperature scale, called the thermodynamic scale. Since the efficiency of a Carnot cycle is a function only of temperature, it follows that

$$\eta_{\text{thermal}} = 1 - \frac{Q_L}{Q_H} = 1 - \psi(T_L, T_H) \tag{7.3}$$

where $\psi$ designates a functional relation.

There are many functional relations that could be chosen to satisfy the relation given in Eq. 7.3. For simplicity, the thermodynamic scale is defined as

$$\frac{Q_H}{Q_L} = \frac{T_H}{T_L} \tag{7.4}$$

Substituting this definition into Eq. 7.3 results in the following relation between thermal efficiency of a Carnot cycle and the absolute temperatures of the two reservoirs:

$$\eta_{\text{thermal}} = 1 - \frac{Q_L}{Q_H} = 1 - \frac{T_L}{T_H} \tag{7.5}$$

It should be noted, however, that the definition of Eq. 7.4 is not complete, in that it does not specify the magnitude of the degree of temperature or a fixed reference point value. In the following section, we will discuss in greater detail the ideal-gas absolute temperature previously introduced in Section 3.6 and show that this scale satisfies the relation as defined by Eq. 7.4.

## 7.8 THE IDEAL-GAS TEMPERATURE SCALE

In this section we reconsider in greater detail the ideal-gas temperature scale introduced in Section 3.6. This scale is based on the observation that as the pressure of a real gas approaches zero, its equation of state approaches that of an ideal gas:

$$Pv = RT$$

It will be shown that the ideal-gas temperature scale satisfies the definition of thermodynamic temperature given in the preceding section by Eq. 7.4, but first let us consider how an ideal gas might be used to measure temperature in a constant-volume gas thermometer, shown schematically in Fig. 7.21.

Let the gas bulb be placed in the location where the temperature is to be measured, and let the mercury column be adjusted so that the level of mercury stands at the reference mark $A$. Thus, the volume of the gas remains constant. Assume that the gas in the capillary tube is at the same temperature as the gas in the bulb. Then the pressure of the gas, which is indicated by the height $L$ of the mercury column, is a measure of the temperature.

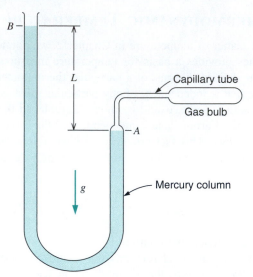

**FIGURE 7.21** Schematic diagram of a constant-volume gas thermometer.

Let the pressure that is associated with the temperature of the triple point of water (273.16 K) also be measured, and let us designate this pressure $P_{t.p.}$. Then, from the definition of an ideal gas, any other temperature $T$ could be determined from a pressure measurement $P$ by the relation

$$T = 273.16 \left( \frac{P}{P_{t.p.}} \right)$$

**EXAMPLE 7.3**

In a certain constant-volume ideal gas thermometer, the measured pressure at the ice point (see Section 2.11) of water, 0°C, is 110.9 kPa and at the steam point, 100°C, is 151.5 kPa. Extrapolating, at what Celsius temperature does the pressure go to zero (i.e., zero absolute temperature)?

**Analysis**

From the ideal gas equation of state $PV = mRT$ at constant mass and volume, the pressure is directly proportional to temperature, as shown in Fig. 7.22,

$$P = CT, \text{ where } T \text{ is the absolute ideal-gas temperature}$$

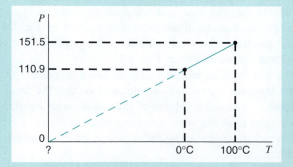

**FIGURE 7.22** Plot for Example 7.3.

**Solution**

Slope $\dfrac{\Delta P}{\Delta T} = \dfrac{151.5 - 110.9}{100 - 0} = 0.406$

Extrapolating from the 0°C point to $P = 0$,

$$T = 0 - \frac{110.9}{0.406} = -273.15°C$$

establishing the relation between absolute ideal-gas Kelvin and Celsius temperature scales.

(Note: Compatible with the subsequent present-day definition of the Kelvin and the Celsius scale in Section 2.11.)

From a practical point of view, we have the problem that no gas behaves exactly like an ideal gas. However, we do know that as the pressure approaches zero, the behavior of all gases approaches that of an ideal gas. Suppose then that a series of measurements is made with varying amounts of gas in the gas bulb. This means that the pressure measured at the triple point, and also the pressure at any other temperature, will vary. If the indicated temperature $T_i$ (obtained by assuming that the gas is ideal) is plotted against the pressure of gas with the bulb at the triple point of water, a curve like the one shown in Fig. 7.23 is obtained. When this curve is extrapolated to zero pressure, the correct ideal-gas temperature is obtained. Different curves might result from different gases, but they would all indicate the same temperature at zero pressure.

We have outlined only the general features and principles for measuring temperature on the ideal-gas scale of temperatures. Precision work in this field is difficult and laborious, and there are only a few laboratories in the world where such work is carried on. The International Temperature Scale, which was mentioned in Chapter 2, closely approximates the thermodynamic temperature scale and is much easier to work with in actual temperature measurement.

We now demonstrate that the ideal-gas temperature scale discussed earlier is, in fact, identical to the thermodynamic temperature scale, which was defined in the discussion of the Carnot cycle and the second law. Our objective can be achieved by using an ideal gas as the working fluid for a Carnot-cycle heat engine and analyzing the four processes that make up the cycle. The four state points, 1, 2, 3, and 4, and the four processes

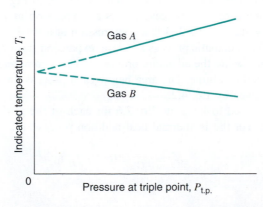

**FIGURE 7.23** Sketch showing how the ideal-gas temperature is determined.

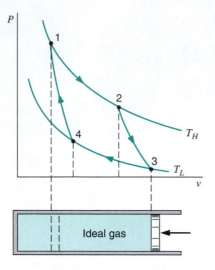

**FIGURE 7.24**   The ideal-gas Carnot cycle.

are as shown in Fig. 7.24. For convenience, let us consider a unit mass of gas inside the cylinder. Now for each of the four processes, the reversible work done at the moving boundary is given by Eq. 4.2:

$$\delta w = P \, dv$$

Similarly, for each process the gas behavior is, from the ideal-gas relation, Eq. 3.5,

$$Pv = RT$$

and the internal energy change, from Eq. 5.20, is

$$du = C_{v0} \, dT$$

Assuming no changes in kinetic or potential energies, the first law is, from Eq. 5.7 at unit mass,

$$\delta q = du + \delta w$$

Substituting the three previous expressions into this equation, we have for each of the four processes

$$\delta q = C_{v0} \, dT + \frac{RT}{v} \, dv \tag{7.6}$$

The shape of the two isothermal processes shown in Fig. 7.23 is known, since $Pv$ is constant in each case. The process 1–2 is an expansion at $T_H$, such that $v_2$ is larger than $v_1$. Similarly, the process 3–4 is a compression at a lower temperature, $T_L$, and $v_4$ is smaller than $v_3$. The adiabatic process 2–3 is an expansion from $T_H$ to $T_L$, with an increase in specific volume, while the adiabatic process 4–1 is a compression from $T_L$ to $T_H$, with a decrease in specific volume. The area below each process line represents the work for that process, as given by Eq. 4.2.

We now proceed to integrate Eq. 7.6 for each of the four processes that make up the Carnot cycle. For the isothermal heat addition process 1–2, we have

$$q_H = {}_1q_2 = 0 + RT_H \ln \frac{v_2}{v_1} \tag{7.7}$$

For the adiabatic expansion process 2–3,

$$0 = \int_{T_H}^{T_L} \frac{C_{v0}}{T} \, dT + R \ln \frac{v_3}{v_2} \tag{7.8}$$

For the isothermal heat rejection process 3–4,

$$q_L = -_3q_4 = -0 - RT_L \ln \frac{v_4}{v_3}$$

$$= +RT_L \ln \frac{v_3}{v_4} \tag{7.9}$$

and for the adiabatic compression process 4–1,

$$0 = \int_{T_L}^{T_H} \frac{C_{v0}}{T} \, dT + R \ln \frac{v_1}{v_4} \tag{7.10}$$

From Eqs. 7.8 and 7.10, we get

$$\int_{T_L}^{T_H} \frac{C_{v0}}{T} \, dT = R \ln \frac{v_3}{v_2} = -R \ln \frac{v_1}{v_4}$$

Therefore,

$$\frac{v_3}{v_2} = \frac{v_4}{v_1}, \qquad \text{or} \qquad \frac{v_3}{v_4} = \frac{v_2}{v_1} \tag{7.11}$$

Thus, from Eqs. 7.7 and 7.9 and substituting Eq. 7.11, we find that

$$\frac{q_H}{q_L} = \frac{RT_H \ln \dfrac{v_2}{v_1}}{RT_L \ln \dfrac{v_3}{v_4}} = \frac{T_H}{T_L}$$

which is Eq. 7.4, the definition of the thermodynamic temperature scale in connection with the second law.

## 7.9 IDEAL VERSUS REAL MACHINES

Following the definition of the thermodynamic temperature scale by Eq. 7.4, it was noted that the thermal efficiency of a Carnot cycle heat engine is given by Eq. 7.5. It also follows that a Carnot cycle operating as a refrigerator or heat pump will have a coefficient of performance expressed as

$$\beta = \frac{Q_L}{Q_H - Q_L} {}_{\text{Carnot}} = \frac{T_L}{T_H - T_L} \tag{7.12}$$

$$\beta' = \frac{Q_H}{Q_H - Q_L} {}_{\text{Carnot}} = \frac{T_H}{T_H - T_L} \tag{7.13}$$

For all three "efficiencies" in Eqs.7.5, 7.12 and 7.13, the first equality sign is the definition with the use of the energy equation and thus is always true. The second equality sign is valid only if the cycle is reversible, that is, a Carnot cycle. Any real heat engine, refrigerator, or heat pump will be less efficient, such that

$$\eta_{\text{real thermal}} = 1 - \frac{Q_L}{Q_H} \leq 1 - \frac{T_L}{T_H}$$

$$\beta_{\text{real}} = \frac{Q_L}{Q_H - Q_L} \leq \frac{T_L}{T_H - T_L}$$

$$\beta'_{\text{real}} = \frac{Q_H}{Q_H - Q_L} \leq \frac{T_H}{T_H - T_L}$$

A final point needs to be made about the significance of absolute-zero temperature in connection with the second law and the thermodynamic temperature scale. Consider a Carnot-cycle heat engine that receives a given amount of heat from a given high-temperature reservoir. As the temperature at which heat is rejected from the cycle is lowered, the net work output increases and the amount of heat rejected decreases. In the limit, the heat rejected is zero, and the temperature of the reservoir corresponding to this limit is absolute zero.

Similarly, for a Carnot-cycle refrigerator, the amount of work required to produce a given amount of refrigeration increases as the temperature of the refrigerated space decreases. Absolute zero represents the limiting temperature that can be achieved, and the amount of work required to produce a finite amount of refrigeration approaches infinity as the temperature at which refrigeration is provided approaches zero.

**EXAMPLE 7.4**

Let us consider the heat engine, shown schematically in Fig. 7.25, that receives a heat-transfer rate of 1 MW at a high temperature of 550°C and rejects energy to the ambient surroundings at 300 K. Work is produced at a rate of 450 kW. We would like to know how much energy is discarded to the ambient surroundings and the engine efficiency and compare both of these to a Carnot heat engine operating between the same two reservoirs.

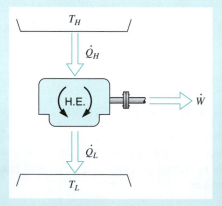

FIGURE 7.25 A heat engine operating between two constant temperature energy reservoirs for Example 7.4.

**Solution**

If we take the heat engine as a control volume, the energy equation gives

$$\dot{Q}_L = \dot{Q}_H - \dot{W} = 1000 - 450 = 550\,\text{kW}$$

and from the definition of the efficiency

$$\eta_{\text{thermal}} = \dot{W}/\dot{Q}_H = 450/1000 = 0.45$$

For the Carnot heat engine, the efficiency is given by the temperature of the reservoirs

$$\eta_{\text{Carnot}} = 1 - \frac{T_L}{T_H} = 1 - \frac{300}{550 + 273} = 0.635$$

The rates of work and heat rejection become

$$\dot{W} = \eta_{\text{Carnot}}\dot{Q}_H = 0.635 \times 1000 = 635 \text{ kW}$$

$$\dot{Q}_L = \dot{Q}_H - \dot{W} = 1000 - 635 = 365 \text{ kW}$$

The actual heat engine thus has a lower efficiency than the Carnot (ideal) heat engine, with a value of 45% typical for a modern steam power plant. This also implies that the actual engine rejects a larger amount of energy to the ambient surroundings (55%) compared with the Carnot heat engine (36%).

---

**EXAMPLE 7.5**

As one mode of operation of an air conditioner is the cooling of a room on a hot day, it works as a refrigerator, shown in Fig. 7.26. A total of 4 kW should be removed from a room at 24°C to the outside atmosphere at 35°C. We would like to estimate the magnitude of the required work. To do this we will not analyze the processes inside the refrigerator, which is deferred to Chapter 11, but we can give a lower limit for the rate of work assuming it is a Carnot-cycle refrigerator.

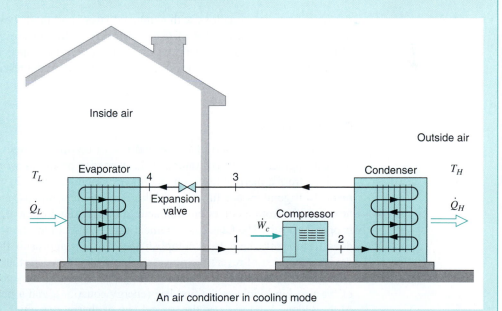

An air conditioner in cooling mode

**FIGURE 7.26** An air conditioner in cooling mode where $T_L$ is the room.

**Solution**

The coefficient of performance (COP) is

$$\beta = \frac{\dot{Q}_L}{\dot{W}} = \frac{\dot{Q}_L}{\dot{Q}_H - \dot{Q}_L} = \frac{T_L}{T_H - T_L} = \frac{273 + 24}{35 - 24} = 27$$

so the rate of work or power input will be

$$\dot{W} = \dot{Q}_L/\beta = 4/27 = 0.15 \text{ kW}$$

Since the power was estimated assuming a Carnot refrigerator, it is the smallest amount possible. Recall also the expressions for heat-transfer rates in Chapter 4. If the refrigerator should push 4.15 kW out to the atmosphere at 35°C, the high-temperature side of it should be at a higher temperature, maybe 45°C, to have a reasonably small-sized heat exchanger. As it cools the room, a flow of air of less than, say, 18°C would be needed. Redoing the COP with a high of 45°C and a low of 18°C gives 10.8, which is more realistic. A real refrigerator would operate with a COP of the order of 5 or less.

In the previous discussion and examples we considered the constant-temperature energy reservoirs and used those temperatures to calculate the Carnot-cycle efficiency. However, if we recall the expressions for the rate of heat transfer by conduction, convection, or radiation in Chapter 4, they can all be shown as

$$\dot{Q} = C\Delta T \qquad (7.14)$$

The constant $C$ depends on the mode of heat transfer as

Conduction: $\qquad C = \dfrac{kA}{\Delta x} \qquad$ Convection: $C = hA$

Radiation: $\qquad C = \epsilon\sigma A(T_s^2 + T_\infty^2)(T_s + T_\infty)$

For more complex situations with combined layers and modes, we also recover the form in Eq. 7.14, but with a value of $C$ that depends on the geometry, materials, and modes of heat transfer. To have a heat transfer, we therefore must have a temperature difference so that the working substance inside a cycle cannot attain the reservoir temperature unless the area is infinitely large.

## SUMMARY

The classical presentation of the second law of thermodynamics starts with the concept of heat engines and refrigerators. A heat engine produces work from a heat transfer obtained from a thermal reservoir, and its operation is limited by the Kelvin–Planck statement. Refrigerators are functionally the same as heat pumps, and they drive energy by heat transfer from a colder environment to a hotter environment, something that will not happen by itself. The Clausius statement says in effect that the refrigerator or heat pump does need work input to accomplish the task. To approach the limit of these cyclic devices, the idea of reversible processes is discussed and further explained by the opposite, namely, irreversible processes and impossible machines. A perpetual motion machine of the first kind violates the first law (energy equation), and a perpetual-motion machine of the second kind violates the second law of thermodynamics.

The limitations for the performance of heat engines (thermal efficiency) and heat pumps or refrigerators (coefficient of performance or COP) are expressed by the corresponding Carnot-cycle device. Two propositions about the Carnot cycle device are another way of expressing the second law of thermodynamics instead of the statements of Kelvin–Planck or Clausius. These propositions lead to the establishment of the thermodynamic absolute temperature, done by Lord Kelvin, and the Carnot-cycle efficiency. We show this temperature to be the same as the ideal-gas temperature introduced in Chapter 3.

You should have learned a number of skills and acquired abilities from studying this chapter that will allow you to

- Understand the concepts of heat engines, heat pumps, and refrigerators.
- Have an idea about reversible processes.
- Know a number of irreversible processes and recognize them.
- Know what a Carnot-cycle is.
- Understand the definition of thermal efficiency of a heat engine.
- Understand the definition of coefficient of performance of a heat pump.
- Know the difference between the absolute and relative temperature.
- Know the limits of thermal efficiency as dictated by the thermal reservoirs and the Carnot-cycle device.
- Have an idea about the thermal efficiency of real heat engines.
- Know the limits of coefficient of performance as dictated by the thermal reservoirs and the Carnot-cycle device.
- Have an idea about the coefficient of performance of real refrigerators.

## KEY CONCEPTS AND FORMULAS

(All $W$, $Q$ can also be rates $\dot{W}$, $\dot{Q}$)

| | |
|---|---|
| Heat engine | $W_{HE} = Q_H - Q_L$; $\quad \eta_{HE} = \dfrac{W_{HE}}{Q_H} = 1 - \dfrac{Q_L}{Q_H}$ |
| Heat pump | $W_{HP} = Q_H - Q_L$; $\quad \beta_{HP} = \dfrac{Q_H}{W_{HP}} = \dfrac{Q_H}{Q_H - Q_L}$ |
| Refrigerator | $W_{REF} = Q_H - Q_L$; $\quad \beta_{REF} = \dfrac{Q_L}{W_{REF}} = \dfrac{Q_L}{Q_H - Q_L}$ |

Factors that make processes irreversible — Friction, unrestrained expansion $(W = 0)$, $Q$ over $\Delta T$, mixing, current through a resistor, combustion, or valve flow (throttle).

Carnot cycle
1–2 Isothermal heat addition $Q_H$ in at $T_H$
2–3 Adiabatic expansion process $T$ goes down
3–4 Isothermal heat rejection $Q_L$ out at $T_L$
4–1 Adiabatic compression process $T$ goes up

| | | |
|---|---|---|
| Proposition I | $\eta_{any} \le \eta_{reversible}$ | Same $T_H$, $T_L$ |
| Proposition II | $\eta_{Carnot\,1} = \eta_{Carnot\,2}$ | Same $T_H$, $T_L$ |

Absolute temperature $\quad \dfrac{T_L}{T_H} = \dfrac{Q_L}{Q_H}$

Real heat engine $\quad \eta_{HE} = \dfrac{W_{HE}}{Q_H} \le \eta_{Carnot\,HE} = 1 - \dfrac{T_L}{T_H}$

Real heat pump $\quad \beta_{HP} = \dfrac{Q_H}{W_{HP}} \le \beta_{Carnot\,HP} = \dfrac{T_H}{T_H - T_L}$

Real refrigerator $\quad \beta_{REF} = \dfrac{Q_L}{W_{REF}} \le \beta_{Carnot\,REF} = \dfrac{T_L}{T_H - T_L}$

Heat-transfer rates $\quad \dot{Q} = C\,\Delta T$

# Homework Problems

## Concept questions

**7.1** Compare two heat engines receiving the same Q, one at 1200 K and the other at 1800 K; they both reject heat at 500 K. Which one is better?

**7.2** Assume we have a refrigerator operating at steady state using 500 W of electric power with a COP of 2.5. What is the net effect on the kitchen air?

**7.3** If the efficiency of a power plant goes up as the low temperature drops, why not let the heat rejection go to a refrigerator at, say, $-10°C$ instead of ambient 20°C?

**7.4** Suppose we forget the model for heat transfer as $\dot{Q} = CA\,\Delta T$. Can we draw some information about direction of $Q$ from the second law?

**7.5** If the efficiency of a power plant goes up as the low temperature drops, why do power plants not just reject energy at say $-40°C$?

**7.6** A coal-fired power plant operates with a high $T$ of 600°C whereas a jet engine has about 1400 K. Does that mean we should replace all power plants with jet engines?

**7.7** A heat transfer requires a temperature difference (see Chapter 4) to push the $\dot{Q}$. What implications does that have for a real heat engine? A refrigerator?

**7.8** Hot combustion gases (air) at 1500 K are used as heat source in a heat engine where the gas is cooled to 750 K and the ambient is at 300 K. This is not a constant T source. How does that affect the efficiency?

**7.9** After you have returned from a car trip, the car engine has cooled down and is thus back to the state in which it started. What happened to all the energy released in the burning of the gasoline? What happened to all the work the engine gave out?

**7.10** Does a reversible heat engine burning coal (which, in practice, cannot be done reversibly) have impacts on our world other than depletion of the coal reserve?

**7.11** A car engine takes atmospheric air in at 20°C, no fuel, and exhausts the air at $-20°C$ producing work in the process. What do the first and the second laws say about that?

## Heat engines and refrigerators

**7.12** A gasoline engine produces 20 hp using 35 kW of heat transfer from burning fuel. What is its thermal efficiency and how much power is rejected to the ambient?

**7.13** Calculate the thermal efficiency of the steam power plant given in Example 6.9.

**7.14** A refrigerator removes 1.5 kJ from the cold space using 1 kJ work input. How much energy goes into the kitchen and what is its coefficient of performance?

**7.15** Calculate the coefficient of performance of the R-134a refrigerator given in Example 6.10.

**7.16** A car engine delivers 25 hp to the driveshaft with a thermal efficiency of 30%. The fuel has a heating value of 40 000 kJ/kg. Find the rate of fuel consumption and the combined power rejected through the radiator and exhaust.

**7.17** In a steam power plant 1 MW is added in the boiler, 0.58 MW is taken out in the condenser, and the pump work is 0.02 MW. Find the plant thermal efficiency. If everything could be reversed, find the coefficient of performance as a refrigerator?

**7.18** A window air-conditioner unit is placed on a laboratory bench and tested in cooling mode using 750 W of electric power with a COP of 1.75. What is the cooling power capacity and what is the net effect on the laboratory?

**7.19** Electric solar cells can produce electricity with 15% efficiency. Compare that to a heat engine driving an electric generator of efficiency 80%. What should the heat engine efficiency be to have the same overall efficiency as the solar cells?

**7.20** A power plant generates 150 MW of electrical power. It uses a supply of 1000 MW from a geothermal source and rejects energy to the atmosphere. Find the power to the air and how much air should be flowed to the cooling tower (kg/s) if its temperature cannot be increased more than 10°C.

**7.21** For each of the cases below, determine if the heat engine satisfies the first law (energy equation) and if it violates the second law.
a. $\dot{Q}_H = 6\,\text{kW}$  $\dot{Q}_L = 4\,\text{kW}$  $\dot{W} = 2\,\text{kW}$
b. $\dot{Q}_H = 6\,\text{kW}$  $\dot{Q}_L = 0\,\text{kW}$  $\dot{W} = 6\,\text{kW}$
c. $\dot{Q}_H = 6\,\text{kW}$  $\dot{Q}_L = 2\,\text{kW}$  $\dot{W} = 5\,\text{kW}$
d. $\dot{Q}_H = 6\,\text{kW}$  $\dot{Q}_L = 6\,\text{kW}$  $\dot{W} = 0\,\text{kW}$

**7.22** For each of the cases in Problem 7.21 determine if a heat pump satisfies the first law (energy equation) and if it violates the second law.

**7.23** A room is heated with a 1500 W electric heater. How much power can be saved if a heat pump with a COP of 2.0 is used instead?

**7.24** Calculate the coefficient of performance of the R-12 heat pump cycle described in Problem 6.87.

**7.25** An air conditioner discards 5.1 kW to the ambient surroundings with a power input of 1.5 kW. Find the rate of cooling and the coefficient of performance.

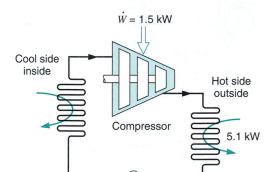

**FIGURE P7.25**

**7.26** A farmer runs a heat pump with a 2 kW motor. It should keep a chicken hatchery at 30°C, which loses energy at a rate of 10 kW to the colder ambient $T_{amb}$. What is the minimum coefficient of performance that will be acceptable for the heat pump?

**7.27** A house needs to be heated by a heat pump, with $\beta' = 2.2$, and maintained at 20°C at all times. It is estimated that it loses 0.8 kW per degree the ambient temperature is lower than the 20°C. Assume an outside temperature of −10°C and find the needed power to drive the heat pump.

**7.28** A large stationary diesel engine produces 15 MW with a thermal efficiency of 40%. The exhaust gas flows out at 800 K, which we assume is air, and the intake air is 290 K. How large a mass flow rate is that, assuming this is the only way we reject heat? Can the exhaust flow energy be used?

**7.29** Refrigerant R-12 at 95°C with $x = 0.1$ flowing at 2 kg/s is brought to saturated vapor in a constant-pressure heat exchanger. The energy is supplied by a heat pump

with a coefficient of performance of $\beta' = 2.5$. Find the required power to drive the heat pump.

## Second law and processes

**7.30** Discuss the factors that would make the power plant cycle described in Problem 6.82 an irreversible cycle.

**7.31** Discuss the factors that would make the heat pump cycle described in Problem 6.87 an irreversible cycle.

**7.32** Prove that a cyclic device that violates the Kelvin–Planck statement of the second law also violates the Clausius statement of the second law.

**7.33** The water in a shallow pond heats up during the day and cools down during the night. Heat transfer by radiation, conduction, and convection with the ambient surroundings thus cycles the water temperature. Is such a cyclic process reversible or irreversible?

**7.34** Assume a cyclic machine that exchanges 6 kW with a 250°C reservoir and has

a. $\dot{Q}_L = 0\ kW$ $\dot{W} = 6\ kW$

b. $\dot{Q}_L = 6\ kW$ $\dot{W} = 0\ kW$

and $\dot{Q}_L$ is exchanged with a 30°C ambient surroundings. What can you say about the processes in the two cases a and b if the machine is a heat engine? Repeat the question for the case of a heat pump.

**7.35** Consider a heat engine and heat pump connected as shown in Fig. P7.35. Assume $T_{H1} = T_{H2} > T_{amb}$ and determine for each of the three cases if the setup satisfies the first law and/or violates the second law.

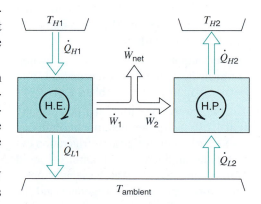

| | $\dot{Q}_{H1}$ | $\dot{Q}_{L1}$ | $\dot{W}_1$ | $\dot{Q}_{H2}$ | $\dot{Q}_{L2}$ | $\dot{W}_2$ |
|---|---|---|---|---|---|---|
| a | 6 | 4 | 2 | 3 | 2 | 1 |
| b | 6 | 4 | 2 | 5 | 4 | 1 |
| c | 3 | 2 | 1 | 4 | 3 | 1 |

**FIGURE P7.35**

**7.36** Consider the four cases of a heat engine in Problem 7.21 and determine if any of those are perpetual-motion machines of the first or second kind.

# Carnot cycles and absolute temperature

**7.37** Calculate the thermal efficiency of a Carnot-cycle heat engine operating between reservoirs at 300°C and 45°C. Compare the result to that of Problem 7.13.

**7.38** Calculate the thermal efficiency of a Carnot-cycle heat pump operating between reservoirs at 0°C and 45°C. Compare the result to that of Problem 7.24.

**7.39** At a few places where the air is very cold in the winter, for example, −30°C, it is possible to find a temperature of 13°C down below ground. What efficiency will a heat engine have operating between these two thermal reservoirs?

**7.40** Differences in surface water and deep water temperature can be utilized for power generation. It is proposed to construct a cyclic heat engine that will operate near Hawaii, where the ocean temperature is 20°C near the surface and 5°C at some depth. What is the possible thermal efficiency of such a heat engine?

**7.41** A refrigerator should remove 500 kJ from some food. Assume the refrigerator works in a Carnot cycle between −10°C and 45°C with a motor-compressor of 500 W. How much time does it take if this is the only cooling load?

**7.42** An air conditioner provides 1 kg/s of air at 15°C cooled from outside atmospheric air at 35°C. Estimate the amount of power needed to operate the air conditioner. Clearly state all assumptions made.

**7.43** Find the power output and the low $T$ heat rejection rate for a Carnot-cycle heat engine that receives 6 kW at 250°C and rejects heat at 30°C as in Problem 7.34.

**7.44** Helium has the lowest normal boiling point of any of the elements at 4.2 K. At this temperature the enthalpy of evaporation is 83.3 kJ/kmol. A Carnot refrigeration cycle is analyzed for the production of 1 kmol of liquid helium at 4.2 K from saturated vapor at the same temperature. What is the work input to the refrigerator and the coefficient of performance for the cycle with an ambient temperature at 300 K?

**7.45** A cyclic machine, shown in Fig. P7.45, receives 325 kJ from a 1000 K energy reservoir. It rejects 125 kJ to a 400 K energy reservoir, and the cycle produces 200 kJ of work as output. Is this cycle reversible, irreversible, or impossible?

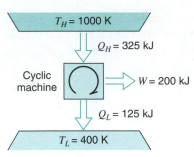

**FIGURE P7.45**

**7.46** In a cryogenic experiment you need to keep a container at −125°C although it gains 100 W due to heat transfer. What is the smallest motor you would need for a heat pump absorbing heat from the container and rejecting heat to the room at 20°C?

**7.47** Find the maximum coefficient of performance for the refrigerator in your kitchen, assuming it runs in a Carnot cycle.

**7.48** Calculate the amount of work input a refrigerator needs to make ice cubes out of a tray of 0.25 kg liquid water at 10°C. Assume the refrigerator works in a Carnot cycle between −8°C and 35°C with a motor-compressor of 750 W. How much time does it take if this is the only cooling load?

**7.49** We propose to heat a house in the winter with a heat pump. The house is to be maintained at 20°C at all times. When the ambient temperature outside drops to −10°C, the rate at which heat is lost from the house is estimated to be 25 kW. What is the minimum electrical power required to drive the heat pump?

**FIGURE P7.49**

**7.50** A thermal storage device is made with a rock (granite) bed of 2 m³ that is heated to 400 K using solar energy. A heat engine receives a $Q_H$ from the bed and rejects heat to the ambient surroundings at 290 K. The rock bed

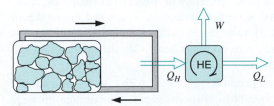

**FIGURE P7.50**

therefore cools down, and as it reaches 290 K the process stops. Find the energy the rock bed can give out. What is the heat engine efficiency at the beginning of the process, and what is it at the end of the process?

**7.51** A salesperson selling refrigerators and deep freezers will guarantee a minimum coefficient of performance of 4.5 year round. How would the performance of these machines compare? Would it be steady throughout the year?

**7.52** Liquid sodium leaves a nuclear reactor at 800°C and is used as the energy source in a steam power plant. The condenser cooling water comes from a cooling tower at 15°C. Determine the maximum thermal efficiency of the power plant. Is it misleading to use the temperatures given to calculate this value?

**7.53** An inventor has developed a refrigeration unit that maintains the cold space at −10°C, while operating in a 25°C room. A coefficient of performance of 8.5 is claimed. How do you evaluate this?

**7.54** A car engine burns 5 kg of fuel (equivalent to adding $Q_H$) at 1500 K and rejects energy to the radiator and exhaust at an average temperature of 750 K. Assume the fuel has a heating value of 40 000 kJ/kg and find the maximum amount of work the engine can provide.

**7.55** A household freezer operates in a room at 20°C. Heat must be transferred from the cold space at a rate of 2 kW to maintain its temperature at −30°C. What is the theoretically smallest (power) motor required to operate this freezer?

**7.56** A certain solar-energy collector produces a maximum temperature of 100°C. The energy is used in a cyclic heat engine that operates in a 10°C environment. What is the maximum thermal efficiency? What is it if the collector is redesigned to focus the incoming light to produce a maximum temperature of 300°C?

**7.57** A large heat pump should upgrade 5 MW of heat at 85°C to be delivered as heat at 150°C. What is the minimum amount of work (power) input that will drive this?

**7.58** A steel bottle of $V = 0.1$ m$^3$ contains R-134a at 20°C and 200 kPa. It is placed in a deep freezer where it is cooled to −20°C. The deep freezer sits in a room with ambient temperature of 20°C and has an inside temperature of −20°C. Find the amount of energy the freezer must remove from the R-134a and the extra amount of work input to the freezer to do the process.

**7.59** Sixty kilograms per hour of water runs through a heat exchanger, entering as saturated liquid at 200 kPa and leaving as saturated vapor. The heat is supplied by a Carnot heat pump operating from a low-temperature reservoir at 16°C. Find the rate of work into the heat pump.

**7.60** It is proposed to build a 1000 MW electric power plant with steam as the working fluid. The condensers are to be cooled with river water (see Fig. P7.60). The maximum steam temperature is 550°C, and the pressure in the condensers will be 10 kPa. Estimate the temperature rise of the river downstream from the power plant.

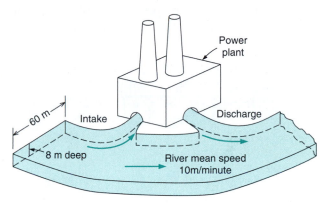

**FIGURE P7.60**

**7.61** Two different fuels can be used in a heat engine operating between the fuel burning temperature and a low temperature of 350 K. Fuel A burns at 2200 K delivering 30 000 kJ/kg and costs $1.50/kg. Fuel B burns at 1200 K, delivering 40 000 kJ/kg and costs $1.30/kg. Which fuel would you buy and why?

**7.62** A heat engine has a solar collector receiving 0.2 kW/m$^2$, inside of which a transfer media is heated to 450 K. The collected energy powers a heat engine that rejects heat at 40°C. If the heat engine should deliver 2.5 kW, what is the minimum size (area) solar collector?

## Finite $\Delta T$ heat transfer

**7.63** A refrigerator keeping 5°C inside is located in a 30°C room. It must have a high temperature $\Delta T$ above room temperature and a low temperature $\Delta T$ below the refrigerated space in the cycle to actually transfer the heat. For a $\Delta T$ of 0, 5, and 10°C, respectively, calculate the COP assuming a Carnot cycle.

**7.64** A refrigerator uses a power input of 2.5 kW to cool a 5°C space with the high temperature in the cycle as 50°C. The $Q_H$ is pushed to the ambient air at 35°C in a heat exchanger where the transfer coefficient is 50 W/m$^2$K. Find the required minimum heat transfer area.

**7.65** A large heat pump should upgrade 5 MW of heat at 85°C to be delivered as heat at 150°C. Suppose the actual heat pump has a COP of 2.5. How much power is required to drive the unit? For the same COP, how high a high temperature would a Carnot heat pump have assuming the same low $T$?

**7.66** A house is heated by a heat pump driven by an electric motor using the outside as the low-temperature reservoir. The house loses energy in direct proportion to the temperature difference as $\dot{Q}_{loss} = K(T_H - T_L)$. Determine the minimum electric power required to drive the heat pump as a function of the two temperatures.

**FIGURE P7.66**

**7.67** A house is heated by an electric heat pump using the outside as the low-temperature reservoir. For several different winter outdoor temperatures, estimate the percent savings in electricity if the house is kept at 20°C instead of 24°C. Assume that the house is losing energy to the outside as in Eq. 7.14.

**7.68** A house is cooled by an electric heat pump using the outside as the high-temperature reservoir. For several different summer outdoor temperatures, estimate the percent savings in electricity if the house is kept at 25°C instead of 20°C. Assume that the house is gaining energy from the outside in direct proportion to the temperature difference, as in Eq. 7.14.

**FIGURE P7.68**

**7.69** A farmer runs a heat pump with a motor of 2 kW. It should keep a chicken hatchery at 30°C, which loses energy at a rate of 0.5 kW per degree difference to the colder ambient $T_{amb}$. The heat pump has a coefficient of performance that is 50% of a Carnot heat pump. What is the minimum ambient temperature for which the heat pump is sufficient?

**7.70** A heat pump has a COP of $\beta' = 0.5\,\beta'_{CARNOT}$ and maintains a house at $T_H = 20°C$, while it leaks energy out as $\dot{Q} = 0.6(T_H - T_L)[kW]$. For a maximum of 1.0 kW

power input, find the minimum outside temperature, $T_L$, for which the heat pump is a sufficient heat source.

**7.71** An air conditioner cools a house at $T_L = 20°C$ with a maximum of 1.2 kW power input. The house gains energy as $\dot{Q} = 0.6(T_H - T_L)[kW]$ and the refrigeration COP is $\beta = 0.6\,\beta_{CARNOT}$. Find the maximum outside temperature, $T_H$, for which the air conditioner unit provides sufficient cooling.

**7.72** A Carnot heat engine, shown in Fig. P7.72 receives energy from a reservoir at $T_{res}$ through a heat exchanger where the heat transferred is proportional to the temperature difference as $\dot{Q}_H = K(T_{res} - T_H)$. It rejects heat at a given low temperature $T_L$. To design the heat engine for maximum work output, show that the high temperature, $T_H$, in the cycle should be selected as $T_H = (T_L T_{res})^{1.2}$.

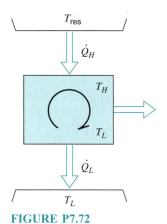

**FIGURE P7.72**

## Ideal-gas Carnot cycles

**7.73** Hydrogen gas is used in a Carnot cycle having an efficiency of 60% with a low temperature of 300 K. During the heat rejection the pressure changes from 90 kPa to 120 kPa. Find the high- and low-temperature heat transfer and the net cycle work per unit mass of hydrogen.

**7.74** An ideal-gas Carnot cycle with air in a piston cylinder has a high temperature of 1200 K and a heat rejection at 400 K. During the heat addition, the volume triples. Find the two specific heat transfers ($q$) in the cycle and the overall cycle efficiency.

**7.75** Air in a piston/cylinder setup goes through a Carnot cycle with the $P$–$v$ diagram shown in Fig. 7.24. The high and low temperatures are 600 K and 300 K, respectively. The heat added at the high temperature is 250 kJ/kg, and the lowest pressure in the cycle is 75 kPa. Find the specific volume and pressure after heat rejection and the net work per unit mass.

# Chapter 8

# Entropy

Up to this point in our consideration of the second law of thermodynamics, we have dealt only with thermodynamic cycles. Although this is a very important and useful approach, we are often concerned with processes rather than cycles. Thus, we might be interested in the second-law analysis of processes we encounter daily, such as the combustion process in an automobile engine, the cooling of a cup of coffee, or the chemical processes that take place in our bodies. It would also be beneficial to be able to deal with the second law quantitatively as well as qualitatively.

In our consideration of the first law, we initially stated the law in terms of a cycle, but we then defined a property, the internal energy, that enabled us to use the first law quantitatively for processes. Similarly, we have stated the second law for a cycle, and we now find that the second law leads to a property, entropy, that enables us to treat the second law quantitatively for processes. Energy and entropy are both abstract concepts that help to describe certain observations. As we noted in Chapter 2, thermodynamics can be described as the science of energy and entropy. The significance of this statement will become increasingly evident.

## 8.1 THE INEQUALITY OF CLAUSIUS

The first step in our consideration of the property we call entropy is to establish the inequality of Clausius, which is

$$\oint \frac{\delta Q}{T} \leq 0$$

The inequality of Clausius is a corollary or a consequence of the second law of thermodynamics. It will be demonstrated to be valid for all possible cycles, including both reversible and irreversible heat engines and refrigerators. Since any reversible cycle can be represented by a series of Carnot cycles, in this analysis we need consider only a Carnot cycle that leads to the inequality of Clausius.

Consider first a reversible (Carnot) heat engine cycle operating between reservoirs at temperatures $T_H$ and $T_L$, as shown in Fig. 8.1. For this cycle, the cyclic integral of the heat transfer, $\oint \delta Q$, is greater than zero.

$$\oint \delta Q = Q_H - Q_L > 0$$

Since $T_H$ and $T_L$ are constant, from the definition of the absolute temperature scale and from the fact this is a reversible cycle, it follows that

$$\oint \frac{\delta Q}{T} = \frac{Q_H}{T_H} - \frac{Q_L}{T_L} = 0$$

If $\oint \delta Q$, the cyclic integral of $\delta Q$, approaches zero (by making $T_H$ approach $T_L$) and the cycle remains reversible, the cyclic integral of $\delta Q/T$ remains zero. Thus, we conclude that for all reversible heat engine cycles

$$\oint \delta Q \geq 0$$

and

$$\oint \frac{\delta Q}{T} = 0$$

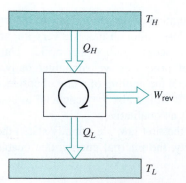

**FIGURE 8.1** Reversible heat engine cycle for demonstration of the inequality of Clausius.

Now consider an irreversible cyclic heat engine operating between the same $T_H$ and $T_L$ as the reversible engine of Fig. 8.1 and receiving the same quantity of heat $Q_H$. Comparing the irreversible cycle with the reversible one, we conclude from the second law that

$$W_{\text{irr}} < W_{\text{rev}}$$

Since $Q_H - Q_L = W$ for both the reversible and irreversible cycles, we conclude that

$$Q_H - Q_{L\,\text{irr}} < Q_H - Q_{L\,\text{rev}}$$

and therefore

$$Q_{L\,\text{irr}} > Q_{L\,\text{rev}}$$

Consequently, for the irreversible cyclic engine,

$$\oint \delta Q = Q_H - Q_{L\,\text{irr}} > 0$$

$$\oint \frac{\delta Q}{T} = \frac{Q_H}{T_H} - \frac{Q_{L\,\text{irr}}}{T_L} < 0$$

Suppose that we cause the engine to become more and more irreversible, but keep $Q_H$, $T_H$, and $T_L$ fixed. The cyclic integral of $\delta Q$ then approaches zero, and that for $\delta Q/T$ becomes a progressively larger negative value. In the limit, as the work output goes to zero,

$$\oint \delta Q = 0$$

$$\oint \frac{\delta Q}{T} < 0$$

Thus, we conclude that for all irreversible heat engine cycles

$$\oint \delta Q \geq 0$$

$$\oint \frac{\delta Q}{T} < 0$$

To complete the demonstration of the inequality of Clausius, we must perform similar analyses for both reversible and irreversible refrigeration cycles. For the reversible refrigeration cycle shown in Fig. 8.2,

$$\oint \delta Q = -Q_H + Q_L < 0$$

and

$$\oint \frac{\delta Q}{T} = -\frac{Q_H}{T_H} + \frac{Q_L}{T_L} = 0$$

As the cyclic integral of $\delta Q$ approaches zero reversibly ($T_H$ approaches $T_L$), the cyclic integral of $\delta Q/T$ remains at zero. In the limit,

$$\oint \delta Q = 0$$

$$\oint \frac{\delta Q}{T} = 0$$

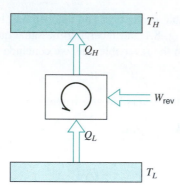

**FIGURE 8.2** Reversible refrigeration cycle for demonstration of the inequality of Clausius.

Thus, for all reversible refrigeration cycles,

$$\oint \delta Q \leq 0$$

$$\oint \frac{\delta Q}{T} = 0$$

Finally, let an irreversible cyclic refrigerator operate between temperatures $T_H$ and $T_L$ and receive the same amount of heat $Q_L$ as the reversible refrigerator of Fig. 8.2. From the second law, we conclude that the work input required will be greater for the irreversible refrigerator, or

$$W_{\text{irr}} > W_{\text{rev}}$$

Since $Q_H - Q_L = W$ for each cycle, it follows that

$$Q_{H\,\text{irr}} - Q_L > Q_{H\,\text{rev}} - Q_L$$

and therefore,

$$Q_{H\,\text{irr}} > Q_{H\,\text{rev}}$$

That is, the heat rejected by the irreversible refrigerator to the high-temperature reservoir is greater than the heat rejected by the reversible refrigerator. Therefore, for the irreversible refrigerator,

$$\oint \delta Q = -Q_{H\,\text{irr}} + Q_L < 0$$

$$\oint \frac{\delta Q}{T} = -\frac{Q_{H\,\text{irr}}}{T_H} + \frac{Q_L}{T_L} < 0$$

As we make this machine progressively more irreversible, but keep $Q_L$, $T_H$, and $T_L$ constant, the cyclic integrals of $\delta Q$ and $\delta Q/T$ both become larger in the negative direction. Consequently, a limiting case as the cyclic integral of $\delta Q$ approaches zero does not exist for the irreversible refrigerator.

Thus, for all irreversible refrigeration cycles,

$$\oint \delta Q < 0$$

$$\oint \frac{\delta Q}{T} < 0$$

Summarizing, we note that, in regard to the sign of $\oint \delta Q$, we have considered all possible reversible cycles (i.e., $\oint \delta Q \gtrless 0$), and for each of these reversible cycles

$$\oint \frac{\delta Q}{T} = 0$$

We have also considered all possible irreversible cycles for the sign of $\oint \delta Q$ (that is, $\oint \delta Q \gtrless 0$), and for all these irreversible cycles

$$\oint \frac{\delta Q}{T} < 0$$

Thus, for all cycles we can write

$$\oint \frac{\delta Q}{T} \leq 0 \tag{8.1}$$

where the equality holds for reversible cycles and the inequality for irreversible cycles. This relation, Eq. 8.1, is known as the inequality of Clausius.

The significance of the inequality of Clausius may be illustrated by considering the simple steam power plant cycle shown in Fig. 8.3. This cycle is slightly different from the usual cycle for steam power plants in that the pump handles a mixture of liquid and vapor in such proportions that saturated liquid leaves the pump and enters the boiler. Suppose that someone reports that the pressure and quality at various points in the cycle are as given in Fig. 8.3. Does this cycle satisfy the inequality of Clausius?

Heat is transferred in two places, the boiler and the condenser. Therefore,

$$\oint \frac{\delta Q}{T} = \int \left(\frac{\delta Q}{T}\right)_{\text{boiler}} + \int \left(\frac{\delta Q}{T}\right)_{\text{condenser}}$$

Since the temperature remains constant in both the boiler and condenser, this may be integrated as follows:

$$\oint \frac{\delta Q}{T} = \frac{1}{T_1}\int_1^2 \delta Q + \frac{1}{T_3}\int_3^4 \delta Q = \frac{{}_1Q_2}{T_1} + \frac{{}_3Q_4}{T_3}$$

Let us consider a 1 kg mass as the working fluid. We have then

$$_1q_2 = h_2 - h_1 = 2066.3 \text{ kJ/kg}, \qquad T_1 = 164.97°C$$

$$_3q_4 = h_4 - h_3 = 463.4 - 2361.8 = -1898.4 \text{ kJ/kg}, \qquad T_3 = 53.97°C$$

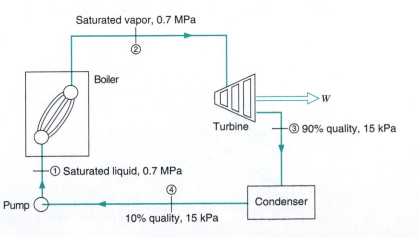

**FIGURE 8.3** A simple steam power plant that demonstrates the inequality of Clausius.

Therefore,

$$\oint \frac{\delta Q}{T} = \frac{2066.3}{164.97 + 273.15} - \frac{1898.4}{53.97 + 273.15} = -1.087 \text{ kJ/kg-K}$$

Thus, this cycle satisfies the inequality of Clausius, which is equivalent to saying that it does not violate the second law of thermodynamics.

---

**In-Text Concept Questions**

a. Does Clausius say anything about the sign for $\oint \delta Q$?

b. Does the statement of Clausius require a constant $T$ for the heat transfer as in a Carnot cycle?

---

## 8.2 ENTROPY—A PROPERTY OF A SYSTEM

By applying Eq. 8.1 and Fig. 8.4, we can demonstrate that the second law of thermodynamics leads to a property of a system that we call entropy. Let a system (control mass) undergo a reversible process from state 1 to state 2 along a path $A$, and let the cycle be completed along path $B$, which is also reversible.

Because this is a reversible cycle, we can write

$$\oint \frac{\delta Q}{T} = 0 = \int_1^2 \left(\frac{\delta Q}{T}\right)_A + \int_2^1 \left(\frac{\delta Q}{T}\right)_B$$

Now consider another reversible cycle, which proceeds first along path $C$ and is then completed along path $B$. For this cycle we can write

$$\oint \frac{\delta Q}{T} = 0 = \int_1^2 \left(\frac{\delta Q}{T}\right)_C + \int_2^1 \left(\frac{\delta Q}{T}\right)_B$$

Subtracting the second equation from the first, we have

$$\int_1^2 \left(\frac{\delta Q}{T}\right)_A = \int_1^2 \left(\frac{\delta Q}{T}\right)_C$$

Since the $\oint \delta Q/T$ is the same for all reversible paths between states 1 and 2, we conclude that this quantity is independent of the path and it is a function of the end states only; it is therefore a property. This property is called entropy and is designated $S$. It follows that entropy may be defined as a property of a substance in accordance with the relation

$$dS \equiv \left(\frac{\delta Q}{T}\right)_{rev} \tag{8.2}$$

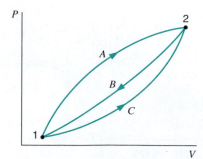

**FIGURE 8.4** Two reversible cycles demonstrating the fact that entropy is a property of a substance.

Entropy is an extensive property, and the entropy per unit mass is designated $s$. It is important to note that entropy is defined here in terms of a reversible process.

The change in the entropy of a system as it undergoes a change of state may be found by integrating Eq. 8.2. Thus,

$$S_2 - S_1 = \int_1^2 \left(\frac{\delta Q}{T}\right)_{rev} \tag{8.3}$$

To perform this integration, we must know the relation between $T$ and $Q$, and illustrations will be given subsequently. The important point is that since entropy is a property, the change in the entropy of a substance in going from one state to another is the same for all processes, both reversible and irreversible, between these two states. Eq. 8.3 enables us to find the change in entropy only along a reversible path. However, once the change has been evaluated, this value is the magnitude of the entropy change for all processes between these two states.

Eq. 8.3 enables us to calculate changes of entropy, but it tells us nothing about absolute values of entropy. From the third law of thermodynamics, which is based on observations of low-temperature chemical reactions, it is concluded that the entropy of all pure substances (in the appropriate structural form) can be assigned the absolute value of zero at the absolute zero of temperature. It also follows from the subject of statistical thermodynamics that all pure substances in the (hypothetical) ideal-gas state at absolute zero temperature have zero entropy.

However, when there is no change of composition, as would occur in a chemical reaction, for example, it is quite adequate to give values of entropy relative to some arbitrarily selected reference state, such as was done earlier when tabulating values of internal energy and enthalpy. In each case, whatever reference value is chosen, it will cancel out when the change of property is calculated between any two states. This is the procedure followed with the thermodynamic tables to be discussed in the following section.

A word should be added here regarding the role of $T$ as an integrating factor. We noted in Chapter 4 that $Q$ is a path function, and therefore $\delta Q$ in an inexact differential. However, since $(\delta Q/T)_{rev}$ is a thermodynamic property, it is an exact differential. From a mathematical perspective, we note that an inexact differential may be converted to an exact differential by the introduction of an integrating factor. Therefore, $1/T$ serves as the integrating factor in converting the inexact differential $\delta Q$ to the exact differential $\delta Q/T$ for a reversible process.

## 8.3 THE ENTROPY OF A PURE SUBSTANCE

Entropy is an extensive property of a system. Values of specific entropy (entropy per unit mass) are given in tables of thermodynamic properties in the same manner as specific volume and specific enthalpy. The units of specific entropy in the steam tables, refrigerant tables, and ammonia tables are kJ/kg K, and the values are given relative to an arbitrary reference state. In the steam tables the entropy of saturated liquid at 0.01°C is given the value of zero. For many refrigerants, the entropy of saturated liquid at −40°C is assigned the value of zero.

In general, we use the term entropy to refer to both total entropy and entropy per unit mass, since the context or appropriate symbol will clearly indicate the precise meaning of the term.

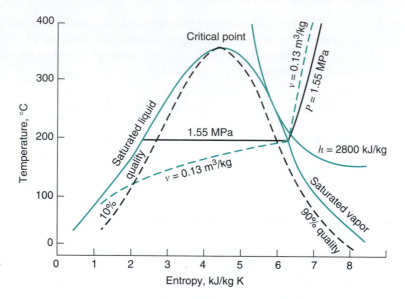

**FIGURE 8.5**
Temperature–entropy diagram for steam.

In the saturation region the entropy may be calculated using the quality. The relations are similar to those for specific volume, internal energy and enthalpy.

$$s = (1 - x)s_f + xs_g$$

$$s = s_f + xs_{fg} \tag{8.4}$$

The entropy of a compressed liquid is tabulated in the same manner as the other properties. These properties are primarily a function of the temperature and are not greatly different from those for saturated liquid at the same temperature. Table 4 of the steam tables, which is summarized in Table B.1.4, give the entropy of compressed liquid water in the same manner as for other properties.

The thermodynamic properties of a substance are often shown on a temperature–entropy diagram and on an enthalpy–entropy diagram, which is also called a Mollier diagram, after Richard Mollier (1863–1935) of Germany. Figures 8.5 and 8.6 show the essential elements of temperature–entropy and enthalpy–entropy diagrams for steam. The

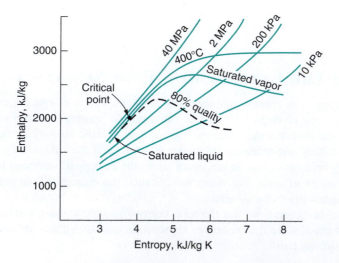

**FIGURE 8.6**  Enthalpy–entropy diagram for steam.

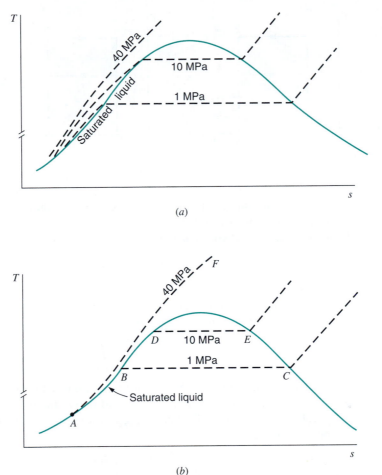

**FIGURE 8.7**
Temperature–entropy diagram to show properties of a compressed liquid, water.

general features of such diagrams are the same for all pure substances. A more complete temperature–entropy diagram for steam is shown in Fig. E.1 in Appendix E.

These diagrams are valuable both because they present thermodynamic data and because they enable us to visualize the changes of state that occur in various processes. As our study progresses, the student should acquire facility in visualizing thermodynamic processes on these diagrams. The temperature–entropy diagram is particularly useful for this purpose.

For most substances the difference in the entropy of a compressed liquid and a saturated liquid at the same temperature is so small that a process in which liquid is heated at constant pressure nearly coincides with the saturated-liquid line until the saturation temperature is reached (Fig. 8.7). Thus, if water at 10 MPa is heated from 0°C to the saturation temperature, it would be shown by line *ABD*, which coincides with the saturated-liquid line.

## 8.4 ENTROPY CHANGE IN REVERSIBLE PROCESSES

Having established that entropy is a thermodynamic property of a system, we now consider its significance in various processes. In this section we will limit ourselves to systems that undergo reversible processes and consider the Carnot cycle, reversible heat-transfer processes, and reversible adiabatic processes.

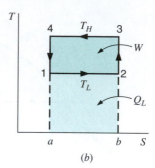

**FIGURE 8.8** The Carnot cycle on the temperature–entropy diagram.

Let the working fluid of a heat engine operating on the Carnot cycle make up the system. The first process is the isothermal transfer of heat to the working fluid from the high-temperature reservoir. For this process we can write

$$S_2 - S_1 = \int_1^2 \left( \frac{\delta Q}{T} \right)_{\text{rev}}$$

Since this is a reversible process in which the temperature of the working fluid remains constant, the equation can be integrated to give

$$S_2 - S_1 = \frac{1}{T_H} \int_1^2 \delta Q = \frac{{}_1 Q_2}{T_H}$$

This process is shown in Fig. 8.8a, and the area under line 1–2, area 1–2–b–a–1, represents the heat transferred to the working fluid during the process.

The second process of a Carnot cycle is a reversible adiabatic one. From the definition of entropy,

$$dS = \left( \frac{\delta Q}{T} \right)_{\text{rev}}$$

it is evident that the entropy remains constant in a reversible adiabatic process. A constant-entropy process is called an isentropic process. Line 2–3 represents this process, and this process is concluded at state 3 when the temperature of the working fluid reaches $T_L$.

The third process is the reversible isothermal process in which heat is transferred from the working fluid to the low-temperature reservoir. For this process we can write

$$S_4 - S_3 = \int_3^4 \left( \frac{\delta Q}{T} \right)_{\text{rev}} = \frac{{}_3 Q_4}{T_L}$$

Because during this process the heat transfer is negative (in regard to the working fluid), the entropy of the working fluid decreases. Moreover, because the final process 4–1, which completes the cycle, is a reversible adiabatic process (and therefore isentropic), it is evident that the entropy decrease in process 3–4 must exactly equal the entropy increase in process 1–2. The area under line 3–4, area 3–4–a–b–3, represents the heat transferred from the working fluid to the low-temperature reservoir.

Since the net work of the cycle is equal to the net heat transfer, then area 1–2–3–4–1 must represent the net work of the cycle. The efficiency of the cycle may also be expressed in terms of areas:

$$\eta_{\text{th}} = \frac{W_{\text{net}}}{Q_H} = \frac{\text{area } 1\text{–}2\text{–}3\text{–}4\text{–}1}{\text{area } 1\text{–}2\text{–}b\text{–}a\text{–}1}$$

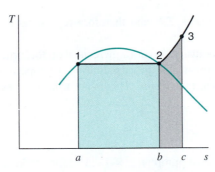

**FIGURE 8.9** A temperature–entropy diagram to show areas that represent heat transfer for an internally reversible process.

Some statements made earlier about efficiencies may now be understood graphically. For example, increasing $T_H$ while $T_L$ remains constant increases the efficiency. Decreasing $T_L$ as $T_H$ remains constant increases the efficiency. It is also evident that the efficiency approaches 100% as the absolute temperature at which heat is rejected approaches zero.

If the cycle is reversed, we have a refrigerator or heat pump. The Carnot cycle for a refrigerator is shown in Fig. 8.8b. Notice that the entropy of the working fluid increases at $T_L$, since heat is transferred to the working fluid at $T_L$. The entropy decreases at $T_H$ because of heat transfer from the working fluid.

Let us next consider reversible heat-transfer processes. Actually, we are concerned here with processes that are internally reversible, that is, processes that have no irreversibilities within the boundary of the system. For such processes the heat transfer to or from a system can be shown as an area on a temperature–entropy diagram. For example, consider the change of state from saturated liquid to saturated vapor at constant pressure. This process would correspond to the process 1–2 on the $T$–$s$ diagram of Fig. 8.9 (note that absolute temperature is required here), and the area 1–2–$b$–$a$–1 represents the heat transfer. Since this is a constant-pressure process, the heat transfer per unit mass is equal to $h_{fg}$. Thus,

$$s_2 - s_1 = s_{fg} = \frac{1}{m}\int_1^2 \left(\frac{\delta Q}{T}\right)_{rev} = \frac{1}{mT}\int_1^2 \delta Q = \frac{{}_1q_2}{T} = \frac{h_{fg}}{T}$$

This relation gives a clue about how $s_{fg}$ is calculated for tabulation in tables of thermodynamic properties. For example, consider steam at 10 MPa. From the steam tables we have

$$h_{fg} = 1317.1 \text{ kJ/kg}$$

$$T = 311.06 + 273.15 = 584.21 \text{ K}$$

Therefore,

$$s_{fg} = \frac{h_{fg}}{T} = \frac{1317.1}{584.21} = 2.2544 \text{ kJ/kg K}$$

This is the value listed for $s_{fg}$ in the steam tables.

If heat is transferred to the saturated vapor at constant pressure, the steam is superheated along line 2–3. For this process we can write

$$_2q_3 = \frac{1}{m}\int_2^3 \delta Q = \int_2^3 T\, ds$$

Since $T$ is not constant, this equation cannot be integrated unless we know a relation between temperature and entropy. However, we do realize that the area under line 2–3, area 2–3–c–b–2, represents $\int_2^3 T\,ds$ and therefore represents the heat transferred during this reversible process.

The important conclusion to draw here is that for processes that are internally reversible, the area underneath the process line on a temperature–entropy diagram represents the quantity of heat transferred. This is not true for irreversible processes, as will be demonstrated later.

---

**EXAMPLE 8.1**

Consider a Carnot-cycle heat pump with R-134a as the working fluid. Heat is absorbed into the R-l34a at 0°C, during which process it changes from a two-phase state to saturated vapor. The heat is rejected from the R-134a at 60°C so that it ends up as saturated liquid. Find the pressure after compression, before the heat rejection process, and determine the coefficient of performance for the cycle.

**Solution**

From the definition of the Carnot cycle we have two constant-temperature (isothermal) processes that involve heat transfer and two adiabatic processes in which the temperature changes. The variation in $s$ follows from Eq. 8.2

$$ds = \delta q / T$$

and the Carnot cycle is shown in Fig. 8.8 and for this case in Fig. 8.10. We therefore have

State 4 Table B.5.1:     $s_4 = s_3 = s_{f@60deg} = 1.2857$ kJ/kg K
State 1 Table B.5.1:     $s_1 = s_2 = s_{g@0deg} = 1.7262$ kJ/kg K
State 2 Table B.5.2:     $60°C, s_2 = s_1 = s_{g@0deg}$

Interpolate between 1400 kPa and 1600 kPa in Table B.5.2:

$$P_2 = 1400 + (1600 - 1400)\frac{1.7262 - 1.736}{1.7135 - 1.736} = 1487.1 \text{ kPa}$$

From the fact that it is a Carnot cycle the COP becomes, from Eq. 7.13

$$\beta' = \frac{q_H}{w_{\text{IN}}} = \frac{T_H}{T_H - T_L} = \frac{333.15}{60} = 5.55$$

*Remark*: Notice how much the pressure varies during the heat rejection process. Because this process is very difficult to accomplish in a real device, no heat pump or refrigerator is designed to attempt to approach a Carnot cycle.

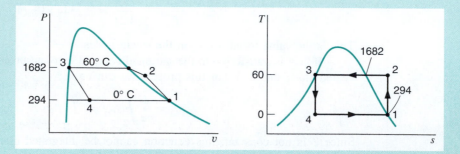

**FIGURE 8.10** Diagram for Example 8.1.

**EXAMPLE 8.2**

A cylinder/piston setup contains 1 L of saturated liquid refrigerant R-12 at 20°C. The piston now slowly expands, maintaining constant temperature to a final pressure of 400 kPa in a reversible process. Calculate the required work and heat transfer to accomplish this process.

**Solution**

C.V. The refrigerant R-12, which is a control mass.

Continuity Eq.: $m_2 = m_1 = m$;

Energy Eq. 5.11: $m(u_2 - u_1) = {}_1Q_2 - {}_1W_2$

Entropy Eq. 8.3: $m(s_2 - s_1) \geq \int \delta Q / T$

Process: $T$ = constant, reversible so equal sign applies in entropy equation.

State 1 $(T, P)$ Table B.3.1: $u_1 = 54.45$ kJ/kg, $s_1 = 0.2078$ kJ/kg K

$$m = V/v_1 = 0.001/0.000752 = 1.33 \text{ kg}$$

State 2 $(T, P)$ Table B.3.2: $u_2 = 180.57$ kJ/kg, $s_2 = 0.7204$ kJ/kg K

As $T$ is constant we have $\int \delta Q / T = {}_1Q_2 / T$, so from the entropy equation

$${}_1Q_2 = mT(s_2 - s_1) = 1.33 \times 293.15 \times (0.7204 - 0.2078) = 200 \text{ kJ}$$

The work is then, from the energy equation,

$${}_1W_2 = m(u_1 - u_2) + {}_1Q_2 = 1.33 \times (54.45 - 180.57) + 200 = 32.3 \text{ kJ}$$

Note from Fig. 8.11 that it would be difficult to calculate the work as the area in the $P$–$v$ diagram due to the shape of the process curve. The heat transfer is the area in the $T$–$s$ diagram.

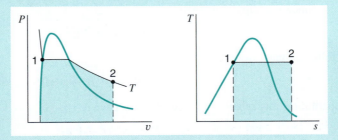

**FIGURE 8.11** Diagram for Example 8.2.

---

**In-Text Concept Questions**

**c.** How can you change $s$ of a substance going through a reversible process?

**d.** A reversible process adds heat to a substance. If $T$ is varying, does that influence the change in $s$?

**e.** Water at 100 kPa, 150°C receives 75 kJ/kg in a reversible process by heat transfer. Which process changes $s$ the most: constant $T$, constant $v$, or constant $P$?

## 8.5 The Thermodynamic Property Relation

At this point we derive two important thermodynamic relations for a simple compressible substance. These relations are

$$T\,dS = dU + P\,dV$$

$$T\,dS = dH - V\,dP$$

The first of these relations can be derived by considering a simple compressible substance in the absence of motion or gravitational effects. The first law for a change of state under these conditions can be written

$$\delta Q = dU + \delta W$$

The equations we are deriving here deal first with the changes of state in which the state of the substance can be identified at all times. Thus, we must consider a quasi-equilibrium process or, to use the term introduced in the last chapter, a reversible process. For a reversible process of a simple compressible substance, we can write

$$\delta Q = T\,dS \qquad \text{and} \qquad \delta W = P\,dV$$

Substituting these relations into the first-law equation, we have

$$T\,dS = dU + P\,dV \tag{8.5}$$

which is the first equation we set out to derive. Note that this equation was derived by assuming a reversible process. This equation can therefore be integrated for any reversible process, for during such a process the state of the substance can be identified at any point during the process. We also note that Eq. 8.5 deals only with properties. Suppose we have an irreversible process taking place between the given initial and final states. The properties of a substance depend only on the state, and therefore the changes in the properties during a given change of state are the same for an irreversible process as for a reversible process. Therefore, Eq. 8.5 is often applied to an irreversible process between two given states, but the integration of Eq. 8.5 is performed along a reversible path between the same two states.

Since enthalpy is defined as

$$H = U + PV$$

it follows that

$$dH = dU + P\,dV + V\,dP$$

Substituting this relation into Eq. 8.5, we have

$$T\,dS = dH - V\,dP \tag{8.6}$$

which is the second relation that we set out to derive. These two expressions, Eqs. 8.5 and 8.6, are two forms of the thermodynamic property relation and are frequently called Gibbs equations.

These equations can also be written for a unit mass,

$$T\,ds = du + P\,dv \tag{8.7}$$

$$T\,ds = dh - v\,dP \tag{8.8}$$

The Gibbs equations will be used extensively in certain subsequent sections of this book.

If we consider substances of fixed composition other than a simple compressible substance, we can write "$T\,dS$" equations other than those just given for a simple compressible substance. In Eq. 4.15 we noted that for a reversible process we can write the following expression for work:

$$\delta W = P\,dV - \mathcal{T}\,dL - \mathcal{S}\,dA - \mathcal{E}\,dZ + \cdots$$

It follows that a more general expression for the thermodynamic property relation would be

$$T\,dS = dU + P\,dV - \mathcal{T}\,dL - \mathcal{S}\,dA - \mathcal{E}\,dZ + \cdots \tag{8.9}$$

## 8.6 Entropy Change of a Solid or Liquid

In Section 5.6 we considered the calculation of the internal energy and enthalpy changes with temperature for solids and liquids and found that, in general, it is possible to express both in terms of the specific heat, in the simple manner of Eq. 5.17, and in most instances in the integrated form of Eq. 5.18. We can now use this result and the thermodynamic property relation, Eq. 8.7, to calculate the entropy change for a solid or liquid. Note that for such a phase the specific volume term in Eq. 8.7 is very small, so that substituting Eq. 5.17 yields

$$ds \simeq \frac{du}{T} \simeq \frac{C}{T}\,dT \tag{8.10}$$

Now, as was mentioned in Section 5.6, for many processes involving a solid or liquid, we may assume that the specific heat remains constant, in which case Eq. 8.10 can be integrated. The result is

$$s_2 - s_1 \simeq C \ln \frac{T_2}{T_1} \tag{8.11}$$

If the specific heat is not constant, then commonly $C$ is known as a function of $T$, in which case Eq. 8.10 can also be integrated to find the entropy change. Equation 8.11 illustrates what happens in a reversible ($ds_{\text{gen}} = 0$) adiabatic ($dq = 0$) process, which therefore is isentropic. In this process, the approximation of constant $v$ leads to constant temperature, which explains why pumping liquid does not change the temperature.

**EXAMPLE 8.3**

One kilogram of liquid water is heated from 20°C to 90°C. Calculate the entropy change, assuming constant specific heat, and compare the result with that found when using the steam tables.

*Control mass*: Water.
*Initial and final states*: Known.
*Model*: Constant specific heat, value at room temperature.

**Solution**

For constant specific heat, from Eq. 8.11,

$$s_2 - s_1 = 4.184 \ln\left(\frac{363.2}{293.2}\right) = 0.8958 \text{ kJ/kg K}$$

Comparing this result with that obtained by using the steam tables, we have

$$s_2 - s_1 = s_{f\,90°C} - s_{f\,20°C} = 1.1925 - 0.2966$$
$$= 0.8959 \text{ kJ/kg K}$$

## 8.7 ENTROPY CHANGE OF AN IDEAL GAS

Two very useful equations for computing the entropy change of an ideal gas can be developed from Eq. 8.7 by substituting Eqs. 5.20 and 5.24:

$$T\,ds = du + P\,dv$$

For an ideal gas

$$du = C_{v0}\,dT \qquad \text{and} \qquad \frac{P}{T} = \frac{R}{v}$$

Therefore,

$$ds = C_{v0}\frac{dT}{T} + \frac{R\,dv}{v} \tag{8.12}$$

$$s_2 - s_1 = \int_1^2 C_{v0}\frac{dT}{T} + R\ln\frac{v_2}{v_1} \tag{8.13}$$

Similarly,

$$T\,ds = dh - v\,dP$$

For an ideal gas

$$dh = C_{p0}\,dT \qquad \text{and} \qquad \frac{v}{T} = \frac{R}{P}$$

Therefore,

$$ds = C_{p0}\frac{dT}{T} - R\frac{dP}{P} \tag{8.14}$$

$$s_2 - s_1 = \int_1^2 C_{p0}\frac{dT}{T} - R\ln\frac{P_2}{P_1} \tag{8.15}$$

To integrate Eqs. 8.13 and 8.15, we must know the temperature dependence of the specific heats. However, if we recall that their difference is always constant as expressed by Eq. 5.27, we realize that we need to examine the temperature dependence of only one of the specific heats.

As in Section 5.7, let us consider the specific heat $C_{p0}$. Again, there are three possibilities to examine, the simplest of which is the assumption of constant specific heat. In this instance it is possible to integrate Eq. 8.15 directly to

$$s_2 - s_1 = C_{p0}\ln\frac{T_2}{T_1} - R\ln\frac{P_2}{P_1} \tag{8.16}$$

Similarly, integrating Eq. 8.13 for constant specific heat, we have

$$s_2 - s_1 = C_{v0}\ln\frac{T_2}{T_1} + R\ln\frac{v_2}{v_1} \tag{8.17}$$

The second possibility for the specific heat is to use an analytical equation for $C_{p0}$ as a function of temperature, for example, one of those listed in Table A.6. The third possibility is to integrate the results of the calculations of statistical thermodynamics from reference temperature $T_0$ to any other temperature $T$ and define the standard entropy

$$s_T^0 = \int_{T_0}^T \frac{C_{p0}}{T}\,dT \tag{8.18}$$

This function can then be tabulated in the single-entry (temperature) ideal-gas table, as for air in Table A.7 or for other gases in Table A.8. The entropy change between any two states 1 and 2 is then given by

$$s_2 - s_1 = (s_{T2}^0 - s_{T1}^0) - R \ln \frac{P_2}{P_1} \tag{8.19}$$

As with the energy functions discussed in Section 5.7, the ideal-gas tables, Tables A.7 and A.8, would give the most accurate results, and the equations listed in Table A.6 would give a close empirical approximation. Constant specific heat would be less accurate, except for monatomic gases and for other gases below room temperature. Again, it should be remembered that all these results are part of the ideal-gas model, which may or may not be appropriate in any particular problem.

| | |
|---|---|
| **EXAMPLE 8.4** | Consider Example 5.7, in which oxygen is heated from 300 to 1500 K. Assume that during this process the pressure dropped from 200 to 150 kPa. Calculate the change in entropy per kilogram. |

**Solution**

The most accurate answer for the entropy change, assuming ideal-gas behavior, would be from the ideal-gas tables. Table A.8. This result is, using Eq. 8.19,

$$s_2 - s_1 = (8.0649 - 6.4168) - 0.2598 \ln \left( \frac{150}{200} \right)$$

$$= 1.7228 \text{ kJ/kg K}$$

The empirical equation from Table A.6 should give a good approximation to this result. Integrating Eq. 8.15, we have

$$s_2 - s_1 = \int_{T_1}^{T_2} C_{p0} \frac{dT}{T} - R \ln \frac{P_2}{P_1}$$

$$s_2 - s_1 = \left[ 0.88 \ln \theta - 0.0001\theta + \frac{0.54}{2} \theta^2 - \frac{0.33}{3} \theta^3 \right]_{\theta_1 = 0.3}^{\theta_2 = 1.5}$$

$$- 0.2598 \ln \left( \frac{150}{200} \right)$$

$$= 1.7058 \text{ kJ/kg K}$$

which is within 1.0% of the previous value. For constant specific heat, using the value at 300 K from Table A.5, we have

$$s_2 - s_1 = 0.922 \ln \left( \frac{1500}{300} \right) - 0.2598 \ln \left( \frac{150}{200} \right)$$

$$= 1.5586 \text{ kJ/kg K}$$

which is too low by 9.5%. If, however, we assume that the specific heat is constant at its value at 900 K, the average temperature, as in Example 5.7, then

$$s_2 - s_1 = 1.0767 \ln \left( \frac{1500}{300} \right) + 0.0747 = 1.8076 \text{ kJ/kg K}$$

which is high by 4.9%.

**EXAMPLE 8.5**

Calculate the change in entropy per kilogram as air is heated from 300 to 600 K while pressure drops from 400 to 300 kPa. Assume:

1. Constant specific heat.
2. Variable specific heat.

**Solution**

1. From Table A.5 for air at 300 K,

$$C_{p0} = 1.004 \text{ kJ/kg K}$$

Therefore, using Eq. 8.16, we have

$$s_2 - s_1 = 1.004 \ln\left(\frac{600}{300}\right) - 0.287 \ln\left(\frac{300}{400}\right) = 0.7785 \text{ kJ/kg K}$$

2. From Table A.7,

$$s_{T1}^0 = 6.8693 \text{ kJ/kg K},$$

$$s_{T2}^0 = 7.5764 \text{ kJ/kg K}$$

Using Eq. 8.19 gives

$$s_2 - s_1 = 7.5764 - 6.8693 - 0.287 \ln\left(\frac{300}{400}\right) = 0.7897 \text{ kJ/kg K}$$

Let us now consider the case of an ideal gas undergoing an isentropic process, a situation that is analyzed frequently. We conclude that Eq. 8.15 with the left side of the equation equal to zero then expresses the relation between the pressure and temperature at the initial and final states, with the specific relation depending on the nature of the specific heat as a function of $T$. As was discussed following Eq. 8.15, there are three possibilities to examine. Of these, the most accurate is the third, that is, the ideal-gas Tables A.7 or A.8 and Eq. 8.19, with the integrated temperature function $s_T^0$ defined by Eq. 8.18. The following Example illustrates the procedure.

**EXAMPLE 8.6**

One kilogram of air is contained in a cylinder fitted with a piston at a pressure of 400 kPa and a temperature of 600 K. The air is expanded to 150 kPa in a reversible, adiabatic process. Calculate the work done by the air.

| | |
|---|---|
| *Control mass*: | Air. |
| *Initial state*: | $P_1$, $T_1$; state 1 fixed. |
| *Final state*: | $P_2$. |
| *Process*: | Reversible and adiabatic. |
| *Model*: | Ideal gas and air tables, Table A.7. |

**Analysis**

From the first law we have

$$0 = u_2 - u_1 + w$$

The second law gives us

$$s_2 = s_1$$

**Solution**

From Table A.7,

$$u_1 = 435.10 \text{ kJ/kg}, \qquad s_{T1}^0 = 7.5764 \text{ kJ/kg K}$$

From Eq. 8.19,

$$s_2 - s_1 = 0 = (s_{T2}^0 - s_{T1}^0) - R \ln \frac{P_2}{P_1}$$

$$= (s_{T2}^0 - 7.5764) - 0.287 \ln \left(\frac{150}{400}\right)$$

$$s_{T2}^0 = 7.2949 \text{ kJ/kg K}$$

From Table A.7,

$$T_2 = 457 \text{ K}, \qquad u_2 = 328.14 \text{ kJ/kg}$$

Therefore,

$$w = 435.10 - 328.14 = 106.96 \text{ kJ/kg}$$

The first of the three possibilities, constant specific heat, is also worth analyzing as a special case. In this instance, the result is Eq. 8.16 with the left side equal to zero, or

$$s_2 - s_1 = 0 = C_{p0} \ln \frac{T_2}{T_1} - R \ln \frac{P_2}{P_1}$$

This expression can also be written as

$$\ln \left(\frac{T_2}{T_1}\right) = \frac{R}{C_{p0}} \ln \left(\frac{P_2}{P_1}\right)$$

or

$$\frac{T_2}{T_1} = \left(\frac{P_2}{P_1}\right)^{R/C_{p0}} \tag{8.20}$$

However,

$$\frac{R}{C_{p0}} = \frac{C_{p0} - C_{v0}}{C_{p0}} = \frac{k - 1}{k} \tag{8.21}$$

where $k$, the ratio of the specific heats, is defined as

$$k = \frac{C_{p0}}{C_{v0}} \tag{8.22}$$

Equation (8.20) is now conveniently written as

$$\frac{T_2}{T_1} = \left(\frac{P_2}{P_1}\right)^{(k-1)/k} \tag{8.23}$$

From this expression and the ideal-gas equation of state, it also follows that

$$\frac{T_2}{T_1} = \left(\frac{v_1}{v_2}\right)^{k-1} \tag{8.24}$$

and

$$\frac{P_2}{P_1} = \left(\frac{v_1}{v_2}\right)^{k} \tag{8.25}$$

From this last expression, we note that for this process

$$Pv^k = \text{constant} \tag{8.26}$$

This is a special case of a polytropic process in which the polytropic exponent $n$ is equal to the specific heat ratio $k$.

## 8.8 THE REVERSIBLE POLYTROPIC PROCESS FOR AN IDEAL GAS

When a gas undergoes a reversible process in which there is heat transfer, the process frequently takes place in such a manner that a plot of log $P$ versus log $V$ is a straight line, as shown in Fig. 8.12. For such a process $PV^n$ is a constant.

A process having this relation between pressure and volume is called a polytropic process. An example is the expansion of the combustion gases in the cylinder of a water-cooled reciprocating engine. If the pressure and volume are measured during the expansion stroke of a polytropic process, as might be done with an engine indicator, and the logarithms of the pressure and volume are plotted, the result would be similar to the straight line in Fig. 8.12. From this figure it follows that

$$\frac{d \ln P}{d \ln V} = -n$$

$$d \ln P + n \, d \ln V = 0$$

If $n$ is a constant (which implies a straight line on the log $P$ versus log $V$ plot), this equation can be integrated to give the following relation:

$$PV^n = \text{constant} = P_1V_1^n = P_2V_2^n \tag{8.27}$$

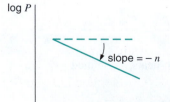

log $P$

slope $= -n$

log $V$  **FIGURE 8.12**  Example of a polytropic process.

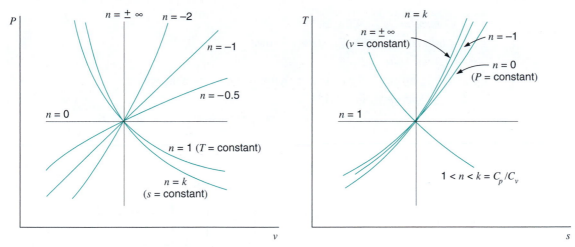

**FIGURE 8.13**   Polytropic process on $P$–$v$ and $T$–$s$ diagrams.

From this equation the following relations can be written for a polytropic process:

$$\frac{P_2}{P_1} = \left(\frac{V_1}{V_2}\right)^n$$

$$\frac{T_2}{T_1} = \left(\frac{P_2}{P_1}\right)^{(n-1)/n} = \left(\frac{V_1}{V_2}\right)^{n-1} \tag{8.28}$$

For a control mass consisting of an ideal gas, the work done at the moving boundary during a reversible polytropic process can be derived from the relations

$$_1W_2 = \int_1^2 P\,dV \qquad \text{and} \qquad PV^n = \text{constant}$$

$$_1W_2 = \int_1^2 P\,dV = \text{constant} \int_1^2 \frac{dV}{V^n}$$

$$= \frac{P_2V_2 - P_1V_1}{1 - n} = \frac{mR(T_2 - T_1)}{1 - n} \tag{8.29}$$

for any value of $n$ except $n = 1$.

The polytropic processes for various values of $n$ are shown in Fig. 8.13 on $P$–$v$ and $T$–$s$ diagrams. The values of $n$ for some familiar processes are

| | | |
|---|---|---|
| Isobaric process: | $n = 0$, | $P = \text{constant}$ |
| Isothermal process: | $n = 1$, | $T = \text{constant}$ |
| Isentropic process: | $n = k$, | $s = \text{constant}$ |
| Isochoric process: | $n = \infty$, | $v = \text{constant}$ |

**EXAMPLE 8.7**   In a reversible process, nitrogen is compressed in a cylinder from 100 kPa and 20°C to 500 kPa. During this compression process, the relation between pressure and volume is $PV^{1.3} = \text{constant}$. Calculate the work and heat transfer per kilogram, and show this process on $P$–$V$ and $T$–$S$ diagrams.

*Control mass*: Nitrogen.
*Initial state*: $P_1, T_1$; state 1 known.
*Final state*: $P_2$.
*Process*: Reversible, polytropic with exponent $n < k$.
*Diagram*: Fig. 8.14
*Model*: Ideal gas, constant specific heat—value at 300 K.

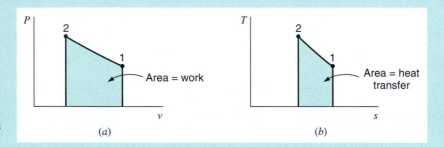

**FIGURE 8.14** Diagram for Example 8.7.

**Analysis**

We need to find the boundary movement work. From Eq. 8.29, we have

$$_1W_2 = \int_1^2 P \, dV = \frac{P_2V_2 - P_1V_1}{1 - n} = \frac{mR(T_2 - T_1)}{1 - n}$$

The first law is

$$_1q_2 = u_2 - u_1 + {}_1w_2 = C_{v0}(T_2 - T_1) + {}_1w_2$$

**Solution**

From Eq. 8.28

$$\frac{T_2}{T_1} = \left(\frac{P_2}{P_1}\right)^{(n-1)/n} = \left(\frac{500}{100}\right)^{(1.3-1)/1.3} = 1.4498$$

$$T_2 = 293.2 \times 1.4498 = 425 \text{ K}$$

Then

$$_1w_2 = \frac{R(T_2 - T_1)}{1 - n} = \frac{0.2968(425 - 293.2)}{(1 - 1.3)} = -130.4 \text{ kJ/kg}$$

and from the first law,

$$_1q_2 = C_{v0}(T_2 - T_1) + {}_1w_2$$
$$= 0.745(425 - 293.2) - 130.4 = -32.2 \text{ kJ/kg}$$

The reversible isothermal process for an ideal gas is of particular interest. In this process

$$PV = \text{constant} = P_1V_1 = P_2V_2 \tag{8.30}$$

The work done at the boundary of a simple compressible mass during a reversible isothermal process can be found by integrating the equation

$$_1W_2 = \int_1^2 P \, dV$$

The integration is

$$_1W_2 = \int_1^2 P \, dV = \text{constant} \int_1^2 \frac{dV}{V} = P_1V_1 \ln \frac{V_2}{V_1} = P_1V_1 \ln \frac{P_1}{P_2} \tag{8.31}$$

or

$$_1W_2 = mRT \ln \frac{V_2}{V_1} = mRT \ln \frac{P_1}{P_2} \tag{8.32}$$

Because there is no change in internal energy or enthalpy in an isothermal process, the heat transfer is equal to the work (neglecting changes in kinetic and potential energy). Therefore, we could have derived Eq. 8.31 by calculating the heat transfer.

For example, using Eq. 8.7, we have

$$\int_1^2 T \, ds = {}_1q_2 = \int_1^2 du + \int_1^2 P \, dv$$

But $du = 0$ and $Pv = \text{constant} = P_1v_1 = P_2v_2$, such that

$$_1q_2 = \int_1^2 P \, dv = P_1v_1 \ln \frac{v_2}{v_1}$$

which yields the same result as Eq. 8.31.

---

**In-Text Concept Questions**

**f.** A liquid is compressed in a reversible adiabatic process. What is the change in $T$?

**g.** An ideal gas goes through a constant $T$ reversible heat addition process. How do the properties ($v$, $u$, $h$, $s$, $P$) change (up, down, or constant)?

**h.** Carbon dioxide is compressed to a smaller volume in a polytropic process with $n = 1.2$. How do the properties ($u$, $h$, $s$, $P$, $T$) change (up, down, or constant)?

---

## 8.9 Entropy Change of a Control Mass During an Irreversible Process

Consider a control mass that undergoes the cycles shown in Fig. 8.15. The cycle made up of the reversible processes $A$ and $B$ is a reversible cycle. Therefore, we can write

$$\oint \frac{\delta Q}{T} = \int_1^2 \left(\frac{\delta Q}{T}\right)_A + \int_2^1 \left(\frac{\delta Q}{T}\right)_B = 0$$

The cycle made of the irreversible process $C$ and the reversible process $B$ is an irreversible cycle. Therefore, for this cycle the inequality of Clausius may be applied, giving the result

$$\oint \frac{\delta Q}{T} = \int_1^2 \left(\frac{\delta Q}{T}\right)_C + \int_2^1 \left(\frac{\delta Q}{T}\right)_B < 0$$

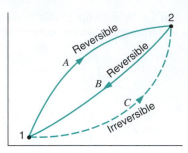

**FIGURE 8.15** Entropy change of a control mass during an irreversible process.

Subtracting the second equation from the first and rearranging, we have

$$\int_1^2 \left(\frac{\delta Q}{T}\right)_A > \int_1^2 \left(\frac{\delta Q}{T}\right)_C$$

Since path $A$ is reversible, and since entropy is a property,

$$\int_1^2 \left(\frac{\delta Q}{T}\right)_A = \int_1^2 dS_A = \int_1^2 dS_C$$

Therefore,

$$\int_1^2 dS_C > \int_1^2 \left(\frac{\delta Q}{T}\right)_C$$

As path $C$ was arbitrary, the general result is

$$dS \geq \frac{\delta Q}{T}$$

$$S_2 - S_1 \geq \int_1^2 \frac{\delta Q}{T} \tag{8.33}$$

In these equations the equality holds for a reversible process and the inequality for an irreversible process.

This is one of the most important equations of thermodynamics. It is used to develop a number of concepts and definitions. In essence, this equation states the influence of irreversibility on the entropy of a control mass. Thus, if an amount of heat $\delta Q$ is transferred to a control mass at temperature $T$ in a reversible process, the change of entropy is given by the relation

$$dS = \left(\frac{\delta Q}{T}\right)_{rev}$$

If any irreversible effects occur while the amount of heat $\delta Q$ is transferred to the control mass at temperature $T$, however, the change of entropy will be greater than for the reversible process. We would then write

$$dS > \left(\frac{\delta Q}{T}\right)_{irr}$$

Equation 8.33 holds when $\delta Q = 0$, when $\delta Q < 0$, and when $\delta Q > 0$. If $\delta Q$ is negative, the entropy will tend to decrease as a result of the heat transfer. However, the influence of irreversibilities is still to increase the entropy of the mass, and from the absolute numerical perspective we can still write for $\delta Q$:

$$dS \geq \frac{\delta Q}{T}$$

**i.** A substance has heat transfer out. Can you say anything about changes in $s$ if the process is reversible? If it is irreversible?

**j.** A substance is compressed adiabatically so $P$ and $T$ go up. Does that change $s$?

## 8.10 Entropy Generation

The conclusion from the previous considerations is that the entropy change for an irreversible process is larger than the change in a reversible process for the same $\delta Q$ and $T$. This can be written out in a common form as an equality

$$dS = \frac{\delta Q}{T} + \delta S_{gen} \tag{8.34}$$

provided the last term is positive,

$$\delta S_{gen} \geq 0 \tag{8.35}$$

The amount of entropy, $\delta S_{gen}$, is the entropy generation in the process due to irreversibilities occurring inside the system, a control mass for now but later extended to the more general control volume. This internal generation can be caused by the processes mentioned in Section 7.4, such as friction, unrestrained expansions, and the internal transfer of energy (redistribution) over a finite temperature difference. In addition to this internal entropy generation, external irreversibilities are possible by heat transfer over finite temperature differences as the $\delta Q$ is transferred from a reservoir or by the mechanical transfer of work.

Eq. 8.35 is then valid with the equal sign for a reversible process and the greater than sign for an irreversible process. Since the entropy generation is always positive and the smallest in a reversible process, namely zero, we may deduce some limits for the heat transfer and work terms.

Consider a reversible process, for which the entropy generation is zero, and the heat transfer and work terms therefore are

$$\delta Q = T\,dS \qquad \text{and} \qquad \delta W = P\,dV$$

For an irreversible process with a nonzero entropy generation, the heat transfer from Eq. 8.34 becomes

$$\delta Q_{irr} = T\,dS - T\,\delta S_{gen}$$

and thus is smaller than that for the reversible case for the same change of state, $dS$. We also note that for the irreversible process, the work is no longer equal to $P\,dV$ but is smaller. Furthermore, since the first law is

$$\delta Q_{irr} = dU + \delta W_{irr}$$

and the property relation is valid,

$$T\,dS = dU + P\,dV$$

it is found that

$$\delta W_{irr} = P\,dV - T\,\delta S_{gen} \tag{8.36}$$

showing that the work is reduced by an amount proportional to the entropy generation. For this reason the term $T\,\delta S_{gen}$ is often called "lost work," although it is not a real work or energy quantity lost but rather a lost opportunity to extract work.

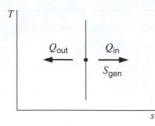

**FIGURE 8.16** Change of entropy due to heat transfer and entropy generation.

Equation 8.34 can be integrated between initial and final states to

$$S_2 - S_1 = \int_1^2 dS = \int_1^2 \frac{\delta Q}{T} + {}_1S_{2\,gen} \qquad (8.37)$$

Thus, we have an expression for the change of entropy for an irreversible process as an equality, whereas in the last section we had an inequality. In the limit of a reversible process, with a zero-entropy generation, the change in $S$ expressed in Eq. 8.37 becomes identical to Eq. 8.33 as the equal sign applies and the work term becomes $\int P\, dV$. Equation 8.37 is now the entropy balance equation for a control mass in the same form as the energy equation in Eq. 5.5, and it could include several subsystems. The equation can also be written in the general form

$$\Delta \text{ Entropy } = +\text{in} - \text{out} + \text{gen}$$

expressing that we can generate but not destroy entropy. This is in contrast to energy which we can neither generate nor destroy.

Some important conclusions can be drawn from Eqs. 8.34 to 8.37. First, there are two ways in which the entropy of a system can be increased—by transferring heat to it and by having an irreversible process. Since the entropy generation cannot be less than zero, there is only one way in which the entropy of a system can be decreased, and that is to transfer heat from the system. These changes are illustrated in a $T$–$s$ diagram in Fig. 8.16 showing the halfplane into which the state moves due to a heat transfer or an entropy generation.

Second, as we have already noted for an adiabatic process, $\delta Q = 0$, and therefore the increase in entropy is always associated with the irreversibilities.

Third, the presence of irreversibilities will cause the work to be smaller than the reversible work. This means less work out in an expansion process and more work into the control mass ($\delta W < 0$) in a compression process.

Finally, it should be emphasized that the change in $s$ associated with the heat transfer is a transfer across the control surface so a gain for the control volume is accompanied by a loss of the same magnitude outside the control volume. This is in contrast to the generation term that expresses all the entropy generated inside the control volume due to any irreversible process.

One other point concerning the representation of irreversible processes on $P$–$V$ and $T$–$S$ diagrams should be made. The work for an irreversible process is not equal to $\int P\, dV$, and the heat transfer is not equal to $\int T\, dS$. Therefore, the area underneath the path does not represent work and heat on the $P$–$V$ and $T$–$S$ diagrams, respectively. In fact, in many situations we are not certain of the exact state through which a system passes when it undergoes an irreversible process. For this reason it is advantageous to show irreversible processes as dashed lines and reversible processes as solid lines. Thus, the area underneath the dashed line will never represent work or heat. Fig. 8.17$a$ shows an irreversible process, and, because the heat transfer and work for this process are zero, the area underneath

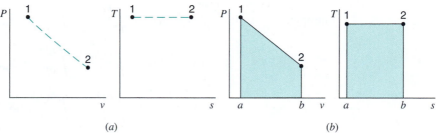

**FIGURE 8.17** Reversible and irreversible processes on pressure-volume and temperature-entropy diagrams.

the dashed line has no significance. Fig. 8.17$b$ shows the reversible process, and area 1–2–$b$–$a$–1 represents the work on the $P$–$V$ diagram and the heat transfer on the $T$–$S$ diagram.

## 8.11 PRINCIPLE OF THE INCREASE OF ENTROPY

In the previous section, we considered irreversible processes in which the irreversibilities occurred inside the system or control mass. We also found that the entropy change of a control mass could be either positive or negative, since entropy can be increased by internal entropy generation and either increased or decreased by heat transfer, depending on the direction of that transfer. Now we would like to emphasize the difference between the energy and entropy equations and point out that energy is conserved, but entropy is not.

Consider two mutually exclusive control volumes $A$ and $B$ with a common surface and their surroundings $C$ such that they collectively include the whole world. Let some processes take place so these control volumes exchange work and heat transfer as indicated in Fig. 8.18. Since a $Q$ or $W$ is transferred from one to another control volume, we only keep one symbol for each term and give the direction with the arrow. We will now write the energy and entropy equations for each control volume and then add them to see what the net effect is. As we write the equations, we do not try to memorize them, but just write them as

$$\text{Change} = +\text{in} - \text{out} + \text{generation}$$

and refer to the figure for the sign. We should know, however, that we cannot generate energy, but only entropy.

Energy:
$$(E_2 - E_1)_A = Q_a - W_a - Q_b + W_b$$
$$(E_2 - E_1)_B = Q_b - W_b - Q_c + W_c$$
$$(E_2 - E_1)_C = Q_c + W_a - Q_a - W_c$$

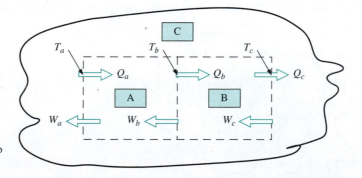

**FIGURE 8.18**
Total world divided into three control volumes.

Entropy:
$$(S_2 - S_1)_A = \int \frac{\delta Q_a}{T_a} - \int \frac{\delta Q_b}{T_b} + S_{\text{gen }A}$$

$$(S_2 - S_1)_B = \int \frac{\delta Q_b}{T_b} - \int \frac{\delta Q_c}{T_c} + S_{\text{gen }B}$$

$$(S_2 - S_1)_C = \int \frac{\delta Q_c}{T_c} - \int \frac{\delta Q_a}{T_a} + S_{\text{gen }C}$$

Now add all the energy equations to get the energy change for the total world

$$(E_2 - E_1)_{\text{total}} = (E_2 - E_1)_A + (E_2 - E_1)_B + (E_2 - E_1)_C$$

$$= Q_a - W_a - Q_b + W_b + Q_b - W_b - Q_c + W_c + Q_c + W_a - Q_a - W_c$$

$$= 0 \tag{8.38}$$

and we see that total energy has not changed, i.e., energy is conserved as all the right-hand side transfer terms pairwise cancel out. The energy is not stored in the same form or place as it was before the process, but the total amount is the same. For entropy we get something slightly different

$$(S_2 - S_1)_{\text{total}} = (S_2 - S_1)_A + (S_2 - S_1)_B + (S_2 - S_1)_C$$

$$= \int \frac{\delta Q_a}{T_a} - \int \frac{\delta Q_b}{T_b} + S_{\text{gen }A} + \int \frac{\delta Q_b}{T_b} - \int \frac{\delta Q_c}{T_c} + S_{\text{gen }B}$$

$$+ \int \frac{\delta Q_c}{T_c} - \int \frac{\delta Q_a}{T_a} + S_{\text{gen }C}$$

$$= S_{\text{gen }A} + S_{\text{gen }B} + S_{\text{gen }C} \geq 0 \tag{8.39}$$

where all the transfer terms cancel leaving only the positive entropy generation terms for each part of the total world. The total entropy increases and is then not conserved. Only if we have reversible processes in all parts of the world will the right-hand side become zero. This concept is referred to as the principle of the increase of entropy. Notice that if we add all the changes in entropy for the whole world from state 1 to state 2 we would get the total generation (increase), but we would not be able to specify where in the world the entropy was made. In order to get this more detailed information, we must make separate control volumes like $A$, $B$, and $C$ and thus also evaluate all the necessary transfer terms so we get the entropy generation by the balance of stored changes and transfers.

As an example of an irreversible process, consider a heat transfer process in which energy flows from a higher temperature domain to a lower temperature domain, as shown in Fig. 8.19. Let control volume $A$ be a control mass at temperature $T$ that receives a heat transfer of $\delta Q$ from a surrounding control volume $C$ at uniform temperature $T_0$. The

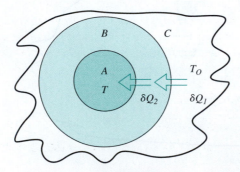

**FIGURE 8.19** The heat transfer through a wall.

transfer goes through the walls, control volume $B$, that separates the domains $A$ and $C$. Let us then analyze the incremental process from the point of view of control volume $B$, the walls, which do not have a change of state in time, but the state is non-uniform in space (it has $T_0$ on the outer side and $T$ on the inner side).

$$\text{Energy Eq.:} \qquad dE = 0 = \delta Q_1 - \delta Q_2 \Rightarrow \quad \delta Q_1 = \delta Q_2 = \delta Q$$

$$\text{Entropy Eq.:} \qquad dS = 0 = \frac{\delta Q}{T_0} - \frac{\delta Q}{T} + \delta S_{\text{gen }B}$$

So from the energy equation we find the two heat transfers to be the same, but realize that they take place at two different temperatures leading to an entropy generation as

$$\delta S_{\text{gen }B} = \frac{\delta Q}{T} - \frac{\delta Q}{T_0} = \delta Q \left( \frac{1}{T} - \frac{1}{T_0} \right) \geq 0 \tag{8.40}$$

Since $T_0 > T$ for the heat transfer to move in the indicated direction, we see that the entropy generation is positive. Suppose the temperatures were reversed so $T_0 < T$. Then the parenthesis would be negative; to have a positive entropy generation, $\delta Q$ must be negative, i.e. move in the opposite direction. The direction of the heat transfer from a higher to a lower temperature domain is thus a logical consequence of the second law.

The principle of the increase of entropy (total entropy generation), Eq. 8.39, is illustrated by the following example.

---

**EXAMPLE 8.8**

Suppose that 1 kg of saturated water vapor at 100°C is condensed to a saturated liquid at 100°C in a constant-pressure process by heat transfer to the surrounding air, which is at 25°C. What is the total increase in entropy of the water plus surroundings?

**Solution**

For the control mass (water), from the steam tables, we obtain

$$\Delta S_{\text{c.m.}} = -m s_{fg} = -1 \times 6.0480 = -6.0480 \text{ kJ/K}$$

Concerning the surroundings, we have

$$Q_{\text{to surroundings}} = m h_{fg} = 1 \times 2257.0 = 2257 \text{ kJ}$$

$$\Delta S_{\text{surr}} = \frac{Q}{T_0} = \frac{2257}{298.15} = 7.5700 \text{ kJ/K}$$

$$\Delta S_{\text{gen total}} = \Delta S_{\text{c.m.}} + \Delta S_{\text{surr}} = -6.0480 + 7.5700 = 1.5220 \text{ kJ/K}$$

This increase in entropy is in accordance with the principle of the increase of entropy and tells us, as does our experience, that this process can take place.

It is interesting to note how this heat transfer from the water to the surroundings might have taken place reversibly. Suppose that an engine operating on the Carnot cycle received heat from the water and rejected heat to the surroundings, as shown in Fig. 8.20. The decrease in the entropy of the water is equal to the increase in the entropy of the surroundings.

$$\Delta S_{\text{c.m.}} = -6.0480 \text{ kJ/K}$$

$$\Delta S_{\text{surr}} = 6.0480 \text{ kJ/K}$$

$$Q_{\text{to surroundings}} = T_0 \Delta S = 298.15(6.0480) = 1803.2 \text{ kJ}$$

$$W = Q_H - Q_L = 2257 - 1803.2 = 453.8 \text{ kJ}$$

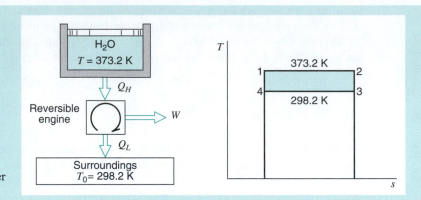

FIGURE 8.20
Reversible heat transfer with the surroundings.

Since this is a reversible cycle, the engine could be reversed and operated as a heat pump. For this cycle the work input to the heat pump would be 453.8 kJ.

## 8.12 Entropy as a Rate Equation

The second law of thermodynamics was used to write the balance of entropy in Eq. 8.34 for a variation and in Eq. 8.37 for a finite change. In some cases the equation is needed in a rate form so that a given process can be tracked in time. The rate form is also the basis for the development of the entropy balance equation in the general control volume analysis for an unsteady situation.

Take the incremental change in $S$ from Eq. 8.34 and divide by $\delta t$. We get

$$\frac{dS}{\delta t} = \frac{1}{T}\frac{\delta Q}{\delta t} + \frac{\delta S_{\text{gen}}}{\delta t} \qquad (8.41)$$

For a given control volume we may have more than one source of heat transfer, each at a certain surface temperature (semidistributed situation). Since we did not have to consider the temperature at which the heat transfer crossed the control surface for the energy equation, all the terms were added into a net heat transfer in a rate form in Eq. 5.31. Using this and a dot to indicate a rate, the final form for the entropy equation in the limit is

$$\frac{dS_{\text{c.m.}}}{dt} = \sum \frac{1}{T}\dot{Q} + \dot{S}_{\text{gen}} \qquad (8.42)$$

expressing the rate of entropy change as due to the flux of entropy into the control mass from heat transfer and an increase due to irreversible processes inside the control mass. If only reversible processes take place inside the control volume, the rate of change of entropy is determined by the rate of heat transfer divided by the temperature terms alone.

EXAMPLE 8.9

Consider an electric space heater that converts 1 kW of electric power into a heat flux of 1 kW delivered at 600 K from the hot wire surface. Let us look at the process of the energy conversion from electricity to heat transfer and find the rate of total entropy generation.

*Control mass*:   The electric heater wire.

*State*:   Constant wire temperature 600 K.

**Analysis**

The first and the second laws of thermodynamics in rate form become

$$\frac{dE_{\text{c.m.}}}{dt} = \frac{dU_{\text{c.m.}}}{dt} = 0 = \dot{W}_{\text{el.in}} - \dot{Q}_{\text{out}}$$

$$\frac{dS_{\text{c.m.}}}{dt} = 0 = -\dot{Q}_{\text{out}}/T_{\text{surface}} + \dot{S}_{\text{gen}}$$

Notice that we neglected kinetic and potential energy changes in going from a rate of $E$ to a rate of $U$. Then the left-hand side is zero since it is steady state and the right-hand side of the energy equation is electric work in minus the heat transfer out. For the entropy equation the left-hand side is zero because of steady state and the right-hand side has a flux of entropy out due to heat transfer, and entropy is generated in the wire.

**Solution**

We now get the entropy generation as

$$\dot{S}_{\text{gen}} = \dot{Q}_{\text{out}}/T = 1/600 = 0.001\,67 \text{ kW/K}$$

## 8.13 How-To Section

### How do I find a change in $s$ for a liquid or solid?

For a liquid or a solid, we have a constant volume ($v = $ constant) and the specific volume is also small. The entropy is then a function of temperature as in Eq. 8.11

$$s_2 - s_1 = C \ln\frac{T_2}{T_1}$$

### How do I find a change in $s$ for an ideal gas?

For an ideal gas, "$u$" and "$h$" are functions of $T$ only, but $s$ is a function of both $T$ and $P$ (or $T$ and $v$). Gibbs relations, Eqs. 8.7–8, and the knowledge of the standard entropy give

$$s_2 - s_1 = s_{T2}^0 - s_{T1}^0 - R \ln\frac{P_2}{P_1} \approx C_p \ln\frac{T_2}{T_1} - R \ln\frac{P_2}{P_1} = C_v \ln\frac{T_2}{T_1} + R \ln\frac{v_2}{v_1}$$

The first most accurate expression, Eq. 8.19 uses Table A.7, A.8, and the approximations. Eqs. 8.16–17 use an average specific heat $C_p$ or $C_v$ listed in Table A.5 for 25°C.

### Are the power relations for an isentropic process always valid?

The power relations in Eq. 8.28 for a polytropic process also covers the isentropic process ($n = k$) if we have constant specific heats. These come from Eqs. 8.16–17 and are shown in Eqs. 8.23–26. If the specific heat varies we can use them as approximations with an **average** value of the specific heat.

### How do I know if a process is reversible?

This is one of the more difficult questions and there is no direct positive answer. To answer it, we try to look for the opposite. Namely, is there any obvious irreversible process taking place inside the chosen *CV*. Is there any heat transfer between different temperature domains, any unrestrained expansions, mixing of masses with different states, electrical current through an ohmic resistor, flow through a restriction, and so forth? Check for the presence of any of the irreversible processes listed in Chapter 7 and if none are recognized, then assume a reversible process.

### How do I deal with an irreversible process?

In an irreversible process, entropy is produced and the only thing we can use the entropy equation for is to find the amount generated. You must therefore have information about the device/process that implicitly tells you how irreversible the process is. Knowing the beginning and end states (and/or the inlet and exit flow states) together with the fluxes of entropy ($\dot{Q}/T$) and energy ($\dot{Q}$, $\dot{W}$), you have all the information you need. Sometimes the information can be given in terms of an efficiency. In that case you need to calculate the ideal device/process first and then get the states and fluxes for the actual device. After that, as the last part, you may calculate the entropy generation in the actual process.

### How do I know which *T* to use in the entropy equation for the term $\dot{Q}/T$?

The term in Eq. 8.41 actually is a summation (integral) over the surface if not all of $\dot{Q}$ comes at the same *T*. The term is evaluated right at the control volume surface where $\dot{Q}$ crosses into the *CV* and you must use the absolute temperature. If you move the *CV* to place it somewhere else, such as from the inside to the outside of a window pane, often $\dot{Q}$ is the same but the temperature is different. There is thus a difference in the flux of *s* ($\dot{Q}/T$) at the two locations. The lower *T* location with the same heat transfer has a higher flux of *s* due to generation of *s* in the space where *T* dropped.

### How do I deal with a heat engine or refrigerator with varying $T_H$ or $T_L$?

If an energy supply/sink is finite such as a thermal storage rock bed or if it is a hot flow of a substance into a heat exchanger, the total heat transfer is not transferred at a single temperature but over a range of temperatures. We then cannot use the efficiency for a Carnot heat engine in Chapter 7, but we must formulate the second law for the control volume(s) involved as done in Chapter 8. Recall that Chapter 7 is the historical classical presentation of the second law, whereas Chapter 8 gives the modern version of it expressed in the entropy balance equation.

Suppose we have a finite source of energy, a constant volume mass at a state 1 being cooled to a state 2 as it gives out heat transfer to a reversible heat engine, as shown in Fig. 8.21. For simplicity, assume the heat engine has a fixed $T_L = T_0$. Take a *CV* that includes the mass and the heat engine and write the energy and entropy equations

$$E_2 - E_1 = m(u_2 - u_1) = -W_{HE} - Q_L$$
$$S_2 - S_1 = m(s_2 - s_1) = -Q_L/T_L \quad \Rightarrow \quad Q_L = -T_L m(s_2 - s_1)$$

Now solve for the work from the energy equation

$$W_{HE} = -Q_L - m(u_2 - u_1) = T_L m(s_2 - s_1) - m(u_2 - u_1)$$

With a *CV* around the heat engine we have a $Q_H = -m(u_2 - u_1)$ into it for the energy and an integral $\int dQ_H/T$ for the entropy, where the *T* is varying between $T_1$ and $T_2$. We

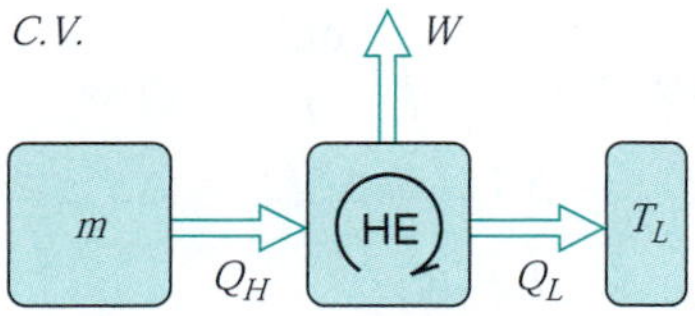

**FIGURE 8.21**

can not evaluate that integral by itself, but must relate it to the change of state in the mass, $\int dQ_H/T = s$ out of mass $= -m(s_2 - s_1)$ assuming a reversible process.

## SUMMARY

The inequality of Clausius and the property entropy ($s$) are modern statements of the second law. The final statement of the second law is the entropy balance equation that includes generation of entropy. All the results that were derived from the classical formulation of the second law in Chapter 7 can be re-derived with the entropy balance equation applied to the cyclic devices. For all reversible processes, entropy generation is zero and all real (irreversible) processes have a positive entropy generation. How large the entropy generation is depends on the actual process.

Thermodynamic property relations for $s$ are derived from consideration of a reversible process and leads to Gibbs relations. Changes in the property $s$ are covered through general tables, approximations for liquids and solids, as well as ideal gases. Changes of entropy in various processes are examined in general together with special cases of polytropic processes. Just as a reversible specific boundary work is the area below the process curve in a $P–v$ diagram, the reversible heat transfer is the area below the process curve in a $T–s$ diagram.

You should have learned a number of skills and acquired abilities from studying this chapter that will allow you to

- Know that Clausius inequality is an alternative statement of the second law.
- Know the relation between the entropy and the reversible heat transfer.
- Locate states in the tables involving entropy.
- Understand how a Carnot cycle looks in a $T–s$ diagram.
- Know how different simple process curves look in a $T–s$ diagram.
- Understand how to apply the entropy balance equation for a control mass.
- Recognize processes that generate entropy and where the entropy is made.
- Evaluate changes in $s$ for liquids, solids, and ideal gases.
- Know the various property relations for a polytropic process in an ideal gas.
- Know the application of the unsteady entropy equation and what a flux of $s$ is.

## KEY CONCEPTS AND FORMULAS

Clausius inequality $\qquad \displaystyle\int \frac{dQ}{T} \leq 0$

Entropy $\qquad ds = \dfrac{dq}{T} + ds_{\text{gen}}; \qquad ds_{\text{gen}} \geq 0$

Rate equation for entropy $\qquad \dot{S}_{\text{c.m.}} = \sum \dfrac{\dot{Q}_{\text{c.m.}}}{T} + \dot{S}_{\text{gen}}$

Entropy equation

$$m(s_2 - s_1) = \int_1^2 \frac{\delta Q}{T} + {}_1S_{2\,\text{gen}}; \qquad {}_1S_{2\,\text{gen}} \geq 0$$

Total entropy change

$$\Delta S_{\text{net}} = \Delta S_{\text{cm}} + \Delta S_{\text{surr}} = \Delta S_{\text{gen}} \geq 0$$

Lost work

$$W_{\text{lost}} = \int T\, dS_{\text{gen}}$$

Actual boundary work

$${}_1W_2 = \int P\, dV - W_{\text{lost}}$$

Gibbs relations

$$T\, ds = du + P\, dv$$
$$T\, ds = dh - v\, dP$$

## Solids, Liquids

$$v = \text{constant}, \qquad dv = 0$$

Change in $s$

$$s_2 - s_1 = \int \frac{du}{T} = \int C \frac{dT}{T} \approx C \ln \frac{T_2}{T_1}$$

## Ideal Gas

Standard entropy

$$s_T^0 = \int_{T_0}^T \frac{C_{p0}}{T}\, dT \qquad\qquad \text{(Function of } T)$$

Change in $s$

$$s_2 - s_1 = s_{T2}^0 - s_{T1}^0 - R \ln \frac{P_2}{P_1} \qquad \text{(Using Table A.7 or A.8)}$$

$$s_2 - s_1 = C_{p0} \ln \frac{T_2}{T_1} - R \ln \frac{P_2}{P_1} \qquad \text{(For constant } C_p, C_v)$$

$$s_2 - s_1 = C_{v0} \ln \frac{T_2}{T_1} + R \ln \frac{v_2}{v_1} \qquad \text{(For constant } C_p, C_v)$$

Ratio of specific heats

$$k = C_{p0}/C_{v0}$$

Polytropic processes

$$Pv^n = \text{constant}; \qquad PV^n = \text{constant}$$

$$\frac{P_2}{P_1} = \left(\frac{V_1}{V_2}\right)^n = \left(\frac{v_1}{v_2}\right)^n = \left(\frac{T_2}{T_1}\right)^{\frac{n}{n-1}}$$

$$\frac{T_2}{T_1} = \left(\frac{v_1}{v_2}\right)^{n-1} = \left(\frac{P_2}{P_1}\right)^{\frac{n-1}{n}}$$

$$\frac{v_2}{v_1} = \left(\frac{P_1}{P_2}\right)^{\frac{1}{n}} = \left(\frac{T_1}{T_2}\right)^{\frac{1}{n-1}}$$

Specific work

$${}_1w_2 = \frac{1}{1-n}(P_2v_2 - P_1v_1) = \frac{R}{1-n}(T_2 - T_1) \qquad n \neq 1$$

$${}_1w_2 = P_1v_1 \ln\frac{v_2}{v_1} = RT_1 \ln\frac{v_2}{v_1} = RT_1 \ln\frac{P_1}{P_2} \qquad n = 1$$

The work is moving boundary work $w = \displaystyle\int P\, dv$

Identifiable processes

$$n = 0; \qquad P = \text{constant}; \qquad \text{Isobaric}$$
$$n = 1; \qquad T = \text{constant}; \qquad \text{Isothermal}$$
$$n = k; \qquad s = \text{constant}; \qquad \text{Isentropic}$$
$$n = \infty; \qquad v = \text{constant}; \qquad \text{Isochoric or isometric}$$

# HOMEWORK PROBLEMS

## Concept problems

**8.1** When a substance has completed a cycle, $v$, $u$, $h$, and $s$ are unchanged. Did anything happen? Explain.

**8.2** Water at 100°C, quality 50% in a rigid box is heated to 110°C. How do the properties ($P$, $v$, $x$, $u$, and $s$) change? (increase, stay about the same, or decrease)

**8.3** Liquid water at 20°C, 100 kPa is compressed in a piston/cylinder without any heat transfer to a pressure of 200 kPa. How do the properties ($T$, $v$, $u$, and $s$) change? (increase, stay about the same, or decrease)

**8.4** A reversible process in a piston/cylinder is shown in Fig. P8.4. Indicate the storage change $u_2 - u_1$ and transfers $_1w_2$ and $_1q_2$ as positive, zero, or negative

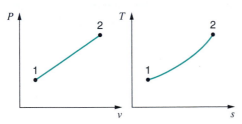

**FIGURE P8.4**

**8.5** A reversible process in a piston/cylinder is shown in Fig. P8.5. Indicate the storage change $u_2 - u_1$ and transfers $_1w_2$ and $_1q_2$ as positive, zero, or negative

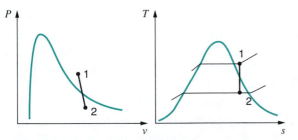

**FIGURE P8.5**

**8.6** Air at 290 K, 100 kPa in a rigid box is heated to 325 K. How do the properties ($P$, $v$, $u$, and $s$) change? (increase, stay about the same, or decrease)

**8.7** Air at 20°C, 100 kPa is compressed in a piston/cylinder without any heat transfer to a pressure of 200 kPa. How do the properties ($T$, $v$, $u$, and $s$) change? (increase, stay about the same, or decrease)

**8.8** Carbon dioxide is compressed to a smaller volume in a polytropic process with $n = 1.4$. How do the properties ($u$, $h$, $s$, $P$, $T$) change (up, down, or constant)?

**8.9** Process A: Air at 300 K, 100 kPa is heated to 310 K at constant pressure. Use the table below to compare this to process B: Heat air at 1300 K to 1310 K at constant 100 kPa. How do the property changes compare?

| Property | $\Delta_A > \Delta_B$ | $\Delta_A \approx \Delta_B$ | $\Delta_A < \Delta_B$ |
|---|---|---|---|
| $\Delta = v_2 - v_1$ | | | |
| $\Delta = h_2 - h_1$ | | | |
| $\Delta = s_2 - s_1$ | | | |

**8.10** A reversible heat pump has a flux of $s$ entering as $\dot{Q}_L/T_L$. What can you say about the exit flux of $s$ at $T_H$?

**8.11** An electric baseboard heater receives 1500 W of electrical power that heats the room air which looses the same amount out through the walls and windows. Specify exactly where entropy is generated in that process.

## Inequality of Clausius

**8.12** Consider the steam power plant in Example 6.9 and assume an average $T$ in the line between 1 and 2. Show that this cycle satisfies the inequality of Clausius.

**8.13** Assume the heat engine in Problem 7.21 has a high temperature of 1200 K and a low temperature of 400 K. What does the inequality of Clausius say about each of the four cases?

**8.14** Let the steam power plant in Problem 7.17 have 700°C in the boiler and 40°C during the heat rejection in the condenser. Does that satisfy the inequality of Clausius? Repeat the question for the cycle operated in reverse as a refrigerator.

**8.15** A heat engine receives 6 kW from a 250°C source and rejects heat at 30°C. Examine each of three cases with respect to the inequality of Clausius.

a. $\dot{W} = 6$ kW

b. $\dot{W} = 0$ kW

c. Carnot cycle

## Entropy of a pure substance

**8.16** Find the missing properties and give the phase of the substance.

a. $H_2O$    $s = 7.70$ kJ/kg K,    $h = ?$ $T = ?$ $x = ?$
   $P = 25$ kPa

b. $H_2O$    $u = 3400$ kJ/kg,    $T = ?$ $x = ?$ $s = ?$
   $P = 10$ MPa

**8.17** Find the missing properties and give the phase of the ammonia, $NH_3$.

a. $T = 65°C, P = 600$ kPa     $s = ?$   $v = ?$

b. $T = 20°C, P = 100$ kPa     $s = ?$   $v = ? x = ?$

c. $T = 50°C, v = 0.1185$ m³/kg   $s = ?$   $P = ? x = ?$

**8.18** Find the entropy for the following water states and indicate each state on a $T$–$s$ diagram relative to the two-phase region.

a. $250°C, v = 0.02$ m³/kg

b. $250°C, 2000$ kPa

c. $-2°C, 100$ kPa

**8.19** Find the missing properties and give the phase of the substance.

a. R-12     $T = 0°C,$        $s = ? x = ?$
            $P = 200$ kPa

b. R-134a   $T = -10°C,$      $v = ? s = ?$
            $x = 0.45$

c. $NH_3$    $T = 20°C,$       $u = ? x = ?$
            $s = 5.50$ kJ/kg K

**8.20** Saturated liquid water at $20°C$ is compressed to a higher pressure with constant temperature. Find the changes in $u$ and $s$ when the final pressure is

a. 500 kPa      b. 2000 kPa      c. 20 000 kPa

**8.21** Saturated vapor water at $150°C$ is expanded to a lower pressure with constant temperature. Find the changes in $u$ and $s$ when the final pressure is

a. 100 kPa      b. 50 kPa      c. 10 kPa

**8.22** Two kg water at $120°C$ with a quality of 25% has its temperature raised $20°C$ in a constant volume process. What are the new quality and specific entropy?

**8.23** Two kg water at 200 kPa with a quality of 25% has its temperature raised $20°C$ in a constant pressure process. What is the change in entropy?

**8.24** Determine the missing property among $P, T, s,$ and $x$ for the following states:

a. R-134a    $5°C, s = 1.7$ kJ/kg K

b. R-134a    $50°C, s = 1.9$ kJ/kg K

c. R-22      $100$ kPa, $v = 0.3$ m³/kg

## Reversible processes

**8.25** Consider a Carnot-cycle heat engine with water as the working fluid. The heat transfer to the water occurs at $300°C$, during which process the water changes from saturated liquid to saturated vapor. The heat is rejected from the water at $40°C$. Show the cycle on a $T$–$s$ diagram and find the quality of the water at the beginning and the end

of the heat rejection process. Determine the net work output per kg water and the cycle thermal efficiency.

**8.26** In a Carnot engine with ammonia as the working fluid the high temperature is $T_H = 60°C$, and as $Q_H$ is received the ammonia changes from saturated liquid to saturated vapor. The ammonia pressure at the low temperature is $P_{low} = 190$ kPa. Find $T_L$, the cycle thermal efficiency, the heat added per kg, and the entropy, $s$, at the beginning of the heat rejection process.

**8.27** Water is used as the working fluid in a Carnot-cycle heat engine, where it changes from saturated liquid to saturated vapor at $200°C$ as heat is added. Heat is rejected in a constant-pressure process (also constant $T$) at 20 kPa. The heat engine powers a Carnot-cycle refrigerator that operates between $-15°C$ and $+20°C$, shown in Fig. P8.27. Find the heat added to the water per kg water. How much heat should be added to the water in the heat engine so the refrigerator can remove 1 kJ from the cold space?

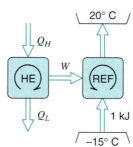

**FIGURE P8.27**

**8.28** Consider a Carnot-cycle heat pump with R-22 as the working fluid. Heat is rejected from the R-22 at $40°C$, during which process the R-22 changes from saturated vapor to saturated liquid. The heat is transferred to the R-22 at $-5°C$.

a. Show the cycle on a $T$–$s$ diagram.

b. Find the quality of the R-22 at the beginning and end of the isothermal heat addition process at $-5°C$.

c. Determine the coefficient of performance for the cycle.

**8.29** Do Problem 8.28 using refrigerant R-134a instead of R-22.

**8.30** Ammonia at 1 MPa and $50°C$ is expanded in a piston cylinder to 500 kPa and $20°C$ in a reversible process. Find the sign for both the work and the heat transfer.

**8.31** Water at 200 kPa with $x = 1.0$ is compressed in a piston cylinder to 1 MPa and $350°C$ in a reversible process. Find the sign for the work and the sign for the heat transfer.

**8.32** Water at 200 kPa with $x = 1.0$ is compressed in a piston cylinder to 1 MPa and 250°C in a reversible process. Find the sign for the work and the sign for the heat transfer.

**8.33** A piston cylinder maintaining constant pressure contains 0.1 kg saturated liquid water at 100°C. It is now boiled to become saturated vapor in a reversible process. Find the work term and then the heat transfer from the energy equation. Find the heat transfer from the entropy equation. Is it the same?

**8.34** One kilogram of ammonia in a piston cylinder at 50°C and 1000 kPa is expanded in a reversible isobaric process to 140°C, shown in Fig. P8.34. Find the work and heat transfer for this process.

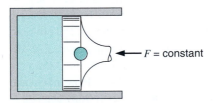

**FIGURE P8.34**

**8.35** One kilogram of ammonia in a piston cylinder at 50°C and 1000 kPa is expanded in a reversible isothermal process to 100 kPa. Find the work and heat transfer for this process.

**8.36** Saturated water vapor at 200 kPa is compressed to 600 kPa in a reversible adiabatic process. Find the change in $v$ and $T$.

**8.37** Compression and heat transfer brings R-134a in a piston cylinder from 500 kPa and 50°C to saturated vapor in an isothermal process. Find the specific heat transfer and the specific work.

**8.38** One kilogram of ammonia in a piston cylinder at 50°C and 1000 kPa is expanded in a reversible adiabatic process to 100 kPa, shown in Fig. P8.38. Find the work and heat transfer for this process.

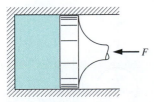

**FIGURE P8.38**

**8.39** A piston cylinder contains 0.25 kg of R-134a at 100 kPa. It will be compressed in an adiabatic reversible process to 400 kPa and should be 70°C. What should the initial temperature be?

**8.40** An insulated cylinder fitted with a piston contains 0.1 kg of water at 100°C with 90% quality. The piston is moved, compressing the water until it reaches a pressure of 1.2 MPa. How much work is required in the process?

**8.41** One kilogram of water at 300°C expands against a piston in a cylinder until it reaches ambient pressure, 100 kPa, at which point the water has a quality of 90.2%. It may be assumed that the expansion is reversible and adiabatic. What was the initial pressure in the cylinder and how much work is done by the water?

**8.42** Water at 1000 kPa and 250°C is brought to saturated vapor in a rigid container, shown in Fig. P8.42. Find the final $T$ and the specific heat transfer in this isometric process.

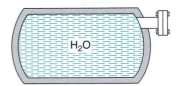

**FIGURE P8.42**

**8.43** Estimate the specific heat transfer from the area in the $T$–$s$ diagram and compare it to the correct value for the states and process in Problem 8.42.

**8.44** A cylinder containing R-134a at 10°C and 150 kPa has an initial volume of 20 L. A piston compresses the R-134a in a reversible, isothermal process until it reaches the saturated vapor state. Calculate the required work and heat transfer to accomplish this process.

**8.45** A piston cylinder contains 0.25 kg of R-134a at 100 kPa, −20°C. It is compressed in an adiabatic reversible process to 400 kPa. How much work is needed?

**8.46** Water in a piston/cylinder device at 400°C and 2000 kPa is expanded in a reversible adiabatic process. The specific work is measured to be 415.72 kJ/kg out. Find the final $P$ and $T$ and show the $P$–$v$ and the $T$–$s$ diagrams for the process.

**8.47** A piston cylinder with 2 kg ammonia at 50°C and 100 kPa is compressed to 1000 kPa. The process happens so slowly that the temperature is constant. Find the heat transfer and the work for the process assuming it to be reversible.

**8.48** A piston/cylinder device with 2 kg water at 1000 kPa and 250°C is cooled with a constant loading on the piston. This isobaric process ends when the water has reached a state of saturated liquid. Find the work and heat transfer and sketch the process in both a $P$–$v$ and a $T$–$s$ diagram.

**8.49** Water at 1000 kPa and 250°C is brought to saturated vapor in a piston/cylinder assembly with an isothermal process. Find the specific work and heat transfer. Estimate the specific work from the area in the P–v diagram and compare it to the correct value.

**8.50** A piston cylinder with R-134a at −20°C and 100 kPa is compressed to 500 kPa in a reversible adiabatic process. Find the final temperature and the specific work.

**8.51** A heavily insulated cylinder/piston contains ammonia at 1200 kPa, 60°C. The piston is moved, expanding the ammonia in a reversible process until the temperature is −20°C. During the process 600 kJ of work is given out by the ammonia. What was the initial volume of the cylinder?

**8.52** A rigid, insulated vessel contains superheated vapor steam at 3 MPa, 400°C. A valve on the vessel is opened, allowing steam to escape as shown in Fig. P8.52. The overall process is irreversible, but the steam remaining inside the vessel goes through a reversible adiabatic expansion. Determine the fraction of steam that has escaped when the final state inside is saturated vapor.

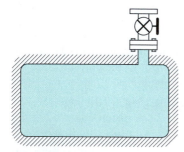

**FIGURE P8.52**

**8.53** A piston/cylinder setup contains 2 kg of water at 200°C and 10 MPa. The piston is slowly moved to expand the water in an isothermal process to a pressure of 200 kPa. Heat transfer takes place with an ambient surrounding at 200°C, and the whole process may be assumed reversible. Sketch the process in a P–V diagram and calculate both the heat transfer and the total work.

## Entropy of a liquid or solid

**8.54** A piston/cylinder setup has constant pressure of 2000 kPa with water at 20°C. It is now heated up to 100°C. Find the heat transfer and the entropy change using the steam tables. Repeat the calculation using constant heat capacity and incompressibility.

**8.55** A large slab of concrete, 5 × 8 × 0.3 m, is used as a thermal storage mass in a solar-heated house. If the slab cools overnight from 23°C to18°C, what is the change in entropy of the concrete slab?

**8.56** Two 5 kg blocks of steel, one at 250°C the other at 25°C, come in thermal contact. Find the final temperature and the change in the entropy of the steel.

**8.57** In a sink, 5 liters of water at 70°C are combined with 1 kg aluminum pots; 1 kg of steel flatware and 1 kg of glass all put in at 20°C. What is the final uniform temperature and the change in stored entropy neglecting any heat loss and work?

**8.58** A foundry form box with 25 kg of 200°C hot sand is dumped into a bucket with 50 L of water at 15°C. Assuming no heat transfer with the surroundings and no boiling away of liquid water, calculate the net entropy change for the process.

**8.59** A 4 L jug of milk at 25°C is placed in your refrigerator where it is cooled down to the refrigerator's inside constant temperature of 5°C. Assume the milk has the property of liquid water and find the entropy change of the milk.

**8.60** A 12-kg steel container has 0.2 kg of superheated water vapor at 1000 kPa and 200°C. The total mass is now cooled to the ambient temperature of 30°C. How much heat transfer occurs? Find the steel and water entropy change.

**8.61** A computer CPU chip consists of 50 g silicon, 20 g copper, and 50 g polyvinyl chloride (plastic). It now heats from 15°C to 70°C as the computer is turned on. How much did the entropy increase?

**8.62** A 5 kg steel container is cured at 500°C. An amount of liquid water at 15°C, 100 kPa is added to the container so a final uniform temperature of the steel and the water becomes 75°C. Neglect any water that might evaporate during the process and any air in the container. How much water should be added and how much was the entropy changed?

**8.63** A pan in an autoshop contains 5 L of engine oil at 20°C, 100 kPa. Now 2 L of hot 100°C oil is mixed into the pan. Neglect any work term and find the final temperature and the change in entropy.

**8.64** Find the total work the heat engine can give out as it receives energy from the rock bed as described in Problem 7.61 (see Fig. P8.64). *Hint:* Write the entropy balance

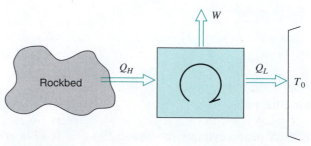

**FIGURE P8.64**

equation for the control volume that is the combination of the rockbed and the heat engine.

**8.65** Two kg of liquid lead initially at 500°C are poured into a form. It then cools at constant pressure down to room temperature of 20°C as heat is transferred to the room. The melting point of lead is 327°C, and the enthalpy change between the phases, $h_{if}$, is 24.6 kJ/kg. The specific heats are in Tables A.3 and A.4. Calculate the net entropy change for this process.

## Entropy of ideal gases

**8.66** Water at 400 kPa is brought from 150°C to 1200°C in a constant pressure process. Evaluate the change in specific entropy using

a. the steam tables

b. the ideal gas Table A.8

c. the specific heat Table A.5

**8.67** Air inside a rigid tank is heated from 300 to 350 K. Find the entropy increase $s_2 - s_1$. What if it is heated from 1300 to 1350 K?

**8.68** Consider a Carnot-cycle heat pump having 1 kg of nitrogen gas in a cylinder/piston arrangement. This heat pump operates between reservoirs at 300 K and 400 K. At the beginning of the low-temperature heat addition, the pressure is 1 MPa. During this process the volume triples. Analyze each of the four processes in the cycle and determine

a. the pressure, volume, and temperature at each point.

b. the work and heat transfer for each process.

**8.69** Oxygen gas in a piston/cylinder assembly at 300 K and 100 kPa with volume 0.1 m³ is compressed in a reversible adiabatic process to a final temperature of 700 K. Find the final pressure and volume using Table A.5.

**8.70** Oxygen gas in a piston/cylinder device at 300 K and 100 kPa with volume 0.1 m³ is compressed in a reversible adiabatic process to a final temperature of 700 K. Find the final pressure and volume using Table A.8.

**8.71** A rigid tank contains 1 kg methane at 500 K, 1500 kPa. It is now cooled down to 300 K. Find the heat transfer and the change in entropy using ideal gas.

**8.72** A mass of 1 kg of air contained in a cylinder at 1.5 MPa and 1000 K expands in a reversible isothermal process to a volume 10 times larger. Calculate the heat transfer during the process and the change of entropy of the air.

**8.73** Consider a small air pistol with a cylinder volume of 1 cm³ at 250 kPa and 27°C. The bullet acts as a piston initially held by a trigger, shown in Fig. P8.73. The

bullet is released so that the air expands in an adiabatic process. If the pressure should be 100 kPa as the bullet leaves the cylinder, find the final volume and the work done by the air.

**FIGURE P8.73**

**8.74** A hydrogen gas in a piston/cylinder assembly at 280 K, 100 kPa with a volume of 0.1 m³ is now compressed to a volume of 0.01 m³ in a reversible adiabatic process. What is the new temperature and how much work is required?

**8.75** An insulated cylinder/piston setup contains carbon dioxide gas at 120 kPa and 400 K. The gas is compressed to 2.5 MPa in a reversible adiabatic process. Calculate the final temperature and the work per unit mass, assuming

a. Variable specific heat (Table A.8).

b. Constant specific heat (value from Table A.5).

**8.76** A hydrogen gas in a piston/cylinder assembly at 300 K, 100 kPa with a volume of 0.1 m³ is now slowly compressed to a volume of 0.01 m³ while being cooled in a reversible isothermal process. What is the final pressure, the heat transfer, and the work required?

**8.77** Two rigid tanks shown in Fig. P8.77 each contain 10 kg of $N_2$ gas at 1000 K and 500 kPa. They are now thermally connected to a reversible heat pump, which heats one and cools the other with no heat transfer to the surroundings. When one tank is heated to 1500 K the process stops. Find the final $(P, T)$ in both tanks and the work input to the heat pump, assuming constant heat capacities.

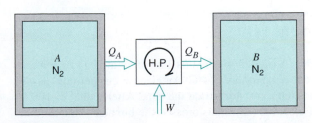

**FIGURE P8.77**

**8.78** We wish to obtain a supply of cold helium gas by applying the following technique. Helium contained in a cylinder at ambient conditions, 100 kPa and 20°C, is compressed in a reversible isothermal process to 600 kPa, after which the gas is expanded back to 100 kPa in a reversible adiabatic process.

a. Show the process on a *T–s* diagram.

b. Calculate the final temperature and the net work per kilogram of helium.

**8.79** A hand-held pump for a bicycle (see Fig. P8.79) has a volume of 25 cm$^3$ when fully extended. You now press the plunger (piston) in while holding your thumb over the exit hole so that an air pressure of 300 kPa is obtained. The outside atmosphere is at $P_0$ and $T_0$. Consider two cases: (1) It is done quickly (~1 s) and (2) it is done very slowly (~1 h).

a. State assumptions about the process for each case.

b. Find the final volume and temperature for both cases.

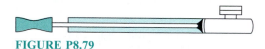

**FIGURE P8.79**

**8.80** A 1 m$^3$ insulated, rigid tank contains air at 800 kPa and 25°C. A valve on the tank is opened, and the pressure inside quickly drops to 150 kPa, at which point the valve is closed. Assuming that the air remaining inside has undergone a reversible adiabatic expansion, calculate the mass withdrawn during the process.

**8.81** A piston/cylinder assembly shown in Fig. P8.81, contains air at 1380 K and 15 MPa, with $V_1 = 10$ cm$^3$ and $A_{cyl} = 5$ cm$^2$. The piston is released, and just before the piston exits the end of the cylinder, the pressure inside is 200 kPa. If the cylinder is insulated, what is its length? How much work is done by the air inside?

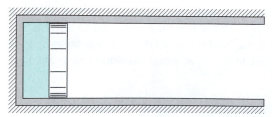

**FIGURE P8.81**

**8.82** A rigid tank contains 2 kg of air at 200 kPa and ambient temperature, 20°C. An electric current now passes through a resistor inside the tank. After a total of 100 kJ of electrical work has crossed the boundary, the air temperature inside is 80°C. Is this possible?

## Polytropic processes

**8.83** A piston/cylinder contains air at 300 K, 100 kPa. A reversible polytropic process with $n = 1.3$ brings the air to 500 K. Sketch the process in a *P–v* and a *T–s* diagram. Find the specific work and specific heat transfer in the process.

**8.84** Neon at 400 kPa and 20°C is brought to 100°C in a polytropic process with $n = 1.4$. Give the sign for the heat transfer and work terms and explain.

**8.85** A mass of 1 kg of air contained in a cylinder at 1.5 MPa and 1000 K expands in a reversible adiabatic process to 100 kPa. Calculate the final temperature and the work done during the process, using

a. Constant specific heat (value from Table A.5).

b. The ideal gas tables (Table A.7).

**8.86** Hot combustion air at 1500 K expands in a polytropic process to a volume 6 times as large with $n = 1.5$. Find the specific boundary work and the specific heat transfer.

**8.87** An ideal gas having a constant specific heat undergoes a reversible polytropic expansion with exponent $n = 1.4$. If the gas is carbon dioxide, will the heat transfer for this process be positive, negative, or zero?

**8.88** A cylinder/piston contains 1 kg methane gas at 100 kPa, 20°C. The gas is compressed reversibly to a pressure of 800 kPa. Calculate the work required if the process is adiabatic.

**8.89** Helium in a piston/cylinder assembly at 20°C and 100 kPa is brought to 400 K in a reversible polytropic process with exponent $n = 1.25$. You may assume helium is an ideal gas with constant specific heat. Find the final pressure and both the specific heat transfer and specific work.

**8.90** The power stroke in an internal combustion engine can be approximated with a polytropic expansion. Consider air in a cylinder volume of 0.2 L at 7 MPa and 1800 K, shown in Fig. P8.90. It now expands in a reversible polytropic process with exponent $n = 1.5$, through a volume ratio of 8:1. Show this process on *P–v* and *T–s* diagrams, and calculate the work and heat transfer for the process.

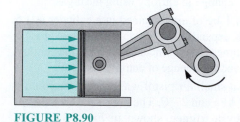

**FIGURE P8.90**

**8.91** A piston/cylinder contains air at 300 K, 100 kPa. It is now compressed in a reversible adiabatic process to a volume 7 times as small. Use constant heat capacity and find the final pressure and temperature, the specific work, and specific heat transfer for the process.

**8.92** A cylinder/piston contains carbon dioxide at 1 MPa, 300°C with a volume of 200 L. The total external force acting on the piston is proportional to $V^3$. This system is allowed to cool to room temperature, 20°C. What is the work and heat transfer for the process?

**8.93** A mass of 2 kg ethane gas at 500 kPa, 100°C, undergoes a reversible polytropic expansion with exponent, $n = 1.3$, to a final temperature of 20°C. Calculate the work and heat transfer for the process.

**8.94** A cylinder/piston contains saturated vapor R-22 at 10°C; the volume is 10 L. The R-22 is compressed to 2 MPa, 60°C in a reversible polytropic process. Find the polytropic exponent $n$ and calculate the work and heat transfer.

## Entropy generation

**8.95** One kg of water at 500°C and 1 kg of saturated water vapor, both at 200 kPa, are mixed in a constant-pressure and adiabatic process. Find the final temperature and the entropy generation for the process.

**8.96** A computer chip dissipates 2 kJ of electric work over time and rejects that as heat transfer from its 50°C surface to 25°C air. How much entropy is generated in the chip? How much, if any, is generated outside the chip?

**8.97** An insulated cylinder/piston arrangement contains R-134a at 1 MPa and 50°C, with a volume of 100 L. The R-134a expands, moving the piston until the pressure in the cylinder has dropped to 100 kPa. It is claimed that the R-134a does 190 kJ of work against the piston during the process. Is that possible?

**8.98** A car uses an average power of 25 hp for a one-hour round trip. With a thermal efficiency of 35%, how much fuel energy was used? What happened to all the energy? What change in entropy took place if we assume ambient at 20°C?

**8.99** A mass- and atmosphere-loaded piston/cylinder device contains 2 kg of water at 5 MPa and 100°C. Heat is added from a reservoir at 700°C to the water until it reaches 700°C. Find the work, heat transfer, and total entropy production for the system and surroundings.

**8.100** A piston/cylinder setup has 2.5 kg of ammonia at 50 kPa and −20°C. Now it is heated to 50°C at constant pressure through the bottom of the cylinder from external hot gas at 200°C. Find the heat transfer to the ammonia and the total entropy generation.

**8.101** Two 5 kg blocks of steel, one at 250°C the other at 25°C, come in thermal contact. Find the final temperature and the total entropy generation in the process.

**8.102** A cylinder fitted with a movable piston contains water at 3 MPa with 50% quality, at which point the volume is 20 L. The water now expands to 1.2 MPa as a result of receiving 600 kJ of heat from a large source at 300°C. It is claimed that the water does 124 kJ of work during this process. Is this possible?

**8.103** A cylinder/piston assembly contains water at 200 kPa and 200°C with a volume of 20 L. The piston is moved slowly, compressing the water to a pressure of 800 kPa. The loading on the piston is such that the product $PV$ is a constant. Assuming that the room temperature is 20°C, show that this process does not violate the second law.

**8.104** A piston/cylinder device keeping a constant pressure of 500 kPa has 1 kg of water at 20°C and 1 kg of water at 100°C separated by a membrane, shown in Fig. P8.104. The membrane is broken and the water comes to a uniform state with no external heat transfer. Find the final temperature and the entropy generation.

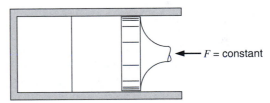

**FIGURE P8.104**

**8.105** A piston/cylinder has ammonia at 2000 kPa, 80°C with a volume of 0.1 m$^3$. The piston is loaded with a linear spring, and the outside ambient is at 20°C, shown in Fig. P8.105. The ammonia now cools down to 20°C at which point it has a quality of 10%. Find the work, heat transfer, and total entropy generation in the process.

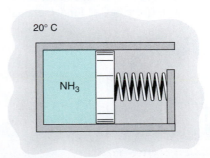

**FIGURE P8.105**

**8.106** A piston/cylinder device loaded so it gives constant pressure has 0.75 kg of saturated vapor water at 200 kPa. It is now cooled so that the volume becomes half the initial volume by heat transfer to the ambient surroundings at 20°C. Find the work, heat transfer, and total entropy generation.

**8.107** One kg of air at 300 K is mixed with one kg air at 400 K in a process at a constant 100 kPa and $Q = 0$. Find the final $T$ and the entropy generation in the process.

**8.108** One kg of carbon dioxide at 100 kPa, 500 K is mixed with two kg carbon dioxide at 200 kPa, 2000 K, in a rigid insulated tank. Find the final state $(P, T)$ and the entropy generation in the process using constant heat capacity from Table A.5.

**8.109** One kg of carbon dioxide at 100 kPa, 500 K is mixed with two kg carbon dioxide at 200 kPa, 2000 K, in a rigid insulated tank. Find the final state $(P, T)$ and the entropy generation in the process using Table A. 8.

**8.110** A piston/cylinder contains 1.5 kg of air at 300 K and 150 kPa. It is now heated up in a two-step process from a source at 1600 K: first, constant volume to 1000 K (state 2) and then a constant pressure process to 1500 K, state 3. Find the heat transfer and the entropy generation for the process.

**8.111** A spring-loaded piston/cylinder setup contains 1.5 kg of air at 27°C and 160 kPa. It is now heated in a process where pressure is linear in volume, $P = A + BV$, to twice the initial volume where it reaches 900 K. Find the work, heat transfer, and total entropy generation assuming a source at 900 K.

**8.112** Argon in a light bulb is at 90 kPa and heated from 20°C to 60°C with electrical power. Do not consider any radiation, or the glass mass. Find the total entropy generation per unit mass of argon.

**8.113** Nitrogen at 600 kPa and 127°C is in a 0.5 m³ insulated tank connected to a pipe with a valve to a second insulated initially empty tank with a volume of 0.5 m³, shown in Fig. P8.113. The valve is opened, and the nitrogen fills both tanks at a uniform state. Find the final

pressure and temperature and the entropy generation this process causes. Why is the process irreversible?

**8.114** A cylinder/piston device contains carbon dioxide at 1 MPa and 300°C with a volume of 200 L. The total external force acting on the piston is proportional to $V^3$. This system is allowed to cool to room temperature, 20°C. What is the total entropy generation for the process?

**8.115** A mass of 2 kg of ethane gas at 500 kPa and 100°C undergoes a reversible polytropic expansion with exponent $n = 1.3$ to a final temperature of the ambient surroundings, 20°C. Calculate the total entropy generation for the process if the heat is exchanged with the ambient surroundings.

**8.116** A cylinder/piston device contains saturated vapor R-22 at 10°C; the volume is 10 L. The R-22 is compressed to 2 MPa at 60°C in a reversible (internally) polytropic process. If all the heat transfer during the process is with the ambient surroundings at 10°C, calculate the net entropy change.

**8.117** A 5-kg aluminum radiator holds 2 kg of liquid R-134a at −10°C. The setup is brought indoors and heated with 220 kJ from a heat source at 100°C. Find the total entropy generation for the process assuming the R-134a remains a liquid.

## Rates or fluxes of entropy

**8.118** A mass of 3 kg nitrogen gas at 2000 K, $V = C$, cools with 500 W. What is $dS/dt$?

**8.119** A window receives 200 W of heat transfer at the inside surface of 20°C and transmits the 200 W from its outside surface at 2°C continuing to ambient air at −5°C. Find the flux of entropy at all three surfaces and the window's rate of entropy generation.

**8.120** A reversible heat pump uses 1 kW of power input to heat a 25°C room, drawing energy from the outside at 15°C. Assuming every process is reversible, what are the total rates of entropy into the heat pump from the outside and from the heat pump to the room?

**8.121** A car engine block receives 2 kW at its surface of 450 K from hot combustion gases at 1500 K. Near the cooling channel the engine block transmits 2 kW out at its 400 K surface to the coolant flowing at 370 K. Finally, in the radiator the coolant at 350 K delivers the 2 kW to air which is at 25°C. Find the rate of entropy generation inside the engine block, inside the coolant, and in the radiator/air combination.

**8.122** Room air at 23°C is heated by a 2000 W space heater with a surface filament temperature of 700 K,

shown in Fig. P8.122. The room at steady state loses heat to the outside, which is at 7°C. Find the rate(s) of entropy generation and specify where it is made.

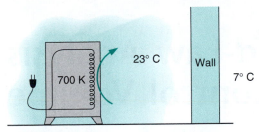

23° C

Wall

700 K

7° C

**FIGURE P8.122**

**8.123** The automatic transmission in a car receives 25 kW shaft work and gives out 24 kW to the drive shaft. The balance is dissipated in the hydraulic fluid and metal casing, all at 45°C, which in turn transmits it to the outer atmosphere at 20°C. What is the rate of entropy generation inside the transmission unit? What is it outside the unit?

# Chapter 9

# Second-Law Analysis for a Control Volume

## TOPICS DISCUSSED

THE SECOND LAW OF THERMODYNAMICS FOR A
CONTROL VOLUME
THE STEADY-STATE PROCESS AND THE TRANSIENT PROCESS
THE STEADY-STATE SINGLE FLOW PROCESS
PRINCIPLE OF THE INCREASE OF ENTROPY
EFFICIENCY
AVAILABLE ENERGY, REVERSIBLE WORK,
AND IRREVERSIBILITY
AVAILABILITY AND SECOND-LAW EFFICIENCY
SOME GENERAL COMMENTS ABOUT ENTROPY AND CHAOS
HOW-TO SECTION

In the preceding two chapters we discussed the second law of thermodynamics and the thermodynamic property entropy. As was done with the first-law analysis, we now consider the more general application of these concepts, the control volume analysis, and a number of cases of special interest. We will also discuss usual definitions of thermodynamic efficiencies.

## 9.1 THE SECOND LAW OF THERMODYNAMICS FOR A CONTROL VOLUME

The second law of thermodynamics can be applied to a control volume by a procedure similar to that used in Section 6.1, where the first law was developed for a control volume.

254

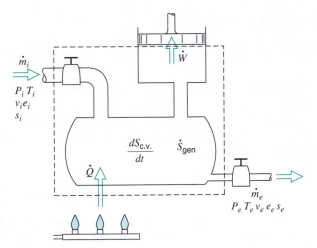

**FIGURE 9.1** The entropy balance for a control volume on a rate form.

We start with the second law expressed as a change of the entropy for a control mass in a rate form from Eq. 8.43,

$$\frac{dS_{\text{c.m.}}}{dt} = \sum \frac{\dot{Q}}{T} + \dot{S}_{\text{gen}} \tag{9.1}$$

to which we now will add the contributions from the mass flow rates in and out of the control volume, A simple example of such a situation is illustrated in Fig. 9.1. The flow of mass does carry an amount of entropy, $s$, per unit mass flowing, but it does not give rise to any other contributions. As a process may take place in the flow, entropy can be generated, but this is attributed to the space it belongs to (i.e., either inside or outside of the control volume).

The balance of entropy as an equation then states that the rate of change in total entropy inside the control volume is equal to the net sum of fluxes across the control surface plus the generation rate. That is,

$$\text{rate of change} = +\text{in} - \text{out} + \text{generation}$$

or

$$\frac{dS_{\text{c.v.}}}{dt} = \sum \dot{m}_i s_i - \sum \dot{m}_e s_e + \sum \frac{\dot{Q}_{\text{c.v.}}}{T} + \dot{S}_{\text{gen}} \tag{9.2}$$

These fluxes are mass flow rates carrying a level of entropy and the rate of heat transfer that takes place at a certain temperature (the temperature right at the control surface). The accumulation and generation terms cover the total control volume and are expressed in the lumped (integral form) so that

$$S_{\text{c.v.}} = \int \rho s \, dV = m_{\text{c.v.}} s = m_A s_A + m_B s_B + m_C s_C + \cdots$$

$$\dot{S}_{\text{gen}} = \int \rho \dot{s}_{\text{gen}} \, dV = \dot{S}_{\text{gen.}A} + \dot{S}_{\text{gen.}B} + \dot{S}_{\text{gen.}C} + \cdots \tag{9.3}$$

If the control volume has several different accumulation units with different fluid states and processes occurring in them, we may have to sum the various contributions over the different domains. If the heat transfer is distributed over the control surface, then an integral has to be done over the total surface area using the local temperature and rate of heat transfer per unit area, $\dot{Q}/A$, as

$$\sum \frac{\dot{Q}_{c.v.}}{T} = \int \frac{\delta \dot{Q}}{T} = \int_{surface} \frac{(\dot{Q}/A)}{T} \, dA \tag{9.4}$$

These distributed cases typically require a much more detailed analysis, which is beyond the scope of the current presentation of the second law.

The generation term(s) in Eq. 9.2 from a summation of individual positive internal-irreversibility entropy-generation terms in Eq. 9.3 is necessarily positive (or zero), such that an inequality is often written as

$$\frac{dS_{c.v.}}{dt} \geq \sum \dot{m}_i s_i - \sum \dot{m}_e s_e + \sum \frac{\dot{Q}_{c.v.}}{T} \tag{9.5}$$

Now the equality applies to internally reversible processes and the inequality to internally irreversible processes. The form of the second law in Eq. 9.2 or 9.5 is general, such that any particular case results in a form that is a subset (simplification) of this form. Examples of various classes of problems are illustrated in the following sections.

If there is no mass flow into or out of the control volume, it simplifies to a control mass and the equation for the total entropy reverts back to Eq. 8.43. Since that version of the second law has been covered in Chapter 8 here we will consider the remaining cases that were done for the first law of thermodynamics in Chapter 6.

## 9.2 THE STEADY-STATE PROCESS AND THE TRANSIENT PROCESS

We now consider in turn the application of the second-law control volume equation, Eq. 9.2 or 9.5, to the two control volume model processes developed in Chapter 6.

### Steady-State Process

For the steady-state process, which has been defined in Section 6.3, we conclude that there is no change with time of the entropy per unit mass at any point within the control volume, and therefore the first term of Eq. 9.2 equals zero. That is,

$$\frac{dS_{c.v.}}{dt} = 0 \tag{9.6}$$

so that, for the steady-state process,

$$\sum \dot{m}_e s_e - \sum \dot{m}_i s_i = \sum_{c.v.} \frac{\dot{Q}_{c.v.}}{T} + \dot{S}_{gen} \tag{9.7}$$

in which the various mass flows, heat transfer and entropy generation rates, and states are all constant with time.

If in a steady-state process there is only one area over which mass enters the control volume at a uniform rate and only one area over which mass leaves the control volume at a uniform rate, we can write

$$\dot{m}(s_e - s_i) = \sum_{c.v.} \frac{\dot{Q}_{c.v.}}{T} + \dot{S}_{gen} \tag{9.8}$$

and dividing the mass flow rate out gives

$$s_e = s_i + \sum \frac{q}{T} + s_{gen}$$

Since $s_{gen}$ is always greater than or equal to zero, for an adiabatic process it follows that

$$s_e = s_i + s_{gen} \geq s_i \qquad (9.9)$$

where the equality holds for a reversible adiabatic process.

**EXAMPLE 9.1**

Steam enters a steam turbine at a pressure of 1 MPa, a temperature of 300°C, and a velocity of 50 m/s. The steam leaves the turbine at a pressure of 150 kPa and a velocity of 200 m/s. Determine the work per kilogram of steam flowing through the turbine, assuming the process to be reversible and adiabatic.

| | |
|---:|:---|
| *Control volume*: | Turbine. |
| *Sketch*: | Fig. 9.2. |
| *Inlet state*: | Fixed (Fig. 9.2). |
| *Exit state*: | $P_e$, $\mathbf{V}_e$ known. |
| *Process*: | Steady state, reversible and adiabatic. |
| *Model*: | Steam tables. |

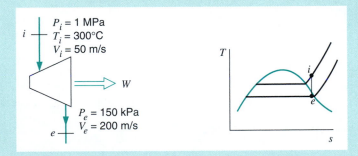

**FIGURE 9.2** Sketch for Example 9.1.

**Analysis**

The continuity equation gives us

$$\dot{m}_e = \dot{m}_i = \dot{m}$$

From the first law we have

$$h_i + \frac{\mathbf{V}_i^2}{2} = h_e + \frac{\mathbf{V}_e^2}{2} + w$$

and the second law is

$$s_e = s_i$$

**Solution**

From the steam tables, we get

$$h_i = 3051.2 \text{ kJ/kg}, \qquad s_i = 7.1228 \text{ kJ/kg K}$$

The two properties known in the final state are pressure and entropy:

$$P_e = 0.15 \text{ MPa}, \qquad s_e = s_i = 7.1228 \text{ kJ/kg K}$$

The quality and enthalpy of the steam leaving the turbine can be determined as follows:

$$s_e = 7.1228 = s_f + x_e s_{fg} = 1.4335 + x_e 5.7897$$

$$x_e = 0.9827$$

$$h_e = h_f + x_e h_{fg} = 467.1 + 0.9827(2226.5)$$

$$= 2655.0 \text{ kJ/kg}$$

Therefore, the work per kilogram of steam for this isentropic process may be found using the equation for the first law:

$$w = 3051.2 + \frac{50 \times 50}{2 \times 1000} - 2655.0 - \frac{200 \times 200}{2 \times 1000} = 377.5 \text{ kJ/kg}$$

**EXAMPLE 9.2**

Consider the reversible adiabatic flow of steam through a nozzle. Steam enters the nozzle at 1 MPa and 300°C, with a velocity of 30 m/s. The pressure of the steam at the nozzle exit is 0.3 MPa. Determine the exit velocity of the steam from the nozzle, assuming a reversible, adiabatic, steady-state process.

| | |
|---|---|
| *Control volume*: | Nozzle. |
| *Sketch*: | Fig. 9.3. |
| *Inlet state*: | Fixed (Fig. 9.3). |
| *Exit State*: | $P_e$ known. |
| *Process*: | Steady state, reversible, and adiabatic. |
| *Model*: | Steam tables. |

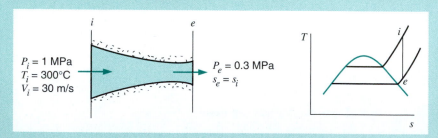

**FIGURE 9.3** Sketch for Example 9.2.

**Analysis**

Because this is a steady-state process in which the work, heat transfer, and changes in potential energy are zero, we can write

Continuity equation: $\dot{m}_e = \dot{m}_i = \dot{m}$

First law: $h_i + \dfrac{\mathbf{V}_i^2}{2} = h_e + \dfrac{\mathbf{V}_e^2}{2}$

Second law: $s_e = s_i$

**Solution**

From the steam tables, we have

$$h_i = 3051.2 \text{ kJ/kg}, \qquad s_i = 7.1228 \text{ kJ/kg K}$$

The two properties that we know in the final state are entropy and pressure:

$$s_e = s_i = 7.1228 \text{ kJ/kg K}, \qquad P_e = 0.3 \text{ MPa}$$

Therefore,

$$T_e = 159.1°C, \qquad h_e = 2780.2 \text{ kJ/kg}$$

Substituting into the equation for the first law, we have

$$\frac{\mathbf{V}_e^2}{2} = h_i - h_e + \frac{\mathbf{V}_i^2}{2}$$

$$= 3051.2 - 2780.2 + \frac{30 \times 30}{2 \times 1000} = 271.5 \text{ kJ/kg}$$

$$\mathbf{V}_e = \sqrt{2000 \times 271.5} = 737 \text{ m/s}$$

---

**EXAMPLE 9.3**

An inventor reports having a refrigeration compressor that receives saturated R-134a vapor at $-20°C$ and delivers the vapor at 1 MPa and 40°C. The compression process is adiabatic. Does the process described violate the second law?

| | |
|---|---|
| *Control volume*: | Compressor. |
| *Inlet state*: | Fixed (saturated vapor at $T_i$). |
| *Exit state*: | Fixed ($P_e$, $T_e$ known). |
| *Process*: | Steady state, adiabatic. |
| *Model*: | R-134a tables. |

**Analysis**

Because this is a steady-state adiabatic process, we can write the second law as

$$s_e \geq s_i$$

**Solution**

From the R-134a tables, we read

$$s_e = 1.7148 \text{ kJ/kg K}, \qquad s_i = 1.7395 \text{ kJ/kg K}$$

Therefore, $s_e < s_i$, whereas for this process the second law requires that $s_e \geq s_i$. The process described involves a violation of the second law and thus would be impossible.

---

**EXAMPLE 9.4**

An air compressor in a gas station, see Fig. 9.4, takes in a flow of ambient air at 100 kPa, 290 K, and compresses it to 1000 kPa in a reversible adiabatic process. We want to know the specific work required and the exit air temperature.

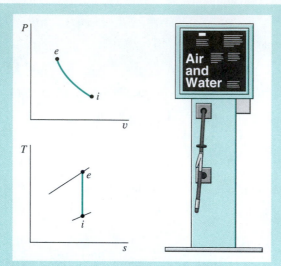

**FIGURE 9.4** Diagram for Example 9.4.

**Solution**

C.V. air compressor, steady state, single flow through it, and we assume adiabatic $\dot{Q} = 0$.

Continuity Eq. 6.11: $\dot{m}_i = \dot{m}_e = \dot{m}$,

Energy Eq. 6.12: $\dot{m}h_i = \dot{m}h_e + \dot{W}_C$,

Entropy Eq. 9.8: $\dot{m}s_i + \dot{S}_{gen} = \dot{m}s_e$

Process: Reversible $\dot{S}_{gen} = 0$

Use constant specific heat from Table A.5, $C_{P0} = 1.004$ kJ/kg K, $k = 1.4$. Entropy equation gives constant $s$, which gives the relation in Eq. 8.23:

$$s_i = s_e \Rightarrow T_e = T_i \left(\frac{P_e}{P_i}\right)^{\frac{k-1}{k}}$$

$$T_e = 290 \left(\frac{1000}{100}\right)^{0.2857} = 559.9 \text{ K}$$

The energy equation per unit mass gives the work term

$$w_c = h_i - h_e = C_{P0}(T_i - T_e) = 1.004(290 - 559.9) = -271 \text{ kJ/kg}$$

---

**EXAMPLE 9.5**

A de-superheater works by injecting liquid water into a flow of superheated steam. With 2 kg/s at 300 kPa, 200°C, steam flowing in, what mass flow rate of liquid water at 20°C should be added to generate saturated vapor at 300 kPa? We also want to know the rate of entropy generation in the process.

**Solution**

C.V. De-superheater, see Fig. 9.5, no external heat transfer, and no work.

Continuity Eq. 6.9: $\dot{m}_1 + \dot{m}_2 = \dot{m}_3$;

Energy Eq. 6.10: $\dot{m}_1 h_1 + \dot{m}_2 h_2 = \dot{m}_3 h_3 = (\dot{m}_1 + \dot{m}_2)h_3$

Entropy Eq. 9.7: $\dot{m}_1 s_1 + \dot{m}_2 s_2 + \dot{S}_{gen} = \dot{m}_3 s_3$

Process: $P = \text{constant}, \dot{W} = 0, \text{and } \dot{Q} = 0$

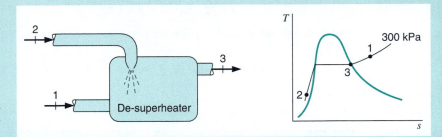

**FIGURE 9.5** Sketch and diagram for Example 9.5.

All the states are specified (approximate state 2 with saturated liquid 20°C)

B.1.3: $h_1 = 2865.54 \dfrac{kJ}{kg}$, $s_1 = 7.3115 \dfrac{kJ}{kg\ K}$; $h_3 = 2725.3 \dfrac{kJ}{kg}$, $s_3 = 6.9918 \dfrac{kJ}{kg\ K}$

B.1.1: $h_2 = 83.94 \dfrac{kJ}{kg}$, $s_2 = 0.2966 \dfrac{kJ}{kg\ K}$

Now we can solve for the flow rate $\dot{m}_2$ from the energy equation, having eliminated $\dot{m}_3$ by the continuity equation

$$\dot{m}_2 = \dot{m}_1 \frac{h_1 - h_3}{h_3 - h_2} = 2 \frac{2865.54 - 2725.3}{2725.3 - 83.94} = 0.1062 \text{ kg/s}$$

$$\dot{m}_3 = \dot{m}_1 + \dot{m}_2 = 2.1062 \text{ kg/s}$$

Generation is from the entropy equation

$$\dot{S}_{gen} = \dot{m}_3 s_3 - \dot{m}_1 s_1 - \dot{m}_2 s_2$$

$$\dot{S}_{gen} = 2.1062 \times 6.9918 - 2 \times 7.3115 - 0.1062 \times 0.2966 = 0.072 \text{ kW/K}$$

## Transient Process

For the transient process, which was described in Section 6.5, the second law for a control volume, Eq. 9.2, can be written in the following form:

$$\frac{d}{dt}(ms)_{c.v.} = \sum \dot{m}_i s_i - \sum \dot{m}_e s_e + \sum \frac{\dot{Q}_{c.v.}}{T} + \dot{S}_{gen} \qquad (9.10)$$

If this is integrated over the time interval $t$, we have

$$\int_0^t \frac{d}{dt}(ms)_{c.v.} \, dt = (m_2 s_2 - m_1 s_1)_{c.v.}$$

$$\int_0^t \left( \sum \dot{m}_i s_i \right) dt = \sum m_i s_i, \qquad \int_0^t \left( \sum \dot{m}_e s_e \right) dt = \sum m_e s_e, \qquad \int_0^t \dot{S}_{gen} \, dt = {}_1 S_{2\,gen}$$

Therefore, for this period of time $t$, we can write the second law for the transient process as

$$(m_2 s_2 - m_1 s_1)_{c.v.} = \sum m_i s_i - \sum m_e s_e + \int_0^t \sum_{c.v.} \frac{\dot{Q}_{c.v.}}{T} \, dt + {}_1 S_{2\,gen} \qquad (9.11)$$

Since in this process the temperature is uniform throughout the control volume at any instant of time, the integral on the right reduces to

$$\int_0^t \sum_{c.v.} \frac{\dot{Q}_{c.v.}}{T}\, dt = \int_0^t \frac{1}{T} \sum_{c.v.} \dot{Q}_{c.v.}\, dt = \int_0^t \frac{\dot{Q}_{c.v.}}{T}\, dt$$

and therefore the second law for the transient process can be written

$$(m_2 s_2 - m_1 s_1)_{c.v.} = \sum m_i s_i - \sum m_e s_e + \int_0^t \frac{\dot{Q}_{c.v.}}{T}\, dt + {}_1 S_{2\,gen} \qquad (9.12)$$

**EXAMPLE 9.6**

Assume an air tank has 40 L of 100 kPa air at ambient temperature 17°C. The adiabatic and reversible compressor is started so that it charges the tank up to a pressure of 1000 kPa and then it shuts off. We want to know how hot the air in the tank gets and the total amount of work required to fill the tank.

**Solution**

C.V. compressor and air tank in Fig. 9.6.

Continuity Eq. 6.15:   $m_2 - m_1 = m_{in}$

Energy Eq. 6.16:   $m_2 u_2 - m_1 u_1 = {}_1Q_2 - {}_1W_2 + m_{in} h_{in}$

Entropy Eq. 9.12:   $m_2 s_2 - m_1 s_1 = \int dQ/T + {}_1S_{2\,gen} + m_{in} s_{in}$

Process:   Adiabatic ${}_1Q_2 = 0$,   Process ideal ${}_1S_{2\,gen} = 0$,   $s_1 = s_{in}$

$\Rightarrow m_2 s_2 = m_1 s_1 + m_{in} s_{in} = (m_1 + m_{in}) s_1 = m_2 s_1 \Rightarrow s_2 = s_1$

Constant $s \Rightarrow$   Eq. 8.19   $s^0_{T2} = s^0_{T1} + R \ln(P_2/P_i)$

$s^0_{T2} = 6.83521 + 0.287 \ln(10) = 7.49605$ kJ/kg K

Interpolate in Table A.7   $\Rightarrow T_2 = 555.7$ K, $u_2 = 401.49$ kJ/kg

$m_1 = P_1 V_1/RT_1 = 100 \times 0.04/(0.287 \times 290) = 0.04806$ kg

$m_2 = P_2 V_2/RT_2 = 1000 \times 0.04/(0.287 \times 555.7) = 0.2508$ kg

$\Rightarrow m_{in} = 0.2027$ kg

${}_1W_2 = m_{in} h_{in} + m_1 u_1 - m_2 u_2$

$= m_{in}(290.43) + m_1(207.19) - m_2(401.49) = -31.9$ kJ

*Remark*: The high final temperature makes the assumption of zero heat transfer poor. The charging process does not happen rapidly so there will be a heat transfer loss. We need to know this to make a better approximation about the real process.

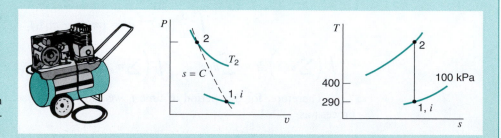

**FIGURE 9.6** Sketch and diagram for Example 9.6.

**In-Text
Concept
Questions**

**a.** A reversible adiabatic flow of liquid water in a pump has increasing $P$. What about $T$?

**b.** A reversible adiabatic flow of air in a compressor has increasing $P$. What about $T$?

**c.** A compressor receives R-134a at $-10°C$, 200 kPa with an exit of 1200 kPa, 50°C. What can you say about the process?

**d.** A flow of water at some velocity out of a nozzle is used to wash a car. The water then falls to the ground. What happens to the water state in terms of $\mathbf{V}$, $T$, and $s$?

## 9.3 THE STEADY-STATE SINGLE FLOW PROCESS

An expression can be derived for the work in a single flow steady-state process that shows how the significant variables influence the work output. We have noted that when a steady-state process involves a single flow of fluid into and out of a control volume, the first law, Eq. 6.13, can be written

$$q + h_i + \frac{1}{2}\mathbf{V}_i^2 + gZ_i = h_e + \frac{1}{2}\mathbf{V}_e^2 + gZ_e + w$$

The second law, Eq. 9.8 and recall Eq. 9.4, is

$$s_i + s_{gen} + \int \frac{\delta q}{T} = s_e$$

which we will write in a differential form as

$$\delta s_{gen} + \delta q/T = ds \qquad \Rightarrow \qquad \delta q = T ds - T\delta s_{gen}$$

To facilitate the integration and find $q$, we use the property relation, Eq. 8.8, and get

$$\delta q = T\,ds - T\,\delta s_{gen} = dh - v\,dP - T\,\delta s_{gen}$$

and we now have

$$q = \int_i^e \delta q = \int_i^e dh - \int_i^e v\,dP - \int_i^e T\,\delta s_{gen} = h_e - h_i - \int_i^e v\,dP - \int_i^e T\,\delta s_{gen}$$

This result is substituted into the energy equation, which we solve for work as

$$w = q + h_i - h_e + \frac{1}{2}(\mathbf{V}_i^2 - \mathbf{V}_e^2) + g(Z_i - Z_e)$$

$$= h_e - h_i - \int_i^e v\,dP - \int_i^e T\,\delta s_{gen} + h_i - h_e + \frac{1}{2}(\mathbf{V}_i^2 - \mathbf{V}_e^2) + g(Z_i - Z_e)$$

The enthalpy terms cancel and the shaft work for a single flow going through an actual process becomes

$$w = -\int_i^e v\,dP + \frac{1}{2}(\mathbf{V}_i^2 - \mathbf{V}_e^2) + g(Z_i - Z_e) - \int_i^e T\,\delta s_{gen} \tag{9.13}$$

Several comments for this expression are in order

**1.** We note that the last term always subtracts ($T > 0$ and $\delta s_{gen} \geq 0$) and we get the maximum work out for a reversible process where this term is zero. This is identical to the conclusion for the boundary work, Eq. 8.36, where it was concluded that any entropy generation reduces the work output. We do not write Eq. 9.13 because we expect to calculate the last integral for a process, but we show it to illustrate the effect of an entropy generation.

2. For a reversible process, the shaft work is associated with changes in pressure, kinetic energy, or potential energy either individually or in combination. When the pressure increases, (pump or compressor) work tends to be negative, that is, we must have shaft work in and when the pressure decreases (turbine), the work tends to be positive. The specific volume does not affect the sign of the work, but its magnitude, so we will have a large amount of work involved when the specific volume is large (the fluid is a gas), whereas the work will be smaller when the specific volume is small (as for a liquid). When the flow reduces its kinetic energy (windmill) or potential energy (dam and a turbine), we can extract the difference as work.

3. If the control volume does not have a shaft ($w = 0$), then the right-hand side terms must balance out to zero. Any change in one of the terms must be accompanied by a net change of opposite sign in the other terms, and notice that the last term can only subtract. As an example, let us briefly look at a pipe flow with no changes in kinetic or potential energy. If the flow is considered reversible, then the last term is zero and the first term must be zero, that is, the pressure must be constant. Realizing the flow has some friction and is therefore irreversible, the first term must be positive (pressure is decreasing) to balance out the last term.

As mentioned in the comment above, Eq. 9.13 is useful to illustrate the work involved in a large class of flow processes such as turbines, compressors, and pumps in which changes in kinetic and potential energies of the working fluid are small. The model process for these machines is then a reversible, steady-state process with no changes in kinetic or potential energy. The process is often also adiabatic but this is not required for this expression which reduces to

$$w = -\int_i^e v \, dP \qquad (9.14)$$

From this result, we conclude that the shaft work associated with this type of process is given by the area shown in Fig. 9.7. It is important to note that this result applies to a very specific situation of a flow device and is very different from the boundary type work $\int_1^2 P \, dv$ in a piston cylinder arrangement. It was also mentioned in the comments that the shaft work involved in this type of process is closely related to the specific volume of the fluid during the process. To amplify this point further, consider the simple steam power plant shown in Fig. 9.8. Suppose that this is a set of ideal components with no pressure drop in the piping, the boiler, or the condenser. Thus, the pressure increase in the pump is equal to the pressure decrease in the turbine. Neglecting kinetic and potential energy changes, the work done in each of these processes is given by Eq. 9.14. Since the pump

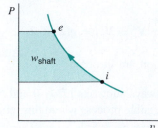

**FIGURE 9.7**  Shaft work from Eq. 9.14.

**FIGURE 9.8**
Simple steam power plant.

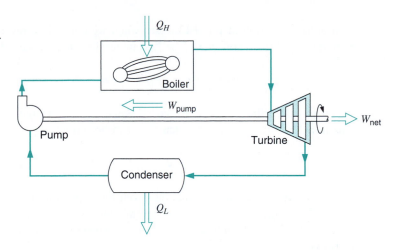

handles liquid, which has a very small specific volume compared to that of the vapor that flows through the turbine, the power input to the pump is much less than the power output of the turbine. The difference is the net power output of the power plant.

This same line of reasoning can be qualitatively applied to actual devices that involve steady-state processes, even though the processes are not exactly reversible and adiabatic.

---

**EXAMPLE 9.7**

Calculate the work per kilogram to pump water isentropically from 100 kPa and 30°C to 5 MPa.

| | |
|---:|:---|
| *Control volume*: | Pump. |
| *Inlet state*: | $P_i, T_i$ known; state fixed. |
| *Exit state*: | $P_e$ known. |
| *Process*: | Steady-state, isentropic. |
| *Model*: | Steam tables. |

**Analysis**

Since the process is steady, state, reversible, and adiabatic, and because changes in kinetic and potential energies can be neglected, we have

$$\text{First law: } h_i = h_e + w$$

$$\text{Second law: } s_e - s_i = 0$$

**Solution**

Since $P_e$ and $s_e$ are known, state $e$ is fixed and therefore $h_e$ is known and $w$ can be found from the first law. However, the process is reversible and steady state, with negligible changes in kinetic and potential energies, so that Eq. 9.14 is also valid. Furthermore, since a liquid is being pumped, the specific volume will change very little during the process.

From the steam tables, $v_i = 0.001\ 004$ m³/kg. Assuming that the specific volume remains constant and using Eq. 9.14, we have

$$-w = \int_1^2 v\ dP = v(P_2 - P_1) = 0.001\ 004(5000 - 100) = 4.92 \text{ kJ/kg}$$

A simplified version of Eq. 9.13 arises when we consider a reversible flow of an incompressible fluid ($v$ = constant). The first integral is then readily done to give

$$w = -v(P_e - P_i) + \frac{1}{2}(\mathbf{V}_i^2 - \mathbf{V}_e^2) + g(Z_i - Z_e) \qquad (9.15)$$

which is called the extended Bernoulli equation after Daniel Bernoulli who wrote the equation for zero work term which then can be written

$$vP_i + \frac{1}{2}\mathbf{V}_i^2 + gZ_i = vP_e + \frac{1}{2}\mathbf{V}_e^2 + gZ_e \qquad (9.16)$$

and known as the Bernoulli equation. From this equation, it follows that the sum of flow work ($Pv$), kinetic energy, and potential energy is constant along a flow line—for instance, as the flow might go up there is a corresponding reduction in the kinetic energy or pressure.

**EXAMPLE 9.8**

Consider a nozzle used to spray liquid water. If the line pressure is 300 kPa and the water is 20°C, how high a velocity can an ideal nozzle generate in the exit flow?

**Analysis**

For this single steady-state flow, we have no work or heat transfer and since it is incompressible and reversible, the Bernoulli equation applies giving

$$vP_i + \frac{1}{2}\mathbf{V}_i^2 + gZ_i = vP_i + 0 + 0 = vP_e + \frac{1}{2}\mathbf{V}_e^2 + gZ = vP_0 + \frac{1}{2}\mathbf{V}_e^2 + 0$$

and the exit kinetic energy becomes

$$\frac{1}{2}\mathbf{V}_e^2 = v(P_i - P_0)$$

**Solution**

We can now solve for the velocity using a value of $v = v_f = 0.001002$ m³/kg at 20°C from the steam tables.

$$\mathbf{V}_e = \sqrt{2v(P_i - P_0)} = \sqrt{2 \times 0.001002(300 - 100)1000} = 20 \text{ m/s}$$

Notice the factor of 1000 to convert from kPa to Pa for proper units.

As a final application of Eq. 9.13, we recall the reversible polytropic process for an ideal gas, discussed in Section 8.8 for a control mass process. For the steady-state process with no change in kinetic and potential energies, we have the relations

$$w = -\int_i^e v\,dP \qquad \text{and} \qquad Pv^n = \text{constant} = C^n$$

$$w = -\int_i^e v\,dP = -C\int_i^e \frac{dP}{P^{1/n}}$$

$$= -\frac{n}{n-1}(P_e v_e - P_i v_i) = -\frac{nR}{n-1}(T_e - T_i) \qquad (9.17)$$

If the process is isothermal, then $n = 1$ and the integral becomes

$$w = -\int_i^e v \, dP = -\text{constant} \int_i^e \frac{dP}{P} = -P_i v_i \ln \frac{P_e}{P_i} \tag{9.18}$$

Note that the $P$–$v$ and $T$–$s$ diagrams of Fig. 8.13 are applicable to represent the slope of polytropic processes in this case as well.

These evaluations of the integral

$$\int_i^e v \, dP$$

may also be used in conjunction with Eq. 9.13 for instances in which kinetic and potential energy changes are not negligibly small.

---

**In-Text Concept Questions**

e. In a steady-state single flow, $s$ is either constant or it increases. Is that true?

f. If a flow-device has the same inlet and exit pressure, can you have shaft work?

g. A polytropic flow process with $n = 0$ might be which device?

## 9.4 PRINCIPLE OF THE INCREASE OF ENTROPY

The principle of the increase of entropy for a control mass analysis was discussed in Section 8.11. The same general conclusion is reached for a control volume analysis. This is demonstrated by the split of the whole world into a control volume $A$ and its surroundings, control volume $B$, as shown in Fig. 9.9. Assume a process takes place in control volume $A$ exchanging mass flows, energy, and entropy transfers with the surroundings. Right where the heat transfer enters control volume $A$, we have a temperature of $T_A$, which is not necessarily equal to the ambient temperature far away from the control volume.

First let us write the entropy balance equation for the two control volumes

$$\frac{dS_{CVA}}{dt} = \dot{m}_i s_i - \dot{m}_e s_e + \frac{\dot{Q}}{T_A} + \dot{S}_{\text{gen }A} \tag{9.19}$$

$$\frac{dS_{CVB}}{dt} = -\dot{m}_i s_i + \dot{m}_e s_e - \frac{\dot{Q}}{T_A} + \dot{S}_{\text{gen }B} \tag{9.20}$$

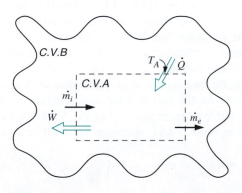

**FIGURE 9.9** Entropy change for a control volume plus surroundings.

and notice that the transfer terms are all evaluated right at the control volume surface. Now we will add the two entropy balance equations to find the net rate of change of $S$ for the total world

$$\frac{dS_{net}}{dt} = \frac{dS_{CVA}}{dt} + \frac{dS_{CVB}}{dt}$$

$$= \dot{m}_i s_i - \dot{m}_e s_e + \frac{\dot{Q}}{T_A} + \dot{S}_{gen\,A} - \dot{m}_i s_i + \dot{m}_e s_e - \frac{\dot{Q}}{T_A} + \dot{S}_{gen\,B}$$

$$= \dot{S}_{gen\,A} + \dot{S}_{gen\,B} \geq 0 \qquad (9.21)$$

Here we notice that all the transfer terms cancel out, leaving only the positive generation terms for each part of the world. If no process takes place in the ambient, that generation term is zero. However, we also notice that for the heat transfer to move in the indicated direction, we must have $T_B \geq T_A$, that is the heat transfer takes place over a finite temperature difference so an irreversible process occurs in the surroundings. Such a situation is called an external irreversible process. This distinguishes it from any generation of $s$ inside the control volume $A$, then called an internal irreversible process.

For this general control volume analysis, we arrive at the same conclusion as for the control mass situation—the entropy for the total world must increase or stay constant, $dS_{net}/dt \geq 0$, from Eq. 9.21. Only those processes that satisfy this equation can possibly take place; any process that would reduce the total entropy is impossible and will not occur.

Some other comments about the principle of the increase of entropy are in order. If we look at and evaluate changes in states for various parts of the world, we can find the net rate by the left-hand side of Eq. 9.21 and thus verify that it is positive for processes we consider. As we do this, we limit the focus to a control volume with a process occurring and the immediate ambient affected by this process. Notice that the left-hand side sums the storage, but it does not explain where the entropy is made. If we want detailed information about where the entropy is made, we must make a number of control volume analyses and evaluate the storage and transfer terms for each control volume. Then the rate of generation is found from the balance, i.e., from an equation like Eq. 9.21 and that must be positive or, at the least, zero. So not only must the total entropy increase by the sum of the generation terms, but we also must have a positive or at least zero entropy generation in every conceivable control volume. This applies to very small (even differential $dV$) control volumes, so only processes that locally generate entropy (or let it stay constant) will happen; any process that locally would destroy entropy cannot take place. Remember this does not preclude that entropy for some mass decreases as long as that is caused by a heat transfer (or net transfer by mass flow) out of that mass, i.e., the negative storage is explained by a negative transfer term.

When we use Eq. 9.21 to check any particular process for a possible violation of the second law, there are situations in which we can calculate the entropy generation terms directly. However, in many cases it is necessary to calculate the entropy changes of the control volume and surroundings separately, and then add them together to see if Eq. 9.21 is satisfied. In these cases, it is preferable to rewrite Eq. 9.20 for the surroundings $B$ to account for the fact that the heat transfer originates at temperature $T_B$. This means that the entropy generation in $B$ in Eq. 9.20 is given by Eq. 8.40 (on a rate basis), as was found in Section 8.11, such that Eq. 9.20 becomes

$$\frac{dS_{CVB}}{dt} = -\dot{m}_i s_i + \dot{m}_e s_e - \frac{\dot{Q}}{T_A} + \left\{ \frac{\dot{Q}}{T_A} - \frac{\dot{Q}}{T_B} \right\} = \dot{m}_e s_e - \dot{m}_i s_i - \frac{\dot{Q}}{T_B} \qquad (9.22)$$

For a steady-state process, we realize that $\frac{dS_{CVA}}{dt} = 0$, so that all of the entropy increase is observed in the surroundings $B$, and that this can be calculated from Eq. 9.22. For a transient process, there are both control volume $A$ and surroundings $B$ terms to evaluate. Each term is integrated over the time $t$ of the process, as was done in Section 9.2. Thus Eq. 9.21 is integrated to

$$\Delta S_{\text{net}} = \Delta S_{CVA} + \Delta S_{\text{surr }B} \tag{9.23}$$

in which the control volume $A$ term is

$$\Delta S_{CVA} = (m_2 s_2 - m_1 s_1)_{CVA} \tag{9.24}$$

The term for the surroundings is, after applying Eq. 9.21 to the surroundings and integrating,

$$\Delta S_{\text{surr }B} = m_e s_e - m_i s_i - \frac{Q}{T_B} \tag{9.25}$$

## 9.5 EFFICIENCY

In Chapter 7 we noted that the second law of thermodynamics led to the concept of thermal efficiency for a heat engine cycle, namely

$$\eta_{\text{th}} = \frac{W_{\text{net}}}{Q_H}$$

where $W_{\text{net}}$ is the net work of the cycle and $Q_H$ is the heat transfer from the high-temperature body.

In this chapter we have extended our consideration of the second law to control volume processes, which leads us now to consider the efficiency of a process. For example, we might be interested in the efficiency of a turbine in a steam power plant or the compressor in a gas turbine engine.

In general, we can say that to determine the efficiency of a machine in which a process takes place, we compare the actual performance of the machine under given conditions to the performance that would have been achieved in an ideal process. It is in the definition of this ideal process that the second law becomes a major consideration. For example, a steam turbine is intended to be an adiabatic machine. The only heat transfer is the unavoidable heat transfer that takes place between the given turbine and the surroundings. We also note that for a given steam turbine operating in a steady-state manner, the state of the steam entering the turbine and the exhaust pressure are fixed. Therefore, the ideal process is a reversible adiabatic process, which is an isentropic process, between the inlet state and the turbine exhaust pressure. In other words, the variables $P_i$, $T_i$, and $P_e$ are the design variables—the first two because the working fluid has been prepared in prior processes to be at these conditions at the turbine inlet, while the exit pressure is fixed by the environment into which the turbine exhausts. Thus, the ideal turbine process would go from state $i$ to state $e_s$, as shown in Fig. 9.10, whereas the real turbine process is irreversible, with the exhaust at a larger entropy at the real exit state $e$. Figure 9.10 shows typical states for a steam turbine, where state $e_s$ is in the two-phase region, and state $e$ may be as well, or may be in the superheated vapor region, depending on the extent of irreversibility of the real process. Denoting the work done in the real process $i$ to $e$ as $w$, and that done in the ideal,

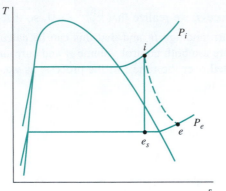

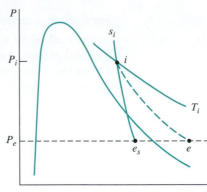

**FIGURE 9.10** The process in a reversible adiabatic steam turbine and an actual turbine.

isentropic process from the same $P_i$, $T_i$ to the same $P_e$ as $w_s$, we define the efficiency of the turbine as

$$\eta_{\text{turbine}} = \frac{w}{w_s} = \frac{h_i - h_e}{h_i - h_{es}} \tag{9.26}$$

The same definition applies to a gas turbine, where all states are in the gaseous phase. Typical turbine efficiencies are 0.70–0.88, with large turbines usually having higher efficiencies than small ones.

---

**EXAMPLE 9.9**

A steam turbine receives steam at a pressure of 1 MPa and a temperature of 300°C. The steam leaves the turbine at a pressure of 15 kPa. The work output of the turbine is measured and is found to be 600 kJ/kg of steam flowing through the turbine. Determine the efficiency of the turbine.

> *Control volume*: Turbine.
> *Inlet state*: $P_i$, $T_i$ known; state fixed.
> *Exit state*: $P_e$ known.
> *Process*: Steady-state.
> *Model*: Steam tables.

**Analysis**

The efficiency of the turbine is given by Eq. 9.26:

$$\eta_{\text{turbine}} = \frac{w_a}{w_s}$$

Thus, to determine the turbine efficiency, we calculate the work that would be done in an isentropic process between the given inlet state and final pressure. For this isentropic process, we have

Continuity equation: $\quad \dot{m}_i = \dot{m}_e = \dot{m}$

First law: $\quad\quad\quad\quad\; h_i = h_{es} + w_s$

Second law: $\quad\quad\quad\; s_i = s_{es}$

**Solution**

From the steam tables, we get

$$h_i = 3051.2 \text{ kJ/kg}, \qquad s_i = 7.1228 \text{ kJ/kg K}$$

Therefore, at $P_e = 15$ kPa,

$$s_{es} = s_i = 7.1228 = 0.7548 + x_{es} 7.2536$$

$$x_{es} = 0.8779$$

$$h_{es} = 225.9 + 0.8779(2373.1) = 2309.3 \text{ kJ/kg}$$

From the first law for the isentropic process,

$$w_s = h_i - h_{es} = 3051.2 - 2309.3 = 741.9 \text{ kJ/kg}$$

But, since

$$w_a = 600 \text{ kJ/kg}$$

we find that

$$\eta_{\text{turbine}} = \frac{w_a}{w_s} = \frac{600}{741.9} = 0.809 = 80.9\%$$

In connection with this example, it should be noted that to find the actual state e of the steam exiting the turbine, we need to analyze the real process taking place. For the real process

$$\dot{m}_i = \dot{m}_e = \dot{m}$$

$$h_i = h_e + w_a$$

$$s_e > s_i$$

Therefore, from the first law for the real process, we have

$$h_e = 3051.2 - 600 = 2451.2 \text{ kJ/kg}$$

$$2451.2 = 225.9 + x_e 2373.1$$

$$x_e = 0.9377$$

It is important to keep in mind that the turbine efficiency is defined in terms of an ideal, isentropic process from $P_i$ and $T_i$ to $P_e$, even when one or more of these variables is unknown. This is illustrated in the following example.

| **EXAMPLE 9.10** | Air enters a gas turbine at 1600 K and exits at 100 kPa and 830 K. The turbine efficiency is estimated to be 85%. What is the turbine inlet pressure? |
|---|---|

|  |  |
|---|---|
| *Control volume*: | Turbine. |
| *Inlet state*: | $T_i$ known. |
| *Exit state*: | $P_e, T_e$ known; state fixed. |
| *Process*: | Steady state. |
| *Model*: | Air tables, Table A.7. |

**Analysis**

The efficiency, which is 85%, is given by Eq. 9.26,

$$\eta_{\text{turbine}} = \frac{w}{w_s}$$

The first law for the real, irreversible process is

$$h_i = h_e + w$$

For the ideal, isentropic process from $P_i$, $T_i$ to $P_e$, the first law is

$$h_i = h_{es} + w_s$$

and the second law is, from Eq. 8.19,

$$s_{es} - s_i = 0 = s_{es}^0 - s_i^0 - R \ln \frac{P_e}{P_i}$$

(Note that this equation is only for the ideal, isentropic process and not for the real process, for which $s_e - s_i > 0$.)

**Solution**

From the air tables, Table A.7, at 1600 K, we get

$$h_i = 1757.3 \text{ kJ/kg}, \qquad s_i^0 = 8.6905 \text{ kJ/kg K}$$

From the air tables at 830 K (the actual turbine exit temperature),

$$h_e = 855.3 \text{ kJ/kg}$$

Therefore, from the first law for the real process.

$$w = 1757.3 - 855.3 = 902.0 \text{ kJ/kg}$$

Using the definition of turbine efficiency,

$$w_s = 902.0/0.85 = 1061.2 \text{ kJ/kg}$$

From the first law for the isentropic process,

$$h_{es} = 1757.3 - 1061.2 = 696.1 \text{ kJ/kg}$$

so that, from the air tables,

$$T_{es} = 683.7 \text{ K}, \qquad s_{es}^0 = 7.7148 \text{ kJ/kg K}$$

and the turbine inlet pressure is determined from

$$0 = 7.7148 - 8.6905 - 0.287 \ln \frac{100}{P_i}$$

or

$$P_i = 2995 \text{ kPa}$$

As was discussed in Section 6.4, unless specifically noted to the contrary, we normally assume compressors or pumps to be adiabatic. In this case the fluid enters the compressor at $P_i$ and $T_i$, the condition at which it exists, and exits at the desired value of $P_e$,

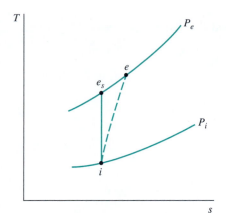 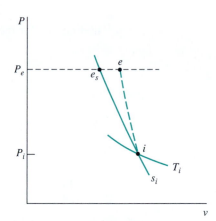

**FIGURE 9.11** The compression process in an ideal and actual adiabatic compressor.

the reason for building the compressor. Thus, the ideal process between the given inlet state $i$ and the exit pressure would be an isentropic process between state $i$ and state $e_s$, as shown in Fig. 9.11 with a work input of $w_s$. The real process, however, is irreversible, and the fluid exits at the real state $e$ with a larger entropy, and a larger amount of work input $w$ is required. The compressor (or pump, in the case of a liquid) efficiency is defined as

$$\eta_{\text{comp}} = \frac{w_s}{w} = \frac{h_i - h_{es}}{h_i - h_e} \tag{9.27}$$

Typical compressor efficiencies are in the range of 0.70–0.88 with large compressors usually having higher efficiencies than small ones.

If an effort is made to cool a gas during compression by using a water jacket or fins, the ideal process is considered a reversible isothermal process, the work input for which is $w_T$, compared to the larger required work $w$ for the real compressor. The efficiency of the cooled compressor is then

$$\eta_{\text{cooled comp}} = \frac{w_T}{w} \tag{9.28}$$

---

**EXAMPLE 9.11**

Air enters an automotive supercharger at 100 kPa and 300 K and is compressed to 150 kPa. The efficiency is 70%. What is the required work input per kg of air? What is the exit temperature?

*Control volume*: Supercharger (compressor).
*Inlet state*: $P_i$, $T_i$ known; state fixed.
*Exit state*: $P_e$ known.
*Process*: Steady-state.
*Model*: Ideal gas, 300 K specific heat, Table A.5.

**Analysis**

The efficiency, which is 70%, is given by Eq. 9.27,

$$\eta_{\text{comp}} = \frac{w_s}{w}$$

The first law for the real, irreversible process is

$$h_i = h_e + w, \qquad w = C_{p0}(T_i - T_e)$$

For the ideal, isentropic process from $P_i$, $T_i$ to $P_e$, the first law is

$$h_i = h_{es} + w_s, \qquad w_s = C_{p0}(T_i - T_{es})$$

and the second law is, from Eq. 8.23

$$\frac{T_{es}}{T_i} = \left(\frac{P_e}{P_i}\right)^{(k-1)/k}$$

**Solution**

Using $C_{p0}$ and $k$ from Table A.5, from the second law, we get

$$T_{es} = 300\left(\frac{150}{100}\right)^{0.286} = 336.9 \text{ K}$$

From the first law for the isentropic process, we have

$$h_{es} = 1.004(300 - 336.9) = -37.1 \text{ kJ/kg}$$

so that, from the efficiency, the real work input is

$$w = -37.1/0.70 = -53.0 \text{ kJ/kg}$$

and from the first law for the real process, the temperature at the supercharger exit is

$$T_e = 300 - \frac{-53.0}{1.004} = 352.8 \text{ K}$$

Our final example is that of nozzle efficiency. As discussed in Section 6.4, the purpose of a nozzle is to produce a high-velocity fluid stream, or in terms of energy, a large kinetic energy, at the expense of the fluid pressure. The design variables are the same as for a turbine, $P_i$, $T_i$, and $P_e$. A nozzle is usually assumed to be adiabatic, such that the ideal process is an isentropic process from state $i$ to state $e_s$, as shown in Fig. 9.12, with the production of velocity $\mathbf{V}_{es}$. The real process is irreversible, with the exit state $e$ having

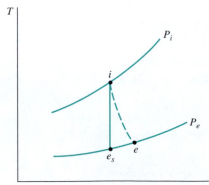

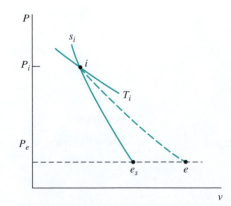

**FIGURE 9.12** The ideal and actual process in an adiabatic nozzle.

a larger entropy, and a smaller exit velocity $\mathbf{V}_e$. The nozzle efficiency is defined in terms of the corresponding kinetic energies,

$$\eta_{\text{nozz}} = \frac{\mathbf{V}_e^2/2}{\mathbf{V}_{es}^2/2} \qquad (9.29)$$

Nozzles are simple devices with no moving parts. As a result, nozzle efficiency may be very high, typically 0.90–0.97.

In summary, to determine the efficiency of a device that carries out a process (rather than a cycle), we compare the actual performance to what would be achieved in a related, but well-defined ideal process.

## 9.6 AVAILABLE ENERGY, REVERSIBLE WORK, AND IRREVERSIBILITY

From the concept of the efficiency of a device, such as a turbine, nozzle, or compressor (perhaps more correctly termed a first-law efficiency, since it is given as the ratio of two energy terms), we proceed now to develop concepts that include more meaningful second-law analysis. Our ultimate goal is to use this analysis to manage our natural resources and environment better.

We first focus our attention on the potential for producing useful work from some source or supply of energy. Consider the simple situation shown in Fig. 9.13a, in which there is an energy source $Q$ in the form of heat transfer from a very large and, there-fore, constant-temperature reservoir at temperature $T$. What is the ultimate potential for producing work?

To answer this question, we imagine that a cyclic heat engine is available, as shown in Fig. 9.13b. To convert the maximum fraction of $Q$ to work requires that the engine be completely reversible, that is, follow a Carnot cycle, and that the lower-temperature reservoir be at the lowest temperature possible, often, but not necessarily, at the ambient temperature. From the first and second laws for the Carnot cycle and the usual consid-eration of all the $Q$s as positive quantities, we find

$$W_{\text{rev H.E.}} = Q - Q_0$$

$$\frac{Q}{T} = \frac{Q_0}{T_0}$$

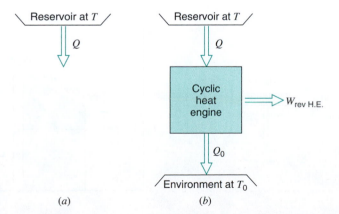

**FIGURE 9.13** Constant-temperature energy source.

(a)            (b)

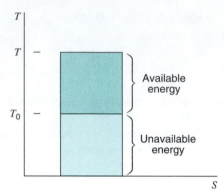

**FIGURE 9.14** *T–S* diagram for constant-temperature energy source.

so that

$$W_{\text{rev H.E.}} = Q\left(1 - \frac{T_0}{T}\right) \tag{9.30}$$

We might say that the fraction of $Q$ given by the right side of Eq. 9.30 is the available portion of the total energy quantity $Q$. To carry this thought one step farther, consider the situation shown on the *T–s* diagram in Fig. 9.14. The total shaded area is $Q$. The portion of $Q$ that is below $T_0$, the environment temperature, cannot be converted into work by the heat engine and must instead be thrown away. This portion is therefore the unavailable portion of total energy $Q$, and the portion lying between the two temperatures $T$ and $T_0$ is the available energy.

Let us next consider the same situation, except that the heat transfer $Q$ is available from a constant-pressure source, for example, a simple heat exchanger as shown in Fig. 9.15*a*. The Carnot cycle must now be replaced by a sequence of such engines, with the result shown in Fig. 9.15*b*. The only difference between the first and second examples is that the second includes an integral, which corresponds to $\Delta S$,

$$\Delta S = \int \frac{\delta Q_{\text{rev}}}{T} = \frac{Q_0}{T_0} \tag{9.31}$$

Substituting into the first law, we have

$$W_{\text{rev H.E.}} = Q - T_0 \, \Delta S \tag{9.32}$$

Note that this $\Delta S$ quantity does not include the standard sign convention. It corresponds to the amount of change of entropy shown in Fig. 9.15*b*. Equation 9.32 specifies the

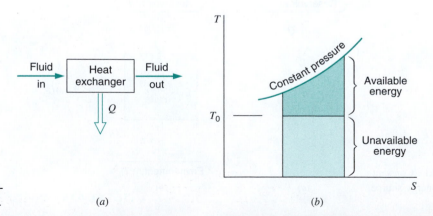

**FIGURE 9.15** Changing-temperature energy source.

(a)

(b)

available portion of the quantity $Q$. The portion unavailable for producing work in this circumstance lies below $T_0$ in Fig. 9.15$b$.

### The Steady-State Process

We now proceed to extend our analysis to real, irreversible processes. In doing so, we will consider only the case of the steady-state control volume process, since the vast majority of applications of this type of analysis refer to components of industrial systems such as power plants or refrigerators. It should be kept in mind that a parallel development for a control mass analysis or for a transient control volume process could also be made giving analogous results.

Consider the real steady-state process shown in Fig. 9.16, in which a control volume has a single fluid stream entering at state $i$, has a single stream exiting at state $e$, receives an amount of specific heat $q$ (per unit mass flow) from a reservoir crossing the control volume surface at temperature $T_H$, and does an amount of specific work $w$. The energy equation is, assuming no changes in kinetic or potential energies,

$$0 = q + h_i - h_e - w \tag{9.33}$$

The real process is irreversible, such that the entropy balance equation is

$$0 = s_i - s_e + \frac{q}{T_H} + s_{\text{gen}} \tag{9.34}$$

with some specific entropy generation inside this control volume. We wish to establish a quantitative measure of the extent to which the process is irreversible and express that in terms of work we potentially could get out.

We now compare the actual control volume with one that has the same inlet and exit states, the same heat transfer $q$ at $T_H$, but we require that it accomplishes this in exclusively reversible processes. We know from our previous considerations of processes that we will get the maximum amount of work out if the process is reversible, recall Eq. 8.36 and Eq. 9.13. The reversible (ideal) control volume has all the same terms as the actual one in Eq. 9.34 except the last term, the entropy generation, which must be zero for reversible processes. We cannot just remove the term, but we must replace it by something else of the exact same value. The only reversible process explanation for a change in entropy is a heat transfer. Since $s_{\text{gen}}$ is positive, the heat transfer must come in and the only available general source is the ambient at $T_0$. We therefore allow the reversible control volume to have a heat transfer of

$$q_0 = T_0 \, s_{\text{gen}} \geq 0 \tag{9.35}$$

so the reversible control volume energy and entropy equations become

$$0 = q + q_0 + h_i - h_e - w^{\text{rev}} \tag{9.36}$$

$$0 = s_i - s_e + \frac{q}{T_H} + \frac{q_0}{T_0} \tag{9.37}$$

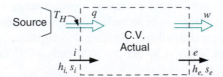

FIGURE 9.16  A real irreversible flow process.

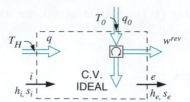

**FIGURE 9.17** An ideal reversible flow process.

As the heat transfer processes inside the control volume must be reversible, they can only occur over infinitesimal temperature differences so we recognize that both the heat transfers may go through reversible heat engines or heat pumps. These are located inside the control volume boundary of Fig. 9.17 and are illustrated explicitly for the ambient heat transfer.

The heat transfer from the ambient is expressed from Eq. 9.37 as

$$q_0 = T_0 s_{\text{gen}} = T_0 (s_e - s_i) - \frac{T_0}{T_H} q \tag{9.38}$$

so the reversible work is

$$w^{\text{rev}} = h_i - h_e + q + q_0$$

$$= h_i - h_e + q + T_0(s_e - s_i) - \frac{T_0}{T_H} q$$

$$= (h_i - T_0 s_i) - (h_e - T_0 s_e) + \left(1 - \frac{T_0}{T_H}\right) q \tag{9.39}$$

This expression establishes the theoretical upper limit for the shaft work given the heat transfer from the source at $T_H$ and the inlet and exit states. Notice that the analysis leading to Eq. 9.39 does not tell you how to accomplish it, only that this is the limit implied by the second law of thermodynamics. We furthermore recognize the separate contributions from the inlet and exit flow terms versus the contribution by the heat transfer from the source at $T_H$ giving a work term as if the $q$ were delivered to a Carnot heat engine.

The difference between the reversible work and the actual work for the real process

$$w = h_i - h_e + q$$

is the extent of the irreversibility $i$ (irreversibility per unit mass flow) of the real process

$$i = w^{\text{rev}} - w$$

$$= h_i - h_e + q + q_0 - (h_i - h_e + q)$$

$$= q_0 = T_0 s_{\text{gen}} \tag{9.40}$$

This specific irreversibility is seen to be directly proportional to the entropy generation and its value depends on the ambient temperature. This is a measure of the lost opportunity to extract work from the process—something the real process failed to do—and the measure is now done in energy units instead of in entropy units.

The expression for reversible work for the steady-state process, Eq. 9.39, was derived without including kinetic and potential energy terms. Whenever necessary, especially in nozzles and diffusers where obtaining kinetic energy change is the reason for building the device, these terms can be included along with the enthalpy terms of the fluid stream in and out of the control volume. Another way of including these terms would be to say that the enthalpies in Eq. 9.39 are the total enthalpies, as used in Eq. 6.8. There are also steady-state processes involving more than one fluid stream entering or exiting the control volume. In such cases, it is necessary to rewrite Eq. 9.39 on a rate basis, including the mass flow rates of the different streams involved in the process.

**EXAMPLE 9.12**

Consider an air compressor that receives ambient air at 100 kPa and 25°C. It compresses the air to a pressure of 1 MPa, where it exits at a temperature of 540 K. Since the air and compressor housing are hotter than the ambient surroundings, 50 kJ per kilogram air flowing through the compressor are lost. Find the reversible work, and the irreversibility in the process.

| | |
|---|---|
| *Control volume*: | The air compressor. |
| *Sketch*: | Fig. 9.18. |
| *Inlet state*: | $P_i$, $T_i$ known; state fixed. |
| *Exit state*: | $P_e$, $T_e$ known; state fixed. |
| *Process*: | Nonadiabatic compression with no change in kinetic or potential energy. |
| *Model*: | Ideal gas. |

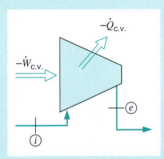

**FIGURE 9.18**   Illustration for Example 9.12.

**Analysis**

This steady-state process has a single inlet and exit flow so all quantities are done on a mass basis as specific quantities. From the ideal gas air tables, we obtain

$$h_i = 298.6 \text{ kJ/kg}, \qquad s^0_{T_i} = 6.8631 \text{ kJ/kg K}$$

$$h_e = 544.7 \text{ kJ/kg}, \qquad s^0_{T_e} = 7.4664 \text{ kJ/kg K}$$

so the energy equation for the actual compressor gives the work as

$$q = -50 \text{ kJ/kg}$$

$$w = h_i - h_e + q = 298.6 - 544.7 - 50 = -296.1 \text{ kJ/kg}$$

The reversible work for the given change of state is, from Eq. 9.39,

$$w^{\text{rev}} = T_0(s_e - s_i) - (h_e - h_i) + q\left(1 - \frac{T_0}{T_H}\right)$$

$$= 298.2(7.4664 - 6.8631 - 0.287 \ln 10) - (544.7 - 298.6) + 0$$

$$= -17.2 - 246.1 = -263.3 \text{ kJ/kg}$$

From Eq. 9.40, we get

$$i = w^{rev} - w$$

$$= -263.3 - (-296.1) = 32.8 \text{ kJ/kg}$$

---

**EXAMPLE 9.13**

A feedwater heater has 5 kg/s water at 5 MPa and 40°C flowing through it, being heated from two sources as shown in Fig. 9.19. One source adds 900 kW from a 100°C reservoir and the other source transfers heat from a 200°C reservoir such that the water exit condition is 5 MPa, 180°C. Find the reversible work and the irreversibility.

| | |
|---|---|
| *Control volume*: | Feedwater heater extending out to the two reservoirs. |
| *Inlet state*: | $P_i$, $T_i$ known; state fixed. |
| *Exit state*: | $P_e$, $T_e$ known; state fixed. |
| *Process*: | Constant-pressure heat addition with no change in kinetic or potential energy. |
| *Model*: | Steam tables. |

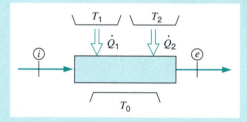

**FIGURE 9.19** The feed water heater for Example 9.13.

**Analysis**

This control volume has a single inlet and exit flow with two heat transfer rates coming from reservoirs different from the ambient surroundings. There is no actual work or actual heat transfer with the surroundings at 25°C. For the actual feedwater heater, the energy equation becomes

$$h_i + q_1 + q_2 = h_e$$

The reversible work for the given change of state is, from Eq. 9.39, with heat transfer $q_1$ from reservoir $T_1$ and heat transfer $q_2$ from reservoir $T_2$,

$$w^{rev} = T_0(s_e - s_i) - (h_e - h_i) + q_1\left(1 - \frac{T_0}{T_1}\right) + q_2\left(1 - \frac{T_0}{T_2}\right)$$

From Eq. 9.40, since the actual work is zero, we have

$$i = w^{rev} - w = w^{rev}$$

**Solution**

From the steam tables the inlet and exit state properties are

$$h_i = 171.97 \text{ kJ/kg}, \qquad s_i = 0.5705 \text{ kJ/kg K}$$

$$h_e = 765.25 \text{ kJ/kg}, \qquad s_e = 2.1341 \text{ kJ/kg K}$$

The second heat transfer is found from the energy equation as

$$q_2 = h_e - h_i - q_1 = 765.25 - 171.97 - 900/5 = 413.28 \text{ kJ/kg}$$

The reversible work is

$$w^{\text{rev}} = T_0(s_e - s_i) - (h_e - h_i) + q_1\left(1 - \frac{T_0}{T_1}\right) + q_2\left(1 - \frac{T_0}{T_2}\right)$$

$$= 298.2(2.1341 - 0.5705) - (765.25 - 171.97)$$

$$+ 180\left(1 - \frac{298.2}{373.2}\right) + 413.28\left(1 - \frac{298.2}{473.2}\right)$$

$$= 466.27 - 593.28 + 36.17 + 152.84 = 62.0 \text{ kJ/kg}$$

The irreversibility is

$$i = w^{\text{rev}} = 62.0 \text{ kJ/kg}$$

**In-Text Concept Questions**

**h.** Can I have any energy transfer as heat transfer that is 100% available?

**i.** Is energy transfer as work 100% available?

**j.** We cannot create or destroy energy. How can we create or destroy available energy?

**k.** Energy can be stored as internal energy, potential energy, or kinetic energy. Are those energy forms all 100% available?

## 9.7 AVAILABILITY AND SECOND-LAW EFFICIENCY

What is the maximum reversible work that can be done by a mass in a given state? The answer to this question depends in part on what type of process it is allowed to undergo. In the previous section, we developed an expression for reversible work for a given change of state in a steady-state process, and we also noted that analogous expressions would result from analysis of other types of processes. In general, for any type of process, when the mass comes into equilibrium with the environment, no spontaneous change of state will occur and the mass will be incapable of doing any work. Therefore, if a mass in a given state undergoes a completely reversible process until it reaches a state in which it is in equilibrium with the environment, the maximum reversible work will have been done by the mass. In this sense, we refer to the availability at the original state in terms of the potential for achieving the maximum possible work by the mass.

If a control mass is in equilibrium with the surroundings, it must certainly be in pressure and temperature equilibrium with the surroundings, that is, at pressure $P_0$ and temperature $T_0$. It must also be in chemical equilibrium with the surroundings, which implies that no further chemical reaction will take place. Equilibrium with the surroundings also requires that the system have zero velocity and minimum potential energy. Similar requirements could be set forth regarding electrical and surface effects if these are relevant to a given problem.

The same general remarks can be made about a quantity of mass that undergoes a steady-state process. With a given state for the mass entering the control volume, the reversible work will be a maximum when this mass leaves the control volume in equilibrium with the surroundings. This means that as the mass leaves the control volume, it must be at the pressure and temperature of the surroundings, in chemical equilibrium with the surroundings, and have minimum potential energy and zero velocity. (The mass leaving the control volume must of necessity have some velocity, but this can be made to approach zero.)

Let us consider the availability associated with a steady-state process. For a control volume with a single-flow stream, the reversible work is given by Eq. 9.39. Including kinetic and potential energies, this expression is rewritten in the form,

$$w^{\text{rev}} = (h_{\text{TOT}\,i} - T_0 s_i) - (h_{\text{TOT}\,e} - T_0 s_e) + q\left(1 - \frac{T_0}{T_H}\right)$$

From the discussion of the heat engine that led to Eq. 9.30, it is clear that the last term in the expression for the reversible work is the contribution to the net reversible work from the heat transfers. These can be viewed as transfer of availability associated with $q$, which gives a potential to do work as in a heat engine. Such contributions are separate from the availability in the flow itself. The steady-state flow reversible work will be maximum, relative to the surroundings, when the mass leaving the control volume is in equilibrium with the surroundings. The state in which the fluid is in equilibrium with the surroundings is designated with subscript 0, and the reversible work will be maximum when $h_e = h_0$, $s_e = s_0$, $\mathbf{V}_e = 0$, and $Z_e = Z_0$. This maximum reversible work per unit mass flow without the additional heat transfers is the flow availability or *exergy* and assigned the symbol $\psi$:

$$\psi = \left(h - T_0 s + \frac{1}{2}\mathbf{V}^2 + gZ\right) - (h_0 - T_0 s_0 + gZ_0) \tag{9.41}$$

It is written without subscript for the inlet state to indicate that this is the flow availability associated with a substance in any state as it enters the control volume in a steady-state process. The reversible work is therefore seen to be equal to the decrease in flow availability plus the reversible work that can be extracted from heat engines operating with the heat transfer at $T_H$ and the ambient temperature.

Irreversibility is as usual defined as the difference between the reversible work and the actual work. If we write this as a general expression on a rate basis, including summations to account for the possibility of more than one flow stream and also more than one heat transfer, the result is

$$\dot{I}_{\text{c.v.}} = \left(\sum \dot{m}_i \psi_i - \sum \dot{m}_e \psi_e\right) + \sum \left(1 - \frac{T_0}{T_j}\right)\dot{Q}_{\text{c.v.},j} - \dot{W}_{\text{c.v.}} \tag{9.42}$$

In this form the irreversibility is equal to the decrease in the availability of the mass flows plus the decrease of availability of each heat transfer rate $j$ at reservoir $T_j$ minus the increase in availability of the surroundings that receive the actual work. The rate of irreversibility is thus seen to be the rate of destruction of availability, which is also directly proportional to the net rate of entropy generation, as noted in Eq. 9.40.

The less the irreversibility associated with a given change of state, the greater the amount of work that will be done (or the smaller the amount of work that will be required). This relation is significant in at least two regards. The first is that availability is one of our natural resources. This availability is found in such forms as oil reserves, coal reserves, and uranium reserves. Suppose we wish to accomplish a given objective that requires a certain amount of work. If this work is produced reversibly while drawing on one of the availability reserves, the decrease in availability is exactly equal to the reversible work. However, since there are irreversibilities in producing this required amount of work, the actual work will be less than the reversible work, and the decrease in availability will be greater (by the amount of the irreversibility) than if this work had been produced reversibly. Thus the more irreversibilities we have in all our processes, the greater will be the decrease in our availability reserves.[1] The conservation and effective use of these availability reserves is an important responsibility for all of us.

The second reason that it is desirable to accomplish a given objective with the smallest irreversibility is an economic one. Work costs money, and in many cases a given objective can be accomplished at less cost when the irreversibility is less. It should be noted, however, that many factors enter into the total cost of accomplishing a given objective, and an optimization process that considers many factors is often necessary to arrive at the most economical design. For example, in a heat-transfer process, the smaller the temperature difference across which the heat is transferred, the less the irreversibility. However, for a given rate of heat transfer, a smaller temperature difference will require a larger (and therefore more expensive) heat exchanger. These various factors must all be considered in developing the optimum and most economical design.

In many engineering decisions other factors, such as the impact on the environment (for example, air pollution and water pollution) and the impact on society must be considered in developing the optimum design.

Along with the increased use of availability analysis in recent years, a term called the "second-law efficiency" has come into more common use. This term refers to comparison of the desired output of a process with the cost, or input, in terms of the thermodynamic availability. Thus, the isentropic turbine efficiency defined by Eq. 9.26 as the actual work output divided by the work for a hypothetical isentropic expansion from the same inlet state to the same exit pressure might well be called a "first-law efficiency," in that it is a comparison of two energy quantities. The "second-law efficiency," as just described, would be the actual work output of the turbine divided by the decrease in availability from the same inlet state to the same exit state. For the turbine shown in Fig. 9.20, the second-law efficiency is

$$\eta_{\text{2nd law}} = \frac{w_a}{\psi_i - \psi_e} \tag{9.43}$$

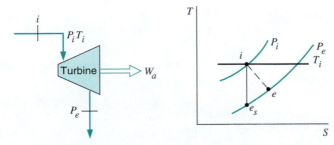

**FIGURE 9.20** Irreversible turbine.

---

[1] In many popular talks reference is made to our energy reserves. From a thermodynamic point of view, availability reserves would be a much more acceptable term. There is much energy in the atmosphere and the ocean, but relatively little availability.

In this sense, this concept provides a rating or measure of the real process in terms of the actual change of state and is simply another convenient way of utilizing the concept of thermodynamic availability. In a similar manner, the second-law efficiency of a pump or compressor is the ratio of the increase in availability to the work input to the device.

---

**EXAMPLE 9.14**

An insulated steam turbine (Fig. 9.21) receives 30 kg of steam per second at 3 MPa and 350°C. At the point in the turbine where the pressure is 0.5 MPa, steam is bled off for processing equipment at the rate of 5 kg/s. The temperature of this steam is 200°C. The balance of the steam leaves the turbine at 15 kPa with a 90% quality. Determine the availability per kilogram of the steam entering and at both points at which steam leaves the turbine, the isentropic efficiency, and the second-law efficiency for this process.

*Control volume*:   Turbine.

*Inlet state*:   $P_1$, $T_1$ known; state fixed.

*Exit state*:   $P_2$, $T_2$ known; $P_3$, $x_3$ known; both states fixed.

*Process*:   Steady-state.

*Model*:   Steam tables.

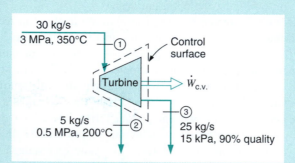

**FIGURE 9.21**   Sketch for Example 9.14.

**Analysis**

The availability at any point for the steam entering or leaving the turbine is given by Eq. 9.41,

$$\psi = (h - h_0) - T_0(s - s_0) + \frac{\mathbf{V}^2}{2} + g(Z - Z_0)$$

Since there are no changes in kinetic and potential energy in this problem, this equation reduces to

$$\psi = (h - h_0) - T_0(s - s_0)$$

For the ideal isentropic turbine,

$$\dot{W}_s = \dot{m}_1 h_1 - \dot{m}_2 h_{2s} - \dot{m}_3 h_{3s}$$

For the actual turbine,

$$\dot{W} = \dot{m}_1 h_1 - \dot{m}_2 h_2 - \dot{m}_3 h_3$$

**Solution**

At the pressure and temperature of the surroundings, 0.1 MPa, 25°C, the water is a slightly compressed liquid, and the properties of the water are essentially equal to those for saturated liquid at 25°C:

$$h_0 = 104.9 \text{ kJ/kg}; \qquad s_0 = 0.3674 \text{ kJ/kg K}$$

From Eq. 9.41, we have

$$\psi_1 = (3115.3 - 104.9) - 298.15(6.7428 - 0.3674) = 1109.6 \text{ kJ/kg}$$

$$\psi_2 = (2855.4 - 104.9) - 298.15(7.0592 - 0.3674) = 755.3 \text{ kJ/kg}$$

$$\psi_3 = (2361.8 - 104.9) - 298.15(7.2831 - 0.3674) = 195.0 \text{ kJ/kg}$$

$$\dot{m}_1\psi_1 - \dot{m}_2\psi_2 - \dot{m}_3\psi_3 = 30(1109.6) - 5(755.3) - 25(195.0) = 24\,637 \text{ kW}$$

For the ideal isentropic turbine,

$$s_{2s} = 6.7428 = 1.8606 + x_{2s} \times 4.9606, \qquad x_{2s} = 0.9842$$

$$h_{2s} = 640.2 + 0.9842 \times 2108.5 = 2715.4 \text{ kJ/kg}$$

$$s_{3s} = 6.7428 = 0.7549 + x_{3s} \times 7.2536, \qquad x_{3s} = 0.8255$$

$$h_{3s} = 225.9 + 0.8255 \times 2373.1 = 2184.9 \text{ kJ/kg}$$

$$\dot{W}_s = 30(3115.3) - 5(2715.4) - 25(2184.9) = 25\,260 \text{ kW}$$

For the actual turbine,

$$\dot{W} = 30(3115.3) - 5(2855.4) - 25(2361.8) = 20\,137 \text{ kW}$$

The isentropic efficiency is

$$\eta_s = \frac{20\,137}{25\,260} = 0.797$$

and the second-law efficiency is

$$\eta_{\text{2nd law}} = \frac{20\,137}{24\,637} = 0.817$$

For a device that does not involve the production or the input of work, the definition of second-law efficiency refers to the accomplishment of the goal of the process relative to the process input, in terms of availability changes or transfers. For example, in a heat exchanger, where energy is transferred from a high-temperature fluid stream to a low-temperature fluid stream, as shown in Fig. 9.22, the second-law efficiency is defined as

$$\eta_{\text{2nd law}} = \frac{\dot{m}_1(\psi_2 - \psi_1)}{\dot{m}_3(\psi_3 - \psi_4)} \tag{9.44}$$

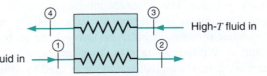

**FIGURE 9.22** A two-fluid heat exchanger.

## 9.8 SOME GENERAL COMMENTS ABOUT ENTROPY AND CHAOS

It is quite possible at this point that a student may have a good grasp of the material that has been covered and yet may have only a vague understanding of the significance of entropy. In fact, the question, "What is entropy?" is frequently raised by students with the implication that no one really knows! This section has been included in an attempt to give insight into the qualitative and philosophical aspects of the concept of entropy, and to illustrate the broad application of entropy to many different disciplines.

First, we recall that the concept of energy arises from the first law of thermodynamics and the concept of entropy from the second law of thermodynamics. Actually it is just as difficult to answer the question, "What is energy?" as it is to answer the question, "What is entropy?" However, since we regularly use the term energy and are able to relate this term to phenomena that we observe every day, the word energy has a definite meaning to us and thus serves as an effective vehicle for thought and communication. The word entropy could serve in the same capacity. If, when we observed a highly irreversible process (such as cooling coffee by placing an ice cube in it), we said, "That surely increases the entropy," we would soon be as familiar with the word *entropy* as we are with the word *energy*. In many cases when we speak about a higher efficiency we are actually speaking about accomplishing a given objective with a smaller total increase in entropy.

A second point to be made regarding entropy is that in statistical thermodynamics, the property entropy is defined in terms of probability. Although this topic will not be examined in detail in this text, a few brief remarks regarding entropy and probability may prove helpful. From this point of view the net increase in entropy that occurs during an irreversible process can be associated with a change of state from a less probable state to a more probable state. For instance, to use a previous example, one is more likely to find gas on both sides of the ruptured membrane in Fig. 7.15 than to find a gas on one side and a vacuum on the other. Thus, when the membrane ruptures, the direction of the process is from a less probable state to a more probable state, and associated with this process is an increase in entropy. Similarly, the more probable state is that a cup of coffee will be at the same temperature as its surroundings than at a higher (or lower) temperature. Therefore, as the coffee cools as the result of a transferring of heat to the surroundings, there is a change from a less probable to a more probable state, and associated with this is an increase in entropy.

To tie entropy a little closer to physics and to the level of disorder or chaos, let us consider a very simple system. Properties like $U$ and $S$ for a substance at a given state are averaged over many particles on the molecular level so they (atoms and molecules) do not all exist in the same detailed quantum state. There is a number of different configurations possible for a given state that constitutes an uncertainty or chaos in the system. The number of possible configurations, $w$, is called the thermodynamic probability and each of these is equally possible; this is used to define the entropy as

$$S = k \ln w$$

where $k$ is Boltzmann constant, and it is from this definition that $S$ is connected to the uncertainty or chaos. The larger the number of possible configurations is, the larger $S$ is. For a given system, we would have to evaluate all the possible quantum states for kinetic energy, rotational energy, vibrational energy, and so forth to find the equilibrium distribution and $w$. Without going into those details, which is the subject of statistical thermodynamics, a very simplistic example is used to illustrate the principle.

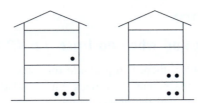

**FIGURE 9.23** Illustration of energy distribution.

Assume we have 4 identical objects that can only posess one kind of energy, namely potential energy associated with elevation (the floor) in a tall building. Let the 4 objects have a combined 2 units of energy (floor height times mass times gravitation). How can this system be configured? We can have one object on the second floor and the remaining 3 on the ground floor, giving a total of 2 energy units. We could also have two objects on the first floor and 2 on the ground floor, again with a total of 2 energy units. These two configurations are equally possible and we could therefore see the system 50% of the time in one and 50% of the time in the other configuration; we have some positive value of $S$.

Now let us add 2 energy units by heat transfer and that is done by giving the objects some energy that they share. Now the total energy is 4 units and we can see the system in the following configurations (a–e):

| Floor number: | | 0 | 1 | 2 | 3 | 4 |
|---|---|---|---|---|---|---|
| Number of objects | a: | 3 | | | | 1 |
| Number of objects | b: | 2 | 1 | | 1 | |
| Number of objects | c: | 2 | | 2 | | |
| Number of objects | d: | 1 | 2 | 1 | | |
| Number of objects | e: | | 4 | | | |

Now we have 5 different configurations ($w = 5$)—each equally possible—so we will observe the system 20% of the time in each one, and we now have a larger value of $S$.

On the other hand, if we increase the energy by 2 units through work, it acts differently. Work is associated with the motion of a boundary, so now we pull in the building to make it higher and stretch it to be twice as tall, i.e., the first floor has 2 energy units per object and so forth as compared with the original state. This means that we simply double the energy per object in the original configuration without altering the number of configurations, which stay at $w = 2$. In effect, $S$ has not changed.

| Floor number: | | 0 | 1 | 2 | 3 | 4 |
|---|---|---|---|---|---|---|
| Number of objects | f: | 3 | | 1 | | |
| Number of objects | g: | 2 | 2 | | | |

So this example illustrates the profound difference between adding energy as a heat transfer changing $S$ versus adding energy through a work term leaving $S$ unchanged. In the first situation, we move a number of particles from lower energy levels to higher energy levels thus changing the distribution and increasing the chaos. In the second situation, we do not move the particles between energy states, but we change the energy level of a given state thus preserving the order and chaos.

## 9.9 HOW-TO SECTION

## When do I use $\int P \, dv$ and when do I use $\int v \, dP$ for work?

A boundary work happens when there is a pressure (or force) on a moving surface that constitutes a force and a displacement. A moving surface also implies that we have a change in volume ($dV = A \, dx$) and the proper expression for a reversible process is

$$_1W_2 = \int F \, dx = \int PA \, dx = \int P \, dV = m \int P \, dv = m_1 w_2$$

The other expression comes from the reversible shaft work in a flow process, Eq. 9.14, where

$$w = -\int v \, dP + \frac{1}{2}(\mathbf{V}_i^2 - \mathbf{V}_e^2) + g(Z_i - Z_e)$$

and there must be changes in pressure or kinetic energy or potential energy to have a shaft work. Examples are a pump for the first term, a windmill for the second term, and a flow of liquid water from a river to a higher elevation irrigation channel for the last term.

## How do I deal with an efficiency?

An efficiency describes the performance of an actual device/process or complete system/cycle relative to a well-known reference. The kind of reference system is implied by the name of the efficiency (isentropic or isothermal or thermal or second law) and the kind of device or system it describes (i.e., is it a single process/device or is it a complete cycle like a heat engine, heat pump, or refrigerator).

By clearly understanding the reference system, we can easily calculate the performance of that and then use the efficiency to get the performance of the actual system. Generally, first law efficiencies are ratios of energy terms, whereas second law efficiencies are ratios of exergy (availability) terms.

Complete cycle efficiencies are often conversion efficiencies (heat engine or COP for a refrigerator) and single process/device efficiency is a relative performance measure (actual work relative to isentropic work or actual kinetic energy relative to isentropic kinetic energy). A second law efficiency is not a conversion efficiency of energy but a comparison between an actual system and a thought of (non-existing) comparable reversible system.

## How do I make a general control volume analysis?

One of the more important subjects to learn is the control volume formulation of the general laws (conservation of mass, momentum and energy, balance of entropy) and the specific laws that in the current presentation are given in an integral (mass averaged) form. The following steps show a systematic way to formulate a thermodynamic problem so it does not become a formula chase, and you will be able to solve general and even unfamiliar problems.

## Formulation Steps

## STEP 1

Make a physical model of the system with components and illustrate all mass flows, heat flows, and work rates. Include also an indication of forces like external pressures and gravitation.

## STEP 2

Define (i.e., choose) a control mass or control volume by placing a control surface that contains the substance you want to analyze. This choice is very important since the formulation will depend on it. Be sure that only those mass flows, heat fluxes, and work terms you want to analyze cross the control surface. Include as much of the system as you can to eliminate flows and fluxes that you don't want to enter the formulation. Number the states of the substance where it enters or leaves the control volume and if it does not have the same state, label different parts of the system with storage.

## STEP 3

Write the general laws down for each of the chosen control volumes. For control volumes that have mass flows or heat and work fluxes between them, make certain that what leaves one control volume enters the other (i.e., have one term in each equation with an opposite sign). When the equations are written down, use the most general form and cancel terms that are not present. Only two forms of the general laws should be used in the formulation: (1) the original rate form (Eq. 9.2 for $S$) and (2) the time integrated form (Eq. 9.11 for $S$), where now terms that are not present are canceled. It is very important to distinguish between storage terms (left-hand side) and flow terms.

## STEP 4

Write the auxiliary or particular laws down for whatever is inside each of the control volumes. The constitution for a substance is either written down or referenced to a table. The equation for a given process is normally easily written down. It is given by the way the system or devices are constructed and often is an approximation to reality. That is, we make a mathematical model of the real-world behavior.

## STEP 5

Finish the formulation by combining all the equations but don't put in numbers yet. At this point, check which quantities are known and which are unknown. Here it is important to be able to find all the states of the substance and determine which two independent properties determine any given state. This task is most easily done by illustrating all the processes and states in a $(P–v)$, $(T–v)$, $(T–s)$ or similar diagram. These diagrams will also show what numbers to look up in the tables to determine where a given state is.

## STEP 6

The equations are now solved for the unknowns by writing all terms with unknown variables on one side and known terms on the other. It is usually easy to do this, but in some cases, it may require an iteration technique to solve the equations (for instance, if you have a combined property of $u$, $P$, $v$, like $u + 1/2\ Pv$ and not $h = u + Pv$). As you find the numerical values for different quantities, make sure they make sense and are within reasonable ranges.

## SUMMARY

The second law of thermodynamics is extended to a general control volume with mass flow rates in or out for steady and transient processes. The vast majority of common devices and complete systems can be treated as nearly steady-state operation even if they have slower transients as in a car engine or jet engine. Simplification of the entropy equation arises when applied to steady-state and single-flow devices like a turbine, nozzle, compressor, or pump. The second law and Gibbs property relation are used to develop a general expression for reversible shaft work in a single flow that is useful in understanding the importance of the specific volume (or density) that influences the magnitude of the work. For a flow with no shaft work, consideration of the reversible process also leads to the derivation of the energy equation for an incompressible fluid as the Bernoulli equation. This covers the flows of liquids such as water or hydraulic fluid as well as airflow at low speeds, which can be considered incompressible for velocities less than a third of the speed of sound.

Many actual devices operate with some irreversibility in the processes that occur, so we also have entropy generation in the flow processes and the total entropy is always increasing. The characterization of performance of actual devices can be done with a comparison to a corresponding ideal device, giving efficiency as the ratio of two energy terms (work or kinetic energy).

Work out of a Carnot-cycle heat engine is the available energy in the heat transfer from the hot source; the heat transfer to the ambient is unavailable. When an actual device is compared to an ideal device with the same flows and states in and out, we get to the concept of reversible work and exergy (availability). The reversible work is the maximum work we can get out of a given set of flows and heat transfers or, alternatively, the minimum work we have to put into the device. The comparison between the actual work and the theoretical maximum work gives a second-law efficiency. When exergy (availability) is used, the second-law efficiency can also be used for devices that do not involve shaft-work such as heat exchangers. In that case, we compare the exergy given out by one flow to the exergy gained by the other flow, giving a ratio of exergies instead of energies used for the first-law efficiency. Any irreversibility (entropy generation) in a process does destroy exergy (availability) and is undesirable.

You should have learned a number of skills and acquired abilities from studying this chapter that will allow you to

- Apply the second law to more general control volumes.
- Analyze steady-state, single-flow devices such as turbines, nozzles, compressors, and pumps, both reversible and irreversible.
- Know how to extend the second law to transient processes.
- Analyze complete systems as a whole or divide them into individual devices.
- Apply the second law to multiple-flow devices such as heat exchangers, mixing chambers, and turbines with several inlets and outlets.
- Recognize when you have an incompressible flow where you can apply the Bernoulli equation or the expression for reversible shaft work.
- Know when you can apply the Bernoulli equation and when you cannot.
- Know how to evaluate the shaft work for a polytropic process.
- Know how to apply the analysis to an actual device using an efficiency and identify the closest ideal approximation to the actual device.
- Know the difference between a cycle efficiency and a device efficiency.
- Have a sense of entropy as a measure of disorder or chaos.
- Understand the concept of available energy.
- Understand that energy and availability are different concepts.

- Be able to conceptualize the ideal counterpart to an actual system and find the reversible work and heat transfer in the ideal system.
- Understand the difference between a first-law and a second-law efficiency.

## KEY CONCEPTS AND FORMULAS

Rate equation for entropy  rate of change $= +$ in $-$ out $+$ generation

$$\dot{S}_{\text{c.v.}} = \sum \dot{m}_i s_i - \sum \dot{m}_e s_e + \sum \frac{\dot{Q}_{\text{c.v.}}}{T} + \dot{S}_{\text{gen}}$$

Steady state single flow

$$s_e = s_i + \int_i^e \frac{\delta q}{T} + s_{\text{gen}}$$

Reversible shaft work

$$w = -\int_i^e v \, dP + \frac{1}{2}\mathbf{V}_i^2 - \frac{1}{2}\mathbf{V}_e^2 + gZ_i - gZ_e$$

Reversible heat transfer

$$q = \int_i^e T \, ds = h_e - h_i - \int_i^e v \, dP \qquad \text{(from Gibbs relation)}$$

Bernoulli equation

$$v(P_i - P_e) + \frac{1}{2}\mathbf{V}_i^2 - \frac{1}{2}\mathbf{V}_e^2 + gZ_i - gZ_e = 0 \quad (v = \text{constant})$$

Polytropic process work

$$w = -\frac{n}{n-1}(P_e v_e - P_i v_i) = -\frac{nR}{n-1}(T_e - T_i) \qquad n \neq 1$$

$$w = -P_i v_i \ln \frac{P_e}{P_i} = -RT_i \ln \frac{P_e}{P_i} = RT_i \ln \frac{v_e}{v_i} \qquad n = 1$$

The work is shaft work $w = -\displaystyle\int_i^e v \, dP$ and for ideal gas

Isentropic efficiencies

$\eta_{\text{turbine}} = w_{Tac}/w_{Ts}$      (Turbine work is out)

$\eta_{\text{compressor}} = w_{Cs}/w_{Cac}$      (Compressor work is in)

$\eta_{\text{pump}} = w_{Ps}/w_{Pac}$      (Pump work is in)

$\eta_{\text{nozzle}} = \dfrac{1}{2}\mathbf{V}_{ac}^2 / \dfrac{1}{2}\mathbf{V}_s^2$      (Kinetic energy is out)

Available work from heat

$$W = Q\left(1 - \frac{T_0}{T_H}\right)$$

Reversible flow work with extra $q_0^{\text{rev}}$

from ambient at $T_0$ and $q$ in at $T_H$    $q_0^{\text{rev}} = T_0(s_e - s_i) - q\dfrac{T_0}{T_H}$

$$w^{\text{rev}} = h_i - h_e - T_0(s_i - s_e) + q\left(1 - \frac{T_0}{T_H}\right)$$

Flow irreversibility

$$i = w^{\text{rev}} - w = q_0^{\text{rev}} = T_0 \dot{S}_{\text{gen}}/\dot{m} = T_0 s_{\text{gen}}$$

Second-law efficiency

$$\eta_{\text{2nd law}} = \frac{\text{wanted output}}{\text{supplied input}} \text{ (availability transfers)}$$

Exergy, flow availability

$$\psi = \left[h - T_0 s + \frac{1}{2}\mathbf{V}^2 + gZ\right] - [h_0 - T_0 s_0 + gZ_0]$$

Exergy destruction

$$\text{Rate of destruction} = T_0 \dot{S}_{\text{gen}} = \dot{I}$$

# HOMEWORK PROBLEMS

## Concept Problems

**9.1** Which process will make the statement in In-Text Concept question *e* (on p. 267) true?

**9.2** Friction in a pipe flow causes a slight pressure decrease and a slight temperature increase. How does that affect entropy?

**9.3** A tank contains air at 400 kPa, 300 K and a valve opens up for flow to the outside, which is at 100 kPa, 300 K. What happens to the air temperature inside?

**9.4** If we follow a mass element going through a reversible adiabatic flow process, what can we say about the change of state?

**9.5** A reversible process in a steady flow with negligible kinetic and potential energy changes is shown in Fig. P9.5. Indicate the change $h_e - h_i$ and transfers $w$ and $q$ as positive, zero, or negative.

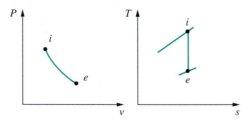

**FIGURE P9.5**

**9.6** A reversible process in a steady flow of air with negligible kinetic and potential energy changes is shown in Fig. P9.6. Indicate the change $h_e - h_i$ and transfers $w$ and $q$ as positive, zero, or negative.

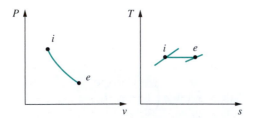

**FIGURE P9.6**

**9.7** To increase the work out of a turbine for a given inlet and exit pressure, how should the inlet state be changed?

**9.8** Liquid water is sprayed into the hot gases before they enter the turbine section of a large gas turbine power plant. It is claimed that the larger mass flow rate produces more work. Is that the reason?

**9.9** An air compressor has a significant heat transfer out. See Example 9.4 for how high $T$ becomes if there is no heat transfer. Is that good, or should it be insulated?

**9.10** The shaft work in a pump to increase the pressure is small compared to the shaft work in an air compressor for the same pressure increase. Why?

**9.11** In a single steady flow what increases the exergy the most?

   **a.** 1 kW of heat transfer

   **b.** 1 kW of shaft work added to the flow

**9.12** A flow of water at room $T$ and $P$ is cooled by a refrigerator to 5°C. Has its energy increased or decreased? Has its exergy increased or decreased?

## Steady-state reversible processes

### Simple devices

**9.13** A compressor brings a hydrogen gas flow at 280 K, 100 kPa up to a pressure of 1000 kPa in a reversible process. How hot is the exit flow and what is the specific work input?

**9.14** A first stage in a turbine receives steam at 10 MPa and 800°C, with an exit pressure of 800 kPa. Assume the stage is adiabatic and neglect kinetic energies. Find the exit temperature and the specific work.

**9.15** Steam enters a turbine at 3 MPa and 450°C, expands in a reversible adiabatic process, and exhausts at 10 kPa. Changes in kinetic and potential energies between the inlet and the exit of the turbine are small. The power output of the turbine is 800 kW. What is the mass flow rate of steam through the turbine?

**9.16** A compressor in a commercial refrigerator receives R-22 at −25°C and $x = 1$. The exit is at 1000 kPa and the process assumed reversible and adiabatic. Neglect kinetic energies and find the exit temperature and the specific work.

**9.17** In a heat pump that uses R-134a as the working fluid, the R-134a enters the compressor at 150 kPa and −10°C at a rate of 0.1 kg/s. In the compressor the R-134a is compressed in an adiabatic process to 1 MPa. Calculate the power input required to the compressor, assuming the process to be reversible.

**9.18** A boiler section boils 3 kg/s saturated liquid water at 2000 kPa to saturated vapor in a reversible constant-pressure process. Assume you do not know that there is no work. Prove that there is no shaftwork using the first and second laws of thermodynamics.

**9.19** Consider the design of a nozzle in which nitrogen gas flowing in a pipe at 500 kPa and 200°C at a velocity of 10 m/s is to be expanded to produce a velocity of 300 m/s. Determine the exit pressure and cross-sectional area of the nozzle if the mass flow rate is 0.15 kg/s and the expansion is reversible and adiabatic.

**9.20** Atmospheric air at −45°C and 60 kPa enters the front diffuser of a jet engine, shown in Fig. P9.20, with a velocity of 900 km/h and frontal area of 1 m². After leaving the adiabatic diffuser, the velocity is 20 m/s. Find the diffuser exit temperature and the maximum pressure possible.

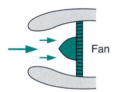

**FIGURE P9.20**

**9.21** A highly cooled compressor brings a hydrogen gas flow at 300 K, 100 kPa up to a pressure of 1000 kPa in an isothermal process. Find the specific work assuming a reversible process.

**9.22** A compressor is surrounded by cold R-134a so it works as an isothermal compressor. The inlet state is 0°C, 100 kPa, and the exit state is saturated vapor. Find the specific heat transfer and specific work.

**9.23** The exit nozzle in a jet engine receives air at 1200 K and 150 kPa with negligible kinetic energy. The exit pressure is 80 kPa, and the process is reversible and adiabatic. Use constant heat capacity at 300 K to find the exit velocity.

**9.24** Do the previous problem using the air tables in Table A.7.

**9.25** An expander receives 0.5 kg/s air at 2000 kPa, 300 K with an exit state of 400 kPa, 300 K. Assume the process is reversible and isothermal. Find the rates of heat transfer and work neglecting kinetic and potential energy changes.

**9.26** A flow of 2 kg/s saturated vapor R-22 at 500 kPa is heated at constant pressure to 60°C. The heat is supplied by a heat pump that receives heat from the ambient at 300 K and work input shown in Fig. P9.26. Assume everything is reversible and find the rate of work input.

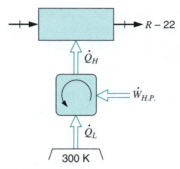

**FIGURE P9.26**

**9.27** A flow of 5 kg/s water at 100 kPa, 20°C should be delivered as steam at 1000 kPa, 350°C to some application. We have a heat source at constant 500°C. If the process should be reversible, how much heat transfer should we have?

**9.28** A reversible steady-state device receives a flow of 1 kg/s air at 400 K and 450 kPa, and the air leaves at 600 K and 100 kPa. Heat transfer of 800 kW is added from a 1000 K reservoir, 100 kW is rejected at 350 K, and some heat transfer takes place at 500 K. (See Fig. P9.28.) Find the heat transferred at 500 K and the rate of work produced.

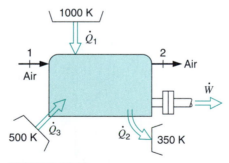

**FIGURE P9.28**

**9.29** A steam turbine in a power plant receives 5 kg/s steam at 3000 kPa, 500°C. Twenty percent of the flow is extracted at 1000 kPa to a feedwater heater and the remainder flows out at 200 kPa. Find the two exit temperatures and the turbine power output.

**9.30** A reversible adiabatic compression of an air flow from 20°C, 100 kPa to 200 kPa is followed by an expansion down to 100 kPa in an ideal nozzle. What are the two processes? How hot does the air get? What is the exit velocity?

**9.31** A small turbine delivers 150 kW and is supplied with steam at 700°C and 2 MPa. The exhaust passes through a heat exchanger where the pressure is 10 kPa and exits as saturated liquid. The turbine is reversible and adiabatic. Find the specific turbine work and the heat transfer in the heat exchanger.

**9.32** Two flows of air are both at 200 kPa; one has 1 kg/s at 400 K, and the other has 2 kg/s at 290 K. The two lines exchange energy through a number of ideal heat engines, taking energy from the hot line and rejecting it to the colder line. The two flows then leave at the same temperature. Assume the whole setup is reversible and find the exit temperature and the total power out of the heat engines.

**9.33** Consider a steam turbine power plant operating near critical pressure, as shown in Fig. P9.33. As a first approximation, it may be assumed that the turbine and the pump processes are reversible and adiabatic. Neglecting any changes in kinetic and potential energies, calculate

  **a.** The specific turbine work output and the turbine exit state.

  **b.** The pump work input and enthalpy at the pump exit state.

  **c.** The thermal efficiency of the cycle.

$$P_4 = P_1 = 20 \text{ MPa} \qquad T_1 = 700°C$$
$$P_2 = P_3 = 20 \text{ kPa} \qquad T_3 = 40°C$$

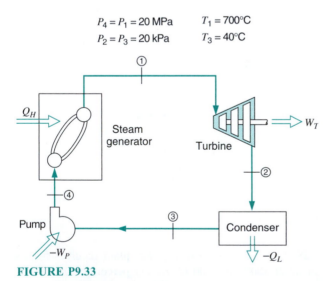

**FIGURE P9.33**

**9.34** A turbocharger boosts the inlet air pressure to an automobile engine. It consists of an exhaust gas-driven turbine directly connected to an air compressor, as shown in Fig. P9.34. For a certain engine load, the conditions are given in the figure. Assume that both the turbine and the compressor are reversible and adiabatic, having also the same mass flow rate. Calculate the turbine exit temperature and power output. Find also the compressor exit pressure and temperature.

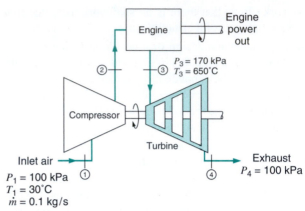

$$P_3 = 170 \text{ kPa}$$
$$T_3 = 650°C$$

$$P_1 = 100 \text{ kPa}$$
$$T_1 = 30°C$$
$$\dot{m} = 0.1 \text{ kg/s}$$

$$P_4 = 100 \text{ kPa}$$

**FIGURE P9.34**

## Reversible shaft work, Bernoulli equation

**9.35** A pump has a 2 kW motor. How much liquid water at 15°C can I pump to 250 kPa from 100 kPa?

**9.36** A garden water hose has liquid water at 200 kPa and 15°C. How high a velocity can be generated in a small ideal nozzle? If you direct the water spray straight up how high will it go?

**9.37** A small pump takes in water at 20°C and 100 kPa and pumps it to 2.5 MPa at a flow rate of 100 kg/min. Find the required pump power input.

**9.38** A large storage tank contains saturated liquid nitrogen at ambient pressure, 100 kPa; it is to be pumped to 500 kPa and fed to a pipeline at the rate of 0.5 kg/s. How much power input is required for the pump, assuming it to be reversible?

**9.39** Liquid water at ambient conditions, 100 kPa and 25°C, enters a pump at the rate of 0.5 kg/s. Power input to the pump is 3 kW. Assuming the pump process to be reversible, determine the pump exit pressure and temperature.

**9.40** An irrigation pump takes water from a river at 10°C, 100 kPa and pumps it up to an open canal at a 100 m higher elevation. The pipe diameter in and out of the pump is 0.1 m and the motor driving the pump is 5 hp. Neglect kinetic energies and friction; find the maximum possible mass flow rate.

**9.41** A small dam has a 0.5-m-diameter pipe carrying liquid water at 150 kPa and 20°C with a flow rate of 2000 kg/s. The pipe runs to the bottom of the dam 15 m lower into a turbine with pipe diameter 0.35 m, shown in Fig. P9.41. Assume no friction or heat transfer in the pipe

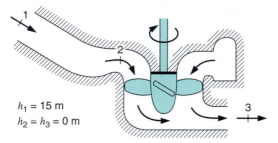

$h_1 = 15$ m
$h_2 \approx h_3 = 0$ m

**FIGURE P9.41**

and find the pressure of the turbine inlet. If the turbine exhausts to 100 kPa with negligible kinetic energy, what is the rate of work?

**9.42** A firefighter on a ladder 25 m above ground should be able to spray water an additional 10 m up with the hose nozzle of exit diameter 2.5 cm. Assume a water pump on the ground and a reversible flow (hose, nozzle included) and find the minimum required power.

**9.43** A wave comes rolling in to the beach at 2 m/s horizontal velocity. Neglect friction and find how high up (elevation) on the beach the wave will reach.

**9.44** A small pump is driven by at 2 kW motor with liquid water at 150 kPa and 10°C entering. Find the maximum water flow rate you can get with an exit pressure of 1 MPa and negligible kinetic energies. The exit flow goes through a small hole in a spray nozzle out to the atmosphere at 100 kPa, shown in Fig. P9.44. Find the spray velocity.

— Nozzle

**FIGURE P9.44**

**9.45** Saturated R-134a at $-10$°C is pumped/compressed to a pressure of 1.0 MPa at the rate of 0.5 kg/s in a reversible adiabatic process. Calculate the power required and the exit temperature for the two cases of inlet state of the R-134a:

    **a.** Quality of 100%    **b.** Quality of 0%.

**9.46** A small water pump on ground level has an inlet pipe down into a well at a depth $H$ with the water at 100 kPa and 15°C. The pump delivers water at 400 kPa to a building. The absolute pressure of the water must be at least twice the saturation pressure to avoid cavitation. What is the maximum depth this setup will allow?

**9.47** Atmospheric air at 100 kPa and 17°C blows at 60 km/h toward the side of a building. Assuming the air is nearly incompressible, find the pressure and the temperature at the stagnation (zero-velocity) point on the wall.

**9.48** You drive on the highway at 120 km/h on a day with 17°C, 100 kPa atmosphere. When you put your hand out of the window flat against the wind you feel the force from the air stagnating (i.e., it comes to relative zero velocity on your skin). Assume that the air is nearly incompressible and find the air temperature and pressure right on your hand.

**9.49** An airflow at 100 kPa, 290 K, and 200 m/s is directed toward a wall. At the wall the flow stagnates (comes to zero velocity) without any heat transfer, as shown in Fig. P9.49. Find the stagnation pressure (a) assuming incompressible flow, (b) assuming an adiabatic compression. *Hint*: $T$ comes from the energy equation.

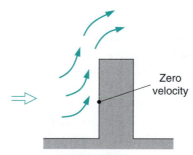

Zero velocity

**FIGURE P9.49**

**9.50** Helium gas enters a steady-flow expander at 800 kPa and 300°C and exits at 120 kPa. The mass flow rate is 0.2 kg/s, and the expansion process can be considered as a reversible polytropic process with exponent $n = 1.3$. Calculate the power output of the expander.

**9.51** Air at 100 kPa and 300 K flows through a device at steady state with the exit at 1000 K during which it went through a polytropic process with $n = 1.3$. Find the exit pressure, the specific work, and heat transfer.

**9.52** An expansion in a gas turbine can be approximated with a polytropic process with exponent $n = 1.25$. The inlet air is at 1200 K, 800 kPa and the exit pressure is 125 kPa with a mass flow rate of 0.75 kg/s. Find the turbine heat transfer and power output.

## Steady-state irreversible processes

**9.53** R-134a at 30°C, 800 kPa is throttled in a steady flow to a lower pressure so it comes out at −10°C. What is the specific entropy generation?

**9.54** Analyze the steam turbine described in Problem 6.64. Is it possible?

**9.55** Methane at 1 MPa, 300 K is throttled through a valve to 100 kPa. Assume no change in the kinetic energy and ideal gas behavior. What is the specific entropy generation?

**9.56** Carbon dioxide at 300 K and 200 kPa flows through a steady device where it is heated to 500 K by a 600 K reservoir in a constant-pressure process. Find the specific work, specific heat transfer, and specific entropy generation.

**9.57** The throttle process described in Example 6.5 is an irreversible process. Find the entropy generation per kg of ammonia in the throttling process.

**9.58** A large condenser in a steam power plant dumps 15 MW at 45°C with an ambient at 25°C. What is the entropy generation rate?

**9.59** Two flowstreams of water, one of saturated vapor at 0.6 MPa, and the other at 0.6 MPa and 600°C, mix adiabatically in a steady flow to produce a single flow out at 0.6 MPa and 400°C. Find the total entropy generation for this process.

**9.60** A heat exchanger that follows a compressor receives 0.1 kg/s air at 1000 kPa and 500 K and cools it in a constant-pressure process to 320 K. The heat is absorbed by ambient air at 300 K. Find the total rate of entropy generation.

**9.61** A compressor in a commercial refrigerator receives R-22 at −25°C and $x = 1$. The exit is at 1000 kPa and 60°C. Neglect kinetic energies and find the specific entropy generation.

**9.62** A condenser in a power plant receives 5 kg/s steam at 15 kPa with a quality of 90% and rejects the heat to cooling water with an average temperature of 17°C. Find the power given to the cooling water in this constant-pressure process, shown in Fig. P9.62, and the total rate of entropy generation when saturated liquid exits the condenser.

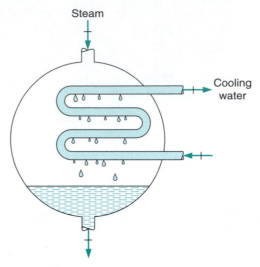

Steam

Cooling water

**FIGURE P9.62**

**9.63** Two flows of air are both at 200 kPa; one has 1 kg/s at 400 K, and the other has 2 kg/s at 290 K. The two flows are mixed together in an insulated box to produce a single exit flow at 200 kPa. Find the exit temperature and the total rate of entropy generation.

**9.64** A steam turbine has an inlet of 2 kg/s water at 1000 kPa and 350°C with velocity of 15 m/s. The exit is at 100 kPa, 150°C and very low velocity. Find the power produced and the rate of entropy generation.

**9.65** Saturated liquid nitrogen at 600 kPa enters a boiler at a rate of 0.005 kg/s and exits as saturated vapor. It then flows into a super heater also at 600 kPa where it exits at 600 kPa, 280 K. Assume the heat transfer comes from a 300 K source and find the rates of entropy generation in the boiler and the super heater.

**9.66** A counterflowing heat exchanger has one line with 2 kg/s air at 125 kPa and 1000 K entering, and the air is leaving at 100 kPa and 400 K. The other line has 0.5 kg/s water coming in at 200 kPa and 20°C and leaving at 200 kPa. (See Fig. P9.66.) What is the exit temperature of the water and the total rate of entropy generation?

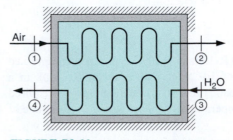

Air

$H_2O$

**FIGURE P9.66**

**9.67** Air at 1000 kPa, 300 K is throttled to 500 kPa. What is the specific entropy generation?

**9.68** A supply of 5 kg/s ammonia at 500 kPa and 20°C is needed. Two sources are available: One is saturated liquid at 20°C, and the other is at 500 kPa and 140°C. Flows from the two sources are fed through valves to an insulated mixing chamber, which then produces the desired output state. Find the two source mass flow rates and the total rate of entropy generation by this setup.

**9.69** A coflowing (same direction) as shown in Fig. P9.69, heat exchanger has one line with 0.25 kg/s oxygen at 17°C and 200 kPa entering, and the other line has 0.6 kg/s nitrogen at 150 kPa and 500 K entering. The heat exchanger is long enough so that the two flows exit at the same temperature. Use constant heat capacities and find the exit temperature and the total rate of entropy generation.

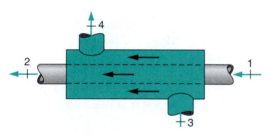

**FIGURE P9.69**

**9.70** A steam turbine in a power plant receives steam at 3000 kPa, 500°C. The turbine has two exit flows, one is 20% of the flow at 1000 kPa, 350°C to a feedwater heater and the remainder flows out at 200 kPa, 200°C. Find the specific turbine work and the specific entropy generation both per kg flow in.

**9.71** Carbon dioxide flows through a device entering at 300 K and 200 kPa and leaving at 500 K. The process is steady-state polytropic with $n = 3.8$, and heat transfer comes from a 600 K source. Find the specific work, specific heat transfer, and specific entropy generation due to this process.

## Transient processes

**9.72** Calculate the specific entropy generated in the filling process given in Example 6.11.

**9.73** Calculate the total entropy generated in the filling process given in Example 6.12.

**9.74** A 1 L can of R-134a is at room temperature 20°C with a quality of 50%. A leak in the top valve allows vapor to escape and heat transfer from the room takes place so we reach final state of 5°C with a quality of 100%.

Find the mass that escaped, the heat transfer, and the entropy generation not including that made in the valve.

**9.75** An initially empty 0.1 m³ canister is filled with R-12 from a line flowing saturated liquid at −5°C. This is done quickly such that the process is adiabatic. Find the final mass, and determine liquid and vapor volumes, if any, in the canister. Is the process reversible?

**9.76** Air in a tank is at 300 kPa and 400 K with a volume of 2 m³. A valve on the tank is opened to let some air escape to the ambient surroundings to leave a final pressure inside of 200 kPa. At the same time the tank is heated so the air remaining has a constant temperature. What is the mass average value (Table A.7 reference) of the $s$ leaving, assuming this is an internally reversible process?

**9.77** An empty canister of 0.002 m³ is filled with R-134a from a line flowing saturated liquid R-134a at 0°C. The filling is done quickly so it is adiabatic. Find the final mass in the canister and the total entropy generation.

**9.78** A 10 m tall 0.1 m diameter pipe is filled with liquid water at 20°C. It is open at the top to the atmosphere, 100 kPa, and a small nozzle is mounted in the bottom. The water is now let out through the nozzle splashing out to the ground until the pipe is empty. Find the water initial exit velocity, the average kinetic energy in the exit flow, and the total entropy generation for the process.

**9.79** Air in a tank is at 300 kPa and 400 K with a volume of 2 m³. A valve on the tank is opened to let some air escape to the ambient surroundings to leave a final pressure inside of 200 kPa. Find the final temperature and mass assuming a reversible adiabatic process for the air remaining inside the tank.

**9.80** A 1 m³ rigid tank contains 100 kg of R-22 at ambient temperature, 15°C shown in Fig. P9.80. A valve on top of the tank is opened, and saturated vapor is throttled to ambient pressure, 100 kPa, and flows to a collector system. During the process, the temperature inside the tank remains at 15°C. The valve is closed when no more liquid remains inside. Calculate the heat transfer to the tank and the total entropy generation in the process.

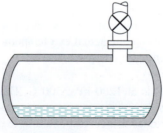

**FIGURE P9.80**

**9.81** A balloon is filled with air from a line at 200 kPa, 300 K to a final state of 110 kPa, 300 K with a mass of 0.1 kg air. Assume the pressure is proportional to the balloon volume as: $P = 100 \text{ kPa} + CV$. Find the heat transfer to/from the ambient at 300 K and the total entropy generation.

**9.82** An old abandoned salt mine, 100 000 m³ in volume, contains air at 290 K and 100 kPa. The mine is used for energy storage so the local power plant pumps it up to 2.1 MPa using outside air at 290 K and 100 kPa. Assume the pump is ideal and the process is adiabatic. Find the final mass and temperature of the air and the required pump work.

**9.83** An insulated 2 m³ tank is to be charged with R-134a from a line flowing the refrigerant at 3 MPa. The tank is initially evacuated, and the valve is closed when the pressure inside the tank reaches 3 MPa. The line is supplied by an insulated compressor that takes in R-134a at 5°C, with a quality of 96.5%, and compresses it to 3 MPa in a reversible process. Calculate the total work input to the compressor to charge the tank.

**9.84** A 0.2 m³ initially empty container is filled with water from a line at 500 kPa and 200°C until there is no more flow. Assume the process is adiabatic and find the final mass, final temperature, and total entropy generation.

**9.85** An initially empty canister with a volume of 0.2 m³ is filled with carbon dioxide from a line at 1000 kPa and 500 K. Assume the process is adiabatic and the flow continues until it stops by itself. Use constant heat capacity to solve for the final mass and temperature of the carbon dioxide in the canister and the total entropy generation by the process.

**9.86** A cook filled a pressure cooker with 3 kg water at 20°C and a small amount of air and forgot about it. The pressure cooker has a vent valve so if $P > 200$ kPa, steam escapes to maintain the pressure at 200 kPa. How much entropy was generated in the throttling of the steam through the vent to 100 kPa when half the original mass has escaped?

## Device efficiency

**9.87** A compressor is used to bring saturated water vapor at 1 MPa up to 17.5 MPa, where the actual exit temperature is 650°C. Find the isentropic compressor efficiency and the entropy generation.

**9.88** A steam turbine inlet is at 1200 kPa, 500°C. The exit is at 200 kPa. What is the lowest possible exit temperature? Which efficiency does that correspond to?

**9.89** A steam turbine inlet is at 1200 kPa, 500°C. The exit is at 200 kPa. What is the highest possible exit temperature? Which efficiency does that correspond to?

**9.90** A steam turbine inlet is at 1200 kPa, 500°C. The exit is at 200 kPa, 275°C. What is the isentropic efficiency?

**9.91** A pump receives water at 100 kPa and 15°C and has a power input of 1.5 kW. The pump has an isentropic efficiency of 75%, and it should flow 1.2 kg/s delivered at 30 m/s exit velocity. How high an exit pressure can the pump produce?

**9.92** A centrifugal compressor takes in ambient air at 100 kPa and 15°C and discharges it at 450 kPa. The compressor has an isentropic efficiency of 80%. What is your best estimate for the discharge temperature?

**9.93** Find the isentropic efficiency of the R-134a compressor in Example 6.10, assuming the ideal compressor is adiabatic.

**9.94** An emergency drain pump, shown in Fig. P9.94, should be able to pump 0.1 m³/s of liquid water at 15°C, 10 m vertically up delivering it with a velocity of 20 m/s. It is estimated that the pump, pipe, and nozzle have a combined isentropic efficiency expressed for the pump as 60%. How much power is needed to drive the pump?

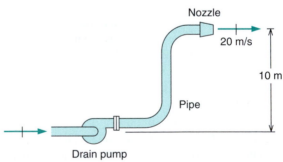

**FIGURE P9.94**

**9.95** A compressor in a commercial refrigerator receives R-22 at −25°C and $x = 1$. The exit is at 1000 kPa and 60°C. Neglect kinetic energies and find the isentropic compressor efficiency.

**9.96** Find the isentropic efficiency of the nozzle in Example 6.4.

**9.97** The exit velocity of a nozzle is 500 m/s. If $\eta_{\text{nozzle}} = 0.88$, what is the ideal exit velocity?

**9.98** A small air turbine with an isentropic efficiency of 80% should produce 270 kJ/kg of work. The inlet temperature is 1000 K, and the turbine exhausts to the atmosphere. Find the required inlet pressure and the exhaust temperature.

**9.99** Steam enters a turbine at 300°C, 600 kPa, and exhausts as saturated vapor at 20 kPa. What is the isentropic efficiency?

**9.100** A compressor in an industrial air-conditioner compresses ammonia from a state of saturated vapor at 150 kPa to a pressure 800 kPa. At the exit, the temperature is measured to be 100°C and the mass flow rate is 0.5 kg/s. What is the required motor size for this compressor and what is its isentropic efficiency?

**9.101** A turbine receives air at 1500 K and 1000 kPa and expands it to 100 kPa. The turbine has an isentropic efficiency of 85%. Find the actual turbine exit air temperature and the specific entropy increase in the actual turbine.

**9.102** The small turbine in Problem 9.31 was ideal. Assume instead that the isentropic turbine efficiency is 88%. Find the actual specific turbine work and the entropy generated in the turbine.

**9.103** Carbon dioxide, $CO_2$, enters an adiabatic compressor at 100 kPa and 300 K and exits at 1000 kPa and 520 K. Find the compressor efficiency and the entropy generation for the process.

**9.104** Air enters an insulated compressor at ambient conditions, 100 kPa and 20°C, at the rate of 0.1 kg/s and exits at 200°C. The isentropic efficiency of the compressor is 70%. Assume the ideal and actual compressor have the same exit pressure. What is the exit pressure? How much power is required to drive the unit?

**9.105** Assume an actual compressor has the same exit pressure and specific heat transfer as the ideal isothermal compressor in Problem 9.22 with an isothermal efficiency of 80%. Find the specific work and exit temperature for the actual compressor.

**9.106** A water-cooled air compressor takes air in at 20°C and 90 kPa and compresses it to 500 kPa. The isothermal efficiency is 80%, and the actual compressor has the same heat transfer as the ideal one. Find the specific compressor work and the exit temperature.

**9.107** A nozzle in a high-pressure liquid water sprayer has an area of 0.5 cm². It receives water at 250 kPa, 20°C, and the exit pressure is 100 kPa. Neglect the inlet kinetic energy and assume a nozzle isentropic efficiency of 85%. Find the ideal nozzle exit velocity and the actual nozzle mass flow rate.

**9.108** A nozzle should produce a flow of air with 200 m/s at 20°C and 100 kPa. It is estimated that the nozzle has an isentropic efficiency of 92%. What nozzle inlet pressure and temperature are required assuming the inlet kinetic energy is negligible?

**9.109** Air flows into an insulated nozzle at 1 MPa and 1200 K with 15 m/s and a mass flow rate of 2 kg/s. It expands to 650 kPa, and the exit temperature is 1100 K. Find the exit velocity and the nozzle efficiency.

# Reversible work, availability, and exergy

**9.110** Find the availability of 110 kW delivered at 500 K when the ambient temperature is 300 K.

**9.111** Consider availability (exergy) associated with a flow. The total exergy is based on the thermodynamic state and the kinetic and potential energies. Can they all be negative?

**9.112** Does a reversible process change the availability if there is no work involved?

**9.113** Are reversible work and availability (exergy) connected?

**9.114** Find the specific reversible work for an R-134a compressor with inlet state of −20°C, 100 kPa and an exit state of 600 kPa, 50°C. Use a 25°C ambient temperature.

**9.115** Find the specific reversible work for a steam turbine with inlet at 4 MPa and 500°C and an actual exit state of 100 kPa, $x = 1.0$ with 25°C ambient surroundings.

**9.116** A heat engine receives 5 kW at 800 K and 10 kW at 1000 K, rejecting energy by heat transfer at 600 K. Assume it is reversible and find the power output. How much power could be produced if it could reject energy at $T_0 = 298$ K?

**9.117** Is the reversible work between two states the same as ideal work for the device?

**9.118** When is the reversible work the same as the isentropic work?

**9.119** If I heat some cold liquid water to $T_0$, do I increase its availability?

**9.120** An airflow of 5 kg/min at 1500 K, 125 kPa, goes through a constant pressure heat exchanger, giving energy to a heat engine shown in Fig. P9.120. The air exits at 500 K, and the ambient is at 298 K, 100 kPa. Find the rate of heat transfer delivered to the engine and the power the engine can produce.

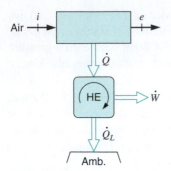

Air

$i$  $e$

$\dot{Q}$

HE  $\dot{W}$

$\dot{Q}_L$

Amb.

**FIGURE P9.120**

**9.121** Calculate the irreversibility for the condenser in Problem 9.62 assuming an ambient temperature at 17°C.

**9.122** A flow of air at 1000 kPa, 300 K is throttled to 500 kPa. What is the irreversibility? What is the drop in flow availability?

**9.123** The throttle process in Example 6.5 is an irreversible process. Find the reversible work and irreversibility assuming an ambient temperature at 25°C.

**9.124** A supply of steam at 100 kPa, 150°C, is needed in a hospital for cleaning purposes at a rate of 15 kg/s. A supply of steam at 150 kPa, 250°C, is available from a boiler, and tap water at 100 kPa, 15°C, is also available. The two sources are then mixed in a mixing chamber to generate the desired state as output. Determine the rate of irreversibility of the mixing process.

**9.125** Two flows of air both at 200 kPa of equal flow rates mix in an insulated mixing chamber. One flow is at 1500 K, and the other is at 300 K. Find the irreversibility in the process per kilogram of air flowing out.

**9.126** A 2 kg/s flow of steam at 1 MPa and 700°C should be brought to 500°C by spraying in liquid water at 1 MPa and 20°C in a steady flow. Find the rate of irreversibility, assuming that the surroundings are at 20°C.

**9.127** A steady combustion of natural gas yields 0.15 kg/s of products (having approximately the same properties as air) at 1100°C and 100 kPa. The products are passed through a heat exchanger and exit at 550°C. What is the maximum theoretical power output from a cyclic heat engine operating on the heat rejected from the combustion products, assuming that the ambient temperature is 20°C?

**9.128** A steady stream of R-22 at ambient temperature, 10°C, and at 750 kPa enters a solar collector. The stream exits at 80°C and 700 kPa. Calculate the change in availability of the R-22 between these two states.

**9.129** Find the availability at all four states in the power plant of Problem 9.33 with an ambient temperature at 298 K.

**9.130** Water at 1000°C, 15 MPa is flowing through a heat exchanger giving off energy to come out saturated liquid water at 15 MPa in a steady flow process. Find the specific heat transfer and the specific flow availability (exergy) the water has delivered.

**9.131** A heat pump has a coefficient of performance of 2 using a power input of 4 kW. Its low temperature is 20°C and the high temperature is 80°C, with an ambient at $T_0$. Find the fluxes of exergy associated with the energy fluxes in and out.

**9.132** Calculate the change in availability (kW) of the two flows in Problem 9.66 assuming $T_0 = 20$°C.

**9.133** Find the change in availability from inlet to exit of the condenser in Problem 9.33.

**9.134** Nitrogen flows in a pipe with a velocity of 300 m/s at 500 kPa and 300°C. What is its availability with respect to ambient surroundings at 100 kPa and 20°C?

**9.135** Air flows at 1500 K and 100 kPa through a constant-pressure heat exchanger giving energy to a heat engine and comes out at 500 K. At what constant temperature should the same heat transfer be delivered to provide the same availability?

**9.136** A steam turbine inlet is at 1200 kPa, 500°C. The actual exit is at 300 kPa with an actual work of 407 kJ/kg. What is its second law efficiency?

**9.137** A heat exchanger increases the availability of 3 kg/s water by 1650 kJ/kg using 10 kg/s air coming in at 1400 K and leaving with 600 kJ/kg less availability. What are the irreversibility and the second law efficiency?

**9.138** Air enters a compressor at ambient conditions, 100 kPa and 300 K, and exits at 800 kPa. If the isentropic compressor efficiency is 85%, what is the second-law efficiency of the compressor process?

**9.139** A heat engine receives 1 kW heat transfer at 1000 K and gives out 600 W as work with the rest as heat transfer to the ambient. What are the fluxes of exergy in and out?

**9.140** A heat engine receives 1 kW heat transfer at 1000 K and gives out 600 W as work with the rest as heat transfer to the ambient. Find its first and second law efficiencies.

**9.141** A steam turbine has inlet at 4 MPa and 500°C and actual exit of 100 kPa with $x = 1.0$. Find its first-law (isentropic) and its second-law efficiencies.

**9.142** A compressor takes in saturated vapor R-134a at $-20$°C and delivers it at 30°C and 0.4 MPa. Assuming that the compression is adiabatic, find the isentropic efficiency and the second-law efficiency.

**9.143** Find the second law efficiency of the heat pump in Problem 9.131.

**9.144** A compressor in a refrigerator receives saturated vapor R-22 at 150 kPa and it brings it up to 600 kPa, 50°C in an adiabatic compression. Find the specific work, entropy generation, and irreversibility.

**9.145** Find the isentropic efficiency and the second law efficiency for the compressor in Problem 9.144.

**9.146** Calculate the second-law efficiency of the counter-flowing heat exchanger in Problem 9.66 with an ambient temperature at 20°C.

**9.147** Air enters a steady-flow turbine at 1600 K and exhausts to the atmosphere at 1000 K. The second-law efficiency is 85%. What is the turbine inlet pressure?

**9.148** Find the isentropic efficiency and the second law efficiency for the steam turbine in Problem 9.70.

**9.149** A flow of 0.1 kg/s hot water at 80°C is mixed with a flow of 0.2 kg/s cold water at 20°C in a shower fixture. What is the rate of exergy destruction (irreversibility) for this process?

**9.150** Consider the car engine in Example 7.1 and assume the fuel energy is delivered at a constant 1500 K. The 70% of the energy that is lost is 40% exhaust flow at 900 K, and the remainder 30% heat transfer to the walls at 450 K goes on to the coolant fluid at 370 K, finally ending up in atmospheric air at ambient 20°C. Find all the energy and exergy flows for this heat engine. Also find the exergy destruction and where that is done.

# Gas Mixtures

Up to this point in our development of thermodynamics we have considered primarily pure substances. A large number of thermodynamic problems involve mixtures of different pure substances. Sometimes these mixtures are referred to as solutions, particularly in the liquid and solid phases.

In this chapter we shall turn our attention to various thermodynamic considerations of gas mixtures. We begin with a consideration of a rather simple problem: mixtures of ideal gases. This leads to a consideration of a simplified but very useful model of certain mixtures, such as air and water vapor, which may involve a condensed (solid or liquid) phase of one of the components.

## 10.1 GENERAL CONSIDERATIONS AND MIXTURES OF IDEAL GASES

Let us consider a general mixture of $N$ components, each a pure substance, so the total mass and the total number of moles are

$$m_{\text{tot}} = m_1 + m_2 + \cdots + m_N = \sum m_i$$

$$n_{\text{tot}} = n_1 + n_2 + \cdots + n_N = \sum n_i$$

The mixture is usually described by a mass fraction (concentration)

$$c_i = \frac{m_i}{m_{tot}} \tag{10.1}$$

or a mole fraction for each component as

$$y_i = \frac{n_i}{n_{tot}} \tag{10.2}$$

which are related through the molecular weight, $M_i$, as $m_i = n_i M_i$. We may then convert from a mole basis to a mass basis as

$$c_i = \frac{m_i}{m_{tot}} = \frac{n_i M_i}{\sum n_j M_j} = \frac{n_i M_i / n_{tot}}{\sum n_j M_j / n_{tot}} = \frac{y_i M_i}{\sum y_j M_j} \tag{10.3}$$

and from a mass basis to a mole basis as

$$y_i = \frac{n_i}{n_{tot}} = \frac{m_i / M_i}{\sum m_j / M_j} = \frac{m_i / (M_i m_{tot})}{\sum m_j / (M_j m_{tot})} = \frac{c_i / M_i}{\sum c_j / M_j} \tag{10.4}$$

The molecular weight for the mixture becomes

$$M_{mix} = \frac{m_{tot}}{n_{tot}} = \frac{\sum n_i M_i}{n_{tot}} = \sum y_i M_i \tag{10.5}$$

**EXAMPLE 10.1**

A mole-basis analysis of a gaseous mixture yields the following results:

CO$_2$    12.0%
O$_2$     4.0
N$_2$     82.0
CO     2.0

Determine the analysis on a mass basis and the molecular weight for the mixture. Assume ideal-gas behavior.

*Control mass*:  Gas mixture.
*State*:  Composition known.

TABLE 10.1

| Constituent | Percent by Mole | Mole Fraction | Molecular Weight | Mass kg per kmol of Mixture | Analysis on Mass Basis, Percent |
|---|---|---|---|---|---|
| CO$_2$ | 12 | 0.12 | × 44.0 | = 5.28 | $\frac{5.28}{30.08}$ = 17.55 |
| O$_2$ | 4 | 0.04 | × 32.0 | = 1.28 | $\frac{1.28}{30.08}$ = 4.26 |
| N$_2$ | 82 | 0.82 | × 28.0 | = 22.96 | $\frac{22.96}{30.08}$ = 76.33 |
| CO | 2 | 0.02 | × 28.0 | = 0.56 | 0.56 = 1.86 |
| | | | | 30.08 | 30.08    100.00 |

**Solution**

It is convenient to set up and solve this problem as shown in Table 10.1. The mass-basis analysis is found using Eq. 10.3, as shown in the table. It is also noted that during this calculation, the molecular weight of the mixture is found to be 30.08.

If the analysis has been given on a mass basis, and the mole fractions or percentages are desired, the procedure shown in Table 10.2 is followed, using Eq. 10.4.

TABLE 10.2

| Constituent | Mass Fraction | | Molecular Weight | | kmol per kg of Mixture | Mole Fraction | Mole Percent |
|---|---|---|---|---|---|---|---|
| $CO_2$ | 0.1755 | ÷ | 44.0 | = | 0.003 99 | 0.120 | 12.0 |
| $O_2$ | 0.0426 | ÷ | 32.0 | = | 0.001 33 | 0.040 | 4.0 |
| $N_2$ | 0.7633 | ÷ | 28.0 | = | 0.027 26 | 0.820 | 82.0 |
| CO | 0.0186 | ÷ | 28.0 | = | 0.000 66 | 0.020 | 2.0 |
| | | | | | 0.033 24 | 1.000 | 100.0 |

Consider a mixture of two gases (not necessarily ideal gases) such as shown in Fig. 10.1. What properties can we experimentally measure for such a mixture? Certainly we can measure the pressure, temperature, volume, and mass of the mixture. We can also experimentally measure the composition of the mixture, and thus determine the mole and mass fractions.

Suppose that this mixture undergoes a process or a chemical reaction and we wish to perform a thermodynamic analysis of this process or reaction. What type of thermodynamic data would we use in performing such an analysis? One possibility would be to have tables of thermodynamic properties of mixtures. However, the number of different mixtures that is possible, both as regards the substances involved and the relative amounts of each, is such that we would need a library full of tables of thermodynamic properties to handle all possible situations. It would be much simpler if we could determine the thermodynamic properties of a mixture from the properties of the pure components. This is in essence the approach that is used in dealing with ideal gases and certain other simplified models of mixtures.

One exception to this procedure is the case where a particular mixture is encountered very frequently, the most familiar being air. Tables and charts of the thermodynamic properties of air are available. However, even in this case it is necessary to define

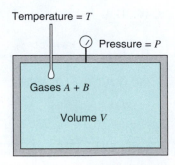

FIGURE 10.1 A mixture of two gases.

the composition of the "air" for which the tables are given, because the composition of the atmosphere varies with altitude, with the number of pollutants, and with other variables at a given location. The composition of air on which air tables are usually based is as follows:

| Component | % on Mole Basis |
|---|---|
| Nitrogen | 78.10 |
| Oxygen | 20.95 |
| Argon | 0.92 |
| $CO_2$ & trace elements | 0.03 |

In this chapter we focus on mixtures of ideal gases. We assume that each component is uninfluenced by the presence of the other components and that each component can be treated as an ideal gas. In the case of a real gaseous mixture at high pressure this assumption would probably not be accurate because of the nature of the interaction between the molecules of the different components. In this text, we will consider only a single model in analyzing gas mixtures, namely, the Dalton model.

## Dalton Model

For the Dalton model of gas mixtures, the properties of each component of the mixture are considered as though each component exists separately and independently at the temperature and volume of the mixture, as shown in Fig. 10.2. We further assume that both the gas mixture and the separated components behave according to the ideal gas model, Eqs. 3.3–3.6. In general, we would prefer to analyze gas mixture behavior on a mass basis. However, in this particular case it is more convenient to use a mole basis, since the gas constant is then the universal gas constant for each component and also for the mixture. Thus, we may write for the mixture (Fig. 10.1)

$$PV = n\bar{R}T$$

$$n = n_A + n_B \tag{10.6}$$

and for the components (Fig. 10.2)

$$P_AV = n_A\bar{R}T$$

$$P_BV = n_B\bar{R}T \tag{10.7}$$

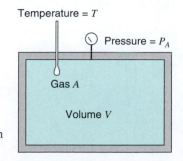
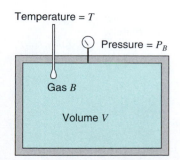

**FIGURE 10.2**  The Dalton model.

On substituting, we have

$$n = n_A + n_B$$

$$\frac{PV}{\overline{R}T} = \frac{P_A V}{\overline{R}T} + \frac{P_B V}{\overline{R}T} \tag{10.8}$$

or

$$P = P_A + P_B \tag{10.9}$$

where $P_A$ and $P_B$ are referred to as partial pressures. Thus, for a mixture of ideal gases, the pressure is the sum of the partial pressures of the individual components, where, using Eqs. 10.6 and 10.7,

$$P_A = y_A P, \qquad P_B = y_B P \tag{10.10}$$

That is, each partial pressure is the product of that component's mole fraction and the mixture pressure.

In determining the internal energy, enthalpy, and entropy of a mixture of ideal gases, the Dalton model proves useful because the assumption is made that each constituent behaves as though it occupies the entire volume by itself. Thus, the internal energy, enthalpy, and entropy can be evaluated as the sum of the respective properties of the constituent gases at the condition at which the component exists in the mixture. Since for ideal gases the internal energy and enthalpy are functions only of temperature, it follows that for a mixture of components $A$ and $B$, on a mass basis,

$$U = mu = m_A u_A + m_B u_B$$

$$= m(c_A u_A + c_B u_B) \tag{10.11}$$

$$H = mh = m_A h_A + m_B h_B$$

$$= m(c_A h_A + c_B h_B) \tag{10.12}$$

In Eqs. 10.11 and 10.12, the quantities $u_A$, $u_B$, $h_A$, and $h_B$ are the ideal-gas properties of the components at the temperature of the mixture. For a process involving a change of temperature, the changes in these values are evaluated by one of the three models discussed in Section 5.7—involving either the ideal-gas tables A.7 or the specific heats of the components.

The ideal-gas mixture equation of state on a mass basis is

$$PV = mR_{\mathrm{mix}}T \tag{10.13}$$

where

$$R_{\mathrm{mix}} = \frac{1}{m}\left(\frac{PV}{T}\right) = \frac{1}{m}(n\overline{R}) = \overline{R}/M_{\mathrm{mix}} \tag{10.14}$$

Alternately,

$$R_{\mathrm{mix}} = \frac{1}{m}(n_A \overline{R} + n_B \overline{R})$$

$$= \frac{1}{m}(m_A R_A + m_B R_B)$$

$$= c_A R_A + c_B R_B \tag{10.15}$$

The entropy of an ideal-gas mixture is expressed as

$$S = ms = m_A s_A + m_B s_B$$
$$= m(c_A s_A + c_B s_B) \tag{10.16}$$

It must be emphasized that the component entropies in Eq. 10.16 must each be evaluated at the mixture temperature and the corresponding partial pressure of the component in the mixture, using Eq. 10.10 in terms of the mole fraction.

To evaluate Eq. 10.16 using the ideal-gas entropy expression 8.24, it is necessary to use one of the specific heat models discussed in Section 8.10. The simplest model is constant specific heat, Eq. 8.25, using an arbitrary reference state $T_0, P_0, s_0$, for each component $i$ in the mixture at $T$ and $P$,

$$s_i = s_{0i} + C_{p0i} \ln\left(\frac{T}{T_0}\right) - R_i \ln\left(\frac{y_i P}{P_0}\right) \tag{10.17}$$

The reference values $s_{0i}, T_0, P_0$ all cancel out when Eq. 10.17 is used to calculate a change of entropy for a process.

An alternative model is to use the $s_T^0$ function defined in Eq. 8.18, in which case each component entropy in Eq. 10.16 is expressed as

$$s_i = s_{T_i}^0 - R_i \ln\left(\frac{y_i P}{P_0}\right) \tag{10.18}$$

---

**EXAMPLE 10.2**

Let a mass $m_A$ of ideal gas $A$ at a given pressure and temperature, $P$ and $T$, be mixed with $m_B$ of ideal gas $B$ at the same $P$ and $T$, such that the final ideal-gas mixture is also at $P$ and $T$. Determine the change in entropy for this process.

> *Control mass*: All gas ($A$ and $B$).
> *Initial states*: $P, T$ known for $A$ and $B$.
> *Final state*: $P, T$ of mixture known.

**Analysis and Solution**

The mixture entropy is given by Eq. 10.16. Therefore, the change of entropy can be grouped into changes for $A$ and for $B$, with each change expressed by Eq. 8.24. Since there is no temperature change for either component, this reduces to

$$\Delta S_{\text{mix}} = m_A \left(0 - R_A \ln\frac{P_A}{P}\right) + m_B \left(0 - R_B \ln\frac{P_B}{P}\right)$$

$$= -m_A R_A \ln y_A - m_B R_B \ln y_B$$

which can also be written in the form

$$\Delta S_{\text{mix}} = -n_A \bar{R} \ln y_A - n_B \bar{R} \ln y_B$$

---

The result of Example 10.2 can readily be generalized to account for the mixing of any number of components at the same temperature and pressure. The result is

$$\Delta S_{\text{mix}} = -\bar{R} \sum_k n_k \ln y_k \tag{10.19}$$

The interesting thing about this equation is that the increase in entropy depends only on the number of moles of component gases and is independent of the composition of the gas. For example, when 1 mol of oxygen and 1 mol of nitrogen are mixed, the increase in entropy is the same as when 1 mol of hydrogen and 1 mol of nitrogen are mixed. But we also know that if 1 mol of nitrogen is "mixed" with another mole of nitrogen there is no increase in entropy. The question that arises is how dissimilar must the gases be in order to have an increase in entropy? The answer lies in our ability to distinguish between the two gases (based on their different molecular weights). The entropy increases whenever we can distinguish between the gases being mixed. When we cannot distinguish between the gases, there is no increase in entropy.

One special case that arises frequently involves an ideal-gas mixture undergoing a process in which there is no change in composition. Let us also assume that the constant-specific heat model is reasonable. For this case, from Eq. 10.11 on a unit mass basis, the internal energy change is

$$u_2 - u_1 = c_A C_{v0\,A}(T_2 - T_1) + c_B C_{v0\,B}(T_2 - T_1)$$
$$= C_{v0\,\text{mix}}(T_2 - T_1) \tag{10.20}$$

where

$$C_{v0\,\text{mix}} = c_A C_{v0\,A} + c_B C_{v0\,B} \tag{10.21}$$

Similarly, from Eq. 10.12, the enthalpy change is

$$h_2 - h_1 = c_A C_{p0\,A}(T_2 - T_1) + c_B C_{p0\,B}(T_2 - T_1)$$
$$= C_{p0\,\text{mix}}(T_2 - T_1) \tag{10.22}$$

where

$$C_{p0\,\text{mix}} = c_A C_{p0\,A} + c_B C_{p0\,B} \tag{10.23}$$

The entropy change is calculated using Eq. 10.17. Since the reference values and the (constant) mole fractions cancel out, and also substituting Eqs. 10.15 and 10.23, we get the result

$$s_2 - s_1 = C_{p0\,\text{mix}} \ln\left(\frac{T_2}{T_1}\right) - R_{\text{mix}} \ln\left(\frac{P_2}{P_1}\right) \tag{10.24}$$

We see that the set of Eqs. 10.20, 10.22, and 10.24 are the same as those for a pure substance, with the gas constant and specific heats for the mixture calculated from Eqs. 10.15, 10.21, and 10.23.

| **In-Text Concept Questions** | |
|---|---|

**a.** Are the mass and mole fractions for a mixture ever the same?

**b.** For a mixture, how many component concentrations are needed?

**c.** Are any of the properties ($P$, $T$, $v$) for oxygen and nitrogen in air the same?

**d.** If I want to heat a flow of a 4-component mixture from 300 to 310 K at constant $P$, how many properties and which properties do I need to know to find the heat transfer?

**e.** To evaluate the change in entropy between two states at different $T$ and $P$s for a given mixture, do I need to find the partial pressures?

## 10.2 A SIMPLIFIED MODEL OF A MIXTURE INVOLVING GASES AND A VAPOR

Let us now consider a simplification, which is often a reasonable one, of the problem involving a mixture of ideal gases that is in contact with a solid or liquid phase of one of the components. The most familiar example is a mixture of air and water vapor in contact with liquid water or ice, such as encountered in air conditioning or in drying. We are all familiar with the condensation of water from the atmosphere when it cools on a summer day.

This problem and a number of similar problems can be analyzed quite simply and with considerable accuracy if the following assumptions are made:

1. The solid or liquid phase contains no dissolved gases.
2. The gaseous phase can be treated as a mixture of ideal gases.
3. When the mixture and the condensed phase are at a given pressure and temperature, the equilibrium between the condensed phase and its vapor is not influenced by the presence of the other component. This means that when equilibrium is achieved, the partial pressure of the vapor will be equal to the saturation pressure corresponding to the temperature of the mixture.

Since this approach is used extensively and with considerable accuracy, let us give some attention to the terms that have been defined and the type of problems for which this approach is valid and relevant. In our discussion we will refer to this as a gas–vapor mixture.

The dew point of a gas–vapor mixture is the temperature at which the vapor condenses or solidifies when it is cooled at constant pressure. This is shown on the $T$–$s$ diagram for the vapor shown in Fig. 10.3. Suppose that the temperature of the gaseous mixture and the partial pressure of the vapor in the mixture are such that the vapor is initially superheated at state 1. If the mixture is cooled at constant pressure, the partial pressure of the vapor remains constant until point 2 is reached, and then condensation begins. The temperature at state 2 is the dew-point temperature. Lines 1–3 on the diagram indicate that if the mixture is cooled at constant volume the condensation begins at point 3, which is slightly lower than the dew-point temperature.

If the vapor is at the saturation pressure and temperature, the mixture is referred to as a saturated mixture, and for an air–water vapor mixture, the term saturated air is used.

The relative humidity $\phi$ is defined as the ratio of the mole fraction of the vapor in the mixture to the mole fraction of vapor in a saturated mixture at the same temperature and total pressure. Since the vapor is considered an ideal gas, the definition reduces to

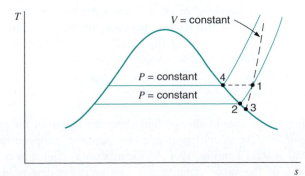

**FIGURE 10.3** Temperature–entropy diagram to show definition of the dew point.

the ratio of the partial pressure of the vapor as it exists in the mixture, $P_v$, to the saturation pressure of the vapor at the same temperature, $P_g$:

$$\phi = \frac{P_v}{P_g}$$

In terms of the numbers on the $T$–$s$ diagram of Fig. 10.3, the relative humidity $\phi$ would be

$$\phi = \frac{P_1}{P_4}$$

Since we are considering the vapor to be an ideal gas, the relative humidity can also be defined in terms of specific volume or density:

$$\phi = \frac{P_v}{P_g} = \frac{\rho_v}{\rho_g} = \frac{v_g}{v_v} \tag{10.25}$$

The humidity ratio $\omega$ of an air–water vapor mixture is defined as the ratio of the mass of water vapor $m_v$ to the mass of dry air $m_a$. The term dry air is used to emphasize that this refers only to air and not to the water vapor. The term specific humidity is used synonymously with humidity ratio.

$$\omega = \frac{m_v}{m_a} \tag{10.26}$$

This definition is identical for any other gas–vapor mixture, and the subscript $a$ refers to the gas, exclusive of the vapor. Since we consider both the vapor and the mixture to be ideal gases, a very useful expression for the humidity ratio in terms of partial pressures and molecular weights can be developed. Writing

$$m_v = \frac{P_v V}{R_v T} = \frac{P_v V M_v}{\overline{R}T}, \qquad m_a = \frac{P_a V}{R_a T} = \frac{P_a V M_a}{\overline{R}T}$$

we have

$$\omega = \frac{P_v V / R_v T}{P_a V / R_a T} = \frac{R_a P_v}{R_v P_a} = \frac{M_v P_v}{M_a P_a} \tag{10.27}$$

For an air–water vapor mixture, this reduces to

$$\omega = 0.622 \frac{P_v}{P_a} \tag{10.28}$$

The degree of saturation is defined as the ratio of the actual humidity ratio to the humidity ratio of a saturated mixture at the same temperature and total pressure.

An expression for the relation between the relative humidity $\phi$ and the humidity ratio $\omega$ can be found by solving Eqs. 10.25 and 10.28 for $P_v$ and equating them. The resulting relation for an air–water vapor mixture is

$$\phi = \frac{\omega P_a}{0.622 \, P_g} \tag{10.29}$$

A few words should also be said about the nature of the process that occurs when a gas–vapor mixture is cooled at constant pressure. Suppose that the vapor is initially superheated at state 1 in Fig. 10.4. As the mixture is cooled at constant pressure, the partial pressure of the vapor remains constant until the dew point is reached at point 2, where the vapor in the mixture is saturated. The initial condensate is at state 4 and is in

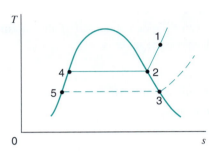

**FIGURE 10.4**  Temperature–entropy diagram to show the cooling of a gas–vapor mixture at a constant pressure.

equilibrium with the vapor at state 2. As the temperature is lowered further, more of the vapor condenses, which lowers the partial pressure of the vapor in the mixture. The vapor that remains in the mixture is always saturated, and the liquid or solid is in equilibrium with it. For example, when the temperature is reduced to $T_3$, the vapor in the mixture is at state 3, and its partial pressure is the saturation pressure corresponding to $T_3$. The liquid in equilibrium with it is at state 5.

---

**EXAMPLE 10.3**

Consider 100 m$^3$ of an air–water vapor mixture at 0.1 MPa, 35°C, and 70% relative humidity. Calculate the humidity ratio, dew point, mass of air, and mass of vapor.

*Control mass*:  Mixture.

*State*:  $P$, $T$, $\phi$ known; state fixed.

**Analysis and Solution**

From Eq. 10.25 and the steam tables, we have

$$\phi = 0.70 = \frac{P_v}{P_g}$$

$$P_v = 0.70(5.628) = 3.94 \text{ kPa}$$

The dew point is the saturation temperature corresponding to this pressure, which is 28.6°C.

The partial pressure of the air is

$$P_a = P - P_v = 100 - 3.94 = 96.06 \text{ kPa}$$

The humidity ratio can be calculated from Eq. 10.28:

$$\omega = 0.622 \times \frac{P_v}{P_a} = 0.622 \times \frac{3.94}{96.06} = 0.0255$$

The mass of air is

$$m_a = \frac{P_a V}{R_a T} = \frac{96.06 \times 100}{0.287 \times 308.2} = 108.6 \text{ kg}$$

The mass of the vapor can be calculated by using the humidity ratio or by using the ideal-gas equation of state:

$$m_v = \omega m_a = 0.0255(108.6) = 2.77 \text{ kg}$$

$$m_v = \frac{3.94 \times 100}{0.4615 \times 308.2} = 2.77 \text{ kg}$$

**EXAMPLE 10.4**

Calculate the amount of water vapor condensed if the mixture of Example 10.3 is cooled to 5°C in a constant-pressure process.

*Control mass*: Mixture.
*Initial state*: Known (Example 10.3).
*Final state*: $T$ known.
*Process*: Constant pressure.

**Analysis**

At the final temperature, 5°C, the mixture is saturated, since this is below the dew-point temperature. Therefore,

$$P_{v2} = P_{g2}, \qquad P_{a2} = P - P_{v2}$$

and

$$\omega_2 = 0.622 \frac{P_{v2}}{P_{a2}}$$

From the conservation of mass, it follows that the amount of water condensed is equal to the difference between the initial and final mass of water vapor, or

$$\text{Mass of vapor condensed} = m_a(\omega_1 - \omega_2)$$

**Solution**

We have

$$P_{v2} = P_{g2} = 0.8721 \text{ kPa}$$

$$P_{a2} = 100 - 0.8721 = 99.128 \text{ kPa}$$

Therefore,

$$\omega_2 = 0.622 \times \frac{0.8721}{99.128} = 0.0055$$

$$\text{Mass of vapor condensed} = m_a(\omega_1 - \omega_2) = 108.6(0.0255 - 0.0055)$$

$$= 2.172 \text{ kg}$$

## 10.3 THE FIRST LAW APPLIED TO GAS–VAPOR MIXTURES

In applying the first law of thermodynamics to gas–vapor mixtures, it is helpful to realize that because of our assumption that ideal gases are involved, the various components can be treated separately when calculating changes of internal energy and enthalpy. Therefore, in dealing with air–water vapor mixtures, the changes in enthalpy of the water vapor can be found from the steam tables and the ideal-gas relations can be applied to the air. This is illustrated by the examples that follow.

**EXAMPLE 10.5**

An air-conditioning unit is shown in Fig. 10.5, with pressure, temperature, and relative humidity data. Calculate the heat transfer per kilogram of dry air, assuming that changes in kinetic energy are negligible.

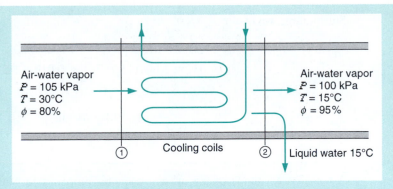

Air-water vapor
$P = 105$ kPa
$T = 30°C$
$\phi = 80\%$

Air-water vapor
$P = 100$ kPa
$T = 15°C$
$\phi = 95\%$

① Cooling coils ② Liquid water 15°C

**FIGURE 10.5** Sketch for Example 10.5.

| | |
|---|---|
| *Control volume*: | Duct, excluding cooling coils. |
| *Inlet state*: | Known (Fig. 10.5). |
| *Exit state*: | Known (Fig. 10.5). |
| *Process*: | Steady state with no kinetic or potential energy changes. |
| *Model*: | Air—ideal gas, constant specific heat, value at 300 K. Water—steam tables. (Since the water vapor at these low pressures is being considered an ideal gas, the enthalpy of the water vapor is a function of the temperature only. Therefore the enthalpy of slightly superheated water vapor is equal to the enthalpy of saturated vapor at the same temperature.) |

**Analysis**

From the continuity equations for air and water, we have

$$\dot{m}_{a1} = \dot{m}_{a2}$$

$$\dot{m}_{v1} = \dot{m}_{v2} + \dot{m}_{l2}$$

The first law gives

$$\dot{Q}_{c.v.} + \sum \dot{m}_i h_i = \sum \dot{m}_e h_e$$

$$\dot{Q}_{c.v.} + \dot{m}_a h_{a1} + \dot{m}_{v1} h_{v1} = \dot{m}_a h_{a2} + \dot{m}_{v2} h_{v2} + \dot{m}_{l2} h_{l2}$$

If we divide this equation by $\dot{m}_a$, introduce the continuity equation for the water, and note that $\dot{m}_v = \omega \dot{m}_a$, we can write the first law in the form

$$\frac{\dot{Q}_{c.v.}}{\dot{m}_a} + h_{a1} + \omega_1 h_{v1} = h_{a2} + \omega_2 h_{v2} + (\omega_1 - \omega_2) h_{l2}$$

**Solution**

We have

$$P_{v1} = \phi_1 P_{g1} = 0.80(4.246) = 3.397 \text{ kPa}$$

$$\omega_1 = \frac{R_a}{R_v} \frac{P_{v1}}{P_{a1}} = 0.622 \times \left(\frac{3.397}{105 - 3.4}\right) = 0.0208$$

$$P_{v2} = \phi_2 P_{g2} = 0.95(1.7051) = 1.620 \text{ kPa}$$

$$\omega_2 = \frac{R_a}{R_v} \times \frac{P_{v2}}{P_{a2}} = 0.622 \times \left(\frac{1.62}{100 - 1.62}\right) = 0.0102$$

Substituting, we obtain

$$\dot{Q}_{c.v.}/\dot{m}_a + h_{a1} + \omega_1 h_{v1} = h_{a2} + \omega_2 h_{v2} + (\omega_1 - \omega_2)h_{l2}$$

$$\dot{Q}_{c.v.}/\dot{m}_a = 1.004(15 - 30) + 0.0102(2528.9)$$

$$-0.0208(2556.3) + (0.0208 - 0.0102)(62.99)$$

$$= -41.76 \text{ kJ/kg dry air}$$

**EXAMPLE 10.6**

A tank has a volume of 0.5 m³ and contains nitrogen and water vapor. The temperature of the mixture is 50°C and the total pressure is 2 MPa. The partial pressure of the water vapor is 5 kPa. Calculate the heat transfer when the contents of the tank are cooled to 10°C.

$$
\begin{array}{rl}
\textit{Control mass}: & \text{Nitrogen and water.} \\
\textit{Initial state}: & P_1, T_1 \text{ known; state fixed.} \\
\textit{Final state}: & T_2 \text{ known.} \\
\textit{Process}: & \text{Constant volume.} \\
\textit{Model}: & \text{Ideal gas mixture; constant specific heat for nitrogen; steam} \\
& \text{tables for water.}
\end{array}
$$

**Analysis**

This is a constant-volume process. Since the work is zero, the first law reduces to

$$Q = U_2 - U_1 = m_{N_2} C_{v(N_2)}(T_2 - T_1) + (m_2 u_2)_v + (m_2 u_2)_l - (m_1 u_1)_v$$

This equation assumes that some of the vapor condensed. This assumption must be checked, however, as shown in the solution:

**Solution**

The mass of nitrogen and water vapor can be calculated using the ideal-gas equation of state:

$$m_{N_2} = \frac{P_{N_2} V}{R_{N_2} T} = \frac{1995 \times 0.5}{0.2968 \times 323.2} = 10.39 \text{ kg}$$

$$m_{v_1} = \frac{P_{v1} V}{R_v T} = \frac{5 \times 0.5}{0.4615 \times 323.2} = 0.016\,76 \text{ kg}$$

If condensation takes place, the final state of the vapor will be saturated vapor at 10°C. Therefore,

$$m_{v2} = \frac{P_{v2} V}{R_v T} = \frac{1.2276 \times 0.5}{0.4615 \times 283.2} = 0.004\,70 \text{ kg}$$

Since this amount is less than the original mass of vapor, there must have been condensation.

The mass of liquid that is condensed, $m_{l2}$, is

$$m_{l2} = m_{v_1} - m_{v_2} = 0.016\,76 - 0.004\,70 = 0.012\,06 \text{ kg}$$

The internal energy of the water vapor is equal to the internal energy of saturated water vapor at the same temperature. Therefore,

$$u_{v_1} = 2443.5 \text{ kJ/kg}$$

$$u_{v_2} = 2389.2 \text{ kJ/kg}$$

$$u_{l2} = 42.0 \text{ kJ/kg}$$

$$Q_{\text{c.v.}} = 10.39 \times 0.745(10 - 50) + 0.0047(2389.2)$$
$$+ \ 0.012\,06(42.0) - 0.016\,76(2443.5)$$
$$= -338.8 \text{ kJ}$$

## 10.4 THE ADIABATIC SATURATION PROCESS

An important process for an air–water vapor mixture is the adiabatic saturation process. In this process, an air–vapor mixture comes in contact with a body of water in a well-insulated duct (Fig. 10.6). If the initial relative humidity is less than 100%, some of the water will evaporate and the temperature of the air–vapor mixture will decrease. If the mixture leaving the duct is saturated and if the process is adiabatic, the temperature of the mixture on leaving is known as the adiabatic saturation temperature. For this to take place as a steady-state process, make-up water at the adiabatic saturation temperature is added at the same rate at which water is evaporated. The pressure is assumed to be constant.

Considering the adiabatic saturation process to be a steady-state process, and neglecting changes in kinetic and potential energy, the first law reduces to

$$h_{a1} + \omega_1 h_{v1} + (\omega_2 - \omega_1)h_{l2} = h_{a2} + \omega_2 h_{v2}$$

$$\omega_1(h_{v1} - h_{l2}) = C_{pa}(T_2 - T_1) + \omega_2(h_{v2} - h_{l2})$$

$$\omega_1(h_{v1} - h_{l2}) = C_{pa}(T_2 - T_1) + \omega_2 h_{fg2} \qquad (10.30)$$

The most significant point to be made about the adiabatic saturation process is that the adiabatic saturation temperature, the temperature of the mixture when it leaves the duct, is a function of the pressure, temperature, and relative humidity of the entering air–vapor mixture and of the exit pressure. Thus, the relative humidity and the humidity ratio of the entering air–vapor mixture can be determined from the measurements of the pressure and temperature of the air–vapor mixture entering and leaving the adiabatic saturator. Since these measurements are relatively easy to make, this is one means of determining the humidity of an air–vapor mixture.

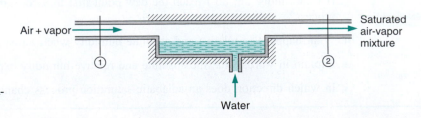

**FIGURE 10.6** The adiabatic saturation process.

**EXAMPLE 10.7**

The pressure of the mixture entering and leaving the adiabatic saturator is 0.1 MPa, the entering temperature is 30°C, and the temperature leaving is 20°C, which is the adiabatic saturation temperature. Calculate the humidity ratio and relative humidity of the air–water vapor mixture entering.

*Control volume*:   Adiabatic saturator.

    *Inlet state*:   $P_1$, $T_1$ known.

    *Exit state*:   $P_2$, $T_2$ known; $\phi_2 = 100\%$; state fixed.

    *Process*:   Steady state, adiabatic saturation (Fig. 10.6).

    *Model*:   Ideal-gas mixture; constant specific heat for air; steam tables for water.

**Analysis**

Use continuity and the first law, Eq. 10.30.

**Solution**

Since the water vapor leaving is saturated, $P_{v2} = P_{g2}$ and $\omega_2$ can be calculated.

$$\omega_2 = 0.622 \times \left( \frac{2.339}{100 - 2.34} \right) = 0.0149$$

$\omega_1$ can be calculated using Eq. 10.30.

$$\omega_1 = \frac{C_{pa}(T_2 - T_1) + \omega_2 h_{fg2}}{(h_{v1} - h_{l2})}$$

$$\omega_1 = \frac{1.004(20 - 30) + 0.0149 \times 2454.1}{2556.3 - 83.96} = 0.0107$$

$$\omega_1 = 0.0107 = 0.622 \times \left( \frac{P_{v1}}{100 - P_{v1}} \right)$$

$$P_{v1} = 1.691 \text{ kPa}$$

$$\phi_1 = \frac{P_{v1}}{P_{g1}} = \frac{1.691}{4.246} = 0.398$$

---

**In-Text Concept Questions**

**f.** What happens to relative and absolute humidity when moist air is heated?

**g.** If I cool moist air, do I reach the dew point first in a constant $P$ or constant $V$ process?

**h.** What happens to relative and absolute humidity when moist air is cooled?

**i.** Explain in words what the absolute and relative humidity express.

**j.** In which direction does an adiabatic saturation process change $\Phi$, $\omega$, and $T$?

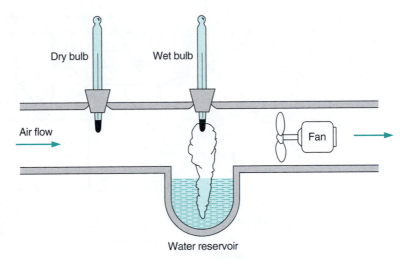

**FIGURE 10.7** Steady-flow apparatus for measuring wet- and dry-bulb temperatures.

## 10.5 WET-BULB AND DRY-BULB TEMPERATURES

The humidity of air–water vapor mixtures has traditionally been measured with a device called a psychrometer, which uses the flow of air past wet-bulb and dry-bulb thermometers. The bulb of the wet-bulb thermometer is covered with a cotton wick that is saturated with water. The dry-bulb thermometer is used simply to measure the temperature of the air. The airflow can be maintained by a fan, as shown in the continuous-flow psychrometer depicted in Fig. 10.7.

The processes that take place at the wet-bulb thermometer are somewhat complicated. First, if the air–water vapor mixture is not saturated, some of the water in the wick evaporates and diffuses into the surrounding air, which cools the water in the wick. As soon as the temperature of the water drops, however, heat is transferred to the water from both the air and the thermometer, with corresponding cooling. A steady state, determined by heat and mass transfer rates, will be reached, in which the wet-bulb thermometer temperature is lower than the dry-bulb temperature.

It can be argued that this evaporative cooling process is very similar, but not identical, to the adiabatic saturation process described and analyzed in Section 10.4. In fact, the adiabatic saturation temperature is often termed the thermodynamic wet-bulb temperature. It is clear, however, that the wet-bulb temperature as measured by a psychrometer is influenced by heat and mass transfer rates, which depend, for example, on the airflow velocity and not simply on thermodynamic equilibrium properties. It does happen that the two temperatures are very close for air–water vapor mixtures at atmospheric temperature and pressure, and they will be assumed to be equivalent in this text.

In recent years, humidity measurements have been made using other phenomena and other devices, primarily electronic devices for convenience and simplicity. For example, some substances tend to change in length, in shape, or in electrical capacitance, or in a number of other ways, when they absorb moisture. They are therefore sensitive to the amount of moisture in the atmosphere. An instrument making use of such a substance can be calibrated to measure the humidity of air–water vapor mixtures. The instrument output can be programmed to furnish any of the desired parameters, such as relative humidity, humidity ratio, or wet-bulb temperature.

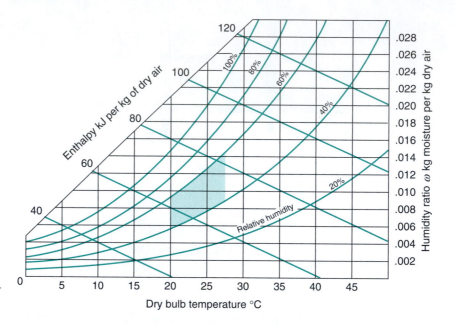

**FIGURE 10.8** Psychrometric chart.

## 10.6 THE PSYCHROMETRIC CHART

Properties of air–water vapor mixtures are given in graphical form on psychrometric charts. These are available in a number of different forms, and only the main features are considered here. It should be recalled that three independent properties—such as pressure, temperature, and mixture composition—will describe the state of this binary mixture.

A simplified version of the chart included in Appendix E, Fig. E.4, is shown in Fig. 10.8. This basic psychrometric chart is a plot of humidity ratio (ordinate) as a function of dry-bulb temperature (abscissa) with relative humidity, wet-bulb temperature, and mixture enthalpy per mass of dry air as parameters. If we fix the total pressure for which the chart is to be constructed (which in our chart is 1 bar, or 100 kPa), lines of constant relative humidity and wet-bulb temperature can be drawn on the chart, because for a given dry-bulb temperature, total pressure, and humidity ratio, the relative humidity and wet-bulb temperature are fixed. The partial pressure of the water vapor is fixed by the humidity ratio and the total pressure, and therefore a second ordinate scale that indicates the partial pressure of the water vapor could be constructed. Likewise it would also be possible to include the mixture specific volume and entropy on the chart.

Most psychrometric charts give the enthalpy of an air–vapor mixture per kilogram of dry air. The values given assume that the enthalpy of the dry air is zero at −20°C, and the enthalpy of the vapor is taken from the steam tables (which are based on the assumption that the enthalpy of saturated liquid is zero at 0°C). The value used in the psychrometric chart is then

$$\tilde{h} \equiv h_a - h_a(-20°C) + \omega h_v$$

This procedure is satisfactory because we are usually concerned only with differences in enthalpy. The fact that the lines of constant enthalpy are essentially parallel to lines of constant wet-bulb temperature is evident from the fact that the wet-bulb temperature is essentially equal to the adiabatic saturation temperature. Thus in Fig. 10.6, if we neglect the enthalpy of the liquid entering the adiabatic saturator, the enthalpy of the air–vapor mixture leaving at a given adiabatic saturation temperature fixes the enthalpy of the mixture entering.

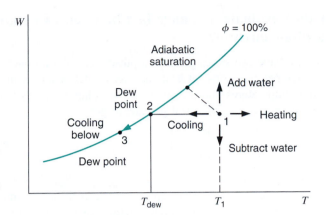

**FIGURE 10.9** Processes on a psychrometric chart.

The chart plotted in Fig. 10.8 also indicates the human comfort zone, as the range of conditions most agreeable for human well being. An air conditioner should then be able to maintain an environment within the comfort zone regardless of the outside atmospheric conditions to be considered adequate. Some charts are available that give corrections for variation from standard atmospheric pressures. Before using a given chart one should fully understand the assumptions made in constructing it and should recognize that it is applicable to the particular problem at hand.

The direction in which various processes proceed for an air–water vapor mixture is shown on the psychrometric chart of Fig. 10.9. For example, a constant-pressure cooling process beginning at state 1 proceeds at constant humidity ratio to the dew point at state 2, with continued cooling below that temperature moving along the saturation line (100% relative humidity) to point 3. Other processes could be traced out in a similar manner.

## 10.7 HOW-TO SECTION

### How do I find a $T$ if I know an $h$ or $u$ for the mixture?

This situation arises if we heat a flow with $q$ at a constant pressure; then the energy equation leads to

$$h_2 - h_1 = q$$

So if we know the starting state and the heat added, then we know $h_2$. To relate this to the temperature, we write the equation as

$$h_2 - h_1 = \sum c_i(h_{i2} - h_{i1}) \approx \sum c_i C_{pi}(T_2 - T_1) = C_{p\,\text{mix}}(T_2 - T_1)$$

The last expression uses the heat capacities from Table A.5 at 300 K and allows an explicit calculation to find $T_2$. If the temperature is high and we want to be accurate, then we must use the first expression and the enthalpies from Table A.8 and do

$$h_2 = \sum c_i h_{i2} = q + h_1 = q + \sum c_i h_{i1} = \text{known value}$$

Now we must guess a $T_2$, get all the $h_{i2}$ from Table A.8, and compute the left-hand side. If it is too small, guess a higher $T_2$ and redo the calculation. Keep doing this until you have bracketed the known value and then do a final linear interpolation. If it is a mass heated in a constant volume tank, we get

$$u_2 - u_1 = q \approx \sum c_i C_{vi}(T_2 - T_1) = C_{v\,\text{mix}}(T_2 - T_1)$$

and the procedure is very similar using the internal energy values instead of enthalpies.

### How do I evaluate a change in $s$ between two states for a given mixture?

Assume we have two states given in a prescribed mixture, and we want to find $s_2 - s_1$. The low temperature constant heat capacity model is shown in Eq. 10.24, assuming we know the temperatures and pressures. If the volume ratio is known, then just solve for the pressure ratio using the ideal gas law to get

$$\frac{P_2}{P_1} = \frac{T_2}{T_1}\frac{V_1}{V_2} \quad \Rightarrow \quad s_2 - s_1 = C_{v0\,\text{mix}}\ln\frac{T_2}{T_1} + R_{\text{mix}}\ln\frac{V_2}{V_1}$$

Recall Eqs. 8.16–17 and $C_{v0\,\text{mix}} = C_{p0\,\text{mix}} - R_{\text{mix}}$. For a high temperature, we do not want to use the constant heat capacity so we evaluate the entropy by Eq. 10.18. When we find the change $s_2 - s_1$, the mole fraction for the pressure correction drops out and the result is

$$s_2 - s_1 = \sum c_i(s^0_{T2i} - s^0_{T1i}) - \sum c_i R_i \ln\frac{P_2}{P_1}$$

$$= \sum c_i(s^0_{T2i} - s^0_{T1i}) - R_{\text{mix}}\ln\frac{P_2}{P_1}$$

Here we use the standard entropies from Table A.8 as a function of temperature.

### How do I find a $T$ if I know an $s$ for the mixture?

If we do an adiabatic compression or expansion process and assume it is reversible, then we have an isentropic ($s$ = constant) process. Knowing the initial (inlet) state and the final (exit) $P$ or $v$, we then want to know the temperature. Now recall that entropy is a function of two properties $(T, P)$ or $(T, v)$ so we get

$$0 = s_2 - s_1 = \sum c_i(s^0_{T2i} - s^0_{T1i}) - R_{\text{mix}}\ln\frac{P_2}{P_1} \approx C_{p\,\text{mix}}\ln\frac{T_2}{T_1} - R_{\text{mix}}\ln\frac{P_2}{P_1}$$

Using the average heat capacity, we use the last expression and solve for $T_2$ explicitly giving Eq. 8.32 or Eq. 8.33 with $k = k_{\text{mix}} = C_{p\,\text{mix}}/C_{v0\,\text{mix}}$. Now with a high temperature and a desire to be accurate, we use the standard entropies and write the equation as

$$\sum c_i s^0_{T2i} = \sum c_i s^0_{T1i} + R_{\text{mix}}\ln\frac{P_2}{P_1}$$

so we evaluate the right-hand side and it becomes trial and error for $T_2$ using Table A.8 for the standard entropies.

### How do I find a $P$ if I know an $s$ for the mixture?

Since the pressure appears explicitly in the entropy variation, we solve for the pressure indirectly as

$$R_{\text{mix}}\ln\frac{P_2}{P_1} = \sum c_i(s^0_{T2i} - s^0_{T1i}) \approx C_{p\,\text{mix}}\ln\frac{T_2}{T_1}$$

and directly as

$$\frac{P_2}{P_1} = \exp\left[\sum c_i(s^0_{T2i} - s^0_{T1i})/R_{\text{mix}}\right] \approx \left(\frac{T_2}{T_1}\right)^{C_{p\,\text{mix}}/R_{\text{mix}}}$$

Again we use the first expression for the high temperature case and recover the power relation Eq. 8.20 for the constant heat capacity approximation.

### How do I find the dew point temperature?

For the given state, you must find the water vapor pressure $P_v$ and then interpolate in the steam tables to find the temperature for which this pressure is the saturated pressure

$$P_g(T_{\text{dew}}) = P_v$$

This amounts to moving from state 1 to state 2 in Fig. 10.3.

### How do I find the adiabatic saturation temperature?

Assume we have the situation as shown in Fig. 10.6, which leads to the energy equation in Eq. 10.30. The adiabatic saturation temperature $T_2$ is now explicit in one term and implicit in these terms

From B.1.1 at $T_2$: $\quad h_{12} = h_{f2} \quad$ and $\quad h_{fg2}; \qquad w_2 = w_{\text{max}} = 0.622 \dfrac{P_{g2}}{P_{\text{tot}} - P_{g2}}$

The easiest way to iterate on the equation is to guess a $T_2$ and evaluate all the implicit terms. Then you solve for the explicit $T_2$, use this as the new guess, and repeat the process.

### How do I use the psychrometric chart together with the steam and air tables?

Offset the air tables enthalpy values as follows

$$\tilde{h} = h_a + \omega h_v = h_{a\,A7} - h_{a\,A7\,\text{at}\,-20C} + \omega h_{vB1} = h_{a\,A7} - 253.45 + \omega h_{vB1}$$

and now they are consistent.

## SUMMARY

A mixture of gases is treated from the definition of the mixture composition of the various components based on mass or based on moles. This leads to the mass fractions and mole fractions, both of which can be called concentrations. The mixture has an overall average molecular weight and other mixture properties on a mass or mole basis. Another simple model includes Dalton's model of ideal mixtures of ideal gases, which leads to partial pressures as the contribution from each component to the total pressure given by the mole fraction. As entropy is sensitive to pressure, the mole fraction enters into the entropy generation by mixing. However, for processes other than mixing of different components, we can treat the mixture as we treat a pure substance by using the mixture properties.

Special treatment and nomenclature is used for moist air as a mixture of air and water vapor. The water content is quantified by the relative humidity (how close the water vapor is to a saturated state) or by the humidity ratio (also called absolute humidity). As moist air is cooled down, it eventually reaches the dew point (relative humidity is 100%), where we have saturated moist air. Vaporizing liquid water without external heat transfer gives an adiabatic saturation process also used in a process called evaporative cooling. In an actual apparatus, we can obtain wet-bulb and dry-bulb temperatures, indirectly measuring the humidity of the incoming air. These property relations are shown in a psychrometric chart.

You should have learned a number of skills and acquired abilities from studying this chapter that will allow you to

- Handle the composition of a multicomponent mixture on a mass or mole basis.
- Convert concentrations from a mass to a mole basis and vice versa.
- Compute average properties for the mixture on a mass or mole basis.
- Know partial pressures and how to evaluate them.
- Know how to treat mixture properties such as $v, u, h, s, C_{p \, \text{mix}}, R_{\text{mix}}$.
- Find entropy generation by a mixing process.
- Formulate the general conservation equations for mass, energy, and entropy for the case of a mixture instead of a pure substance.
- Know how to use the simplified formulation of the energy equation using the frozen heat capacities for the mixture.
- Deal with a polytropic process when the substance is a mixture of ideal gases.
- Know the special properties ($\phi, \omega$) describing humidity in moist air.
- Have a sense of what changes relative humidity and humidity ratio and recognize that you can change one and not the other in a given process.

## KEY CONCEPTS AND FORMULAS

### Composition

Mass concentration
$$c_i = \frac{m_i}{m_{\text{tot}}} = \frac{y_i M_j}{\sum y_i M_j}$$

Mole concentration
$$y_i = \frac{n_i}{n_{\text{tot}}} = \frac{c_i / M_j}{\sum c_i / M_j}$$

Molecular weight
$$M_{\text{mix}} = \sum y_i M_i$$

### Properties

| | | |
|---|---|---|
| Internal energy | $u_{\text{mix}} = \sum c_i u_i$; | $\bar{u}_{\text{mix}} = \sum y_i \bar{u}_i = u_{\text{mix}} M_{\text{mix}}$ |
| Enthalpy | $h_{\text{mix}} = \sum c_i h_i$; | $\bar{h}_{\text{mix}} = \sum y_i \bar{h}_i = h_{\text{mix}} M_{\text{mix}}$ |
| Gas constant | $R_{\text{mix}} = \bar{R}/M_{\text{mix}} = \sum c_i R_i$ | |
| Heat capacity frozen | $C_{v \, \text{mix}} = \sum c_i C_{v \, i}$; | $\bar{C}_{v \, \text{mix}} = \sum y_i \bar{C}_{v \, i}$ |
| | $C_{v \, \text{mix}} = C_{p \, \text{mix}} - R_{\text{mix}}$; | $\bar{C}_{v \, \text{mix}} = \bar{C}_{p \, \text{mix}} - \bar{R}$ |
| | $C_{p \, \text{mix}} = \sum c_i C_{p \, i}$; | $\bar{C}_{p \, \text{mix}} = \sum y_i \bar{C}_{p \, i}$ |
| Ratio of specific heats | $k_{\text{mix}} = C_{p \, \text{mix}}/C_{v \, \text{mix}}$ | |
| Dalton model | $P_i = y_i P_{\text{tot}}$ & | $V_i = V_{\text{tot}}$ |
| Entropy | $s_{\text{mix}} = \sum c_i s_i$; | $\bar{s}_{\text{mix}} = \sum y_i \bar{s}_i$ |
| Component entropy | $s_i = s_{Ti}^0 - R_i \ln\left[y_i P/P_0\right]$ | $\bar{s}_i = \bar{s}_{Ti}^0 - \bar{R} \ln\left[y_i P/P_0\right]$ |

### Air–Water Mixtures

Relative humidity  $\phi = \dfrac{P_v}{P_g}$

Humidity ratio  $\omega = \dfrac{m_v}{m_a} = 0.622\,\dfrac{P_v}{P_a} = 0.622\,\dfrac{\phi P_g}{P_{tot} - \phi P_g}$

Enthalpy per kg dry air  $\tilde{h} = h_a + \omega h_v$

# HOMEWORK PROBLEMS

## Concept-study guide problems

**10.1**  An ideal mixture at $T$, $P$ is made from ideal gases at $T$, $P$ by charging them into a steel tank. Assume heat is transferred so $T$ stays the same as the supply. How do the properties ($P$, $v$ and $u$) for each component change up, down, or constant?

**10.2**  An ideal mixture at $T$, $P$ is made from ideal gases at $T$, $P$ by flow into a mixing chamber without any external heat transfer and an exit at $P$. How do the properties ($P$, $v$, and $h$) for each component change up, down, or constant?

**10.3**  If a certain mixture is used in a number of different processes, do I need to consider partial pressures?

**10.4**  Why is it that I can use a set of tables for air, which is a mixture, without dealing with its composition?

**10.5**  Can moist air below the freezing point, say $-5°C$, have a dew point?

**10.6**  Why does a car with an air-conditioner running often have water dripping out?

**10.7**  I want to bring air at $35°C$, $\Phi = 40\%$ to a state of $25°C$, $\omega = 0.01$. Do I need to add or subtract water?

## Mixture composition and Properties

**10.8**  A gas mixture at $120°C$ and $125$ kPa is 50% $N_2$, 30% $H_2O$, and 20% $O_2$ on a mole basis. Find the mass fractions, the mixture gas constant, and the volume for 5 kg of mixture.

**10.9**  A mixture of 60% $N_2$, 30% argon, and 10% $O_2$ on a mass basis is in a cylinder at $250$ kPa and $310$ K with a volume of $0.5$ m$^3$. Find the mole fractions and the mass of argon.

**10.10**  If oxygen is 21% by mole of air, what is the oxygen state ($P$, $T$, $v$) in a room at $300$ K, $100$ kPa of total volume $60$ m$^3$?

**10.11**  A new refrigerant R-407 is a mixture of 23% R-32, 25% R-125, and 52% R-134a on a mass basis. Find the mole fractions, the mixture gas constant, and the mixture heat capacities for this new refrigerant.

**10.12**  A carbureted internal combustion engine is converted to run on methane gas (natural gas). The air–fuel ratio in the cylinder is to be 20 to 1 on a mass basis. How many moles of oxygen per mole of methane are there in the cylinder?

**10.13**  Weighing of masses gives a mixture at $60°C$, $225$ kPa with $0.5$ kg $O_2$, $1.5$ kg $N_2$, and $0.5$ kg $CH_4$. Find the partial pressures of each component, the mixture specific volume (mass basis), mixture molecular weight, and the total volume.

**10.14**  A flow of oxygen and one of nitrogen, both $300$ K, are mixed to produce 1 kg/s air at $300$ K, $100$ kPa. What are the mass and volume flow rates of each line?

**10.15**  A 2-kg mixture of 25% $N_2$, 50% $O_2$, and 25% $CO_2$ by mass is at $150$ kPa and $300$ K. Find the mixture gas constant and the total volume.

**10.16**  A $100$ m$^3$ storage tank with fuel gases is at $20°C$ and $100$ kPa containing a mixture of acetylene $C_2H_2$, propane $C_3H_8$, and butane $C_4H_{10}$. A test shows the partial pressure of the $C_2H_2$ is $15$ kPa and that of $C_3H_8$ is $65$ kPa. How much mass is there of each component?

**10.17**  A new refrigerant R-410a is a mixture of R-32 and R-125 in a 1:1 mass ratio. What are the overall molecular weight, the gas constant, and the ratio of specific heats for such a mixture?

**10.18**  A pipe, of cross-sectional area $0.1$ m$^2$, carries a flow of 75% $O_2$ and 25% $N_2$ by mole with a velocity of $-25$ m/s at $200$ kPa and $290$ K. To install and operate a mass flowmeter, it is necessary to know the mixture density and the gas constant. What are they? What mass flow rate should the meter then show?

## Simple processes

**10.19** A rigid container has 1 kg $CO_2$ at 300 K and 1 kg argon at 400 K both at 150 kPa. Now they are allowed to mix without any heat transfer. What is final $T$, $P$?

**10.20** The mixture in Problem 10.15 is heated to 500 K with constant volume. Find the final pressure and the total heat transfer needed using Table A.5.

**10.21** The mixture in problem 10.15 is heated up to 500 K in a constant-pressure process. Find the final volume and the total heat transfer using Table A.5.

**10.22** A flow of 1 kg/s argon at 300 K and another flow of 1 kg/s $CO_2$ at 1600 K both at 150 kPa are mixed without any heat transfer. What is the exit $T$, $P$?

**10.23** A pipe flows 1.5 kg/s of a mixture with mass fractions of 40% $CO_2$ and 60% $N_2$ at 400 kPa and 300 K, shown in Fig. P10.23. Heating tape is wrapped around a section of pipe with insulation added, and 2 kW of electrical power is heating the pipe flow. Find the mixture exit temperature.

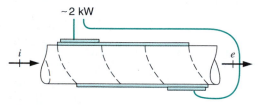

**FIGURE P10.23**

**10.24** A rigid insulated vessel contains 12 kg of oxygen at 200 kPa and 280 K separated by a membrane from 26 kg of carbon dioxide at 400 kPa and 360 K. The membrane is removed, and the mixture comes to a uniform state. Find the final temperature and pressure of the mixture.

**10.25** A steady flow of 0.1 kg/s carbon dioxide at 1000 K in one line is mixed with 0.2 kg/s of nitrogen at 400 K from another line, both at 100 kPa. The mixing chamber is insulated and has constant pressure at 100 kPa. Use constant heat capacity to find the exit temperature.

**10.26** A mixture of 40% water and 60% carbon dioxide by mass is heated from 400 K to 1000 K at a constant pressure of 120 kPa. Find the total change in enthalpy and entropy using Table A.5 values.

**10.27** Do Problem 10.26 but with variable heat capacity using values from Table A.8.

**10.28** An insulated gas turbine receives a mixture of 10% $CO_2$, 10% $H_2O$, and 80% $N_2$ on a mass basis at 1000 K and 500 kPa. The inlet volume flow rate is 2 $m^3$/s, and the exhaust is at 700 K and 100 kPa. Find the power output in kW using constant specific heat from Table A.5 at 300 K.

**10.29** Solve Problem 10.28 using values of enthalpy from Table A.8.

**10.30** A piston/cylinder device contains 0.1 kg of a mixture of 40% methane and 60% propane by mass at 300 K and 100 kPa. The gas is now slowly compressed in an isothermal ($T$ = constant) process to a final pressure of 250 kPa. Show the process in a $P$–$V$ diagram and find both the work and heat transfer in the process.

**10.31** A mixture of carbon dioxide and oxygen is made by having a rigid tank with two sections separated by a membrane. In $A$ we have 1 kg $CO_2$ at 1000 K and in $B$ we have 2 kg $O_2$ at 300 K both at 200 kPa. The membrane is broken and the mixture comes to a uniform state without any external heat transfer. Find the final $T$ and $P$.

**10.32** A mixture of 0.5 kg of nitrogen and 0.5 kg of oxygen is at 100 kPa and 300 K in a piston cylinder keeping constant pressure. Now 800 kJ is added by heating. Find the final temperature and the increase in entropy of the mixture using Table A.5 values.

**10.33** Repeat Problem 10.32, but solve using values from Table A.8.

**10.34** New refrigerant R-410a is a mixture of R-32 and R-25 in a 1:1 mass ratio. A process brings 0.5 kg R-410a from 270 K to 320 K at a constant pressure of 250 kPa in a piston cylinder. Find the work and heat transfer.

**10.35** A piston cylinder contains 0.5 kg of argon and 0.5 kg of hydrogen at 300 K and 100 kPa. The mixture is compressed in an adiabatic process to 400 kPa by an external force on the piston. Find the final temperature, the work, and the heat transfer in the process.

**10.36** Natural gas as a mixture of 75% methane and 25% ethane by mass is flowing to a compressor at 17°C and 100 kPa. The reversible adiabatic compressor brings the flow to 250 kPa. Find the exit temperature and the needed work per kg flow.

**10.37** The substance R-410a (see Problem 10.34) is at 100 kPa and 290 K. It is now brought to 250 kPa and 400 K in a reversible polytropic process. Find the change in specific volume, specific enthalpy, and specific entropy for the process.

**10.38** A steady flow of 0.1 kg/s carbon dioxide at 1000 K in one line is mixed with 0.2 kg/s of nitrogen at 400 K from another line, both at 100 kPa. The exit mixture at 100 kPa is compressed by an reversible adiabatic compressor to 500 kPa. Use constant heat capacity to find the mixing chamber exit temperature and the needed compressor power.

**10.39** Two insulated tanks $A$ and $B$ are connected by a valve, shown in Fig. P10.39. Tank $A$ has a volume of 1 m³ and initially contains argon at 300 kPa and 10°C. Tank $B$ has a volume of 2 m³ and initially contains ethane at 200 kPa and 50°C. The valve is opened and remains open until the resulting gas mixture comes to a uniform state. Determine the final pressure and temperature.

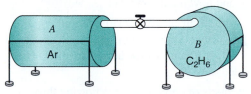

**FIGURE P10.39**

**10.40** A compressor brings R-410a (see Problem 10.34) from −10°C and 125 kPa up to 500 kPa in an adiabatic reversible compression. Assume ideal-gas behavior and find the exit temperature and the specific work.

**10.41** A mixture of 50% carbon dioxide and 50% water by mass is brought from 1500 K and 1 MPa to 500 K and 200 kPa in a polytropic process through a steady-state device. Find the necessary heat transfer and work involved using values from Table A.5.

**10.42** A 50/50 (by mass) gas mixture of methane $CH_4$ and ethylene $C_2H_4$ is contained in a cylinder piston at the initial state of 480 kPa, 330 K, and 1.05 m³. The piston is now moved, compressing the mixture in a reversible, polytropic process to the final state of 260 K and 0.03 m³. Calculate the final pressure, the polytropic exponent, the work and heat transfer, and entropy change for the mixture.

**10.43** A piston cylinder has 0.1 kg mixture of 25% argon, 25% nitrogen and 50% carbon dioxide by mass at total pressure 100 kPa and 290 K. Now the piston compresses the gases to a volume 7 times smaller in a polytropic process with $n = 1.3$. Find the final pressure and temperature, the work, and the heat transfer in the process.

**10.44** A mixture of 2 kg of oxygen and 2 kg of argon is in an insulated piston–cylinder arrangement at 100 kPa and 300 K. The piston now compresses the mixture to half its initial volume. Find the final pressure, final temperature, and the piston work.

## Entropy generation

**10.45** A flow of gas $A$ and a flow of gas $B$ are mixed in a 1:1 mole ratio with the same $T$. What is the entropy generation per kmole flow out?

**10.46** What is the rate of entropy increase in problem 10.22?

**10.47** What is the entropy generation in Problem 10.19?

**10.48** A flow of 2 kg/s mixture of 50% $CO_2$ and 50% $O_2$ by mass is heated in a constant-pressure heat exchanger from 400 K to 1000 K by a radiation source at 1400 K. Find the rate of heat transfer and the entropy generation in the process shown in Fig. P10.48.

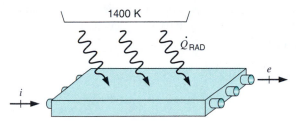

**FIGURE P10.48**

**10.49** Carbon dioxide gas at 320 K is mixed with nitrogen at 280 K in an insulated mixing chamber. Both flows are at 100 kPa, and the mass ratio of carbon dioxide to nitrogen is 2:1. Find the exit temperature and the total entropy generation per kg of the exit mixture.

**10.50** A rigid container has 1 kg argon at 300 K and 1 kg argon at 400 K both at 150 kPa. Now they are allowed to mix without any external heat transfer. What is final $T, P$? Is any $s$ generated?

**10.51** Repeat Problem 10.49 with inlet temperatures of 1400 K for the carbon dioxide and 300 K for the nitrogen. First estimate the exit temperature with the specific heats from Table A.5 and use this to start iterations with values from Table A.8.

**10.52** A flow of 1 kg/s carbon dioxide at 1600 K, 100 kPa is mixed with a flow of 2 kg/s water at 800 K, 100 kPa. After the mixing it goes through a heat exchanger where it is cooled to 500 K by a 400 K ambient, as shown in Fig. P10.52. How much heat transfer is taken out in the heat exchanger? What is the entropy generation rate for the whole process?

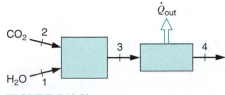

**FIGURE P10.52**

**10.53** A mixture of 60% helium and 40% nitrogen by mass enters a turbine at 1 MPa and 800 K at a rate of

2 kg/s. The adiabatic turbine has an exit pressure of 100 kPa and an isentropic efficiency of 85%. Find the turbine work.

**10.54** A large air separation plant takes in ambient air (79% $N_2$, 21% $O_2$ by mole) at 100 kPa and 20°C at a rate of 25 kg/s. It discharges a stream of pure $O_2$ gas at 200 kPa and 100°C and a stream of pure $N_2$ gas at 100 kPa and 20°C. The plant operates on an electrical power input of 2000 kW, shown in Fig. P10.54. Calculate the net rate of entropy change for the process.

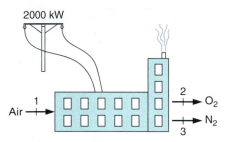

**FIGURE P10.54**

**10.55** A steady flow of 0.3 kg/s of 50% carbon dioxide and 50% water by mass at 1200 K and 200 kPa is used in a heat exchanger where 300 kW is extracted from the flow. Find the flow exit temperature and the rate of change of entropy using Table A.8.

**10.56** A flow of 1.8 kg/s steam at 400 kPa, 400°C, is mixed with 3.2 kg/s oxygen at 400 kPa, 400 K, in a steady flow mixing-chamber without any heat transfer. Find the exit temperature and the rate of entropy generation.

**10.57** Three steady flows are mixed in an adiabatic chamber at 150 kPa. Flow one is 2 kg/s of $O_2$ at 340 K, flow two is 4 kg/s of $N_2$ at 280 K, and flow three is 3 kg/s of $CO_2$ at 310 K. All flows are at 150 kPa, the same as the total exit pressure. Find the exit temperature and the rate of entropy generation in the process.

**10.58** Reconsider Problem 10.39, but let the tanks have a small amount of heat transfer so the final mixture is at 400 K. Find the final pressure, the heat transfer, and the entropy change for the process.

## Air-water vapor mixtures

**10.59** If I have air at 100 kPa and a) −10°C b) 45°C and c) 110°C what is the maximum absolute humidity I can have?

**10.60** If I have air at 200 kPa and a) −10°C b) 45°C and c) 110°C what is the maximum absolute humidity I can have?

**10.61** Atmospheric air is at 100 kPa and 25°C with a relative humidity of 75%. Find the absolute humidity and the dew point of the mixture. If the mixture is heated to 30°C, what is the new relative humidity?

**10.62** A 1 kg/s flow of saturated moist air (relative humidity 100%) at 100 kPa and 10°C goes through a heat exchanger and comes out at 25°C. What is the exit relative humidity and how much power is needed?

**10.63** A room with air at 40% relative humidity, 20°C having 50 kg of dry air is made moist by boiling water to a final state of 20°C and 100% humidity. How much water was added to the air?

**10.64** The products of combustion are flowing through a heat exchanger with 12% $CO_2$, 13% $H_2O$, and 75% $N_2$ on a volume basis at the rate 0.1 kg/s and 100 kPa. What is the dew-point temperature? If the mixture is cooled 10°C below the dew-point temperature, how long will it take to collect 10 kg of liquid water?

**10.65** Consider 100 $m^3$ of atmospheric air, which is an air–water vapor mixture at 100 kPa, 15°C, and 40% relative humidity. Find the mass of water and the humidity ratio. What is the dew point of the mixture?

**10.66** A flow of 0.2 kg/s liquid water at 80°C is sprayed into a chamber together with 16 kg/s dry air at 80°C. All the water evaporates and the air leaves at 40°C. What is the exit relative humidity and the heat transfer.

**10.67** A new high-efficiency home heating system includes an air-to-air heat exchanger, which uses energy from outgoing stale air to heat the fresh incoming air. If the outside ambient temperature is −10°C and the relative humidity is 30%, how much water will have to be added to the incoming air, if it flows in at the rate of 1 $m^3$/s and must eventually be conditioned to 20°C and 40% relative humidity?

**10.68** Consider a 1 $m^3$/s flow of atmospheric air at 100 kPa, 25°C, and 80% relative humidity. Assume this flows into a basement room where it cools to 15°C at 100 kPa. How much liquid water will condense out?

**10.69** A 2 kg/s flow of completely dry air at $T_1$ and 100 kPa is cooled down to 10°C by spraying liquid water at 10°C and 100 kPa into it so it becomes saturated moist air at 10°C. The process is steady state with no external heat transfer or work. Find the exit moist air humidity ratio and the flow rate of liquid water. Find also the dry air inlet temperature $T_1$.

**10.70** A piston cylinder has 100 kg of saturated moist air at 100 kPa and 5°C. If it is heated to 45°C in an isobaric process, find $_1q_2$ and the final relative humidity.

**10.71** Ambient moist air enters a steady-flow air-conditioning unit at 102 kPa and 30°C with a 60% relative humidity. The volume flow rate entering the unit is 100 L/s. The moist air leaves the unit at 95 kPa and 15°C with a relative humidity of 100%. Liquid condensate also leaves the unit at 15°C. Determine the rate of heat transfer for this process.

**10.72** To make dry coffee powder, we spray 0.2 kg/s coffee (assume liquid water) at 80°C into a chamber where we add 8 kg/s dry air at $T$. All the water should evaporate and the air should leave with a minimum 40°C and we neglect the powder. How high should $T$ in the inlet air flow be?

**10.73** Consider at 500 L rigid tank containing an air–water vapor mixture at 100 kPa and 35°C with a 70% relative humidity. The system is cooled until the water just begins to condense. Determine the final temperature in the tank and the heat transfer for the process.

**10.74** A 300 L rigid vessel initially contains moist air at 150 kPa and 40°C with a relative humidity of 10%. A supply line connected to this vessel by a valve carries steam at 600 kPa and 200°C. The valve is opened, and steam flows into the vessel until the relative humidity of the resultant moist air mixture is 90%. Then the valve is closed. Sufficient heat is transferred from the vessel so that the temperature remains at 40°C during the process. Determine the heat transfer for the process, the mass of steam entering the vessel, and the final pressure inside the vessel.

**10.75** A rigid container, 10 m³ in volume, contains moist air at 45°C and 100 kPa with $\Phi = 40\%$. The container is now cooled to 5°C. Neglect the volume of any liquid that might be present and find the final mass of water vapor, final total pressure, and the heat transfer.

# Tables and formulas or psychrometric chart

**10.76** A flow of moist air at 100 kPa, 40°C, and 40% relative humidity is cooled to 15°C in a constant-pressure device. Find the humidity ratio of the inlet and the exit flow and the heat transfer in the device per kg dry air.

**10.77** A flow, 0.2 kg/s dry air, of moist air at 40°C and 50% relative humidity flows from the outside state 1 down into a basement where it cools to 16°C, state 2. Then it flows up to the living room where it is heated to 25°C, state 3. Find the dew point for state 1, any amount of liquid that may appear, the heat transfer that takes place in the basement, and the relative humidity in the living room at state 3.

**10.78** The discharge moist air from a clothes dryer is at 35°C, 80% relative humidity. The flow is guided through a pipe up through the roof and a vent to the atmosphere shown in Fig. P10.78. Due to heat transfer in the pipe, the flow is cooled to 24°C by the time it reaches the vent. Find the humidity ratio in the flow out of the clothes dryer and at the vent. Find the heat transfer and any amount of liquid that may be forming per kg dry air for the flow.

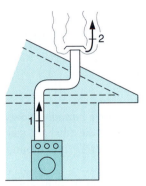

**FIGURE P10.78**

**10.79** Use the formulas and the steam tables to find the missing property of: $\Phi$, $\omega$, and $T_{dry}$, for a total pressure of 100 kPa; repeat the answers using the psychrometric chart.

a. $\Phi = 50\%$, $\omega = 0.010$

b. $T_{dry} = 25°C$, $T_{wet} = 21°C$

**10.80** A steady supply of 1.0 m³/s air at 25°C, 100 kPa, and 50% relative humidity is needed to heat a building in the winter. The ambient outdoors is at 10°C, 100 kPa, and 50% relative humidity. What are the required liquid water input and heat transfer rates for this purpose?

**10.81** A combination air cooler and dehumidification unit receives outside ambient air at 35°C, 100 kPa, and 90% relative humidity. The moist air is first cooled to a low temperature $T_2$ to condense the proper amount of water; assume all the liquid leaves at $T_2$. The moist air is then heated and leaves the unit at 20°C, 100 kPa, and 30% relative humidity with a volume flow rate of 0.01 m³/s. Find the temperature $T_2$, the mass of liquid per kilogram of dry air, and the overall heat transfer rate.

**10.82** A flow of moist air from a domestic furnace, state 1, is at 45°C, 10% relative humidity with a flow rate of 0.05 kg/s dry air. A small electric heater adds steam at 100°C, 100 kPa, generated from tap water at 15°C shown in Fig. P10.82. Up in the living room, the flow comes out at state 4: 30°C, 60% relative humidity. Find the power needed for the electric heater and the heat transfer to the flow from state 1 to state 4.

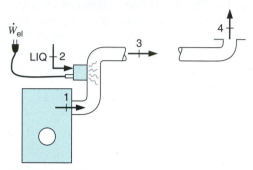

**FIGURE P10.82**

**10.83** Moist air at 31°C and 50% relative humidity flows over a large surface of liquid water. Find the adiabatic saturation temperature by trial and error. *Hint:* it is around 22.5°C.

**10.84** A water-cooling tower for a power plant cools 45°C liquid water by evaporation. The tower receives air at 19.5°C, $\Phi = 30\%$, and 100 kPa that is blown through/over the water such that it leaves the tower at 25°C and $\Phi = 70\%$. The remaining liquid water flows back to the condenser at 30°C having given off 1 MW. Find the mass flow rate of air, and determine the amount of water that evaporates.

**10.85** A flow of air at 5°C, $\Phi = 90\%$, is brought into a house, where it is conditioned to 25°C, 60% relative humidity. This is done with a combined heater-evaporator where any liquid water is at 10°C. Find any flow of liquid and the necessary heat transfer, both per kilogram dry air flowing. Find the dew point for the final mixture.

**10.86** In a car's defrost/defog system atmospheric air at 21°C and 80% relative humidity is taken in and cooled such that liquid water drips out. The now dryer air is heated to 41°C and then blown onto the windshield, where it should have a maximum of 10% relative humidity to remove water from the windshield. Find the dew point of the atmospheric air, specific humidity of air onto the windshield, the lowest temperature, and the specific heat transfer in the cooler.

**10.87** Atmospheric air at 35°C with a relative humidity of 10%, is too warm and also too dry. An air conditioner should deliver air at 21°C and 50% relative humidity in the amount of 3600 m³/h. Sketch a setup to accomplish this. Find any amount of liquid (at 20°C) that is needed or discarded and any heat transfer.

**10.88** A flow of moist air at 45°C, 10% relative humidity with a flow rate of 0.2 kg/s dry air is mixed with a flow of moist air at 25°C, and absolute humidity of $w = 0.018$ with a rate of 0.3 kg/s dry air. The mixing takes place in an air duct at 100 kPa with some heat transfer to reach a final temperature of 40°C. Find the heat transfer and the final exit relative humidity.

**10.89** An indoor pool evaporates 1.512 kg/h of water, which is removed by a dehumidifier to maintain 21°C, $\Phi = 70\%$ in the room. The dehumidifier, shown in Fig. P10.89, is a refrigeration cycle in which air flowing over the evaporator cools such that liquid water drops out, and the air continues flowing over the condenser. For an airflow rate of 0.1 kg/s the unit requires 1.4 kW input to a motor driving a fan and the compressor, and it has a coefficient of performance, $\beta = \dot{Q}_L/\dot{W}_c = 2.0$. Find the state of the air as it returns to the room and the compressor work input.

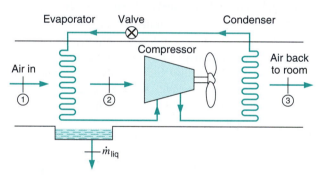

**FIGURE P10.89**

# Psychrometric chart only

**10.90** Use the psychrometric chart to find the missing property of: $\Phi, \omega, T_{wet}, T_{dry}$.
a. $T_{dry} = 25°C$, $\Phi = 80\%$
b. $T_{dry} = 15°C$, $\Phi = 100\%$
c. $T_{dry} = 20°C$, $\omega = 0.008$
d. $T_{dry} = 25°C$, $T_{wet} = 23°C$

**10.91** Use the psychrometric chart to find the missing property of: $\Phi, \omega, T_{wet}, T_{dry}$.
a. $\Phi = 50\%$, $\omega = 0.012$
b. $T_{wet} = 15°C$, $\Phi = 60\%$
c. $\omega = 0.008$, $T_{wet} = 17°C$
d. $T_{dry} = 10°C$, $\omega = 0.006$

**10.92** For each of the states in Problem 10.91 find the dew-point temperature.

**10.93** Compare the weather in two places where it is cloudy and breezy. At beach $A$ the temperature is 20°C, the pressure is 103.5 kPa, and the relative humidity is 90%; beach $B$ has 25°C, 99 kPa, and 20% relative humidity. Suppose you just took a swim and came out of the water. Where would you feel more comfortable, and why?

**10.94** An air-conditioner should cool a flow of ambient moist air at 40°C, 40% relative humidity having 0.2 kg/s flow of dry air. The exit temperature should be 25°C and the pressure is 100 kPa. Find the rate of heat transfer needed and check for the formation of liquid water.

**10.95** Ambient air at 100 kPa, 30°C, and 40% relative humidity goes through a constant-pressure heat exchanger as a steady flow. In one case it is heated to 45°C, and in another case it is cooled until it reaches saturation. For both cases find the exit relative humidity and the amount of heat transfer per kilogram of dry air.

**10.96** Two moist air streams with 85% relative humidity, both flowing at a rate of 0.1 kg/s of dry air, are mixed in a steady-flow setup. One inlet stream is at 32.5°C and the other at 16°C. Find the exit relative humidity.

**10.97** One means of air-conditioning hot summer air is by evaporative cooling, which is a process similar to the adiabatic saturation process. Consider outdoor ambient air at 35°C, 100 kPa, 30% relative humidity. What is the maximum amount of cooling that can be achieved by such a technique? What disadvantage is there to this approach?

**10.98** A flow of moist air at 100 kPa, 35°C, 40 relative humidity is cooled by adiabatic evaporation of liquid 20°C water to reach a saturated state. Find the amount of water added per kg dry air and the exit temperature.

**10.99** A flow of moist air at 21°C with 60% relative humidity should be produced from mixing two different moist airflows. Flow 1 is at 10°C and 80% relative

humidity; flow 2 is at 32°C and has $T_{wet} = 27$°C. The mixing chamber can be followed by a heater or a cooler, as shown in Fig. P10.99. No liquid water is added, and $P = 100$ kPa. Find the two controls—one is the ratio of the two mass flow rates $\dot{m}_{a1}/\dot{m}_{a2}$ and the other is the heat transfer in the heater/cooler per kg dry air.

**10.100** In a hot and dry climate, air enters an air-conditioner unit at 100 kPa, 40°C, and 5% relative humidity, at the steady rate of 1.0 m³/s. Liquid water at 20°C is sprayed into the air in the AC unit at the rate of 20 kg/h, and heat is rejected from the unit at the rate 20 kW. The exit pressure is 100 kPa. What are the exit temperature and relative humidity?

**10.101** An insulated tank has an air inlet, $\omega_1 = 0.0084$ and an outlet, $T_2 = 22$°C, $\Phi_2 = 90\%$, both at 100 kPa. A third line sprays 0.25 kg/s of water at 80°C and 100 kPa, as shown in Fig. P10.101. For steady operation, find the outlet specific humidity, the mass flow rate of air needed, and the required air inlet temperature, $T_1$.

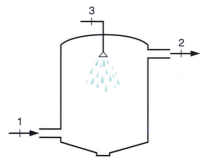

**FIGURE P10.101**

**10.102** Consider two states of atmospheric air (1) 35°C, $T_{wet} = 18$°C and (2) 26.5°C, $\Phi = 60\%$. Suggest a system of devices that will allow air in a steady flow to change from (1) to (2) and from (2) to (1). Heaters, coolers, (de)humidifiers, liquid traps, and the like are available, and any liquid/solid flowing is assumed to be at the lowest temperature seen in the process. Find the specific and relative humidity for state 1, dew point for state 2, and the heat transfer per kilogram dry air in each component in the systems.

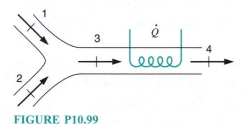

**FIGURE P10.99**

# Chapter 11

# Power and Refrigeration Systems

## TOPICS DISCUSSED

Some power plants, such as the simple steam power plant, which we have considered several times, operate in a cycle. That is, the working fluid undergoes a series of processes and finally returns to the initial state. In other power plants, such as the internal-combustion engine and the gas turbine, the working fluid does not go through a thermodynamic cycle, even though the engine itself may operate in a mechanical cycle. In this instance the working fluid has a different composition or is in a different state at the conclusion of the process than it had or was at the beginning. Such equipment is sometimes said to operate on the open cycle (the word cycle is really a misnomer), whereas the steam power plant operates on a closed cycle. The same distinction between open and closed cycles can be made regarding refrigeration devices. For both the open- and closed-cycle apparatus, however, it is advantageous to analyze the performance of an idealized closed cycle similar to the actual cycle. Such a procedure is particularly advantageous for determining the influence of certain variables on performance. For example, the spark-ignition internal-combustion engine is usually approximated by the Otto cycle. From an analysis of the Otto cycle we conclude that increasing the compression ratio increases the efficiency. This is also true for the actual engine, even though the Otto-cycle efficiencies may deviate significantly from the actual efficiencies.

This chapter is concerned with these idealized cycles for both power and refrigeration apparatus. Both vapors and ideal gases are considered as working fluids. An attempt will be made to point out how the processes in actual apparatus deviate from the ideal. Consideration is also given to certain modifications of the basic cycles that are intended to improve performance. These modifications include the use of devices such as regenerators, multistage compressors and expanders, and intercoolers. Various combinations of these types of systems and also special applications, such as cogeneration of electrical power and energy, combined cycles, topping and bottoming cycles, and binary cycle systems, are also discussed in this chapter and in the problems.

## 11.1 INTRODUCTION TO POWER SYSTEMS

In introducing the second law of thermodynamics in Chapter 7, we considered cyclic heat engines consisting of four separate processes. We noted there that it is possible to have these engines operate as steady-state devices involving shaft work, as shown in Fig. 7.16, or instead as cylinder/piston devices involving boundary-movement work, as shown in Fig. 7.17. The former may have a working fluid that changes phase during the processes in the cycle, or may have a single-phase working fluid throughout. The latter type would normally have a gaseous working fluid throughout the cycle.

For a reversible steady-state process involving negligible kinetic and potential energy changes, the shaft work per unit mass is given by Eq. 9.19,

$$w = -\int v \, dP$$

For a reversible process involving a simple compressible substance, the boundary movement work per unit mass is given by Eq. 4.3,

$$w = \int P \, dv$$

The areas represented by these two integrals are shown in Fig. 11.1. It is of interest to note that, in the former case, there is no work involved in a constant-pressure process, while in the latter case, there is no work involved in a constant-volume process.

Let us now consider a power system consisting of four steady-state processes, as in Fig. 7.16. We assume that each process is internally reversible and has negligible changes

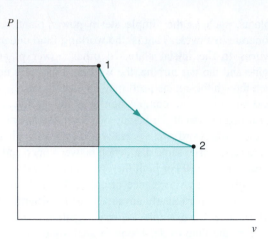

**FIGURE 11.1** Comparison of shaft work and boundary-movement work.

in kinetic and potential energies, which results in the work for each process being given by Eq. 9.19. For convenience of operation, we will make the two heat-transfer processes (boiler and condenser) constant-pressure processes, such that those are simple heat exchangers involving no work. Let us also assume that the turbine and pump processes are both adiabatic, such that they are therefore isentropic processes. Thus, the four processes comprising the cycle are as shown in Fig. 11.2. Note that if the entire cycle takes place inside the two-phase liquid–vapor dome, the resulting cycle is the Carnot cycle, since the two constant-pressure processes are also isothermal. Otherwise, this cycle is not a Carnot cycle. In either case, we find that the net work output for this power system is given by

$$w_{\text{net}} = -\int_1^2 v\,dP + 0 - \int_3^4 v\,dP + 0 = -\int_1^2 v\,dP + \int_4^3 v\,dP$$

and, since $P_2 = P_3$ and also $P_1 = P_4$, we find that the system produces a net work output because the specific volume is larger during the expansion from 3 to 4 than it is during the compression from 1 to 2. This result is also evident from the areas $-\int v\,dP$ in Fig. 11.2. We conclude that it would be advantageous to have this difference in specific volume be as large as possible, as, for example, the difference between a vapor and a liquid.

If the four-process cycle shown in Fig. 11.2 were accomplished in a cylinder/piston system involving boundary-movement work, then the net work output for this power system is given by

$$w_{\text{net}} = \int_1^2 P\,dv + \int_2^3 P\,dv + \int_3^4 P\,dv + \int_4^1 P\,dv$$

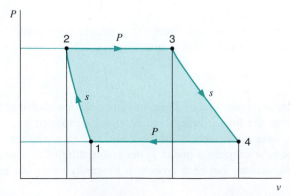

**FIGURE 11.2** Four-process power cycle.

and from these four areas on Fig. 11.2, we note that the pressure is higher during any given change in volume in the two expansion processes than in the two compression processes, resulting in a net positive area and a net work output.

For either of the two cases just analyzed, it is noted from Fig. 11.2 that the net work output of the cycle is equal to the area enclosed by the process lines 1–2–3–4–1, and this area is the same for both cases, even though the work terms for the four individual processes are different for the two cases.

In the next several sections, we consider the Rankine cycle, which is the ideal, four-steady-state process cycle shown in Fig. 11.2, utilizing a phase change between vapor and liquid to maximize the difference in specific volume during expansion and compression. This is the idealized model for a steam power plant system.

## 11.2 THE RANKINE CYCLE

We now consider the idealized four-steady-state-process cycle shown in Fig. 11.2, in which state 1 is saturated liquid and state 3 either saturated vapor or superheated vapor. This system is termed the Rankine cycle and is the model for the simple steam power plant. It is convenient to show the states and processes on a $T$–$s$ diagram, as given in Fig. 11.3. The four processes are:

**1–2:** Reversible adiabatic pumping process in the pump

**2–3:** Constant-pressure transfer of heat in the boiler

**3–4:** Reversible adiabatic expansion in the turbine (or other prime mover such as a steam engine)

**4–1:** Constant-pressure transfer of heat in the condenser

As mentioned above, the Rankine cycle also includes the possibility of superheating the vapor, as cycle 1–2–3′–4′–1.

If changes of kinetic and potential energy are neglected, heat transfer and work may be represented by various areas on the $T$–$s$ diagram. The heat transferred to the working fluid is represented by area $a$–2–2′–3–$b$–$a$, and the heat transferred from the working fluid by area $a$–1–4–$b$–$a$. From the first law we conclude that the area representing the work is the difference between these two areas—area 1–2–2′–3–4–1. The thermal efficiency is defined by the relation

$$\eta_{\text{th}} = \frac{w_{\text{net}}}{q_H} = \frac{\text{area } 1\text{--}2\text{--}2'\text{--}3\text{--}4\text{--}1}{\text{area } a\text{--}2\text{--}2'\text{--}3\text{--}b\text{--}a} \tag{11.1}$$

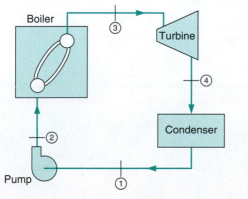
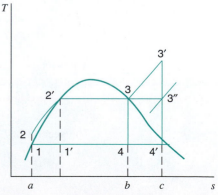

**FIGURE 11.3** Simple steam power plant that operates on the Rankine cycle.

For analyzing the Rankine cycle, it is helpful to think of efficiency as depending on the average temperature at which heat is supplied and the average temperature at which heat is rejected. Any changes that increase the average temperature at which heat is supplied or decrease the average temperature heat is rejected will increase the Rankine-cycle efficiency.

In analyzing the ideal cycles in this chapter, the changes in kinetic and potential energies from one point in the cycle to another are neglected. In general, this is a reasonable assumption for the actual cycles.

It is readily evident that the Rankine cycle has a lower efficiency than a Carnot cycle with the same maximum and minimum temperatures as a Rankine cycle, because the average temperature between 2 and 2′ is less than the temperature during evaporation. We might well ask, why choose the Rankine cycle as the ideal cycle? Why not select the Carnot cycle 1′–2′–3–4–1′? At least two reasons can be given. The first reason concerns the pumping process. State 1′ is a mixture of liquid and vapor. Great difficulties are encountered in building a pump that will handle the mixture of liquid and vapor at 1′ and deliver saturated liquid at 2′. It is much easier to condense the vapor completely and handle only liquid in the pump: The Rankine cycle is based on this fact. The second reason concerns superheating the vapor. In the Rankine cycle the vapor is superheated at constant pressure, process 3–3′. In the Carnot cycle all the heat transfer is at constant temperature, and therefore the vapor is superheated in process 3–3″. Note, however, that during this process the pressure is dropping, which means that the heat must be transferred to the vapor as it undergoes an expansion process in which work is done. This heat transfer is also very difficult to achieve in practice. Thus, the Rankine cycle is the ideal cycle that can be approximated in practice. In the following sections we will consider some variations on the Rankine cycle that enable it to approach more closely the efficiency of the Carnot cycle.

Before we discuss the influence of certain variables on the performance of the Rankine cycle, an example is given.

| | |
|---|---|
| **EXAMPLE 11.1** | Determine the efficiency of a Rankine cycle using steam as the working fluid in which the condenser pressure is 10 kPa. The boiler pressure is 2 MPa. The steam leaves the boiler as saturated vapor. |

In solving Rankine-cycle problems, we let $w_p$ denote the work into the pump per kilogram of fluid flowing and $q_L$ denote the heat rejected from the working fluid per kilogram of fluid flowing.

To solve this problem we consider, in succession, a control surface around the pump, the boiler, the turbine, and the condenser. For each the thermodynamic model is the steam tables, and the process is steady state with negligible changes in kinetic and potential energies. First consider the pump:

> *Control volume*:  Pump.
> *Inlet state*:  $P_1$ known, saturated liquid; state fixed.
> *Exit state*:  $P_2$ known

**Analysis**

From the first law, we have

$$w_p = h_2 - h_1$$

The second law gives

$$s_2 = s_1$$

and so

$$h_2 - h_1 = \int_1^2 v\, dP$$

**Solution**

Assuming the liquid to be incompressible, we have

$$w_p = v(P_2 - P_1) = (0.001\ 01)(2000 - 10) = 2.0\ \text{kJ/kg}$$

$$h_2 = h_1 + w_p = 191.8 + 2.0 = 193.8\ \text{kJ/kg}$$

Now consider the boiler:

> *Control volume*:  Boiler.
> *Intel state*:  $P_2, h_2$ known; state fixed.
> *Exit state*:  $P_3$ known, saturated vapor; state fixed.

**Analysis**

From the first law, we write

$$q_H = h_3 - h_2$$

**Solution**

Substituting, we obtain

$$q_H = h_3 - h_2 = 2799.5 - 193.8 = 2605.7\ \text{kJ/kg}$$

Turning to the turbine next, we have:

> *Control volume*:  Turbine.
> *Inlet state*:  State 3 known (above).
> *Exit state*:  $P_4$ known.

**Analysis**

The first law gives

$$w_t = h_3 - h_4$$

The second law gives

$$s_3 = s_4$$

**Solution**

We can determine the quality at state 4 as follows:

$$s_3 = s_4 = 6.3409 = 0.6493 + x_4 7.5009, \qquad x_4 = 0.7588$$

$$h_4 = 191.8 + 0.7588(2392.8) = 2007.5\ \text{kJ/kg}$$

$$w_t = 2799.5 - 2007.5 = 792.0\ \text{kJ/kg}$$

Finally, we consider the condenser.

> *Control volume*: Condenser.
> *Inlet state*: State 4 known (as determined).
> *Exit state*: State 1 known (as given).

**Analysis**

The first law gives

$$q_L = h_4 - h_1$$

**Solution**

Substituting, we obtain

$$q_L = h_4 - h_1 = 2007.5 - 191.8 = 1815.7 \text{ kJ/kg}$$

We can now calculate the thermal efficiency:

$$\eta_{th} = \frac{w_{net}}{q_H} = \frac{q_H - q_L}{q_H} = \frac{w_t - w_p}{q_H} = \frac{792.0 - 2.0}{2605.7} = 30.3\%$$

We could also write an expression for thermal efficiency in terms of properties at various points in the cycle:

$$\eta_{th} = \frac{(h_3 - h_2) - (h_4 - h_1)}{h_3 - h_2} = \frac{(h_3 - h_4) - (h_2 - h_1)}{h_3 - h_2}$$

$$= \frac{2605.7 - 1815.7}{2605.7} = \frac{792.0 - 2.0}{2605.7} = 30.3\%$$

Let us now consider the effects of pressure and temperature on the performance of the Rankine cycle. The variables are the exhaust pressure (and temperature), the maximum cycle temperature, and the maximum (boiler) cycle pressure. In Fig. 11.4, we realize that the net cycle work output of the basic Rankine cycle is given by the area enclosed by the process lines 1–2–3–4–1, as was discussed in Sect. 11.1. We note that whatever is changed to expand the size of this process envelope will increase the net cycle work.

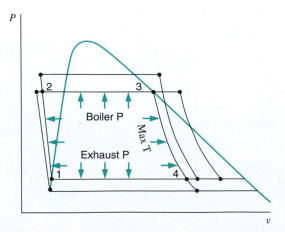

**FIGURE 11.4** Effect of pressure and temperature on Rankine-cycle work

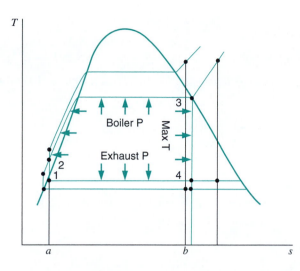

**FIGURE 11.5** Effect of pressure and temperature on Rankine-cycle efficiency

That is, a decrease in exhaust pressure, an increase in maximum temperature (by superheating the steam), or an increase in maximum pressure will all increase this area.

In the *T–s* diagram of the Rankine cycle shown in Fig. 11.5, we recall that the net cycle work output again is given by the area enclosed by the process lines 1–2–3–4–1, since that is the difference between the heat input area *a*–2–3–*b*–*a* and the heat rejected *a*–1–4–*b*–1. The cycle efficiency is given by the ratio of the net work to the heat input, or the ratio of those two areas on the *T–s* diagram. Again, it is seen that this ratio increases with a decrease in exhaust pressure, an increase in maximum temperature, or an increase in maximum pressure.

---

**EXAMPLE 11.2**

In a Rankine cycle steam leaves the boiler and enters the turbine at 4 MPa and 400°C. The condenser pressure is 10 kPa. Determine the cycle efficiency.

To determine the cycle efficiency, we must calculate the turbine work, the pump work, and the heat transfer to the steam in the boiler. We do this by considering a control surface around each of these components in turn. In each case the thermodynamic model is the steam tables, and the process is steady state with negligible changes in kinetic and potential energies.

> *Control volume*:  Pump.
> *Inlet state*:  $P_1$ known, saturated liquid; state fixed.
> *Exit state*:  $P_2$ known.

**Analysis**

From the first law, we have

$$w_p = h_2 - h_1$$

The second law gives

$$s_2 = s_1$$

Since $s_2 = s_1$,

$$h_2 - h_1 = \int_1^2 v \, dP = v(P_2 - P_1)$$

**Solution**

Substituting, we obtain

$$w_p = v(P_2 - P_1) = (0.001\ 01)(4000 - 10) = 4.0 \text{ kJ/kg}$$

$$h_1 = 191.8 \text{ kJ/kg}$$

$$h_2 = 191.8 + 4.0 = 195.8 \text{ kJ/kg}$$

For the turbine we have:

*Control volume*: Turbine.
*Inlet state*: $P_3$, $T_3$ known; state fixed.
*Exit state*: $P_4$ known.

**Analysis**

The first law is

$$w_t = h_3 - h_4$$

and the second law is

$$s_4 = s_3$$

**Solution**

Upon substitution we get

$$h_3 = 3213.6 \text{ kJ/kg}, \qquad s_3 = 6.7690 \text{ kJ/kg K}$$

$$s_3 = s_4 = 6.7690 = 0.6493 + x_4 7.5009, \qquad x_4 = 0.8159$$

$$h_4 = 191.8 + 0.8159(2392.8) = 2144.1 \text{ kJ/kg}$$

$$w_t = h_3 - h_4 = 3213.6 - 2144.1 = 1069.5 \text{ kJ/kg}$$

$$w_{\text{net}} = w_t - w_p = 1069.5 - 4.0 = 1065.5 \text{ kJ/kg}$$

Finally, for the boiler we have:

*Control volume*: Boiler.
*Inlet state*: $P_2$, $h_2$ known; state fixed.
*Exit state*: State 3 fixed (as given).

**Analysis**

The first law is

$$q_H = h_3 - h_2$$

**Solution**

Substituting gives

$$q_H = h_3 - h_2 = 3213.6 - 195.8 = 3017.8 \text{ kJ/kg}$$

$$\eta_{\text{th}} = \frac{w_{\text{net}}}{q_H} = \frac{1065.5}{3017.8} = 35.3\%$$

The net work could also be determined by calculating the heat rejected in the condenser, $q_L$, and noting, from the first law, that the net work for the cycle is equal to the net heat transfer. Considering a control surface around the condenser, we have

$$q_L = h_4 - h_1 = 2144.1 - 191.8 = 1952.3 \text{ kJ/kg}$$

Therefore,

$$w_{\text{net}} = q_H - q_L = 3017.8 - 1952.3 = 1065.5 \text{ kJ/kg}$$

## 11.3 THE REGENERATIVE CYCLE

An important variation from the Rankine cycle is the regenerative cycle, which uses feedwater heaters. The basic concepts of this cycle can be demonstrated by considering the Rankine cycle without superheat as shown in Fig. 11.6. During the process between states 2 and 2′, the working fluid is heated while in the liquid phase, and the average temperature of the working fluid is much lower than during the vaporization process 2′–3. The process between states 2 and 2′ causes the average temperature at which heat is supplied in the Rankine cycle to be lower than in the Carnot cycle 1′–2′–3–4–1′. Consequently, the efficiency of the Rankine cycle is lower than that of the corresponding Carnot cycle. In the regenerative cycle the working fluid enters the boiler at some state between 2 and 2′, and consequently the average temperature at which heat is supplied is higher.

In an ideal regenerative cycle, the liquid exiting the pump at state 2 in Fig. 11.6 would be preheated to state 2′ by circulating the liquid around the turbine casing in a counter-flowing direction to that of vapor flow through the turbine. This would require a large heat transfer between the two flow streams. Since this is not practical, a more reasonable approach is to extract a portion of the (partially expanded) turbine vapor to mix with the liquid in an open feedwater heater, as shown in the schematic diagram of Fig. 11.7.

Superheated steam enters the turbine at state 5. After expansion to state 6, some of the steam is extracted and enters the feedwater heater. The steam that is not extracted is expanded in the turbine to state 7 and is then condensed in the condenser. This condensate is pumped into the feedwater heater where it mixes with the steam extracted from the turbine. The proportion of steam extracted is just sufficient to cause the liquid

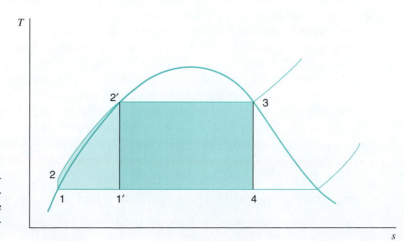

**FIGURE 11.6**
Temperature–entropy diagram showing the relationship between Carnot-cycle efficiency and Rankine-cycle efficiency.

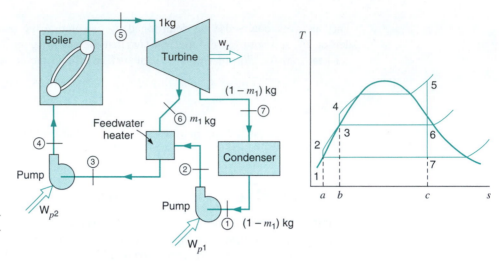

**FIGURE 11.7** Regenerative cycle with open feedwater heater.

leaving the feedwater heater to be saturated at state 3. Note that the liquid has not been pumped to the boiler pressure, but only to the intermediate pressure corresponding to state 6. Another pump is required to pump the liquid leaving the feedwater heater to boiler pressure. The significant point is that the average temperature at which heat is supplied has been increased.

This cycle is somewhat difficult to show on a $T$–$s$ diagram because the masses of steam flowing through the various components vary. The $T$–$s$ diagram of Fig. 11.7 simply shows the state of the fluid at the various points.

Area 4–5–$c$–$b$–4 in Fig. 11.7 represents the heat transferred per kilogram of working fluid. Process 7–1 is the heat rejection process, but since not all the steam passes through the condenser, area 1–7–$c$–$a$–1 represents the heat transfer per kilogram flowing through the condenser, which does not represent the heat transfer per kilogram of working fluid entering the turbine. Between states 6 and 7 only part of the steam is flowing through the turbine. The example that follows illustrates the calculations for the regenerative cycle.

| **EXAMPLE 11.3** | Consider a regenerative cycle using steam as the working fluid. Steam leaves the boiler and enters the turbine at 4 MPa and 400°C. After expansion to 400 kPa, some of the steam is extracted from the turbine for the purpose of heating the feedwater in an open feedwater heater. The pressure in the feedwater heater is 400 kPa and the water leaving it is saturated liquid at 400 kPa. The steam not extracted expands to 10 kPa. Determine the cycle efficiency. |
|---|---|

The line diagram and $T$–$s$ diagram for this cycle are shown in Fig. 11.7.

As in previous examples, the model for each control volume is the steam tables, the process is steady state, and kinetic and potential energy changes are negligible.

From Example 11.2 we have the following properties:

$$h_5 = 3213.6 \text{ kJ/kg}, \qquad h_7 = 2144.1 \text{ kJ/kg}, \qquad h_1 = 191.8 \text{ kJ/kg}$$

Also, at $P_6 = 400$ kPa, from the second law,

$$s_6 = s_5 = 6.7690 = 1.7766 + x_6 5.1193, \qquad x_6 = 0.9752$$

$$h_6 = 604.7 + 0.9752(2133.8) = 2685.6 \text{ kJ/kg}$$

First consider the low-pressure pump.

$\quad$ *Control volume*: Low-pressure pump.
$\qquad$ *Inlet state*: $P_1$ known, saturated liquid; state fixed.
$\qquad$ *Exit state*: $P_2$ known.

**Analysis**

The first law is

$$w_{p1} = h_2 - h_1$$

The second law is

$$s_2 = s_1$$

Therefore,

$$h_2 - h_1 = \int_1^2 v \, dP = v(P_2 - P_1)$$

**Solution**

We have

$$w_{p1} = v(P_2 - P_1) = (0.001\,01)(400 - 10) = 0.4 \text{ kJ/kg}$$

$$h_2 = h_1 + w_p = 191.8 + 0.4 = 192.2 \text{ kJ/kg}$$

For the turbine we have:

$\quad$ *Control volume*: Turbine.
$\qquad$ *Inlet state*: $P_5, T_5$ known; state fixed.
$\qquad$ *Exit state*: $P_6$ known; $P_7$ known.

**Analysis**

The first and second laws give

$$w_t = (h_5 - h_6) + (1 - m_1)(h_6 - h_7)$$

$$s_5 = s_6 = s_7$$

**Solution**

From the second law, the values for $h_6$ and $h_7$ were given previously. Next consider the feedwater heater.

$\quad$ *Control volume*: Feedwater heater.
$\qquad$ *Inlet states*: States 2 and 6 both known (as given).
$\qquad$ *Exit state*: $P_3$ known, saturated liquid; state fixed.

**Analysis**

The first law gives us

$$m_1(h_6) + (1 - m_1)h_2 = h_3$$

**Solution**

Substituting, we obtain

$$m_1(2685.6) + (1 - m_1)(192.2) = 604.7$$

$$m_1 = 0.1654$$

We can now calculate the turbine work:

$$w_t = (h_5 - h_6) + (1 - m_1)(h_6 - h_7)$$

$$= (3213.6 - 2685.6) + (1 - 0.1654)(2685.6 - 2144.1)$$

$$= 979.9 \text{ kJ/kg}$$

Let us turn now to the high-pressure pump.

> *Control volume*:  High-pressure pump.
> *Inlet state*:  State 3 known (as given).
> *Exit state*:  $P_4$ known.

**Analysis**

The first and second laws are

$$w_{p2} = h_4 - h_3$$

$$s_4 = s_3$$

**Solution**

Substitution leads to

$$w_{p2} = v(P_4 - P_3) = (0.001\ 084)(4000 - 400) = 3.9 \text{ kJ/kg}$$

$$h_4 = h_3 + w_{p2} = 604.7 + 3.9 = 608.6 \text{ kJ/kg}$$

Therefore,

$$w_{\text{net}} = w_t - (1 - m_1)w_{p1} - w_{p2}$$

$$= 979.9 - (1 - 0.1654)(0.4) - 3.9 = 975.7 \text{ kJ/kg}$$

Finally, for the boiler:

> *Control volume*:  Boiler.
> *Inlet state*:  $P_4$, $h_4$ known (as given); state fixed.
> *Exit state*:  State 5 known (as given).

**Analysis**

The first law is

$$q_H = h_5 - h_4$$

**Solution**

Substitution gives

$$q_H = h_5 - h_4 = 3213.6 - 608.6 = 2605.0 \text{ kJ/kg}$$

Therefore,

$$\eta_{th} = \frac{w_{net}}{q_H} = \frac{975.7}{2605.0} = 37.5\%$$

Note the increase in efficiency over the efficiency of the Rankine cycle of Example 11.2.

The above discussion and example have tacitly assumed that the extraction steam and feedwater are mixed in the feedwater heater. Another much-used type of feedwater heater, known as a closed heater, is one in which the steam and feedwater do not mix; rather heat is transferred from the extracted steam as it condenses on the outside of tubes while the feedwater flows through the tubes. In a closed heater, the steam and feedwater may be at considerably different pressures. The condensate may be pumped into the feedwater line, or it may be removed through a trap to a lower-pressure heater or to the condenser. (A trap is a device that permits liquid but not vapor to flow to a region of lower pressure.)

Open feedwater heaters have the advantage of being less expensive and having better heat-transfer characteristics compared to closed feedwater heaters. They have the disadvantage of requiring a pump to handle the feedwater between each heater.

In many power plants, several stages of extraction and feedwater heating are used, with the number and type being determined by economics. Using more stages results in somewhat improved efficiency, but the benefit may be offset by the cost of additional equipment and increased system complexity.

## 11.4 DEVIATION OF ACTUAL CYCLES FROM IDEAL CYCLES

Before we leave the matter of vapor power cycles, a few comments are in order regarding the ways in which an actual cycle deviates from an ideal cycle. The most important of these losses are due to the turbine, the pump(s), the pipes, and the condenser. These losses are discussed next.

### Turbine Losses

Turbine losses, as described in Section 9.5, represent by far the largest discrepancy between the performance of a real cycle and a corresponding ideal Rankine-cycle power plant. The large positive turbine work is the principal number in the numerator of the cycle thermal efficiency and is directly reduced by the factor of the isentropic turbine efficiency. Turbine losses are primarily those associated with the flow of the working fluid through the turbine blades and passages, with heat transfer to the surroundings also being a loss, but of secondary importance. The turbine process might be represented as shown in Fig. 11.8, where state $4_s$ is the state after an ideal isentropic turbine expansion, and state 4 is the actual state leaving the turbine following an irreversible process. The turbine governing procedures may also cause a loss in the turbine, particularly if a throttling process is used to govern the turbine operation.

### Pump Losses

The losses in the pump are similar to those of the turbine and are due primarily to the irreversibilities with the fluid flow. Pump efficiency was discussed in Section 9.5, and

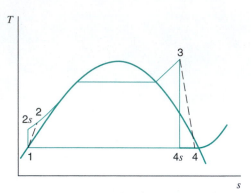

**FIGURE 11.8** Temperature–entropy diagram showing effect of turbine and pump inefficiencies on cycle performance.

the ideal exit state $2_s$ and real exit state 2 are shown in Fig. 11 .8. Pump losses are much smaller than those of the turbine, since the associated work is very much smaller.

## Piping Losses

Pressure drops caused by frictional effects and heat transfer to the surroundings are the most important piping losses. Consider, for example, the pipe connecting the turbine to the boiler. If only frictional effects occur, states *a* and *b* in Fig. 11.9 would represent the states of the steam leaving the boiler and entering the turbine, respectively. Note that the frictional effects cause an increase in entropy. Heat transferred to the surroundings at constant pressure can be represented by process *bc*. This effect decreases entropy. Both the pressure drop and heat transfer decrease the availability of the steam entering the turbine. The irreversibility of this process can be calculated by the methods outlined in Chapter 9.

A similar loss is the pressure drop in the boiler. Because of this pressure drop, the water entering the boiler must be pumped to a higher pressure than the desired steam pressure leaving the boiler, and this requires additional pump work.

## Condenser Losses

The losses in the condenser are relatively small. One of these minor losses is the cooling below the saturation temperature of the liquid leaving the condenser. This represents a loss because additional heat transfer is necessary to bring the water to its saturation temperature.

The influence of these losses on the cycle is illustrated in the following example, which should be compared to Example 11.2.

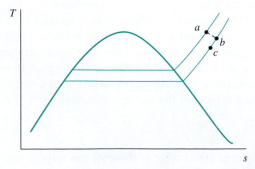

**FIGURE 11.9** Temperature–entropy diagram showing effect of losses between boiler and turbine.

**EXAMPLE 11.4**

A steam power plant operates on a cycle with pressures and temperatures as designated in Fig. 11.10. The efficiency of the turbine is 86% and the efficiency of the pump is 80%. Determine the thermal efficiency of this cycle.

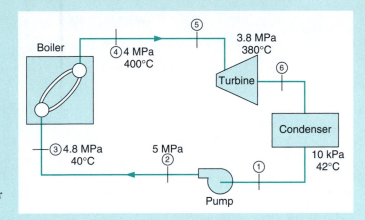

**FIGURE 11.10**
Schematic diagram for Example 11.4.

As in previous examples, for each control volume the model used is the steam tables, and each process is steady state with no changes in kinetic or potential energy. This cycle is shown on the *T–s* diagram of Fig. 11.11.

*Control volume*: Turbine.
*Inlet state*: $P_5$, $T_5$ known; state fixed.
*Exit state*: $P_6$ known.

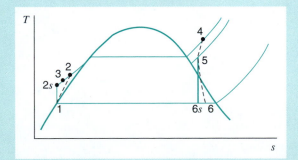

**FIGURE 11.11**
Temperature–entropy diagram for Example 11.4.

**Analysis**

From the first law, we have

$$w_t = h_5 - h_6$$

The second law is

$$s_{6s} = s_5$$

The efficiency is

$$\eta_t = \frac{w_t}{h_5 - h_{6s}} = \frac{h_5 - h_6}{h_5 - h_{6s}}$$

**Solution**

From the steam tables, we get

$$h_5 = 3169.1 \text{ kJ/kg}, \qquad s_5 = 6.7235$$

$$s_{6s} = s_5 = 6.7235 = 0.6493 + x_{6s}7.5009, \qquad x_{6s} = 0.8098$$

$$h_{6s} = 191.8 + 0.8098(2392.8) = 2129.5 \text{ kJ/kg}$$

$$w_t = \eta_t(h_5 - h_{6s}) = 0.86(3169.1 - 2129.5) = 894.1 \text{ kJ/kg}$$

For the pump, we have:

> *Control volume*: Pump.
> *Inlet state*: $P_1$, $T_1$ known; state fixed.
> *Exit state*: $P_2$ known.

**Analysis**

The first law gives

$$w_p = h_2 - h_1$$

The second law gives

$$s_{2s} = s_1$$

The pump efficiency is

$$\eta_p = \frac{h_{2s} - h_1}{w_p} = \frac{h_{2s} - h_1}{h_2 - h_1}$$

Since $s_{2s} = s_1$,

$$h_{2s} - h_1 = v(P_2 - P_1)$$

Therefore,

$$w_p = \frac{h_{2s} - h_1}{\eta_p} = \frac{v(P_2 - P_1)}{\eta_p}$$

**Solution**

Substituting, we obtain

$$w_p = \frac{v(P_2 - P_1)}{\eta_p} = \frac{(0.001\,009)(5000 - 10)}{0.80} = 6.3 \text{ kJ/kg}$$

Therefore,

$$w_{\text{net}} = w_t - w_p = 894.1 - 6.3 = 887.8 \text{ kJ/kg}$$

Finally, for the boiler:

> *Control volume*: Boiler.
> *Inlet state*: $P_3$, $T_3$ known; state fixed.
> *Exit state*: $P_4$, $T_4$ known; state fixed.

**Analysis**

The first law is

$$q_H = h_4 - h_3$$

**Solution**

Substition gives

$$q_H = h_4 - h_3 = 3213.6 - 171.8 = 3041.8 \text{ kJ/kg}$$

$$\eta_{\text{th}} = \frac{887.8}{3041.8} = 29.2\%$$

This result compares to the Rankine efficiency of 35.3% for the similar cycle of Example 11.2.

## 11.5 COGENERATION

There are many occasions in industrial settings where the need arises for a specific source or supply of energy within the environment in which a steam power plant is being used to generate electricity. In such cases, it is appropriate to consider supplying this source of energy in the form of steam that has already been expanded through the high-pressure section of the turbine in the power plant cycle, thereby eliminating the construction and use of a second boiler or other energy source. Such an arrangement is shown in Fig. 11.12, in which the turbine is tapped at some intermediate pressure to furnish the necessary amount of process steam required for the particular energy need—perhaps to operate a special process in the plant, or in many cases simply for the purpose of space heating the facilities. This type of application is termed cogeneration, and if the system is designed as a package with both the electrical and the process steam requirements in

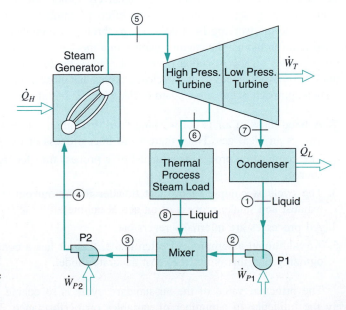

**FIGURE 11.12** Example of a cogeneration system.

mind, it is possible to achieve a substantial savings in capital cost of equipment and also in the operating cost, through careful consideration of all the requirements and optimization of the various parameters involved. Specific examples of cogeneration systems are considered in the problems at the end of the chapter.

---

**In-Text Concept Questions**

**a.** Consider a Rankine cycle without superheat. How many single properties are needed to determine the cycle? Repeat the answer for a cycle with superheat.

**b.** Which component determines the high pressure in a Rankine cycle? What factor determines the low pressure?

**c.** What is the difference between an open and a closed feedwater heater?

**d.** In a cogenerating power plant, what is cogenerated?

---

## 11.6 AIR-STANDARD POWER CYCLES

In Section 11.1, we considered idealized four-process cycles, including both steady-state-process and cylinder/piston boundary-movement cycles. The question of phase-change cycles and single-phase cycles was also mentioned. We then proceeded to examine the Rankine power plant cycle in detail, the idealized model of a phase-change power cycle. However, many work-producing devices (engines) utilize a working fluid that is always a gas. The spark-ignition automotive engine is a familiar example, and so are the diesel engine and the conventional gas turbine. In all these engines there is a change in the composition of the working fluid, because during combustion it changes from air and fuel to combustion products. For this reason these engines are called internal-combustion engines. In contrast, the steam power plant may be called an external-combustion engine, because heat is transferred from the products of combustion to the working fluid. External-combustion engines using a gaseous working fluid (usually air) have been built. To date they have had only limited application, but the use of the gas-turbine cycle in conjunction with a nuclear reactor has been investigated extensively. Other external-combustion engines are currently receiving serious attention in an effort to combat air pollution.

Because the working fluid does not go through a complete thermodynamic cycle in the engine (even though the engine operates in a mechanical cycle) the internal-combustion engine operates on the so-called open cycle. However, for analyzing internal-combustion engines, it is advantageous to devise closed cycles that closely approximate the open cycles. One such approach is the air-standard cycle, which is based on the following assumptions:

1. A fixed mass of air is the working fluid throughout the entire cycle, and the air is always an ideal gas. Thus, there is no inlet process or exhaust process.

2. The combustion process is replaced by a process transferring heat from an external source.

3. The cycle is completed by heat transfer to the surroundings (in contrast to the exhaust and intake process of an actual engine).

4. All processes are internally reversible.

5. An additional assumption is often made, that air has a constant specific heat, recognizing that this is not the most accurate model.

The principal value of the air-standard cycle is to enable us to examine qualitatively the influence of a number of variables on performance. The quantitative results obtained from the air-standard cycle, such as efficiency and mean effective pressure, will differ from those of the actual engine. Our emphasis, therefore, in our consideration of the air-standard cycle, will be primarily on the qualitative aspects.

The term "mean effective pressure" (mep), which is used in conjunction with reciprocating engines, is defined as the pressure that, if it acted on the piston during the entire power stroke, would do an amount of work equal to that actually done on the piston. The work for one cycle is found by multiplying this mean effective pressure by the area of the piston (minus the area of the rod on the crank end of a double-acting engine) and by the stroke.

## 11.7 THE BRAYTON CYCLE

In discussing idealized four-steady-state-process power cycles in Section 11.1, a cycle involving two constant-pressure and two isentropic processes was examined, and the results shown in Fig. 11.2. This cycle used with a condensing working fluid is the Rankine cycle, but when used with a single-phase, gaseous working fluid it is termed the Brayton cycle. The air-standard Brayton cycle is the ideal cycle for the simple gas turbine. The simple open-cycle gas turbine utilizing an internal-combustion process and the simple closed-cycle gas turbine, which utilizes heat-transfer processes, are both shown schematically in Fig. 11.13. The air-standard Brayton cycle is shown on the $P$–$v$ and $T$–$s$ diagrams of Fig. 11.14.

The efficiency of the air-standard Brayton cycle is found as follows:

$$\eta_{\text{th}} = 1 - \frac{Q_L}{Q_H} = 1 - \frac{C_p(T_4 - T_1)}{C_p(T_3 - T_2)} = 1 - \frac{T_1(T_4/T_1 - 1)}{T_2(T_3/T_2 - 1)}$$

We note, however, that

$$\frac{P_3}{P_4} = \frac{P_2}{P_1}$$

$$\frac{P_2}{P_1} = \left(\frac{T_2}{T_1}\right)^{k/(k-1)} = \frac{P_3}{P_4} = \left(\frac{T_3}{T_4}\right)^{k/(k-1)}$$

$$\frac{T_3}{T_4} = \frac{T_2}{T_1} \quad \therefore \frac{T_3}{T_2} = \frac{T_4}{T_1} \quad \text{and} \quad \frac{T_3}{T_2} - 1 = \frac{T_4}{T_1} - 1$$

$$\eta_{\text{th}} = 1 - \frac{T_1}{T_2} = 1 - \frac{1}{(P_2/P_1)^{(k-1)/k}} \qquad (11.2)$$

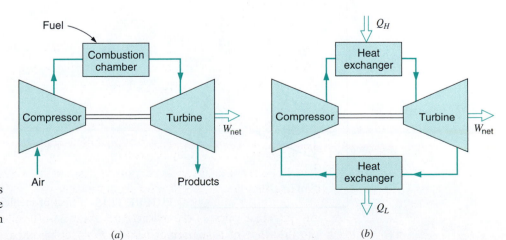

**FIGURE 11.13** A gas turbine operating on the Brayton cycle. (*a*) Open cycle. (*b*) Closed cycle.

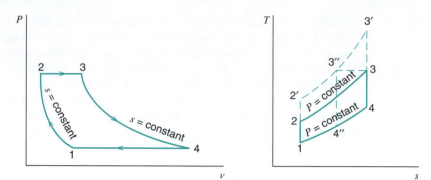

**FIGURE 11.14** The air-standard Brayton cycle.

The efficiency of the air-standard Brayton cycle is therefore a function of the isentropic pressure ratio. The fact that efficiency increases with pressure ratio is evident from the $T$–$s$ diagram of Fig. 11.14 because increasing the pressure ratio changes the cycle from 1–2–3–4–1 to 1–2′–3′–4–1. The latter cycle has a greater heat supply and the same heat rejected as the original cycle; therefore, it has a greater efficiency. Note that the latter cycle has a higher maximum temperature, $T_3'$, than the original cycle, $T_3$. In the actual gas turbine the maximum temperature of the gas entering the turbine is fixed by material considerations. Therefore, if we fix the temperature $T_3$ and increase the pressure ratio, the resulting cycle is 1–2′–3″–4″–1. This cycle would have a higher efficiency than the original cycle, but the work per kilogram of working fluid is thereby changed.

With the advent of nuclear reactors, the closed-cycle gas turbine has become more important. Heat is transferred, either directly or via a second fluid, from the fuel in the nuclear reactor to the working fluid in the gas turbine. Heat is rejected from the working fluid to the surroundings.

The actual gas-turbine engine differs from the ideal cycle primarily because of irreversibilities in the compressor and turbine, and because of pressure drop in the flow passages and combustion chamber (or in the heat exchanger of a closed-cycle turbine). Thus, the state points in a simple open-cycle gas turbine might be as shown in Fig. 11.15.

The efficiencies of the compressor and turbine are defined in relation to isentropic processes. With the states designated as in Fig. 11.15, the definitions of compressor and

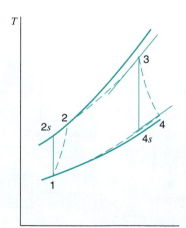

**FIGURE 11.15** Effect of inefficiencies on the gas-turbine cycle.

turbine efficiencies are

$$\eta_{comp} = \frac{h_{2s} - h_1}{h_2 - h_1} \qquad (11.3)$$

$$\eta_{turb} = \frac{h_3 - h_4}{h_3 - h_{4s}} \qquad (11.4)$$

Another important feature of the Brayton cycle is the large amount of compressor work (also called back work) compared to turbine work. Thus, the compressor might require from 40 to 80% of the output of the turbine. This is particularly important when the actual cycle is considered, because the effect of the losses is to require a larger amount of compression work from a smaller amount of turbine work, and thus, the overall efficiency drops very rapidly with a decrease in the efficiencies of the compressor and turbine. In fact, if these efficiencies drop below about 60%, all the work of the turbine will be required to drive the compressor, and the overall efficiency will be zero. This is in sharp contrast to the Rankine cycle, where only 1 or 2% of the turbine work is required to drive the pump. This demonstrates the inherent advantage of the cycle utilizing a condensing working fluid, such that a much larger difference in specific volume between the expansion and compression processes is utilized effectively.

**EXAMPLE 11.5**

In an air-standard Brayton cycle the air enters the compressor at 0.1 MPa and 15°C. The pressure leaving the compressor is 1.0 MPa, and the maximum temperature in the cycle is 1100°C. Determine

1. The pressure and temperature at each point in the cycle.
2. The compressor work, turbine work, and cycle efficiency.

For each of the control volumes analyzed, the model is ideal gas with constant specific heat, at 300 K, and each process is steady state with no kinetic or potential energy changes. The diagram for this example is Fig. 11.14.

We consider the compressor, the turbine, and the high-temperature and low-temperature heat exchangers in turn.

*Control volume*: Compressor.
*Inlet state*: $P_1$, $T_1$ known; state fixed.
*Exit state*: $P_2$ known.

**Analysis**

The first law gives

$$w_c = h_2 - h_1$$

(Note that the compressor work $w_c$ is here defined as work input to the compressor.) The second law is

$$s_2 = s_1$$

so that

$$\frac{T_2}{T_1} = \left(\frac{P_2}{P_1}\right)^{(k-1)/k}$$

**Solution**

Solving for $T_2$ we get

$$\left(\frac{P_2}{P_1}\right)^{(k-1)/k} = 10^{0.286} = 1.932, \qquad T_2 = 556.8 \text{ K}$$

Therefore,

$$w_c = h_2 - h_1 = C_p(T_2 - T_1)$$
$$= 1.004(556.8 - 288.2) = 269.5 \text{ kJ/kg}$$

Consider the turbine next.

> *Control volume*:   Turbine.
> *Inlet state*:   $P_3(= P_2)$ known, $T_3$ known; state fixed.
> *Exit state*:   $P_4(= P_1)$ known.

**Analysis**

The first law gives

$$w_t = h_3 - h_4$$

The second law is

$$s_3 = s_4$$

so that

$$\frac{T_3}{T_4} = \left(\frac{P_3}{P_4}\right)^{(k-1)/k}$$

**Solution**

Solving for $T_4$ we get

$$\left(\frac{P_3}{P_4}\right)^{(k-1)/k} = 10^{0.286} = 1.932, \qquad T_4 = 710.8 \text{ K}$$

Therefore,

$$w_t = h_3 - h_4 = c_p(T_3 - T_4)$$
$$= 1.004(1373.2 - 710.8) = 664.7 \text{ kJ/kg}$$

$$w_{\text{net}} = w_t - w_c = 664.7 - 269.5 = 395.2 \text{ kJ/kg}$$

Now we turn to the heat exchangers.

> *Control volume*:   High-temperature heat exchanger.
> *Inlet state*:   State 2 fixed (as determined).
> *Exit state*:   State 3 fixed (as given).

**Analysis**

The first law is

$$q_H = h_3 - h_2 = C_p(T_3 - T_2)$$

**Solution**

Substitution gives

$$q_H = h_3 - h_2 = C_p(T_3 - T_2) = 1.004(1373.2 - 556.8) = 819.3 \text{ kJ/kg}$$

| | |
|---|---|
| *Control volume*: | Low-temperature heat exchanger. |
| *Inlet state*: | State 4 fixed (above). |
| *Exit state*: | State 1 fixed (above). |

**Analysis**

The first law is

$$q_L = h_4 - h_1 = C_p(T_4 - T_1)$$

**Solution**

Upon substitution we have

$$q_L = h_4 - h_1 = C_p(T_4 - T_1) = 1.004(710.8 - 288.2) = 424.1 \text{ kJ/kg}$$

Therefore,

$$\eta_{\text{th}} = \frac{w_{\text{net}}}{q_H} = \frac{395.2}{819.3} = 48.2\%$$

This may be checked by using Eq. 11.2.

$$\eta_{\text{th}} = 1 - \frac{1}{(P_2/P_1)^{(k-1)/k}} = 1 - \frac{1}{10^{0.286}} = 48.2\%$$

---

**EXAMPLE 11.6**

Consider a gas turbine with air entering the compressor under the same conditions as in Example 11.5 and leaving at a pressure of 1.0 MPa. The maximum temperature is 1100°C. Assume a compressor efficiency of 80%, a turbine efficiency of 85%, and a pressure drop between the compressor and turbine of 15 kPa. Determine the compressor work, turbine work, and cycle efficiency.

As in the previous example, for each control volume the model is ideal gas with constant specific heat, at 300 K, and each process is steady state with no kinetic or potential energy changes. In this example the diagram is Fig. 11.15.

We consider the compressor, the turbine and the high-temperature heat exchanger in turn.

| | |
|---|---|
| *Control volume*: | Compressor. |
| *Inlet state*: | $P_1$, $T_1$ known; state fixed. |
| *Exit state*: | $P_2$ known. |

**Analysis**

The first law for the real process is

$$w_c = h_2 - h_1$$

The second law for the ideal process is

$$s_{2s} = s_1$$

so that

$$\frac{T_{2s}}{T_1} = \left(\frac{P_2}{P_1}\right)^{(k-1)/k}$$

In addition,

$$\eta_c = \frac{h_{2s} - h_1}{h_2 - h_1} = \frac{T_{2s} - T_1}{T_2 - T_1}$$

**Solution**

Solving for $T_{2s}$, we get

$$\left(\frac{P_2}{P_1}\right)^{(k-1)/k} = \frac{T_{2s}}{T_1} = 10^{0.286} = 1.932, \qquad T_{2s} = 556.8 \text{ K}$$

The efficiency is

$$\eta_c = \frac{h_{2s} - h_1}{h_2 - h_1} = \frac{T_{2s} - T_1}{T_2 - T_1} = \frac{556.8 - 288.2}{T_2 - T_1} = 0.80$$

Therefore,

$$T_2 - T_1 = \frac{556.8 - 288.2}{0.80} = 335.8, \qquad T_2 = 624.0 \text{ K}$$

$$w_c = h_2 - h_1 = C_p(T_2 - T_1)$$

$$= 1.004(624.0 - 288.2) = 337.0 \text{ kJ/kg}$$

For the turbine, we have.

> *Control volume*:   Turbine.
> *Inlet state*:   $P_3(P_2 - \text{drop})$ known, $T_3$ known; state fixed.
> *Exit state*:   $P_4$ known.

**Analysis**

The first law for the real process is

$$w_t = h_3 - h_4$$

The second law for the ideal process is

$$s_{4s} = s_3$$

so that

$$\frac{T_3}{T_{4s}} = \left(\frac{P_3}{P_4}\right)^{(k-1)/k}$$

In addition,

$$\eta_t = \frac{h_3 - h_4}{h_3 - h_{4s}} = \frac{T_3 - T_4}{T_3 - T_{4s}}$$

**Solution**

Substituting numerical values, we obtain

$$P_3 = P_2 - \text{pressure drop} = 1.0 - 0.015 = 0.985 \text{ MPa}$$

$$\left(\frac{P_3}{P_4}\right)^{(k-1)/k} = \frac{T_3}{T_{4s}} = 9.85^{0.286} = 1.9236, \qquad T_{4s} = 713.9 \text{ K}$$

$$\eta_t = \frac{h_3 - h_4}{h_3 - h_{4s}} = \frac{T_3 - T_4}{T_3 - T_{4s}} = 0.85$$

$$T_3 - T_4 = 0.85(1373.2 - 713.9) = 560.4 \text{ K}$$

$$T_4 = 812.8 \text{ K}$$

$$w_t = h_3 - h_4 = C_p(T_3 - T_4)$$

$$= 1.004(1373.2 - 812.8) = 562.4 \text{ kJ/kg}$$

$$w_{\text{net}} = w_t - w_c = 562.4 - 337.0 = 225.4 \text{ kJ/kg}$$

Finally, for the heat exchanger:

*Control volume*:   High-temperature heat exchanger.
*Inlet state*:   State 2 fixed (as determined).
*Exit state*:   State 3 fixed (as given).

**Analysis**

The first law is

$$q_H = h_3 - h_2$$

**Solution**

Substituting, we have

$$q_H = h_3 - h_2 = C_p(T_3 - T_2)$$

$$= 1.004(1373.2 - 624.0) = 751.8 \text{ kJ/kg}$$

so that

$$\eta_{\text{th}} = \frac{w_{\text{net}}}{q_H} = \frac{225.4}{751.8} = 30.0\%$$

The following comparisons can be made between Examples 11.5 and 11.6:

|  | $w_c$ | $w_t$ | $w_{\text{net}}$ | $q_H$ | $\eta_{\text{th}}$ |
|---|---|---|---|---|---|
| Example 11.5 (Ideal) | 269.5 | 664.7 | 395.2 | 819.3 | 48.2 |
| Example 11.6 (Actual) | 337.0 | 562.4 | 225.4 | 751.8 | 30.0 |

As stated previously, the irreversibilities decrease the turbine work and increase the compressor work. Since the net work is the difference between these two, it decreases very rapidly as compressor and turbine efficiencies decrease. The development of compressors and turbines of high efficiency is therefore an important aspect of the development of gas turbines.

Note that in the ideal cycle (Example 11.5) about 41% of the turbine work is required to drive the compressor and 59% is delivered as net work. In the actual turbine (Example 11.6) 60% of the turbine work is required to drive the compressor and 40% is delivered as net work. Thus, if the net power of this unit is to be 10 000 kW, a 25 000 kW turbine and a 15 000 kW compressor are required. This result demonstrates that a gas turbine has a high back-work ratio.

## 11.8 THE SIMPLE GAS-TURBINE CYCLE WITH A REGENERATOR

The efficiency of the gas-turbine cycle may be improved by introducing a regenerator. The simple open-cycle gas-turbine cycle with a regenerator is shown in Fig. 11.16, and the corresponding ideal air-standard cycle with a regenerator is shown on the $P$–$v$ and $T$–$s$ diagrams. In cycle 1–2–$x$–3–4–$y$–1, the temperature of the exhaust gas leaving the turbine in state 4 is higher than the temperature of the gas leaving the compressor. Therefore, heat can be transferred from the exhaust gases to the high-pressure gases leaving the compressor. If this is done in a counterflow heat exchanger (a regenerator), the temperature of the high-pressure gas leaving the regenerator, $T_x$, may, in the ideal case, have a temperature equal to $T_4$, the temperature of the gas leaving the turbine. Heat transfer from the external source is necessary only to increase the temperature from $T_x$ to $T_3$. Area $x$–3–$d$–$b$–$x$ represents the heat transferred, and area $y$–1–$a$–$c$–$y$ represents the heat rejected.

The influence of pressure ratio on the simple gas-turbine cycle with a regenerator is shown by considering cycle 1–2′–3′–4–1. In this cycle the temperature of the exhaust gas leaving the turbine is just equal to the temperature of the gas leaving the compressor; therefore, utilizing a regenerator is not possible. This can be shown more exactly by determining the efficiency of the ideal gas-turbine cycle with a regenerator.

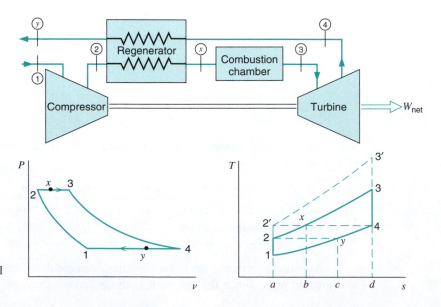

**FIGURE 11.16** The ideal regenerative cycle.

The efficiency of this cycle with regeneration is found as follows, where the states are as given in Fig. 11.16:

$$\eta_{th} = \frac{w_{net}}{q_H} = \frac{w_t - w_c}{q_H}$$

$$q_H = C_p(T_3 - T_x)$$

$$w_t = C_p(T_3 - T_4)$$

But for an ideal regenerator, $T_4 = T_x$, and therefore $q_H = w_t$. Consequently,

$$\eta_{th} = 1 - \frac{w_c}{w_t} = 1 - \frac{C_p(T_2 - T_1)}{C_p(T_3 - T_4)}$$

$$= 1 - \frac{T_1(T_2/T_1 - 1)}{T_3(1 - T_4/T_3)} = 1 - \frac{T_1}{T_3}\frac{\left[(P_2/P_1)^{(k-1)/k} - 1\right]}{\left[1 - (P_1/P_2)^{(k-1)/k}\right]}$$

$$\eta_{th} = 1 - \frac{T_1}{T_3}\left(\frac{P_2}{P_1}\right)^{(k-1)/k}$$

Thus, for the ideal cycle with regeneration the thermal efficiency depends not only on the pressure ratio but also on the ratio of the minimum to maximum temperature. We note that, in contrast to the Brayton cycle, the efficiency decreases with an increase in pressure ratio.

The effectiveness or efficiency of a regenerator is given by the regenerator efficiency, which can best be defined by reference to Fig. 11.17. State $x$ represents the high-pressure gas leaving the regenerator. In the ideal regenerator there would be only an infinitesimal temperature difference between the two streams, and the high-pressure gas would leave the regenerator at temperature $T_x'$, and $T_x' = T_4$. In an actual regenerator, which must operate with a finite temperature difference $T_x$, the actual temperature leaving the regenerator is therefore less than $T_x'$. The regenerator efficiency is defined by

$$\eta_{reg} = \frac{h_x - h_2}{h_x' - h_2} \tag{11.5}$$

If the specific heat is assumed to be constant, the regenerator efficiency is also given by the relation

$$\eta_{reg} = \frac{T_x - T_2}{T_x' - T_2}$$

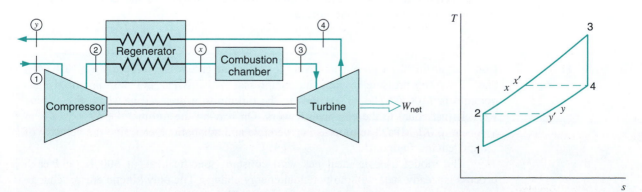

**FIGURE 11.17** Temperature–entropy diagram to illustrate the definition of regenerator efficiency.

It should be pointed out that a higher efficiency can be achieved by using a regenerator with a greater heat-transfer area. However, this also increases the pressure drop, which represents a loss, and both the pressure drop and the regenerator efficiency must be considered in determining which regenerator gives maximum thermal efficiency for the cycle. From an economic point of view, the cost of the regenerator must be weighed against the savings that can be effected by its use.

---

**EXAMPLE 11.7**

If an ideal regenerator is incorporated into the cycle of Example 11.5, determine the thermal efficiency of the cycle.

The diagram for this example is Fig. 11. 17. Values are from Example 11.5. Therefore, for the analysis of the high-temperature heat exchanger (combustion chamber), from the first law, we have

$$q_H = h_3 - h_x$$

so that the solution is

$$T_x = T_4 = 710.8 \text{ K}$$

$$q_H = h_3 - h_x = C_p(T_3 - T_x) = 1.004(1373.2 - 710.8) = 664.7 \text{ kJ/kg}$$

$$w_{\text{net}} = 395.2 \text{ kJ/kg} \quad \text{(from Example 11.5)}$$

$$\eta_{\text{th}} = \frac{395.2}{664.7} = 59.5\%$$

---

## 11.9 THE AIR-STANDARD CYCLE FOR JET PROPULSION

The next air-standard power cycle we consider is utilized in jet propulsion. In this cycle the work done by the turbine is just sufficient to drive the compressor. The gases are expanded in the turbine to a pressure for which the turbine work is just equal to the compressor work. The exhaust pressure of the turbine will then be greater than that of the surroundings, and the gas can be expanded in a nozzle to the pressure of the surroundings. Since the gases leave at a high velocity, the change in momentum that the gases undergo gives a thrust to the aircraft in which the engine is installed. A jet engine was shown in Fig. 1.11, and the air-standard cycle for this situation is shown in Fig. 11.18. The principles governing this cycle follow from the analysis of the Brayton cycle plus that for a reversible, adiabatic nozzle.

---

**EXAMPLE 11.8**

Consider an ideal jet propulsion cycle in which air enters the compressor at 0.1 MPa and 15°C. The pressure leaving the compressor is 1.0 MPa, and the maximum temperature is 1100°C. The air expands in the turbine to a pressure at which the turbine work is just equal to the compressor work. On leaving the turbine, the air expands in a nozzle to 0.1 MPa. The process is reversible and adiabatic. Determine the velocity of the air leaving the nozzle.

The model used is ideal gas with constant specific heat, at 300 K, and each process is steady state with no potential energy change. The only kinetic energy change occurs in the nozzle. The diagram is shown in Fig. 11.18.

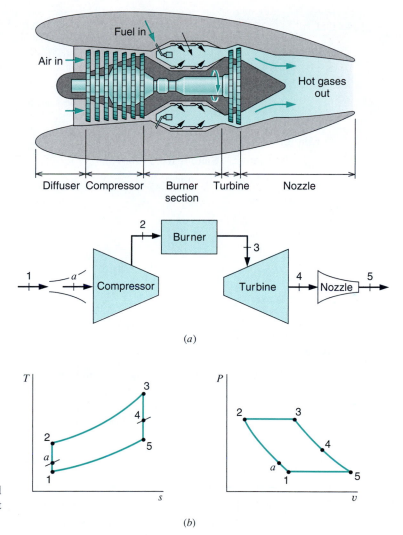

**FIGURE 11.18** The ideal gas-turbine cycle for a jet engine.

(a)

(b)

The compressor analysis is the same as in Example 11.5. From the results of that solution, we have

$$P_1 = 0.1 \text{ MPa}, \qquad T_1 = 288.2 \text{ K}$$

$$P_2 = 1.0 \text{ MPa}, \qquad T_2 = 556.8 \text{ K}$$

$$w_c = 269.5 \text{ kJ/kg}$$

The turbine analysis is also the same as in Example 11.5. Here, however,

$$P_3 = 0.1 \text{ MPa}, \qquad T_3 = 1373.2 \text{ K}$$

$$w_c = w_t = C_p(T_3 - T_4) = 269.5 \text{ kJ/kg}$$

$$T_3 - T_4 = \frac{269.5}{1.004} = 268.6, \qquad T_4 = 1104.6 \text{ K}$$

so that

$$\frac{T_3}{T_4} = \left(\frac{P_3}{P_4}\right)^{(k-1)/k} = \frac{1373.2}{1104.6} = 1.2432$$

$$\frac{P_3}{P_4} = 2.142, \qquad P_4 = 0.4668 \text{ MPa}$$

> *Control volume*:  Nozzle.
> *Inlet state*:  State 4 fixed (above).
> *Exit state*:  $P_5$ known.

**Analysis**

The first law gives

$$h_4 = h_5 + \frac{\mathbf{V}_5^2}{2}$$

The second law is

$$s_4 = s_5$$

**Solution**

Since $P_5$ is 0.1 MPa, from the second law we find that $T_5 = 710.8$ K. Then

$$\mathbf{V}_5^2 = 2C_{p0}(T_4 - T_5)$$

$$\mathbf{V}_5^2 = 2 \times 1000 \times 1.004(1104.6 - 710.8)$$

$$\mathbf{V}_5 = 889 \text{ m/s}$$

---

**In-Text Concept Questions**

**e.** The Brayton cycle has the same 4 processes as the Rankine cycle, but the $T$–$s$ and $P$–$v$ diagrams look very different; why is that?

**f.** Is it always possible to add a regenerator to the Brayton cycle? What happens when the pressure ratio is increased?

**g.** Why would you use an intercooler between compressor stages?

**h.** The jet engine does not produce shaft work; how is power produced?

## 11.10 RECIPROCATING ENGINE POWER CYCLES

In Section 11.1, we discussed power cycles incorporating either steady-state processes or cylinder/piston boundary-work processes. In that section, it was noted that for the steady-state process, there is no work in a constant-pressure process. Each of the steady-state power cycles presented in subsequent sections of this chapter incorporated two constant-pressure heat-transfer processes. It was further noted in Section 11.1 that in a boundary-movement work process, there is no work in a constant-volume process. In the next three sections, we will present ideal air-standard power cycles for cylinder/piston boundary-movement work processes, each example of which includes either one or two constant-volume heat-transfer processes.

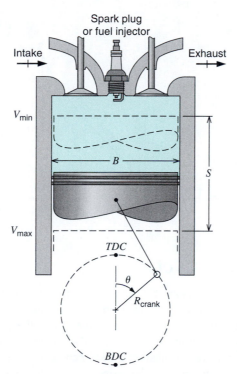

**FIGURE 11.19** The piston-cylinder configuration for an internal combustion engine.

Before we describe the reciprocating engine cycles, we want to present a few common definitions and terms. Car engines typically have 4, 6, or 8 cylinders, each with a diameter called bore B. The piston is connected to a crankshaft, as shown in Figure 11.19, and as it rotates changing the crank angle, $\theta$, the piston moves up or down with a stroke.

$$S = 2R_{\text{crank}} \tag{11.6}$$

This gives a displacement for all cylinders as

$$V_{\text{displ}} = N_{\text{cyl}}(V_{\text{max}} - V_{\text{min}}) = N_{\text{cyl}}A_{\text{cyl}}S \tag{11.7}$$

which is the main characterization of the engine size. The ratio of the largest to the smallest volume is the compression ratio

$$r_v = CR = V_{\text{max}}/V_{\text{min}} \tag{11.8}$$

and both of these characteristics are fixed with the engine geometry. The net specific work in a complete cycle is used to define a mean effective pressure

$$w_{\text{net}} = \oint P \, dv \equiv P_{\text{meff}}(v_{\text{max}} - v_{\text{min}}) \tag{11.9}$$

or net work per cylinder per cycle

$$W_{\text{net}} = m w_{\text{net}} = P_{\text{meff}}(V_{\text{max}} - V_{\text{min}}) \tag{11.10}$$

We now use this to find the rate of work (power) for the whole engine as

$$\dot{W} = N_{\text{cyl}} \, m w_{\text{net}} \frac{\text{RPM}}{60} = P_{\text{meff}} V_{\text{displ}} \frac{\text{RPM}}{60} \tag{11.11}$$

where RPM is revolutions per minute. This result should be corrected with a factor $\frac{1}{2}$ for a four-stroke engine, where 2 revolutions are needed for a complete cycle to also accomplish the intake and exhaust strokes.

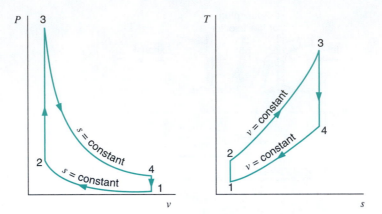

**FIGURE 11.20** The air-standard Otto cycle.

## 11.11 The Otto Cycle

The air-standard Otto cycle is an ideal cycle that approximates a spark-ignition internal-combustion engine. This cycle is shown on the *P–v* and *T–s* diagrams of Fig. 11.20. Process 1–2 is an isentropic compression of the air as the piston moves from bottom dead center (BDC) to top dead center. Heat is then added at constant volume while the piston is momentarily at rest at head-end dead center. (This process corresponds to the ignition of the fuel–air mixture by the spark and the subsequent burning in the actual engine.) Process 3–4 is an isentropic expansion, and process 4–1 is the rejection of heat from the air while the piston is at bottom dead center.

The thermal efficiency of this cycle is found as follows, assuming constant specific heat of air:

$$
\eta_{\text{th}} = \frac{Q_H - Q_L}{Q_H} = 1 - \frac{Q_L}{Q_H} = 1 - \frac{mC_v(T_4 - T_1)}{mC_v(T_3 - T_2)}
$$

$$
= 1 - \frac{T_1(T_4/T_1 - 1)}{T_2(T_3/T_2 - 1)}
$$

We note further that

$$
\frac{T_2}{T_1} = \left(\frac{V_1}{V_2}\right)^{k-1} = \left(\frac{V_4}{V_3}\right)^{k-1} = \frac{T_3}{T_4}
$$

Therefore,

$$
\frac{T_3}{T_2} = \frac{T_4}{T_1}
$$

and

$$
\eta_{\text{th}} = 1 - \frac{T_1}{T_2} = 1 - (r_v)^{1-k} = 1 - \frac{1}{r_v^{k-1}} \qquad (11.12)
$$

where

$$
r_v = \text{compression ratio} = \frac{V_1}{V_2} = \frac{V_4}{V_3}
$$

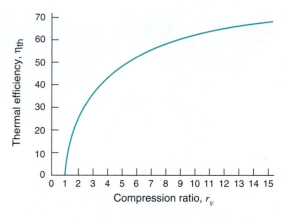

**FIGURE 11.21** Thermal efficiency of the Otto cycle as a function of compression ratio.

It is important to note that the efficiency of the air-standard Otto cycle is a function only of the compression ratio and that the efficiency is increased by increasing the compression ratio. Figure 11.21 shows a plot of the air-standard cycle thermal efficiency versus compression ratio. It is also true of an actual spark-ignition engine that the efficiency can be increased by increasing the compression ratio. The trend toward higher compression ratios is prompted by the effort to obtain higher thermal efficiency. In the actual engine there is an increased tendency for the fuel to detonate as the compression ratio is increased. After detonation the fuel burns rapidly, and strong pressure waves present in the engine cylinder give rise to the so-called spark knock. Therefore, the maximum compression ratio that can be used is fixed by the fact that detonation must be avoided. Advances in compression ratios over the years in actual engines were originally made possible by developing fuels with better antiknock characteristics, primarily through the addition of tetraethyl lead. More recently, however, nonleaded gasolines with good antiknock characteristics have been developed in an effort to reduce atmospheric contamination.

Some of the most important ways in which the actual open-cycle spark-ignition engine deviates from the air-standard cycle are as follows:

1. The specific heats of the actual gases increase with an increase in temperature.
2. The combustion process replaces the heat-transfer process at high temperature, and combustion may be incomplete.
3. Each mechanical cycle of the engine involves an inlet and an exhaust process and, because of the pressure drop through the valves, a certain amount of work is required to charge the cylinder with air and exhaust the products of combustion.
4. There will be considerable heat transfer between the gases in the cylinder and the cylinder walls.
5. There will be irreversibilities associated with pressure and temperature gradients.

**EXAMPLE 11.9**

The compression ratio in an air-standard Otto cycle is 10. At the beginning of the compression stroke the pressure is 0.1 MPa and the temperature is 15°C. The heat transfer to the air per cycle is 1800 kJ/kg air. Determine

1. The pressure and temperature at the end of each process of the cycle.
2. The thermal efficiency.
3. The mean effective pressure.

*Control mass*: Air inside cylinder.

*Diagram*: Fig. 11.21.

*State information*: $P_1 = 0.1$ Mpa, $T_1 = 288.2$ K.

*Process information*: Four processes known (Fig. 11.21). Also $r_v = 10$ and $q_H = 1800$ kJ/kg.

*Model*: Ideal gas, constant specific heat, value at 300 K.

**Analysis**

The second law for compression process 1–2 is

$$s_2 = s_1$$

so that

$$\frac{T_2}{T_1} = \left(\frac{V_1}{V_2}\right)^{k-1}$$

$$\frac{P_2}{P_1} = \left(\frac{V_1}{V_2}\right)^{k}$$

The first law for heat addition process 2–3 is

$$q_H = {}_2q_3 = u_3 - u_2 = C_v(T_3 - T_2)$$

The second law for expansion process 3–4 is

$$s_4 = s_3$$

so that

$$\frac{T_3}{T_4} = \left(\frac{V_4}{V_3}\right)^{k-1}$$

$$\frac{P_3}{P_4} = \left(\frac{V_4}{V_3}\right)^{k}$$

In addition,

$$\eta_{th} = 1 - \frac{1}{r_v^{k-1}}, \qquad \text{mep} = \frac{w_{net}}{v_1 - v_2}$$

**Solution**

Substitution yields the following:

$$v_1 = \frac{0.287 \times 288.2}{100} = 0.827 \text{ m}^3/\text{kg}$$

$$\frac{T_2}{T_1} = \left(\frac{V_1}{V_2}\right)^{k-1} = 10^{0.4} = 2.5119, \qquad T_2 = 723.9 \text{ K}$$

$$\frac{P_2}{P_1} = \left(\frac{V_1}{V_2}\right)^{k} = 10^{1.4} = 25.12, \qquad P_2 = 2.512 \text{ MPa}$$

$$v_2 = \frac{0.827}{10} = 0.0827 \text{ m}^3/\text{kg}$$

$$_2q_3 = C_v(T_3 - T_2) = 1800 \text{ kJ/kg}$$

$$T_3 - T_2 = \frac{1800}{0.717} = 2510 \text{ K}, \qquad T_3 = 3234 \text{ K}$$

$$\frac{T_3}{T_2} = \frac{P_3}{P_2} = \frac{3234}{723.9} = 4.467, \qquad P_3 = 11.222 \text{ MPa}$$

$$\frac{T_3}{T_4} = \left(\frac{V_4}{V_3}\right)^{k-1} = 10^{0.4} = 2.5119, \qquad T_4 = 1287.5 \text{ K}$$

$$\frac{P_3}{P_4} = \left(\frac{V_4}{V_3}\right)^{k} = 10^{1.4} = 25.12, \qquad P_4 = 0.4467 \text{ MPa}$$

$$\eta_{th} = 1 - \frac{1}{r_v^{k-1}} = 1 - \frac{1}{10^{0.4}} = 0.602 = 60.2\%$$

This can be checked by finding the heat rejected:

$$_4q_1 = C_v(T_1 - T_4) = 0.717(288.2 - 1287.5) = -716.5 \text{ kJ/kg}$$

$$\eta_{th} = 1 - \frac{716.5}{1800} = 0.602 = 60.2\%$$

$$w_{net} = 1800 - 716.5 = 1083.5 \text{ kJ/kg} = (v_1 - v_2)\text{ mep}$$

$$\text{mep} = \frac{1083.5}{(0.827 - 0.0827)} = 1456 \text{ kPa}$$

This is a high value for mean effective pressure, largely because the two constant-volume heat transfer processes keep the total volume change to a minimum (compared with a Brayton cycle, for example). Thus, the Otto cycle is a good model to emulate in the cylinder–piston internal-combustion engine. At the other extreme, a low mean effective pressure means a large piston displacement for a given power output, which in turn means high frictional losses in an actual engine.

## 11.12 THE DIESEL CYCLE

The air-standard diesel cycle is shown in Fig. 11.22. This is the ideal cycle for the diesel engine, which is also called the compression-ignition engine.

In this cycle the heat is transferred to the working fluid at constant pressure. This process corresponds to the injection and burning of the fuel in the actual engine. Since the gas is expanding during the heat addition in the air-standard cycle, the heat transfer must be just sufficient to maintain constant pressure. When state 3 is reached the heat addition ceases and the gas undergoes an isentropic expansion, process 3–4, until the piston reaches crank-end dead center. As in the air-standard Otto cycle, a constant-volume rejection of heat at crank-end dead center replaces the exhaust and intake processes of the actual engine.

The efficiency of the diesel cycle is given by the relation

$$\eta_{th} = 1 - \frac{Q_L}{Q_H} = 1 - \frac{C_v(T_4 - T_1)}{C_p(T_3 - T_2)} = 1 - \frac{T_1(T_4/T_1 - 1)}{kT_2(T_3/T_2 - 1)} \tag{11.13}$$

It is important to note that the isentropic compression ratio is greater than the isentropic expansion ratio in the diesel cycle. In addition, for a given state before compression

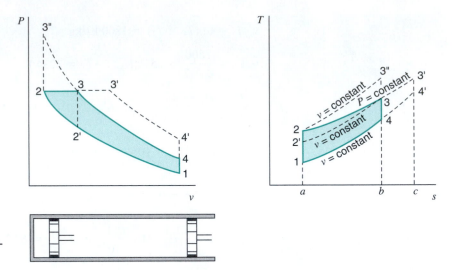

**FIGURE 11.22** The air-standard diesel cycle.

and a given compression ratio (that is, given states 1 and 2), the cycle efficiency decreases as the maximum temperature increases. This is evident from the $T$–$s$ diagram, because the constant-pressure and constant-volume lines converge, and increasing the temperature from 3 to $3'$ requires a large addition of heat (area $3$–$3'$–$c$–$b$–$3$) and results in a relatively small increase in work (area $3$–$3'$–$4'$–$4$–$3$).

There are a number of comparisons between the Otto cycle and the diesel cycle, but here we will note only two. Consider Otto cycle $1$–$2$–$3''$–$4$–$1$ and diesel cycle $1$–$2$–$3$–$4$–$1$, which have the same state at the beginning of the compression stroke and the same piston displacement and compression ratio. From the $T$–$s$ diagram we see that the Otto cycle has the higher efficiency. In practice, however, the diesel engine can operate on a higher compression ratio than the spark-ignition engine. The reason is that in the spark-ignition engine an air–fuel mixture is compressed, and detonation (spark knock) becomes a serious problem if too high a compression ratio is used. This problem does not exist in the diesel engine because only air is compressed during the compression stroke.

Therefore, we might compare an Otto cycle with a diesel cycle and in each case select a compression ratio that might be achieved in practice. Such a comparison can be made by considering Otto cycle $1$–$2'$–$3$–$4$–$1$ and diesel cycle $1$–$2$–$3$–$4$–$1$. The maximum pressure and temperature are the same for both cycles, which means that the Otto cycle has a lower compression ratio than the diesel cycle. It is evident from the $T$–$s$ diagram that in this case the diesel cycle has the higher efficiency. Thus, the conclusions drawn from a comparison of these two cycles must always be related to the basis on which the comparison has been made.

The actual compression-ignition open cycle differs from the air-standard diesel cycle in much the same way that the spark-ignition open cycle differs from the air-standard Otto cycle.

| EXAMPLE 11.10 | An air-standard diesel cycle has a compression ratio of 20, and the heat transferred to the working fluid per cycle is 1800 kJ/kg. At the beginning of the compression process the pressure is 0.1 MPa and the temperature is 15°C. Determine |
| --- | --- |

1. The pressure and temperature at each point in the cycle.
2. The thermal efficiency.
3. The mean effective pressure.

*Control mass*: Air inside cylinder.

*Diagram*: Fig. 11.22.

*State information*: $P_1 = 0.1$ MPa, $T_1 = 288.2$ K.

*Process information*: Four processes known (Fig. 11.22). Also $r_v = 20$ and $q_H = 1800$ kJ/kg.

*Model*: Ideal gas, constant specific heat, value at 300 K.

**Analysis**

The second law for compression process 1–2 is

$$s_2 = s_1$$

so that

$$\frac{T_2}{T_1} = \left(\frac{V_1}{V_2}\right)^{k-1}$$

$$\frac{P_2}{P_1} = \left(\frac{V_1}{V_2}\right)^{k}$$

The first law for heat addition process 2–3 is

$$q_H = {}_2q_3 = C_p(T_3 - T_2)$$

and the second law for expansion process 3–4 is

$$s_4 = s_3$$

so that

$$\frac{T_3}{T_4} = \left(\frac{V_4}{V_3}\right)^{k-1}$$

In addition,

$$\eta_{\text{th}} = \frac{w_{\text{net}}}{q_H}, \qquad \text{mep} = \frac{w_{\text{net}}}{v_1 - v_2}$$

**Solution**

Substitution gives

$$v_1 = \frac{0.287 \times 288.2}{100} = 0.827 \text{ m}^3/\text{kg}$$

$$v_2 = \frac{v_2}{20} = \frac{0.827}{20} = 0.04135 \text{ m}^3/\text{kg}$$

$$\frac{T_2}{T_1} = \left(\frac{V_1}{V_2}\right)^{k-1} = 20^{0.4} = 3.3145, \qquad T_2 = 955.2 \text{ K}$$

$$\frac{P_2}{P_1} = \left(\frac{V_1}{V_2}\right)^{k} = 20^{1.4} = 66.29, \qquad P_2 = 6.629 \text{ MPa}$$

$$q_H = {}_2q_3 = C_P(T_3 - T_2) = 1800 \text{ kJ/kg}$$

$$T_3 - T_2 = \frac{1800}{1.004} = 1793 \text{ K}, \qquad T_3 = 2748 \text{ K}$$

$$\frac{V_3}{V_2} = \frac{T_3}{T_2} = \frac{2748}{955.2} = 2.8769, \qquad v_3 = 0.118\ 96 \text{ m}^3/\text{kg}$$

$$\frac{T_3}{T_4} = \left(\frac{V_4}{V_3}\right)^{k-1} = \left(\frac{0.827}{0.118\ 96}\right)^{0.4} = 2.1719, \qquad T_4 = 1265 \text{ K}$$

$$q_L = {}_4q_1 = C_v(T_1 - T_4) = 0.717(288.2 - 1265) = -700.4 \text{ kJ/kg}$$

$$w_{\text{net}} = 1800 - 700.4 = 1099.6 \text{ kJ/kg}$$

$$\eta_{\text{th}} = \frac{w_{\text{net}}}{q_H} = \frac{1099.6}{1800} = 61.1\%$$

$$\text{mep} = \frac{w_{\text{net}}}{v_1 - v_2} = \frac{1099.6}{0.827 - 0.041\ 35} = 1400 \text{ kPa}$$

**In-Text Concept Questions**

**i.** How is the compression in the Otto cycle different from the Brayton cycle?

**j.** How many parameters do you need to know to completely describe the Otto cycle? How about the diesel cycle?

**k.** The exhaust and inlet flow processes are not included in the Otto or diesel cycles. How do these necessary processes affect the cycle performance?

## 11.13 INTRODUCTION TO REFRIGERATION SYSTEMS

In Section 11.1, we discussed cyclic heat engines consisting of four separate processes, either steady-state or cylinder–piston boundary-movement work devices. We further allowed for a working fluid that changes phase or for one that remains in a single phase throughout the cycle. We then considered a power system comprised of four reversible steady-state processes, two of which were constant-pressure heat transfer processes, for simplicity of equipment requirements, since these two processes involve no work. It was further assumed that the other two work-involved processes were adiabatic and therefore isentropic. The resulting power cycle appeared as in Fig. 11.2.

We now consider the basic ideal refrigeration system cycle in exactly the same terms as those described above, except that each process is the reverse of that in the power cycle. The result is the ideal cycle shown in Fig. 11.23. Note that if the entire cycle takes place inside the two-phase liquid–vapor dome, the resulting cycle is, as with the power cycle, the Carnot cycle, since the two constant-pressure processes are also isothermal. Otherwise, this cycle is not a Carnot cycle. It is also noted, as before, that the net work input to the cycle is equal to the area enclosed by the process lines 1–2–3–4–1, independently of whether the individual processes are steady state or cylinder/piston boundary movement.

In the next section, we make one modification to this idealized basic refrigeration system cycle in presenting and applying the model of refrigeration and heat pump systems.

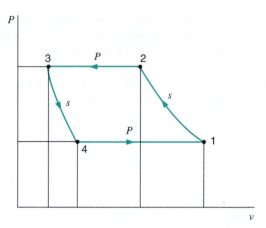

**FIGURE 11.23**  Four-process refrigeration cycle.

## 11.14 THE VAPOR-COMPRESSION REFRIGERATION CYCLE

In this section, we consider the ideal refrigeration cycle for a working substance that changes phase during the cycle, in a manner equivalent to that done with the Rankine power cycle in Section 11.2. In doing so, we note that state 3 in Fig. 11.23 is saturated liquid at the condenser temperature and state 1 is saturated vapor at the evaporator temperature. This means that the isentropic expansion process from 3–4 will be in the two-phase region, and the substance there will be mostly liquid. As a consequence, there will be very little work output from this process, such that it is not worth the cost of including this piece of equipment in the system. We therefore replace the turbine with a throttling device, usually a valve or a length of small-diameter tubing, by which the working fluid is throttled from the high-pressure to the low-pressure side. The resulting cycle become the ideal model for a vapor-compression refrigeration system, which is shown in Fig. 11.24. Saturated vapor at low pressure enters the compressor and undergoes a reversible adiabatic compression, process 1–2. Heat is then rejected at constant pressure in process 2–3, and the working fluid exits the condenser as saturated liquid. An adiabatic throttling process, 3–4, follows, and the working fluid is then evaporated at constant pressure, process 4–1, to complete the cycle.

The similarity of this cycle to the reverse of the Rankine cycle has already been noted. We also note the difference between this cycle and the ideal Carnot cycle, in which

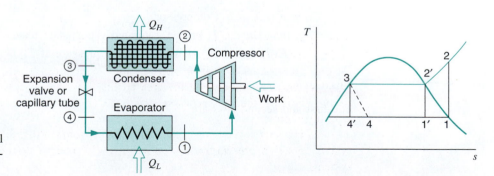

**FIGURE 11.24**  The ideal vapor-compression refrigeration cycle.

the working fluid always remains inside the two-phase region, $1'$–$2'$–$3$–$4'$–$1'$. It is much more expedient to have a compressor handle only vapor than a mixture of liquid and vapor, as would be required in process $1'$–$2'$ of the Carnot cycle. It is virtually impossible to compress, at a reasonable rate, a mixture such as that represented by state $1'$ and still maintain equilibrium between liquid and vapor. The other difference, that of replacing the turbine by the throttling process, has already been discussed.

It should be pointed out that the system described in Fig. 11.24 can be used for either of two purposes. The first use is as a refrigeration system, in which case it is desired to maintain a space at a low temperature $T_1$ relative to the ambient temperature $T_3$. (In a real system, it would be necessary to allow a finite temperature difference in both the evaporator and condenser to provide a finite rate of heat transfer in each.) Thus, the reason for building the system in this case is the quantity $q_L$. The measure of performance of a refrigeration system is given in terms of the coefficient of performance, $\beta$, which was defined in Chapter 7 as

$$\beta = \frac{q_L}{w_c} \tag{11.14}$$

The second use of the system described in Fig. 11.24 is as a heat pump system, in which case it is desired to maintain a space at a temperature $T_3$ above that of the ambient (or other source) $T_1$. In this case, the reason for building the system is the quantity $q_H$, and the coefficient of performance for the heat pump, $\beta'$, is now

$$\beta' = \frac{q_H}{w_c} \tag{11.15}$$

Refrigeration systems and heat pump systems are, of course, different in terms of design variables, but the analysis of the two is the same. When we discuss refrigerators in this and the following two sections, it should be kept in mind that the same comments generally apply to heat pump systems, as well.

---

**EXAMPLE 11.11**

Consider an ideal refrigeration cycle that uses R-134a as the working fluid. The temperature of the refrigerant in the evaporator is $-20°C$ and in the condenser it is $40°C$. The refrigerant is circulated at the rate of 0.03 kg/s. Determine the coefficient of performance and the capacity of the plant in rate of refrigeration.

The diagram for this example is as shown in Fig. 11.24. For each control volume analyzed, the thermodynamic model is the R-134a tables. Each process is steady state with no changes in kinetic or potential energy.

> *Control volume*: Compressor.
> *Inlet state*: $T_1$ known, saturated vapor; state fixed.
> *Exit state*: $P_2$ known (saturation pressure at $T_3$).

**Analysis**

The first and second laws are

$$w_c = h_2 - h_1$$

$$s_2 = s_1$$

**Solution**

At $T_3 = 40°C$,

$$P_g = P_2 = 1017 \text{ kPa}$$

From the R-134a tables, we get

$$h_1 = 386.1 \text{ kJ/kg}, \qquad s_1 = 1.7395 \text{ kJ/kg K}$$

Therefore,

$$s_2 = s_1 = 1.7395 \text{ kJ/kg K}$$

so that

$$T_2 = 47.7°C \qquad \text{and} \qquad h_2 = 428.4 \text{ kJ/kg}$$

$$w_c = h_2 - h_1 = 428.4 - 386.1 = 42.3 \text{ kJ/kg}$$

| | |
|---:|:---|
| *Control volume*: | Expansion valve. |
| *Inlet state*: | $T_3$ known, saturated liquid; state fixed. |
| *Exit state*: | $T_4$ known. |

**Analysis**

The first law is

$$h_3 = h_4$$

**Solution**

Numerically, we have

$$h_4 = h_3 = 256.5 \text{ kJ/kg}$$

| | |
|---:|:---|
| *Control volume*: | Evaporator. |
| *Inlet state*: | State 4 known (as given). |
| *Exit state*: | State 1 known (as given). |

**Analysis**

The first law is

$$q_L = h_1 - h_4$$

**Solution**

Substituting, we have

$$q_L = h_1 - h_4 = 386.1 - 256.5 = 129.6 \text{ kJ/kg}$$

Therefore,

$$\beta = \frac{q_L}{w_c} = \frac{129.6}{42.3} = 3.064$$

Refrigeration capacity $= 129.6 \times 0.03 = 3.89 \text{ kW}$

## 11.15 Working Fluids for Vapor-Compression Refrigeration Systems

A much larger number of different working fluids (refrigerants) are utilized in vapor-compression refrigeration systems than in vapor power cycles. Ammonia and sulfur dioxide were important in the early days of vapor-compression refrigeration, but both are highly toxic and therefore dangerous substances. For many years now, the principal refrigerants have been the halogenated hydrocarbons, which are marketed under the trade names of Freon and Genatron. For example, dichlorodifluoromethane ($CCl_2F_2$) is known as Freon-12 and Genatron-12, and therefore as refrigerant-12 or R-12. This group of substances, known commonly as chlorofluorocarbons or CFCs, are chemically very stable at ambient temperature, especially those lacking any hydrogen atoms. This characteristic is necessary for a refrigerant working fluid. This same characteristic, however, has devastating consequences if the gas, having leaked from an appliance into the atmosphere, spends many years slowly diffusing upward into the stratosphere. There it is broken down, releasing chlorine, which destroys the protective ozone layer of the stratosphere. It is therefore of overwhelming importance to us all to eliminate completely the widely used but life-threatening CFCs, particularly R-11 and R-12, and to develop suitable and acceptable replacements. The CFCs containing hydrogen (often termed HCFCs), such as R-22, have shorter atmospheric lifetimes and therefore are not as likely to reach the stratosphere before being broken up and rendered harmless. The most desirable fluids, called HFCs, contain no chlorine atoms at all.

There are two important considerations when selecting refrigerant working fluids: the temperature at which refrigeration is needed and the type of equipment to be used.

As the refrigerant undergoes a change of phase during the heat transfer process, the pressure of the refrigerant will be the saturation pressure during the heat supply and heat rejection processes. Low pressures mean large specific volumes and correspondingly large equipment. High pressures mean smaller equipment, but it must be designed to withstand higher pressure. In particular, the pressures should be well below the critical pressure. For extremely low temperature applications a binary fluid system may be used by cascading two separate systems.

The type of compressor used has a particular bearing on the refrigerant. Reciprocating compressors are best adapted to low specific volumes, which means higher pressures, whereas centrifugal compressors are most suitable for low pressures and high specific volumes.

It is also important that the refrigerants used in domestic appliances be nontoxic. Other beneficial characteristics, in addition to being environmentally acceptable, are miscibility with compressor oil, dielectric strength, stability, and low cost. Refrigerants, however, have an unfortunate tendency to cause corrosion. For given temperatures during evaporation and condensation, not all refrigerants have the same coefficient of performance for the ideal cycle. It is, of course, desirable to use the refrigerant with the highest coefficient of performance, other factors permitting.

## 11.16 Deviation of the Actual Vapor-Compression Refrigeration Cycle from the Ideal Cycle

The actual refrigeration cycle deviates from the ideal cycle primarily because of pressure drops associated with fluid flow and heat transfer to or from the surroundings. The actual cycle might approach the one shown in Fig. 11.25.

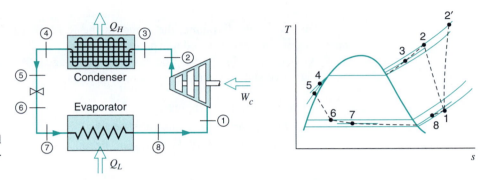

**FIGURE 11.25** The actual vapor-compression refrigeration cycle.

The vapor entering the compressor will probably be superheated. During the compression process there are irreversibilities and heat transfer either to or from the surroundings, depending on the temperature of the refrigerant and the surroundings. Therefore, the entropy might increase or decrease during this process, for the irreversibility and the heat transferred to the refrigerant cause an increase in entropy, and the heat transferred from the refrigerant causes a decrease in entropy. These possibilities are represented by the two dashed lines 1–2 and 1–2′. The pressure of the liquid leaving the condenser will be less than the pressure of the vapor entering, and the temperature of the refrigerant in the condenser will be somewhat higher than that of the surroundings to which heat is being transferred. Usually the temperature of the liquid leaving the condenser is lower than the saturation temperature. It might drop somewhat more in the piping between the condenser and expansion valve. This represents a gain, however, because as a result of this heat transfer the refrigerant enters the evaporator with a lower enthalpy, which permits more heat to be transferred to the refrigerant in the evaporator.

There is some drop in pressure as the refrigerant flows through the evaporator. It may be slightly superheated as it leaves the evaporator, and through heat transferred from the surroundings its temperature will increase in the piping between the evaporator and the compressor. This heat transfer represents a loss, because it increases the work of the compressor, since the fluid entering it has an increased specific volume.

**EXAMPLE 11.12**

A refrigeration cycle utilizes R-12 as the working fluid. The following are the properties at various points of the cycle designated in Fig. 11.25.

$$P_1 = 125 \text{ kPa}, \qquad T_1 = -10°C$$
$$P_2 = 1.2 \text{ MPa}, \qquad T_2 = 100°C$$
$$P_3 = 1.19 \text{ MPa}, \qquad T_3 = 80°C$$
$$P_4 = 1.16 \text{ MPa}, \qquad T_4 = 45°C$$
$$P_5 = 1.15 \text{ MPa}, \qquad T_5 = 40°C$$
$$P_6 = P_7 = 140 \text{ kPa}, \qquad x_6 = x_7$$
$$P_8 = 130 \text{ kPa}, \qquad T_8 = -20°C$$

The heat transfer from R-12 during the compression process is 4 kJ/kg. Determine the coefficient of performance of this cycle.

For each control volume, the model is the R-12 tables. Each process is steady state with no changes in kinetic or potential energy.

As before, we break the process down into stages, treating the compressor, the throttling valve and line, and the evaporator in turn.

*Control volume*:  Compressor.
*Inlet state*:  $P_1$, $T_1$ known; state fixed.
*Exit state*:  $P_2$, $T_2$ known; state fixed.

**Analysis**

From the first law, we have

$$q + h_1 = h_2 + w$$
$$w_c = -w = h_2 - h_1 - q$$

**Solution**

From the R-12 tables, we read

$$h_1 = 185.16 \text{ kJ/kg}, \qquad h_2 = 245.52 \text{ kJ/kg}$$

Therefore,

$$w_c = 245.52 - 185.16 - (-4) = 64.36 \text{ kJ/kg}$$

*Control volume*:  Throttling valve plus line.
*Inlet state*:  $P_5$, $T_5$ known; state fixed.
*Exit state*:  $P_7 = P_6$ known, $x_7 = x_6$.

**Analysis**

The first law is

$$h_5 = h_6$$

Since $x_7 = x_6$, it follows that $h_7 = h_6$.

**Solution**

Numerically, we obtain

$$h_5 = h_6 = h_7 = 74.53 \text{ kJ/kg}$$

*Control volume*:  Evaporator.
*Inlet state*:  $P_7$, $h_7$ known (above).
*Exit state*:  $P_8$, $T_8$ known; state fixed.

**Analysis**

The first law is

$$q_L = h_8 - h_7$$

**Solution**

Substitution gives

$$q_L = h_8 - h_7 = 179.12 - 74.53 = 104.59 \text{ kJ/kg}$$

Therefore,

$$\beta = \frac{q_L}{w_c} = \frac{104.59}{64.36} = 1.625$$

## 11.17 THE AIR-STANDARD REFRIGERATION CYCLE

If we consider the original ideal four-process refrigeration cycle of Fig. 11.23 with a non-condensing (gaseous) working fluid, then the work output during the isentropic expansion process is not negligibly small, as was the case with a condensing working fluid. Therefore, we retain the turbine in the four-steady-state process ideal air-standard refrigeration cycle shown in Fig. 11.26. This cycle is seen to be the reverse Brayton cycle, and it is used in practice in the liquefaction of air and other gases and also in certain special situations that require refrigeration, such as aircraft cooling systems. After compression from states 1 to 2, the air is cooled as heat is transferred to the surroundings at temperature $T_0$. The air is then expanded in process 3–4 to the pressure entering the compressor, and the temperature drops to $T_4$ in the expander. Heat may then be transferred to the air until temperature $T_L$ is reached. The work for this cycle is represented by area 1–2–3–4–1, and the refrigeration effect is represented by area 4–1–b–a–4. The coefficient of performance is the ratio of these two areas.

In practice, this cycle has been used to cool aircraft in an open cycle. A simplified form is shown in Fig. 11.27. Upon leaving the expander, the cool air is blown directly into the cabin, thus providing the cooling effect where needed.

When counterflow heat exchangers are incorporated, very low temperatures can be obtained. This is essentially the cycle used in low-pressure air liquefaction plants and in other liquefaction devices such as the Collins helium liquefier. The ideal cycle is as shown in Fig. 11.28. Because the expander operates at very low temperature, the designer is faced with unique problems in providing lubrication and choosing materials.

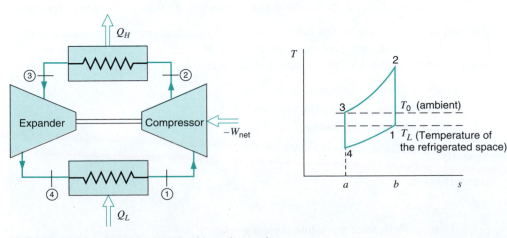

**FIGURE 11.26**  The air-standard refrigeration cycle.

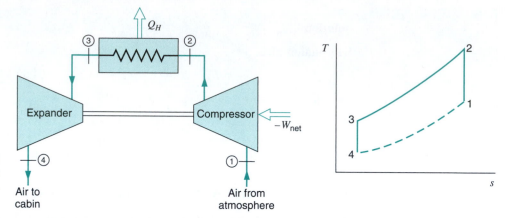

**FIGURE 11.27** An air refrigeration cycle that might be utilized for aircraft cooling.

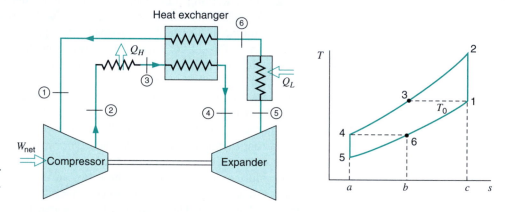

**FIGURE 11.28** The air refrigeration cycle utilizing a heat exchanger.

**EXAMPLE 11.13**

Consider the simple air-standard refrigeration cycle of Fig. 11.26. Air enters the compressor at 0.1 MPa and −20°C and leaves at 0.5 MPa. Air enters the expander at 15°C. Determine

1. The coefficient of performance for this cycle.
2. The rate at which air must enter the compressor to provide 1 kW of refrigeration.

For each control volume in this example, the model is ideal gas with constant specific heat, at 300 K, and each process is steady state with no kinetic or potential energy changes. The diagram for this example is Fig. 11.26.

> *Control volume*: Compressor.
> *Inlet state*: $P_1$, $T_1$ known; state fixed.
> *Exit state*: $P_2$ known.

**Analysis**

The first law is

$$w_c = h_2 - h_1$$

(Here $w_c$ designates work into the compressor.)

The second law gives

$$s_1 = s_2$$

so that

$$\frac{T_2}{T_1} = \left(\frac{P_2}{P_1}\right)^{(k-1)/k}$$

**Solution**

Substituting, we obtain

$$\frac{T_2}{T_1} = \left(\frac{P_2}{P_1}\right)^{(k-1)/k} = 5^{0.286} = 1.5845, \qquad T_2 = 401.2 \text{ K}$$

$$w_c = h_2 - h_1 = C_p(T_2 - T_1)$$

$$= 1.004(401.2 - 253.2) = 148.5 \text{ kJ/kg}$$

> *Control volume*: Expander.
> *Inlet state*: $P_3(= P_2)$ known, $T_3$ known; state fixed.
> *Exit state*: $P_4(= P_1)$ known.

**Analysis**

From the first law, we have

$$w_t = h_3 - h_4$$

The second law gives

$$s_3 = s_4$$

so that

$$\frac{T_3}{T_4} = \left(\frac{P_3}{P_4}\right)^{(k-1)/k}$$

**Solution**

Therefore,

$$\frac{T_3}{T_4} = \left(\frac{P_3}{P_4}\right)^{(k-1)/k} = 5^{0.286} = 1.5845, \qquad T_4 = 181.9 \text{ K}$$

$$w_t = h_3 - h_4 = 1.004(288.2 - 181.9) = 106.7 \text{ kJ/kg}$$

> *Control volume*: High-temperature heat exchanger.
> *Inlet state*: State 2 known (as given).
> *Exit state*: State 3 known (as given).

**Analysis**

The first law is

$$q_H = h_2 - h_3 \quad \text{(heat rejected)}$$

**Solution**

Substitution gives

$$q_H = h_2 - h_3 = C_p(T_2 - T_3) = 1.004(401.2 - 288.2) = 113.4 \text{ kJ/kg}$$

> *Control volume*: Low-temperature heat exchanger.
> *Inlet state*: State 4 known (as given).
> *Exit state*: State 1 known (as given).

**Analysis**

The first law is

$$q_L = h_1 - h_4$$

**Solution**

Substituting, we obtain

$$q_L = h_1 - h_4 = C_p(T_1 - T_4) = 1.004(253.2 - 181.9) = 71.6 \text{ kJ/kg}$$

Therefore,

$$w_{net} = w_c - w_t = 148.5 - 106.7 = 41.8 \text{ kJ/kg}$$

$$\beta = \frac{q_L}{w_{net}} = \frac{71.6}{41.8} = 1.713$$

To provide 1 kW of refrigeration capacity, we have

$$\dot{m} = \frac{\dot{Q}_L}{q_L} = \frac{1}{71.6} = 0.014 \text{ kg/s}$$

**In-Text Concept Questions**

l. A refrigerator in my 20°C kitchen uses R-12 and I want to make ice cubes at −5°C. What is the minimum high $P$ and the maximum low $P$ it can use?

m. How many parameters are needed to completely determine a standard vapor compression refrigeration cycle?

## 11.18 COMBINED-CYCLE POWER AND REFRIGERATION SYSTEMS

There are many situations in which it is desirable to combine two cycles in series, either power systems or refrigeration systems, to take advantage of a very wide temperature range or to utilize what would otherwise be waste heat to improve efficiency. One combined power cycle, shown in Fig. 11.29 as a simple steam cycle with a liquid metal topping cycle, is often referred to as a binary cycle. The advantage of this combined system is that the liquid metal has a very low vapor pressure relative to that for water; therefore it is possible for an isothermal boiling process in the liquid metal to take place at a high temperature, much higher than the critical temperature of water, but still at a moderate pressure.

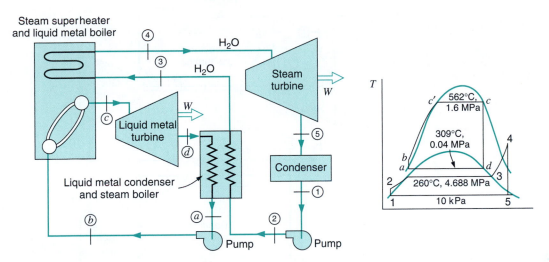

**FIGURE 11.29** Liquid metal–water binary power system.

The liquid metal condenser then provides an isothermal heat source as input to the steam boiler, such that the two cycles can be closely matched by proper selection of the cycle variables, with the resulting combined cycle then having a high thermal efficiency. Saturation pressures and temperatures for a typical liquid metal–water binary cycle are shown in the $T$–$s$ diagram of Fig. 11.29.

A different type of combined cycle that has seen considerable attention is to use the "waste heat" exhaust from a Brayton cycle gas-turbine engine (or another combustion engine such as a diesel engine) as the heat source for a steam or other vapor power cycle, in which case the vapor cycle acts as a bottoming cycle for the gas engine, in order to improve the overall thermal efficiency of the combined power system. Such a system, utilizing a gas turbine and a steam Rankine cycle, is shown in Fig. 11.30. In such a combination, there is a natural mismatch using the cooling of a noncondensing gas as the energy source to effect an isothermal boiling process plus superheating the vapor, and careful design is required to avoid a pinch point, a condition at which the gas has cooled to the vapor boiling temperature without having provided sufficient energy to complete the boiling process.

One way to take advantage of the cooling exhaust gas in the Brayton-cycle portion of the combined system is to utilize a mixture as the working fluid in the Rankine cycle. An example of this type of application is the Kalina cycle, which uses ammonia–water mixtures as the working fluid in the Rankine-type cycle. Such a cycle can be made very efficient, inasmuch as the temperature differences between the two fluid streams can be controlled through careful design of the combined system.

Combined cycles are used in refrigeration systems in cases in which there is a very large temperature difference between the ambient surroundings and the refrigerated space. Such a refrigeration system is often called a cascade system, an example of which is shown in Fig. 11.31. In this case, the refrigerant R-22 is used in the refrigeration system rejecting heat to the ambient surroundings, while its evaporator picks up the heat rejected in the low-temperature system condenser, the low temperature working fluid in this case being R-23, whose thermodynamic properties are suited to work as a refrigerant in this low-temperature range. As with the other combined-cycle systems, the working fluids and design variables must be considered very carefully to optimize the performance of each unit.

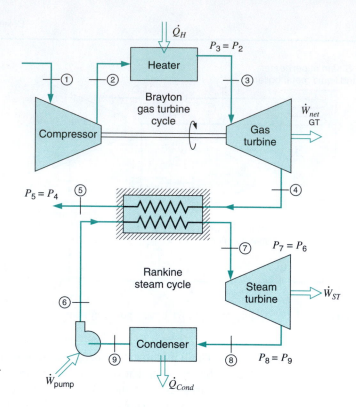

**FIGURE 11.30**
Combined Brayton/Rankine cycle power system.

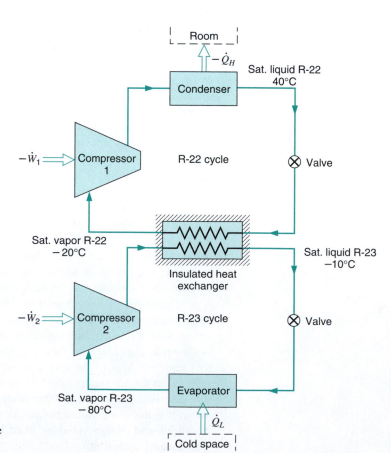

**FIGURE 11.31**
Combined-cycle cascade refrigeration system.

We have described only a few combined-cycle systems here, as examples of the types of applications that can be dealt with, and the resulting improvement in overall performance that can occur. Obviously, there are many other combinations of power and refrigeration systems. Some of these are discussed in the problems at the end of the chapter.

## SUMMARY

A number of standard power-producing cycles and refrigeration cycles are presented. First we cover a number of stationary and mobile power-producing heat engines. The Rankine cycle and its variations represent a steam power plant, which produces most of the world production of electricity. The heat input can come from combustion of fossil fuels, a nuclear reactor, solar radiation, or any other heat source that can generate a temperature high enough to boil water at a high pressure. In low- or very high-temperature applications, substances other than water can be used. Modifications to the basic cycle such as closed and open feedwater heaters are covered together with applications where the electricity is cogenerated with a base demand for process steam.

A Brayton cycle is a gas turbine producing electricity and, with a modification, a jet engine producing thrust. This is a high-power, low-mass, and low-volume device and is used where space and weight are at a premium cost. A high back-work ratio makes this cycle sensitive to the compressor efficiency.

Piston/cylinder devices are shown for the Otto and Diesel cycles modeling the gasoline and diesel engines, which can be two- or four-stroke engines. Cold air properties are used to show the influence of compression ratio on the thermal efficiency, and the mean effective pressure is used to relate the engine size to total power output.

Standard refrigeration systems are covered by the vapor-compression refrigeration cycle. This applies to household refrigerators, air conditioners, and heat pumps as well as commercial units to lower temperature ranges. The air-standard refrigeration cycle is also covered in detail.

The chapter is completed with a short description of combined cycle applications. This covers stacked or cascade systems for large temperature spans and combinations of different kinds of cycles where one can be added as a topping cycle or a bottoming cycle. Often a Rankine cycle uses exhaust energy from a Brayton cycle in larger stationary applications.

You should have learned a number of skills and acquired abilities from studying this chapter that will allow you to

- Apply the general laws to control volumes with several devices forming a complete system.
- Have a knowledge of how many common power-producing devices work.
- Have a knowledge of how simple refrigerators and heat pumps work.
- Know that most cycle devices do not operate in Carnot cycles.
- Know that real devices have lower efficiencies/COP than ideal cycles.
- Have a sense of the most influential parameters for each type of cycle.
- Have an idea about the importance of the component efficiency for the overall cycle efficiency or COP.
- Know that most real cycles have modifications to the basic cycle setup.
- Know that many of these devices affect our environment.
- Know the principle of combining different cycles.

## KEY CONCEPTS AND FORMULAS

### Rankine Cycle

| | |
|---|---|
| Open feedwater heater | Feedwater mixed with extraction steam, exit as saturated liquid |
| Closed feedwater heater | Feedwater heated by extraction steam, no mixing |
| Cogeneration | Turbine power is cogenerated with a desired steam supply |

### Brayton Cycle

| | |
|---|---|
| Compression ratio | Pressure ratio $r_p = P_{high}/P_{low}$ |
| Regenerator | Dual fluid heat exchanger, uses exhaust flow energy |
| Intercooler | Cooler between compressor stages, reduces work input |
| Jet engine | No shaftwork out, kinetic energy generated in exit nozzle |
| Thrust | $F = \dot{m}(\mathbf{V}_e - \mathbf{V}_i)$ momentum equation |
| Propulsive power | $\dot{W} = F\,\mathbf{V}_{aircraft} = \dot{m}(\mathbf{V}_e - \mathbf{V}_i)\mathbf{V}_{aircraft}$ |

### Piston Cylinder Power Cycles

| | |
|---|---|
| Compression ratio | Volume ratio $r_v = CR = V_{max}/V_{min}$ |
| Displacement (1 cyl.) | $\Delta V = V_{max} - V_{min} = m(v_{max} - v_{min}) = SA_{cyl}$ |
| Stroke | $S = 2\,R_{crank}$, piston travel in compression or expansion. |
| Mean effective pressure | $P_{meff} = w_{net}/(v_{max} - v_{min}) = W_{net}/(V_{max} - V_{min})$ |
| Power by 1 cylinder | $\dot{W} = mw_{net}\dfrac{RPM}{60}$ (times $\frac{1}{2}$ for four-stroke cycle) |
| Total power | $\dot{W} = N_{cyl}\,mw_{net}\dfrac{RPM}{60} = P_{meff}V_{displ}\dfrac{RPM}{60}$ |

### Refrigeration Cycle

| | |
|---|---|
| Coefficient of performance | $\text{COP}_{REF} = \beta_{REF} = \dfrac{\dot{Q}_L}{\dot{W}_C} = \dfrac{q_L}{w_c}$ |
| | $\text{COP}_{HP} = \beta_{HP} = \dfrac{\dot{Q}_H}{\dot{W}_C} = \dfrac{q_H}{w_c}$ |

### Combined Cycles

| | |
|---|---|
| Topping, bottoming cycle | The high- and low-temperature cycles |
| Cascade system | Stacked refrigeration cycles |

# HOMEWORK PROBLEMS

## Rankine cycles, power plants

### Simple cycles

**11.1** Is a steam power plant running in a Carnot cycle? Name the four processes in the cycle.

**11.2** Can the energy removed in a power plant condenser be useful?

**11.3** A steam power plant, as shown in Fig. 11.3, operating in a Rankine cycle has saturated vapor at 3 MPa leaving the boiler. The turbine exhausts to the condenser operating at 10 kPa. Find the specific work and heat transfer in each of the ideal components and the cycle efficiency.

**11.4** The power plant in Problem 11.3 is modified to have a superheater section following the boiler so that the steam leaves the superheater at 3 MPa and 400°C. Find the specific work and heat transfer in each of the ideal components and the cycle efficiency.

**11.5** Consider a solar-energy-powered ideal Rankine cycle that uses water as the working fluid. Saturated vapor leaves the solar collector at 175°C, and the condenser pressure is 10 kPa. Determine the thermal efficiency of this cycle.

**11.6** A utility runs a Rankine cycle with a water boiler at 3 MPa, and the cycle has the highest and lowest temperatures of 450°C and 45°C, respectively. Find the plant efficiency and the efficiency of a Carnot cycle with the same temperatures.

**11.7** A power plant for a polar expedition uses ammonia that is heated to 80°C at 1000 kPa in the boiler and the condenser is maintained at −15°C. Find the cycle efficiency.

**11.8** A Rankine cycle uses ammonia as the working substance and powered by solar energy. It heats the ammonia to 140°C at 5000 kPa in the boiler/superheater. The condenser is water cooled, and the exit is kept at 25°C. Find ($T$, $P$, and $x$ if applicable) for all four states in the cycle.

**11.9** A steam power plant operating in an ideal Rankine cycle has a high pressure of 5 MPa and a low pressure of 15 kPa. The turbine exhaust state should have a quality of at least 95%, and the turbine power generated should be 7.5 MW. Find the necessary boiler exit temperature and the total mass flow rate.

**11.10** For a special low temperature heat source, we want to run a power plant with R-134a so it is supercritical with high pressure of 6 MPa. How much should the R-134a be heated so the turbine exit is minimum quality of 0.9 at a low temperature in the condenser of 40°C.

**11.11** A supply of geothermal hot water is to be used as the energy source in an ideal Rankine cycle, with R-134a as the cycle working fluid. Saturated vapor R-134a leaves the boiler at a temperature of 85°C, and the condenser temperature is 40°C. Calculate the thermal efficiency of this cycle.

**11.12** Do Problem 11.11 with R-22 as the working fluid.

**11.13** Do Problem 11.11 with ammonia as the working fluid.

**11.14** Consider the ammonia Rankine-cycle power plant shown in Fig. P11.14, a plant that was designed to operate in a location where the ocean water temperature is 25°C near the surface and 5°C at some greater depth. The mass flow rate of the working fluid is 1000 kg/s.

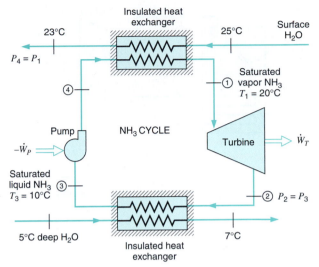

**FIGURE P11.14**

**a.** Determine the turbine power output and the pump power input for the cycle.

**b.** Determine the mass flow rate of water through each heat exchanger.

**c.** What is the thermal efficiency of this power plant?

**11.15** A smaller power plant produces 25 kg/s steam at 3 MPa, 600°C, in the boiler. It cools the condenser with ocean water coming in at 12°C and returned at 15°C so the condenser exit is at 45°C. Find the net power output and the required mass flow rate of ocean water.

**11.16** A steam power plant has a steam generator exit at 4 MPa and 500°C and a condenser exit temperature of 45°C. Assume all components are ideal and find the cycle efficiency and the specific work and heat transfer in the components.

**11.17** Consider an ideal Rankine cycle using water with a high-pressure side of the cycle at a supercritical pressure. Such a cycle has a potential advantage of minimizing local temperature differences between the fluids in the steam generator, such as the instance in which the high-temperature energy source is the hot exhaust gas from a gas-turbine engine. Calculate the thermal efficiency of the cycle if the state entering the turbine is 30 MPa, 550°C, and the condenser pressure is 5 kPa. What is the steam quality at the turbine exit?

**11.18** Consider an ideal steam reheat cycle as shown in Fig. P11.18, where steam enters the high-pressure turbine at 3 MPa and 400°C and then expands to 0.8 MPa. It is then reheated at constant pressure 0.8 MPa to 400°C and expands to 10 kPa in the low-pressure turbine. Calculate

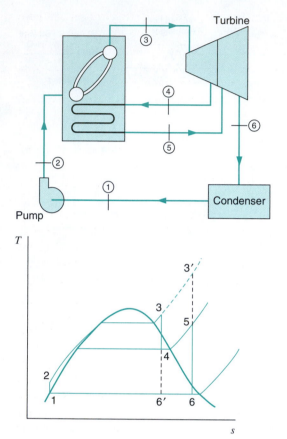

**FIGURE P11.18**

the thermal efficiency and the moisture content of the steam leaving the low-pressure turbine.

**11.19** Consider an ideal steam reheat cycle as shown in Fig. P11.18, where steam enters the high-pressure turbine at 4 MPa and 450°C with a mass flow rate of 20 kg/s. After expansion to 400 kPa it is reheated to $T_5$ flowing through the low-pressure turbine out to the condenser operating at 10 kPa. Find $T_5$ so the turbine exit quality is at least 95%.

## Open feedwater heaters

**11.20** An open feedwater heater in a regenerative steam power cycle receives 20 kg/s of water at 100°C and 2 MPa. The extraction steam from the turbine enters the heater at 2 MPa and 275°C, and all the feedwater leaves as saturated liquid. What is the required mass flow rate of the extraction steam?

**11.21** Consider an ideal steam regenerative cycle in which steam enters the turbine at 3 MPa and 400°C and exhausts to the condenser at 10 kPa. Steam is extracted from the turbine at 0.8 MPa for an open feedwater heater. The

feedwater leaves the heater as saturated liquid. The appropriate pumps are used for the water leaving the condenser and the feedwater heater. Calculate the thermal efficiency of the cycle and the net work per kilogram of steam.

**11.22** A power plant with one open feedwater heater has a condenser temperature of 45°C, a maximum pressure of 5 MPa, and boiler exit temperature of 900°C. Extraction steam at 1 MPa to the feedwater heater is mixed with the feedwater line so that the exit is saturated liquid into the second pump. Find the fraction of extraction steam flow and the two specific pump work inputs.

**11.23** A Rankine cycle operating with ammonia is heated by some low-temperature source so that the highest $T$ is 120°C at a pressure of 5000 kPa. Its low pressure is 1003 kPa, and it operates with one open feedwater heater at 2033 kPa. The total flow rate is 5 kg/s. Find the extraction flow rate to the feedwater heater assuming its outlet state is saturated liquid at 2033 kPa. Find the total power to the two pumps.

**11.24** The power plant in Problem 11.7 is modified to have an open "feedwater" heater operating at 400 kPa with an exit state of saturated liquid. Find the fraction of mass flow extracted in the turbine and the turbine work per kg flowing into it.

**11.25** A steam power plant has high and low pressures of 20 MPa and 10 kPa, and one open feedwater heater operating at 1 MPa with the exit as saturated liquid. The maximum temperature is 800°C, and the turbine has a total power output of 5 MW. Find the fraction of the flow for extraction to the feedwater and the total condenser heat transfer rate.

## Closed feedwater heaters

**11.26** A closed feedwater heater in a regenerative steam power cycle, as shown in Fig. P11.26, heats 20 kg/s of

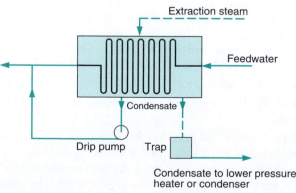

**FIGURE P11.26**

water from 100°C and 20 MPa to 250°C and 20 MPa. The extraction steam from the turbine enters the heater at 4 MPa and 275°C and leaves as saturated liquid. What is the required mass flow rate of the extraction steam?

**11.27** Repeat Problem 11.21, but assume a closed instead of an open feedwater heater. A single pump is used to pump the water leaving the condenser up to the boiler pressure of 3 MPa. Condensate from the feedwater heater is drained through a trap to the condenser.

**11.28** Do Problem 11.25 with a closed feedwater heater instead of an open heater and a drip pump to add the extraction flow to the feedwater line at 20 MPa. Assume the temperature is 175°C after the drip pump flow is added to the line. One main pump brings the water to 20 MPa from the condenser.

**11.29** Assume the power plant in Problem 11.23 has one closed feedwater heater (FWH) instead of the open FWH. The extraction flow out of the FWH is saturated liquid at 2033 kPa being dumped into the condenser and the feedwater is heated to 50°C. Find the extraction flow rate and the total turbine power output.

## Nonideal cycles

**11.30** Steam enters the turbine of a power plant at 5 MPa and 400°C and exhausts to the condenser at 10 kPa. The turbine produces a power output of 20 000 kW with an isentropic efficiency of 85%. What is the mass flow rate of steam around the cycle and the rate of heat rejection in the condenser? Find the thermal efficiency of the power plant. How does this compare with a Carnot cycle?

**11.31** A steam power plant has a high pressure of 5 MPa and maintains 50°C in the condenser. The boiler exit temperature is 600°C. All the components are ideal except the turbine, which has an actual exit state of saturated vapor at 50°C. Find the cycle efficiency with the actual turbine and the turbine isentropic efficiency.

**11.32** Consider the power plant in Problem 11.24. Assume the high temperature source is a flow of liquid water at 120°C into a heat exchanger at constant pressure 300 kPa and that the water leaves at 90°C as shown in Fig. P11.32. Assume the condenser rejects heat to the

ambient, which is at −20°C. List all the places that have entropy generation and find the entropy generated in the boiler heat exchanger per kg ammonia flowing.

**11.33** A steam power cycle has a high pressure of 3 MPa and a condenser exit temperature of 45°C. The turbine efficiency is 85%, and other cycle components are ideal. If the boiler superheats to 800°C, find the cycle thermal efficiency.

**11.34** A steam power plant operates with a high pressure of 5 MPa and has a boiler exit temperature of 600°C receiving heat from a 700°C source. The ambient air at 20°C provides cooling for the condenser so it can maintain 45°C inside. All the components are ideal except for the turbine, which has an exit state with a quality of 97%. Find the work and heat transfer in all components per kg of water and the turbine isentropic efficiency. Find the rate of entropy generation per kg of water in the boiler/heat source setup.

**11.35** A steam power plant operates with a high pressure of 4 MPa and has a boiler exit of 600°C receiving heat from a 700°C source. The ambient at 20°C provides cooling to maintain the condenser at 60°C. All components are ideal except for the turbine which has an isentropic efficiency of 92%. Find the ideal and the actual turbine exit qualities. Find the actual specific work and specific heat transfer in all four components.

**11.36** For Problem 11.35 find also the specific entropy generation in the boiler heat source setup.

**11.37** A small steam power plant has a boiler exit of 3 MPa and 400°C, while it maintains 50 kPa in the condenser. All the components are ideal except the turbine, which has an isentropic efficiency of 80%, and it should deliver a shaft power of 9.0 MW to an electric generator. Find the specific turbine work, the needed flow rate of steam, and the cycle efficiency.

**11.38** Repeat Problem 11.25 assuming the turbine has an isentropic efficiency of 85%.

## Cogeneration

**11.39** A cogenerating steam power plant, as in Fig. 11.12 operates with a boiler output of 25 kg/s steam at 7 MPa and 500°C. The condenser operates at 7.5 kPa, and the process heat is extracted at 5 kg/s from the turbine at 500 kPa, state 6, and after use is returned as saturated liquid at 100 kPa, state 8. Assume all components are ideal and find the temperature after pump 1, the total turbine output, and the total process heat transfer.

**11.40** A boiler delivers steam at 10 MPa, 550°C to a two-stage turbine as shown in Fig 11.12. After the first stage,

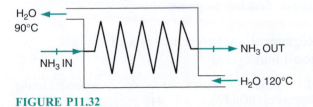

**FIGURE P11.32**

25% of the steam is extracted at 1.4 MPa for a process application and returned at 1 MPa, 90°C, to the feedwater line. The remainder of the steam continues through the low-pressure turbine stage, which exhausts to the condenser at 10 kPa. One pump brings the feedwater to 1 MPa, and a second pump brings it to 10 MPa, Assume all components are ideal. If the process application requires 5 MW of power, how much power can then be cogenerated by the turbine?

**11.41** A 10-kg/s steady supply of saturated-vapor steam at 500 kPa is required for drying a wood pulp slurry in a paper mill (see Fig. P11.41). It is decided to supply this steam by cogeneration; that is, the steam supply will be the exhaust from a steam turbine. Water at 20°C and 100 kPa is pumped to a pressure of 5 MPa and then fed to a steam generator with an exit at 400°C. What is the additional heat-transfer rate to the steam generator beyond what would have been required to produce only the desired steam supply? What is the difference in net power?

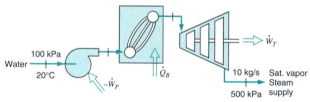

**FIGURE P11.41**

**11.42** In a cogenerating steam power plant the turbine receives steam from a high-pressure steam drum and a low-pressure steam drum, as shown in Fig. P11.42. The condenser is made as two closed heat exchangers used to heat water running in a separate loop for district heating.

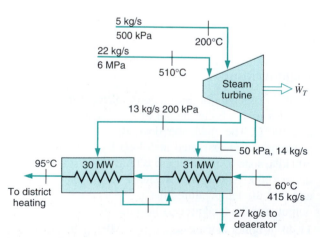

**FIGURE P11.42**

The high-temperature heater adds 30 MW, and the low-temperature heater adds 31 MW to the district heating water flow. Find the power cogenerated by the turbine and the temperature in the return line to the deaerator.

# Brayton cycles, gas turbines

**11.43** Why is the back work ratio in the Brayton cycle much higher than in the Rankine cycle?

**11.44** A Brayton cycle produces 14 MW with an inlet state of 17°C, 100 kPa, and a compression ratio of 16:1. The heat added in the combustion is 960 kJ/kg. What is the highest temperature and the mass flow rate of air, assuming cold air properties?

**11.45** A Brayton cycle has inlet at 290 K, 90 kPa and the combustion adds 1000 kJ/kg. How high can the compression ratio be so the highest temperature is below 1700 K? Use cold air properties to solve.

**11.46** Consider an ideal air-standard Brayton cycle in which the air into the compressor is at 100 kPa and 20°C, and the pressure ratio across the compressor is 12:1. The maximum temperature in the cycle is 1100°C, and the air-flow rate is 10 kg/s. Assume constant specific heat for the air (from Table A.5). Determine the compressor work, the turbine work, and the thermal efficiency of the cycle.

**11.47** A large stationary Brayton-cycle gas turbine power plant delivers a power output of 100 MW to an electric generator. The minimum temperature in the cycle is 300 K, and the maximum temperature is 1600 K. The minimum pressure in the cycle is 100 kPa, and the compressor pressure ratio is 14 to 1. Calculate the power output of the turbine. What fraction of the turbine output is required to drive the compressor? What is the thermal efficiency of the cycle?

**11.48** Repeat Problem 11.46, but assume variable specific heat for the air (Table A.7).

**11.49** A Brayton-cycle inlet is at 300 K and 100 kPa, and the combustion adds 670 kJ/kg. The maximum temperature is 1200 K due to material considerations. What is the maximum allowed compression ratio? For this ratio, calculate the net work and cycle efficiency assuming variable specific heat for the air (Table A.7).

**11.50** Do Problem 11.44 with properties from Table A.7.1 instead of cold air properties.

# Regenerators, intercoolers and non-ideal cycles

**11.51** Would it be beneficial to add a regenerator to the Brayton cycle in Problem 11.44?

**11.52** An ideal regenerator is incorporated into the ideal air-standard Brayton cycle of Problem 11.46. Find the thermal efficiency of the cycle with this modification.

**11.53** Consider an ideal gas-turbine cycle with a pressure ratio across the compressor of 12 to 1. The compressor inlet is at 300 K and 100 kPa, and the cycle has a maximum temperature of 1600 K. An ideal regenerator is also incorporated in the cycle. Find the thermal efficiency of the cycle using cold air (298 K) properties. If the compression ratio is raised, $T_4 - T_2$ goes down. At what compression ratio is $T_2 = T_4$ so the regenerator cannot be used?

**11.54** Repeat Problem 11.47, but include a regenerator with 75% efficiency in the cycle.

**11.55** A two-stage air compressor has an intercooler between the two stages, as shown in Fig. P11.55. The inlet state is 100 kPa, 290 K, and the final exit pressure is 1.6 MPa. Assume that the constant-pressure intercooler cools the air to the inlet temperature, $T_3 = T_1$. It can be shown that the optimal pressure is $P_2 = (P_1P_4)^{1/2}$, for minimum total compressor work. Find the specific compressor works and the intercooler heat transfer for the optimal $P_2$.

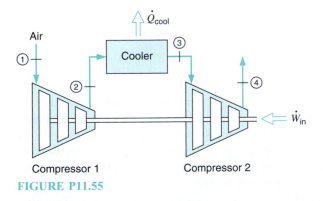

**FIGURE P11.55**

**11.56** Suppose a compressor in Problem 11.44 has an intercooler that cools the air to 330 K operating at 500 kPa followed by the second stage of compression to 1600 kPa. Find the specific heat transfer in the intercooler and the total compression work required.

**11.57** A two-stage compressor in a gas turbine brings atmospheric air at 100 kPa and 17°C to 500 kPa, and then cools it in an intercooler to 27°C at constant $P$. The second stage brings the air to 1000 kPa. Assume that both stages are adiabatic and reversible. Find the combined specific work to the compressor stages. Compare that to the specific work for the case of no intercooler (i.e., one compressor from 100 to 1000 kPa).

**11.58** Repeat Problem 11.47, but assume that the compressor has an isentropic efficiency of 85% and the turbine an isentropic efficiency of 88%.

**11.59** A gas turbine has two stages of compression, with an intercooler between the stages (see Fig. P11.55). Air enters the first stage at 100 kPa and 300 K. The pressure ratio across each compressor stage is 5 to 1, and each stage has an isentropic efficiency of 82%. Air exits the intercooler at 330 K. Calculate the exit temperature from each compressor stage and the total specific work required.

# Jet engine cycles

**11.60** The Brayton cycle in Problem 11.47 is changed to be a jet engine. Find the exit velocity using cold air properties.

**11.61** Consider an ideal air-standard cycle for a gas-turbine, jet propulsion unit, such as that shown in Fig. 11.18. The pressure and temperature entering the compressor are 90 kPa and 290 K. The pressure ratio across the compressor is 14 to 1, and the turbine inlet temperature is 1500 K. When the air leaves the turbine, it enters the nozzle and expands to 90 kPa. Determine the velocity of the air leaving the nozzle.

**11.62** The turbine section in a jet engine shown in Fig. P11. 62 receives gas (assume air) at 1200 K, 800 kPa followed by a nozzle open to the atmosphere at 80 kPa and all the turbine work drives the compressor. Find the turbine exit pressure so the nozzle has an exit velocity of 800 m/s. Hint: take the CV around both turbine and nozzle.

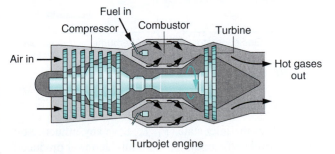

Turbojet engine

**FIGURE P11.62**

**11.63** The turbine in a jet engine receives air at 1250 K and 1.5 MPa. It exhausts to a nozzle at 250 kPa, which in turn exhausts to the atmosphere at 100 kPa. The isentropic efficiency of the turbine is 85%, and the nozzle efficiency is 95%. Find the nozzle inlet temperature and the nozzle exit velocity. Assume negligible kinetic energy out of the turbine and cold air properties.

**11.64** Consider an air-standard jet engine cycle operating in a 280-K, 100-kPa environment. The compressor requires a shaft power input of 4000 kW. Air enters the turbine state 3 at 1600 K and 2 MPa, at the rate of 9 kg/s, and the isentropic efficiency of the turbine is 85%. Determine the pressure and temperature entering the nozzle using cold air properties.

**11.65** A jet aircraft is flying at an altitude of 4900 m, where the ambient pressure is approximately 55 kPa and the ambient temperature is −18°C. The velocity of the aircraft is 280 m/s, the pressure ratio across the compressor is 14:1, and the cycle maximum temperature is 1450 K. Assume that the inlet flow goes through a diffuser to zero relative velocity at state a, Fig. 11.18. Find the temperature and pressure at state a.

**11.66** An afterburner in a jet engine, as shown in Fig. P11.66 adds fuel after the turbine, thus raising the pressure and temperature via the energy of combustion. Assume a standard condition of 800 K and 250 kPa after the turbine into the nozzle that exhausts at 95 kPa. Assume the afterburner adds 450 kJ/kg to that state with a rise in pressure for the same specific volume, and neglect any upstream effects on the turbine. Find the nozzle exit velocity before and after the afterburner is turned on.

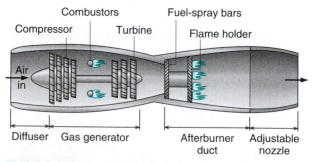

**FIGURE P11.66**

## Otto cycles, gasoline engines

**11.67** Does the inlet state $(P_1, T_1)$ have any influence on the Otto cycle efficiency? How about the power produced by a real car engine?

**11.68** Air flows into a gasoline engine at 95 kPa and 300 K. The air is then compressed with a volumetric compression ratio of 8:1. The combustion process releases 1300 kJ/kg of energy as the fuel burns. Find the temperature and pressure after combustion using cold air properties.

**11.69** A gasoline engine has a volumetric compression ratio of 9. The state before compression is 290 K, 90 kPa, and the peak cycle temperature is 1800 K. Find the pressure after expansion, the cycle net work, and the cycle efficiency using properties from Table A.5.

**11.70** A 4 stroke gasoline engine with a 3.0 L displacement and a compression ratio of 10:1 as shown in Fig. P11.70 runs at 1500 RPM with the air at 75 kPa, 280 K after the intake is done. The mean effective pressure is 1000 kPa. Find the power produced, the engine efficiency, and the specific energy the combustion adds.

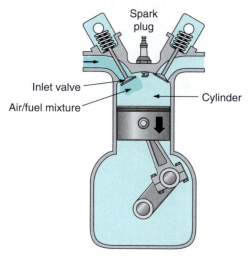

**FIGURE P11.70**

**11.71** A 4 stroke gasoline engine of displacement 4.2 L has inlet state of 280 K and 85 kPa. The maximum temperature is 2000 K and the highest pressure is 5 MPa and it is running at 2000 RPM. Using cold air properties, find the compression ratio, the mean effective pressure, and the total power output.

**11.72** A gasoline engine has a volumetric compression ratio of 10 and before compression has air at 290 K, 85 kPa, in the cylinder. The combustion peak pressure is 6000 kPa. Assume cold air properties. What is the highest temperature in the cycle? Find the temperature at the beginning of the exhaust (heat rejection) and the overall cycle efficiency.

**11.73** A four-stroke gasoline engine has a compression ratio of 10:1 with 4 cylinders of total displacement 2.3 L shown in Fig. P11.73. The inlet state is 280 K, 70 kPa, and the engine is running at 2100 RPM with the fuel adding 1800 kJ/kg in the combustion process. What is the net work in the cycle, and how much power is produced?

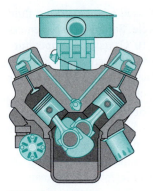

**FIGURE P11.73**

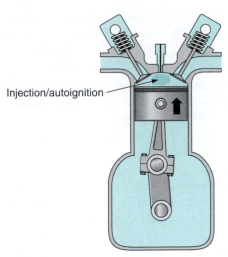

Injection/autoignition

**FIGURE P11.79**

**11.74** A gasoline engine takes air in at 290 K and 90 kPa and then compresses it. The combustion adds 1000 kJ/kg to the air, after which the temperature is 2050 K. Use the cold air properties (i.e., constant heat capacities at 300 K) and find the compression ratio, the compression specific work, and the highest pressure in the cycle.

**11.75** Answer the same three questions for the previous problem, but use variable heat capacities (use Table A. 7).

**11.76** When methanol is used as a fuel, it provides a slightly smaller energy release compared to gasoline. Assume an engine with compression ratio of 10:1 and an inlet at 10°C and 90 kPa with 1700 kJ/kg added in the combustion process. Find the maximum temperature and pressure, the cycle efficiency, and the mean effective pressure.

**11.77** It is found experimentally that the power stroke expansion in an internal combustion engine can be approximated, with a polytropic process with a value of the polytropic exponent $n$ somewhat larger than the specific heat ratio $k$. Repeat Problem 11.71, but assume that the expansion process is reversible and polytropic (instead of the isentropic expansion in the Otto cycle) with $n$ equal to 1.50.

## Diesel cycles

**11.78** A diesel engine has a state before compression of 95 kPa, 290 K, a peak pressure of 6000 kPa, and a maximum temperature of 2400 K. Find the volumetric compression ratio and the thermal efficiency using Table A.5.

**11.79** A diesel engine shown in Fig.P11.79 has a bore of 0.1 m, a stroke of 0.11 m, and a compression ratio of 19:1 running at 2000 RPM (revolutions per minute). Each cycle takes two revolutions and has a mean effective pressure of 1400 kPa. With a total of 6 cylinders, find the engine power in kW and horsepower, hp.

**11.80** A diesel engine has a compression ratio of 20:1 with an inlet state of 290 K, 80 kPa. The combustion adds 1200 kJ/kg to the charge. Find the highest $P$ and $T$ in the cycle and the mean effective pressure.

**11.81** A diesel engine has a compression ratio of 20:1 with an inlet of 95 kPa and 290 K, state 1, with volume 0.5 L. The maximum cycle temperature is 1800 K. Find the maximum pressure, the net specific work, and the thermal efficiency.

**11.82** At the beginning of compression in a diesel cycle $T = 300$ K and $P = 200$ kPa; after combustion (heat addition) is complete $T = 1500$ K and $P = 7.0$ MPa. Find the compression ratio, the thermal efficiency, and the mean effective pressure.

**11.83** Do Problem 11.78 but use the properties from Table A.7 and not the cold air properties.

**11.84** A diesel engine has air before compression at 280 K and 85 kPa. The highest temperature is 2200 K, and the highest pressure is 6 MPa. Find the volumetric compression ratio and the mean effective pressure using cold air properties at 300 K.

## Refrigeration cycles

**11.85** A refrigerator with R-12 as the working fluid has a minimum temperature of $-10°C$ and a maximum pressure of 1 MPa. Assume an ideal refrigeration cycle as in Fig. 11.24. Find the specific heat transfer from the cold space and that to the hot space, and determine the coefficient of performance.

**11.86** An old cooling system runs with R-12 with a high pressure of 1500 kPa and a low pressure of 200 kPa. What is the coefficient of performance as a heat pump?

**11.87** Consider an ideal refrigeration cycle that has a condenser temperature of 45°C and an evaporator temperature of −15°C. Determine the coefficient of performance of this refrigerator for the working fluids R-12 and R-22.

**11.88** The environmentally safe refrigerant R-134a is one of the replacements for R-12 in refrigeration systems. Repeat Problem 11.87 using R-134a.

**11.89** A refrigerator in a meat warehouse must keep a low temperature of −15°C. It uses R-12 as the refrigerant, which must remove 5 kW from the cold space. Assume that the outside temperature is 20°C. Find the flow rate of the R-12 needed, assuming a standard vapor compression refrigeration cycle with a condenser at 20°C.

**11.90** A steady flow of R-22 in an ideal refrigeration cycle as saturated vapor at −20°C goes into the adiabatic compressor that brings it to 1000 kPa. Find the compressor specific work and the cycle coefficient of performance.

**11.91** A small heat pump unit is used to heat water for a hot-water supply. Assume that the unit uses R-22 and operates on the ideal refrigeration cycle. The evaporator temperature is 15°C, and the condenser temperature is 60°C. If the amount of hot water needed is 0.1 kg/s, determine the amount of energy saved by using the heat pump instead of directly heating the water from 15 to 60°C.

**11.92** A heat pump for heat upgrade uses ammonia with a low temperature of 25°C and a high pressure of 5000 kPa. If it receives 1 MW of shaft work, what is the rate of heat transfer at the high temperature?

**11.93** Reconsider the heat pump in Problem 11.92. Assume the compressor is split into two —first compress to 2000 kPa, then take heat transfer out at constant $P$ to reach saturated vapor, then compress to the 5000 kPa. Find the two rates of heat transfer, at 2000 kPa and at 5000 kPa, for a total of 1 MW shaft work input.

**11.94** Consider an ideal heat pump that has a condenser temperature of 50°C and an evaporator temperature of 0°C. Determine the coefficient of performance of this heat pump for the working fluids R-12, R-22, and ammonia.

**11.95** The air conditioner in a car uses R-134a, and the compressor power input is 1.5 kW, bringing the R-134a from 201.7 kPa to 1200 kPa by compression. The cold space is a heat exchanger that cools 30°C atmospheric air from the outside down to 10°C and blows it into the car. What is the mass flow rate of the R-134a, and what is the low-temperature heat-transfer rate? What is the mass flow rate of air at 10°C?

**11.96** The refrigerant R-22 is used as the working fluid in a conventional heat pump cycle. Saturated vapor enters the compressor of this unit at 10°C; its exit temperature from the compressor is measured and found to be 85°C. If the compressor exit is 2 MPa, what is the compressor isentropic efficiency and the cycle COP?

**11.97** A refrigerator has a steady flow of R-22 as saturated vapor at −20°C into the adiabatic compressor that brings it to 1000 kPa. After the compressor, the temperature is measured to be 60°C. Find the actual compressor work and the actual cycle coefficient of performance.

**11.98** A refrigerator in a laboratory uses R-22 as the working substance. The high pressure is 1200 kPa, the low pressure is 201 kPa, and the compressor is reversible. It should remove 500 W from a specimen currently at −20°C (not equal to $T_L$ in the cycle) that is inside the refrigerated space. Find the cycle COP and the electrical power required.

**11.99** Consider the previous problem with a laboratory at 20°C and find the three rates of entropy generation in the whole setup and where they occur.

# Air-standard refrigeration cycles

**11.100** The formula for the coefficient of performance when we use cold air properties is not given in the text. Derive the expression for COP as function of the compression ratio similar to how the Brayton-cycle efficiency was found.

**11.101** A standard air refrigeration cycle has −10°C, 100 kPa into the compressor, and the ambient cools the air down to 35°C at 400 kPa. Find the lowest temperature in the cycle, the low $T$ specific heat transfer, and the specific compressor work.

**11.102** A heat exchanger is incorporated into an ideal air-standard refrigeration cycle, as shown in Fig. P11.102. It

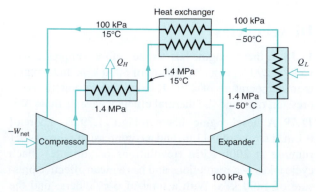

**FIGURE P11.102**

may be assumed that both the compression and the expansion are reversible adiabatic processes in this ideal case. Determine the coefficient of performance for the cycle.

# Combined cycles

**11.103** Why would one consider a combined cycle system for a power plant? For a heat pump or refrigerator?

**11.104** A refrigerator using R-22 is powered by a small natural gas-fired heat engine with a thermal efficiency of 25%, as shown in Fig. P11.104. The R-22 condenses at 40°C, it evaporates at −20°C, and the cycle is standard. Find the two specific heat transfers in the refrigeration cycle. What is the overall coefficient of performance as $Q_L/Q_1$?

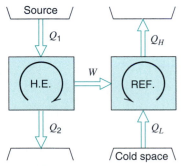

**FIGURE P11.104**

**11.105** A binary system power plant uses mercury for the high-temperature cycle and water for the low-temperature cycle, as shown in Fig. 11.29. The temperatures and pressures are shown in the corresponding $T–s$ diagram. The maximum temperature in the steam cycle is where the steam leaves the superheater at point 4 where it is 500°C. Determine the ratio of the mass flow rate of mercury to the mass flow rate of water in the heat exchanger that condenses mercury and boils the water and the thermal efficiency of this ideal cycle.

The following saturation properties for mercury are known:

| $P$, MPa | $T_g$, °C | $h_f$, kJ/kg | $h_g$, kJ/kg | $s_f$,kJ/ kg-K | $s_g$, kJ/ kg-K |
|---|---|---|---|---|---|
| 0.04 | 309 | 42.21 | 335.64 | 0.1034 | 0.6073 |
| 1.60 | 562 | 75.37 | 364.04 | 0.1498 | 0.4954 |

**11.106** A cascade system is composed of two ideal refrigeration cycles, as shown in Fig. 11.31. The high-temperature cycle uses R-22. Saturated liquid leaves the con-

denser at 40°C, and saturated vapor leaves the heat exchanger at −20°C. The low-temperature cycle uses a different refrigerant, R-23. Saturated vapor leaves the evaporator at −80°C with $h = 330$ kJ/kg, and saturated liquid leaves the heat exchanger at −10°C with $h = 185$ kJ/kg. R-23 out of the compressor has $h = 405$ kJ/kg. Calculate the ratio of the mass flow rates through the two cycles an the COP of the total system.

**11.107** Consider an ideal dual-loop heat-powered refrigeration cycle using R-12 as the working fluid, as shown in Fig. P11.107. Saturated vapor at 105°C leaves the boiler and expands in the turbine to the condenser pressure. Saturated vapor at −15°C leaves the evaporator and is compressed to the condenser pressure. The ratio of the flows through the two loops is such that the turbine produces just enough power to drive the compressor. The two exiting streams mix together and enter the condenser. Saturated liquid leaving the condenser at 45°C is then separated into two streams in the necessary proportions. Determine the ratio of mass flow rate through the power loop to that through the refrigeration loop. Find also the performance of the cycle, in terms of the ratio $Q_L/Q_H$.

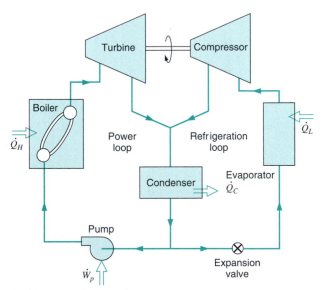

**FIGURE P11.107**

# Availability or exergy concepts

**11.108** Since any heat transfer is driven by a temperature difference, how does that affect all the real cycles relative to the ideal cycles?

**11.109** If we neglect the external irreversibilities due to the heat transfers over finite temperature differences in a power plant, how would you define its second law efficiency?

**11.110** How would you define a second law efficiency for a heat pump?

**11.111** Find the flows and fluxes of exergy in the condenser of Problem 11.15. Use those to determine the second law efficiency.

**11.112** Find the availability of the water at all four states in the Rankine cycle described in Problem 11.4. Assume that the high-temperature source is 500°C and the low-temperature reservoir is at 25°C. Determine the flow of availability in or out of the reservoirs per kilogram of steam flowing in the cycle. What is the overall cycle second-law efficiency?

**11.113** Find the flows of exergy into and out of the feed-water heater in Problem 11.23.

**11.114** Consider the Brayton cycle in Problem 11.44. Find all the flows and fluxes of exergy and find the overall cycle second-law efficiency. Assume the heat transfers are internally reversible processes, and we then neglect any external irreversibility.

**11.115** The power plant using ammonia in Problem 11.32 has a flow of liquid water at 120°C, 300 kPa as a heat source, and the water leaves the heat exchanger at 90°C. Find the second law efficiency of this heat exchanger.

**11.116** What is the second law efficiency of the heat pump in Problem 11.92?

# Chapter 12

# Chemical Reactions

## TOPICS DISCUSSED

Many thermodynamic problems involve chemical reactions. Among the most familiar of these is the combustion of hydrocarbon fuels, for this process is utilized in most of our power-generating devices. However, we can all think of a host of other processes involving chemical reactions, including those that occur in the human body.

This chapter considers a first- and second-law analysis of systems undergoing a chemical reaction. In many respects, this chapter is simply an extension of our previous consideration of the first and second laws. However, a number of new terms are introduced, and it will also be necessary to introduce the third law of thermodynamics.

In this chapter the combustion process is considered in detail. There are two reasons for this emphasis. First, the combustion process is of great significance in many problems and devices with which the engineer is concerned. Second, the combustion process provides an excellent vehicle for teaching the basic principles of the thermodynamics of chemical reactions. The student should keep both of these objectives in mind as the study of this chapter progresses.

Two other topics, fuel cells and chemical equilibrium (as applied to the dissociation of combustion products), are also considered in this chapter.

## 12.1 FUELS

A thermodynamics textbook is not the place for a detailed treatment of fuels. However, some knowledge of them is a prerequisite to a consideration of combustion, and this section is therefore devoted to a brief discussion of some of the hydrocarbon fuels. Most fuels fall into one of three categories—coal, liquid hydrocarbons, or gaseous hydrocarbons.

Coal consists of the remains of vegetation deposits of past geologic ages, after subjection of biochemical actions, high pressure, temperature, and submersion. The characteristics of coal vary considerably with location, and even within a given mine there is some variation in composition.

The analysis of a sample of coal is given on one of two bases: the proximate analysis specifies, on a mass basis, the relative amounts of moisture, volatile matter, fixed carbon, and ash; the ultimate analysis specifies, on a mass basis, the relative amounts of carbon, sulfur, hydrogen, nitrogen, oxygen, and ash. The ultimate analysis may be given on an "as-received" basis or on a dry basis. In the latter case, the ultimate analysis does not include the moisture as determined by the proximate analysis.

A number of other properties of coal are important in evaluating a coal for a given use. Some of these are the fusibility of the ash, the grindability or ease of pulverization, the weathering characteristics, and size.

Most liquid and gaseous hydrocarbon fuels are a mixture of many different hydrocarbons. For example, gasoline consists primarily of a mixture of about 40 hydrocarbons, with many others present in very small quantities. In discussing hydrocarbon fuels, therefore, brief consideration should be given to the most important families of hydrocarbons, which are summarized in Table 12.1.

Three concepts should be defined. The first pertains to the structure of the molecule. The important types are the ring and chain structures; the difference between the two is illustrated in Fig. 12.1. The same figure illustrates the definition of saturated and unsaturated hydrocarbons. An unsaturated hydrocarbon has two or more adjacent carbon atoms joined by a double or triple bond, whereas in a saturated hydrocarbon all the carbon

TABLE 12.1
*Characteristics of Some of the Hydrocarbon Families*

| Family | Formula | Structure | Saturated |
|---|---|---|---|
| Paraffin | $C_nH_{2n+2}$ | Chain | Yes |
| Olefin | $C_nH_{2n}$ | Chain | No |
| Diolefin | $C_nH_{2n-2}$ | Chain | No |
| Naphthene | $C_nH_{2n}$ | Ring | Yes |
| Aromatic | | | |
| Benzene | $C_nH_{2n-6}$ | Ring | No |
| Naphthalene | $C_nH_{2n-12}$ | Ring | No |

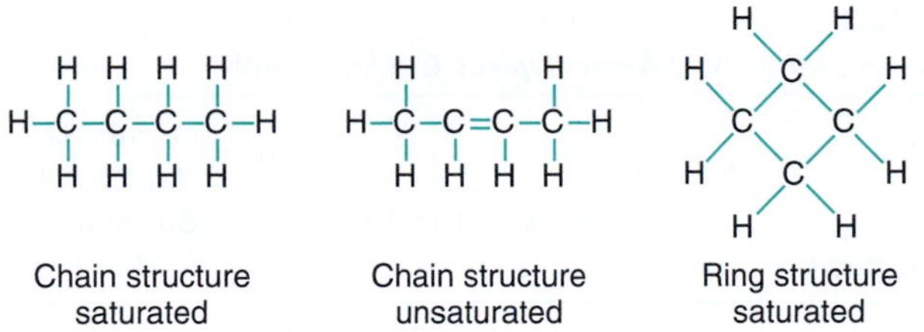

FIGURE 12.1 Molecular structure of some hydrocarbon fuels.

Chain structure saturated

Chain structure unsaturated

Ring structure saturated

atoms are joined by a single bond. The third term to be defined is an isomer. Two hydrocarbons with the same number of carbon and hydrogen atoms and different structures are called isomers. Thus, there are several different octanes ($C_8H_{18}$), each having 8 carbon atoms and 18 hydrogen atoms, but each with a different structure.

The various hydrocarbon families are identified by a common suffix. The compounds comprising the paraffin family all end in "-ane" (as propane and octane). Similarly, the compounds comprising the olefin family end in "-ylene" or "-ene" (as propene and octene), and the diolefin family ends in "-diene" (as butadiene). The naphthene family has the same chemical formula as the olefin family but has a ring rather than chain structure. The hydrocarbons in the naphthene family are named by adding the prefix "cyclo-" (as cyclopentane).

The aromatic family includes the benzene series ($C_nH_{2n-6}$) and the naphthalene series ($C_nH_{2n-12}$). The benzene series has a ring structure and is unsaturated.

Alcohols are sometimes used as a fuel in internal combustion engines. Characteristically, in the alcohol family, one of the hydrogen atoms is replaced by an OH radical. Thus, methyl alcohol, also called methanol, is $CH_3OH$.

Most liquid hydrocarbon fuels are mixtures of hydrocarbons that are derived from crude oil through distillation and cracking processes. Thus, from a given crude oil, a variety of different fuels can be produced, some of the common ones being gasoline, kerosene, diesel fuel, and fuel oil. Within each of these classifications there is a wide variety of grades, and each is made up of a large number of different hydrocarbons. The important distinction between these fuels is the distillation curve, Fig. 12.2. The distillation curve is obtained by slowly heating a sample of fuel so that it vaporizes. The vapor

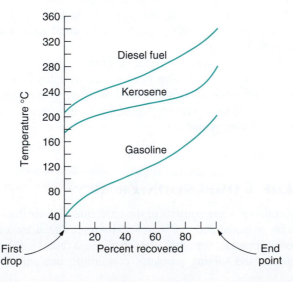

FIGURE 12.2 Typical distillation curves of some hydrocarbon fuels.

**TABLE 12.2**
*Volumetric Analyses of Some Typical Gaseous Fuels*

| Constituent | Various Natural Gases | | | | Producer Gas from Bituminous Coal | Carbureted Water Gas | Coke- Oven Gas |
|---|---|---|---|---|---|---|---|
| | A | B | C | D | | | |
| Methane | 93.9 | 60.1 | 67.4 | 54.3 | 3.0 | 10.2 | 32.1 |
| Ethane | 3.6 | 14.8 | 16.8 | 16.3 | | | |
| Propane | 1.2 | 13.4 | 15.8 | 16.2 | | | |
| Butanes plus[a] | 1.3 | 4.2 | | 7.4 | | | |
| Ethene | | | | | | 6.1 | 3.5 |
| Benzene | | | | | | 2.8 | 0.5 |
| Hydrogen | | | | | 14.0 | 40.5 | 46.5 |
| Nitrogen | | 7.5 | | 5.8 | 50.9 | 2.9 | 8.1 |
| Oxygen | | | | | 0.6 | 0.5 | 0.8 |
| Carbon monoxide | | | | | 27.0 | 34.0 | 6.3 |
| Carbon dioxide | | | | | 4.5 | 3.0 | 2.2 |

[a]This includes butane and all heavier hydrocarbons

is then condensed and the amount measured. The more volatile hydrocarbons are vaporized first, and thus the temperature of the nonvaporized fraction increases during the process. The distillation curve, which is a plot of the temperature of the nonvaporized fraction versus the amount of vapor condensed, is an indication of the volatility of the fuel.

For the combustion of liquid fuels, it is convenient to express the composition in terms of a single hydrocarbon, even though it is a mixture of many hydrocarbons. Thus, gasoline is usually considered to be octane, $C_8H_{18}$, and diesel fuel is considered to be dodecane, $C_{12}H_{26}$. The composition of a hydrocarbon fuel may also be given in terms of percentage of carbon and hydrogen.

The two primary sources of gaseous hydrocarbon fuels are natural gas wells and certain chemical manufacturing processes. Table 12.2 gives the composition of a number of gaseous fuels. The major constituent of natural gas is methane, which distinguishes it from manufactured gas.

At the present time, considerable effort is being devoted to develop more economical processes for producing gaseous and also liquid hydrocarbon fuels from coal, oil shale, and tar sands deposits. Several alternative techniques have been demonstrated to be feasible, and these resources promise to provide an increasing proportion of our fuel supply in future years.

## 12.2 THE COMBUSTION PROCESS

The combustion process consists of the oxidation of constituents in the fuel that are capable of being oxidized and can therefore be represented by a chemical equation. During a combustion process, the mass of each element remains the same. Thus, writing chemical equations and solving problems concerning quantities of the various constituents basically involve the conservation of mass of each element. This chapter presents a brief review of this subject, particularly as it applies to the combustion process.

Consider first the reaction of carbon with oxygen.

$$\text{Reactants} \qquad \text{Products}$$
$$C + O_2 \;\rightarrow\; CO_2$$

This equation states that 1 kmol of carbon reacts with 1 kmol of oxygen to form 1 kmol of carbon dioxide. This also means that 12 kg of carbon react with 32 kg of oxygen to form 44 kg of carbon dioxide. All the initial substances that undergo the combustion process are called the reactants, and the substances that result from the combustion process are called the products.

When a hydrocarbon fuel is burned, both the carbon and the hydrogen are oxidized. Consider the combustion of methane as an example.

$$CH_4 + 2O_2 \rightarrow CO_2 + 2H_2O \qquad (12.1)$$

Here the products of combustion include both carbon dioxide and water. The water may be in the vapor, liquid, or solid phases, depending on the temperature and pressure of the products of combustion.

In the combustion process, many intermediate products are formed during the chemical reaction. In this book we are concerned with the initial and final products and not with the intermediate products, but this aspect is very important in a detailed consideration of combustion.

In most combustion processes, the oxygen is supplied as air rather than as pure oxygen. The composition of air on a molal basis is approximately 21% oxygen, 78% nitrogen, and 1% argon. We assume that the nitrogen and the argon do not undergo chemical reaction (except for dissociation, which will be considered later). They do leave at the same temperature as the other products, however, and therefore undergo a change of state if the products are at a temperature other than the original air temperature. At the high temperatures achieved in internal-combustion engines, there is actually some reaction between the nitrogen and oxygen, and this gives rise to the air pollution problem associated with the oxides of nitrogen in the engine exhaust.

In combustion calculations concerning air, the argon is usually neglected, and the air is considered to be composed of 21% oxygen and 79% nitrogen by volume. When this assumption is made, the nitrogen is sometimes referred to as atmospheric nitrogen. Atmospheric nitrogen has a molecular weight of 28.16 (which takes the argon into account) as compared to 28.013 for pure nitrogen. This distinction will not be made in this text, and we will consider the 79% nitrogen to be pure nitrogen.

The assumption that air is 21.0% oxygen and 79.0% nitrogen by volume leads to the conclusion that for each mole of oxygen, $79.0/21.0 = 3.76$ moles of nitrogen are involved. Therefore, when the oxygen for the combustion of methane is supplied as air, the reaction can be written

$$CH_4 + 2O_2 + 2(3.76)N_2 \rightarrow CO_2 + 2H_2O + 7.52N_2 \qquad (12.2)$$

The minimum amount of air that supplies sufficient oxygen for the complete combustion of all the carbon, hydrogen, and any other elements in the fuel that may oxidize is called the theoretical air. When complete combustion is achieved with theoretical air, the products contain no oxygen. A general combustion reaction with a hydrocarbon fuel and air is thus written

$$C_xH_y + \nu_{O_2}(O_2 + 3.76\,N_2) \rightarrow \nu_{CO_2}CO_2 + \nu_{H_2O}H_2O + \nu_{N_2}N_2 \qquad (12.3)$$

with the coefficients to the substances called stoichiometric coefficients. The balance of atoms yields the theoretical amount of air as

$$C: \quad \nu_{CO_2} = x$$

$$H: \quad 2\nu_{H_2O} = y$$

$$N_2: \quad \nu_{N_2} = 3.76 \times \nu_{O_2}$$

$$O_2: \quad \nu_{O_2} = \nu_{CO_2} + \nu_{H_2O}/2 = x + y/4$$

and the total number of moles of air for 1 mole of fuel becomes

$$n_{air} = \nu_{O_2} \times 4.76 = 4.76(x + y/4)$$

This amount of air is equal to 100% theoretical air. In practice, complete combustion is not likely to be achieved unless the amount of air supplied is somewhat greater than the theoretical amount. Two important parameters often used to express the ratio of fuel and air are the air–fuel ratio (designed AF) and its reciprocal, the fuel–air ratio (designated FA). These ratios are usually expressed on a mass basis, but a mole basis is used at times.

$$AF_{mass} = \frac{m_{air}}{m_{fuel}} \tag{12.4}$$

$$AF_{mole} = \frac{n_{air}}{n_{fuel}} \tag{12.5}$$

They are related through the molecular weights as

$$AF_{mass} = \frac{m_{air}}{m_{fuel}} = \frac{n_{air} M_{air}}{n_{fuel} M_{fuel}} = AF_{mole} \frac{M_{air}}{M_{fuel}}$$

and a subscript $s$ is used to indicate the ratio for 100% theoretical air, also called a stoichiometric mixture. In an actual combustion process, an amount of air is expressed as a fraction of the theoretical amount, called percent theoretical air. A similar ratio named the equivalence ratio equals the actual fuel–air ratio divided by the theoretical fuel–air ratio as

$$\Phi = FA/FA_s = AF_s/AF \tag{12.6}$$

the reciprocal of percent theoretical air. Since the percent theoretical air and the equivalence ratio are both ratios of the stoichiometric air–fuel ratio and the actual air–fuel ratio, the molecular weights cancel out and they are the same whether a mass basis or a mole basis is used.

Thus, 150% theoretical air means that the air actually supplied is 1.5 times the theoretical air and the equivalence ratio is $^2/_3$. The complete combustion of methane with 150% theoretical air is written

$$CH_4 + 1.5 \times 2(O_2 + 3.76 N_2) \rightarrow CO_2 + 2H_2O + O_2 + 11.28 N_2 \tag{12.7}$$

having balanced all the stoichiometric coefficients from conservation of all the atoms.

The amount of air actually supplied may also be expressed in terms of percent excess air. The excess air is the amount of air supplied over and above the theoretical air. Thus, 150% theoretical air is equivalent to 50% excess air. The terms "theoretical air," "excess air," and "equivalence ratio" are all in current usage and give an equivalent information about the reactant mixture of fuel and air.

When the amount of air supplied is less than the theoretical air required, the combustion is incomplete. If there is only a slight deficiency of air, the usual result is that some of the carbon unites with the oxygen to form carbon monoxide (CO) instead of carbon dioxide ($CO_2$). If the air supplied is considerably less than the theoretical air, there may also be some hydrocarbons in the products of combustion.

Even when some excess air is supplied, small amounts of carbon monoxide may be present, the exact amount depending on a number of factors including the mixing and turbulence during combustion. Thus, the combustion of methane with 110% theoretical air might be as follows:

$$CH_4 + 2(1.1)O_2 + 2(1.1)3.76N_2 \rightarrow$$
$$+0.95\,CO_2 + 0.05\,CO + 2H_2O + 0.225\,O_2 + 8.27\,N_2 \qquad (12.8)$$

The material covered so far in this section is illustrated by the following examples.

---

**EXAMPLE 12.1**

Calculate the theoretical air–fuel ratio for the combustion of octane, $C_8H_{18}$.

**Solution**

The combustion equation is

$$C_8H_{18} + 12.5\,O_2 + 12.5(3.76)N_2 \rightarrow 8CO_2 + 9H_2O + 47.0\,N_2$$

The air–fuel ratio on a mole basis is

$$AF = \frac{12.5 + 47.0}{1} = 59.5 \text{ kmol air/kmol fuel}$$

The theoretical air–fuel ratio on a mass basis is found by introducing the molecular weight of the air and fuel.

$$AF = \frac{59.5(28.97)}{114.2} = 15.0 \text{ kg air/kg fuel}$$

---

**EXAMPLE 12.2**

Determine the molal analysis of the products of combustion when octane, $C_8H_{18}$, is burned with 200% theoretical air, and determine the dew point of the products if the pressure is 0.1 MPa.

**Solution**

The equation for the combustion of octane with 200% theoretical air is

$$C_8H_{18} + 12.5(2)O_2 + 12.5(2)(3.76)N_2 \rightarrow 8CO_2 + 9H_2O + 12.5\,O_2 + 94.0\,N_2$$

Total kmols of product = $8 + 9 + 12.5 + 94.0 = 123.5$
Molal analysis of products:

$$CO_2 = 8/123.5 \quad = \quad 6.47\%$$
$$H_2O = 9/123.5 \quad = \quad 7.29$$
$$O_2 = 12.5/123.5 = \quad 10.12$$
$$N_2 = 94/123.5 \quad = \quad \underline{76.12}$$
$$100.00\%$$

The partial pressure of the water is $100(0.0729) = 7.29$ kPa.

The saturation temperature corresponding to this pressure is 39.7°C, which is also the dew-point temperature.

The water condensed from the products of combustion usually contains some dissolved gases and therefore may be quite corrosive. For this reason the products of combustion are often kept above the dew point until discharged to the atmosphere.

**EXAMPLE 12.3**

Producer gas from bituminous coal (see Table 12.2) is burned with 20% excess air. Calculate the air–fuel ratio on a volumetric basis and on a mass basis.

**Solution**

To calculate the theoretical air requirement, let us write the combustion equation for the combustible substances in 1 kmol of fuel.

$$0.14\,H_2 + 0.070\,O_2 \rightarrow 0.14\,H_2O$$

$$0.27\,CO + 0.135\,O_2 \rightarrow 0.27\,CO_2$$

$$0.03\,CH_4 + \underline{0.06\,O_2} \;\rightarrow 0.03\,CO_2 + 0.06\,H_2O$$

$$0.265 = \text{kmol oxygen required/kmol fuel}$$

$$\underline{-0.006} = \text{oxygen in fuel/kmol fuel}$$

$$0.259 = \text{kmol oxygen required from air/kmol fuel}$$

Therefore, the complete combustion equation for 1 kmol of fuel is

$$\overbrace{0.14\,H_2 + 0.27\,CO + 0.03\,CH_4 + 0.006\,O_2 + 0.509\,N_2 + 0.045\,CO_2}^{\text{fuel}}$$

$$\overbrace{+0.259\,O_2 + 0.259(3.76)\,N_2}^{\text{air}} \rightarrow 0.20\,H_2O + 0.345\,CO_2 + 1.482\,N_2$$

$$\left(\frac{\text{kmol air}}{\text{kmol fuel}}\right)_{\text{theo}} = \frac{0.259 \times 4.76}{1} = 1.233$$

If the air and fuel are at the same pressure and temperature, this also represents the ratio of the volume of air to the volume of fuel.

$$\text{For 20\% excess air, } \frac{\text{kmol air}}{\text{kmol fuel}} = 1.233 \times 1.200 = 1.48$$

The air–fuel ratio on a mass basis is

$$AF = \frac{1.48(28.97)}{0.14(2) + 0.27(28) + 0.03(16) + 0.006(32) + 0.509(28) + 0.045(44)}$$

$$= \frac{1.48(28.97)}{24.74} = 1.73 \text{ kg air/kg fuel}$$

An analysis of the products of combustion affords a very simple method for calculating the actual amount of air supplied in a combustion process. There are various experimental methods by which such an analysis can be made. Some yield results on a "dry" basis, that is, the fractional analysis of all the components, except for water vapor. Other experimental procedures give results that include the water vapor. In this presentation we are not concerned with the experimental devices and procedures, but rather with the use of such information in a thermodynamic analysis of the chemical reaction. The following examples illustrate how an analysis of the products can be used to determine the chemical reaction and the composition of the fuel.

The basic principle in using the analysis of the products of combustion to obtain the actual fuel–air ratio is conservation of the mass of each of the elements. Thus, in changing from reactants to products, we can make a carbon balance, hydrogen balance, oxygen balance, and nitrogen balance (plus any other elements that may be involved). Furthermore, we recognize that there is a definite ratio between the amounts of some of these elements. Thus, the ratio between the nitrogen and oxygen supplied in the air is fixed, as well as the ratio between carbon and hydrogen if the composition of a hydrocarbon fuel is known.

Often the combustion of a fuel uses atmospheric air as the oxidizer, in which case the reactants also hold some water vapor. Assuming we know the humidity ratio for the moist air, $\omega$, then we would like to know the composition of air per mole of oxygen as

$$1\,O_2 + 3.76\,N_2 + x\,H_2O$$

Since the humidity ratio is, $\omega = m_v/m_a$, the number of moles of water is

$$n_v = \frac{m_v}{M_v} = \frac{\omega m_a}{M_v} = \omega n_a \frac{M_a}{M_v}$$

and the number of moles of dry air per mole of oxygen is $(1 + 3.76)/1$, so we get

$$x = \frac{n_v}{n_{\text{oxygen}}} = \omega 4.76 \frac{M_a}{M_v} = 7.655\omega \qquad (12.9)$$

This amount of water is found in the products together with the water produced by the oxidation of the hydrogen in the fuel.

---

**EXAMPLE 12.4**

Methane ($CH_4$) is burned with atmospheric air. The analysis of the products on a dry basis is as follows:

| | |
|---|---|
| $CO_2$ | 10.00% |
| $O_2$ | 2.37 |
| CO | 0.53 |
| $N_2$ | 87.10 |
| | 100.00% |

Calculate the air–fuel ratio and the percent theoretical air, and determine the combustion equation.

**Solution**

The solution consists of writing the combustion equation for 100 kmol of dry products, introducing letter coefficients for the unknown quantities, and then solving for them.

From the analysis of the products, the following equation can be written, keeping in mind that this analysis is on a dry basis.

$$aCH_4 + bO_2 + cN_2 \rightarrow 10.0\,CO_2 + 0.53\,CO + 2.37\,O_2 + dH_2O + 87.1\,N_2$$

A balance for each of the elements will enable us to solve for all the unknown coefficients:

Nitrogen balance: $c = 87.1$

Since all the nitrogen comes from the air,

$$\frac{c}{b} = 3.76 \qquad b = \frac{87.1}{3.76} = 23.16$$

Carbon balance: $a = 10.00 + 0.53 = 10.53$

Hydrogen balance: $d = 2a = 21.06$

Oxygen balance: All the unknown coefficients have been solved for, and therefore the oxygen balance provides a check on the accuracy. Thus, $b$ can also be determined by an oxygen balance

$$b = 10.00 + \frac{0.53}{2} + 2.37 + \frac{21.06}{2} = 23.16$$

Substituting these values for $a$, $b$, $c$, and $d$, we have

$$10.53\,CH_4 + 23.16\,O_2 + 87.1\,N_2 \rightarrow$$

$$10.0\,CO_2 + 0.53\,CO + 2.37\,O_2 + 21.06\,H_2O + 87.1\,N_2$$

Dividing through by 10.53 yields the combustion equation per kmol of fuel.

$$CH_4 + 2.2\,O_2 + 8.27\,N_2 \rightarrow 0.95\,CO_2 + 0.05\,CO + 2H_2O + 0.225\,O_2 + 8.27\,N_2$$

The air–fuel ratio on a mole basis is

$$2.2 + 8.27 = 10.47 \text{ kmol air/kmol fuel}$$

The air–fuel ratio on a mass basis is found by introducing the molecular weights.

$$AF = \frac{10.47 \times 28.97}{16.0} = 18.97 \text{ kg air/kg fuel}$$

The theoretical air–fuel ratio is found by writing the combustion equation for theoretical air.

$$CH_4 + 2O_2 + 2(3.76)N_2 \rightarrow CO_2 + 2H_2O + 7.52\,N_2$$

$$AF_{theo} = \frac{(2 + 7.52)28.97}{16.0} = 17.23 \text{ kg air/kg fuel}$$

The percent theoretical air is $\dfrac{18.97}{17.23} = 110\%$

**EXAMPLE 12.5**    Coal from Jenkin, Kentucky, has the following ultimate analysis on a dry basis, percent by mass:

| Component | Percent by Mass |
|-----------|-----------------|
| Sulfur    | 0.6             |
| Hydrogen  | 5.7             |
| Carbon    | 79.2            |
| Oxygen    | 10.0            |
| Nitrogen  | 1.5             |
| Ash       | 3.0             |

This coal is to be burned with 30% excess air. Calculate the air–fuel ratio on a mass basis.

**Solution**

One approach to this problem is to write the combustion equation for each of the combustible elements per 100 kg of fuel. The molal composition per 100 kg of fuel is found first.

$$\text{kmol S/100 kg fuel} = \frac{0.6}{32} = 0.02$$

$$\text{kmol H}_2\text{/100 kg fuel} = \frac{5.7}{2} = 2.85$$

$$\text{kmol C/100 kg fuel} = \frac{79.2}{12} = 6.60$$

$$\text{kmol O}_2\text{/100 kg fuel} = \frac{10}{32} = 0.31$$

$$\text{kmol N}_2\text{/100 kg fuel} = \frac{1.5}{28} = 0.05$$

The combustion equations for the combustible elements are now written, which enables us to find the theoretical oxygen required.

$$0.02\,\text{S} + 0.02\,\text{O}_2 \rightarrow 0.02\,\text{SO}_2$$

$$2.85\,\text{H}_2 + 1.42\,\text{O}_2 \rightarrow 2.85\,\text{H}_2\text{O}$$

$$6.60\,\text{C} + 6.60\,\text{O}_2 \rightarrow 6.60\,\text{CO}_2$$

$$8.04 \text{ kmol O}_2 \text{ required/100 kg fuel}$$

$$-0.31 \text{ kmol O}_2 \text{ in fuel/100 kg fuel}$$

$$7.73 \text{ kmol O}_2 \text{ from air/100 kg fuel}$$

$$AF_{\text{theo}} = \frac{[7.73 + 7.73(3.76)]28.97}{100} = 10.63 \text{ kg air/kg fuel}$$

For 30% excess air the air–fuel ratio is

$$AF = 1.3 \times 10.63 = 13.82 \text{ kg air/kg fuel}$$

**a.** How many kmoles of air are needed to burn 1 kmol of carbon?

**b.** If I burn 1 kmol of hydrogen $H_2$ with 6 kmol air, what is $A/F$ ratio on a mole basis and what is the percent theoretical air?

**c.** For the 110% theoretical air in Eq. 12.8, what is the equivalence ratio? Is that mixture rich or lean?

**d.** In most cases, combustion products are exhausted above the dew point. Why?

## 12.3 Enthalpy of Formation

In the first eleven chapters of this book, the problems always concerned a fixed chemical composition and never a change of composition through a chemical reaction. Therefore, in dealing with a thermodynamic property, we used tables of thermodynamic properties for the given substance, and in each of these tables the thermodynamic properties were given relative to some arbitrary base. In the steam tables, for example, the internal energy of saturated liquid at 0.01°C is assumed to be zero. This procedure is quite adequate when there is no change in composition because we are concerned with the changes in the properties of a given substance. The properties at the condition of the reference state cancel out in the calculation. However, now that we are to include the possibility of a chemical reaction, it will become necessary to choose these reference state values on a common and consistent basis. We will use as our reference state a temperature of 25°C, a pressure of 0.1 MPa, and a hypothetical ideal-gas condition for those substances that are gases.

Consider the simple steady-state combustion process shown in Fig. 12.3. This idealized reaction involves the combustion of solid carbon with gaseous (ideal gas) oxygen, each of which enters the control volume at the reference state, 25°C and 0.1 MPa. The carbon dioxide (ideal gas) formed by the reaction leaves the chamber at the reference state, 25°C and 0.1 MPa. If the heat transfer could be accurately measured, it would be found to be $-393\,522$ kJ/kmol of carbon dioxide formed. The chemical reaction can be written

$$C + O_2 \rightarrow CO_2$$

Applying the first law to this process we have

$$Q_{\text{c.v.}} + H_R = H_P \tag{12.10}$$

where the subscripts $R$ and $P$ refer to the reactants and products, respectively. We will find it convenient to also write the first law for such a process in the form

$$Q_{\text{c.v.}} + \sum_R n_i \bar{h}_i = \sum_P n_e \bar{h}_e \tag{12.11}$$

where the summations refer, respectively, to all the reactants or all the products.

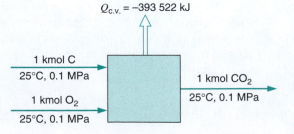

$Q_{\text{c.v.}} = -393\,522$ kJ

1 kmol C
25°C, 0.1 MPa

1 kmol $O_2$
25°C, 0.1 MPa

1 kmol $CO_2$
25°C, 0.1 MPa

**FIGURE 12.3**  Example of combustion process.

Thus, a measurement of the heat transfer would give us the difference between the enthalpy of the products and the reactants, where each is at the reference state condition. Suppose, however, that we assign the value of zero to the enthalpy of all the elements at the reference state. In this case, the enthalpy of the reactants is zero, and

$$Q_{c.v.} = H_P = -393\,522 \text{ kJ/kmol}$$

The enthalpy of (hypothetical) ideal-gas carbon dioxide at 25°C, 0.1 MPa pressure (with reference to this arbitrary base in which the enthalpy of the elements is chosen to be zero), is called the enthalpy of formation. We designate this with the symbol $\overline{h}_f$. Thus, for carbon dioxide

$$\overline{h}_f^\circ = -393\,522 \text{ kJ/kmol}$$

The enthalpy of carbon dioxide in any other sate, relative to this base in which the enthalpy of the elements is zero, would be found by adding the change of enthalpy between ideal gas at 25°C, 0.1 MPa, and the given state to the enthalpy of formation. That is, the enthalpy at any temperature and pressure, $h_{T,P}$, is

$$\overline{h}_{T,P} = (\overline{h}_f^\circ)_{298,0.1\text{MPa}} + (\Delta\overline{h})_{298,0.1\text{MPa}\to T,P} \tag{12.12}$$

where the term $(\Delta\overline{h})_{298,0.1\text{MPa}\to T,P}$ represents the difference in enthalpy between any given state and the enthalpy of ideal gas at 298.15 K, 0.1 MPa. For convenience we usually drop the subscripts in the examples that follow.

The procedure that we have demonstrated for carbon dioxide can be applied to any compound.

Table A.10 gives values of the enthalpy of formation for a number of substances in the units kJ/kmol (or Btu/lb mol in G.11).

Three further observations should be made in regard to enthalpy of formation.

1. We have demonstrated the concept of enthalpy of formation in terms of the measurement of the heat transfer in an idealized chemical reaction in which a compound is formed from the elements. Actually, the enthalpy of formation is usually found by the application of statistical thermodynamics, using observed spectroscopic data.

2. The justification of this procedure of arbitrarily assigning the value of zero to the enthalpy of the elements at 25°C, 0.1 MPa, rests on the fact that in the absence of nuclear reactions the mass of each element is conserved in a chemical reaction. No conflicts or ambiguities arise with this choice of reference state, and it proves to be very convenient in studying chemical reactions from a thermodynamic point of view.

3. In certain cases an element or compound can exist in more than one state at 25°C, 0.1 MPa. Carbon, for example, can be in the form of graphite or diamond. It is essential that the state to which a given value is related be clearly identified. Thus, in Table A.10, the enthalpy of formation of graphite is given the value of zero, and the enthalpy of each substance that contains carbon is given relative to this base. Another example is that oxygen may exist in the monatomic or diatomic form, and also as ozone, $O_3$. The value chosen as zero is for the form that is chemically stable at the reference state, which in the case of oxygen is the diatomic form. Then each of the other forms must have an enthalpy of formation consistent with the chemical reaction and heat transfer for the reaction that produces that form of oxygen.

It will be noted from Table A.10 that two values are given for the enthalpy of formation for water; one is for liquid water and the other for gaseous (hypothetical ideal gas) water, both at the reference state of 25°C, 0.1 MPa. It is convenient to use the hypothetical ideal-gas reference in connection with the ideal-gas table property changes given in Table A.9, and to use the real liquid reference in connection with real water property changes as given in the steam tables, Table B.1.

The real liquid property value is found from the hypothetical ideal-gas value at 0.1 MPa by making an ideal-gas calculation from that state to a very low pressure (say 3.169 kPa, at which real water behaves nearly as an ideal gas) and then using the steam tables, B.1, to calculate the property change from there to the real liquid state at 0.1 MPa.

Frequently, students are bothered by the minus sign when the enthalpy of formation is negative. For example, the enthalpy of formation of carbon dioxide is negative. This is quite evident because the heat transfer is negative during the steady-flow chemical reaction, and the enthalpy of the carbon dioxide must be less than the sum of enthalpy of the carbon and oxygen initially, both of which are assigned the value of zero. This is quite analogous to the situation we would have in the steam tables if we let the enthalpy of saturated vapor be zero at 0.1 MPa pressure. In this case the enthalpy of the liquid would be negative, and we would simply use the negative value for the enthalpy of the liquid when solving problems.

## 12.4 FIRST-LAW ANALYSIS OF REACTING SYSTEMS

The significance of the enthalpy of formation is that it is most convenient in performing a first-law analysis of a reacting system, for the enthalpies of different substances can be added or subtracted, since they are all given relative to the same base.

In such problems, we will write the first law for a steady-state, steady-flow process in the form

$$Q_{c.v.} + H_R = W_{c.v.} + H_P$$

or

$$Q_{c.v.} + \sum_R n_i \bar{h}_i = W_{c.v.} + \sum_P n_e \bar{h}_e$$

where $R$ and $P$ refer to the reactants and products, respectively. In each problem it is necessary to choose one parameter as the basis of the solution. Usually this is taken as 1 kmol of fuel.

---

**EXAMPLE 12.6**

Consider the following reaction, which occurs in a steady-state, steady-flow process.

$$CH_4 + 2O_2 \rightarrow CO_2 + 2H_2O(l)$$

The reactants and products are each at a total pressure of 0.1 MPa and 25°C. Determine the heat transfer per kilomole of fuel entering the combustion chamber.

| | |
|---:|:---|
| *Control volume*: | Combustion chamber. |
| *Inlet state*: | P and T known; state fixed. |
| *Exit state*: | P and T known, state fixed. |
| *Process*: | Steady state. |
| *Model*: | Three gases ideal gases; real liquid water. |

**Analysis**

First law:

$$Q_{\text{c.v.}} + \sum_R n_i \bar{h}_i = \sum_P n_e \bar{h}_e$$

**Solution**

Using values from Table A.10, we have

$$\sum_R n_i \bar{h}_i = (\bar{h}_f^0)_{\text{CH}_4} = -74\,873 \text{ kJ}$$

$$\sum_P n_e \bar{h}_e = (\bar{h}_f^0)_{\text{CO}_2} + 2(\bar{h}_f^0)_{\text{H}_2\text{O}(l)}$$

$$= -393\,522 + 2(-285\,830) = -965\,182 \text{ kJ}$$

$$Q_{\text{c.v.}} = -965\,182 - (-74\,873) = -890\,309 \text{ kJ}$$

In most instances, however, the substances that comprise the reactants and products in a chemical reaction are not at a temperature of 25°C and a pressure of 0.1 MPa (the state at which the enthalpy of formation is given). Therefore, the change of enthalpy between 25°C and 0.1 MPa and the given state must be known. For a solid or liquid, this change of enthalpy can usually be found from a table of thermodynamic properties or from specific heat data. For gases, the change of enthalpy can usually be found by one of the following procedures.

1. Assume ideal-gas behavior between 25°C, 0.1 MPa, and the given state. In this case, the enthalpy is a function of the temperature only and can be found by an equation of $\overline{C}_{p0}$ or from tabulated values of enthalpy as a function of temperature (which assumes ideal-gas behavior). Table A.6 gives an equation for $\overline{C}_{p0}$ for a number of substances and Table A.9 gives values of $\bar{h}^\circ - \bar{h}_{298}^\circ$ (that is, the $\Delta\bar{h}$ of Eq. 12.11) in kJ/kmol, ($\bar{h}_{298}^\circ$ refers to 25°C or 298.15 K. For simplicity this is designated $\bar{h}_{298}^\circ$.) The superscript 0 is used to designate that this is the enthalpy at 0.1 MPa pressure, based on ideal-gas behavior, that is, the standard-state enthalpy.

2. If a table of thermodynamic properties is available, $\Delta\bar{h}$ can be found directly from these tables if a real substance behavior reference state is being used, such as that described above for liquid water. If a hypothetical ideal-gas reference state is being used, then it is necessary to account for the real substance correction to properties at that state to gain entry to the tables.

Thus, in general, for applying the first law to a steady-state process involving a chemical reaction and negligible changes in kinetic and potential energy, we can write

$$Q_{\text{c.v.}} + \sum_R n_i(\bar{h}_f^\circ + \Delta\bar{h})_i = W_{\text{c.v.}} + \sum_P n_e(\bar{h}_f^\circ + \Delta\bar{h})_e \tag{12.13}$$

**EXAMPLE 12.7**

Calculate the enthalpy of water (on a kilomole basis) at 3.5 MPa, 300°C, relative to the 25°C and 0.1 MPa base, using the following procedures.

1. Assume the steam to be an ideal gas with the value of $\overline{C}_{p0}$ given in the Appendix, Table A.6.

2. Assume the steam to be an ideal gas with the value for $\Delta \bar{h}$ as given in the Appendix, Table A.9.

3. The steam tables.

**Solution**

For each of these procedures, we can write

$$\bar{h}_{T,P} = (\bar{h}_f^{\circ} + \Delta \bar{h})$$

The only difference is in the procedure by which we calculate $\Delta \bar{h}$. From Table A.10 we note that

$$(\bar{h}_f^{\circ})_{H_2O(g)} = -241\ 826\ \text{kJ/kmol}$$

1. Using the specific heat equation for $H_2O(g)$ from Table A.6,

$$C_{p0} = 1.79 + 0.107\theta + 0.586\theta^2 - 0.20\theta^3, \theta = T/1000$$

The specific heat at the average temperature

$$T_{\text{avg}} = \frac{298.15 + 573.15}{2} = 435.65\ \text{K}$$

is

$$C_{p0} = 1.79 + 0.107(0.43565) + 0.586(0.43565)^2 - 0.2(0.43565)^3$$

$$= 1.9313\ \frac{\text{kJ}}{\text{kg K}}$$

Therefore,

$$\Delta \bar{h} = MC_{p0}\ \Delta T$$

$$= 18.015 \times 1.9313(573.15 - 298.15) = 9568\ \frac{\text{kJ}}{\text{kmol}}$$

$$\bar{h}_{T,P} = -241\ 826 + 9568 = -232\ 258\ \text{kJ/kmol}$$

2. Using Table A.9 for $H_2O(g)$,

$$\Delta \bar{h} = 9539\ \text{kJ/kmol}$$

$$\bar{h}_{T,P} = -241\ 826 + 9539 = -232\ 287\ \text{kJ/kmol}$$

3. Using the steam tables, either the liquid reference or the gaseous reference state may be used.

For the liquid,

$$\Delta \bar{h} = 18.015(2977.5 - 104.9) = 51\ 750\ \text{kJ/kmol}$$

$$\bar{h}_{T,P} = -285\ 830 + 51\ 750 = -234\ 080\ \text{kJ/kmol}$$

For the gas,

$$\Delta \bar{h} = 18.015(2977.5 - 2547.2) = 7752\ \text{kJ/kmol}$$

$$\bar{h}_{T,P} = -241\ 826 + 7752 = -234\ 074\ \text{kJ/kmol}$$

The very small difference results from using the enthalpy of saturated vapor at 25°C (which is almost but not exactly an ideal gas) in calculating the $\Delta \bar{h}$.

The particular approach that is used in a given problem will depend on the data available for the given substance.

**Example 12.8**

A small gas-turbine uses $C_8H_{18}(l)$ for fuel and 400% theoretical air. The air and fuel enter at 25°C, and the products of combustion leave at 900 K. The output of the engine and the fuel consumption are measured, and it is found that the specific fuel consumption is 0.25 kg/s of fuel per megawatt output. Determine the heat transfer from the engine per kilomole of fuel. Assume complete combustion.

| | |
|---:|:---|
| *Control volume*: | Gas-turbine engine. |
| *Inlet states*: | $T$ known for fuel and air. |
| *Exit state*: | $T$ known for combustion products. |
| *Process*: | Steady state. |
| *Model*: | All gases ideal gases, Table A.9; liquid octane, Table A.10. |

**Analysis**

The combustion equation is

$$C_8H_{18}(l) + 4(12.5)O_2 + 4(12.5)(3.76)N_2 \rightarrow 8CO_2 + 9H_2O + 37.5O_2 + 188.0N_2$$

First law:

$$Q_{c.v.} + \sum_R n_i(\bar{h}_f^{\circ} + \Delta\bar{h})_i = W_{c.v.} + \sum_P n_e(\bar{h}_f^{\circ} + \Delta\bar{h})_e$$

**Solution**

Since the air is composed of elements and enters at 25°C, the enthalpy of the reactants is equal to that of the fuel,

$$\sum_R n_i(\bar{h}_f^0 + \Delta\bar{h})_i = (\bar{h}_f^0)_{C_8H_{18}(l)} = -250\ 105 \text{ kJ/kmol fuel}$$

Considering the products, we have

$$\sum_P n_e(\bar{h}_f^0 + \Delta\bar{h})_e = n_{CO_2}(\bar{h}_f^0 + \Delta\bar{h})_{CO_2} + n_{H_2O}(\bar{h}_f^0 + \Delta\bar{h})_{H_2O}$$

$$+ n_{O_2}(\Delta\bar{h})_{O_2} + n_{N_2}(\Delta\bar{h})_{N_2}$$

$$= 8(-393\ 522 + 28\ 030) + 9(-241\ 826 + 21\ 937)$$

$$+ 37.5(19\ 241) + 188(18\ 223)$$

$$= -755\ 476 \text{ kJ/kmol fuel}$$

$$W_{c.v.} = \frac{1000 \text{ kJ/s}}{0.25 \text{ kg/s}} \times \frac{114.23 \text{ kg}}{\text{kmol}} = 456\ 920 \text{ kJ/kmol fuel}$$

Therefore, from the first law,

$$Q_{c.v.} = -755\ 476 + 456\ 920 - (-250\ 105)$$

$$= -48\ 451 \text{ kJ/kmol fuel}$$

**Example 12.9**

A mixture of 1 kmol of gaseous ethene and 3 kmol of oxygen at 25°C reacts in a constant-volume bomb. Heat is transferred until the products are cooled to 600 K. Determine the amount of heat transfer from the system.

| | |
|---:|:---|
| *Control mass*: | Constant-volume bomb. |
| *Initial state*: | $T$ known. |
| *Final state*: | $T$ known. |
| *Process*: | Constant volume. |
| *Model*: | Ideal-gas mixtures, Tables A.9, A.10. |

**Analysis**

The chemical reaction is

$$C_2H_4 + 3O_2 \rightarrow 2CO_2 + 2H_2O(g)$$

First law:

$$Q + U_R = U_P$$

$$Q + \sum_R n(\bar{h}_f^0 + \Delta\bar{h} - \bar{R}T) = \sum_P n(\bar{h}_f^0 + \Delta\bar{h} - \bar{R}T)$$

**Solution**

Using values from Tables A.9 and A.10, gives

$$\sum_R n(\bar{h}_f^0 + \Delta\bar{h} - \bar{R}T) = (\bar{h}_f^0 - \bar{R}T)_{C_2H_4} - n_{O_2}(\bar{R}T)_{O_2} = (\bar{h}_f^0)_{C_2H_4} - 4\bar{R}T$$

$$= 52\,467 - 4 \times 8.3145 \times 298.2 = 42\,550 \text{ kJ}$$

$$\sum_P n(\bar{h}_f^0 + \Delta\bar{h} - \bar{R}T) = 2[(\bar{h}_f^0)_{CO_2} + \Delta\bar{h}_{CO_2}] + 2[(\bar{h}_f^0)_{H_2O(g)} + \Delta\bar{h}_{H_2O(g)}] - 4\bar{R}T$$

$$= 2(-393\,522 + 12\,906) + 2(-241\,826 + 10\,499)$$

$$- 4 \times 8.3145 \times 600$$

$$= -1\,243\,841 \text{ kJ}$$

Therefore,

$$Q = -1\,243\,841 - 42\,550 = -1\,286\,391 \text{ kJ}$$

## 12.5 Enthalpy of Combustion

The enthalpy of combustion, $h_{RP}$, is defined as the difference between the enthalpy of the products and the enthalpy of the reactants when complete combustion occurs at a given temperature and pressure. That is,

$$\bar{h}_{RP} = H_P - H_R$$

$$\bar{h}_{RP} = \sum_P n_e(\bar{h}_f^0 + \Delta\bar{h})_e - \sum_R n_i(\bar{h}_f^0 + \Delta\bar{h})_i \qquad (12.14)$$

The usual parameter for expressing the enthalpy of combustion is a unit mass of fuel, such as a kilogram ($h_{RP}$) or a kilomole ($\bar{h}_{RP}$) of fuel.

As the enthalpy of formation is fixed, we can separate the terms as

$$H = H^0 + \Delta H$$

where

$$H_R^0 = \sum_R n_i \bar{h}_{fi}^0; \qquad \Delta H_R = \sum_R n_i \Delta \bar{h}_i$$

and

$$H_P^0 = \sum_P n_i \bar{h}_{fi}^0; \qquad \Delta H_P = \sum_P n_i \Delta \bar{h}_i$$

Now the difference in enthalpies is written

$$H_P - H_R = H_P^0 - H_R^0 + \Delta H_P - \Delta H_R$$
$$= \bar{h}_{RP0} + \Delta H_P - \Delta H_R \qquad (12.15)$$

explicitly showing the reference enthalpy of combustion, $\bar{h}_{RP0}$, and the two departure terms $\Delta H_P$ and $\Delta H_R$. The latter two terms for the products and reactants are nonzero if they exist at a state other than the reference state.

The tabulated values of the enthalpy of combustion of fuels are usually given for a temperature of 25°C and a pressure of 0.1 MPa. The enthalpy of combustion for a number of hydrocarbon fuels at this temperature and pressure, which we designate $h_{RP0}$, is given in Table 12.3.

Frequently the term "heating value" or "heat of reaction" is used. This represents the heat transferred from the chamber during combustion or reaction at constant temperature. In the case of a constant pressure or steady-flow process, we conclude from the first law of thermodynamics that it is equal to the negative of the enthalpy of combustion. For this reason, this heat transfer is sometimes designated the constant-pressure heating value for combustion processes.

When the term heating value is used, the terms "higher" and "lower" heating value are used. The higher heating value is the heat transfer with liquid water in the products, and the lower heating value is the heat transfer with vapor water in the products.

---

**EXAMPLE 12.10**

Calculate the enthalpy of combustion of propane at 25°C on both a kilomole and kilogram basis under the following conditions.

1. Liquid propane with liquid water in the products.
2. Liquid propane with gaseous water in the products.
3. Gaseous propane with liquid water in the products.
4. Gaseous propane with gaseous water in the products.

This example is designed to show how the enthalpy of combustion can be determined from enthalpies of formation. The enthalpy of evaporation of propane is 370 kJ/kg.

**Analysis and Solution**

The basic combustion equation is

$$C_3H_8 + 5O_2 \rightarrow 3CO_2 + 4H_2O$$

From Table A.10 $(\bar{h}_f^0)_{C_3H_8(g)} = -103\,900$ kJ/kmol. Therefore,

$$(\bar{h}_f^0)_{C_3H_8(l)} = -103\,900 - 44.097(370) = -120\,216 \text{ kJ/kmol}$$

**TABLE 12.3**

*Enthalpy of Combustion of Some Hydrocarbons at 25°C*

| Hydrocarbon | UNITS: $\dfrac{kJ}{kg}$ Formula | LIQUID $H_2O$ IN PRODUCTS | | GAS $H_2O$ IN PRODUCTS | |
|---|---|---|---|---|---|
| | | Liq. HC | Gas HC | Liq. HC | Gas HC |
| **Paraffins** | $C_nH_{2n+2}$ | | | | |
| Methane | $CH_4$ | | −55 496 | | −50 010 |
| Ethane | $C_2H_6$ | | −51 875 | | −47 484 |
| Propane | $C_3H_8$ | −49 973 | −50 343 | −45 982 | −46 352 |
| n-Butane | $C_4H_{10}$ | −49 130 | −49 500 | −45 344 | −45 714 |
| n-Pentane | $C_5H_{12}$ | −48 643 | −49 011 | −44 983 | −45 351 |
| n-Hexane | $C_6H_{14}$ | −48 308 | −48 676 | −44 733 | −45 101 |
| n-Heptane | $C_7H_{16}$ | −48 071 | −48 436 | −44 557 | −44 922 |
| n-Octane | $C_8H_{18}$ | −47 893 | −48 256 | −44 425 | −44 788 |
| n-Decane | $C_{10}H_{22}$ | −47 641 | −48 000 | −44 239 | −44 598 |
| n-Dodecane | $C_{12}H_{26}$ | −47 470 | −47 828 | −44 109 | −44 467 |
| n-Cetane | $C_{16}H_{34}$ | −47 300 | −47 658 | −44 000 | −44 358 |
| **Olefins** | $C_nH_{2n}$ | | | | |
| Ethene | $C_2H_4$ | | −50 296 | | −47 158 |
| Propene | $C_3H_6$ | | −48 917 | | −45 780 |
| Butene | $C_4H_8$ | | −48 453 | | −45 316 |
| Pentene | $C_5H_{10}$ | | −48 134 | | −44 996 |
| Hexene | $C_6H_{12}$ | | −47 937 | | −44 800 |
| Heptene | $C_7H_{14}$ | | −47 800 | | −44 662 |
| Octene | $C_8H_{16}$ | | −47 693 | | −44 556 |
| Nonene | $C_9H_{18}$ | | −47 612 | | −44 475 |
| Decene | $C_{10}H_{20}$ | | −47 547 | | −44 410 |
| **Alkylbenzenes** | $C_{6+n}H_{6+2n}$ | | | | |
| Benzene | $C_6H_6$ | −41 831 | −42 266 | −40 141 | −40 576 |
| Methylbenzene | $C_7H_8$ | −42 437 | −42 847 | −40 527 | −40 937 |
| Ethylbenzene | $C_8H_{10}$ | −42 997 | −43 395 | −40 924 | −41 322 |
| Propylbenzene | $C_9H_{12}$ | −43 416 | −43 800 | −41 219 | −41 603 |
| Butylbenzene | $C_{10}H_{14}$ | −43 748 | −44 123 | −41 453 | −41 828 |
| **Other fuels** | | | | | |
| Gasoline | $C_7H_{17}$ | −48 201 | −48 582 | −44 506 | −44 886 |
| Diesel T-T | $C_{14.4}H_{24.9}$ | −45 700 | −46 074 | −42 934 | −43 308 |
| JP8 jet fuel | $C_{13}H_{23.8}$ | −45 707 | −46 087 | −42 800 | −43 180 |
| Methanol | $CH_3OH$ | −22 657 | −23 840 | −19 910 | −21 093 |
| Ethanol | $C_2H_5OH$ | −29 676 | −30 596 | −26 811 | −27 731 |
| Nitromethane | $CH_3NO_2$ | −11 618 | −12 247 | −10 537 | −11 165 |
| Phenol | $C_6H_5OH$ | −32 520 | −33 176 | −31 117 | −31 774 |
| Hydrogen | $H_2$ | | −141 781 | | −119 953 |

**1.** Liquid propane–liquid water:

$$\overline{h}_{RP_0} = 3(\overline{h}_f^0)_{CO_2} + 4(\overline{h}_f^0)_{H_2O(l)} - (\overline{h}_f^0)_{C_3H_8(l)}$$

$$= 3(-393\ 522) + 4(-285\ 830) - (-120\ 216)$$

$$= -2\ 203\ 670\ \text{kJ/kmol} = -\frac{2\ 203\ 670}{44.097} = -49\ 973\ \text{kJ/kg}$$

The higher heating value of liquid propane is 49 973 kJ/kg.

**2.** Liquid propane–gaseous water:

$$\overline{h}_{RP_0} = 3(\overline{h}_f^0)_{CO_2} + 4(\overline{h}_f^0)_{H_2O(g)} - (\overline{h}_f^0)_{C_3H_8(l)}$$

$$= 3(-393\ 522) + 4(-241\ 826) - (-120\ 216)$$

$$= -2\ 027\ 654\ \text{kJ/kmol} = -\frac{2\ 027\ 654}{44.097} = -45\ 982\ \text{kJ/kg}$$

The lower heating value of liquid propane is 45 982 kJ/kg.

**3.** Gaseous propane–liquid water:

$$\overline{h}_{RP_0} = 3(\overline{h}_f^0)_{CO_2} + 4(\overline{h}_f^0)_{H_2O(l)} - (\overline{h}_f^0)_{C_3H_8(g)}$$

$$= 3(-393\ 522) + 4(-285\ 830) - (-103\ 900)$$

$$= -2\ 219\ 986\ \text{kJ/kmol} = -\frac{2\ 219\ 986}{44.097} = -50\ 343\ \text{kJ/kg}$$

The higher heating value of gaseous propane is 50 343 kJ/kg.

**4.** Gaseous propane–gaseous water:

$$\overline{h}_{RP_0} = 3(\overline{h}_f^0)_{CO_2} + 4(\overline{h}_f^0)_{H_2O(g)} - (\overline{h}_f^0)_{C_3H_8(g)}$$

$$= 3(-393\ 522) + 4(-241\ 826) - (-103\ 900)$$

$$= -2\ 043\ 970\ \text{kJ/kmol} = -\frac{2\ 043\ 970}{44.097} = -46\ 352\ \text{kJ/kg}$$

The lower heating value of gaseous propane is 46 352 kJ/kg.

Each of the four values calculated in this example corresponds to the appropriate value given in Table 12.3.

## 12.6 ADIABATIC FLAME TEMPERATURE

Consider a given combustion process that takes place adiabatically and with no work or changes in kinetic or potential energy involved. For such a process the temperature of the products is referred to as the adiabatic flame temperature. With the assumptions of no work and no changes in kinetic or potential energy, this is the maximum temperature that can be achieved for the given reactants because any heat transfer from the reacting substances and any incomplete combustion would tend to lower the temperature of the products.

For a given fuel and given pressure and temperature of the reactants, the maximum adiabatic flame temperature that can be achieved is with a stoichiometric mixture. The adiabatic flame temperature can be controlled by the amount of excess air that is used. This is important, for example, in gas turbines, where the maximum permissible temperature is determined by metallurgical considerations in the turbine, and close control of the temperature of the products is essential.

Example 12.11 shows how the adiabatic flame temperature may be found. The dissociation that takes place in the combustion products, which has a significant effect on the adiabatic flame temperature, will be considered later in the chapter.

**EXAMPLE 12.11**

Liquid octane at 25°C is burned with 400% theoretical air at 25°C in a steady-state process. Determine the adiabatic flame temperature.

| | |
|---:|:---|
| *Control volume*: | Combustion chamber. |
| *Inlet states*: | $T$ known for fuel and air. |
| *Process*: | Steady state. |
| *Model*: | Gases ideal gases, Table A.9; liquid octane, Table A.10. |

**Analysis**

The reaction is

$$C_8H_{18}(l) + 4(12.5)O_2 + 4(12.5)(3.76)N_2 \rightarrow 8CO_2 + 9H_2O(g) + 37.5O_2 + 188.0N_2$$

First law: Since the process is adiabatic,

$$H_R = H_P$$

$$\sum_R n_i(\bar{h}_f^0 + \Delta\bar{h})_i = \sum_P n_e(\bar{h}_f^0 + \Delta\bar{h})_e$$

where $\Delta\bar{h}_e$ refers to each constituent in the products at the adiabatic flame temperature.

**Solution**

From Tables A.9 and A.10,

$$H_R = \sum_R n_i(\bar{h}_f^0 + \Delta\bar{h})_i = (\bar{h}_f^0)_{C_8H_{18}(l)} = -250\ 105 \text{ kJ/kmol fuel}$$

$$H_P = \sum_P n_e(\bar{h}_f^0 + \Delta\bar{h})_e$$

$$= 8(-393\ 522 + \Delta\bar{h}_{CO_2}) + 9(-241\ 826 + \Delta\bar{h}_{H_2O}) + 37.5\ \Delta\bar{h}_{O_2} + 188.0\ \Delta\bar{h}_{N_2}$$

By trial-and-error solution, a temperature of the products is found that satisfies this equation. Assume that

$$T_P = 900 \text{ K}$$

$$H_P = \sum_P n_e(\bar{h}_f^0 + \Delta\bar{h})_e$$

$$= 8(-393\ 522 + 28\ 030) + 9(-241\ 826 + 21\ 892)$$

$$+ 37.5(19\ 249) + 188(18\ 222)$$

$$= -755\ 769 \text{ kJ/kmol fuel}$$

Assume that

$$T_p = 1000 \text{ K}$$

$$H_P = \sum_P n_e(\bar{h}_f^0 + \Delta\bar{h})_e$$

$$= 8(-393\ 522 + 33\ 400) + 9(-241\ 826 + 25\ 956)$$

$$+ 37.5(22\ 710) + 188(21\ 461)$$

$$= 62\ 487 \text{ kJ/kmol fuel}$$

Since $H_P = H_R = -250\ 105$ kJ/kmol, we find by linear interpolation that the adiabatic flame temperature is 962 K. Because the ideal-gas enthalpy is not really a linear function of temperature, the true answer will be slightly different from this value.

---

**In-Text Concept Questions**

**e.** How is a fuel enthalpy of combustion connected to its enthalpy of formation?

**f.** What are the higher and lower heating values HHV, LHV of $n$-Butane?

**g.** What is the value of $h_{fg}$ for $n$-Octane?

**h.** What happens to the adiabatic flame temperature when I burn rich and when I burn lean?

## 12.7 THE THIRD LAW OF THERMODYNAMICS AND ABSOLUTE ENTROPY

As we consider a second-law analysis of chemical reactions, we face the same problem we had with the first law: What base should be used for the entropy of the various substances? This problem leads directly to a consideration of the third law of thermodynamics.

The third law of thermodynamics was formulated during the early part of the twentieth century. The initial work was done primarily by W. H. Nernst (1864–1941) and Max Planck (1858–1947). The third law deals with the entropy of substances at the absolute zero of temperature and in essence states that the entropy of a perfect crystal is zero at absolute zero. From a statistical point of view, this means that the crystal structure has the maximum degree of order. Furthermore, because the temperature is absolute zero, the thermal energy is minimum. It also follows that a substance that does not have a perfect crystalline structure at absolute zero, but instead has a degree of randomness, such as a solid solution or a glassy solid, has a finite value of entropy at absolute zero. The experimental evidence on which the third law rests is primarily data on chemical reactions at low temperatures and measurements of heat capacity at temperatures approaching absolute zero. In contrast to the first and second laws, which lead, respectively, to the properties of internal energy and entropy, the third law deals only with the question of entropy at absolute zero. However, the implications of the third law are quite profound, particularly in respect to chemical equilibrium.

The particular relevance of the third law is that it provides an absolute base from which to measure the entropy of each substance. The entropy relative to this base is termed the absolute entropy. The increase in entropy between absolute zero and any given state can be found either from calorimetric data or by procedures based on statistical thermodynamics. The calorimetric method gives precise measurements of specific-heat data over the temperature range, as well as of the energy associated with phase transformations. These measurements are in agreement with the calculations based on statistical thermodynamics and observed molecular data.

Table A.10 gives the absolute entropy at 25°C and 0.1-MPa pressure for a number of substances. Table A.9 gives the absolute entropy for a number of gases at 0.1-MPa pressure and various temperatures. For gases the numbers in all these tables are the hypothetical ideal-gas values. The pressure $P^0$ of 0.1 MPa is termed the standard-state pressure, and the absolute entropy as given in these tables is designated $\bar{s}^0$. The temperature is designated in kelvins with a subscript such as $\bar{s}^0_{1000}$.

If the value of the absolute entropy is known at the standard-state pressure of 0.1 MPa and a given temperature, it is a straightforward procedure to calculate the entropy change at a different pressure $P$, if the substance can be assumed to be an ideal gas at that state. If the substance is listed in Table A.9, the result is, on a kmol basis,

$$\bar{s}_{T,P} = \bar{s}^0_T - \bar{R} \ln \frac{P}{P^0} \qquad (12.16)$$

In this expression, the first term on the right side is the value from Table A.9, and the second one is the ideal-gas term to account for a change in pressure from $P^0$ to $P$. If the substance is not an ideal gas at pressure $P$, and a table of properties for that substance is available, the ideal-gas term for the change in pressure should be made to a low pressure $P^*$, at which ideal-gas behavior is a reasonable assumption, but it is also listed in the tables. Then,

$$\bar{s}_{T,P} = \bar{s}^0_T - \bar{R} \ln \frac{P^*}{P^0} + (\bar{s}_{T,P} - \bar{s}^*_{T,P*}) \qquad (12.17)$$

If the substance is not one of those listed in Table A.9, and the absolute entropy is known only at one temperature $T_0$, as given in Table A.10 for example, then it will be necessary to calculate $\bar{s}^0_T$ from

$$\bar{s}^0_T = \bar{s}^0_{T_0} + \int_{T_0}^{T} \frac{\bar{C}_{p0}}{T} \, dT \qquad (12.18)$$

and then proceed with the calculation of Eq. 12.16 or 12.17.

For calculation of the absolute entropy of a mixture of ideal gases at $T$, $P$, the mixture entropy is given in terms of the component entropies as

$$\bar{s}_{\text{mix}} = \sum_i y_i \bar{s}_i \qquad (12.19)$$

where each component entropy $\bar{s}_i$ is calculated from Eq. 10.18, written on a kmol basis,

$$\bar{s}_i = \bar{s}^0_{Ti} - \bar{R} \ln \frac{y_i P}{P^0} \qquad (12.20)$$

## 12.8 SECOND-LAW ANALYSIS OF REACTING SYSTEMS

The concepts of reversible work, irreversibility, and availability (exergy) were introduced in Chapter 9. These concepts included both the first and second laws of thermodynamics. We proceed now to develop this matter further, and we will be particularly concerned with determining the maximum work (availability) that can be done through a combustion process and with examining the irreversibilities associated with such processes.

The reversible work for a steady-state process in which there is no heat transfer with reservoirs other than the surroundings, and also in the absence of changes in kinetic and potential energy is, from Eq. 9.39 on a total mass basis,

$$W^{\text{rev}} = \sum m_i(h_i - T_0 s_i) - \sum m_e(h_e - T_0 s_e)$$

Applying this equation to a steady-state process that involves a chemical reaction, and introducing the symbols from this chapter, we have

$$W^{\text{rev}} = \sum_R n_i(\overline{h}_f^0 + \Delta\overline{h} - T_0\overline{s})_i - \sum_P n_e(\overline{h}_f^0 + \Delta\overline{h} - T_0\overline{s})_e \qquad (12.21)$$

Similarly, the irreversibility for such a process can be written as

$$I = W^{\text{rev}} - W = \sum_P n_e T_0 \overline{s}_e - \sum_R n_i T_0 \overline{s}_i - Q_{\text{c.v.}} \qquad (12.22)$$

The availability, $\psi$, for a steady-flow process, in the absence of kinetic and potential energy changes, is given by Eq. 9.41 as

$$\psi = (h - T_0 s) - (h_0 - T_0 s_0)$$

We further note that if a steady-state chemical reaction takes place in such a manner that both the reactants and products are in temperature equilibrium with the surroundings, then the combination $(h - Ts)$ becomes significant, being itself a thermodynamic property. This property is the Gibbs function, defined as

$$g \equiv h - Ts \qquad (12.23)$$

For such a process, in the absence of changes in kinetic and potential energy, the reversible work is given by the relation

$$W^{\text{rev}} = \sum_R n_i \overline{g}_i - \sum_R n_e \overline{g}_e = -\Delta G \qquad (12.24)$$

in which

$$\Delta G = \Delta H - T\Delta S \qquad (12.25)$$

We should keep in mind that Eq. 12.24 is a special case and that the reversible work is given by Eq. 12.21 if the reactants and products are not in temperature equilibrium with the surroundings.

Let us now consider the second-law analysis of a combustion process that does not occur at constant temperature. This matter is taken up in the following example.

---

| EXAMPLE 12.12 | Ethene gas at 25°C and 200 kPa enters a steady-state adiabatic combustion chamber along with 400% theoretical air at 25°C, 200 kPa., as shown in Fig. 12.4. The product gas mixture exits at the adiabatic flame temperature and 200 kPa. Calculate the irreversibility per kmol of ethene for this process. |

*Control volume*: Combustion chamber

*Inlet states*: $P$, $T$ known for each component gas stream

*Exit state*: $P$, $T$ known

*Model*: All ideal gases, Tables A.9 and A.10

*Sketch*: Fig. 12.4

**FIGURE 12.4** Sketch for Example 12.12

### Analysis

The combustion equation is

$$C_2H_4(g) + 12O_2 + 12(3.76)N_2 \rightarrow 2CO_2 + 2H_2O(g) + 9O_2 + 45.1N_2$$

The adiabatic flame temperature is determined first.

First law:

$$H_R = H_P$$

$$\sum_R n_i(\bar{h}_f^0)_i = \sum_P n_e(\bar{h}_f^0 + \Delta\bar{h})_e$$

### Solution

$$52\,467 = 2(-393\,522 + \Delta\bar{h}_{CO_2}) + 2(-241\,826 + \Delta\bar{h}_{H_2O(g)}) + 9\Delta\bar{h}_{O_2} + 45.1\Delta\bar{h}_{N_2}$$

By a trial-and-error solution we find the adiabatic flame temperature to be 1016 K.

We now proceed to find the change in entropy during this adiabatic combustion process.

$$S_R = S_{C_2H_4} + S_{air}$$

From Eq. 12.16,

$$S_{C_2H_4} = 1\left(219.330 - 8.3145 \ln\frac{200}{100}\right) = 213.567 \text{ kJ/K}$$

From Eqs. 12.19 and 12.20,

$$S_{air} = 12\left(205.147 - 8.3145 \ln\frac{0.21 \times 200}{100}\right)$$

$$+ 45.1\left(191.610 - 8.3145 \ln\frac{0.79 \times 200}{100}\right)$$

$$= 12(212.360) + 45.1(187.807) = 11\,018.416 \text{ kJ/K}$$

$$S_R = 213.567 + 11\,018.416 = 11\,231.983 \text{ kJ/K}$$

For a multicomponent product gas mixture, it is convenient to set up a table, as follows:

| Comp | $n_i$ | $y_i$ | $\bar{R}\ln\dfrac{y_iP}{P^0}$ | $\bar{s}_{Ti}^0$ | $\bar{s}_i$ |
|------|-------|-------|------------------|------------|-----------|
| $CO_2$ | 2 | 0.0344 | −22.254 | 270.194 | 292.448 |
| $H_2O$ | 2 | 0.0344 | −22.254 | 233.355 | 255.609 |
| $O_2$ | 9 | 0.1549 | −9.743 | 244.135 | 253.878 |
| $N_2$ | 45.1 | 0.7763 | +3.658 | 228.691 | 225.033 |

Then, with values from this table for $n_i$ and $\bar{s}_i$ for each component $i$,

$$S_P = \sum_i n_i \bar{s}_i = 13\,530.004 \text{ kJ/K}$$

Since this is an adiabatic process, the irreversibility is, from Eq. 12.28,

$$I = T_0(S_P - S_R) = 298.15\,(13530.004 - 11\,231.983) = 685\,155 \text{ kJ/kmol C}_2\text{H}_4$$

## 12.9 FUEL CELLS

The previous discussion raises the question of the possibility of a reversible chemical reaction. Some reactions can be made to approach reversibility by having them take place in an electrolytic cell, as described in Chapter 1. When a potential exactly equal to the electromotive force of the cell is applied, no reaction takes place. When the applied potential is increased slightly, the reaction proceeds in one direction, and if the applied potential is decreased slightly, the reaction proceeds in the opposite direction. The work done is the electrical energy supplied or delivered.

Consider a reversible reaction occurring at constant temperature equal to that of its environment. The work output of the fuel cell is

$$W = -\left(\sum n_e \bar{g}_e - \sum n_i \bar{g}_i\right) = -\Delta G$$

where $\Delta G$ is the change in Gibbs function for the overall chemical reaction. We also realize that the work is given in terms of the charged electrons flowing through an electrical potential $\mathscr{E}$ as

$$W = \mathscr{E} n_e N_0 e$$

in which $n_e$ is the number of kilomoles of electrons flowing through the external circuit and

$$N_0 e = 6.022\,136 \times 10^{26} \text{ elec/kmol} \times 1.602\,177 \times 10^{-22} \text{ kJ/elec V}$$

$$= 96\,485 \text{ kJ/kmol V}$$

Thus, for a given reaction, the maximum (reversible reaction) electrical potential $\mathscr{E}^0$ of a fuel cell at a given temperature is

$$\mathscr{E}^0 = \frac{-\Delta G}{96\,485 n_e} \tag{12.26}$$

**EXAMPLE 12.13**

Calculate the reversible electromotive force (EMF) at 25°C for the hydrogen–oxygen fuel cell described in Section 1.2.

**Solution**

The anode side reaction was stated to be

$$2\text{H}_2 \rightarrow 4\text{H}^+ + 4e^-$$

and the cathode side reaction is

$$4\text{H}^+ + 4e^- + \text{O}_2 \rightarrow 2\text{H}_2\text{O}$$

Therefore, the overall reaction is, in kilomoles,

$$2\text{H}_2 + \text{O}_2 \rightarrow 2\text{H}_2\text{O}$$

for which 4 kmol of electrons flow through the external circuit. Let us assume that each component is at its standard-state pressure of 0.1 MPa and that the water formed is liquid. Then

$$\Delta H^0 = 2\bar{h}^0_{f_{H_2O(l)}} - 2\bar{h}^0_{f_{H_2}} - \bar{h}^0_{f_{O_2}}$$

$$= 2(-285\ 830) - 2(0) - 1(0) = -571\ 660\ \text{kJ}$$

$$\Delta S^0 = 2\bar{s}^0_{H_2O(l)} - 2\bar{s}^0_{H_2} - \bar{s}^0_{O_2}$$

$$= 2(69.950) - 2(130.678) - 1(205.148) = -326.604\ \text{kJ/K}$$

$$\Delta G^0 = -571\ 660 - 298.15(-326.604) = -474\ 283\ \text{kJ}$$

Therefore, from Eq. 12.28,

$$\mathscr{E}^0 = \frac{-(-474\ 283)}{96\ 485 \times 4} = 1.229\ \text{V}$$

In Example 12.13, we found the shift in the Gibbs function and the reversible EMF at 25°C. In practice, however, many fuel cells operate at an elevated temperature where the water leaves as a gas and not as a liquid; thus, it carries away more energy. The computations can be done for a range of temperatures, leading to lower EMF as the temperature increases. This behavior is shown in Fig. 12.5.

A variety of fuel cells are being investigated for use in stationary as well as mobile power plants. The low-temperature fuel cells use hydrogen as the fuel, whereas the higher temperature cells can use methane and carbon dioxide that are then internally reformed into hydrogen and carbon monoxide. The most important fuel cells are listed in Table 12.4 with their main characteristics.

The low-temperature fuel cells are very sensitive to being poisoned by CO gas so that they require an external reformer and purifier to deliver hydrogen gas. The higher temperature fuel cells can reform natural gas, mainly methane, but also ethane and propane as shown in Table 12.2, into hydrogen gas and carbon monoxide inside the cell. The latest research is being done with gasified coal as a fuel and operating the cell at higher pressures like 15 atmospheres. As the fuel cell has exhaust gas with a small amount of fuel in it, additional combustion can occur and then combine the fuel cell

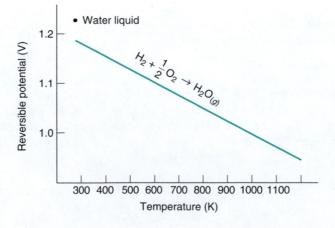

**FIGURE 12.5**
Hydrogen–oxygen fuel cell ideal EMF as a function of temperature.

**TABLE 12.4**

*Fuel Cell Types*

| FUEL CELL | PEC | PAC | MCC | SOC |
|---|---|---|---|---|
| | Polymer Electrolyte | Phosphoric Acid | Molten Carbonate | Solid Oxide |
| T | 80°C | 200°C | 650°C | 900°C |
| Fuel | Hydrogen, $H_2$ | Hydrogen, $H_2$ | CO, Hydrogen | Natural Gas |
| Carrier | $H^+$ | $H^+$ | $CO_3^{--}$ | $O^{--}$ |
| Charge, $n_e$ | $2e^-$ per $H_2$ | $2e^-$ per $H_2$ | $2e^-$ per $H_2$ $2e^-$ per CO | $8e^-$ per $CH_4$ |
| Catalyst | Pt | Pt | Ni | $ZrO_2$ |
| Poison | CO | CO | | |

with a gas turbine or steam power plant to utilize the exhaust gas energy. These combined cycle power plants strive to have an efficiency of up to 60%.

**In-Text Concept Questions**

**i.** Is the irreversibility in a combustion process significant? Explain your answer.

**j.** If the A/F ratio is larger than stoichiometric, is it more or less reversible?

**k.** What makes the fuel cell attractive from a power generating point of view?

## 12.10 CHEMICAL EQUILIBRIUM

The concept of thermodynamic equilibrium was first discussed in Section 2.3, including equilibrium of temperature and of pressure. The requirement for equilibrium for a given $T, P$, is that the reversible work for a possible change approach zero, as given by Eq. 12.26 in terms of the Gibbs function. If we have the possibility of a chemical reaction occurring among the constituents in a system, then it is also possible to have a condition of chemical equilibrium. Consider the reaction

$$\nu_A A + \nu_B B \rightleftharpoons \nu_C C + \nu_D D \tag{12.27}$$

In this expression, the stoichiometric coefficients $\nu_A$, $\nu_B$, $\nu_C$, and $\nu_D$ represent the relative proportions in which the components $A$, $B$, $C$, and $D$ take part in the reaction. The change in number of kmols of each component $i$ present is then given as

$$dn_i = \pm \nu_i \, d\varepsilon \tag{12.28}$$

where $\varepsilon$ is termed the degree of reaction. The components on the same side of the reaction equation have the same sign, opposite to that for those on the other side. The requirement for chemical equilibrium is that the Gibbs function of the overall system be a minimum, as shown in Fig. 12.6 with respect to amount of component $A$ present and as given by the expression

$$d\,G_{TP} = 0 \tag{12.29}$$

This can be represented in terms of the component properties as

$$d\,G_{TP} = \sum_i \bar{g}_i \, dn_i = (+\nu_C \bar{g}_C + \nu_D \bar{g}_D - \nu_A \bar{g}_A - \nu_B \bar{g}_B)\,d\varepsilon \tag{12.30}$$

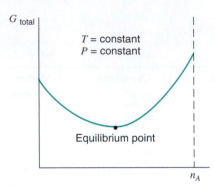

**FIGURE 12.6** Illustration of the requirement for chemical equilibrium.

The component Gibbs functions in Eq. 12.30 can be expressed in terms of the pure-substance standard-state Gibbs functions at the same temperature using Eq. 12.20 for the component entropies. This results in expressions of the form

$$\bar{g}_i = \bar{g}_{Ti}^0 + \bar{R}T \ln \frac{y_i P}{P^0} \tag{12.31}$$

Substituting this expression for each component into Eq. 12.30 and collecting and rearranging terms,

$$dG_{T,P} = \left\{ \Delta G^0 + \bar{R}T \ln \left[ \frac{y_C^{\nu_C} y_D^{\nu_D}}{y_A^{\nu_A} y_B^{\nu_B}} \left( \frac{P}{P^0} \right)^{\nu_C + \nu_D - \nu_A - \nu_B} \right] \right\} d\varepsilon \tag{12.32}$$

in which

$$\Delta G^0 = \nu_C \bar{g}_C^0 + \nu_D \bar{g}_D^0 - \nu_A \bar{g}_A^0 - \nu_B \bar{g}_B^0 \tag{12.33}$$

Note that for a given reaction, $\Delta G^0$ is a function only of temperature, as it is comprised of the pure-substance standard-state Gibbs functions and the fixed stoichiometric coefficients. At equilibrium $dG_{TP} = 0$, and since $d\varepsilon$ is arbitrary,

$$\ln \left[ \frac{y_C^{\nu_C} y_D^{\nu_D}}{y_A^{\nu_A} y_B^{\nu_B}} \left( \frac{P}{P^0} \right)^{\nu_C + \nu_D - \nu_A - \nu_B} \right] = -\frac{\Delta G^0}{\bar{R}T} = \ln K \tag{12.34}$$

in which $K$ is termed the equilibrium constant. Therefore,

$$K = \frac{y_C^{\nu_C} y_D^{\nu_D}}{y_A^{\nu_A} y_B^{\nu_B}} \left( \frac{P}{P^0} \right)^{\nu_C + \nu_D - \nu_A - \nu_B} \tag{12.35}$$

which is the chemical equilibrium equation corresponding to the reaction equation 12.27.

From the equilibrium constant definition in Eqs. 12.34 and 12.35 we can draw a few conclusions. If the shift in the Gibbs function is large and positive, ln $K$ is large and negative leading to a very small value of $K$. At a given $P$ in Eq. 12.35 this leads to relatively small values of the RHS (component $C$ and $D$) concentrations relative to the LHS component concentrations; the reaction is shifted to the left. The opposite is the case of a shift in the Gibbs function that is large and negative, giving a large value of $K$ and the reaction is shifted to the right as shown in Fig. 12.7. If the shift in Gibbs function is zero, then ln $K$ is zero, and $K$ is exactly equal to 1, the reaction is in the middle with all concentrations of the same order of magnitude, unless the stoichiometric coefficients are extreme.

The other trends we can see are the influences of the temperature and pressure. For a higher temperature, but same shift in Gibbs function, the absolute value of ln $K$ is

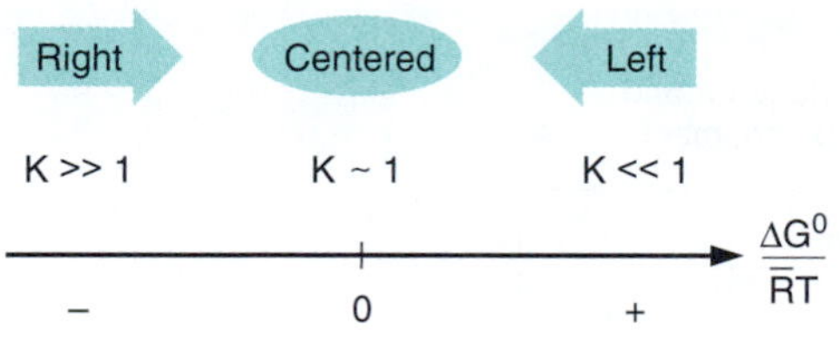

| | | |
|---|---|---|
| Right ▶ | ⬭ Centered | ◀ Left |
| K >> 1 | K ~ 1 | K << 1 |

$$\xrightarrow{\hspace{6cm}} \dfrac{\Delta G^0}{RT}$$

|—————————————|————————|————————| 
| − | 0 | + |

**FIGURE 12.7** The shift in the reaction with the change in Gibbs function.

smaller, which means $K$ is closer to 1 and the reaction is more centered. For low temperatures, the reaction is shifted toward the side with the smallest Gibbs function $G^0$. The pressure has an influence only if the power in Eq. 12.35 is different from zero. That is so when the number of moles on the RHS $(v_C + v_D)$ is different from the number of moles on the LHS $(v_A + v_B)$. Assuming we have more moles on the RHS, then we see that the power is positive. So if the pressure is larger than the reference pressure, the whole pressure factor is larger than 1, which reduces the RHS concentrations as $K$ is fixed for a given temperature. You can argue all the other combinations, and the result is that a higher pressure pushes the reaction toward the side with fewer moles, and a lower pressure pushes the reaction toward the side with more moles. The reaction tries to counteract the externally imposed pressure variation.

Table A.11 gives the values of the equilibrium constant (log basis) for a number of reactions as a function of temperature. For other reactions, the value for $\Delta G^0$ (and $K$) can be calculated from combinations of the reactions given in A.11 or else from the expression

$$\Delta G_T^0 = \Delta H_T^0 - T \, \Delta S_T^0 \tag{12.36}$$

in which the standard-state enthalpy and entropy changes for the reaction at temperature $T$ can be calculated separately.

The following example illustrates the procedure for determining the equilibrium composition for a given dissociation reaction equation.

---

**EXAMPLE 12.14**

One kilomole of carbon at 25°C and 0.1 MPa pressure reacts with 1 kmol of oxygen at 25°C and 0.1 MPa pressure to form an equilibrium mixture of $CO_2$, $CO$, and $O_2$ at 3000 K, 0.1 MPa pressure, in a steady-state process. Determine the equilibrium composition and the heat transfer for this process.

| | |
|---|---|
| *Control volume*: | Combustion chamber. |
| *Inlet states*: | $P$, $T$ known for carbon and for oxygen. |
| *Exit state*: | $P$, $T$ known. |
| *Process*: | Steady-state. |
| *Sketch*: | Figure 12.8. |
| *Model*: | Table A.10 for carbon; ideal gases, Tables A.9 and A.10. |

**Analysis and Solution**

It is convenient to view the overall process as though it occurs in two separate steps, a combustion process followed by a heating and dissociation of the combustion product carbon dioxide, as indicated in Fig. 12.8. This two-step process is represented as

Combustion:   $C + O_2 \rightarrow CO_2$

Dissociation reaction:   $2CO_2 \rightleftharpoons 2CO + O_2$

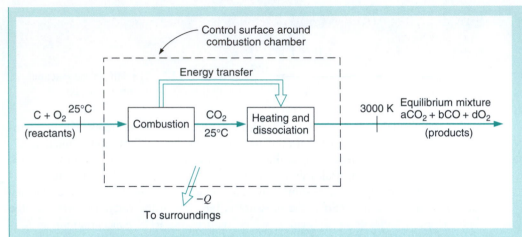

**FIGURE 12.8**   Sketch for Example 12.14.

That is, the energy released by the combustion of C and $O_2$ heats the $CO_2$ formed to high temperature, which causes dissociation of part of the $CO_2$ to CO and $O_2$. Thus, the overall reaction can be written

$$C + O_2 \rightarrow aCO_2 + bCO + dO_2$$

where the unknown coefficients $a$, $b$, and $d$ must be found by solution of the equilibrium equation associated with the dissociation reaction. Once this is accomplished, we can write the first law for a control volume around the combustion chamber to calculate the heat transfer.

From the combustion equation we find that the initial composition for the dissociation reaction is 1 kmol $CO_2$. Therefore, letting $2z$ be the number of kilomoles of $CO_2$ dissociated, we find

|  | $2CO_2 \rightleftharpoons 2CO + O_2$ | | |
|---|---|---|---|
| Initial: | 1 | 0 | 0 |
| Change: | $-2z$ | $+2z$ | $+z$ |
| At equilibrium: | $(1 - 2z)$ | $2z$ | $z$ |

Therefore, the overall reaction is

$$C + O_2 \rightarrow (1 - 2z)CO_2 + 2zCO + zO_2$$

and the total number of kilomoles at equilibrium is

$$n = (1 - 2z) + 2z + z = 1 + z$$

The equilibrium mole fractions are

$$y_{CO_2} = \frac{1 - 2z}{1 + z} \qquad y_{CO} = \frac{2z}{1 + z} \qquad y_{O_2} = \frac{z}{1 + z}$$

From Table A.11 we find that the value of the equilibrium constant at 3000 K for the dissociation reaction considered here is

$$\ln K = -2.217 \qquad K = 0.1089$$

Substituting these quantities along with $P = 0.1$ MPa into Eq. 12.35, we have the equilibrium equation,

$$K = 0.1089 = \frac{y_{CO}^2 y_{O_2}}{y_{CO_2}^2}\left(\frac{P}{P^0}\right)^{2+1-2} = \frac{\left(\dfrac{2z}{1+z}\right)^2\left(\dfrac{z}{1+z}\right)}{\left(\dfrac{1-2z}{1+z}\right)^2} \quad (1)$$

or, in more convenient form,

$$\frac{K}{P/P^0} = \frac{0.1089}{1} = \left(\frac{2z}{1-2z}\right)^2\left(\frac{z}{1+z}\right)$$

To obtain the physically meaningful root of this mathematical relation, we note that the number of moles of each component must be greater than zero. Thus, the root of interest to us must lie in the range

$$0 \le z \le 0.5$$

Solving the equilibrium equation by trial and error, we find

$$z = 0.2189$$

Therefore, the overall process is

$$C + O_2 \rightarrow 0.5622CO_2 + 0.4378CO + 0.2189O_2$$

where the equilibrium mole fractions are

$$y_{CO_2} = \frac{0.5622}{1.2189} = 0.4612$$

$$y_{CO} = \frac{0.4378}{1.2189} = 0.3592$$

$$y_{O_2} = \frac{0.2189}{1.2189} = 0.1796$$

The heat transfer from the combustion chamber to the surroundings can be calculated using the enthalpies of formation and Table A.9. For this process

$$H_R = (\overline{h}_f^0)_C + (\overline{h}_f^0)_{O_2} = 0 + 0 = 0$$

The equilibrium products leave the chamber at 3000 K. Therefore,

$$\begin{aligned}
H_P &= n_{CO_2}(\overline{h}_f^0 + \overline{h}_{3000}^0 - \overline{h}_{298}^0)_{CO_2} \\
&\quad + n_{CO}(\overline{h}_f^0 + \overline{h}_{3000}^0 - \overline{h}_{298}^0)_{CO} \\
&\quad + n_{O_2}(\overline{h}_f^0 + \overline{h}_{3000}^0 - \overline{h}_{298}^0)_{O_2} \\
&= 0.5622(-393\,522 + 152\,853) \\
&\quad + 0.4378(-110\,527 + 93\,504) \\
&\quad + 0.2189(98\,013) \\
&= -121\,302 \text{ kJ}
\end{aligned}$$

Substituting into the first law gives

$$Q_{C.V.} = H_P - H_G$$

$$= -121\ 302\ \text{kJ/kmol C burned}$$

It is of interest to note that in the above example it is not necessary to have 100% theoretical oxygen entering the combustion chamber. For example, if 200% oxygen is supplied, then the initial composition for the dissociation is $1\ CO_2 + 1\ O_2$, such that at equilibrium the number of kmol $O_2$ is $(1 + z)$ and the total number of kmol of mixture is now $(2 + z)$. Proceeding to a solution, it is found that in this case $z = 0.1553$, with a resulting mixture containing more $CO_2$ and less $CO$ than in Example 12.14. It is also noted that a mixture pressure other than 0.1 MPa would affect the solution of the equilibrium equation and therefore the mixture composition, as would the presence of a gas inert to the reaction, such as $N_2$, had air been supplied initially instead of oxygen.

Cases in which more than one independent reaction equation of the type of Eq. 12.27 occur simultaneously result in a mixture where there is an equilibrium equation of the type of Eq. 12.35 for each reaction equation with a corresponding number of unknown variables. Solution of such complex problems will not be considered in this text.

**In-Text Concept Questions**

l. For a mixture of $O_2$ and O the pressure is increased at constant $T$; what happens to the composition?

m. For a mixture of $O_2$ and O the temperature is increased at constant $P$; what happens to the composition?

n. For a mixture of $O_2$ and O I add some argon, keeping constant $T$, $P$; what happens to the moles of O?

## 12.11 HOW-TO SECTION

### How do I find a fuel enthalpy of formation if I know the heating value?

The two numbers are connected by the definition

$$HV = -H_{RP}^0 = -M_{\text{fuel}} \times H_{RP\ \text{TBL 12.3}}^0 = H_R^0 - H_P^0$$

$$= \sum_R n_i \bar{h}_{fi}^0 - \sum_P n_i \bar{h}_{fi}^0 = \bar{h}_{f\,\text{fuel}}^0 - \nu_{CO_2}\bar{h}_{f\,CO_2}^0 - \nu_{H_2O}\bar{h}_{f\,H_2O}^0$$

where the fuel enthalpy of formation is normally the only non-zero term for the reactants and the product summation contains the carbon dioxide and water terms. Notice that all these must be at the reference state to result in a single number for the fuel, and even then we have values for the fuel as a liquid or vapor and the water as a liquid or vapor.

### How do I know which product species are present?

In the present introduction to combustion, we assume that we burn the carbon to carbon dioxide and the hydrogen to water and leave the nitrogen unaltered. If we have extra oxygen (lean) or extra fuel (rich), we will also find the excess in the products unaltered. Reality is more complicated and we need to know what the detailed product list is to treat it more accurately than the above description. If the chemical equilibrium is included, the list of the product species can be very long. The list can have 50–100 different species for heavier fuels.

### How do I find an adiabatic flame temperature?

The adiabatic (or non-adiabatic) flame temperature is found from the knowledge of the combined product energy. Assume we have a flow of the reactants that burns and converts to a flow of products, then the energy equation (written for 1 kmol fuel) is

$$H_R + Q = H_P + W$$

After separation as $(H = H^0 + \Delta H)$ we then solve for the enthalpy, $\Delta H_P$, as

$$\Delta H_P = H_R^0 - H_P^0 + \Delta H_R + Q - W = HV + \Delta H_R + Q - W$$

Dependent upon the actual case, you may or may not have the last three terms. If you want the standard adiabatic flame temperature, we assume all the last three terms are zero. Now we need to express the LHS from the species as

$$\Delta H_P = \sum_P \nu_i \Delta \bar{h}_i$$

and the enthalpies are from Table A.9 so it becomes a function of temperature. Guess a $T$ and compute $\Delta H_P$, compare with what it should be (RHS = $HV + \Delta H_R + Q - W$), and if too low, guess a higher $T$ and so on until you have a match. This procedure requires that you do know the RHS as the energy equation can only solve for one ($T_{\text{ad.}}$) unknown.

### How does a dilution with an inert gas affect the chemical equilibrium?

The presence of a dilution gas like argon (or any other inert gas) does not enter the chemical reaction equation directly. It does enter into the computation of the mole fractions because its presence will lower all the other mole fractions and generally push the reaction towards the side with more moles.

### How do I know which reactions can take place?

Generally you really do not know this before you have tried to find the concentrations of the species. However, you have an idea from the magnitude of the equilibrium constant. If it is large, the reaction is pushed to the right and if it is small, the reaction is pushed to the left, so you would know if the reaction is important (somewhere in the middle). For larger systems, the treatment of many species with multiple and simultaneous reactions does require a slightly more elaborate formulation and solution procedure.

### SUMMARY

An introduction to combustion of hydrocarbon fuels and chemical reactions in general is given. A simple oxidation of a hydrocarbon fuel with pure oxygen or air burns the hydrogen to water and the carbon to carbon dioxide. We apply the continuity equation for each kind of atom to balance the stoichiometric coefficients of the species in the reactants and products. The reactant mixture composition is described by the air–fuel ratio on a mass or mole basis or percent theoretical air or equivalence ratio, according to the practice of the particular area of use. The product of a given fuel for a stoichiometric mixture and complete combustion is unique, whereas actual combustion can lead to incomplete combustion and more complex products described by measurements on a dry or wet basis. As water is part of the products, they have a dew point, so it is possible to see water condensing out from the products as they are cooled.

Because of the chemical changes from the reactants to the products, we need to measure energy from a common reference. Chemically pure substances (not compounds like CO) in their ground state (graphite for carbon, not diamond form) are assigned a value of 0 for the formation enthalpy at reference temperature and pressure (25°C, 100 kPa). Stable compounds have a negative formation enthalpy, and unstable compounds have a positive formation enthalpy. The shift in the enthalpy from the reactants to the products is the enthalpy of combustion, which is also the negative of the heating value HV. When a combustion process takes place without any heat transfer, the resulting product temperature is the adiabatic flame temperature. The enthalpy of combustion, the heating value (lower or higher), and the adiabatic flame temperature depend on the mixture (fuel and A/F ratio) and the reactant's supply temperature. When a single unique number for these properties is used, it is understood to be for a stoichiometric mixture at the reference conditions.

Similarly to the enthalpy, an absolute value of entropy is needed for the application of the second law. The absolute entropy is zero for a perfect crystal at 0 K, which is the third law of thermodynamics. The combustion process is an irreversible process, and there is thus a loss of availability (exergy) associated with it. This irreversibility is increased by mixtures different from stoichiometric and by dilution of the oxygen (i.e., nitrogen in air), which lowers the adiabatic flame temperature. From the concept of flow exergy, we apply the second law to find the reversible work given by the change in Gibbs function. A process that has less irreversibility than combustion at high temperature is the chemical conversion in a fuel cell where we approach a chemical equilibrium process. Here the energy release is directly converted into an electrical power output, a system that is currently under intense study and development for future energy conversion systems.

From previous analysis with the second law we have found the reversible shaft work as the change in Gibbs function. This is extended to give the equilibrium state as the one with minimum Gibbs function at a given $T$, $P$.

Chemical equilibrium is formulated for a single-equilibrium reaction assuming the components are all ideal gases. This leads to an equilibrium equation tying together the mole fractions of the components, the pressure, and the reaction constant. The reaction constant is related to the shift in Gibbs function from the reactants (LHS) to the products (RHS) at a temperature $T$. As temperature or pressure changes, the equilibrium composition will shift according to its sensitivity to $T$ and $P$. For very large equilibrium constants, the reaction is shifted toward the RHS, and for very small ones it is shifted toward the LHS.

You should have learned a number of skills and acquired abilities from studying this chapter that will allow you to:

- Write the combustion equation for the stoichiometric reaction of any fuel.
- Balance the stoichiometric coefficients for a reaction with a set of products measured on a dry basis.
- Handle the combustion of fuel mixtures as well as moist air oxidicer.

- Apply the energy equation with common-base values of enthalpy.
- Use the proper tables for high-temperature products.
- Deal with condensation of water in low-temperature products of combustion.
- Calculate the adiabatic flame temperature for a given set of reactants.
- Know the difference between the enthalpy of formation and the enthalpy of combustion.
- Know the definition of the higher and lower heating values.
- Apply the second law to a combustion problem and find irreversibilities.
- Calculate the change in Gibbs function and the reversible work.
- Know how a fuel cell operates and how to find its electrical potential.
- Understand that the chemical equilibrium is written for ideal-gas mixtures.
- Understand the meaning of the shift in Gibbs function due to the reaction.
- Know when the absolute pressure has an influence on the composition.
- Know the connection between the reaction scheme and the equilibrium constant.
- Understand that all species are present and influence the mole fractions.
- Know that a dilution with an inert gas has an effect.
- Understand the coupling between the chemical equilibrium and the energy equation.
- Intuitively know that most problems must be solved by iterations.
- Be able to treat a dissociation added on to a combustion process.

## KEY CONCEPTS AND FORMULAS

| | |
|---|---|
| Reaction | Fuel + Oxidizer $\Rightarrow$ Products |
| | hydrocarbon + air $\Rightarrow$ carbon dioxide + water + nitrogen |
| Stoichiometric ratio | No excess fuel, no excess oxygen |
| Stoichiometric coefficients | Factors to balance atoms between reactants and products |
| Stoichiometric reaction | $C_xH_y + \nu_{O_2}(O_2 + 3.76\,N_2)$ |
| | $\qquad \Rightarrow \nu_{CO_2}CO_2 + \nu_{H_2O}H_2O + \nu_{N_2}N_2$ |
| | $\nu_{O_2} = x + y/4; \quad \nu_{CO_2} = x; \quad \nu_{H_2O} = y/2; \quad \nu_{N_2} = 3.76\nu_{O_2}$ |

Air–fuel ratio
$$AF_{mass} = \frac{m_{air}}{m_{fuel}} = AF_{mole}\frac{M_{air}}{M_{fuel}}$$

Equivalence ratio
$$\Phi = \frac{FA}{FA_s} = \frac{AF_s}{AF}$$

| | |
|---|---|
| Enthalpy of formation | $\overline{h}_f^0$, zero for chemically pure substance elements, ground state |
| Enthalpy of combustion | $h_{RP} = H_P - H_R$ |
| Heating value HV | $HV = -h_{RP}$ |
| Adiabatic flame temperature | $H_P = H_R$ if flow |
| Reversible work | $W^{rev} = G_R - G_P = -\Delta G = -(\Delta H - T\Delta S)$ |
| | This requires that any $Q$ is transferred at the local $T$. |
| Gibbs function | $G = H - TS$ |
| Irreversibility | $i = w^{rev} - w = T_0\dot{S}_{gen}/\dot{m} = T_0 s_{gen}$ |
| | $I = W^{rev} - W = T_0\dot{S}_{gen}/\dot{n} = T_0 S_{gen}$ for 1 kmole fuel |
| Equilibrium | Minimum $G$ for given $T, P \Rightarrow dG_{T,P} = 0$ |

| Equilibrium reaction | $\nu_A A + \nu_B B \Leftrightarrow \nu_C C + \nu_D D$ |
| Change in Gibbs function | $\Delta G^0 = \nu_C \bar{g}_C^0 + \nu_D \bar{g}_D^0 - \nu_A \bar{g}_A^0 - \nu_B \bar{g}_B^0$ evaluate at $T$, $P^0$ |
| Equilibrium constant | $K = e^{-\Delta G^0/RT}$ |

$$K = \frac{y_C^{\nu_C} y_D^{\nu_D}}{y_A^{\nu_A} y_B^{\nu_B}} \left(\frac{P}{P^0}\right)^{\nu_C + \nu_D - \nu_A - \nu_B}$$

# HOMEWORK PROBLEMS

## Concept study guide problems

**12.1** Why would I sometimes need $A/F$ on a mole basis? on a mass basis?

**12.2** What is the dew point of hydrogen burned with stoichiometric pure oxygen? air?

**12.3** Why does combustion contribute to global warming?

**12.4** What is the enthalpy of formation for oxygen as $O_2$? If O? For $CO_2$?

**12.5** Is a heating value a fixed number for a fuel?

**12.6** Why do some fuels not have entries for liquid fuel in Table 12.3?

**12.7** Does it make a difference for the enthalpy of combustion whether I burn with pure oxygen or air? What about the adiabatic flame temperature?

**12.8** Is an adiabatic flame temperature a fixed number for a fuel?

**12.9** I burn a fuel with air in a steady flow burner. I now add a flow of an inert gas such as argon or helium to the reactant flow. What happens with the flame temperature?

**12.10** A welder uses a bottle with acetylene and a bottle with oxygen. Why does he use the oxygen bottle instead of air?

## Fuels and the combustion process

**12.11** Calculate the theoretical air–fuel ratio on a mass and mole basis for the combustion of ethanol, $C_2H_5OH$.

**12.12** A certain fuel oil has the composition $C_{10}H_{22}$. If this fuel is burned with 150% theoretical air, what is the composition of the products of combustion?

**12.13** In a picnic grill, gaseous propane is fed to a burner together with stoichiometric air. Find the air-fuel ratio on a mass basis and the total amount of reactant mass for 1 kg propane burned.

**12.14** Methane is burned with 200% theoretical air. Find the composition and the dew point of the products.

**12.15** Natural gas B from Table 12.2 is burned with 20% excess air. Determine the composition of the products.

**12.16** For complete stoichiometric combustion of gasoline, $C_7H_{17}$, determine the fuel molecular weight, the combustion products, and the mass of carbon dioxide produced per kg of fuel burned.

**12.17** Liquid propane is burned with dry air. A volumetric analysis of the products of combustion yields the following volume percent composition on a dry basis: 8.6% $CO_2$, 0.6% CO, 7.2% $O_2$, and 83.6% $N_2$. Determine the percent of theoretical air used in this combustion process.

**12.18** For the combustion of methane, 150% theoretical air is used at 25°C, 100 kPa, and relative humidity of 70%. Find the composition and dew point of the products.

**12.19** Butane is burned with dry air at 40°C, 100 kPa, with $AF = 26$ on a mass basis. For complete combustion find the equivalence ratio, % theoretical air, and the dew point of the products. How much water (kg/kg fuel) is condensed out, if any, when the products are cooled down to ambient temperature 40°C?

**12.20** Methanol, $CH_3OH$, is burned with 200% theoretical air in an engine, and the products are brought to 100 kPa, 30°C. How much water is condensed per kilogram of fuel?

**12.21** Pentane is burned with 120% theoretical air in a constant-pressure process at 100 kPa. The products are cooled to ambient temperature, 20°C. How much mass of water is condensed per kilogram of fuel? Repeat the answer, assuming that the air used in the combustion has a relative humidity of 90%.

## Energy equation, enthalpy of formation

**12.22** A rigid vessel initially contains 2 kmol of carbon and 2 kmol of oxygen at 25°C, 200 kPa. Combustion occurs, and the resulting products consist of 1 kmol of carbon dioxide, 1 kmol of carbon monoxide, and excess oxygen at a temperature of 1000 K. Determine the final pressure in the vessel and the heat transfer from the vessel during the process.

**12.23** The combustion of heptane $C_7H_{16}$ takes place in a steady-flow burner where fuel and air are added as gases at $P_0$, $T_0$. The mixture has 125% theoretical air, and the products pass through a heat exchanger where they are cooled to 600 K. Find the heat transfer from the heat exchanger per kmol of heptane burned.

**12.24** Butane gas and 200% theoretical air, both at 25°C, enter a steady-flow combustor. The products of combustion exit at 1000 K. Calculate the heat transfer from the combustor per kmol of butane burned.

**12.25** One alternative to using petroleum or natural gas as fuels is ethanol ($C_2H_5OH$), which is commonly produced from grain by fermentation. Consider a combustion process in which liquid ethanol is burned with 120% theoretical air in a steady-flow process. The reactants enter the combustion chamber at 25°C, and the products exit at 60°C, 100 kPa. Calculate the heat transfer per kilomole of ethanol.

**12.26** Do the previous problem with the ethanol fuel delivered as a vapor.

**12.27** As an alternative fuel, consider liquid methanol burned with stoichiometric air both supplied at $P_0$ and $T_0$ in a constant pressure process exhausting the products at 900 K. What is the heat transfer per kmol of fuel?

**12.28** Another alternative fuel to be seriously considered is hydrogen. It can be produced from water by various techniques that are under extensive study. Its biggest problems at the present time are cost, storage, and safety. Repeat Problem 12.25 using hydrogen gas as the fuel instead of ethanol.

**12.29** Pentene, $C_5H_{10}$, is burned with pure stoichiometric oxygen in a steady-flow process. The products at one point are brought to 700 K and used in a heat exchanger, where they are cooled to 25°C. Find the specific heat transfer in the heat exchanger.

**12.30** Methane, $CH_4$, is burned in a steady-flow adiabatic process with two different oxidizers: Case **A**: Pure oxygen, $O_2$, and case **B**: A mixture of $O_2 + x$Ar. The reactants are supplied at $T_0$, $P_0$ and the products for both cases should be at 1800 K. Find the required equivalence ratio in case **A** and the amount of argon, $x$, for a stoichiometric ratio in case **B**.

## Enthalpy of combustion and heating value

**12.31** Phenol has an entry in Table 12.3, but it does not have a corresponding value of the enthalpy of formation in Table A.10. Can you calculate it?

**12.32** In a picnic grill, gaseous propane and stoichiometric air are mixed and fed to the burner both at ambient $P_0$ and $T_0$. After combustion the products cool down and eventually reaches ambient $T_0$. How much heat transfer was given out for 1 kg of propane and how much heat transfer per kg reactant mixture?

**12.33** Solve Problem 12.23 using Table 12.3 instead of Table A.10 for the solution.

**12.34** Wet biomass waste from a food-processing plant is fed to a catalytic reactor, where in a steady-flow process it is converted into a low-energy fuel gas suitable for firing the processing plant boilers. The fuel gas has a composition of 50% methane, 45% carbon dioxide, and 5% hydrogen on a volumetric basis. Determine the lower heating value of this fuel gas mixture per unit volume.

**12.35** The enthalpy of combustion in Table 12.3 is per kg fuel. In the cycle analysis for the Brayton and Diesel cycles, we need a $q_H$ which is per kg mixture. Find this $q_H$ for methane and gaseous octane in both cases with water vapor in the products and stoichiometric mixture.

**12.36** Solve Problem 12.25 using Table 12.3 instead of Table A.10 for the solution.

**12.37** Propylbenzene, $C_9H_{12}$, is listed in Table 12.3, but not in Table A.10. No molecular weight is listed in the book. Find the molecular weight, the enthalpy of formation for the liquid fuel, and the enthalpy of evaporation.

**12.38** Solve Problem 12.27 using Table 12.3 instead of Table A.10 for the solution.

**12.39** A burner receives a mixture of two fuels with mass fraction 40% n-butane and 60% methanol, both vapor. The fuel is burned with stoichiometric air. Find the product composition and the lower heating value of this fuel mixture (kJ/kg fuel mix).

**12.40** Natural gas, we assume methane, is burned with 200% theoretical air, shown in Fig. P12.40, and the reactants are supplied as gases at the reference temperature and pressure. The products are flowing through a heat exchanger where they give off energy to some water flowing

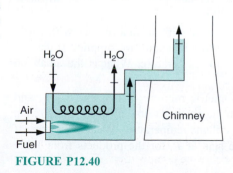

**FIGURE P12.40**

in at 20°C, 500 kPa, and out at 700°C, 500 kPa. The products exit at 400 K to the chimney. How much energy per kmole fuel can the products deliver, and how many kg water per kg fuel can they heat?

**12.41** A constant pressure burner receives a flow of gaseous benzene $C_6H_6$ and a stoichiometric amount of pure oxygen both supplied at $P_0$, $T_0$. To limit the adiabatic flame temperature at 2000 K, we spray liquid water in after the combustion. Find the kmol of liquid water added per kmol of fuel and the dew point of the combined products.

**12.42** Liquid nitromethane is added to the air in a carburetor to make a stoichiometric mixture where both fuel and air are added at 298 K, 100 kPa. After combustion, a constant-pressure heat exchanger brings the products to 600 K before being exhausted. Assume the nitrogen in the fuel becomes $N_2$ gas. Find the total heat transfer per kmol fuel in the whole process.

## Adiabatic flame temperature

**12.43** Hydrogen gas is burned with pure oxygen in a steady-flow burner, shown in Fig. P12.43, where both reactants are supplied in a stoichiometric ratio at the reference pressure and temperature. What is the adiabatic flame temperature?

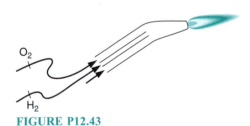

**FIGURE P12.43**

**12.44** In a rocket, hydrogen is burned with air, both reactants supplied as gases at $P_0$, $T_0$. The combustion is adiabatic, and the mixture is stoichiometric (100% theoretical air). Find the products' dew point and the adiabatic flame temperature (~2500 K).

**12.45** Carbon is burned with air in a furnace with 150% theoretical air, and both reactants are supplied at the reference pressure and temperature. What is the adiabatic flame temperature?

**12.46** A stoichiometric mixture of benzene, $C_6H_6$, and air is mixed from the reactants flowing at 25°C, 100 kPa. Find the adiabatic flame temperature. What is the error if constant-specific heat at $T_0$ for the products from Table A.5 is used?

**12.47** Hydrogen gas is burned with 200% theoretical air in a steady-flow burner where both reactants are supplied at the reference pressure and temperature. What is the adiabatic flame temperature?

**12.48** A gas-turbine burns natural gas (assume methane) where the air is supplied to the combustor at 1000 kPa, 500 K, and the fuel is at 298 K, 1000 kPa. What is the percent theoretical air if the adiabatic flame temperature should be limited to 1800 K?

**12.49** Liquid $n$-butane at $T_0$, is sprayed into a gas turbine, as in Fig. P12.49, with primary air flowing at 1.0 MPa, 400 K, in a stoichiometric ratio. After complete combustion, the products are at the adiabatic flame temperature, which is too high, so secondary air at 1.0 MPa, 400 K, is added, with the resulting mixture being at 1400 K. Show that $T_{ad} > 1400$ K and find the ratio of secondary to primary airflow.

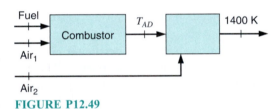

**FIGURE P12.49**

**12.50** Butane gas at 25°C is mixed with 150% theoretical air at 600 K and is burned in an adiabatic steady-flow combustor. What is the temperature of the products exiting the combustor?

**12.51** Natural gas, we assume methane, is burned with 200% theoretical air, and the reactants are supplied as gases at the reference temperature and pressure. The products are flowing through a heat exchanger and then out the exhaust, as in Fig. P12.51. What is the adiabatic flame temperature right after combustion before the heat exchanger?

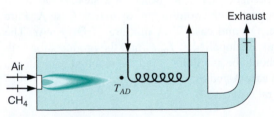

**FIGURE P12.51**

**12.52** Liquid butane at 25°C is mixed with 150% theoretical air at 600 K and is burned in a steady flow burner. Use the enthalpy of combustion from Table 12.3 to find the adiabatic flame temperature out of the burner.

**12.53** Acetylene gas at $P_0$, $T_0$ is fed to the head of a cutting torch. Find the adiabatic flame temperature if it is burned with stoichiometric pure oxygen at $P_0$, $T_0$.

**12.54** Acetylene gas at $P_0$, $T_0$ is fed to the head of a cutting torch. Find the adiabatic flame temperature if it is burned with stoichiometric air at $P_0$, $T_0$.

**12.55** Ethene, $C_2H_4$, burns with 150% theoretical air in a steady-flow constant-pressure process with reactants entering at $P_0$, $T_0$. Find the adiabatic flame temperature.

**12.56** Gaseous ethanol, $C_2H_5OH$, is burned with pure oxygen in a constant-volume combustion bomb. The reactants are charged in a stoichiometric ratio at the reference condition. Assume no heat transfer and find the final temperature ($>5000$ K).

## Second law for the combustion process

**12.57** Consider the combustion of hydrogen with pure oxygen in a stoichiometric ratio under steady-flow adiabatic conditions. The reactants enter separately at 298 K, 100 kPa, and the product(s) exit at a pressure of 100 kPa. What is the exit temperature, and what is the irreversibility?

**12.58** Methane is burned with air, both of which are supplied at the reference conditions. There is enough excess air to give a flame temperature of 1800 K. What are the percent theoretical air and the irreversibility in the process?

**12.59** Consider the combustion of methanol, $CH_3OH$, with 25% excess air. The combustion products are passed through a heat exchanger and exit at 200 kPa, 400 K. Calculate the absolute entropy of the products exiting the heat exchanger assuming all the water is vapor.

**12.60** Consider the combustion of methanol, $CH_3OH$, with 25% excess air. The combustion products are passed through a heat exchanger and exit at 200 kPa, 40°C. Calculate the absolute entropy of the products exiting the heat exchanger per kilomole of methanol burned, using proper amounts of liquid and vapor water.

**12.61** An inventor claims to have built a device that will take 0.001 kg/s of water from the faucet at 10°C, 100 kPa, and produce separate streams of hydrogen and oxygen gas, each at 400 K, 175 kPa. It is stated that this device operates in a 25°C room on 10-kW electrical power input. How do you evaluate this claim?

**12.62** Two kilomoles of ammonia are burned in a steady-flow process with $x$ kmol of oxygen. The products, consisting of $H_2O$, $N_2$, and the excess $O_2$, exit at 200°C, 7 MPa.

a. Calculate $x$ if half the water in the products is condensed.

b. Calculate the absolute entropy of the products at the exit conditions.

**12.63** A flow of hydrogen gas is mixed with a flow of oxygen in a stoichiometric ratio, both at 298 K and 50 kPa. The mixture burns without any heat transfer in complete combustion. Find the adiabatic flame temperature and the amount of entropy generated per kmole hydrogen in the process.

## Fuel cells

**12.64** In Example 12.13, a basic hydrogen–oxygen fuel cell reaction was analyzed at 25°C, 100 kPa. Repeat this calculation, assuming that the fuel cell operates on air at 25°C, 100 kPa, instead of on pure oxygen at this state.

**12.65** Assume the basic hydrogen-oxygen fuel cell operates at 600 K instead of 298 K as in Example 12.13. Find the change in the Gibbs function and the reversible EMF it can generate.

**12.66** Consider a methane–oxygen fuel cell in which the reaction at the anode is

$$CH_4 + 2H_2O \rightarrow CO_2 + 8e^- + 8H^+$$

The electrons produced by the reaction flow through the external load, and the positive ions migrate through the electrolyte to the cathode, where the reaction is

$$8e^- + 8H^+ + 2O_2 \rightarrow 4H_2O$$

Calculate the reversible work and the reversible EMF for the fuel cell operating at 25°C, 100 kPa.

**12.67** Redo the previous problem, but assume that the fuel cell operates at 1200 K instead of at room temperature.

## Chemical equilibrium, equilibrium constant

**12.68** Is the dissociation of water (see Table A.11) pressure sensitive?

**12.69** At 298 K, $K = \exp(-184)$ for the water dissociation. What does that imply?

**12.70** In a combustion process, is the adiabatic flame temperature affected by reactions?

**12.71** When dissociations occur after combustion, does $T$ go up or down?

**12.72** In equilibrium, Gibbs function of the reactants and the products is the same; how about the energy (is $H_R = H_P$)?

**12.73** Calculate the equilibrium constant for the reaction $O_2 \rightleftharpoons 2O$ at temperatures of 298 K and 6000 K. Verify the result with Table A.11.

**12.74** Calculate the equilibrium constant for the reaction $H_2 \rightleftharpoons 2H$ at a temperature of 2000 K, using properties from Table A.9. Compare the result with the value listed in Table A.11.

**12.75** Consider the dissociation of oxygen, $O_2 \Leftrightarrow 2O$, starting with 1 kmol oxygen at 298 K and heating it at constant pressure 100 kPa. At which temperature will we reach a concentration of monatomic oxygen of 10%?

**12.76** Pure oxygen is heated from 25°C to 3200 K in a steady-state process at a constant pressure of 200 kPa. Find the exit composition and the heat transfer.

**12.77** Nitrogen gas, $N_2$, is heated to 4000 K, 10 kPa. What fraction of the $N_2$ is dissociated to N at this state?

**12.78** Hydrogen gas is heated from room temperature to 4000 K, 500 kPa, at which state the diatomic species has partially dissociated to the monatomic form. Determine the equilibrium composition at this state.

**12.79** One kilomole Ar and one kilomole $O_2$ are heated at a constant pressure of 100 kPa to 3200 K, where they come to equilibrium. Find the final mole fractions for Ar, $O_2$, and O.

**12.80** Consider the reaction $2CO_2 \Leftrightarrow 2CO + O_2$ obtained after heating 1 kmol $CO_2$ to 3000 K. Find the equilibrium mole fraction of CO at 3000 K, 200 kPa.

**12.81** Air (assumed to be 79% nitrogen and 21% oxygen) is heated in a steady-state process at a constant pressure of 100 kPa, and some NO is formed (disregard dissociations of $N_2$ and $O_2$). At what temperature will the mole fraction of NO be 0.001?

**12.82** The combustion products from burning pentane, $C_5H_{12}$, with pure oxygen in a stoichiometric ratio exit at 2400 K, 100 kPa. Consider the dissociation of only $CO_2$ and find the equilibrium mole fraction of CO.

**12.83** A mixture of 1 kmol carbon dioxide, 2 kmol carbon monoxide, and 2 kmol oxygen, at 25°C, 150 kPa, is heated in a constant-pressure steady-state process to 3000 K. Assuming that only these same substances are present in the exiting chemical equilibrium mixture, determine the composition of that mixture.

**12.84** Water from the combustion of hydrogen and pure oxygen is at 3800 K and 50 kPa. Assume we only have $H_2O$, $O_2$ and $H_2$ as gases, find the equilibrium composition.

**12.85** A piston/cylinder contains 0.1 kmol hydrogen and 0.1 kmol Ar gas at 25°C, 200 kPa. It is heated in a constant-pressure process so the mole fraction of atomic hydrogen is 10%. Find the final temperature and the heat transfer needed.

**12.86** A tank contains 0.1 kmol hydrogen and 0.1 kmol of argon gas at 25°C, 200 kPa, and the tank keeps constant volume. To what $T$ should it be heated to have a mole fraction of atomic hydrogen, H, of 10%?

**12.87** A gas mixture of 1 kmol carbon monoxide, 1 kmol nitrogen, and 1 kmol oxygen at 25°C, 150 kPa, is heated in a constant-pressure steady-state process. The exit mixture can be assumed to be in chemical equilibrium with $CO_2$, CO, $O_2$, and $N_2$ present. The mole fraction of $CO_2$ at this point is 0.176. Calculate the heat transfer for the process.

# Chapter **13**

# Introduction to Heat Transfer

This chapter will give an introduction to analysis of the heat transfer, a topic that is important for many applications. In the preceeding chapters, heat transfer has been dealt with as a mode of energy transfer either as a rate for steady or transient processes or in a finite amount. The details for the predictions of the rate or the elapsed time to transfer a given amount was not elaborated on, so a short introduction to heat transfer with an emphasis on conduction is given in this chapter. Some coverage of radiation and convection is also presented but the details of convection are left to be studied after or together with fluid mechanics. As a special application, we have included a section about cooling of electronic equipment, which is important for further increases in density and speed of junctions in chips like a CPU.

## 13.1 HEAT TRANSFER MODES

A brief introduction to the three modes of heat transfer is given in Section 4.7 as the energy transport due to a temperature difference. Physically there are two modes of heat transfer: conduction is energy exchange on an atomic scale due to various motions of atoms, molecules, and electrons; and radiation is energy transfer by electromagnetic

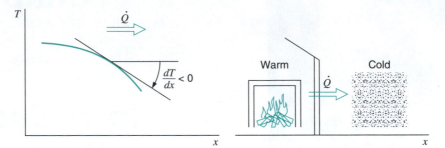

**FIGURE 13.1** Conduction heat transfer to lower temperature region.

waves. When there is a flow of a substance, it has a dominant influence on the local conduction; because of this, a third mode is called convection heat transfer.

## Conduction

The conduction of energy is accomplished through interactions among atoms, molecules, and electrons and thus depends on the detailed structure of the substance mass. With small distances between atoms and relatively strong bonds, solids and liquids show the highest ability to conduct energy. A significant amount of this conduction is done by electrons, so metals are the best conductors owing to the high mobility of the electrons. The opposite is the case with gases, where the intermolecular distances are large and the interactions and forces between them are weak, becoming strong only during short-time collisions where momentum and energy are exchanged.

Heat transfer by conduction is energy flowing from a higher temperature region to a lower temperature region, as shown in Fig. 13.1. The conduction heat transfer rate is modeled according to Fourier's law

$$\dot{Q} = -kA\frac{dT}{dx} \qquad (13.1)$$

where $A$ is the cross sectional area normal to the $x$-direction of the heat flux. The minus sign indicates that heat is transferred from the higher temperature region to the lower temperature region, opposite the temperature gradient. Typical values for the conductivity, $k$, of substances given in Tables A.3–A.5 are shown in Table 13.1.

In the following sections we will look at physical situations where we can solve for the temperature gradient and relate the heat flux, $\dot{Q}$, to finite temperature differences.

## Convection

Heat transfer by convection is determined by the detailed flow in which conduction takes place locally. A solution to the heat transfer must then be accompanied by a solution for the flow. This makes the description of these situations very complex and often difficult to solve mathematically. An average can be performed over the heat transfer layers to express the heat transfer by the overall temperature difference in Newton's law of cooling

$$\dot{Q} = hA(T_s - T_\infty) \qquad (13.2)$$

Here $A$ is the surface area with temperature $T_s$ and $h$ is the convective heat transfer coefficient to the fluid at temperature $T_\infty$ far away from the surface. Variation of the temperature with the distance normal to the surface is indicated in Fig. 13.2. The magnitude of the convective heat transfer coefficient, $h$, depends on the flow, its velocity, and the overall geometry. Some very general estimates for these values are presented in Table 13.2.

**TABLE 13.1**

*Conductivity of Selected Substances from Tables A.3, A.4, and A.5. Properties at 300 K, 100 kPa (or saturation pressure).*

| Solids from A.3 | k [W/m K] | Liquids from A.4 | k [W/m K] |
|---|---|---|---|
| Asphalt | 0.7 | Ammonia | 0.51 |
| Brick, common | 0.7 | Gasoline | 0.12 |
| Carbon diamond | 1350 | Glycerine | 0.29 |
| Carbon graphite | 155 | Methanol | 0.19 |
| Concrete | 1.28 | Oil engine | 0.14 |
| Glass plate | 1.40 | Propane | 0.095 |
| Glass wool | 0.037 | R-12 | 0.07 |
| Granite | 2.9 | R-134a | 0.08 |
| Ice (0°C) | 2.25 | Water | 0.60 |
| Plaster | 0.79 | Lead | 16.3 |
| Plexiglas | 0.18 | **Gases from A.5** | **k [W/m K]** |
| Polystyrene | 0.35 | | |
| Rock salt | 7 | Air | 0.026 |
| Snow, firm | 0.46 | Argon | 0.018 |
| Wood, hard | 0.16 | Carbon dioxide | 0.017 |
| Wood, soft | 0.12 | Helium | 0.155 |
| Wool | 0.036 | Neon | 0.05 |
| **Metals** | | Nitrogen | 0.026 |
| Aluminum | 164 | Oxygen | 0.027 |
| Copper, pure | 372 | Propane | 0.018 |
| Copper, commercial | 52 | R-12 | 0.01 |
| Iron, cast | 52 | R-134a | 0.013 |
| Iron, steel | 15 | Water steam | 0.02 |

Further studies of fluid mechanics and heat transfer will show correlations between a dimensionless form of *h*, the Nusselt number

$$Nu = \frac{hL}{k}$$

and a dimensionless velocity, the Reynolds number

$$Re = \frac{\rho \mathbf{V} L}{\mu}$$

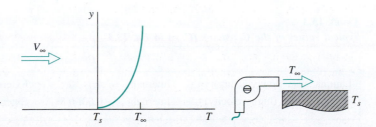

**FIGURE 13.2** Convection heat transfer layer.

**TABLE 13.2**

*Convection Heat Transfer Coefficient*

| Heat Transfer Coefficient $h$ [W/m$^2$ K] | Gas | Liquid |
|---|---|---|
| Natural convection | 5–25 | 50–1000 |
| Forced convection | 25–250 | 50–20 000 |
| Boiling/condensation | | 2500–100 000 |

with the absolute viscosity $\mu$, density $\rho$, and a characteristic length $L$. Such correlations for different situations are shown in technical handbooks typically of the form

$$Nu = C\,Re^n Pr^{1/3} \tag{13.3}$$

with the dimensionless Prandtl number, $Pr = \mu C_p / k$. Some characteristic values of the constants $C$ and $n$ are given in Table 13.3 for flows inside pipes or ducts (internal) or outside (external or free flow) and relates to the average value over the length $L$. Additional explanations and details for different flows are presented in Section 13.6.

## Radiation

Radiation is the transport of energy by electromagnetic waves, which takes place over all wavelengths. The distribution in wavelength depends on the temperature characteristic of the source, which for moderate temperatures (cool to warm) has a major part of the flux in the range invisible to the human eye (infrared). For higher temperatures, a larger fraction of the energy is transmitted in the visible part of the spectrum, for example, the light from the sun or light bulbs. The total flow of energy (a rate) is summed over all the wavelengths, the details of which are beyond the current presentation.

Emission of energy from a surface depends on its surface temperature to the fourth power, and for an ideal black body, the Stefan Boltzmann law expresses the total emissive power as

$$\dot{Q} = \sigma A T_s^4 \tag{13.4}$$

where $T_s$ is the absolute temperature in Kelvins and $\sigma$ is the Stefan-Boltzmann constant

$$\sigma = 5.67051 \times 10^{-8} \text{ W/m}^2 \text{ K}^4$$

Any real surface emits an amount of energy that is a fraction of the ideal black body radiation; this fraction is the emissivity $\varepsilon$. Typical values for various surfaces are shown in Table 13.4 and represent an average over wavelength and solid substances.

Radiation hitting a surface is partly reflected, with the remainder transmitted through or absorbed by the surface material. Assuming the absorption coefficient is the same as the emissivity, $\varepsilon$, for such a gray surface the net heat flux exchanged with the

**TABLE 13.3**

*Typical Values of the Constants (C, n) in Eq. 13.3*

| | | | $C$ | $n$ | | $C$ | $n$ |
|---|---|---|---|---|---|---|---|
| Laminar flow (low velocity): | Internal | | 4, | 0 | External | 0.7, | 0.5 |
| Turbulent flow (high velocity): | Internal | | 0.03, | 0.8 | External | 0.03, | 0.8 |

**TABLE 13.4**

*Typical Surface Emissivities, ε*

| Perfect Black body | 1 | Metal surface | 0.6–0.9 |
| Non-metallic surface | 0.9 | Polished metal | 0.1 |

surroundings can be written

$$\dot{Q} = \varepsilon \sigma A (T_s^4 - T_\infty^4) \tag{13.5}$$

For cases where radiation is exchanged between multiple surfaces with emission, reflection and absorption, a more extensive analysis and treatment is required. Such analysis is beyond the scope of the current introduction.

## Surface Energy Balance

In various situations we often have a control volume for the analysis that is placed on the surface (inside or outside) of some solid material such as the steel wall of a container. This surface usually has a media as a fluid next to it that may be flowing, for example, ambient air or steam inside a steel pipe. Inside the solid material, the heat transfer takes place by conduction, whereas in the fluid, it is convection and possibly radiation.

The connection between the various energy flows (heat transfer) is established with an infinitely thin control volume on the surface, with one side on the solid and the other side to the fluid (see Fig. 13.3). Since this control volume does not have any mass or volume, it cannot store any energy, and the energy equation becomes

$$\frac{dE_{cv}}{dt} = 0 = \dot{Q}_s - \dot{Q}_{Fluid} = (\dot{q}_s'' - \dot{q}_{Fluid}'') A \tag{13.6}$$

Locally there is a balance between the two sides as

$$\dot{Q}_s = \dot{Q}_{Fluid} = \dot{Q}_{Conv} + \dot{Q}_{Rad} \tag{13.7}$$

Substituting the expressions for the conduction in the solid (Eq. 13.1), the convection (Eq. 13.2) and the radiation (Eq. 13.5) in the fluid, and dividing by the area, we get

$$-k_s \left[ \frac{dT}{dx} \right]_s = h(T_s - T_\infty) + \varepsilon \sigma (T_s^4 - T_\infty^4)$$

Notice how this couples the temperature gradient in the solid to the surface temperature, and if radiation is present, this coupling is non-linear. If the fluid does not have any motion (forced or natural convection), then the fluid convection should be replaced by conduction in the fluid.

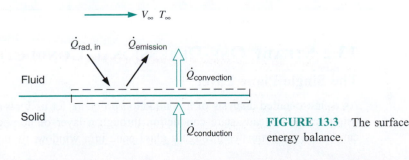

**FIGURE 13.3** The surface energy balance.

**EXAMPLE 13.1**

A steady heat transfer takes place through a 5-mm-thick plastic casing, $k = 0.35$ W/m K, of a power transistor with surface temperature $T_1$ on the left side and an area of 4 cm$^2$ as shown in Fig. 13.4. The right side has $T_s = 85°C$ exposed to a fluid at 25°C with a convective heat transfer coefficient of $h = 25$ W/m$^2$ K, and the surface has an emissivity of 0.9. For a linear temperature variation in the plastic casing, find the convection and radiation heat transfer rates in the fluid and the necessary $T_1$.

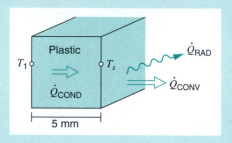

**FIGURE 13.4**  Steady heat transfer balance at a transistor surface.

**Analysis**

The steady state heat transfer has convection from Eq. 13.2, radiation from Eq. 13.5 leading to the surface energy balance in Eq. 13.7 with conduction from Eq. 13.1.

**Solution**

We can find the convection heat transfer rate from Eq. 13.2

$$\dot{Q}_{\text{Conv}} = hA(T_s - T_\infty) = 25 \times 4 \times 10^{-4}(85 - 25) = 0.6 \text{ W}$$

The radiation from Eq. 13.5 is (remember to use absolute $T$)

$$\dot{Q}_{\text{Rad}} = \varepsilon\sigma A(T_s^4 - T_\infty^4) = 0.9 \times 5.67 \times 10^{-8} \times 4 \times 10^{-4}(358.15^4 - 298.15^4)$$
$$= 0.174 \text{ W}$$

From the surface energy balance in Eq. 13.7 the conduction heat transfer is

$$\dot{Q}_{\text{Cond}} = \dot{Q}_{\text{Conv}} + \dot{Q}_{\text{Rad}} = 0.774 \text{ W}$$

Assuming a linear variation of $T$ in the plastic, Eq. 13.1 is evaluated as

$$\dot{Q}_{\text{Cond}} = -kA\frac{dT}{dx} = -kA\frac{T_s - T_1}{L} = kA\frac{T_1 - T_s}{L}$$

Solving for $T_1$ gives

$$T_1 = T_s + \dot{Q}_{\text{Cond}} L/kA = 85 + 0.774 \times 0.005/(0.35 \times 4 \times 10^{-4})$$
$$= 113°C$$

# 13.2 STEADY ONE-DIMENSIONAL CONDUCTION

## The Single Plane Layer

As a first detailed analysis of a conduction problem, let us look at the temperature distribution for a steady-state conduction through a layer of homogeneous material. This can be the heat transfer through a glass pane in a window, through the thin metal wall

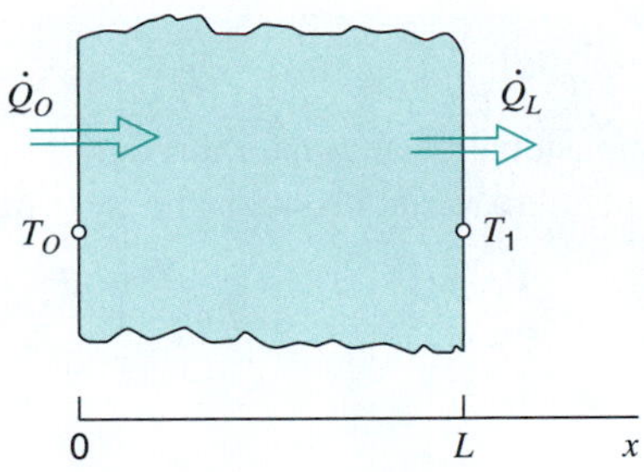

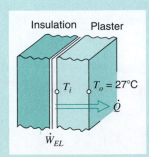

**FIGURE 13.5** A plane steady-state conduction layer.

of a radiator, the view window in an oven, or a layer of substrate in a computer chip. The energy is flowing in a single direction normal to the surface of the layer and there is no generation of energy within the layer. The situation is depicted in Fig. 13.5.

The energy equation for a plane layer of thickness $L$ becomes

$$\frac{dE_{CV}}{dt} = 0 = \dot{Q}_0 - \dot{Q}_L$$

Restating the balance of the two rates and substituting Fourier's law of conduction gives

$$\dot{Q}_0 = \dot{Q}_L = -kA\frac{dT}{dx} = \text{constant} \tag{13.8}$$

For a constant cross-sectional area, $A$, and constant conductivity $k$, we have a constant temperature gradient resulting in a linear temperature variation through the layer. The rate of heat transfer is

$$\dot{Q} = -kA\frac{dT}{dx} = -kA\frac{T_1 - T_0}{L} \tag{13.9}$$

which relates the finite temperature difference to the rate of heat transfer. An example anticipating this result is given in Example 4.4 for a glass pane, and in Example 13.1 for conduction through a brick layer. A different situation is given in the following example.

---

**EXAMPLE 13.2**

Consider an electric heating mat that dissipates 500 W uniformly over an area of 0.5 m$^2$ with a perfectly insulated backing, as shown in Fig. 13.6. On the front side is a l-cm-thick plaster board with an outer surface temperature of 27°C. Find the plaster temperature on the side facing the heater.

**FIGURE 13.6**

**Analysis**

Assume steady-state, one-dimensional heat transfer through the plaster. The constant flux of energy gives a linear temperature variation, with the heat transfer rate as Eq. 13.9:

$$\dot{Q} = \dot{W}_{el} = -kA\frac{dT}{dx} = kA\frac{T_i - T_o}{L}$$

**Solution**

The conductivity from Table 13.1 is $k = 0.79$ W/m K, so we can solve for $T_i$

$$T_i = T_o + \dot{Q}L/(kA) = 27 + 500 \times 0.01/(0.79 \times 0.5)$$

$$= 27 + 13.7 = 39.7°C$$

## Thermal Resistance

The concept of a thermal resistance is introduced by writing the rate of heat transfer from Eq. 13.9 in a different form as

$$\dot{Q} = -kA\frac{T_1 - T_0}{L} = \frac{T_0 - T_1}{(L/kA)} = \Delta T/R_{cond} \tag{13.10}$$

where

$$R_{cond} = \frac{L}{kA} \tag{13.11}$$

is the conduction thermal resistance with units [K/W]. From Eq. 13.10 we notice that a high thermal resistance means a small rate of heat transfer for a given temperature difference, and a small thermal resistance gives a larger rate of heat transfer. This is similar to an electrical current through an ohmic resistance for a given voltage drop, an analogy that is further explained later.

The expression in Eq. 13.11 is for a single conduction layer deduced from Eq. 13.10 so with a similar expression for convection in Eq. 13.2

$$\dot{Q} = hA(T_s - T_\infty) = \Delta T/R_{conv}$$

we have the definition of the single convection layer resistance

$$R_{conv} = \frac{1}{hA} \tag{13.12}$$

It is also possible to use this for the radiation heat transfer, although it is non-linear. We do this by factoring out the temperature difference in Eq. 13.5:

$$T_s^4 - T_\infty^4 = (T_s^2 + T_\infty^2)(T_s^2 - T_\infty^2)$$

$$= (T_s^2 + T_\infty^2)(T_s + T_\infty)(T_s - T_\infty)$$

The radiation heat transfer from Eq. 13.5 is then written as

$$\dot{Q}_{Rad} = \varepsilon\sigma A(T_s^4 - T_\infty^4)$$

$$= \varepsilon\sigma A(T_s^2 + T_\infty^2)(T_s + T_\infty)(T_s - T_\infty) \tag{13.13}$$

$$= (T_s - T_\infty)/R_{Rad}$$

and the radiation thermal resistance is

$$R_{\text{Rad}} = 1/[\varepsilon \sigma A (T_s^2 + T_\infty^2)(T_s + T_\infty)] \tag{13.14}$$

The thermal resistance is given by the mode of the heat transfer, the properties of the media and flow, and the geometry. Once the thermal resistance is evaluated for a given situation, it is used to find the rate of heat transfer as

$$\dot{Q} = \Delta T/R \tag{13.15}$$

or the temperature difference

$$\Delta T = R\dot{Q} \tag{13.16}$$

---

**EXAMPLE 13.3**

Calculate the thermal resistance for the transistor casing plastic, the convective layer, and the radiation in Example 13.1.

**Analysis**

Steady heat transfer in a plane layer has resistances as in Eq. 13.11–12 and 13.14.

**Solution**

The thermal resistance due to the conductivity is from Eq. 13.11:

$$R_{\text{cond}} = \frac{L}{kA} = \frac{0.005}{0.35 \times 4 \times 10^{-4}} = 35.7 \text{ K/W}$$

The thermal convective resistance is from Eq. 13.12:

$$R_{\text{conv}} = 1/hA = 1/(25 \times 4 \times 10^{-4}) = 100 \text{ K/W}$$

Finally, the thermal radiation resistance is from Eq. 13.14:

$$R_{\text{Rad}} = 1/[\varepsilon \sigma A (T_s^2 + T_\infty^2)(T_s + T_\infty)]$$

$$= 1/[0.9 \times 5.67 \times 10^{-8} \times 4 \times 10^{-4}(358.15^2 + 298.15^2)(358.15 + 298.15)]$$

$$= 343 \text{ K/W}$$

---

## Composite Layers

In most constructions that involve heat transfer, the media is not a single uniform layer but often several layers of different substances. This is the case of the heat transfer out through a metal pipe wall wrapped with insulation and a protective cover or the wall in a common building that consists of an inside plasterboard, fiber glass insulation, a vapor barrier, and the outer wall of either brick, wood, or aluminum siding. A household refrigerator typically has an inside plastic wall, a layer of insulation, and an outside metal side with a plastic coating. These are examples of several layers sandwiched together, forming a composite layer.

The analysis of a composite layer follows the single layer analysis with a matching condition where the layers are joined together. Let us look at the combination of two plane layers of different materials under steady conditions and a perfect thermal contact between the two layers so the temperature at the contact is $T_1$ in both materials, as illustrated in Fig. 13.7. Imagine a control volume of zero thickness around the position where

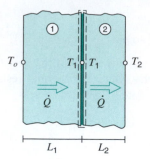

**FIGURE 13.7**  A two-layer composite configuration.

the two layers are joined, as with the surface control volume in Section 13.1.4. Since the control volume does not store energy, the heat transfer at one side equals the heat transfer at the other side,

$$\dot{Q} = -k_1 A \frac{T_1 - T_0}{L_1} = -k_2 A \frac{T_2 - T_1}{L_2} \qquad (13.17)$$

The temperature differences are expressed from the rate of heat transfer as Eq. 13.10:

$$T_0 - T_1 = \dot{Q} L_1 / (k_1 A) = \dot{Q} R_1$$

$$T_1 - T_2 = \dot{Q} L_2 / (k_2 A) = \dot{Q} R_2$$

where we have introduced the thermal resistances for the two layers. If the two temperature differences are added, we get the total temperature difference across the combined two layers as

$$T_0 - T_1 = T_0 - T_1 + T_1 - T_2 = \dot{Q} R_1 + \dot{Q} R_2 = \dot{Q}(R_1 + R_2) \qquad (13.18)$$

Comparing this expression with the form in Eq. 13.10 for the combined layers,

$$\Delta T_{\text{TOT}} = T_0 - T_2 = \dot{Q} R_{\text{TOT}} \qquad (13.19)$$

we have

$$R_{\text{TOT}} = R_1 + R_2 \qquad (13.20)$$

The total thermal resistance of the two layers in series ($\dot{Q}$ goes through both) is then the sum of the two individual layers' resistances.

   If the two layers do not have a perfect thermal contact, the temperature at the two sides facing each other can be different. The contact plane is not perfectly smooth if we look at it under a microscope and see that each surface has mountains and valleys touching only at a few locations, maybe trapping air or some other material in the voids between the two surfaces. Such a situation appears as a temperature jump at the contact location, when the very thin layer is assigned a zero thickness as shown in Fig. 13.8. This contact layer has a temperature difference across it and a rate of heat transfer through it so there is an associated contact resistance

$$\Delta T_C = T_{S1} - T_{S2} = \dot{Q} R_C \qquad (13.21)$$

similar to the expression in Eq. 13.16. This resistance generally depends on the surface preparations, how tightly the two surfaces are pressed together, and if any glue, creme, or insulating sheet is used. How this is done depends on the desire for a maximum resistance (small heat flux or large $\Delta T$) or a small resistance (large heat flux or small $\Delta T$) in the design. Deposits of material on the surface of pipes or ducts are generally

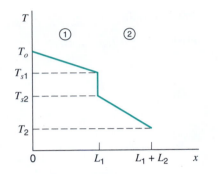

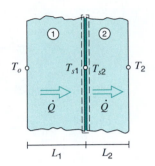

**FIGURE 13.8** A contact resistance between two layers.

undesirable because they cause fouling and represent a resistance to heat transfer as well as the flow inside the pipe or duct.

This result can be generalized from the two-layer composite to any number of layers of different types. A composite layer typically separates two fluids, say steam inside a pipe and air outside, such that there are also two convective layers besides conduction and contact layers.

Assuming we have multiple layers with a steady heat transfer through all the layers, then for each layer $i$

$$\Delta T_i = \dot{Q} R_i \tag{13.22}$$

so the total temperature difference is

$$\Delta T_{\text{TOT}} = \sum \Delta T_i = \sum \dot{Q} R_i = \dot{Q} \sum R_i = \dot{Q} R_{\text{TOT}} \tag{13.23}$$

The total thermal resistance is summed over all the layers in series as

$$R_{\text{TOT}} = \sum R_i = \sum \frac{L}{kA} + \sum \frac{1}{hA} + \sum R_C \tag{13.24}$$

Dividing Eq. 13.22 by the total temperature difference, Eq. 13.23, we obtain

$$\Delta T_i / \Delta T_{\text{TOT}} = \dot{Q} R_i / \dot{Q} R_{\text{TOT}} = R_i / R_{\text{TOT}} \tag{13.25}$$

We now see that each layer's temperature difference is equal to a fraction of the total temperature difference, and that the fraction equals the layer's thermal resistance divided by the total resistance. Notice that this is true only when each layer has the same rate of heat transfer through it.

**EXAMPLE 13.4**

Consider 2 m$^2$ of plane composite walls forming the bottom in a 70°C hot water tank. The walls have 2 mm of steel insulated with 8 cm of glass wool covered by a plastic casing 5 mm thick as shown in Fig. 13.9. The thermal conductivity of the plastic is $k = 0.2$ W/m K. Convection to the outside ambient air at 15°C occurs with a heat transfer coefficient of $h = 10$ W/m$^2$ K. We want to find the heat transfer loss from the water through the bottom of the tank.

**Analysis**

The physical setup gives a steady-state heat transfer through a one-dimensional plane composite layer. We can use the total resistance from Eq. 13.24 to find the heat transfer from the temperatures.

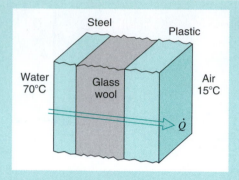

**FIGURE 13.9**  Water tank wall.

**Solution**

$$R_{\text{TOT}} = R_{\text{steel}} + R_{\text{ins}} + R_{\text{plastic}} + R_{\text{conv}}$$

$$= \frac{L_{\text{st}}}{k_{\text{st}} A} + \frac{L_{\text{ins}}}{k_{\text{ins}} A} + \frac{L_{\text{plastic}}}{k_{\text{plastic}} A} + \frac{1}{h_{\text{conv}} A}$$

$$= \frac{0.002}{15 \times 2} + \frac{0.08}{0.037 \times 2} + \frac{0.005}{0.2 \times 2} + \frac{1}{10 \times 2}$$

$$= 0.000067 + 1.081 + 0.0125 + 0.05$$

$$= 1.144 \text{ K/W}$$

The heat transfer from Eq. 13.23 is

$$\dot{Q} = \Delta T_{\text{TOT}}/R_{\text{TOT}} = (70 - 15)/1.144 = 48.1 \text{ W}$$

The temperature difference over each layer could be found from Eq. 13.25.

## Cylindrical Layers

There are numerous situations where heat transfer occurs in a radial direction with cylindrical geometry. Flow inside a pipe has heat transfer through the pipe wall, and an electrical wire that dissipates some power has energy leaving through the insulation layer. A hot water storage tank of cylindrical shape has heat transfer through relatively thin metal walls and a thick layer of insulation to the ambient air.

The solution for the steady conduction of energy through a homogeneous radial layer is similar to the plane-layer solution in Section 13.2.1. Assume a hollow cylinder (pipe) with inner radius $r_0$, surface temperature $T_0$, and a total heat flux of $\dot{Q}_0$ in the positive radial direction as shown in Fig. 13.10.

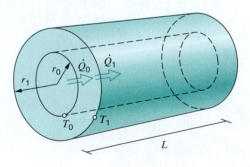

**FIGURE 13.10**  Heat transfer in cylindrical geometry.

If we select a control volume of total length $L$ from $r_0$ to $r_1$ at which surface we have $T_1$ and $\dot{Q}_1$, the energy equation becomes

$$\frac{dE_{cv}}{dt} = 0 = \dot{Q}_0 - \dot{Q}_1 \tag{13.26}$$

Restating the balance of the two rates and substituting Fourier's law of conduction gives

$$\dot{Q}_0 = \dot{Q}_1 = -kA\frac{dT}{dr} = -k\,2\pi rL\,\frac{dT}{dr} = \text{constant} \tag{13.27}$$

where $r$ and the temperature gradient are evaluated at the desired location. If the location $r_0$ is kept fixed and $r_1$ is varied, the energy equation stays the same, so the total heat flux is constant as indicated. Equation (13.27) is now integrated to find the function $T(r)$ as

$$dT = -\frac{\dot{Q}_0}{2\pi kL}\,r^{-1}\,dr \Rightarrow \int_{T_0}^{T} dT = -\int_{r_0}^{r}\frac{\dot{Q}_0}{2\pi kL}\,r^{-1}dr$$

The complete solution is

$$T - T_0 = -\frac{\dot{Q}_0}{2\pi kL}\ln\left(\frac{r}{r_0}\right) \tag{13.28}$$

showing a variation of temperature with radius as a $\ln(r)$ function. With this solution we can relate the total heat flux to the overall temperature difference and solve for the heat transfer rate as

$$\dot{Q}_0 = \frac{2\pi kL}{\ln(r_1/r_0)}(T_0 - T_1) \tag{13.29}$$

By comparison with Eq. 13.10, the thermal conduction resistance of a cylindrical layer is

$$R_{\text{cond cyl}} = \frac{\ln(r_1/r_0)}{2\pi kL} \tag{13.30}$$

The rate of heat transfer is constant, as shown in Eq. 13.27, so with the surface area through which the energy flows increasing with radius, the heat flux per unit area is not constant as in the plane geometry. For very thin pipe walls, the limit gives $\ln(r_1/r_0) = r_1/r_0 - 1$ and with the thickness being $r_1 - r_0$ and area $A = 2\pi r_0 L$, we recover the plane layer result in Eq. 13.11.

---

**EXAMPLE 13.5**

An aerial electric wire shown in Fig. 13.11 carries a current, so it generates 2 W for one meter length. The wire is copper of 5 mm diameter with 2 mm thick insulation around it ($k = 0.2$ W/m K) and it has an outside convection coefficient of $h = 25$ W/m$^2$ K to ambient air at 22°C. Find the copper surface temperature.

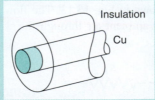

Insulation

Cu

**FIGURE 13.11**   A wire.

**Analysis**

A control volume around the copper wire gives

$$0 = \dot{W}_{el} - \dot{Q}_{out}$$

All the energy generated must leave the surface first as conduction through the insulation, then as convection.

$$\dot{Q}_{out} = \dot{Q}_{cond} = \dot{Q}_{conv}$$

**Solution**

The process happens in radial geometry conduction and convection layers for which we can calculate the resistances. These are:

$$R_{cond\ cyl} = \frac{\ln(r_1/r_0)}{2\pi kL} = \ln\left[0.0045/0.0025\right]/[2\pi \times 0.2 \times 1] = 0.4677\ \text{K/W}$$

$$R_{conv} = 1/[hA] = 1/[25 \times 2\pi \times 0.0045 \times 1] = 1.4147\ \text{K/W}$$

$$R_{TOT} = R_{cond\ cyl} + R_{conv} = 1.882\ \text{K/W}$$

Then from the total resistance Eq. 13.23

$$\Delta T = \dot{Q}\,R_{TOT} = 2\ \text{W} \times 1.882\ \text{K/W} = 3.8\ \text{K}$$

$$T_s = T_\infty + \Delta T = 22 + 3.8 = 25.8°\text{C}$$

## Electrical Analogy

If we write the temperature difference in terms of the rate of heat transfer and the thermal resistance as in Eq. 13.16, $\Delta T = R\dot{Q}$, we have an equation similar to Ohm's law

$$E = RI \tag{13.31}$$

where the current $I$ (analogous to $\dot{Q}$) is driven by the voltage difference $E$ (analogous to $\Delta T$) through the resistance $R$ (analogous to the thermal resistance $R$). All the circuit analysis that applies to electrical circuits is then also directly applicable to the equivalent thermal resistance network. A situation with resistances in series was illustrated with a composite wall leading to Eq. 13.32, where the same rate of heat transfer goes through several different layers. In that case, the total resistance is equal to the sum of all the resistances in series as shown in Fig. 13.12:

$$\dot{Q}_{TOT} = \dot{Q}_i \quad \text{and} \quad \Delta T_{TOT} = \sum \Delta T_i$$

$$\text{SERIES:} \quad R_{TOT} = \sum R_i \tag{13.32}$$

Resistances in parallel can be illustrated with several paths for the heat transfer from one domain to another, as shown in Fig. 13.13. The flow of energy from the warm air in a room can go to the colder ambient air through the windows, the side walls, or

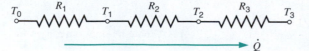

**FIGURE 13.12** Thermal resistance layers in series.

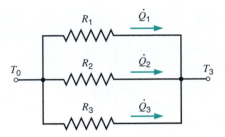

**FIGURE 13.13** Thermal resistance layers in parallel.

the ceiling at different rates. For resistances in parallel, we have

$$\dot{Q}_{TOT} = \sum \dot{Q}_i \quad \text{and} \quad \Delta T_{TOT} = \Delta T_i$$

$$\text{PARALLEL:} \quad \frac{1}{R_{TOT}} = \sum \frac{1}{R_i} \tag{13.33}$$

To create a simple electrical analogy to a physical system, each part of the mass with a near uniform state is assigned as a node with a temperature and then all nodes are connected by thermal resistances. Nodes can represent sources or sinks such as an electrical heater giving out a prescribed rate of heat transfer. The energy equation for the node contains the sum of all the heat transfers to the node and then implicitly determines the temperature.

**EXAMPLE 13.6**

Consider a composite plane wall that separates 100°C steam from the outside air at 5°C. Assume a construction with a 2-mm-thick steel plate containing the steam as shown in Fig. 13.14. Over an area of 2 m² the steel is insulated with 8-cm of glass wool and an outside 1-cm-thick softwood board having an outside convection coefficient of $h = 25$ W/m² K. The remaining area of 0.5 m² has no glass wool, just the wood board with an outside convection coefficient of $h = 20$ W/m² K. We want to find the total rate of heat loss from the steam.

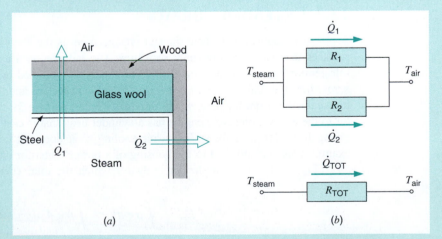

**FIGURE 13.14** (a) The physical system. (b) The electrical equivalent.

### Analysis

In this situation we have a rate of $\dot{Q}_1$ from the steam to the air through wall 1 and a $\dot{Q}_2$ through wall 2. Each one can be written from the temperature difference and the resistance as Eq. 13.15 and each resistance is a series combination from Eq. 13.32.

**Solution**

$$R_1 = \left[ \frac{L_{st}}{k_{st}} + \frac{L_{ins}}{k_{ins}} + \frac{L_{wood}}{k_{wood}} + \frac{1}{h_{conv}} \right] / A_1$$

$$= [(0.002/15) + (0.08/0.037) + (0.01/0.12) + (1/25)]/2 = 1.143 \text{ K/W}$$

$$R_2 = \left[ \frac{L_{st}}{k_{st}} + \frac{L_{wood}}{k_{wood}} + \frac{1}{h_{conv}} \right] / A_2$$

$$= [(0.002/15) + (0.01/0.12) + (1/20)]/0.5 = 0.267 \text{ K/W}$$

Now

$$\dot{Q}_1 = \Delta T/R_1 = (100 - 5)/1.143 = 83.1 \text{ W}$$

$$\dot{Q}_2 = \Delta T/R_2 = (100 - 5)/0.267 = 355.8 \text{ W}$$

and the total is the sum of them ($\cong 439$ W). We could also write

$$\dot{Q}_{TOT} = \dot{Q}_1 + \dot{Q}_2 = (\Delta T/R_1) + (\Delta T/R_2) = \Delta T/R_{TOT}$$

and so by inspection

$$\frac{1}{R_{TOT}} = \frac{1}{R_1} + \frac{1}{R_2} \qquad \text{or} \qquad R_{TOT} = \frac{R_1 R_2}{R_1 + R_2}$$

a familiar result for two resistances in parallel as in Eq. 13.33. In this case

$$R_{TOT} = 1.143 \times 0.267/[1.143 + 0.267] = 0.2164 \text{ K/W}$$

$$\dot{Q}_{TOT} = (100 - 5)/0.2164 = 439 \text{ W}$$

which is the same result, but now obtained by the electrical analogy.

## 13.3 EXTENDED SURFACES

A practical application of heat transfer enhancement is the attachment of fins or extended surfaces to a wall of a container or a pipe. The radiator on a car or the condenser on a refrigerator has a pipe carrying a hot fluid inside being cooled by atmospheric air flowing across the pipe. Thin plates are connected to the pipe surface and therefore become warmer by conduction from the pipe, which provides a larger surface for the convection heat transfer to the air. A motorcycle engine has a cylinder head that is cast with a number of plates, giving it a large area for sufficient air cooling so it does not need cooling water. Other examples are fins attached to the housing of a power transistor or the CPU in a computer.

Fig. 13.15 shows a slender body for which the total energy is summed over the volume:

$$E_{cv} = \int \rho e \, dV = \int \rho e A_c \, dx$$

The total surface convective heat transfer is summed over the surface area, where the perimeter $P$ is the area per unit length $P \, dx = dA_s$, as

$$\dot{Q}_s = \int h(T_s - T_\infty) \, dA_s = \int hP(T_s - T_\infty) \, dx$$

Let us restrict the discussion to a situation where there is no work term (no change in volume), typical for a solid or liquid substance. The energy equation for the control

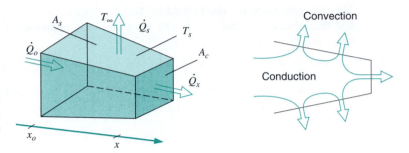

**FIGURE 13.15** Axial conduction and surface convection in a slender body.

volume shown becomes

$$\frac{dE_{cv}}{dt} = 0 = \dot{Q}_0 - \dot{Q}_x - \dot{Q}_s \tag{13.34}$$

Now differentiate the energy equation with respect to $x$, keeping $\dot{Q}_0$ a constant to obtain

$$\frac{d\,\rho e A_c}{dt} = \frac{d}{dx}\left[ kA_c \frac{dT}{dx} \right] - h\,P\,(T_s - T_\infty) \tag{13.35}$$

Assume also that we are looking at steady state, so there are no changes in properties with time, and that the temperature is uniform across the area $A_c$, so the surface temperature $T_s$ is equal to the single temperature $T$, a function of $x$ only. Under these circumstances, the energy equation reduces to

$$0 = \frac{d}{dx}\left[ kA_c \frac{dT}{dx} \right] - h\,P\,(T - T_\infty)$$

stating the balance of conduction of energy in the cross-sectional area with the surface heat transfer.

## Constant Cross-Sectional Area Fin

For a constant cross-sectional area and constant conductivity, the energy equation further simplifies to

$$0 = \frac{d^2T}{dx^2} - a^2\,(T - T_\infty) \tag{13.36}$$

with a parameter

$$a^2 = hP/kA_c \tag{13.37}$$

The solution is found by first changing variable to

$$\theta = T - T_\infty \quad \Rightarrow \quad d\theta = dT$$

The differential equation now becomes

$$\frac{d^2\theta}{dx^2} - a^2\theta = 0 \tag{13.38}$$

This is a linear (in $T$), constant-coefficient, second-order differential equation to determine the temperature as a function of $x$. As the right-hand side is also zero, the differential equation is homogeneous. There are two independent solution functions with the solution as a linear combination

$$\theta = T - T_\infty = C_1\,e^{ax} + C_2\,e^{-ax} \tag{13.39}$$

The two constants are determined by two boundary conditions, one at the base $x = 0$ and the other at the end $x = L$. The most common condition at $x = 0$ is a known base temperature

$$x = 0: \qquad \theta(0) = \theta_b = T_b - T_\infty = C_1 + C_2 \qquad (13.40)$$

giving the first equation to determine the two constants $C_1$ and $C_2$. The boundary condition at the tip of the fin $(x = L)$ can be approximated by one of several types:

**a.** $T = \text{given} = T_L$

**b.** $\dot{q}'' = -k \left. \dfrac{dT}{dx} \right|_{x=L} = \text{given}$

**c.** $-k \left. \dfrac{dT}{dx} \right|_{x=L} = h\left(T_L - T_\infty\right)$

which are the Dirichlet, Neuman, and mixed kind of boundary conditions, respectively. The boundary condition forms the second equation for $C_1$ and $C_2$.

Which kind of boundary condition should be used? The most common situation is that the tip also "sees" ambient fluid with the axial conduction to the tip equal to the surface convection heat transfer expressed in type $c$. If the tip is connected to another larger solid with a given temperature, $T_L$, it is type $a$, often also used for a very long fin, where the fin temperature approaches ambient temperature $T_\infty$ in the limit $L \to \infty$. Insulating the tip or having the location $x = L$ as a symmetry plane gives a zero heat transfer in type $b$.

## Infinitely Long Fin

The simplest solution is obtained in the case of the infinitely long fin where the condition $\theta_L \to 0$ as $L \to \infty$ in Eq. 13.39 is satisfied only if $C_1 = 0$. Then Eq. 13.40 yields the other constant and the solution from Eq. 13.39 becomes

$$\theta = T - T_\infty = \left(T_b - T_\infty\right)e^{-ax} \qquad (13.41)$$

The conduction heat transfer in the fin is

$$\dot{Q} = -k A_c \frac{dT}{dx} = k A_c\, a\left(T_b - T_\infty\right)e^{-ax}$$

so both the temperature difference and the heat transfer drop exponentially with the distance from the base $x$. From this solution, we observe that farther out the fin area is less useful and the cost and space it occupies does not justify the benefits, so it should be made with some reasonable length. The total rate of heat transfer from the fin equals the conduction heat transfer rate at the base, $x = 0$

$$\dot{Q}_b = -k A_c \frac{dT}{dx} = k A_c\, a\left(T_b - T_\infty\right) = \left(hPkA_c\right)^{1/2}\left(T_b - T_\infty\right) \qquad (13.42)$$

proportional to the temperature difference. By comparison with Eq. 13.10, we can find a thermal resistance for the total fin as $R_{\text{fin}} = \left(hPkA_c\right)^{-1/2}$.

## Adiabatic Tip

The solution for the adiabatic tip uses boundary condition type $b$ and thus leads to a zero temperature gradient. From this together with Eq. 13.40, the values of the constants $C_1$

and $C_2$ can be obtained. The solution after some rewriting is expressed as

$$\theta = T - T_\infty = (T_b - T_\infty)\frac{\cosh a(L - x)}{\cosh aL} \qquad (13.43)$$

The total rate of heat transfer from the fin is evaluated from the base, $x = 0$:

$$\dot{Q}_b = -k\, A_c \frac{dT}{dx} = (hPkA_c)^{1/2} \tanh (aL)(T_b - T_\infty) \qquad (13.44)$$

and differs only from the previous result by the factor tanh($aL$). This factor approaches 1 as $aL$ increases and is practically 1 for values of $aL$ larger than 3.

The solution with boundary condition of type $c$ can also be solved for leading to a slightly more complicated and longer solution with the *sinh* and *cosh* functions. This and other solutions can be seen in heat transfer textbooks and handbooks.

---

**EXAMPLE 13.7**

A rectangular aluminum fin, 0.1 m wide, 0.05 m in length and 2 mm thick, is attached to a base at 100°C, as shown in Fig. 13.16. Air at 20°C flows over the fin, resulting in an average convective heat transfer coefficient $h = 50$ W/m$^2$ K. We want to find the fin heat transfer and the tip temperature assuming an adiabatic tip condition.

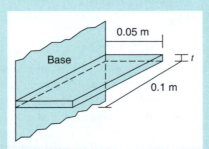

**FIGURE 13.16**   Rectangular fin.

**Analysis**

The adiabatic tip gives the solution in Eq. 13.43.

**Solution**

From the geometry and the dimensions we have

$$A_c = w\,t = 0.1 \times 0.002 = 0.0002 \text{ m}^2$$

$$P = 2w + 2t = 0.2 + 0.004 = 0.204 \text{ m}$$

$$a = (hP/kA_c)^{1/2} = [50 \times 0.204/(164 \times 0.0002)]^{1/2}$$

$$= 17.63 \text{ m}^{-1}$$

$$aL = 17.63 \times 0.05 = 0.8817$$

We calculate the heat transfer for the fin from Eq. 13.44

$$\dot{Q} = (hPkA_c)^{1/2} \tanh (aL)(T_b - T_\infty)$$

$$= [50 \times 0.204 \times 164 \times 0.0002]^{1/2} \times 0.707 \times (100 - 20)$$

$$= 32.7 \text{ W}$$

The tip temperature is evaluated from Eq. 13.43 at location $x = L$

$$T_L = T_\infty + \theta_{tip} = T_\infty + (T_b - T_\infty) \frac{\cosh 0}{\cosh aL} = 20 + 80 \frac{1}{1.414} = 76.6°C$$

From $\tanh(aL) = 0.707$ and $T_L$ we note that this fin cannot be assumed infinitely long.

### Fin Efficiency and Effectiveness

As real fins may not have constant cross-sectional area or perimeter and the convective heat transfer coefficient often varies along the fin surface, our mathematical solution is only approximate. It is possible to solve cases other than the two shown for known variation in the geometry with distance $x$, but rather than showing these more complicated formulas, we will present some results graphically instead.

For a given fin, we may compare the actual fin heat transfer as the conduction through the base area, which equals the sum of the convection from the fin surface to two other well-defined cases. The maximum possible heat transfer occurs if the conductivity of the solid is very large so that the surface temperature is $T_b$ everywhere. Then

$$\dot{Q}_{max} = hA_{surf\,fin}(T_b - T_\infty) \tag{13.45}$$

and the efficiency is defined by

$$\dot{Q}_{fin} = \eta_{fin}\dot{Q}_{max} \tag{13.46}$$

Figure 13.17 shows the fin efficiency for four different geometries, where the first two are assumed infinitely wide, and the tip heat transfer is accounted for by the use of a corrected length. The parameter on the abscissa is close to being $(aL)$ for the constant cross-section area cases, and it is observed that efficiency drops with increase in length.

If the fin is not present, the base area it covered would have a heat transfer as

$$\dot{Q}_{base\,no\,fin} = hA_{c\,b}(T_b - T_\infty) \tag{13.47}$$

which is used to define the fin effectiveness

$$\varepsilon = \dot{Q}_{fin}/\dot{Q}_{base\,no\,fin} \tag{13.48}$$

This measures the effective increase in the base area, so it appears that $A_{c\,b}$ is $\varepsilon$ times larger. It relates to the efficiency by substitution of Eqs. 13.45–13.47 into Eq. 13.48:

$$\varepsilon = \eta_{fin}\dot{Q}_{max}/\dot{Q}_{base\,no\,fin} = \eta_{fin}A_{surf\,fin}/A_{c\,b} \tag{13.49}$$

Once a single fin performance has been evaluated, the heat transfer over several fins and the remainder of the base area can be calculated. If we separate the total surface area, assuming a total of $N$ fins, into

$$A_{total} = NA_{c\,b} + A_{base\,no\,fins}$$

then the heat transfer is

$$\begin{aligned}\dot{Q}_{total} &= \dot{Q}_{all\,fins} + \dot{Q}_{base\,no\,fins} \\ &= N\dot{Q}_{fin} + hA_{base\,no\,fins}(T_b - T_\infty) \\ &= (N\varepsilon A_{c\,b} + A_{base\,no\,fins})h(T_b - T_\infty)\end{aligned} \tag{13.50}$$

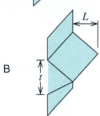

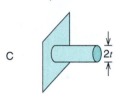

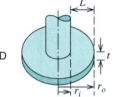

**FIGURE 13.17**  Fin efficiency for various geometries non-adiabatic tip.

---

**EXAMPLE 13.8**

Find the heat transfer from the fin in Example 13.7, and determine its efficiency and effectiveness using the graph in Fig. 13.17.

**Analysis**

The solution is from case $A$ in the Fig. 13.17 graph, which is for a non-adiabatic tip.

**Solution**

For this we compute the corrected length, the fin surface area, and a parameter to be used in the graph

$$L_c = L + t/2 = 0.05 + 0.002/2 = 0.051 \text{ m}$$

$$A_{\text{surf fin}} = 2\,w\,L + 2\,t\,L + w\,t$$

$$= 2 \times 0.1 \times 0.05 + 2 \times 0.002 \times 0.05 + 0.1 \times 0.002 = 0.0104 \text{ m}^2$$

$$L_c\sqrt{\frac{2h}{kt}} = 0.051\sqrt{\frac{2 \times 50}{164 \times 0.002}} = 0.89$$

Entering this result for case $A$ in Figure 13.17 gives an efficiency of $\eta = 0.85$. Then

$$\dot{Q}_{max} = h\, A_{surf\ fin}(T_b - T_\infty) = 50 \times 0.0104 \times (100 - 20) = 41.6\ \text{W}$$

$$\dot{Q}_{fin} = \eta_{fin}\, \dot{Q}_{max} = 0.85 \times 41.6 = 35.4\ \text{W}$$

$$\dot{Q}_{base\ no\ fin} = h\, A_{cb}\,(T_b - T_\infty) = 50\,[0.1 \times 0.002]\,(100 - 20) = 0.8\ \text{W}$$

$$\varepsilon = \dot{Q}_{fin}/\dot{Q}_{base\ no\ fin} = 35.4/0.8 = 44$$

The heat transfer is found to be a little larger than the previous calculation due to the non-adiabatic edges, the use of the corrected length, and the calculation of $A_{surf\ fin}$.

**EXAMPLE 13.9**

An interesting example of fins used to enhance the heat transfer is from the cooling of the permafrost around the base of the structural support for the Alaskan oil pipeline. A heat pipe or thermosyphon contains ammonia with liquid at the bottom being boiled by heat transfer from the permafrost so the vapor rises to the top where it is cooled by the cold air as shown in Fig. 13.18a. As the vapor cools, it condenses and falls to the bottom of the pipe. This is a passive system (no work input) and it only works when the air is colder than the bottom of the pipe. Let us look at the ammonia to air heat transfer and find the expression for the total thermal resistance. The pipe length with fins is 1 m, having 20 aluminum fins 2 mm thick and 6 cm long. The 2 mm thick steel

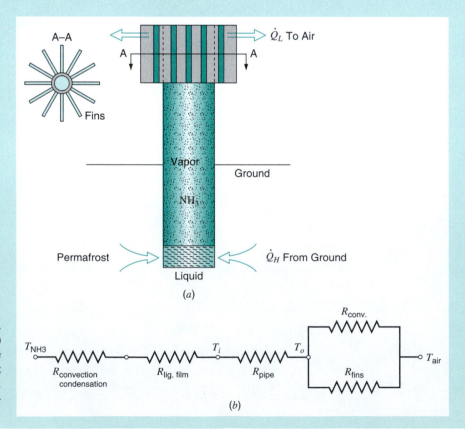

**FIGURE 13.18** (a) A heat pipe (thermosyphon) on the Alaskan pipeline to keep the permafrost under the support cold. (b) The electrical equivalent network.

pipe has outer radius of 2.5 cm, and we estimate a convection heat transfer coefficient of 75 W/m$^2$ K for a relative low velocity wind.

### Analysis

This is assumed to be a steady constant rate heat transfer from the ammonia vapor to a liquid film of ammonia on the inner pipe surface, through the pipe wall. From the outer pipe surface, the heat transfer is partly to the air and partly through the fins to the air. The equivalent resistive network is show in Fig. 13.18b.

### Solution

Since we do not cover the condensation heat transfer, let us assume we asked an expert and were told $h = 2500$ W/m$^2$ K (see Table 13.2) for the vapor-to-pipe inner surface heat transfer.

$$R_i = R_{\text{condensation}} = \frac{1}{h_i A_i} = \frac{1}{2500 \times 2\pi \times 0.023 \times 1} = 0.0028 \text{ K/W}$$

The resistance through the pipe is from Eq. 13.30

$$R_{\text{pipe}} = \frac{\ln(r_0/r_i)}{2\pi k_{\text{steel}} L} = \frac{\ln(0.025/0.023)}{2\pi \times 15 \times 1} = 0.00088 \text{ K/W}$$

and the convection from the pipe outer surface directly to the air is

$$R_{\text{convection}} = \frac{1}{h_0 A_{0 \text{ no fins}}} = \frac{1}{75(2\pi \times 0.025 - 20 \times 0.002) \times 1} = 0.114 \text{ K/W}$$

To calculate the fin, we use Fig. 13.17 for which the parameters are

$$L_c = 0.06 + 0.002/2 = 0.061 \text{ m}, \quad L_c \sqrt{2h/kt} = 0.061 \sqrt{2 \times 75/164 \times 0.002} = 1.3$$

so case A gives

$$\eta = 0.74 \quad \text{and} \quad \varepsilon = \eta \, A_{\text{surf fin}}/A_{c\,b} = \eta \frac{(2L + t)H}{H t} = 0.74 \times \frac{0.122}{0.002} = 45$$

The resistance for 20 fins in parallel becomes

$$R_{\text{fins}} = R_{\text{fin}}/N_{\text{fins}} = \frac{1}{h_0 N_{\text{fins}} \varepsilon A_{c\,b}} = \frac{1}{75 \times 20 \times 45 \times 0.002 \times 1} = 0.0074 \text{ K/W}$$

The fins and the convection resistances are in parallel so the combination is

$$R_{\text{convection}} \parallel R_{\text{fins}} = \frac{0.114 \times 0.0074}{0.114 + 0.0074} = 0.00695 \text{ K/W}$$

All the resistances in series then give

$$R_{\text{fins}} = R_i + R_{\text{pipe}} + R_{\text{convection}} \parallel R_{\text{fins}}$$

$$= 0.0028 + 0.00088 + 0.00695 = 0.0106 \text{ K/W}$$

That is, we get close to 100 W per degree temperature difference between the vapor and air.

## 13.4 INTERNAL ENERGY GENERATION

In the preceeding sections we have dealt with situations where there is no internal energy generation throughout the volume. There are, however, a few situations where such generation is present. These include cases in which chemical reactions release energy and in which electrical currents deposit energy through ohmic resistances (wire, pipe, heating mats, transformer coils and kernels, and so forth). For these and similar cases, the energy equation becomes

$$\frac{dE_{CV}}{dt} = \dot{Q}_{in} - \dot{W}_{out} + \dot{Q}_{gen}$$

To make the presentation reasonably simple, we will assume that the volume is constant so there is no work involved and that there is a uniform energy generation, $\dot{q}'''$, which is constant per unit volume. Finally we will assume a steady state with no rate of energy storage and the energy equation reduces to

$$0 = \dot{Q}_{in} + \dot{Q}_{gen} \tag{13.51}$$

### Energy Generation in Plane Geometry

We will now do the analysis of the plane steady-state conduction from Section 13.2 that includes an internal uniform energy generation as shown in Fig. 13.19. With the previously stated simplifying assumptions Eq. 13.51 becomes

$$0 = \dot{Q}_0 - \dot{Q}_x + \dot{q}'''V$$

$$= \dot{Q}_0 + kA\frac{dT}{dx} + \dot{q}'''Ax \tag{13.52}$$

where we substituted the expression for the conduction at location $x$ and the volume for the domain. Eq. 13.52 is now integrated to solve for $T(x)$ as $\dot{Q}_0$ is a constant, just as was done for Eq. 13.8.

$$\frac{dT}{dx} = -\frac{\dot{Q}_0}{kA} - \frac{\dot{q}'''}{k}x \tag{13.53}$$

Multiplying by $dx$ and integrating, we have

$$\int_{T_0}^{T} dT = \int_{0}^{x}\left[-\frac{\dot{Q}_0}{kA} - \frac{\dot{q}'''}{k}x\right]dx$$

leading to

$$T - T_0 = -\frac{\dot{Q}_0}{kA}x - \frac{\dot{q}'''}{2k}x^2 \tag{13.54}$$

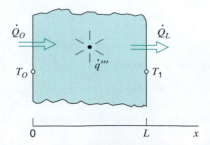

**FIGURE 13.19** Plane conduction layer with generation of energy.

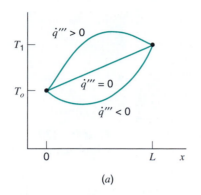

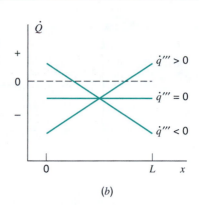

**FIGURE 13.20** (a) Temperature profile. (b) Variation of $\dot{Q}$.

(a)

(b)

This shows the solution as a parabola in $x$ with the left-hand side temperature, the left-hand side rate of heat transfer, and the generation rate as parameters. If we introduce the temperature $T_1$ at $x = L$, we have

$$T_1 - T_0 = -\frac{\dot{Q}_0}{kA}L - \frac{\dot{q}'''}{2k}L^2 \tag{13.55}$$

from which we can solve for $\dot{Q}_0$ and substitute it into Eq. 13.54 to give

$$T = T_0 + \frac{T_1 - T_0}{L}x + \frac{\dot{q}'''}{2k}(L - x)x \tag{13.56}$$

This solution is illustrated in Fig. 13.20 for different signs of the energy-generation term. The rate of heat transfer at any given location $x$ is

$$\dot{Q}_x = -kA\frac{dT}{dx} = (T_0 - T_1)\frac{kA}{L} + \frac{1}{2}\dot{q}'''A(2x - L) \tag{13.57}$$

$$\dot{Q}_x = \dot{Q}_0 + \dot{q}'''Ax \tag{13.58}$$

with the first expression from Eq. 13.56 and the second expression from Eq. 13.52. Notice how the temperature varies as a parabola and the heat transfer varies linearly in $x$. The solution in Eq. 13.56 is used if the temperatures at the boundaries are chosen as variables, and Eq. 13.54 is used if one temperature and one heat transfer rate are used as variables. These are not independent but connected as shown, so that the heat transfer at the right-hand side could be used from Eq. 13.58 with $x = L$:

$$\dot{Q}_L = \dot{Q}_0 + \dot{q}'''AL \tag{13.59}$$

confirming the overall energy equation balance.

**EXAMPLE 13.10**

During the curing of a slab of poured concrete, the chemical reactions release 2.5 kW/m³ of power per unit volume. The slab, shown in Fig. 13.21, is very large with a thickness of 0.25 m, and its top surface is at 25°C exposed to ambient air and the bottom is in contact with the colder ground so it is at 15°C. We want to find the maximum temperature in the concrete assuming steady state and a conductivity of $k = 1$ W/m K.

**Analysis**

Take as a control volume the concrete, which for steady state has no rate of change in $T$ and all the energy generated inside must leave through the two surfaces. Let $x = 0$

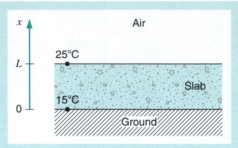

**FIGURE 13.21** A concrete slab with energy release during curing.

be the bottom surface and $x = L$ the top surface, so with the two temperatures and the generation rate we have the temperature profile in Eq. 13.56 as the solution.

**Solution**

From the solution in Eq. 13.56, we find the peak of the temperature by $dT/dx = 0$ as

$$dT/dx = (T_1 - T_0)/L + \frac{\dot{q}'''}{2K}(L - 2x) = 0$$

Solve for the location

$$x = L/2 + (T_1 - T_0)\frac{k}{L\dot{q}'''}$$

$$= 0.125 + (25 - 15)\frac{1}{0.25 \times 2500} = 0.141 \text{ m}$$

The solution Eq. 13.56 gives the temperature as

$$T = T_0 + (T_1 - T_0)\frac{x}{L} + \frac{\dot{q}'''}{2k}(L - x)x$$

$$= 15 + 10 \times \frac{0.141}{0.25} + \frac{2500}{2 \times 1}(0.25 - 0.141) \times 0.141$$

$$= 39.9°C$$

## 13.5 TRANSIENT HEAT TRANSFER

All the previous analyses and examples dealt with steady-state heat transfer. However, when energy is added to or removed from a finite size domain, its state cannot be constant and thus the temperature often changes with time. Placing food in a refrigerator, some matter in an oven or furnace, or flowing a mass of cold water through a pipe in a boiler are all examples of processes where a rate of heat transfer causes a transient in the state of the substance. Since the rate of heat transfer is sensitive to the temperature difference, it changes as the process proceeds.

Assume that we want to heat air inside a rigid container by heating from a hot source attached to a part of the container surface. From our knowledge of the construction and geometry of the container wall and possible convective layers, the previous analysis in Section 13.2 allows us to estimate a total thermal resistance, $R_{TOT}$, from the outside source at $T_\infty$ to the inside air at temperature $T$. The instantaneous rate of heat transfer is

$$\dot{Q} = \frac{\Delta T}{R_{\text{TOT}}} = \frac{T_\infty - T}{R_{\text{TOT}}} \tag{13.60}$$

measured as positive into the air, where we have assumed the steady-state solution from Eq. 13.10 to be valid. As the process proceeds, the air next to the heated wall becomes hotter and it in turns heats the air further away from the wall. A slight non-uniformity of the temperature in the air develops, and this is kept small only if the energy can be redistributed in the air fast enough.

## Lumped Analysis

For the subsequent analysis, we will assume that the temperature of the control volume is uniform over the volume, such that only one temperature describes the state. The complete energy equation for the control mass becomes

$$\frac{dE_{CV}}{dt} = \dot{Q} = \frac{T_\infty - T}{R_{\text{TOT}}} \tag{13.61}$$

with no phase changes

$$dE_{CV} = d(\rho\, e\, V) = d(mu) = m\, C\, dT$$

and so we have

$$\frac{dT}{dt} = \frac{T_\infty - T}{m\, C\, R_{\text{TOT}}} \tag{13.62}$$

To facilitate the integration, we change variable to

$$\theta = T - T_\infty \qquad \Rightarrow \qquad d\theta = dT$$

$$\frac{d\theta}{dt} = -\frac{\theta}{m\, C\, R_{\text{TOT}}} \tag{13.63}$$

Multiplying by $dt$ and dividing by $\theta$ gives

$$\theta^{-1}\, d\theta = -\frac{1}{m\, C\, R_{\text{TOT}}}\, dt \tag{13.64}$$

The equation is integrated from the beginning of the process at $t = 0$ where $\theta = \theta_i$ as

$$\int_{\theta_i}^{\theta} \theta^{-1}\, d\theta = -\int_0^t \frac{1}{m\, C\, R_{\text{TOT}}}\, dt$$

with the result

$$\ln \frac{\theta}{\theta_i} = -\frac{t}{m\, C\, R_{\text{TOT}}} \tag{13.65}$$

The final solution is obtained by taking the inverse of the ln function (an exp) as

$$\frac{\theta}{\theta_i} = \frac{T - T_\infty}{T_i - T_\infty} = \exp\left[-\frac{t}{m\, C\, R_{\text{TOT}}}\right] = \exp\left[-\frac{t}{\tau_t}\right] \tag{13.66}$$

showing the exponential decay of the initial temperature difference $\theta_i = T_i - T_\infty$ falling to zero as time goes toward infinity. Here we have introduced the thermal time constant

$$\tau_t = m\, C\, R_{\text{TOT}} \tag{13.67}$$

with solutions for various values of the thermal time constant shown in Fig. 13.22.

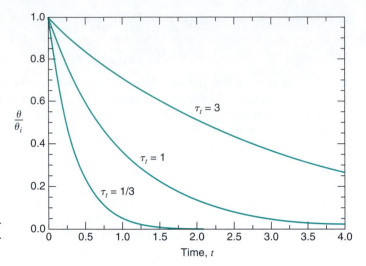

**FIGURE 13.22** The exponential decay of the temperature difference.

For very small values of $\tau_t$ (small mass or heat capacity or thermal resistance), the decay is fast and the substance temperature approaches the ambient temperature in a very short time. The opposite is true for large values of $\tau_t$ (large mass or heat capacity or thermal resistance) where the decay is very slow. For an elapsed time equal to $\tau_t$, the temperature difference has dropped to roughly 37% $[\exp(-1) = 0.3678]$ of the initial difference and for an elapsed time of $3\tau_t$ it has dropped to about 5%. From this solution, the value of the thermal time constant gives some idea about the speed of the process.

## The Biot Number

The simple lumped analysis using an average temperature for the mass requires the temperature differences in the control volume to be small compared to the temperature difference that drives the heat transfer. A similar approximation was needed to develop the mathematical solution to the fin heat transfer, assuming a uniform state over the cross-sectional area.

We want to look at the conditions for which the averaging process is a reasonable assumption. Compare a $\Delta T$ by conduction heat transfer to a surface as

$$\Delta T_{\text{cond}} = \dot{Q} R_{\text{cond}} = \dot{Q} L / k_{\text{solid}} A \tag{13.68}$$

with the $\Delta T$ due to convection heat transfer

$$\Delta T_{\text{conv}} = \dot{Q} R_{\text{conv}} = \dot{Q} / h_{\text{fluid}} A \tag{13.69}$$

The ratio of the two temperature differences defines the dimensionless Biot number

$$\frac{\Delta T_{\text{cond}}}{\Delta T_{\text{conv}}} = \frac{R_{\text{cond}}}{R_{\text{conv}}} = \frac{h_{\text{fluid}} L}{k_{\text{solid}}} \equiv Bi \tag{13.70}$$

The Biot number is based on the same combination of properties as the Nusselt number defined in Section 13.1 but here the heat transfer coefficient is for the fluid and the conductivity is for the solid media. For a small Biot number, we may neglect the temperature differences by conduction relative to those by convection. In practice this is taken with a characteristic length $L_c = V/A_s$, so if

$$Bi = \frac{h_{\text{fluid}} L_c}{k_{\text{solid}}} \leq 0.1 \tag{13.71}$$

the lumped analysis is reasonable and we can assume an averaged temperature over some domain for the conduction media. If the Biot number is much larger, then the temperature distribution inside the conducting media cannot be neglected. In that case a multidimensional conduction analysis may be needed, making the solution mathematically complex.

**EXAMPLE 13.11**

A 20°C cube of granite, measuring 3 cm on a side, is heated by hot gases at 200°C on all outside surfaces by convection with $h = 75$ W/m$^2$ K. How long time will it take for the granite to reach 175°C? Is a lumped analysis valid?

**Analysis**

First calculate the Biot number based on $L_c = V/A_s = (3 \times 3 \times 3)/(6 \times 3 \times 3) = 0.5$ cm:

$$Bi = \frac{hL_c}{k} = \frac{75\ (\text{W/m}^2\ \text{K}) \times 0.5\ \text{m}}{2.9\ (\text{W/m K})} = 0.129$$

which is slightly over the limit for neglecting the $\Delta T$ inside the granite. The analysis leading to Eq. 13.66 is then reasonable with some error.

**Solution**

The convection resistance is

$$R_{\text{TOT}} = 1/hA = 1/(75 \times 6 \times 3 \times 3 \times 10^{-4}) = 2.47\ \text{K/W}$$

and the thermal time constant, with $C$ from Table A.3, is

$$\tau_t = mCR_{\text{TOT}} = \rho V C R_{\text{TOT}}$$

$$= 2750 \times 3^3 \times 10^{-6} \times 0.89 \times 1000 \times 2.47 = 163\ \text{s}$$

From Eq. 13.87 and Eq. 13.89 we can find the time

$$t = \tau_t \ln\left(\frac{\theta}{\theta_i}\right) = 163 \ln\left(\frac{175 - 200}{20 - 200}\right) = 322\ \text{s}$$

## 13.6 CONVECTION HEAT TRANSFER

The convection of a fluid has a strong influence on the heat transfer process as mentioned in the introduction of convection heat transfer in Section 13.1. The following section will examine some more details of the flow situations and the correlations for the heat transfer expressed in the average value of the Nusselt number. We will look at external flows, which could be flows over the outside surface of a pipe, over the top of an electronic chip, or the air flowing over a rooftop. The other class is internal flow such as a flow inside a pipe, a coolant fluid inside an engine, or air in a duct to mention a few common examples.

### External Flow

Consider a free stream flow of fluid that approaches a flat surface where the fluid and the surface have different temperatures as shown in Fig. 13.23. A heat transfer and a momentum transfer takes place between the fluid and the solid surface that we assume

Free stream

$V_\infty$ $T_\infty$

$\delta_T$
Thermal
boundary
layer

$\delta$ Velocity
boundary
layer

$x$          $T_S$

**FIGURE 13.23** The momentum and thermal boundary layers.

has a constant temperature $T_s$ and a zero velocity. Viscous friction slows the fluid down so the velocity, essentially parallel to the surface, changes from the free stream value $\mathbf{V}_\infty$ far from the surface to zero at the surface. The distance measured normal to the surface over which the velocity changes is called the momentum (or velocity) boundary layer and defined more precisely as the distance, $\delta$, where $\mathbf{V}$ has reached 99% of the free stream value $\mathbf{V}_\infty$. The temperature shows a similar variation from the free stream value $T_\infty$ to the surface temperature $T_s$ over a distance corresponding to the thermal boundary layer, $\delta_T$, where both of these boundary layers are shown in Fig. 13.24.

The local heat flux $\dot{q}''$ evaluated in the fluid at the surface, $y = 0$, is

$$\dot{q}'' = -k\left(\frac{\partial T}{\partial y}\right)_{y=0} \tag{13.72}$$

with a direction in the positive y-axis. Looking across the whole thermal boundary layer Newtons law of cooling, Eq. 13.2, is written

$$\dot{q}'' = h(T_s - T_\infty) \tag{13.73}$$

using the same direction. By equating the two heat fluxes in Eqs. 13.72 and 13.73, the heat transfer coefficient $h$ is expressed as

$$h = -k\left(\frac{\partial T}{\partial y}\right)_{y=0} /(T_s - T_\infty) \tag{13.74}$$

The overall heat transfer coefficient can thus be found if the temperature $T(x, y)$ is known so the gradient can be evaluated. This is a local value often varying over the surface area and as the thermal boundary layer grows following the direction of the free stream, $h$ generally decreases with $x$. The average rate of heat transfer from a given surface area becomes

$$\dot{Q} = \int \dot{q}'' \, dA_s = (T_s - T_\infty) \int h \, dA_s$$

which we can express with an average value of $h$ as

$$\dot{Q} = h_{\text{avg}} A_s (T_s - T_\infty) \tag{13.75}$$

**FIGURE 13.24**
The velocity profile $V(y)$ and the temperature profile $T(y)$ near the wall.

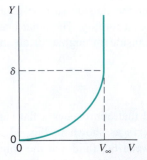

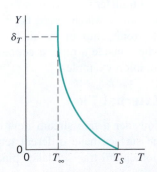

where

$$h_{\text{avg}} = \frac{1}{A_s} \int h \, dA_s$$

For a situation where it is uniform in the $z$-direction (normal to the $x$–$y$ plane), the average becomes over a length in the $x$-domain as

$$h_{\text{avg}} = \frac{1}{L} \int h \, dx \tag{13.76}$$

The Nusselt number as defined in Section 13.1.2 can be restated in terms of the temperature gradient using Eq. 13.74

$$Nu \equiv \frac{hL}{k} = -\left(\frac{\partial T}{\partial y}\right)_{y=0} \frac{L}{T_s - T_\infty} \tag{13.77}$$

and is seen to represent a dimensionless temperature gradient. The local value of $Nu$ and the average value of $Nu$ are related to the corresponding values of $h$ so

$$Nu_{\text{avg}} = \frac{h_{\text{avg}} L}{k} \tag{13.78}$$

A laminar (low velocity) flow over a flat plate leads to a boundary layer development so the local Nusselt number, $Nu = hx/k$, varies proportional to the square root of $x$

$$Nu = \frac{hx}{k} \propto x^{1/2}$$

The local heat transfer coefficient $h$ then behaves as

$$h_x = C x^{-1/2} \tag{13.79}$$

Integration according to Eq. 13.76 forms the average from $x = 0$ to $x = L$ as

$$h_{\text{avg}} = \frac{1}{L} \int C x^{-1/2} \, dx$$

$$= \frac{1}{L} 2C \left[x^{1/2}\right]_0^L = 2C \frac{1}{L} \left[x^{1/2} - 0\right]$$

$$= 2C L^{-1/2} = 2h_L \tag{13.80}$$

showing it is twice the local value at $x = L$. The two functions $h(x)$ and $h_{\text{avg}}(x)$ are shown in Fig. 13.25.

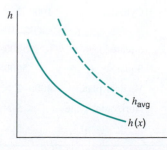

**FIGURE 13.25** The local and the average heat transfer coefficient.

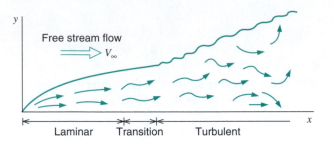

**FIGURE 13.26**
The boundary layer development over a flat plate.

The boundary layer over a flat plate shows several regions in Fig. 13.26. It has a leading section of a laminar boundary layer followed by a transition region and a fully turbulent boundary layer. The location of the transition happens after a critical Reynolds number is reached

$$Re_c = \frac{\rho V x_c}{\mu} = 5 \times 10^5 \tag{13.81}$$

At this point the flow becomes unstable and random fluctuations occur on top of the mean motion. These fluctuations, vortices of various sizes and strength, enhance mixing of fluid near the wall with fluid away from the wall. This increases the transport (exchange) of momentum and energy so the value of $h$ is larger than it otherwise would be. Boundary layer growth rate with distance $x$ is more rapid for a turbulent flow, typically $x^{4/5}$, as compared to the laminar flow growth proportional to $x^{1/2}$ and is accompanied with a more gradual decrease in the local $h$ value as compared to the laminar case.

The two cases for a flat plate with $D$ as the distance from the leading edge follows the correlation for the average Nusselt number as in Eq. 13.3

$$Nu_{avg} = C\, Re_D^n\, Pr^{1/3} \tag{13.82}$$

where the constant $C$ and the power $n$ are listed in Table 13.5 as case $E$.

For smaller values of $Re$ the laminar and turbulent contributions should be used to form the proper average $h$ in Eq. 13.76 as

$$h_{avg} = \frac{1}{L}\left[ \int_0^{xc} h_{laminar}\, dx + \int_{xc}^{L} h_{turbulent}\, dx \right] \tag{13.83}$$

**TABLE 13.5**
***The value of the constants in Eq. 13.82 for several flows.***

| CASE | V | D | Geometry | $Re_D$ | C | n |
|------|---|---|----------|--------|---|---|
| | | | Circular | 40–1000 | 0.51 | 0.5 |
| A | ⇒ | ⊕ | Cylinder | $1000$–$2 \times 10^5$ | 0.26 | 0.6 |
| | | | | $2 \times 10^5$–$10^6$ | 0.076 | 0.7 |
| B | ⇒ | ▯ | Square | $5000$–$10^5$ | 0.1 | 0.68 |
| C | ⇒ | ◈ | Square | $5000$–$10^5$ | 0.25 | 0.6 |
| D | ⇒ | ▯ | Vertical plate | $4000$–$2 \times 10^4$ | 0.23 | 0.73 |
| E | ⇒ | ↔ | Horizontal plate | $< 5 \times 10^5$ | 0.68 | 0.5 |
| | | | | $> 5 \times 10^5$ | 0.03 | 0.8 |

However for larger $Re$, the laminar contribution is small enough so the fully turbulent form for the average $Nu$ is of sufficient accuracy. It should be kept in mind that uncertainty in the actual geometry, the flow velocity and direction, and the surface temperature will lead to heat transfer calculations that are accurate within $\pm 25\%$.

Flows over other surface geometries are also included in Table 13.5 such as cross flow over pipes and ducts with cross sections as circles and squares. Some have a correlation for different ranges of the Reynolds number due to changes in the flow pattern and in general these are sensitive to the details of the approach flow and geometry. Many other flow configurations and multiple arrays of these are correlated with forms similar to Eq. 13.82 and available in the literature.

**EXAMPLE 13.12**

Air at 25°C flows with a velocity of 10 m/s over the flat top surface of a computer chip that is a square 3 cm by 3 cm. If the power dissipated through the top surface is 5 W, how high is the surface temperature? Assume the air dynamic viscosity $v = \mu/\rho = 1.6 \times 10^{-5}\,\text{m}^2/\text{s}$, $Pr = 0.7$ and the velocity is parallel to one edge.

**Analysis**

This is a steady-state convection from a flat surface with a convection correlation as case $E$ in Table 13.5. We must decide which correlation form to use for $Nu$ to find a suitable value of $h$.

**Solution**

First calculate the flow Reynolds number based on the length of flow over the top

$$Re = \rho VL/\mu = VL/v = 10 \times 0.03/1.6 \times 10^{-5} = 18\,750$$

which gives laminar flow over the whole top. Now the $Nu$ number from Table 13.5 is

$$Nu = 0.68\,Re^{0.5}Pr^{1/3} = 0.68 \times 18\,750^{0.5} \times 0.7^{1/3} = 82.7$$

and the average heat transfer coefficient is

$$h = Nu\,k/L = 82.7 \times 0.026/0.03 = 71.7\,\text{W/m}^2\,\text{K}$$

Now the rate of heat transfer is $\dot{Q} = hA\,\Delta T$, so the temperature difference is

$$\Delta T = \dot{Q}/hA = 5/(71.7 \times 0.03 \times 0.03) = 77.5°C$$

and the surface temperature becomes

$$T_s = T_{\text{air}} + \Delta T = 25 + 77.5 = 102.5°C$$

When there is a large temperature difference between the free stream flow and the surface, it may influence the value of properties such as $\rho$, $\mu$ and $k$. Special correlations to account for this effect have been developed as extensions of Eq. 13.82 or similar equations. A quite satisfactory method is to evaluate the fluid properties at the film temperature which is defined as

$$T_f = \frac{1}{2}(T_s + T_\infty) \tag{13.84}$$

This temperature is used for the evaluation of properties needed to compute the Reynolds number and the Nusselt number correlation in Eq. 13.82.

**Example 13.13**

Hot combustion gases at 1500 K flow with 44 m/s across steel pipes to boil water inside the pipes at 225°C. The pipes have $D = 10$ cm and a surface temperature of 227°C. We wish to find the rate of heat transfer per meter pipe length.

**Analysis**

A steady-state external flow over a circular pipe for which Newton's law of cooling applies with geometry as case $A$ in Table 13.5. As the gases are at a high temperature, we will evaluate properties at the film temperature and assume we can use air properties.

**Solution**

The film temperature is

$$T_f = \frac{1}{2}(T_s + T_\infty) = \frac{1}{2}(227 + 273 + 1500) = 1000 \text{ K}$$

At this temperature we may find the properties for air in a handbook as

$$\mu = 42 \times 10^{-6} \text{ kg/s m}, \qquad k = 0.074 \text{ W/m K}, \qquad Pr = 0.73$$

and compute the density from ideal gas law

$$\rho = P/RT = 101.3/(0.287 \times 1000) = 0.353 \text{ kg/m}^3$$

Now the Reynolds number becomes

$$Re = \rho \mathbf{V}D/\mu = 0.353 \times 44 \times 0.1/42 \times 10^{-6} = 36\,981$$

This value falls in the middle range for case $A$ in Table 13.5 so the Nusselt number is

$$Nu = 0.26\, Re^{0.5}\, Pr^{1/3} = 0.26 \times 36\,981^{0.6} \times 0.73^{1/3}$$

$$= 0.26 \times 550.5 \times 0.90 = 128.8$$

and

$$h = Nu\, k/D = 128.8 \times 0.074/0.1 = 95.3 \text{ W/m}^2\text{ K}$$

The rate of heat transfer is

$$\dot{Q} = hA_s\, \Delta T = h\pi D\, L\Delta T$$

so per unit length of pipe

$$\dot{q}' = \dot{Q}/L = h\pi D\Delta T$$

$$= 95.3\, \pi\, 0.1(1500 - 500)$$

$$= 29\,939 \text{ W/m} = 29.9 \text{ kW/m}$$

## Internal Flow

A flow from a tank into a pipe or duct goes through a transition flow in the entry region, as shown in Fig. 13.27, before it turns to a fully developed flow. The presence of the walls are felt through the boundary layer that grows until all the fluid is included. During this transition region, the velocity across the flow changes from a constant called a top hat profile to a smooth function of distance from the centerline. The velocity profile

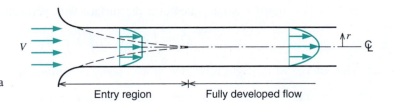

**FIGURE 13.27**
Flow development in a pipe.

Entry region | Fully developed flow

for the fully developed flow has a maximum at the centerline furthest away from the walls and the velocity is zero at the wall. For a laminar flow in a pipe, the profile ends up as a parabola whereas the turbulent flow profile has a steeper gradient near the wall and is more flat in the center—something between the parabola and the top hat profile.

The mass flow rate through the pipe or duct is found as explained in Section 6.1 by integration over the cross sectional area

$$\dot{m} = \int \rho \mathbf{V} \, dA_c = \rho \mathbf{V}_m A_c \tag{13.85}$$

The mean velocity $\mathbf{V}_m$ is used to characterize the flow

$$\mathbf{V}_m = \dot{m}/\rho A_c = \frac{1}{\rho A_c} \int \rho \mathbf{V} \, dA_c \tag{13.86}$$

so for a pipe flow, $A_c = \pi D^2/4$, with constant density and $\mathbf{V}(r)$ we have

$$\mathbf{V}_m = \frac{4}{\pi D^2} \int \mathbf{V}(r) \, 2\pi r \, dr = \frac{8}{D^2} \int_0^{D/2} \mathbf{V}(r) r \, dr \tag{13.87}$$

This velocity is used for computing the Reynolds number in the evaluation of friction and heat transfer. Just as a mean velocity is useful a mean temperature characterizes the energy level of the flow. To express the flux of energy by the flow in terms of temperature, we will assume a single phase flow and use the heat capacity as

$$\dot{m} h = \dot{m} \, C_p \, T_m = \int \rho \mathbf{V} \, C_p \, T \, dA_c$$

$$= \rho \mathbf{V}_m \, C_p \, T_m \, A_c \tag{13.88}$$

Consider a control volume as a section of pipe or duct shown in Fig. 13.28, where the energy equation is

$$\frac{dE_{CV}}{dt} = 0 = \dot{m} h_o + \dot{Q} - \dot{m} h_x \tag{13.89}$$

The enthalpy flow terms are expressed by the mean temperature from Eq. 13.88

$$\dot{m} h_o = \dot{m} \, C_p \, T_{m\,o}; \qquad \dot{m} h_x = \dot{m} \, C_p \, T_m$$

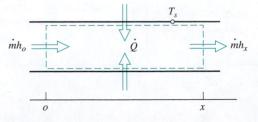

**FIGURE 13.28**  Flow in a section of pipe or duct.

and the heat transfer is integrated over the surface with perimeter P as

$$\dot{Q} = \int \dot{q}'' \, dA_s = \int \dot{q}'' P \, dx$$

The balance of the right hand side in the energy equation, Eq. 13.89, becomes

$$\dot{m} C_p \, T_m = \dot{m} C_p \, T_{mo} + \int \dot{q}'' P \, dx \qquad (13.90)$$

which we will differentiate with respect to $x$ as

$$\frac{d}{dx} \dot{m} C_p \, T_m = \dot{m} C_p \frac{dT_m}{dx} = 0 + \dot{q}'' P \qquad (13.91)$$

The differential equation for the mean temperature is

$$\frac{dT_m}{dx} = \frac{\dot{q}'' P}{\dot{m} C_p} \qquad (13.92)$$

from which the axial variation $T_m(x)$ will be determined. The solution to Eq. 13.92 will be done for two surface conditions, namely one of constant surface heat flux, $\dot{q}''$, and the other of constant surface temperature.

With a constant surface heat flux, constant perimeter $P$ and constant heat capacity the right-hand side of Eq. 13.92 is constant. The equation is then integrated to

$$T_m = T_{mo} + \frac{\dot{q}'' P}{\dot{m} C_p} x \qquad (13.93)$$

and $T$ varies linearly with $x$. With Newton's law of cooling

$$\dot{q}'' = h(T_s - T_m) \qquad (13.94)$$

it is noted that $T_s$ also varies with $x$ as $\dot{q}''$ is a constant. In the entrance region where the boundary layer develops, $h$ decreases so the temperature difference increases, reaching a constant for the fully developed flow, as shown in Fig. 13.29a.

The number of transfer units is defined as

$$\text{NTU} = \frac{h_{\text{avg}} A_s}{\dot{m} C_p} = \frac{1}{\dot{m} C_p R} \qquad (13.95)$$

and expresses the ratio of the heat transfer capacity ($hA$) or ($1/R$) to the flow capacity $\dot{m} C_p$. Each of these is an energy flux (power) per degree temperature difference and the

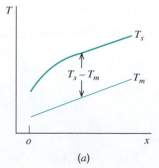

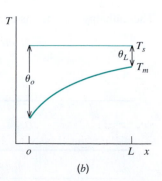

**FIGURE 13.29** Axial variation of the mean temperature (a) constant surface heat flux (b) constant surface temperature.

(a)

(b)

ratio thus relates to the overall magnitude of the heat exchange. For a finite length of pipe, the solution in Eq. 13.93 with $PL = A_s$ using Eq. 13.94 for $\dot{q}''$ becomes

$$T_{mL} - T_{mo} = \text{NTU}(T_s - T_m) \tag{13.96}$$

If the surface temperature is constant, Eq. 13.92 can be integrated by a change of variable to $\theta = T_m - T_s$ so it becomes

$$\frac{d\theta}{dx} = -\frac{hP}{\dot{m}C_p}\theta$$

Integration after separating variables yields the result

$$\ln\frac{\theta_L}{\theta_o} = -\frac{P}{\dot{m}C_p}\int_o^L h\,dx = -\frac{PL}{\dot{m}C_p}h_{\text{avg}} = -\text{NTU} \tag{13.97}$$

Here the average value of $h$ is used from Eq. 13.76 and $PL = A_s$ as the total surface area. With the number of transfer units from Eq. 13.95, the solution can also be written

$$\theta_L = \theta_o\,e^{-\text{NTU}} \tag{13.98}$$

and shown in Fig. 13.29b. The total heat transfer is

$$\dot{Q} = \dot{m}C_p(T_{mL} - T_{mo}) = \dot{m}C_p(\theta_L - \theta_o)$$

$$= \Delta T_{lm}/R \tag{13.99}$$

where the log mean temperature difference is

$$\Delta T_{lm} = \dot{m}C_p R(\theta_L - \theta_o) = (\theta_L - \theta_o)/NTU$$

$$= \frac{\theta_L - \theta_o}{\ln(\theta_o/\theta_L)} = \frac{T_{mL} - T_{mo}}{\ln[(T_o - T_s)/(T_L - T_s)]} \tag{13.100}$$

This made use of the solution in Eq. 13.97 to express NTU as $\ln(\theta_0/\theta_L)$. Using the log mean temperature difference, Eq. 13.99 writes the heat transfer rate as previously shown in Eq. 13.10.

The correlations for the heat transfer coefficient expressed in the Nusselt number are very similar to Eq. 13.82 and already shown in Eq. 13.3. For a laminar flow the solution for the two surface conditions can be obtained mathematically. The Nusselt number based on the pipe diameter becomes a constant as

$$Nu_D = \begin{cases} 4.36 & \dot{q}'' \text{ constant} \\ 3.66 & T_s \text{ constant} \end{cases} \tag{13.101}$$

with a smaller difference between the two cases for non-circular cross sections. In the latter cases a hydraulic diameter is used as

$$D_h = \frac{4A_c}{P} \tag{13.102}$$

with values of $Nu_{Dh}$ in the order of 4.

When the Reynolds number exceed a critical value

$$Re_c = 2300$$

the flow makes a transition to a turbulent flow. A typical correlation used for all geometries based on the hydraulic diameter is

$$Nu_D = 0.027\, Re_D^{0.8} Pr^{1/3} \tag{13.103}$$

Numerous other correlations for special geometries and temperature effects on the properties can be found in the literature.

---

**EXAMPLE 13.14**

A section of 5 cm diameter steel pipe in a steam superheater has a surface temperature of 500°C. Steam at 800 kPa, 300°C enters with $Re = 20\,000$, $Pr = 1$, $k = 0.042$ W/m K, $C_p = 2.1$ kJ/kg K and a massflow rate of 0.15 kg/s. What is the log-mean temperature difference and the length of pipe required to bring the steam to 450°C.

**Analysis**

The section of pipe has a steady flow with assumed constant properties. These conditions give the solution in Eq. 13.97 and 13.98.

**Solution**

The log-mean temperature difference from Eq. 13.100

$$\Delta T_{lm} = \frac{450 - 300}{\ln \dfrac{300 - 500}{450 - 500}} = 108.2°C$$

To proceed we must find $h_{avg}$ that enters in Eq. 13.96–13.97 or 13.95 before we can solve for the length $L$.

$$Re_0 = 20\,000 > 2300 \quad \text{so turbulent flow}$$

From Eq. 13.102 we get

$$Nu = 0.027 \times 20\,000^{0.8} \times 1^{1/3} = 74.5$$

and

$$h = \frac{k}{D} Nu = \frac{0.042 \times 74.5}{0.05} = 62.6 \text{ W/m}^2 \text{ K}$$

$$P = \pi D = 0.157 \text{ m}$$

so from Eq. 13.96 and 13.95 (or equivalently Eq. 13.95)

$$\text{NTU} = \frac{hP}{\dot{m}C_P} \times L = -\ln \frac{\theta_L}{\theta_o}$$

$$= (\theta_L - \theta_o)/\Delta T_{lm} = \frac{-50 + 200}{108.2} = 1.386$$

Now the length becomes

$$L = \frac{\dot{m}C_p}{h P} \times \text{NTU} = \frac{0.15 \times 2.1 \times 1000}{62.6 \times 0.157} \times 1.386$$

$$= 44.4 \text{ m}$$

A more common boundary condition than constant $T_s$ is a constant ambient $T_{amb}$ that may be another fluid. In that case we substitute Eq. 13.94 with

$$\dot{q}' = \dot{q}'' \, P = (T_{amb} - T_m)/R' \tag{13.104}$$

where $R'$ is the thermal resistance between $T_m$ and $T_{amb}$ per unit length of pipe or duct. Now $\theta = T_m - T_{amb}$ and Eq. 13.92 becomes

$$\frac{dT_m}{dx} = \frac{\dot{q}'}{\dot{m}C_p} = \frac{1}{\dot{m}C_p R'} (T_{amb} - T_m) \tag{13.105}$$

or

$$\frac{d\theta}{dx} = -\frac{\theta}{\dot{m}C_p R'} \tag{13.106}$$

and Eq. 13.97 is still valid with NTU as in Eq. 13.96 and

$$R = R'/L = \sum \left( \frac{1}{hA} + \frac{t}{kA} + \cdots \right)$$

$R'$ is then evaluated from the inside convection layer, one or more conduction layers and an external convection layer in the ambient fluid.

## 13.7 ELECTRONICS COOLING

Cooling of electronic equipment was recognized as important from the beginning of the technology that started out with amplifier tubes containing heated wires where a significant amount of energy must be removed. Due to the bulkiness of the components, the cabinets and enclosures were large and left sufficient space for natural or forced convection of air to carry the energy away. As the technology moved towards transistors, the same cooling could be used because each transistor had its own housing and thus a substantial surface-to-volume ratio. Further advances started to use multitransistor chips though the density of the circuits on the chip was low and the packaging of several chips on a printed circuit board (PCB) was also modest. Having a few openings in the frame or box for the equipment to allow a natural air flow was, in many cases, sufficient.

Rapid development in the technology that decreased chip size coupled with increased demands of functionality and performance have raised the density of the circuits in the chip. The number of components on a PCB have also dramatically increased over time so the need for a higher rate of heat removal has become critical. High-end server central processor units (CPU's) now have a local conversion of electrical energy into internal energy of up to 100 W/cm$^2$ that must be removed by heat transfer. This should be done so the maximum temperature in the substrate is no more than 105°C to keep the very small leads from melting and breaking the connections. Reliability is also an issue—for every 10°C the temperature increases, the failure rate doubles so careful consideration of the management of heat and its removal must be designed into the product from the very beginning.

Furthermore, the industry moves towards CMOS (complementary metal oxide semiconductor) technology and the cooler the electronics is the faster it can operate. A reduction of the temperature of 100°C can provide an improvement of 10% to 30% in performance together with an increase in reliability. To achieve such a large reduction in temperature and to reach a temperature below ambient, an active cooling by a refrigeration cycle is necessary. Newer rack-mounted servers can have built-in refrigeration units

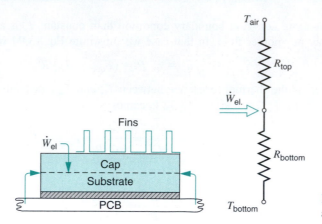

**FIGURE 13.30**  A single chip on a printed circuit board.

using R-134a or external units providing chilled water or other transfer fluid to actively cool the most critical components.

## Passive Cooling Systems

The most used cooling system is a passive system with no moving parts or devices requiring power, this reduces both the cost and increases the reliability. These configurations consist mainly of heat transfer enhancements that will allow a certain amount of heat transfer to occur at a lower temperature difference. Let us look at a single chip with a layer of circuits that is packaged in several layers as illustrated in Figure 13.30.

An amount of electrical power is dissipated in the circuits that is leaving as heat transfer to the top and bottom of the chip at steady state. The electrical analogy to the heat transfer problem is also shown in the figure assuming uniform conditions in the plane of the circuit and disregarding edge effects. From this we can state the energy equation

$$\dot{Q}_{tot} = \dot{W}_{el.} = \dot{Q}_{top} + \dot{Q}_{bottom} = (T - T_{air})/R_{top} + (T - T_{bottom})/R_{bottom} \quad (13.107)$$

where the circuit layer is at temperature $T$, which we like to find. The solution is

$$T = (\dot{W}_{el.} + T_{air} R_{top}^{-1} + T_{bottom} R_{bottom}^{-1})/(R_{top}^{-1} + R_{bottom}^{-1}) \quad (13.108)$$

from which we see that the temperature becomes large if the resistances are large, and for small resistance the temperature approches some average of $T_{air}$ and $T_{bottom}$. The two resistances are then found from the detailed layered construction, serial combinations, and multiple paths such as several fins, parallel connections, or pins for the leads to the circuit board. For the top section we get the resistance as

$$R_{top} = \sum \frac{t}{kA} + R_{fins} \quad (13.109)$$

where the combination of $N$ fins and free surface area are found from Eq. 13.50 as

$$R_{fins} = \frac{1}{h(N \varepsilon A_{cb} + A_{base\ no\ fins})} \quad (13.110)$$

The bottom section may be connected to the circuit board with a heat conducting electrically insulating paste/grease and the leads soldered to the board circuits giving

$$R_{bottom} = \left[ \sum \frac{t}{kA} + R_c \right] \| R_{leads} \quad (13.111)$$

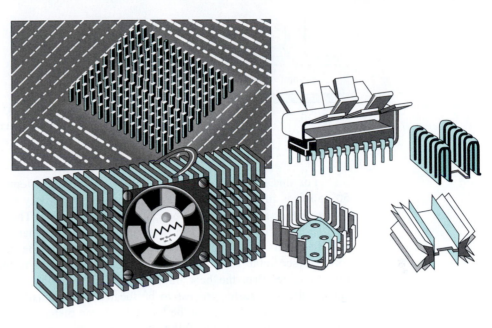

**FIGURE 13.31** Fins used for cooling of chips.

where the contact resistance $R_c$ is such that

$$R_c A \approx 0.1 \frac{\text{cm}^2\,°\text{C}}{\text{W}} \qquad (13.112)$$

and the $N_{\text{leads}}$ leads are conducting through the length giving

$$R_{\text{leads}} = \frac{L_{\text{leads}}}{kA_c\,N_{\text{leads}}} \qquad (13.113)$$

A normal PCB does not have a large conductivity to even out the temperature so there can be local hot spots and $T_{\text{bottom}}$ may be high. To alleviate this problem, a metal base plate can be used to ensure a high conductivity that will reduce local peak temperatures. The drawback of this is higher cost and weight for the whole assembly.

Notice how the resistances for the conduction layers and also the size of the fins scales with the available surface area. To remove a larger amount of heat from a small area, the use of fins increases the effective area exposed to the cooling air. The size and shape of these fins varies greatly as shown in Fig. 13.31 and to have high conductivity they are made of metal often aluminum. Plates or pins in various staggered forms are used so the actual correlations for the heat transfer coefficient must be examined for each case and typically will be supplied by the manufactor of the fins.

A similar effect of boosting the effective heat transfer area can be achieved with a thermosyhon or heat pipe. An example of this was shown in Example 13.9, and an open chamber configuration mounted on two chips is shown in Fig. 13.32.

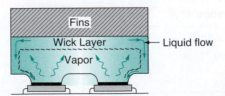

**FIGURE 13.32** A heat pipe or thermosyphon with a wick structure to guide the liquid towards the surface to be cooled.

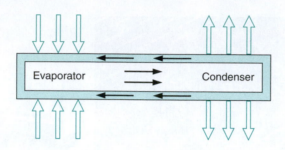

**FIGURE 13.33** A straight heat pipe with a wick structure to transport the liquid back to the evaporator.

For this situation, a transfer fluid with a boiling temperature in the range of 50°–80°C is in a closed volume with a wick or capillary tube to transfer the liquid to the surface in case gravitation cannot be relied upon. The air-cooled fins condense the vapor over an area that can be several times larger than the area where the liquid is boiled, which is right over the surface that should be cooled. It should also be recalled (see Table 13.2) that the heat transfer coefficient to the boiling liquid is magnitudes larger than the heat transfer coefficient to the air. This system allows the enhanced heat transfer process to be located at the smaller area, and the less effective heat transfer process can be located at some distance away where the area can be larger. As a guideline, the following resistances can be used for the phase change processes

$$R_{\text{evap.}}\, A_{\text{evap.}} = R_{\text{cond}}\, A_{\text{cond}} \approx 0.2 \,\frac{\text{cm}^2\,\text{K}}{\text{W}} \tag{13.114}$$

whereas the vapor flow through the pipe section is evaluated as

$$R_{\text{vapor convection}}\, A_{\text{pipe}} \approx 0.02 \,\frac{\text{cm}^2\,\text{K}}{\text{W}} \tag{13.115}$$

If the transfer of liquid should be against the gravitation, the flow is limited by the capillary pressure and smaller structures are used such as grooves, screens, and sintered powder metal. Creating a larger capillary pressure increases the frictional losses and limits the heat flux that can be obtained. For very high heat fluxes, the liquid can evaporate at such a high rate that the surface sees a vapor resulting in much higher resistances than indicated above.

For a straight pipe with a wick structure on the surface as shown in Fig.13.33, the heat flux limits depends on the fluid used, the pipe and the wick material arrangements with some typical axial heat fluxes listed in Table 13.6. As the heat pipe in its various configurations has only minor pressure differences between the evaporator section and the condenser section, the temperatures are not very different. Recall the very low convection resistance in Eq. 13.115.

**TABLE 13.6**

*Typical Heat Fluxes for a Heat Pipe.*

| Fluid | Temperature°C | Flux W/cm² |
|-------|---------------|------------|
| Ammonia | −70 to +50 | 300 |
| Methanol | 80 to 100 | 400 |
| Water | 150 | 500 |

**EXAMPLE 13.15**

Let us compare a typical heat pipe resistance with a similar sized aluminum rod. Assume we have a 1 cm² round pipe with an evaporator and condenser end being 1 cm long each and a total length of pipe of 10 cm.

For the aluminum rod the resistance is from Eq. 13.11

$$R_{\text{rod}} = \frac{L}{kA} = \frac{0.1}{164 \times 1} \frac{\text{m}^2\,\text{K}}{\text{cm}^2\,\text{W}} = 6.1 \frac{\text{K}}{\text{W}}$$

and for the heat pipe the two heat transfer areas are

$$A_{\text{evap.}} = A_{\text{cond}} = L\,\pi\,D = 1\,\text{cm} \times \pi \times 1.13\,\text{cm} = 3.5\,\text{cm}^2$$

so we get from Eq. 13.114 and 13.115

$$R = (0.2/3.5) + (0.02/1) + (0.2/3.5) = 0.134 \frac{\text{K}}{\text{W}}$$

which is 45 times smaller. A reasonable goal for a heat pipe is $R = 0.2$–1 K/W.

## Active Cooling Systems

The cooling systems with active components use fans to give forced convection of air or perhaps a different cooling media, such as a liquid, transferring the heat to a remote location where larger heat sinks to air can be placed. To achieve temperatures below ambient, complete refrigeration cycle systems can be employed as either conventional vapor compression systems or as thermoelectric coolers (TEC) using Peltier elements.

Two types of fans are used, depending on the situation. A larger fan or blower is used to move air through a cabinet with PCBs drawing air in through a filter if dust should be removed. The air is passed over the most critical components first and then out of the cabinet with the largest heat generating parts close to the exit flow. Smaller fans can be mounted directly over components with high heat release, such as a CPU, drawing air in over the fins and then away from the unit. To lessen the cooling load for the rest of the components, the warm exit flow can be guided in a pipe directly out of the cabinet so that flow does not contribute to a temperature increase inside the cabinet.

The thermo electric cooler, TEC, shown in Fig. 1.8a works similar to a thermocouple except a current is the input and it creates a temperature difference between the two junctions so it can achieve cooling below ambient temperature. The maximum temperature difference between the hot and the cold side is about 70°C but it drops linearly with the amount of heat transfer that takes place, becoming zero for some maximum rate.

$$\Delta T_{\text{TEC}} = \Delta T_{\text{max}} - \alpha \dot{Q}_L \tag{13.116}$$

It requires a significant amount of power to operate, so its coefficient of performance is quite low, COP = 1/3 to about 1, and for the higher COP there is very small temperature difference between the two junctions. From the energy equation we have the relation

$$\dot{Q}_H = \dot{Q}_L + \dot{W} = \dot{Q}_L + \dot{Q}_L/\text{COP}$$

as the rate of energy that must be discarded. The hot junction is cooled with air, typically through some fins, leading to a temperature difference to the sink (air) as

$$\Delta T_{\text{sink}} = T_H - T_{\text{air}} = R_{\text{sink}}\,\dot{Q}_H = R_{\text{sink}} \frac{\text{COP} + 1}{\text{COP}}\,\dot{Q}_L$$

The cold junction temperature becomes

$$T_L = T_H - \Delta T_{TEC} = T_{air} + R_{sink} \frac{COP + 1}{COP} \dot{Q}_L - (\Delta T_{max} - \alpha \dot{Q}_L)$$

$$= T_{air} + \left( R_{sink} \frac{COP + 1}{COP} + \alpha \right) \dot{Q}_L - \Delta T_{max} \qquad (13.117)$$

This should be compared with the temperature that can be achieved without the use of the TEC, but just the sink-fins to air resulting in

$$T_L = T_{air} + R_{sink} \dot{Q}_L \qquad (13.118)$$

For small COP and large values of $\alpha$, it is better not to use the TEC whereas large values of COP and smaller values of $\alpha$ makes the TEC beneficial. Because of the limited range $(\dot{Q}_L)$ where the TEC is useful, it is not used as a major part of cooling, but due to its possible small size and very high reliability, it is used for spot cooling of critical components.

A standard vapor compression refrigeration cycle is capable of operating with much higher COP than the TEC and it can also remove a much larger amount of heat transfer at lower temperatures. However, it does require a significant amount of space and increases the level of noise, the weight of the system, and decreases the reliability. The cooling provided by the evaporator can be done by a direct thermal contact to the PCBs or by the use of a transfer fluid that flows to the PCBs. The cold fluid flows in small channels, creating a cold plate or it can flow over the circuits if enclosed in a chamber. These systems require much more extensive mechanical connections and arrangements between the cooling system and the electrical circuits, and to keep the reliability high, two refrigeration systems are used in parallel. Due to cost and size, only high-end computer servers and other performance critical components can justify the use of the refrigeration systems.

# Homework Problems

## Introductory problems and single plane layers

Problems 4.83 to 4.91 are also suitable for this section

**13.1** Air at 20°C, 100 kPa flows over the roof of a building with 5 m/s. Its absolute viscosity is $19 \times 10^{-6}$ kg/m-s and the roof side is 1 m long. Use Eq. 13.3 with $Pr = 0.7$ and assume external turbulent flow. Find the characteristic value of $h$.

**13.2** Air at 17°C, 100 kPa flows over a window 1 m by 1 m, leading to a Reynolds number of 50 000. The Prandtl number is 0.7 for air. What is the characteristic value of the heat transfer coefficient, $h$, if (a) laminar flow or (b) turbulent flow is assumed?

**13.3** Liquid water at 80°C, 100 kPa flows inside a pipe of diameter 0.04 m with a low velocity so the Reynolds number is 400. Use $Pr = 6$ and estimate a suitable $h$ for the flow.

**13.4** Water at 15°C, 100 kPa flows in a square duct 10 cm by 10 cm, leading to a Reynolds number of 2300. The Prandtl number is 6.2 for liquid water. What is the characteristic value of the heat transfer coefficient, $h$, if (a) laminar flow or (b) turbulent flow is assumed?

**13.5** The solar radiation is roughly 1000 W/m² on a clear day. What temperature should a perfect black body have to emit that amount of energy flux?

**13.6** A side of a furnace is 0.75 m² with a surface temperature of 60°C, emissivity of 0.8. The ambient air and walls are at 20°C with a convective heat transfer coefficient of 10 W/m² K. Find the net rate of heat transfer by convection and radiation from the side.

**13.7** An open tray contains 25 kg of boiling liquid water at 100 kPa and is heated from below by hot gases at 200°C through a bottom plate of steel 1 cm thick with a cross-sectional heating area of 0.25 m². The process stops when all the liquid is gone. Find the total heat transfer needed and the time the process takes.

**13.8** A transistor in a stereo amplifier dissipates 0.4 W within a housing that is cylindrical, radius 8 mm and height 8 mm. The top and sides are exposed to room air at 20°C with a convection heat transfer coefficient $h = 40$ W/m$^2$ K. Neglect the bottom and find the transistor surface temperature.

**13.9** Two thermocouples measure the temperature on each side of a 5 cm thick material through which an electric heater pushes 150 W over an area of 0.75 m$^2$. If the temperature difference is measured as 4.5°C, what is the material conductivity? What is the layer thermal resistance?

**13.10** A rigid tank of volume 4 m$^3$ contains argon at 500 kPa, 30°C. It is connected to a piston cylinder (initially empty) with a line and a closed valve. The pressure in the cylinder should be 300 kPa to float the piston. Now the valve is slowly opened and heat is transferred so the argon reaches a final state of 30°C with the valve open. The heat transfer takes place from a 60°C space through a conduction layer 1 cm thick with conductivity $k = 0.1$ W/m K and a surface area of 1.5 m$^2$. Find the final volume of the argon and the total work for the process. Find the total heat transfer needed. How long will the process take?

**13.11** A rectangular chip has a top surface of 0.5 cm by 2 cm cooled by air at 20°C with $h = 150$ W/m K. There are also 16 connector pins, $k = 60$ W/m K, 1 mm wide, 0.5 mm thick, and 6 mm long attached to a circuit board at 30°C. If the surface temperature should not exceed 80°C, how much power can be dissipated through the top surface and the pins together?

## Composite plane layers

**13.12** A heat engine receives 2 kW from a 750°C source through a heat exchanger where it has to go through a 2 mm thick steel plate to a fluid with a convective heat transfer coefficient of 40 W/m$^2$ K. Find the needed cross-sectional area if a maximum of 25°C is allowed for the temperature drop.

**13.13** A water heater tank is cylindrical, 1 m tall, and diameter 0.5 m. It is made of a thin steel sheet covered by a 1.5 cm thick foam insulation $k = 0.025$ W/m K. The combined convection and radiation effects are expressed in a convection coefficient of $h = 12$ W/m$^2$ K. The hot water inside is at 72°C and the outside room air is at 20°C. Neglect the steel sheet, the top and bottom surfaces, and assume plane geometry. Determine the rate of heat loss and the outside surface temperature.

**13.14** A double pane (each 4 mm thick) glass window has an air gap of 1 cm. The outside is at $-10$°C with a convection heat transfer coefficient of 175 W/m$^2$ K. The inside glass surface temperature of the inner pane is at 298 K. Find the rate of heat transfer and the lowest temperature in the air gap.

**13.15** A house has 2000 m$^2$ walls that consist of 1 cm thick plaster plates, 6 cm insulation with $k = 0.04$ W/m K, and outside brick layer 10 cm thick. There are, in addition, 200 m$^2$ windows with two 2 mm thick glass panes separated by a 2 cm air gap. Assume no inside convective layer but an outside convective coefficient of 20 W/m$^2$ K, both windows and brick. The inside is 20°C and the outside is 35°C. Find the thermal resistance of the walls and the windows both per square meter. Find the total heat transfer rate to the inside.

**13.16** An industrial food processing plant boils milk at a constant 100 kPa. The heating takes place from a 900°C source through a 0.5 cm thick brick tile and a 2 mm thick steel wall between which is a contact resistance of 0.05 K/W. For a heating area of 25 cm$^2$, find the rate of heat transfer.

**13.17** A cold room has 200 m$^2$ walls of 1 cm thick plasterboard, 10 cm glasswool, and another 1 cm plaster board facing a room at 25°C. Another wall of total 100 m$^2$ has 1 cm plasterboard, 15 cm glass wool, and 5 cm brick to the outside at 15°C. Inside walls have $h = 15$ and outside wall surfaces have $h = 40$ W/m$^2$ K. How much heat transfer is needed to keep the room at $-20$°C?

**13.18** A plane of electrical circuits have on top a 1 mm layer of substrate, $k = 0.95$ W/m K, then a 1.5 mm layer of plastic, $k = 0.4$ W/m K, outside which is a convective coefficient of $h = 35$ W/m$^2$ K to cooling oil at 25°C. If the electrical circuits generates 1000 W/m$^2$, how hot will they be?

**13.19** A 2 cm thick piece of plexiglass is glued to a 2 cm thick piece of softwood as shown in Fig. P13.19. The glue layer releases 100 W/m$^2$ as it hardens. The outer sides of the plexiglass and the wood sees ambient air at 22°C for

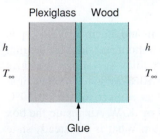

**FIGURE P13.19**

which we assume convection has $h = 5$ W/m² K. Find the steady state glue temperature.

**13.20** A power transistor junction is on a 2 mm thick substrate, $k = 1.25$ W/m K, a 2 mm plastic layer, $k = 0.4$ W/m K, and a 1 mm steel housing. The metal has a contact resistance of $R_cA = 0.0002$ m² K/W to a printed circuit board that is at 35°C. If the transistor junction is at 45°C, how much power per m² goes through these layers?

**13.21** A cubic tank has R-134a at −25°C with a volume of 0.343 m³. The tank has 3 mm steel walls with a 5 cm thick glass-wool insulation except at the bottom surface that stands on a concrete floor giving a contact resistance of 0.25 K/W from steel to the 15°C floor. The other 5 sides are exposed to air at 20°C with a convection coefficient of 50 W/m² K. What is the rate of heat transfer?

**13.22** Consider a 0.5 m² rear car window, 5 mm thick glass. There are convective heat transfer coefficients of inside $h = 10$ W/m² K and outside $h = 20$ W/m² K with an inside air at 20°C and the outside air is at −20°C. Find the thermal resistance of every layer present, and find the outside glass surface temperature.

**13.23** A very thin heating mat dissipates 100 W/m². On the left side it has 2 cm thick soft wood with a 2 mm glass pane exposed to air at 60°C with a convective $h = 25$ W/m² K as shown in Fig. P13.23. The right side has a 2 mm still air gap and then a 1 mm steel plate exposed to a flow of water at 200 kPa, 100°C with an $h = 50$ W/m² K. It is plane geometry and steady state. Find the heating mat temperature and the rate of heat transfer to the right and left.

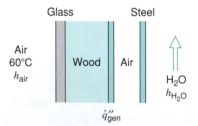

**FIGURE P13.23**

**13.24** A box is 0.5 m by 0.5 m and 25 cm tall. The top is 1 cm thick plexiglass and the bottom and sides are 0.5 mm steel plates covered by 1 cm thick softwood. The sides have $h = 15$ and the top $h = 10$ W/m² K to ambient air at 20°C. The bottom outside has contact to ground 10°C through a resistance of 0.05 m² K/W. Air inside the box is heated with a 200 W heater. What is the steady state temperature?

**13.25** Reconsider the rear car window in Problem 13.22. A thin film deposits $x$ W/m² on the inside glass surface. What should $x$ be so the outside surface is at +2°C?

## Cylindrical geometry

**13.26** Saturated water vapor at 15 kPa runs in a steel pipe inside radius 0.2 m with wall thickness of 3 mm. Outside the convective coefficient is 25 W/m² K to ambient air at 25°C. Find the temperature drop across the steel pipe and the rate of heat transfer per meter length.

**13.27** Saturated water vapor at 15 kPa runs in a steel pipe inside radius 0.2 m with wall thickness of 3 mm. Insulation is wrapped around the pipe, 0.1 m thick with a conductivity of 0.07 W/m K, outside of which the convective coefficient is 25 W/m² K to ambient air at 25°C. Find the temperature drop across the insulation and the rate of heat transfer.

**13.28** Consider the thick foam insulation for the water heater in Problem 13.13. Find its thermal resistance for the proper cylindrical geometry and repeat the calculation for a 1.5 cm plane layer as was done for Problem 13.13.

**13.29** An aerial electrical copper wire carries a current so it generates 2 W per meter length. The wire has diameter 5 mm and 2 mm thick insulation, $k = 0.2$ W/m K with an outside $h = 25$ W/m² K to ambient air at 22°C. Find the metal wire surface temperature.

**13.30** Saturated liquid nitrogen at 77.3 K is stored in a cylindrical container of 5 mm steel $(k = 15)$. Outside the steel of diameter 0.75 m is a 25 cm thick layer of insulation $(k = 0.05)$ covered by a sheet of metal, neglect $k$. The outside convection coefficient is 30 W/m² K to ambient air at 15°C. The cylinder is 3 m tall. Find the rate of heat transfer to the nitrogen, neglecting the top and bottom.

**13.31** A water pipe of 2 mm thick cast iron with inner radius 1 cm is insulated with 3 cm thick glass wool. The pipe is 5 m long and carries water at 15°C, 110 kPa from one building to another as shown Fig. P13.31. On a winter day, the outside air is −20°C giving an external convection coefficient of $h = 75$ W/m² K. Neglect the change in water temperature and find the total rate of heat transfer loss.

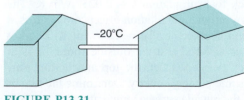

**FIGURE P13.31**

**13.32** Saturated liquid water at 500 kPa flows in a metal pipe, 5 cm inner diameter and $k = 40$ W/m K, with wall thickness 3 mm at a rate of 2 kg/s. The pipe is exposed to ambient gas at 500°C with $h = 100$ W/m² K. The water is being boiled to come out as saturated vapor at 500 kPa. Assume the water temperature is uniform and there is no inside convective layer. Find the total rate of heat transfer that must take place and the needed length of pipe.

**13.33** A 1 cm thick plexiglass cylinder, inner radius 5 cm flows air at 15°C. Outside the cylinder some steam at 100°C flows to give a convective heat transfer coefficient of $h = 50$ W/m² K. Find the total rate of heat transfer for 1 m length cylinder.

**13.34** Take the same situation as in Problem 13.33, but add a 2 cm thick insulation layer on the plexiglass with $k = 0.05$ W/m K outside of which you have the convection layer. Find the rate of heat transfer for a 1 m long cylinder.

**13.35** A 2 m long baseboard heater deposits 1500 W over the surface of a 1 cm diameter 2 m long rod surrounded by a 1 cm thick layer of sand, $k = 0.5$ W/m K, with the whole combination inside a 1 mm thick steel pipe inner diameter 3 cm. The pipe has an outside convection coefficient of 35 W/m² K to ambient air at 20°C. Find the highest temperature in the sand.

**13.36** As refrigerant, R-12 evaporates inside a metal tube ($k = 10$ W/m K) of 2 mm inner diameter and 0.5 mm thickness, the temperature is $-20$°C. A frost layer ($k \approx 0.75$ W/m K) of thickness 2 mm has formed on the tube, outside of which is convection ($h = 45$ W/m² K) to refrigerator air at $+5$°C. Find the total heat transfer for a 2 m long tube.

**13.37** Consider the previous problem. A fan inside the refrigerator blows the air over the tube so $h$ is higher. How high should it be to start to melt the frost layer?

**13.38** A steam drum 6 m long, 1 m inner diameter steel cylinder with 5 mm thickness is shown in Fig. P13.38. It flows water inside at 250°C. It is insulated with 3 cm thick glasswool and the outside convective heat transfer coefficient to air at 20°C is $h = 25$ W/m² K. Find the total rate of heat loss from the drum including ends.

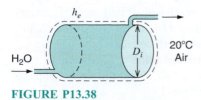

**FIGURE P13.38**

**13.39** A power transistor is in a cylindrical metal ($k = 5$ W/m K) housing of thickness 2 mm with outside radius of 8 mm, height 10 mm as shown in Fig. P13.39. The outside convection coefficient is $h = 50$ W/m² K to air at 25°C. The power of 2 W is dissipated inside an inner layer of 1 mm thick substrate with $k = 1$ W/m K. Assume no heat transfer from the bottom mounted on a board, neglect corner effects. Find the thermal resistance going through the top (substrate, metal, and convection) and through the side (substrate, metal, and convection). Find the temperature on the inside of the metal casing.

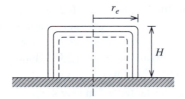

**FIGURE P13.39**

# Extended surfaces, fins

**13.40** A rectangular steel fin is 3 mm thick, 5 cm long and 0.5 m wide attached to a base at 125°C. The fin is exposed to hot air at 250°C flow so the convective coefficient is $h = 100$ W/m² K and we assume the fin tip is adiabatic. Find the rate of heat transfer to the fin. Is this an infinitely long fin?

**13.41** A 4 cm long, 40 cm wide, 2 mm thick steel plate, $k = 15$ W/m K, is used as a fin on a 40 cm wide base at 100°C. The free stream air is at 15°C and $h = 50$ W/m² K and assume the tip is adiabatic. Find the heat transfer for this fin and compare to the case where the plate is cut in half, now 2 cm long, 40 cm wide and used as two identical but shorter fins.

**13.42** A rectangular 1 mm thick aluminum plate, 16 mm by 16 mm is attached on edge (see Fig. P13.42) to the housing of a CPU unit 16 mm by 8 mm. Neglect the sides and bottom of the CPU unit. The top surface plus the fin should remove a total of 1 W having a convective heat transfer coefficient of 40 W/m² K to air at 25°C. How hot is the CPU top surface?

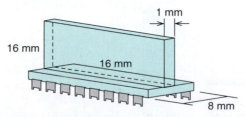

**FIGURE P13.42**

**13.43** A 10 cm long steel rod is square 8 mm by 8 mm cross-section and attached to a base plate at 100°C. The surrounding gas is at 250°C and flows to give a convective heat transfer coefficient of 60 W/m² K. Assume the tip is adiabatic. Is this an infinitely long fin? What is the total rate of heat transfer?

**13.44** A cylindrical rod of aluminum 5 mm in diameter is 30 cm long and placed between two thick plates at 120°C with ambient air at 20°C. The convection coefficient from the rod to air is 60 W/m² K. Find the fin efficiency for the rod and the total rate of heat transfer from the rod.

**13.45** The surface of a steel pipe outer radius 0.25 m is at 54°C. Annular fins of thickness 2 mm and outer radius 0.30 m are mounted on the pipe, and the outside convective coefficient is 25 W/m² K to ambient air at 25°C. Find the heat transfer from each of these fins.

**13.46** An electric baseboard heater has a 2 m long 4 cm diameter metal pipe on which are mounted a number of disks as 1 mm thick annular fins, outer diameter 20 cm. The fin metal has $k = 20$ W/m K and the convection coefficient is 25 W/m² K. What is the fin efficiency and its effectiveness.

**13.47** The baseboard heater in Problem 13.46 has a convective coefficient of $h = 25$ W/m² K on the metal pipe surface and the fin surfaces to ambient air at 18°C. If there is one fin for every cm length of heater, how hot does the 4 cm metal pipe become for a total power dissipation of 1500 W?

**13.48** Annular stainless steel fins of rectangular profile are attached to a tube, outside diameter 100 mm, surface at 400°C. The fins are 2 mm thick and 20 mm long (i.e., outer diameter is 140 mm) and exposed to air at 20°C with a convective heat transfer coefficient of $h = 25$ W/m² K. Find the fin efficiency and effectiveness.

**13.49** Consider the annular fin in the previous problem. Assume we have 100 such fins per meter length of the tube. What is then the rate of heat transfer for 1 m length?

**13.50** A steel plate, 1 m tall and 0.5 m wide, has 15 rectangular fins in the manner described for one fin mounted on it, in each fin is described in Problem 13.40 and shown in Fig P13.50. What is the total rate of heat transfer to the base?

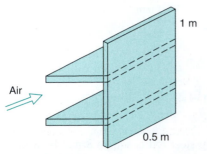

**FIGURE P13.50**

**13.51** A pure copper triangular fin has a base width of 10 cm, and a base thickness of $t = 8$ mm. Find the effectiveness with an outside heat transfer coefficient of 100 W/m² K. What is it if the width is 20 cm?

**13.52** A cylindrical steel rod is used as a fin; it is 2 cm long with diameter 1 cm. The base is 0.5 m² with temperature 200°C and exposed to liquid water at 60°C, 800 kPa flowing so $h = 250$ W/m² K. With 200 rods mounted evenly on the base, find the total rate of heat transfer to the water.

**13.53** A heat exchanger in an air conditioner has to remove 5 kW from a flow of condensing R-12 at 65°C. The R-12 flows behind a flat plate 1 m tall and 0.5 m wide that has some rectangular steel plates as fins on the front similar to Fig. P13.50. Each plate is 5 cm long and 0.5 m wide (along the base plate) with a thickness of 2 mm. A fan blows air at 20°C over the fins to give a convection coefficient $h = 100$ W/m² K. How many fins do we need? If there were no fins, how much heat transfer would take place?

## Energy generation

**13.54** A newly poured concrete slab measures 8 m by 10 m and is 0.12 m thick. The curing process releases a power of 2.5 W/kg. Assume adiabatic bottom and only convective heat transfer through the top surface to atmospheric air at 15°C with $h = 15$ W/m² K and use properties of solid concrete. Find the top surface temperature when equilibrium is reached (steady state).

**13.55** A plate of polystyrene 2 cm thick is placed in a microwave oven tuned so it appears to have an energy generation of 25 000 W/m³. The sides have a convection heat transfer coefficient of 25 W/m² K to ambient air at 25°C. Find the maximum temperature inside the plate at steady state.

**13.56** A 2.5 mm thick layer has electrical generation of 40 kW/m³ and a conductivity of $k = 0.5$ W/m K. One side is perfectly insulated and the other side is connected to 25°C air through a total resistance of $RA = 0.5$ m² K/W. Find the temperature on both sides of the layer.

**13.57** A 20 cm deep pond with liquid water is heated uniformly by solar radiation in the amount of 600 W/m³. Assume the bottom is perfectly insulated and that the top surface has a convective heat transfer coefficient $h = 25$ W/m² K to ambient air at 15°C. Find the water temperature at the bottom of the pond.

**13.58** A 25 cm thick poured concrete slab releases 2500 W/m³ as it hardens. The top surface is at 30°C, the bottom is at 22°C, and the average conductivity is 0.9 W/m K.

What is the maximum and minimum temperature inside the slab?

**13.59** An aerial electrical metal wire ($k$ = 23 W/m K) carries a current so it generates 2 W per meter length. The wire has diameter 5 mm and 2 mm thick insulation, $k$ = 0.4 W/m K with an outside $h$ = 15 W/m² K to ambient air at 25°C. Find the metal wire surface temperature.

**13.60** A 4 mm thick layer with $k$ = 0.4 W/m K has a generation of 1 MW/m³. Both sides have 2 mm plastic layers ($k$ = 0.35 W/m K) and then convection with $h$ = 50 W/m² K to ambient air at 25°C. Find the surface temperature of the generating layer and the maximum temperature in the layer.

**13.61** A cylindrical thorium ($k$ = 50 W/m K) fuel element of a nuclear reactor is inside a steel tubing ($k$ = 15 W/m K) inner radius 25 mm, thickness 5 mm. Energy is uniformly generated in the thorium at a rate of $\dot{q}'''$ = $10^7$ W/m³. Outside the steel tube, water at 100°C flows with a convective heat transfer coefficient of $h$ = 1000 W/m² K. Find the heat transfer rate to the water per meter tube and the surface temperature of the fuel rod.

**13.62** Two plane layers are joined without contact resistance as shown in Fig. P13.62. Layer $A$, 4 cm wide and $k$ = 10 W/m K, has a heat generation of 200 000 W/m³ and layer $B$, 3 cm wide and $k$ = 20 W/m K, has no generation. Ambient air on both outer surfaces is at 40°C and $h$ = 200 W/m² K. Find the three temperatures and the rate of heat transfer per square meter from the left and the right outer surfaces.

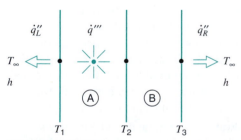

**FIGURE P13.62**

## Transient problems

**13.63** A 20 kg cube of copper at 80°C is dropped into a lake with $T$ = 15°C. The convective heat transfer coefficient is $h$ = 100 W/m² K. Can you assume the temperature inside the copper is uniform? Determine the time it takes for the copper to reach 30°C.

**13.64** A small metal sphere of steel ($k$ = 15 W/m K) is at 20°C. It is immersed in a water bath maintained at

90°C. After 5 seconds, the sphere has a uniform temperature of 85°C due to convection with $h$ = 100 W/m² K. What is the radius of the sphere?

**13.65** A spherical steel ball of 25 kg is at 25°C and heated by 1200 K hot gases uniformly across the lower half surface. The convective heat transfer coefficient is 50 W/m² K, and we assume the steel is at a uniform temperature. How long does it take to bring the steel to 1000 K?

**13.66** Recall the cast iron pipe with water inside in Problem 13.31. Assume the iron has the same temperature as the water at all times. By mistake, the water flow is stopped; neglect energy storage in the insulation and end effects. How long will it take to bring the water to the freezing point?

**13.67** A thin walled steel container contains 200 kg liquid water at 80°C, 110 kPa. It now cools by heat transfer to the ambient 20°C through a 5 cm thick glasswool layer and an outside convective heat transfer coefficient of 15 W/m² K with a total surface area of 2 m². Neglect energy in the steel and insulation and estimate how long it takes to bring the water to 50°C? How about 30°C?

**13.68** A thermocouple is a small metal junction ($k$ = 20 W/m K, $\rho$ = 8000 kg/m³, $C_p$ = 0.4 kJ/kg K) of spherical shape. It is estimated that it will have a convection heat transfer coefficient of $h$ = 150 W/m² K when immersed in a fluid. If the junction has a diameter of 0.001 m, how long does it take to follow 90% of a change in fluid temperature?

**13.69** A 0.1 kg cube of steel is at 20°C. It is immersed in a water bath maintained at 90°C. After 4 seconds, the cube has a uniform temperature of 75°C due to heat transfer. What is the convection heat transfer coefficient?

**13.70** A cold winter has −20°C outside a house with a total of 1500 kg air currently at 20°C. The heating system fails and during a 4-hour period the air cools down to 10°C. Assume all other mass in the house is represented by 16 000 kg solid hard wood of the same temperature as the air inside. What is the thermal resistance between the inside and outside? (Neglect any change in mass, volume, or pressure.)

**13.71** A liquid copper film, $k$ = 370 W/m K, with thickness 4 mm at 1500°C is poured on to a wood plate, $k$ = 0.15 W/m K, with thickness 1 cm. Air at 20°C flows over the top of the film and below the wood plate with a convection coefficient of $h$ = 50 W/m² K. The copper now cools to the melting point at 1083°C. Is the temperature in the copper uniform? Determine the thermal resistance between the copper and the air for 1 m² area. How long does it take to reach the final $T$?

**13.72** A shallow pond, 25 cm deep and an area of 1000 m², contains water at 15°C. During the morning and afternoon it receives a net solar flux of 1000 W/m² for a total of 8 hours. If we neglect heat transfer, how hot does it become? Assume the same pond at 30°C starts to cool overnight with a convective heat transfer coefficient of 20 W/m² K to ambient air at 15°C and no radiation nor is there heat transfer through the bottom. How cold is it after 4 hours?

**13.73** A cubic container, side $L$, contains mass at 20°C, 100 kPa. It is used for energy storage so it is heated up from the outside with $h = 75$ W/m² K. What is the length $L$ so the Biot number is 0.1 if the mass is (a) air, (b) steel, and (c) liquid water.

**13.74** Consider the pond in Problem 13.72. How deep should it be to have $Bi = 0.2$?

**13.75** How thick should the copper liquid film in Problem 13.71 be to have $Bi = 0.1$?

**13.76** Calculate the Biot number for the fin in Problem 13.43. How wide should the side be to have Biot number of 0.1 ?

# Convection

**13.77** Water flows in a pipe at 80°C, 110 kPa from one building 5 m across to another building. The pipe of cast iron has inner diameter 1 cm, 2mm thick with a 2.5 cm glass wool insulation layer. The external convection heat transfer coefficient is $h = 60$ W/m² K to the ambient air at −20°C see Fig. P13.31. Find the thermal resistance between the water and the air. What should the mass flow rate of the water be so the temperature drops 0.5°C along the 5 m pipe?

**13.78** Air flows with 24 m/s and 300 K over a flat plate that is at 400 K. Air has $Pr = 0.7$ and $v = \mu/\rho = 1.6 \times 10^{-5}$ m²/s. Determine the local heat flux at $x = 0.15$ downstream from the leading edge and the average heat flux $\dot{q}''$ up to that location.

**13.79** Assume the same flow as in Problem 13.78 and determine the location $x_c$ of the transition to turbulent flow. Evaluate the local heat flux $\dot{q}''$ and the temperature gradient $(\partial T/\partial y)_{y=0}$ in the laminar flow at $x_c$.

**13.80** The top surface of an electronic chip is at 85°C and a fan blows air at 20°C over it with 10 m/s, along the surface of length 4 cm. The chip is 2 cm wide and air properties are $Pr = 0.7$ and $v = \mu/\rho = 1.6 \times 10^{-5}$ m²/s. Find the total heat transfer rate. How much larger should the velocity be to double the heat transfer rate?

**13.81** Atmospheric air at 0°C, 20 m/s blows over a roof top of 2 cm thick hardwood below which is stagnant air at

5°C. Assume air has $Pr = 0.7$, $v = 1.4 \times 10^{-5}$ m²/s and that the inside hardwood surface is at 5°C. The roof is 10 m by 10 m. Evaluate the total heat loss from the attic air through the roof.

**13.82** Liquid water at 15°C, $v = \mu/\rho = 1.1 \times 10^{-6}$ m²/s, and $Pr = 7$ flows at 10 m/s over a 50°C flat surface in a transformer. The total length is 0.25 m over a width of 0.1 m. Find the total rate of heat transfer.

**13.83** Air at 25°C flows at 15 m/s over a cylindrical rod $D = 2.5$ cm having an outside surface temperature of 75°C. The rod is 1 m long and assume air has $Pr = 0.7$ and $v = \mu/\rho = 1.6 \times 10^{-5}$ m²/s. Find the rate of heat transfer from the rod.

**13.84** Liquid water at 15°C, $v = \mu/\rho = 1.1 \times 10^{-6}$ m²/s, and $Pr = 7$ flows at 1.0 m/s across a duct of square cross section with side of 0.1 m. The duct has surface temperature of 50°C. Determine the maximum heat loss from the duct and the direction the flow should have.

**13.85** Feedwater at 40°C and 400 kPa, 3 m/s flows into a 2 cm diameter pipe heated so it has a surface temperature of 120°C. The water has $Pr = 4$, $v = \mu/\rho = 0.7 \times 10^{-6}$ m²/s, $k = 0.6$ W/m K, and $C_p = 4.18$ kJ/ kg K. Find the heat transfer rate per meter length of pipe assuming fully developed flow.

**13.86** Assume the some physical setup as in the previous problem. How long a pipe is needed to heat the water to 100°C?

**13.87** The boiler in a steam power plant has 5 kg/s water flowing inside a 5 cm inner diameter steel pipe with thickness 3 mm. Assume an inside $h = 2500$ W/m² K from the pipe wall to the water and an outside $h = 250$ W/m² K from 1000 K combustion air to the pipe surface. Consider just the pipe section that brings water from saturated liquid 5 MPa to saturated vapor at 5 MPa. Find the thermal resistance from the water to the combustion air for 1 m pipe length and find the total length needed.

**13.88** Liquid water at 15°C and 100 kPa flows in a square duct, with side 0.05 m and velocity of 0.04 m/s. A constant heat flux of $\dot{q}'' = 4000$ W/m² is applied to the duct. The water has $v = 1 \times 10^{-6}$ m²/s and $Pr = 8$. How long a duct is needed to heat the water to 85°C? What is the duct surface temperature at the beginning assuming fully developed flow?

**13.89** Assume the same physical setup as in the previous problem but change surface condition to be $T_s = 120$°C. How long a duct is needed to heat the water to 85°C. What is the local heat flux $\dot{q}''$ at the beginning assuming fully developed flow?

**13.90** Engine oil flows through a 3 mm diameter tube 15 m long at a rate of 0.02 kg/s. The oil enters at 50°C with a tube surface temperature of 100°C. Oil has $\nu = \mu/\rho = 1.5 \times 10^{-4}$ m²/s, $C_p = 1.9$ kJ/kg K and $Pr = 600$. Find the oil exit temperature.

**13.91** Air at 25°C with $\mathbf{V} = 50$ m/s flows over a 5 cm diameter pipe of steel with thickness 2 mm. Inside the pipe there is a flow of steam at 100 kPa, 250°C with $\mathbf{V} = 40$ m/s. For air: $Pr = 0.7$, $\nu = \mu/\rho = 1.6 \times 10^{-5}$ m²/s and for steam $Pr = 1$, $\nu = \mu/\rho = 4.0 \times 10^{-5}$ m²/s and $C_p = 2$ kJ/kg K. Find the heat loss per unit length of the pipe.

**13.92** Assume the same physical setup as in the previous problem. Find the length of pipe that will cool the steam to 175°C.

## Electronics cooling

**13.93** Assume a rectangular chip of 2 by 4 cm with a uniform power dissipation of 2 W/cm². The top layers are 2 mm epoxy, $k = 0.5$ W/m K and a thin casing with the top surface $h = 60$ W/m² K to air at 25°C. The bottom is 2 mm substrate $k = 0.35$ W/m K and a contact resistance $RA = 1$ cm² °C/W to the board at 40°C. For this case, neglect the leads and find the circuit layer temperature.

**13.94** What is the circuit layer temperature in the previous problem if the chip has no power to it?

**13.95** A circuit on a chip is 2 cm by 2 cm and it dissipates 5 W/cm². On top of the circuit is a 1 mm layer of plastic $k = 0.4$ W/m K and convection to air at 25°C with $h = 50$ W/m² K. The bottom has a combined $RA = 16$ cm² °C/W to 25°C. What is the circuit layer temperature?

**13.96** For the previous problem we want to reduce the temperature so fins are added to the top of the chip. Assume we add 10 fins 2 cm wide and 1 mm thick with an effectiveness of $\varepsilon = 40$. What is the circuit layer temperature then?

**13.97** A chip dissipates 5 W and the top layers have a resistance of 6 K/W conduction plus 40 K/W convection from the top surface to 30°C air. The bottom layers have 6 K/W by conduction/contact to the board in parallel with 2 K/W by the leads to the 40°C board. Find the circuit layer temperature and the split of the heat transfer between the top and the board.

**13.98** For the previous problem assume we add fins to the top surface so the convection resistance is reduced to half (20 K/W). Find the circuit layer temperature.

**13.99** A heat pipe with ammonia is piggy-backed on to a CPU unit and the combined resistances between the ammonia vapor and the top CPU surface is 0.4 K/W. The top

surface of the CPU should be kept at 70°C while it dissipates 50 W. At what pressure is the heat pipe operating?

**13.100** A straight heat pipe is used in the previous problem with a cross section area of 2 cm² equal to both the condenser and evaporator heat transfer areas. What is the temperature difference between the two ends of the heat pipe?

**13.101** The straight heat pipe used in Problem 13.99 with a cross section area of 2 cm² is enhanced with an enlarged condenser section having fins to give a resistance from the condensing ammonia to the air at 25°C of 0.4 K/W. How cold can the air then keep the condensing ammonia?

**13.102** A heat pipe has an evaporator section of 4 cm² with a 1 mm, $k = 0.8$ W/m K layer to the heat generating circuit layer and we neglect the bottom heat transfer. The condenser section also 4 cm² has wall conduction resistance of 0.1 K/W and an effective outer surface area due to fins of 40 cm² with $h = 100$ W/m² K to air at 25°C. Find all the resistances and the circuit layer maximum power dissipation allowed if the layer $T = 80$°C.

**13.103** A TEC unit has a maximum $\Delta T = 70$°C and its size gives $\alpha = 2$ K/W. Assume we designed a heat sink with $R_{sink} = 1$ K/W to air at 20°C. How cold will the cold side be if it operates with $COP = 0.5$ and should remove 20 W from a chip. How cold will the chip surface be if we used the heat sink only without the TEC unit.

**13.104** Assume the same TEC unit as in Problem 13.103. How high should the $\dot{Q}_L$ be so the TEC unit gives the same chip surface temperature as the heat sink alone?

**13.105** A TEC unit with a large junction area has maximum $\Delta T = 70$°C and $\alpha = 1$ K/W. With the larger area we can design a heat sink with $R_{sink} = 0.5$ K/W to air at 25°C. Find the cold junction temperature for a heat removal rate of 25 W and 50 W assuming we have $COP = 0.6$.

**13.106** A refrigeration unit operating with R-134a has a high pressure of 1200 kPa and a low pressure of 200 kPa. It should remove 200 W from a high powered electronics package. The evaporator has a total resistance to the power generating layer of $R = 0.1$ K/W and the condenser is cooled by 25°C air through a resistance of $R = 0.05$ K/W. How cold will the electronics be? How much heat must be given to the air? Can the air cool the R-134a enough?

**13.107** Take the previous problem and assume we were trying to find the maximum allowed thermal resistance of the condenser to air for which the unit will operate. For this assume we need to move 250 W to the air from the condensing R-134a (neglect the temperature variation of the process that is in the superheated vapor region). Find maximum $R$ so the process is still possible.

**13.108** To remove 250 W from a condenser in a refrigeration cycle we blow 25°C air in over a 45°C surface with a number of fins on it. The air therefore heats up to say 35°C and we should use an expression as in Eq. 13.99 and 13.100. How different is the heat transfer using the log mean temperature difference versus using a $\Delta T = 45 - 25 = 20$°C temperature difference for the same $R$.

**13.109** To remove 250 W from a condenser in a refrigeration cycle we blow 25°C air in over a 45°C surface giving $h = 100$ W/m$^2$ K. If the air should not come out hotter than 35°C how much air flow do we need? If the average $\Delta T = 15$°C how large a surface area should there be?

**13.110** A refrigeration unit keeps a server CPU chip at 0°C by flowing refrigerant around the unit in a enclosure. To prevent condensation on the outside the of the 0°C box we put a layer of insulation $k = 0.05$ W/m K outside which is convection to air at 25°C with $h = 100$ W/m$^2$ K. How thick an insulation layer should we have to prevent the outside surface of the insulation to be colder than 15°C?

# Contents of Appendix

# Appendix A

# SI Units: Single-State Properties

**TABLE A.1**
*Conversion Factors*

**Area ($A$)**

$1 \text{ mm}^2 = 1.0 \times 10^{-6} \text{ m}^2$     $1 \text{ ft}^2 = 144 \text{ in.}^2$

$1 \text{ cm}^2 = 1.0 \times 10^{-4} \text{ m}^2 = 0.1550 \text{ in.}^2$     $1 \text{ in.}^2 = 6.4516 \text{ cm}^2 = 6.4516 \times 10^{-4} \text{ m}^2$

$1 \text{ m}^2 = 10.7639 \text{ ft}^2$     $1 \text{ ft}^2 = 0.092\,903 \text{ m}^2$

**Conductivity ($k$)**

$1 \text{ W/m-K} = 1 \text{ J/s-m-K}$

$\qquad\qquad = 0.577\,789 \text{ Btu/h-ft-}^\circ\text{R}$     $1 \text{ Btu/h-ft-R} = 1.730\,735 \text{ W/m-K}$

**Density ($\rho$)**

$1 \text{ kg/m}^3 = 0.06242797 \text{ lbm/ft}^3$     $1 \text{ lbm/ft}^3 = 16.018\,46 \text{ kg/m}^3$

$1 \text{ g/cm}^3 = 1000 \text{ kg/m}^3$

$1 \text{ g/cm}^3 = 1 \text{ kg/L}$

**Energy ($E$, $U$)**

$1 \text{ J} \quad = 1 \text{ N-m} = 1 \text{ kg-m}^2/\text{s}^2$

$1 \text{ J} \quad = 0.737\,562 \text{ lbf-ft}$     $1 \text{ lbf-ft} \quad = 1.355\,818 \text{ J}$

$1 \text{ cal (Int.)} = 4.1868 \text{ J}$     $\qquad\qquad = 1.28507 \times 10^{-3} \text{ Btu}$

    $1 \text{ Btu (Int.)} = 1.055\,056 \text{ kJ}$

$1 \text{ erg} \quad = 1.0 \times 10^{-7} \text{ J}$     $\qquad\qquad = 778.1693 \text{ lbf-ft}$

$1 \text{ eV} \quad = 1.602\,177\,33 \times 10^{-19} \text{ J}$

**Force ($F$)**

$1 \text{ N} = 0.224809 \text{ lbf}$     $1 \text{ lbf} = 4.448\,222 \text{ N}$

$1 \text{ kp} = 9.80665 \text{ N (1 kgf)}$

**Gravitation**

$g = 9.80665 \text{ m/s}^2$     $g = 32.17405 \text{ ft/s}^2$

**Heat capacity ($C_p$, $C_r$, $C$), specific entropy ($s$)**

$1 \text{ kJ/kg-K} = 0.238\,846 \text{ Btu/lbm-}^\circ\text{R}$     $1 \text{ Btu/lbm-}^\circ\text{R} = 4.1868 \text{ kJ/kg-K}$

**Heat flux (per unit area)**

$1 \text{ W/m}^2 = 0.316\,998 \text{ Btu/h-ft}^2$     $1 \text{ Btu/h-ft}^2 = 3.15459 \text{ W/m}^2$

**Table A.1** (*continued*)
*Conversion Factors*

**Heat-transfer coefficient (*h*)**
    $1 \text{ W/m}^2\text{-K} = 0.176\ 11 \text{ Btu/h-ft}^2\text{-}°\text{R}$

**Length (*L*)**
    1 mm = 0.001 m = 0.1 cm
    1 cm  = 0.01 m = 10 mm = 0.3970 in.
    1 m   = 3.28084 ft = 39.370 in.
    1 km  = 0.621 371 mi
    1 mi   = 1609.3 m (US statute)

**Mass (*m*)**
    1 kg    = 2.204 623 lbm
    1 tonne = 1000 kg
    1 grain = $6.47989 \times 10^{-5}$ kg

**Moment (torque, *T*)**
    1 N-m = 0.737 562 lbf-ft

**Momentum (*mV*)**
    1 kg-m/s = 7.232 94 lbm-ft/s
                = 0.224809 lbf-s

**Power ($\dot{Q}, \dot{W}$)**
    1 W         = 1 J/s = 1 N-m/s
                 = 0.737 562 lbf-ft/s
    1 kW       = 3412.14 Btu/h
    1 hp (metric) = 0.735 499 kW

    1 ton of
  refrigeration  = 3.516 85 kW

**Pressure (*P*)**
    1 Pa         = $1 \text{ N/m}^2 = 1 \text{ kg/m-s}^2$
    1 bar       = $1.0 \times 10^5 \text{ Pa} = 100 \text{ kPa}$
    1 atm      = 101.325 kPa
                 = 1.01325 bar
                 = 760 mm Hg $[0°\text{C}]$
                 = $10.332\ 56 \text{ m H}_2\text{O} [4°\text{C}]$
    1 torr      = 1 mm Hg $[0°\text{C}]$
    1 mm Hg $[0°\text{C}]$ = 0.133 322 kPa
    $1 \text{ m H}_2\text{O} [4°\text{C}]$ = 9.806 38 kPa

**Specific energy (*e*, *u*)**
1 kJ/kg = 0.42992 Btu/lbm
          = 334.55 lbf-ft/lbm

$1 \text{ Btu/h-ft}^2\text{-}°\text{R} = 5.67826 \text{ W/m}^2\text{-K}$

1 ft  = 12 in.
1 in. = 2.54 cm = 0.0254 m
1 ft  = 0.3048 m
1 mi = 1.609344 km
1 yd = 0.9144 m

1 lbm = 0.453 592 kg
1 slug = 14.5939 kg
1 ton  = 2000 lbm

1 lbf-ft = 1.355 818 N-m

1 lbm-ft/s = 0.138 256 kg-m/s

1 lbf-ft/s   = 1.355 818 W
              = 4.626 24 Btu/h
1 Btu/s     = 1.055 056 kW
1 hp (UK) = 0.7457 kW
              = 550 lbf-ft/s
              = 2544.43 Btu/h
1 ton of
refrigeration = 12 000 Btu/h

$1 \text{ lbf/in.}^2$      = 6.894 757 kPa

1 atm        = $14.695\ 94 \text{ lbf/in.}^2$
              = 29.921 in. Hg $[32°\text{F}]$
              = $33.899\ 5 \text{ ft H}_2\text{O} [4°\text{C}]$

1 in. Hg $[0°\text{C}]$  = $0.49115 \text{ lbf/in.}^2$
$1 \text{ in. H}_2\text{O} [4°\text{C}]$ = $0.036126 \text{ lbf/in.}^2$

1 Btu/lbm  = 2.326 kJ/kg
1 lbf-ft/lbm = $2.98907 \times 10^{-3}$ kJ/kg
             = $1.28507 \times 10^{-3}$ Btu/lbm

**TABLE A.1** (*continued*)
**Conversion Factors**

**Specific kinetic energy ($\frac{1}{2}\mathbf{V}^2$)**

$1 \text{ m}^2/\text{s}^2 = 0.001 \text{ kJ/kg}$
$1 \text{ kJ/kg} = 1000 \text{ m}^2/\text{s}^2$

$1 \text{ ft}^2/\text{s}^2 \quad = 3.9941 \times 10^{-5} \text{ Btu/lbm}$
$1 \text{ Btu/lbm} = 25037 \text{ ft}^2/\text{s}^2$

**Specific potential energy ($Zg$)**

$1 \text{ m-}g_{\text{std}} = 9.80665 \times 10^{-3} \text{ kJ/kg}$
$\quad\quad\quad = 4.21607 \times 10^{-3} \text{ Btu/lbm}$

$1 \text{ ft-}g_{\text{std}} = 1.0 \text{ lbf-ft/lbm}$
$\quad\quad\quad = 0.001285 \text{ Btu/lbm}$
$\quad\quad\quad = 0.002989 \text{ kJ/kg}$

**Specific volume ($v$)**

$1 \text{ cm}^3/\text{g} = 0.001 \text{ m}^3/\text{kg}$
$1 \text{ cm}^3/\text{g} = 1 \text{ L/kg}$
$1 \text{ m}^3/\text{kg} = 16.018\ 46 \text{ ft}^3/\text{lbm}$

$1 \text{ ft}^3/\text{lbm} = 0.062\ 428 \text{ m}^3/\text{kg}$

**Temperature ($T$)**

$1 \text{ K} = 1°\text{C} = 1.8 \text{ R} = 1.8 \text{ F}$
$\text{TC} = \text{TK} - 273.15$
$\quad\quad = (\text{TF} - 32)/1.8$
$\text{TK} = \text{TR}/1.8$

$1 \text{ R} = (5/9) \text{ K}$
$\text{TF} = \text{TR} - 459.67$
$\quad\quad = 1.8 \text{ TC} + 32$
$\text{TR} = 1.8 \text{ TK}$

**Universal Gas Constant**

$\overline{R} = N_0 k = 8.31451 \text{ kJ/kmol-K}$
$\quad = 1.98589 \text{ kcal/kmol-K}$
$\quad = 82.0578 \text{ atm-L/kmol-K}$

$\overline{R} = 1.98589 \text{ Btu/lbmol-R}$
$\quad = 1545.36 \text{ lbf-ft/lbmol-R}$
$\quad = 0.73024 \text{ atm-ft}^3/\text{lbmol-R}$
$\quad = 10.7317 \text{ (lbf/in.}^2\text{)-ft}^3/\text{lbmol-R}$

**Velocity ($\mathbf{V}$)**

$1 \text{ m/s} \quad = 3.6 \text{ km/h}$
$\quad\quad\quad = 3.28084 \text{ ft/s}$
$\quad\quad\quad = 2.23694 \text{ mi/h}$
$1 \text{ km/h} = 0.27778 \text{ m/s}$
$\quad\quad\quad = 0.91134 \text{ ft/s}$
$\quad\quad\quad = 0.62137 \text{ mi/h}$

$1 \text{ ft/s} \quad = 0.681818 \text{ mi/h}$
$\quad\quad\quad = 0.3048 \text{ m/s}$
$\quad\quad\quad = 1.09728 \text{ km/h}$
$1 \text{ mi/h} = 1.46667 \text{ ft/s}$
$\quad\quad\quad = 0.44704 \text{ m/s}$
$\quad\quad\quad = 1.609344 \text{ km/h}$

**Volume ($V$)**

$1 \text{ m}^3 \quad\quad = 35.3147 \text{ ft}^3$
$1 \text{ L} \quad\quad = 1 \text{ dm}^3 = 0.001 \text{ m}^3$
$1 \text{ Gal (US)} = 3.785\ 412 \text{ L}$
$\quad\quad\quad\quad = 3.785\ 412 \times 10^{-3} \text{ m}^3$

$1 \text{ ft}^3 \quad\quad = 2.831\ 685 \times 10^{-2} \text{ m}^3$
$1 \text{ in.}^3 \quad\quad = 1.6387 \times 10^{-5} \text{ m}^3$
$1 \text{ Gal (UK)} = 4.546\ 090 \text{ L}$
$1 \text{ Gal (US)} = 231.00 \text{ in.}^3$

**Table A.2**
*Critical Constants*

| Substance | Formula | Molec. Mass | Temp. (K) | Press. (MPa) | Vol. ($m^3/kg$) |
|---|---|---|---|---|---|
| Ammonia | $NH_3$ | 17.031 | 405.5 | 11.35 | 0.00426 |
| Argon | Ar | 39.948 | 150.8 | 4.87 | 0.00188 |
| Bromine | $Br_2$ | 159.808 | 588 | 10.30 | 0.000796 |
| Carbon dioxide | $CO_2$ | 44.01 | 304.1 | 7.38 | 0.00212 |
| Carbon monoxide | CO | 28.01 | 132.9 | 3.50 | 0.00333 |
| Chlorine | $Cl_2$ | 70.906 | 416.9 | 7.98 | 0.00175 |
| Fluorine | $F_2$ | 37.997 | 144.3 | 5.22 | 0.00174 |
| Helium | He | 4.003 | 5.19 | 0.227 | 0.0143 |
| Hydrogen (normal) | $H_2$ | 2.016 | 33.2 | 1.30 | 0.0323 |
| Krypton | Kr | 83.80 | 209.4 | 5.50 | 0.00109 |
| Neon | Ne | 20.183 | 44.4 | 2.76 | 0.00206 |
| Nitric oxide | NO | 30.006 | 180 | 6.48 | 0.00192 |
| Nitrogen | $N_2$ | 28.013 | 126.2 | 3.39 | 0.0032 |
| Nitrogen dioxide | $NO_2$ | 46.006 | 431 | 10.1 | 0.00365 |
| Nitrous oxide | $N_2O$ | 44.013 | 309.6 | 7.24 | 0.00221 |
| Oxygen | $O_2$ | 31.999 | 154.6 | 5.04 | 0.00229 |
| Sulfur dioxide | $SO_2$ | 64.063 | 430.8 | 7.88 | 0.00191 |
| Water | $H_2O$ | 18.015 | 647.3 | 22.12 | 0.00317 |
| Xenon | Xe | 131.30 | 289.7 | 5.84 | 0.000902 |
| Acetylene | $C_2H_2$ | 26.038 | 308.3 | 6.14 | 0.00433 |
| Benzene | $C_6H_6$ | 78.114 | 562.2 | 4.89 | 0.00332 |
| *n*-Butane | $C_4H_{10}$ | 58.124 | 425.2 | 3.80 | 0.00439 |
| Chlorodifluoroethane (142b) | $CH_3CClF_2$ | 100.495 | 410.3 | 4.25 | 0.00230 |
| Chlorodifluoromethane (22) | $CHClF_2$ | 86.469 | 369.3 | 4.97 | 0.00191 |
| Dichlorofluoroethane (141) | $CH_3CCl_2F$ | 116.95 | 481.5 | 4.54 | 0.00215 |
| Dichlorotrifluoroethane (123) | $CHCl_2CF_3$ | 152.93 | 456.9 | 3.66 | 0.00182 |
| Difluoroethane (152a) | $CHF_2CH_3$ | 66.05 | 386.4 | 4.52 | 0.00272 |
| Difluoromethane (32) | $CF_2H_2$ | 52.024 | 351.3 | 5.78 | 0.00236 |
| Ethane | $C_2H_6$ | 30.070 | 305.4 | 4.88 | 0.00493 |
| Ethyl alcohol | $C_2H_5OH$ | 46.069 | 513.9 | 6.14 | 0.00363 |
| Ethylene | $C_2H_4$ | 28.054 | 282.4 | 5.04 | 0.00465 |
| *n*-Heptane | $C_7H_{16}$ | 100.205 | 540.3 | 2.74 | 0.00431 |
| *n*-Hexane | $C_6H_{14}$ | 86.178 | 507.5 | 3.01 | 0.00429 |
| Methane | $CH_4$ | 16.043 | 190.4 | 4.60 | 0.00615 |
| Methyl alcohol | $CH_3OH$ | 32.042 | 512.6 | 8.09 | 0.00368 |
| *n*-Octane | $C_8H_{18}$ | 114.232 | 568.8 | 2.49 | 0.00431 |
| Pentafluoroethane (125) | $CHF_2CF_3$ | 120.022 | 339.2 | 3.62 | 0.00176 |
| *n*-Pentane | $C_5H_{12}$ | 72.151 | 469.7 | 3.37 | 0.00421 |
| Propane | $C_3H_8$ | 44.094 | 369.8 | 4.25 | 0.00454 |
| Propene | $C_3H_6$ | 42.081 | 364.9 | 4.60 | 0.00430 |
| Tetrafluoroethane (134a) | $CF_3CH_2F$ | 102.03 | 374.2 | 4.06 | 0.00197 |

## TABLE A.3
*Properties of Selected Solids at 25°C*

| Substance | $\rho$ (kg/m³) | $C_p$ (kJ/kg-K) |
|---|---|---|
| Asphalt | 2120 | 0.92 |
| Brick, common | 1800 | 0.84 |
| Carbon, diamond | 3250 | 0.51 |
| Carbon, graphite | 2000–2500 | 0.61 |
| Coal | 1200–1500 | 1.26 |
| Concrete | 2200 | 0.88 |
| Glass, plate | 2500 | 0.80 |
| Glass, wool | 20 | 0.66 |
| Granite | 2750 | 0.89 |
| Ice (0°C) | 917 | 2.04 |
| Paper | 700 | 1.2 |
| Plexiglass | 1180 | 1.44 |
| Polystyrene | 920 | 2.3 |
| Polyvinyl chloride | 1380 | 0.96 |
| Rubber, soft | 1100 | 1.67 |
| Sand, dry | 1500 | 0.8 |
| Salt, rock | 2100–2500 | 0.92 |
| Silicon | 2330 | 0.70 |
| Snow, firm | 560 | 2.1 |
| Wood, hard (oak) | 720 | 1.26 |
| Wood, soft (pine) | 510 | 1.38 |
| Wool | 100 | 1.72 |
| **Metals** | | |
| Aluminum | 2700 | 0.90 |
| Brass, 60-40 | 8400 | 0.38 |
| Copper, commercial | 8300 | 0.42 |
| Gold | 19300 | 0.13 |
| Iron, cast | 7272 | 0.42 |
| Iron, 304 St Steel | 7820 | 0.46 |
| Lead | 11340 | 0.13 |
| Magnesium, 2% Mn | 1778 | 1.00 |
| Nickel, 10% Cr | 8666 | 0.44 |
| Silver, 99.9% Ag | 10524 | 0.24 |
| Sodium | 971 | 1.21 |
| Tin | 7304 | 0.22 |
| Tungsten | 19300 | 0.13 |
| Zinc | 7144 | 0.39 |

## TABLE A.4
*Properties of Some Liquids at 25°C\**

| Substance | $\rho$ (kg/m³) | $C_p$ (kJ/kg-K) |
|---|---|---|
| Ammonia | 604 | 4.84 |
| Benzene | 879 | 1.72 |
| Butane | 556 | 2.47 |
| $CCl_4$ | 1584 | 0.83 |
| $CO_2$ | 680 | 2.9 |
| Ethanol | 783 | 2.46 |
| Gasoline | 750 | 2.08 |
| Glycerine | 1260 | 2.42 |
| Kerosene | 815 | 2.0 |
| Methanol | 787 | 2.55 |
| *n*-octane | 692 | 2.23 |
| Oil engine | 885 | 1.9 |
| Oil light | 910 | 1.8 |
| Propane | 510 | 2.54 |
| R-12 | 1310 | 0.97 |
| R-22 | 1190 | 1.26 |
| R-32 | 961 | 1.94 |
| R-125 | 1191 | 1.41 |
| R-134a | 1206 | 1.43 |
| Water | 997 | 4.18 |
| **Liquid metals** | | |
| Bismuth, Bi | 10040 | 0.14 |
| Lead, Pb | 10660 | 0.16 |
| Mercury, Hg | 13580 | 0.14 |
| NaK (56/44) | 887 | 1.13 |
| Potassium, K | 828 | 0.81 |
| Sodium, Na | 929 | 1.38 |
| Tin, Sn | 6950 | 0.24 |
| Zinc, Zn | 6570 | 0.50 |

\*Or $T_{melt}$ if higher.

**Table A.5**
*Properties of Various Ideal Gases at 25°C, 100 kPa\* (SI Units)*

| Gas | Chemical Formula | Molecular Mass | $R$ (kJ/Kg-K) | $\rho$ (kg/m$^3$) | $C_{p0}$ (kJ/kg-K) | $C_{v0}$ (kJ/kg-K) | $k = \dfrac{C_p}{C_v}$ |
|---|---|---|---|---|---|---|---|
| Steam | $H_2O$ | 18.015 | 0.4615 | 0.0231 | 1.872 | 1.410 | 1.327 |
| Acetylene | $C_2H_2$ | 26.038 | 0.3193 | 1.05 | 1.699 | 1.380 | 1.231 |
| Air | — | 28.97 | 0.287 | 1.169 | 1.004 | 0.717 | 1.400 |
| Ammonia | $NH_3$ | 17.031 | 0.4882 | 0.694 | 2.130 | 1.642 | 1.297 |
| Argon | Ar | 39.948 | 0.2081 | 1.613 | 0.520 | 0.312 | 1.667 |
| Butane | $C_4H_{10}$ | 58.124 | 0.1430 | 2.407 | 1.716 | 1.573 | 1.091 |
| Carbon dioxide | $CO_2$ | 44.01 | 0.1889 | 1.775 | 0.842 | 0.653 | 1.289 |
| Carbon monoxide | CO | 28.01 | 0.2968 | 1.13 | 1.041 | 0.744 | 1.399 |
| Ethane | $C_2H_6$ | 30.07 | 0.2765 | 1.222 | 1.766 | 1.490 | 1.186 |
| Ethanol | $C_2H_5OH$ | 46.069 | 0.1805 | 1.883 | 1.427 | 1.246 | 1.145 |
| Ethylene | $C_2H_4$ | 28.054 | 0.2964 | 1.138 | 1.548 | 1.252 | 1.237 |
| Helium | He | 4.003 | 2.0771 | 0.1615 | 5.193 | 3.116 | 1.667 |
| Hydrogen | $H_2$ | 2.016 | 4.1243 | 0.0813 | 14.209 | 10.085 | 1.409 |
| Methane | $CH_4$ | 16.043 | 0.5183 | 0.648 | 2.254 | 1.736 | 1.299 |
| Methanol | $CH_3OH$ | 32.042 | 0.2595 | 1.31 | 1.405 | 1.146 | 1.227 |
| Neon | Ne | 20.183 | 0.4120 | 0.814 | 1.03 | 0.618 | 1.667 |
| Nitric oxide | NO | 30.006 | 0.2771 | 1.21 | 0.993 | 0.716 | 1.387 |
| Nitrogen | $N_2$ | 28.013 | 0.2968 | 1.13 | 1.042 | 0.745 | 1.400 |
| Nitrous oxide | $N_2O$ | 44.013 | 0.1889 | 1.775 | 0.879 | 0.690 | 1.274 |
| $n$-octane | $C_8H_{18}$ | 114.23 | 0.07279 | 0.092 | 1.711 | 1.638 | 1.044 |
| Oxygen | $O_2$ | 31.999 | 0.2598 | 1.292 | 0.922 | 0.662 | 1.393 |
| Propane | $C_3H_8$ | 44.094 | 0.1886 | 1.808 | 1.679 | 1.490 | 1.126 |
| R-12 | $CCl_2F_2$ | 120.914 | 0.06876 | 4.98 | 0.616 | 0.547 | 1.126 |
| R-22 | $CHClF_2$ | 86.469 | 0.09616 | 3.54 | 0.658 | 0.562 | 1.171 |
| R-32 | $CF_2H_2$ | 52.024 | 0.1598 | 2.125 | 0.822 | 0.662 | 1.242 |
| R-125 | $CHF_2CF_3$ | 120.022 | 0.06927 | 4.918 | 0.791 | 0.722 | 1.097 |
| R-134a | $CF_3CH_2F$ | 102.03 | 0.08149 | 4.20 | 0.852 | 0.771 | 1.106 |
| Sulfur dioxide | $SO_2$ | 64.059 | 0.1298 | 2.618 | 0.624 | 0.494 | 1.263 |
| Sulfur trioxide | $SO_3$ | 80.053 | 0.10386 | 3.272 | 0.635 | 0.531 | 1.196 |

\*Or saturation pressure if it is less than 100 kPa.

**TABLE A.6**

*Constant-Pressure Specific Heats of Various Ideal Gases*[†]

| Gas | Formula | $C_{p0} = C_0 + C_1\theta + C_2\theta^2 + C_3\theta^3$ (kJ/kg K) | | $\theta = T(\text{Kelvin})/1000$ | |
|-----|---------|--------|--------|--------|--------|
| | | $C_0$ | $C_1$ | $C_2$ | $C_3$ |
| Steam | $H_2O$ | 1.79 | 0.107 | 0.586 | −0.20 |
| Acetylene | $C_2H_2$ | 1.03 | 2.91 | −1.92 | 0.54 |
| Air | — | 1.05 | −0.365 | 0.85 | −0.39 |
| Ammonia | $NH_3$ | 1.60 | 1.4 | 1.0 | −0.7 |
| Argon | Ar | 0.52 | 0 | 0 | 0 |
| Butane | $C_4H_{10}$ | 0.163 | 5.70 | −1.906 | −0.049 |
| Carbon dioxide | $CO_2$ | 0.45 | 1.67 | −1.27 | 0.39 |
| Carbon monoxide | CO | 1.10 | −0.46 | 1.0 | −0.454 |
| Ethane | $C_2H_6$ | 0.18 | 5.92 | −2.31 | 0.29 |
| Ethanol | $C_2H_5OH$ | 0.2 | −4.65 | −1.82 | 0.03 |
| Ethylene | $C_2H_4$ | 1.036 | 5.58 | −3.0 | 0.63 |
| Helium | He | 5.193 | 0 | 0 | 0 |
| Hydrogen | $H_2$ | 13.46 | 4.6 | −6.85 | 3.79 |
| Methane | $CH_4$ | 1.2 | 3.25 | 0.75 | −0.71 |
| Methanol | $CH_3OH$ | 0.66 | 2.21 | 0.81 | −0.89 |
| Neon | Ne | 1.03 | 0 | 0 | 0 |
| Nitric oxide | NO | 0.98 | −0.031 | 0.325 | −0.14 |
| Nitrogen | $N_2$ | 1.11 | −0.48 | 0.96 | −0.42 |
| Nitrous oxide | $N_2O$ | 0.49 | 1.65 | −1.31 | 0.42 |
| n-octane | $C_8H_{18}$ | −0.053 | 6.75 | −3.67 | 0.775 |
| Oxygen | $O_2$ | 0.88 | −0.0001 | 0.54 | −0.33 |
| Propane | $C_3H_8$ | −0.096 | 6.95 | −3.6 | 0.73 |
| R-12* | $CCl_2F_2$ | 0.26 | 1.47 | −1.25 | 0.36 |
| R-22* | $CHClF_2$ | 0.2 | 1.87 | −1.35 | 0.35 |
| R-32* | $CF_2H_2$ | 0.227 | 2.27 | −0.93 | 0.041 |
| R-125* | $CHF_2CF_3$ | 0.305 | 1.68 | −0.284 | 0 |
| R-134a* | $CF_3CH_2F$ | 0.165 | 2.81 | −2.23 | 1.11 |
| Sulfur dioxide | $SO_2$ | 0.37 | 1.05 | −0.77 | 0.21 |
| Sulfur trioxide | $SO_3$ | 0.24 | 1.7 | −1.5 | 0.46 |

[†]Approximate forms valid from 250 K to 1200 K.

*Formula limited to maximum 500 K.

Table A7.1
*Ideal-Gas Properties of Air, Standard Entropy at 0.1-MPa (1-bar) Pressure*

| T (K) | u (kJ/kg) | h (kJ/kg) | $s_T^0$ (kJ/kg-K) | T (K) | u (kJ/kg) | h (kJ/kg) | $s_T^0$ (kJ/kg-K) |
|---|---|---|---|---|---|---|---|
| 200 | 142.77 | 200.17 | 6.46260 | 1100 | 845.45 | 1161.18 | 8.24449 |
| 220 | 157.07 | 220.22 | 6.55812 | 1150 | 889.21 | 1219.30 | 8.29616 |
| 240 | 171.38 | 240.27 | 6.64535 | 1200 | 933.37 | 1277.81 | 8.34596 |
| 260 | 185.70 | 260.32 | 6.72562 | 1250 | 977.89 | 1336.68 | 8.39402 |
| 280 | 200.02 | 280.39 | 6.79998 | 1300 | 1022.75 | 1395.89 | 8.44046 |
| 290 | 207.19 | 290.43 | 6.83521 | 1350 | 1067.94 | 1455.43 | 8.48539 |
| 298.15 | 213.04 | 298.62 | 6.86305 | 1400 | 1113.43 | 1515.27 | 8.52891 |
| 300 | 214.36 | 300.47 | 6.86926 | 1450 | 1159.20 | 1575.40 | 8.57111 |
| 320 | 228.73 | 320.58 | 6.93413 | 1500 | 1205.25 | 1635.80 | 8.61208 |
| 340 | 243.11 | 340.70 | 6.99515 | 1550 | 1251.55 | 1696.45 | 8.65185 |
| 360 | 257.53 | 360.86 | 7.05276 | 1600 | 1298.08 | 1757.33 | 8.69051 |
| 380 | 271.99 | 381.06 | 7.10735 | 1650 | 1344.83 | 1818.44 | 8.72811 |
| 400 | 286.49 | 401.30 | 7.15926 | 1700 | 1391.80 | 1879.76 | 8.76472 |
| 420 | 301.04 | 421.59 | 7.20875 | 1750 | 1438.97 | 1941.28 | 8.80039 |
| 440 | 315.64 | 441.93 | 7.25607 | 1800 | 1486.33 | 2002.99 | 8.83516 |
| 460 | 330.31 | 462.34 | 7.30142 | 1850 | 1533.87 | 2064.88 | 8.86908 |
| 480 | 345.04 | 482.81 | 7.34499 | 1900 | 1581.59 | 2126.95 | 8.90219 |
| 500 | 359.84 | 503.36 | 7.38692 | 1950 | 1629.47 | 2189.19 | 8.93452 |
| 520 | 374.73 | 523.98 | 7.42736 | 2000 | 1677.52 | 2251.58 | 8.96611 |
| 540 | 389.69 | 544.69 | 7.46642 | 2050 | 1725.71 | 2314.13 | 8.99699 |
| 560 | 404.74 | 565.47 | 7.50422 | 2100 | 1774.06 | 2376.82 | 9.02721 |
| 580 | 419.87 | 586.35 | 7.54084 | 2150 | 1822.54 | 2439.66 | 9.05678 |
| 600 | 435.10 | 607.32 | 7.57638 | 2200 | 1871.16 | 2502.63 | 9.08573 |
| 620 | 450.42 | 628.38 | 7.61090 | 2250 | 1919.91 | 2565.73 | 9.11409 |
| 640 | 465.83 | 649.53 | 7.64448 | 2300 | 1968.79 | 2628.96 | 9.14189 |
| 660 | 481.34 | 670.78 | 7.67717 | 2350 | 2017.79 | 2692.31 | 9.16913 |
| 680 | 496.94 | 692.12 | 7.70903 | 2400 | 2066.91 | 2755.78 | 9.19586 |
| 700 | 512.64 | 713.56 | 7.74010 | 2450 | 2116.14 | 2819.37 | 9.22208 |
| 720 | 528.44 | 735.10 | 7.77044 | 2500 | 2165.48 | 2883.06 | 9.24781 |
| 740 | 544.33 | 756.73 | 7.80008 | 2550 | 2214.93 | 2946.86 | 9.27308 |
| 760 | 560.32 | 778.46 | 7.82905 | 2600 | 2264.48 | 3010.76 | 9.29790 |
| 780 | 576.40 | 800.28 | 7.85740 | 2650 | 2314.13 | 3074.77 | 9.32228 |
| 800 | 592.58 | 822.20 | 7.88514 | 2700 | 2363.88 | 3138.87 | 9.34625 |
| 850 | 633.42 | 877.40 | 7.95207 | 2750 | 2413.73 | 3203.06 | 9.36980 |
| 900 | 674.82 | 933.15 | 8.01581 | 2800 | 2463.66 | 3267.35 | 9.39297 |
| 950 | 716.76 | 989.44 | 8.07667 | 2850 | 2513.69 | 3331.73 | 9.41576 |
| 1000 | 759.19 | 1046.22 | 8.13493 | 2900 | 2563.80 | 3396.19 | 9.43818 |
| 1050 | 802.10 | 1103.48 | 8.19081 | 2950 | 2613.99 | 3460.73 | 9.46025 |
| 1100 | 845.45 | 1161.18 | 8.24449 | 3000 | 2664.27 | 3525.36 | 9.48198 |

**TABLE A7.2**

*The Isentropic Relative Pressure and Relative Volume Functions*   FOR AIR!

| $T[K]$ | $P_r$ | $v_r$ | $T[K]$ | $P_r$ | $v_r$ | $T[K]$ | $P_r$ | $v_r$ |
|---|---|---|---|---|---|---|---|---|
| 200 | 0.2703 | 493.47 | 700 | 23.160 | 20.155 | 1900 | 1327.5 | 0.95445 |
| 220 | 0.3770 | 389.15 | 720 | 25.742 | 18.652 | 1950 | 1485.8 | 0.87521 |
| 240 | 0.5109 | 313.27 | 740 | 28.542 | 17.289 | 2000 | 1658.6 | 0.80410 |
| 260 | 0.6757 | 256.58 | 760 | 31.573 | 16.052 | 2050 | 1847.1 | 0.74012 |
| 280 | 0.8756 | 213.26 | 780 | 34.851 | 14.925 | 2100 | 2052.1 | 0.68242 |
| 290 | 0.9899 | 195.36 | 800 | 38.388 | 13.897 | 2150 | 2274.8 | 0.63027 |
| 298.15 | 1.0907 | 182.29 | 850 | 48.468 | 11.695 | 2200 | 2516.2 | 0.58305 |
| 300 | 1.1146 | 179.49 | 900 | 60.520 | 9.9169 | 2250 | 2777.5 | 0.54020 |
| 320 | 1.3972 | 152.73 | 950 | 74.815 | 8.4677 | 2300 | 3059.9 | 0.50124 |
| 340 | 1.7281 | 131.20 | 1000 | 91.651 | 7.2760 | 2350 | 3364.6 | 0.46576 |
| 360 | 2.1123 | 113.65 | 1050 | 111.35 | 6.2885 | 2400 | 3693.0 | 0.43338 |
| 380 | 2.5548 | 99.188 | 1100 | 134.25 | 5.4641 | 2450 | 4046.2 | 0.40378 |
| 400 | 3.0612 | 87.137 | 1150 | 160.73 | 4.7714 | 2500 | 4425.8 | 0.37669 |
| 420 | 3.6373 | 77.003 | 1200 | 191.17 | 4.1859 | 2550 | 4833.0 | 0.35185 |
| 440 | 4.2892 | 68.409 | 1250 | 226.02 | 3.6880 | 2600 | 5269.5 | 0.32903 |
| 460 | 5.0233 | 61.066 | 1300 | 265.72 | 3.2626 | 2650 | 5736.7 | 0.30805 |
| 480 | 5.8466 | 54.748 | 1350 | 310.74 | 2.8971 | 2700 | 6236.2 | 0.28872 |
| 500 | 6.7663 | 49.278 | 1400 | 361.62 | 2.5817 | 2750 | 6769.7 | 0.27089 |
| 520 | 7.7900 | 44.514 | 1450 | 418.89 | 2.3083 | 2800 | 7338.7 | 0.25443 |
| 540 | 8.9257 | 40.344 | 1500 | 483.16 | 2.0703 | 2850 | 7945.1 | 0.23921 |
| 560 | 10.182 | 36.676 | 1550 | 554.96 | 1.8625 | 2900 | 8590.7 | 0.22511 |
| 580 | 11.568 | 33.436 | 1600 | 634.97 | 1.6804 | 2950 | 9277.2 | 0.21205 |
| 600 | 13.092 | 30.561 | 1650 | 723.86 | 1.52007 | 3000 | 10007. | 0.19992 |
| 620 | 14.766 | 28.001 | 1700 | 822.33 | 1.37858 | | | |
| 640 | 16.598 | 25.713 | 1750 | 931.14 | 1.25330 | | | |
| 660 | 18.600 | 23.662 | 1800 | 1051.05 | 1.14204 | | | |
| 680 | 20.784 | 21.818 | 1850 | 1182.9 | 1.04294 | | | |
| 700 | 23.160 | 20.155 | 1900 | 1327.5 | 0.95445 | | | |

The relative pressure and relative volume are temperature functions calculated with two scaling constants $A_1$, $A_2$.

$$P_r = \exp\left[s_T^0/R - A_1\right]; \qquad v_r = A_2 T/P_r$$

such that for an isentropic process ($s_1 = s_2$)

$$\frac{P_2}{P_1} = \frac{P_{r2}}{P_{r1}} = \frac{e^{s_{T2}^0/R}}{e^{s_{T1}^0/R}} \approx \left(\frac{T_2}{T_1}\right)^{C_p/R} \quad \text{and} \quad \frac{v_2}{v_1} = \frac{v_{r2}}{v_{r1}} \approx \left(\frac{T_2}{T_1}\right)^{C_v/R}$$

where the near equalities are for the constant heat capacity approximation.

Table A.8

*Ideal-Gas Properties of Various Substances, Entropies at 0.1-MPa (1-bar) Pressure, Mass Basis*

| | Nitrogen, Diatomic ($N_2$) $R = 0.2968$ kJ/kg-K $M = 28.013$ | | | Oxygen, Diatomic ($O_2$) $R = 0.2598$ kJ/kg-K $M = 31.999$ | | |
|---|---|---|---|---|---|---|
| $T$ (K) | $u$ (kJ/kg) | $h$ (kJ/kg) | $s_T^0$ (kJ/kg-K) | $u$ (kJ/kg) | $h$ (kJ/kg) | $s_T^0$ (kJ/kg-K) |
| 200 | 148.39 | 207.75 | 6.4250 | 129.84 | 181.81 | 6.0466 |
| 250 | 185.50 | 259.70 | 6.6568 | 162.41 | 227.37 | 6.2499 |
| 300 | 222.63 | 311.67 | 6.8463 | 195.20 | 273.15 | 6.4168 |
| 350 | 259.80 | 363.68 | 7.0067 | 228.37 | 319.31 | 6.5590 |
| 400 | 297.09 | 415.81 | 7.1459 | 262.10 | 366.03 | 6.6838 |
| 450 | 334.57 | 468.13 | 7.2692 | 296.52 | 413.45 | 6.7954 |
| 500 | 372.35 | 520.75 | 7.3800 | 331.72 | 461.63 | 6.8969 |
| 550 | 410.52 | 573.76 | 7.4811 | 367.70 | 510.61 | 6.9903 |
| 600 | 449.16 | 627.24 | 7.5741 | 404.46 | 560.36 | 7.0768 |
| 650 | 488.34 | 681.26 | 7.6606 | 441.97 | 610.86 | 7.1577 |
| 700 | 528.09 | 735.86 | 7.7415 | 480.18 | 662.06 | 7.2336 |
| 750 | 568.45 | 791.05 | 7.8176 | 519.02 | 713.90 | 7.3051 |
| 800 | 609.41 | 846.85 | 7.8897 | 558.46 | 766.33 | 7.3728 |
| 850 | 650.98 | 903.26 | 7.9581 | 598.44 | 819.30 | 7.4370 |
| 900 | 693.13 | 960.25 | 8.0232 | 638.90 | 872.75 | 7.4981 |
| 950 | 735.85 | 1017.81 | 8.0855 | 679.80 | 926.65 | 7.5564 |
| 1000 | 779.11 | 1075.91 | 8.1451 | 721.11 | 980.95 | 7.6121 |
| 1100 | 867.14 | 1193.62 | 8.2572 | 804.80 | 1090.62 | 7.7166 |
| 1200 | 957.00 | 1313.16 | 8.3612 | 889.72 | 1201.53 | 7.8131 |
| 1300 | 1048.46 | 1434.31 | 8.4582 | 975.72 | 1313.51 | 7.9027 |
| 1400 | 1141.35 | 1556.87 | 8.5490 | 1062.67 | 1426.44 | 7.9864 |
| 1500 | 1235.50 | 1680.70 | 8.6345 | 1150.48 | 1540.23 | 8.0649 |
| 1600 | 1330.72 | 1805.60 | 8.7151 | 1239.10 | 1654.83 | 8.1389 |
| 1700 | 1426.89 | 1931.45 | 8.7914 | 1328.49 | 1770.21 | 8.2088 |
| 1800 | 1523.90 | 2058.15 | 8.8638 | 1418.63 | 1886.33 | 8.2752 |
| 1900 | 1621.66 | 2185.58 | 8.9327 | 1509.50 | 2003.19 | 8.3384 |
| 2000 | 1720.07 | 2313.68 | 8.9984 | 1601.10 | 2120.77 | 8.3987 |
| 2100 | 1819.08 | 2442.36 | 9.0612 | 1693.41 | 2239.07 | 8.4564 |
| 2200 | 1918.62 | 2571.58 | 9.1213 | 1786.44 | 2358.08 | 8.5117 |
| 2300 | 2018.63 | 2701.28 | 9.1789 | 1880.17 | 2477.79 | 8.5650 |
| 2400 | 2119.08 | 2831.41 | 9.2343 | 1974.60 | 2598.20 | 8.6162 |
| 2500 | 2219.93 | 2961.93 | 9.2876 | 2069.71 | 2719.30 | 8.6656 |
| 2600 | 2321.13 | 3092.81 | 9.3389 | 2165.50 | 2841.07 | 8.7134 |
| 2700 | 2422.66 | 3224.03 | 9.3884 | 2261.94 | 2963.49 | 8.7596 |
| 2800 | 2524.50 | 3355.54 | 9.4363 | 2359.01 | 3086.55 | 8.8044 |
| 2900 | 2626.62 | 3487.34 | 9.4825 | 2546.70 | 3210.22 | 8.8478 |
| 3000 | 2729.00 | 3619.41 | 9.5273 | 2554.97 | 3334.48 | 8.8899 |

**TABLE A.8** (*continued*)
*Ideal-Gas Properties of Various Substances, Entropies at 0.1-MPa (1-bar) Pressure, Mass Basis*

| | CARBON DIOXIDE ($CO_2$) $R = 0.1889$ kJ/kg-K $M = 44.010$ | | | WATER ($H_2O$) $R = 0.4615$ kJ/kg-K $M = 18.015$ | | |
|---|---|---|---|---|---|---|
| *T* (K) | *u* (kJ/kg) | *h* (kJ/kg) | $s_T^0$ (kJ/kg-K) | *u* (kJ/kg) | *h* (kJ/kg) | $s_T^0$ (kJ/kg-K) |
| 200 | 97.49 | 135.28 | 4.5439 | 276.38 | 368.69 | 9.7412 |
| 250 | 126.21 | 173.44 | 4.7139 | 345.98 | 461.36 | 10.1547 |
| 300 | 157.70 | 214.38 | 4.8631 | 415.87 | 554.32 | 10.4936 |
| 350 | 191.78 | 257.90 | 4.9972 | 486.37 | 647.90 | 10.7821 |
| 400 | 228.19 | 303.76 | 5.1196 | 557.79 | 742.40 | 11.0345 |
| 450 | 266.69 | 351.70 | 5.2325 | 630.40 | 838.09 | 11.2600 |
| 500 | 307.06 | 401.52 | 5.3375 | 704.36 | 935.12 | 11.4644 |
| 550 | 349.12 | 453.03 | 5.4356 | 779.79 | 1033.63 | 11.6522 |
| 600 | 392.72 | 506.07 | 5.5279 | 856.75 | 1133.67 | 11.8263 |
| 650 | 437.71 | 560.51 | 5.6151 | 935.31 | 1235.30 | 11.9890 |
| 700 | 483.97 | 616.22 | 5.6976 | 1015.49 | 1338.56 | 12.1421 |
| 750 | 531.40 | 673.09 | 5.7761 | 1097.35 | 1443.49 | 12.2868 |
| 800 | 579.89 | 731.02 | 5.8508 | 1180.90 | 1550.13 | 12.4244 |
| 850 | 629.35 | 789.93 | 5.9223 | 1266.19 | 1658.49 | 12.5558 |
| 900 | 676.69 | 849.72 | 5.9906 | 1353.23 | 1768.60 | 12.6817 |
| 950 | 730.85 | 910.33 | 6.0561 | 1442.03 | 1880.48 | 12.8026 |
| 1000 | 782.75 | 971.67 | 6.1190 | 1532.61 | 1994.13 | 12.9192 |
| 1100 | 888.55 | 1096.36 | 6.2379 | 1719.05 | 2226.73 | 13.1408 |
| 1200 | 996.64 | 1223.34 | 6.3483 | 1912.42 | 2466.25 | 13.3492 |
| 1300 | 1106.68 | 1352.28 | 6.4515 | 2112.47 | 2712.46 | 13.5462 |
| 1400 | 1218.38 | 1482.87 | 6.5483 | 2318.89 | 2965.03 | 13.7334 |
| 1500 | 1331.50 | 1614.88 | 6.6394 | 2531.28 | 3223.57 | 13.9117 |
| 1600 | 1445.85 | 1748.12 | 6.7254 | 2749.24 | 3487.69 | 14.0822 |
| 1700 | 1561.26 | 1882.43 | 6.8068 | 2972.35 | 3756.95 | 14.2454 |
| 1800 | 1677.61 | 2017.67 | 6.8841 | 3200.17 | 4030.92 | 14.4020 |
| 1900 | 1794.78 | 2153.73 | 6.9577 | 3432.28 | 4309.18 | 14.5524 |
| 2000 | 1912.67 | 2290.51 | 7.0278 | 3668.24 | 4591.30 | 14.6971 |
| 2100 | 2031.21 | 2427.95 | 7.0949 | 3908.08 | 4877.29 | 14.8366 |
| 2200 | 2150.34 | 2565.97 | 7.1591 | 4151.28 | 5166.64 | 14.9712 |
| 2300 | 2270.00 | 2704.52 | 7.2206 | 4397.56 | 5459.08 | 15.1012 |
| 2400 | 2390.14 | 2843.55 | 7.2798 | 4646.71 | 5754.37 | 15.2269 |
| 2500 | 2510.74 | 2983.04 | 7.3368 | 4898.49 | 6052.31 | 15.3485 |
| 2600 | 2631.73 | 3122.93 | 7.3917 | 5152.73 | 6352.70 | 15.4663 |
| 2700 | 2753.10 | 3263.19 | 7.4446 | 5409.24 | 6655.36 | 15.5805 |
| 2800 | 2874.81 | 3403.79 | 7.4957 | 5667.86 | 6960.13 | 15.6914 |
| 2900 | 2996.84 | 3544.71 | 7.5452 | 5928.44 | 7266.87 | 15.7990 |
| 3000 | 3119.18 | 3685.95 | 7.5931 | 6190.86 | 7575.44 | 15.9036 |

**Table A.9**

*Ideal-Gas Properties of Various Substances (SI Units), Entropies at 0.1-MPa (1-bar) Pressure, Mole Basis*

| T K | Nitrogen, Diatomic ($N_2$) $\bar{h}_{f,298}^0 = 0$ kJ/kmol $M = 28.13$ | | Nitrogen, Monatomic (N) $\bar{h}_{f,298}^0 = 472\ 680$ kJ/kmol $M = 14.007$ | |
|---|---|---|---|---|
| | $(\bar{h} - \bar{h}_{298}^0)$ kJ/kmol | $\bar{s}_T^0$ kJ/kmol K | $(\bar{h} - \bar{h}_{298}^0)$ kJ/kmol | $\bar{s}_T^0$ kJ/kmol |
| 0 | −8670 | 0 | −6197 | 0 |
| 100 | −5768 | 159.812 | −4119 | 130.593 |
| 200 | −2857 | 179.985 | −2040 | 145.001 |
| 298 | 0 | 191.609 | 0 | 153.300 |
| 300 | 54 | 191.789 | 38 | 153.429 |
| 400 | 2971 | 200.181 | 2117 | 159.409 |
| 500 | 5911 | 206.740 | 4196 | 164.047 |
| 600 | 8894 | 212.177 | 6274 | 167.837 |
| 700 | 11937 | 216.865 | 8353 | 171.041 |
| 800 | 15046 | 221.016 | 10431 | 173.816 |
| 900 | 18223 | 224.757 | 12510 | 176.265 |
| 1000 | 21463 | 228.171 | 14589 | 178.455 |
| 1100 | 24760 | 231.314 | 16667 | 180.436 |
| 1200 | 28109 | 234.227 | 18746 | 182.244 |
| 1300 | 31503 | 236.943 | 20825 | 183.908 |
| 1400 | 34936 | 239.487 | 22903 | 185.448 |
| 1500 | 38405 | 241.881 | 24982 | 186.883 |
| 1600 | 41904 | 244.139 | 27060 | 188.224 |
| 1700 | 45430 | 246.276 | 29139 | 189.484 |
| 1800 | 48979 | 248.304 | 31218 | 190.672 |
| 1900 | 52549 | 250.234 | 33296 | 191.796 |
| 2000 | 56137 | 252.075 | 35375 | 192.863 |
| 2200 | 63362 | 255.518 | 39534 | 194.845 |
| 2400 | 70640 | 258.684 | 43695 | 196.655 |
| 2600 | 77963 | 261.615 | 47860 | 198.322 |
| 2800 | 85323 | 264.342 | 52033 | 199.868 |
| 3000 | 92715 | 266.892 | 56218 | 201.311 |
| 3200 | 100134 | 269.286 | 60420 | 202.667 |
| 3400 | 107577 | 271.542 | 64646 | 203.948 |
| 3600 | 115042 | 273.675 | 68902 | 205.164 |
| 3800 | 122526 | 275.698 | 73194 | 206.325 |
| 4000 | 130027 | 277.622 | 77532 | 207.437 |
| 4400 | 145078 | 281.209 | 86367 | 209.542 |
| 4800 | 160188 | 284.495 | 95457 | 211.519 |
| 5200 | 175352 | 287.530 | 104843 | 213.397 |
| 5600 | 190572 | 290.349 | 114550 | 215.195 |
| 6000 | 205848 | 292.984 | 124590 | 216.926 |

**TABLE A.9** (*continued*)
*Ideal-Gas Properties of Various Substances (SI Units), Entropies at 0.1-MPa (1-bar) Pressure,*
*Mole Basis*

| T K | OXYGEN, DIATOMIC ($O_2$) $\bar{h}_{f,298}^0 = 0$ kJ/kmol $M = 31.999$ | | OXYGEN, MONATOMIC (O) $\bar{h}_{f,298}^0 = 249\,170$ kJ/kmol $M = 16.00$ | |
|---|---|---|---|---|
| | $(\bar{h} - \bar{h}_{298}^0)$ kJ/kmol | $\bar{s}_T^0$ kJ/kmol K | $(\bar{h} - \bar{h}_{298}^0)$ kJ/kmol | $\bar{s}_T^0$ kJ/kmol K |
| 0 | −8683 | 0 | −6725 | 0 |
| 100 | −5777 | 173.308 | −4518 | 135.947 |
| 200 | −2868 | 193.483 | −2186 | 152.153 |
| 298 | 0 | 205.148 | 0 | 161.059 |
| 300 | 54 | 205.329 | 41 | 161.194 |
| 400 | 3027 | 213.873 | 2207 | 167.431 |
| 500 | 6086 | 220.693 | 4343 | 172.198 |
| 600 | 9245 | 226.450 | 6462 | 176.060 |
| 700 | 12499 | 231.465 | 8570 | 179.310 |
| 800 | 15836 | 235.920 | 10671 | 182.116 |
| 900 | 19241 | 239.931 | 12767 | 184.585 |
| 1000 | 22703 | 243.579 | 14860 | 186.790 |
| 1100 | 26212 | 246.923 | 16950 | 188.783 |
| 1200 | 29761 | 250.011 | 19039 | 190.600 |
| 1300 | 33345 | 252.878 | 21126 | 192.270 |
| 1400 | 36958 | 255.556 | 23212 | 193.816 |
| 1500 | 40600 | 258.068 | 25296 | 195.254 |
| 1600 | 44267 | 260.434 | 27381 | 196.599 |
| 1700 | 47959 | 262.673 | 29464 | 197.862 |
| 1800 | 51674 | 264.797 | 31547 | 199.053 |
| 1900 | 55414 | 266.819 | 33630 | 200.179 |
| 2000 | 59176 | 268.748 | 35713 | 201.247 |
| 2200 | 66770 | 272.366 | 39878 | 203.232 |
| 2400 | 74453 | 275.708 | 44045 | 205.045 |
| 2600 | 82225 | 278.818 | 48216 | 206.714 |
| 2800 | 90080 | 281.729 | 52391 | 208.262 |
| 3000 | 98013 | 284.466 | 56574 | 209.705 |
| 3200 | 106022 | 287.050 | 60767 | 211.058 |
| 3400 | 114101 | 289.499 | 64971 | 212.332 |
| 3600 | 122245 | 291.826 | 69190 | 213.538 |
| 3800 | 130447 | 294.043 | 73424 | 214.682 |
| 4000 | 138705 | 296.161 | 77675 | 215.773 |
| 4400 | 155374 | 300.133 | 86234 | 217.812 |
| 4800 | 172240 | 303.801 | 94873 | 219.691 |
| 5200 | 189312 | 307.217 | 103592 | 221.435 |
| 5600 | 206618 | 310.423 | 112391 | 223.066 |
| 6000 | 224210 | 313.457 | 121264 | 224.597 |

**Table A.9** (*continued*)
*Ideal-Gas Properties of Various Substances (SI Units), Entropies at 0.1-MPa (1-bar) Pressure, Mole Basis*

| T<br>K | Carbon Dioxide ($CO_2$)<br>$\bar{h}_{f,298}^0 = -393\ 522$ kJ/kmol<br>$M = 44.01$<br>$(\bar{h} - \bar{h}_{298}^0)$<br>kJ/kmol | $\bar{s}_T^0$<br>kJ/kmol K | Carbon Monoxide (CO)<br>$\bar{h}_{f,298}^0 = -110\ 527$ kJ/kmol<br>$M = 28.01$<br>$(\bar{h} - \bar{h}_{298}^0)$<br>kJ/kmol | $\bar{s}_T^0$<br>kJ/kmol K |
|---|---|---|---|---|
| 0 | −9364 | 0 | −8671 | 0 |
| 100 | −6457 | 179.010 | −5772 | 165.852 |
| 200 | −3413 | 199.976 | −2860 | 186.024 |
| 298 | 0 | 213.794 | 0 | 197.651 |
| 300 | 69 | 214.024 | 54 | 197.831 |
| 400 | 4003 | 225.314 | 2977 | 206.240 |
| 500 | 8305 | 234.902 | 5932 | 212.833 |
| 600 | 12906 | 243.284 | 8942 | 218.321 |
| 700 | 17754 | 250.752 | 12021 | 223.067 |
| 800 | 22806 | 257.496 | 15174 | 227.277 |
| 900 | 28030 | 263.646 | 18397 | 231.074 |
| 1000 | 33397 | 269.299 | 21686 | 234.538 |
| 1100 | 38885 | 274.528 | 25031 | 237.726 |
| 1200 | 44473 | 279.390 | 28427 | 240.679 |
| 1300 | 50148 | 283.931 | 31867 | 243.431 |
| 1400 | 55895 | 288.190 | 35343 | 246.006 |
| 1500 | 61705 | 292.199 | 38852 | 248.426 |
| 1600 | 67569 | 295.984 | 42388 | 250.707 |
| 1700 | 73480 | 299.567 | 45948 | 252.866 |
| 1800 | 79432 | 302.969 | 49529 | 254.913 |
| 1900 | 85420 | 306.207 | 53128 | 256.860 |
| 2000 | 91439 | 309.294 | 56743 | 258.716 |
| 2200 | 103562 | 315.070 | 64012 | 262.182 |
| 2400 | 115779 | 320.384 | 71326 | 265.361 |
| 2600 | 128074 | 325.307 | 78679 | 268.302 |
| 2800 | 140435 | 329.887 | 86070 | 271.044 |
| 3000 | 152853 | 334.170 | 93504 | 273.607 |
| 3200 | 165321 | 338.194 | 100962 | 276.012 |
| 3400 | 177836 | 341.988 | 108440 | 278.279 |
| 3600 | 190394 | 345.576 | 115938 | 280.422 |
| 3800 | 202990 | 348.981 | 123454 | 282.454 |
| 4000 | 215624 | 352.221 | 130989 | 284.387 |
| 4400 | 240992 | 358.266 | 146108 | 287.989 |
| 4800 | 266488 | 363.812 | 161285 | 291.290 |
| 5200 | 292112 | 368.939 | 176510 | 294.337 |
| 5600 | 317870 | 373.711 | 191782 | 297.167 |
| 6000 | 343782 | 378.180 | 207105 | 299.809 |

TABLE A.9 (*continued*)
*Ideal-Gas Properties of Various Substances (SI Units), Entropies at 0.1-MPa (1-bar) Pressure,*
*Mole Basis*

| T<br>K | WATER ($H_2O$)<br>$\bar{h}_{f,298}^0 = -241\ 826$ kJ/kmol<br>$M = 18.015$<br>$(\bar{h} - \bar{h}_{298}^0)$<br>kJ/kmol | $\bar{s}_T^0$<br>kJ/kmol K | HYDROXYL (OH)<br>$\bar{h}_{f,298}^0 = 38\ 987$ kJ/kmol<br>$M = 17.007$<br>$(\bar{h} - \bar{h}_{298}^0)$<br>kJ/kmol | $\bar{s}_T^0$<br>kJ/kmol K |
|---|---|---|---|---|
| 0 | −9904 | 0 | −9172 | 0 |
| 100 | −6617 | 152.386 | −6140 | 149.591 |
| 200 | −3282 | 175.488 | −2975 | 171.592 |
| 298 | 0 | 188.835 | 0 | 183.709 |
| 300 | 62 | 189.043 | 55 | 183.894 |
| 400 | 3450 | 198.787 | 3034 | 192.466 |
| 500 | 6922 | 206.532 | 5991 | 199.066 |
| 600 | 10499 | 213.051 | 8943 | 204.448 |
| 700 | 14190 | 218.739 | 11902 | 209.008 |
| 800 | 18002 | 223.826 | 14881 | 212.984 |
| 900 | 21937 | 228.460 | 17889 | 216.526 |
| 1000 | 26000 | 232.739 | 20935 | 219.735 |
| 1100 | 30190 | 236.732 | 24024 | 222.680 |
| 1200 | 34506 | 240.485 | 27159 | 225.408 |
| 1300 | 38941 | 244.035 | 30340 | 227.955 |
| 1400 | 43491 | 247.406 | 33567 | 230.347 |
| 1500 | 48149 | 250.620 | 36838 | 232.604 |
| 1600 | 52907 | 253.690 | 40151 | 234.741 |
| 1700 | 57757 | 256.631 | 43502 | 236.772 |
| 1800 | 62693 | 259.452 | 46890 | 238.707 |
| 1900 | 67706 | 262.162 | 50311 | 240.556 |
| 2000 | 72788 | 264.769 | 53763 | 242.328 |
| 2200 | 83153 | 269.706 | 60751 | 245.659 |
| 2400 | 93741 | 274.312 | 67840 | 248.743 |
| 2600 | 104520 | 278.625 | 75018 | 251.614 |
| 2800 | 115463 | 282.680 | 82268 | 254.301 |
| 3000 | 126548 | 286.504 | 89585 | 256.825 |
| 3200 | 137756 | 290.120 | 96960 | 259.205 |
| 3400 | 149073 | 293.550 | 104388 | 261.456 |
| 3600 | 160484 | 296.812 | 111864 | 263.592 |
| 3800 | 171981 | 299.919 | 119382 | 265.625 |
| 4000 | 183552 | 302.887 | 126940 | 267.563 |
| 4400 | 206892 | 308.448 | 142165 | 271.191 |
| 4800 | 230456 | 313.573 | 157522 | 274.531 |
| 5200 | 254216 | 318.328 | 173002 | 277.629 |
| 5600 | 278161 | 322.764 | 188598 | 280.518 |
| 6000 | 302295 | 326.926 | 204309 | 283.227 |

**TABLE A.9** (*continued*)
*Ideal-Gas Properties of Various Substances (SI Units), Entropies at 0.1-MPa (1-bar) Pressure, Mole Basis*

| $T$ K | Hydrogen ($H_2$) $\overline{h}_{f,298}^0 = 0$ kJ/kmol $M = 2.016$ | | Hydrogen, Monatomic (H) $\overline{h}_{f,298}^0 = 217\ 999$ kJ/kmol $M = 1.008$ | |
|---|---|---|---|---|
| | $(\overline{h} - \overline{h}_{298}^0)$ kJ/kmol | $\overline{s}_T^0$ kJ/kmol K | $(\overline{h} - \overline{h}_{298}^0)$ kJ/kmol | $\overline{s}_T^0$ kJ/kmol K |
| 0 | −8467 | 0 | −6197 | 0 |
| 100 | −5467 | 100.727 | −4119 | 92.009 |
| 200 | −2774 | 119.410 | −2040 | 106.417 |
| 298 | 0 | 130.678 | 0 | 114.716 |
| 300 | 53 | 130.856 | 38 | 114.845 |
| 400 | 2961 | 139.219 | 2117 | 120.825 |
| 500 | 5883 | 145.738 | 4196 | 125.463 |
| 600 | 8799 | 151.078 | 6274 | 129.253 |
| 700 | 11730 | 155.609 | 8353 | 132.457 |
| 800 | 14681 | 159.554 | 10431 | 135.233 |
| 900 | 17657 | 163.060 | 12510 | 137.681 |
| 1000 | 20663 | 166.225 | 14589 | 139.871 |
| 1100 | 23704 | 169.121 | 16667 | 141.852 |
| 1200 | 26785 | 171.798 | 18746 | 143.661 |
| 1300 | 29907 | 174.294 | 20825 | 145.324 |
| 1400 | 33073 | 176.637 | 22903 | 146.865 |
| 1500 | 36281 | 178.849 | 24982 | 148.299 |
| 1600 | 39533 | 180.946 | 27060 | 149.640 |
| 1700 | 42826 | 182.941 | 29139 | 150.900 |
| 1800 | 46160 | 184.846 | 31218 | 152.089 |
| 1900 | 49532 | 186.670 | 33296 | 153.212 |
| 2000 | 52942 | 188.419 | 35375 | 154.279 |
| 2200 | 59865 | 191.719 | 39532 | 156.260 |
| 2400 | 66915 | 194.789 | 43689 | 158.069 |
| 2600 | 74082 | 197.659 | 47847 | 159.732 |
| 2800 | 81355 | 200.355 | 52004 | 161.273 |
| 3000 | 88725 | 202.898 | 56161 | 162.707 |
| 3200 | 96187 | 205.306 | 60318 | 164.048 |
| 3400 | 103736 | 207.593 | 64475 | 165.308 |
| 3600 | 111367 | 209.773 | 68633 | 166.497 |
| 3800 | 119077 | 211.856 | 72790 | 167.620 |
| 4000 | 126864 | 213.851 | 76947 | 168.687 |
| 4400 | 142658 | 217.612 | 85261 | 170.668 |
| 4800 | 158730 | 221.109 | 93576 | 172.476 |
| 5200 | 175057 | 224.379 | 101890 | 174.140 |
| 5600 | 191607 | 227.447 | 110205 | 175.681 |
| 6000 | 208332 | 230.322 | 118519 | 177.114 |

TABLE A.9 (*continued*)
*Ideal-Gas Properties of Various Substances (SI Units), Entropies at 0.1-MPa (1-bar) Pressure, Mole Basis*

| | Nitric Oxide (NO) $\bar{h}_{f,298}^0 = 90\ 291$ kJ/kmol $M = 30.006$ | | Nitrogen Dioxide ($NO_2$) $\bar{h}_{f,298}^0 = 33\ 100$ kJ/kmol $M = 46.005$ | |
|---|---|---|---|---|
| $T$ K | $(\bar{h} - \bar{h}_{298}^0)$ kJ/kmol | $\bar{s}_T^0$ kJ/kmol K | $(\bar{h} - \bar{h}_{298}^0)$ kJ/kmol | $\bar{s}_T^0$ kJ/kmol K |
| 0 | −9192 | 0 | −10186 | 0 |
| 100 | −6073 | 177.031 | −6861 | 202.563 |
| 200 | −2951 | 198.747 | −3495 | 225.852 |
| 298 | 0 | 210.759 | 0 | 240.034 |
| 300 | 55 | 210.943 | 68 | 240.263 |
| 400 | 3040 | 219.529 | 3927 | 251.342 |
| 500 | 6059 | 226.263 | 8099 | 260.638 |
| 600 | 9144 | 231.886 | 12555 | 268.755 |
| 700 | 12308 | 236.762 | 17250 | 275.988 |
| 800 | 15548 | 241.088 | 22138 | 282.513 |
| 900 | 18858 | 244.985 | 27180 | 288.450 |
| 1000 | 22229 | 248.536 | 32344 | 293.889 |
| 1100 | 25653 | 251.799 | 37606 | 298.904 |
| 1200 | 29120 | 254.816 | 42946 | 303.551 |
| 1300 | 32626 | 257.621 | 48351 | 307.876 |
| 1400 | 36164 | 260.243 | 53808 | 311.920 |
| 1500 | 39729 | 262.703 | 59309 | 315.715 |
| 1600 | 43319 | 265.019 | 64846 | 319.289 |
| 1700 | 46929 | 267.208 | 70414 | 322.664 |
| 1800 | 50557 | 269.282 | 76008 | 325.861 |
| 1900 | 54201 | 271.252 | 81624 | 328.898 |
| 2000 | 57859 | 273.128 | 87259 | 331.788 |
| 2200 | 65212 | 276.632 | 98578 | 337.182 |
| 2400 | 72606 | 279.849 | 109948 | 342.128 |
| 2600 | 80034 | 282.822 | 121358 | 346.695 |
| 2800 | 87491 | 285.585 | 132800 | 350.934 |
| 3000 | 94973 | 288.165 | 144267 | 354.890 |
| 3200 | 102477 | 290.587 | 155756 | 358.597 |
| 3400 | 110000 | 292.867 | 167262 | 362.085 |
| 3600 | 117541 | 295.022 | 178783 | 365.378 |
| 3800 | 125099 | 297.065 | 190316 | 368.495 |
| 4000 | 132671 | 299.007 | 201860 | 371.456 |
| 4400 | 147857 | 302.626 | 224973 | 376.963 |
| 4800 | 163094 | 305.940 | 248114 | 381.997 |
| 5200 | 178377 | 308.998 | 271276 | 386.632 |
| 5600 | 193703 | 311.838 | 294455 | 390.926 |
| 6000 | 209070 | 314.488 | 317648 | 394.926 |

**Table A.10**

*Enthalpy of Formation and Absolute Entropy of Various Substances at 25°C, 100 kPa Pressure*

| Substance | Formula | $M$ | State | $\bar{h}_f^0$ kJ/kmol | $\bar{s}_f^0$ kJ/kmol K |
|---|---|---|---|---|---|
| Water | $H_2O$ | 18.015 | gas | −241 826 | 188.834 |
| Water | $H_2O$ | 18.015 | liq | −285 830 | 69.950 |
| Hydrogen peroxide | $H_2O_2$ | 34.015 | gas | −136 106 | 232.991 |
| Ozone | $O_3$ | 47.998 | gas | +142 674 | 238.932 |
| Carbon (graphite) | C | 12.011 | solid | 0 | 5.740 |
| Carbon monoxide | CO | 28.011 | gas | −110 527 | 197.653 |
| Carbon dioxide | $CO_2$ | 44.010 | gas | −393 522 | 213.795 |
| Methane | $CH_4$ | 16.043 | gas | −74 873 | 186.251 |
| Acetylene | $C_2H_2$ | 26.038 | gas | +226 731 | 200.958 |
| Ethene | $C_2H_4$ | 28.054 | gas | +52 467 | 219.330 |
| Ethane | $C_2H_6$ | 30.070 | gas | −84 740 | 229.597 |
| Propene | $C_3H_6$ | 42.081 | gas | +20 430 | 267.066 |
| Propane | $C_3H_8$ | 44.094 | gas | −103 900 | 269.917 |
| *n*-Butane | $C_4H_{10}$ | 58.124 | gas | −126 200 | 306.647 |
| Pentane | $C_5H_{12}$ | 72.151 | gas | −146 500 | 348.945 |
| Benzene | $C_6H_6$ | 78.114 | gas | +82 980 | 269.562 |
| Hexane | $C_6H_{14}$ | 86.178 | gas | −167 300 | 387.979 |
| Heptane | $C_7H_{16}$ | 100.205 | gas | −187 900 | 427.805 |
| *n*-Octane | $C_8H_{18}$ | 114.232 | gas | −208 600 | 466.514 |
| *n*-Octane | $C_8H_{18}$ | 114.232 | liq | −250 105 | 360.575 |
| Methanol | $CH_3OH$ | 32.042 | gas | −201 300 | 239.709 |
| Methanol | $CH_3OH$ | 32.042 | liq | −239 220 | 126.809 |
| Ethanol | $C_2H_5OH$ | 46.069 | gas | −235 000 | 282.444 |
| Ethanol | $C_2H_5OH$ | 46.069 | liq | −277 380 | 160.554 |
| Ammonia | $NH_3$ | 17.031 | gas | −45 720 | 192.572 |
| *T-T*-Diesel | $C_{14.4}H_{24.9}$ | 198.06 | liq | −174 000 | 525.90 |
| Sulfur | S | 32.06 | solid | 0 | 32.056 |
| Sulfur dioxide | $SO_2$ | 64.059 | gas | −296 842 | 248.212 |
| Sulfur trioxide | $SO_3$ | 80.058 | gas | −395 765 | 256.769 |
| Nitrogen oxide | $N_2O$ | 44.013 | gas | +82 050 | 219.957 |
| Nitromethane | $CH_3NO_2$ | 61.04 | liq | −113 100 | 171.80 |

**TABLE A.11**

*Logarithms to the Base e of the Equilibrium Constant K*

For the reaction $v_A A + v_B B \rightleftharpoons v_C C + v_D D$, the equilibrium constant $K$ is defined as

$$K = \frac{y_C^{v_C} y_D^{v_D}}{y_A^{v_A} y_B^{v_B}} \left(\frac{P}{P^0}\right)^{v_C + v_D - v_A - v_B}, \quad P^0 = 0.1 \text{ MPa}$$

| Temp K | $H_2 \rightleftharpoons 2H$ | $O_2 \rightleftharpoons 2O$ | $N_2 \rightleftharpoons 2N$ | $2H_2O \rightleftharpoons 2H_2 + O_2$ | $2H_2O \rightleftharpoons H_2 + 2OH$ | $2CO_2 \rightleftharpoons 2CO + O_2$ | $N_2 + O_2 \rightleftharpoons 2NO$ | $N_2 + 2O_2 \rightleftharpoons 2NO_2$ |
|---|---|---|---|---|---|---|---|---|
| 298 | −164.003 | −186.963 | −367.528 | −184.420 | −212.075 | −207.529 | −69.868 | −41.355 |
| 500 | −92.830 | −105.623 | −213.405 | −105.385 | −120.331 | −115.234 | −40.449 | −30.725 |
| 1000 | −39.810 | −45.146 | −99.146 | −46.321 | −51.951 | −47.052 | −18.709 | −23.039 |
| 1200 | −30.878 | −35.003 | −80.025 | −36.363 | −40.467 | −35.736 | −15.082 | −21.752 |
| 1400 | −24.467 | −27.741 | −66.345 | −29.222 | −32.244 | −27.679 | −12.491 | −20.826 |
| 1600 | −19.638 | −22.282 | −56.069 | −23.849 | −26.067 | −21.656 | −10.547 | −20.126 |
| 1800 | −15.868 | −18.028 | −48.066 | −19.658 | −21.258 | −16.987 | −9.035 | −19.577 |
| 2000 | −12.841 | −14.619 | −41.655 | −16.299 | −17.406 | −13.266 | −7.825 | −19.136 |
| 2200 | −10.356 | −11.826 | −36.404 | −13.546 | −14.253 | −10.232 | −6.836 | −18.773 |
| 2400 | −8.280 | −9.495 | −32.023 | −11.249 | −11.625 | −7.715 | −6.012 | −18.470 |
| 2600 | −6.519 | −7.520 | −28.313 | −9.303 | −9.402 | −5.594 | −5.316 | −18.214 |
| 2800 | −5.005 | −5.826 | −25.129 | −7.633 | −7.496 | −3.781 | −4.720 | −17.994 |
| 3000 | −3.690 | −4.356 | −22.367 | −6.184 | −5.845 | −2.217 | −4.205 | −17.805 |
| 3200 | −2.538 | −3.069 | −19.947 | −4.916 | −4.401 | −0.853 | −3.755 | −17.640 |
| 3400 | −1.519 | −1.932 | −17.810 | −3.795 | −3.128 | 0.346 | −3.359 | −17.496 |
| 3600 | −0.611 | −0.922 | −15.909 | −2.799 | −1.996 | 1.408 | −3.008 | −17.369 |
| 3800 | 0.201 | −0.017 | −14.205 | −1.906 | −0.984 | 2.355 | −2.694 | −17.257 |
| 4000 | 0.934 | 0.798 | −12.671 | −1.101 | −0.074 | 3.204 | −2.413 | −17.157 |
| 4500 | 2.483 | 2.520 | −9.423 | 0.602 | 1.847 | 4.985 | −1.824 | −16.953 |
| 5000 | 3.724 | 3.898 | −6.816 | 1.972 | 3.383 | 6.397 | −1.358 | −16.797 |
| 5500 | 4.739 | 5.027 | −4.672 | 3.098 | 4.639 | 7.542 | −0.980 | −16.678 |
| 6000 | 5.587 | 5.969 | −2.876 | 4.040 | 5.684 | 8.488 | −0.671 | −16.588 |

*Source:* Consistent with thermodynamic data in *JANAF Thermochemical Tables*, third edition, Thermal Group, Dow Chemical U.S.A., Midland, MI, 1985.

# Appendix B

## SI Units: Thermodynamic Tables

# TABLE B.1
*Thermodynamic Properties of Water*

## TABLE B.1.1
*Saturated Water*

| Temp. (°C) | Press. (kPa) | Specific Volume, m³/kg | | | Internal Energy, kJ/kg | | |
|---|---|---|---|---|---|---|---|
| | | Sat. Liquid $v_f$ | Evap. $v_{fg}$ | Sat. Vapor $v_g$ | Sat. Liquid $u_f$ | Evap. $u_{fg}$ | Sat. Vapor $u_g$ |
| 0.01 | 0.6113 | 0.001000 | 206.131 | 206.132 | 0 | 2375.33 | 2375.33 |
| 5 | 0.8721 | 0.001000 | 147.117 | 147.118 | 20.97 | 2361.27 | 2382.24 |
| 10 | 1.2276 | 0.001000 | 106.376 | 106.377 | 41.99 | 2347.16 | 2389.15 |
| 15 | 1.705 | 0.001001 | 77.924 | 77.925 | 62.98 | 2333.06 | 2396.04 |
| 20 | 2.339 | 0.001002 | 57.7887 | 57.7897 | 83.94 | 2318.98 | 2402.91 |
| 25 | 3.169 | 0.001003 | 43.3583 | 43.3593 | 104.86 | 2304.90 | 2409.76 |
| 30 | 4.246 | 0.001004 | 32.8922 | 32.8932 | 125.77 | 2290.81 | 2416.58 |
| 35 | 5.628 | 0.001006 | 25.2148 | 25.2158 | 146.65 | 2276.71 | 2423.36 |
| 40 | 7.384 | 0.001008 | 19.5219 | 19.5229 | 167.53 | 2262.57 | 2430.11 |
| 45 | 9.593 | 0.001010 | 15.2571 | 15.2581 | 188.41 | 2248.40 | 2436.81 |
| 50 | 12.350 | 0.001012 | 12.0308 | 12.0318 | 209.30 | 2234.17 | 2443.47 |
| 55 | 15.758 | 0.001015 | 9.56734 | 9.56835 | 230.19 | 2219.89 | 2450.08 |
| 60 | 19.941 | 0.001017 | 7.66969 | 7.67071 | 251.09 | 2205.54 | 2456.63 |
| 65 | 25.03 | 0.001020 | 6.19554 | 6.19656 | 272.00 | 2191.12 | 2463.12 |
| 70 | 31.19 | 0.001023 | 5.04114 | 5.04217 | 292.93 | 2176.62 | 2469.55 |
| 75 | 38.58 | 0.001026 | 4.13021 | 4.13123 | 313.87 | 2162.03 | 2475.91 |
| 80 | 47.39 | 0.001029 | 3.40612 | 3.40715 | 334.84 | 2147.36 | 2482.19 |
| 85 | 57.83 | 0.001032 | 2.82654 | 2.82757 | 355.82 | 2132.58 | 2488.40 |
| 90 | 70.14 | 0.001036 | 2.35953 | 2.36056 | 376.82 | 2117.70 | 2494.52 |
| 95 | 84.55 | 0.001040 | 1.98082 | 1.98186 | 397.86 | 2102.70 | 2500.56 |
| 100 | 101.3 | 0.001044 | 1.67185 | 1.67290 | 418.91 | 2087.58 | 2506.50 |
| 105 | 120.8 | 0.001047 | 1.41831 | 1.41936 | 440.00 | 2072.34 | 2512.34 |
| 110 | 143.3 | 0.001052 | 1.20909 | 1.21014 | 461.12 | 2056.96 | 2518.09 |
| 115 | 169.1 | 0.001056 | 1.03552 | 1.03658 | 482.28 | 2041.44 | 2523.72 |
| 120 | 198.5 | 0.001060 | 0.89080 | 0.89186 | 503.48 | 2025.76 | 2529.24 |
| 125 | 232.1 | 0.001065 | 0.76953 | 0.77059 | 524.72 | 2009.91 | 2534.63 |
| 130 | 270.1 | 0.001070 | 0.66744 | 0.66850 | 546.00 | 1993.90 | 2539.90 |
| 135 | 313.0 | 0.001075 | 0.58110 | 0.58217 | 567.34 | 1977.69 | 2545.03 |
| 140 | 361.3 | 0.001080 | 0.50777 | 0.50885 | 588.72 | 1961.30 | 2550.02 |
| 145 | 415.4 | 0.001085 | 0.44524 | 0.44632 | 610.16 | 1944.69 | 2554.86 |
| 150 | 475.9 | 0.001090 | 0.39169 | 0.39278 | 631.66 | 1927.87 | 2559.54 |
| 155 | 543.1 | 0.001096 | 0.34566 | 0.34676 | 653.23 | 1910.82 | 2564.04 |
| 160 | 617.8 | 0.001102 | 0.30596 | 0.30706 | 674.85 | 1893.52 | 2568.37 |
| 165 | 700.5 | 0.001108 | 0.27158 | 0.27269 | 696.55 | 1875.97 | 2572.51 |
| 170 | 791.7 | 0.001114 | 0.24171 | 0.24283 | 718.31 | 1858.14 | 2576.46 |
| 175 | 892.0 | 0.001121 | 0.21568 | 0.21680 | 740.16 | 1840.03 | 2580.19 |
| 180 | 1002.2 | 0.001127 | 0.19292 | 0.19405 | 762.08 | 1821.62 | 2583.70 |
| 185 | 1122.7 | 0.001134 | 0.17295 | 0.17409 | 784.08 | 1802.90 | 2586.98 |
| 190 | 1254.4 | 0.001141 | 0.15539 | 0.15654 | 806.17 | 1783.84 | 2590.01 |

**TABLE B.1.1** (*continued*)
*Saturated Water*

| Temp. (°C) | Press. (kPa) | ENTHALPY, kJ/kg | | | ENTROPY, kJ/kg-K | | |
|---|---|---|---|---|---|---|---|
| | | Sat. Liquid $h_f$ | Evap. $h_{fg}$ | Sat. Vapor $h_g$ | Sat. Liquid $s_f$ | Evap. $s_{fg}$ | Sat. Vapor $s_g$ |
| 0.01 | 0.6113 | 0.00 | 2501.35 | 2501.35 | 0 | 9.1562 | 9.1562 |
| 5 | 0.8721 | 20.98 | 2489.57 | 2510.54 | 0.0761 | 8.9496 | 9.0257 |
| 10 | 1.2276 | 41.99 | 2477.75 | 2519.74 | 0.1510 | 8.7498 | 8.9007 |
| 15 | 1.705 | 62.98 | 2465.93 | 2528.91 | 0.2245 | 8.5569 | 8.7813 |
| 20 | 2.339 | 83.94 | 2454.12 | 2538.06 | 0.2966 | 8.3706 | 8.6671 |
| 25 | 3.169 | 104.87 | 2442.30 | 2547.17 | 0.3673 | 8.1905 | 8.5579 |
| 30 | 4.246 | 125.77 | 2430.48 | 2556.25 | 0.4369 | 8.0164 | 8.4533 |
| 35 | 5.628 | 146.66 | 2418.62 | 2565.28 | 0.5052 | 7.8478 | 8.3530 |
| 40 | 7.384 | 167.54 | 2406.72 | 2574.26 | 0.5724 | 7.6845 | 8.2569 |
| 45 | 9.593 | 188.42 | 2394.77 | 2583.19 | 0.6386 | 7.5261 | 8.1647 |
| 50 | 12.350 | 209.31 | 2382.75 | 2592.06 | 0.7037 | 7.3725 | 8.0762 |
| 55 | 15.758 | 230.20 | 2370.66 | 2600.86 | 0.7679 | 7.2234 | 7.9912 |
| 60 | 19.941 | 251.11 | 2358.48 | 2609.59 | 0.8311 | 7.0784 | 7.9095 |
| 65 | 25.03 | 272.03 | 2346.21 | 2618.24 | 0.8934 | 6.9375 | 7.8309 |
| 70 | 31.19 | 292.96 | 2333.85 | 2626.80 | 0.9548 | 6.8004 | 7.7552 |
| 75 | 38.58 | 313.91 | 2321.37 | 2635.28 | 1.0154 | 6.6670 | 7.6824 |
| 80 | 47.39 | 334.88 | 2308.77 | 2643.66 | 1.0752 | 6.5369 | 7.6121 |
| 85 | 57.83 | 355.88 | 2296.05 | 2651.93 | 1.1342 | 6.4102 | 7.5444 |
| 90 | 70.14 | 376.90 | 2283.19 | 2660.09 | 1.1924 | 6.2866 | 7.4790 |
| 95 | 84.55 | 397.94 | 2270.19 | 2668.13 | 1.2500 | 6.1659 | 7.4158 |
| 100 | 101.3 | 419.02 | 2257.03 | 2676.05 | 1.3068 | 6.0480 | 7.3548 |
| 105 | 120.8 | 440.13 | 2243.70 | 2683.83 | 1.3629 | 5.9328 | 7.2958 |
| 110 | 143.3 | 461.27 | 2230.20 | 2691.47 | 1.4184 | 5.8202 | 7.2386 |
| 115 | 169.1 | 482.46 | 2216.50 | 2698.96 | 1.4733 | 5.7100 | 7.1832 |
| 120 | 198.5 | 503.69 | 2202.61 | 2706.30 | 1.5275 | 5.6020 | 7.1295 |
| 125 | 232.1 | 524.96 | 2188.50 | 2713.46 | 1.5812 | 5.4962 | 7.0774 |
| 130 | 270.1 | 546.29 | 2174.16 | 2720.46 | 1.6343 | 5.3925 | 7.0269 |
| 135 | 313.0 | 567.67 | 2159.59 | 2727.26 | 1.6869 | 5.2907 | 6.9777 |
| 140 | 361.3 | 589.11 | 2144.75 | 2733.87 | 1.7390 | 5.1908 | 6.9298 |
| 145 | 415.4 | 610.61 | 2129.65 | 2740.26 | 1.7906 | 5.0926 | 6.8832 |
| 150 | 475.9 | 632.18 | 2114.26 | 2746.44 | 1.8417 | 4.9960 | 6.8378 |
| 155 | 543.1 | 653.82 | 2098.56 | 2752.39 | 1.8924 | 4.9010 | 6.7934 |
| 160 | 617.8 | 675.53 | 2082.55 | 2758.09 | 1.9426 | 4.8075 | 6.7501 |
| 165 | 700.5 | 697.32 | 2066.20 | 2763.53 | 1.9924 | 4.7153 | 6.7078 |
| 170 | 791.7 | 719.20 | 2049.50 | 2768.70 | 2.0418 | 4.6244 | 6.6663 |
| 175 | 892.0 | 741.16 | 2032.42 | 2773.58 | 2.0909 | 4.5347 | 6.6256 |
| 180 | 1002.2 | 763.21 | 2014.96 | 2778.16 | 2.1395 | 4.4461 | 6.5857 |
| 185 | 1122.7 | 785.36 | 1997.07 | 2782.43 | 2.1878 | 4.3586 | 6.5464 |
| 190 | 1254.4 | 807.61 | 1978.76 | 2786.37 | 2.2358 | 4.2720 | 6.5078 |

TABLE B.1.1  (*continued*)
**Saturated Water**

| Temp. (°C) | Press. (kPa) | SPECIFIC VOLUME, m³/kg | | | INTERNAL ENERGY, kJ/kg | | |
|---|---|---|---|---|---|---|---|
| | | Sat. Liquid $v_f$ | Evap. $v_{fg}$ | Sat. Vapor $v_g$ | Sat. Liquid $u_f$ | Evap. $u_{fg}$ | Sat. Vapor $u_g$ |
| 195 | 1397.8 | 0.001149 | 0.13990 | 0.14105 | 828.36 | 1764.43 | 2592.79 |
| 200 | 1553.8 | 0.001156 | 0.12620 | 0.12736 | 850.64 | 1744.66 | 2595.29 |
| 205 | 1723.0 | 0.001164 | 0.11405 | 0.11521 | 873.02 | 1724.49 | 2597.52 |
| 210 | 1906.3 | 0.001173 | 0.10324 | 0.10441 | 895.51 | 1703.93 | 2599.44 |
| 215 | 2104.2 | 0.001181 | 0.09361 | 0.09479 | 918.12 | 1682.94 | 2601.06 |
| 220 | 2317.8 | 0.001190 | 0.08500 | 0.08619 | 940.85 | 1661.49 | 2602.35 |
| 225 | 2547.7 | 0.001199 | 0.07729 | 0.07849 | 963.72 | 1639.58 | 2603.30 |
| 230 | 2794.9 | 0.001209 | 0.07037 | 0.07158 | 986.72 | 1617.17 | 2603.89 |
| 235 | 3060.1 | 0.001219 | 0.06415 | 0.06536 | 1009.88 | 1594.24 | 2604.11 |
| 240 | 3344.2 | 0.001229 | 0.05853 | 0.05976 | 1033.19 | 1570.75 | 2603.95 |
| 245 | 3648.2 | 0.001240 | 0.05346 | 0.05470 | 1056.69 | 1546.68 | 2603.37 |
| 250 | 3973.0 | 0.001251 | 0.04887 | 0.05013 | 1080.37 | 1522.00 | 2602.37 |
| 255 | 4319.5 | 0.001263 | 0.04471 | 0.04598 | 1104.26 | 1496.66 | 2600.93 |
| 260 | 4688.6 | 0.001276 | 0.04093 | 0.04220 | 1128.37 | 1470.64 | 2599.01 |
| 265 | 5081.3 | 0.001289 | 0.03748 | 0.03877 | 1152.72 | 1443.87 | 2596.60 |
| 270 | 5498.7 | 0.001302 | 0.03434 | 0.03564 | 1177.33 | 1416.33 | 2593.66 |
| 275 | 5941.8 | 0.001317 | 0.03147 | 0.03279 | 1202.23 | 1387.94 | 2590.17 |
| 280 | 6411.7 | 0.001332 | 0.02884 | 0.03017 | 1227.43 | 1358.66 | 2586.09 |
| 285 | 6909.4 | 0.001348 | 0.02642 | 0.02777 | 1252.98 | 1328.41 | 2581.38 |
| 290 | 7436.0 | 0.001366 | 0.02420 | 0.02557 | 1278.89 | 1297.11 | 2575.99 |
| 295 | 7992.8 | 0.001384 | 0.02216 | 0.02354 | 1305.21 | 1264.67 | 2569.87 |
| 300 | 8581.0 | 0.001404 | 0.02027 | 0.02167 | 1331.97 | 1230.99 | 2562.96 |
| 305 | 9201.8 | 0.001425 | 0.01852 | 0.01995 | 1359.22 | 1195.94 | 2555.16 |
| 310 | 9856.6 | 0.001447 | 0.01690 | 0.01835 | 1387.03 | 1159.37 | 2546.40 |
| 315 | 10547 | 0.001472 | 0.01539 | 0.01687 | 1415.44 | 1121.11 | 2536.55 |
| 320 | 11274 | 0.001499 | 0.01399 | 0.01549 | 1444.55 | 1080.93 | 2525.48 |
| 325 | 12040 | 0.001528 | 0.01267 | 0.01420 | 1474.44 | 1038.57 | 2513.01 |
| 330 | 12845 | 0.001561 | 0.01144 | 0.01300 | 1505.24 | 993.66 | 2498.91 |
| 335 | 13694 | 0.001597 | 0.01027 | 0.01186 | 1537.11 | 945.77 | 2482.88 |
| 340 | 14586 | 0.001638 | 0.00916 | 0.01080 | 1570.26 | 894.26 | 2464.53 |
| 345 | 15525 | 0.001685 | 0.00810 | 0.00978 | 1605.01 | 838.29 | 2443.30 |
| 350 | 16514 | 0.001740 | 0.00707 | 0.00881 | 1641.81 | 776.58 | 2418.39 |
| 355 | 17554 | 0.001807 | 0.00607 | 0.00787 | 1681.41 | 707.11 | 2388.52 |
| 360 | 18651 | 0.001892 | 0.00505 | 0.00694 | 1725.19 | 626.29 | 2351.47 |
| 365 | 19807 | 0.002011 | 0.00398 | 0.00599 | 1776.13 | 526.54 | 2302.67 |
| 370 | 21028 | 0.002213 | 0.00271 | 0.00493 | 1843.84 | 384.69 | 2228.53 |
| 374.1 | 22089 | 0.003155 | 0 | 0.00315 | 2029.58 | 0 | 2029.58 |

TABLE B.1.1 (*continued*)
*Saturated Water*

| Temp. (°C) | Press. (kPa) | ENTHALPY, kJ/kg | | | ENTROPY, kJ/kg-K | | |
|---|---|---|---|---|---|---|---|
| | | Sat. Liquid $h_f$ | Evap. $h_{fg}$ | Sat. Vapor $h_g$ | Sat. Liquid $s_f$ | Evap. $s_{fg}$ | Sat. Vapor $s_g$ |
| 195 | 1397.8 | 829.96 | 1959.99 | 2789.96 | 2.2835 | 4.1863 | 6.4697 |
| 200 | 1553.8 | 852.43 | 1940.75 | 2793.18 | 2.3308 | 4.1014 | 6.4322 |
| 205 | 1723.0 | 875.03 | 1921.00 | 2796.03 | 2.3779 | 4.0172 | 6.3951 |
| 210 | 1906.3 | 897.75 | 1900.73 | 2798.48 | 2.4247 | 3.9337 | 6.3584 |
| 215 | 2104.2 | 920.61 | 1879.91 | 2800.51 | 2.4713 | 3.8507 | 6.3221 |
| 220 | 2317.8 | 943.61 | 1858.51 | 2802.12 | 2.5177 | 3.7683 | 6.2860 |
| 225 | 2547.7 | 966.77 | 1836.50 | 2803.27 | 2.5639 | 3.6863 | 6.2502 |
| 230 | 2794.9 | 990.10 | 1813.85 | 2803.95 | 2.6099 | 3.6047 | 6.2146 |
| 235 | 3060.1 | 1013.61 | 1790.53 | 2804.13 | 2.6557 | 3.5233 | 6.1791 |
| 240 | 3344.2 | 1037.31 | 1766.50 | 2803.81 | 2.7015 | 3.4422 | 6.1436 |
| 245 | 3648.2 | 1061.21 | 1741.73 | 2802.95 | 2.7471 | 3.3612 | 6.1083 |
| 250 | 3973.0 | 1085.34 | 1716.18 | 2801.52 | 2.7927 | 3.2802 | 6.0729 |
| 255 | 4319.5 | 1109.72 | 1689.80 | 2799.51 | 2.8382 | 3.1992 | 6.0374 |
| 260 | 4688.6 | 1134.35 | 1662.54 | 2796.89 | 2.8837 | 3.1181 | 6.0018 |
| 265 | 5081.3 | 1159.27 | 1634.34 | 2793.61 | 2.9293 | 3.0368 | 5.9661 |
| 270 | 5498.7 | 1184.49 | 1605.16 | 2789.65 | 2.9750 | 2.9551 | 5.9301 |
| 275 | 5941.8 | 1210.05 | 1574.92 | 2784.97 | 3.0208 | 2.8730 | 5.8937 |
| 280 | 6411.7 | 1235.97 | 1543.55 | 2779.53 | 3.0667 | 2.7903 | 5.8570 |
| 285 | 6909.4 | 1262.29 | 1510.97 | 2773.27 | 3.1129 | 2.7069 | 5.8198 |
| 290 | 7436.0 | 1289.04 | 1477.08 | 2766.13 | 3.1593 | 2.6227 | 5.7821 |
| 295 | 7992.8 | 1316.27 | 1441.78 | 2758.05 | 3.2061 | 2.5375 | 5.7436 |
| 300 | 8581.0 | 1344.01 | 1404.93 | 2748.94 | 3.2533 | 2.4511 | 5.7044 |
| 305 | 9201.8 | 1372.33 | 1366.38 | 2738.72 | 3.3009 | 2.3633 | 5.6642 |
| 310 | 9856.6 | 1401.29 | 1325.97 | 2727.27 | 3.3492 | 2.2737 | 5.6229 |
| 315 | 10547 | 1430.97 | 1283.48 | 2714.44 | 3.3981 | 2.1821 | 5.5803 |
| 320 | 11274 | 1461.45 | 1238.64 | 2700.08 | 3.4479 | 2.0882 | 5.5361 |
| 325 | 12040 | 1492.84 | 1191.13 | 2683.97 | 3.4987 | 1.9913 | 5.4900 |
| 330 | 12845 | 1525.29 | 1140.56 | 2665.85 | 3.5506 | 1.8909 | 5.4416 |
| 335 | 13694 | 1558.98 | 1086.37 | 2645.35 | 3.6040 | 1.7863 | 5.3903 |
| 340 | 14586 | 1594.15 | 1027.86 | 2622.01 | 3.6593 | 1.6763 | 5.3356 |
| 345 | 15525 | 1631.17 | 964.02 | 2595.19 | 3.7169 | 1.5594 | 5.2763 |
| 350 | 16514 | 1670.54 | 893.38 | 2563.92 | 3.7776 | 1.4336 | 5.2111 |
| 355 | 17554 | 1713.13 | 813.59 | 2526.72 | 3.8427 | 1.2951 | 5.1378 |
| 360 | 18651 | 1760.48 | 720.52 | 2481.00 | 3.9146 | 1.1379 | 5.0525 |
| 365 | 19807 | 1815.96 | 605.44 | 2421.40 | 3.9983 | 0.9487 | 4.9470 |
| 370 | 21028 | 1890.37 | 441.75 | 2332.12 | 4.1104 | 0.6868 | 4.7972 |
| 374.1 | 22089 | 2099.26 | 0 | 2099.26 | 4.4297 | 0 | 4.4297 |

**Table B.1.2**

*Saturated Water Pressure Entry*

| Press. (kPa) | Temp. (°C) | Specific Volume, $m^3$/kg | | | Internal Energy, kJ/kg | | |
|---|---|---|---|---|---|---|---|
| | | Sat. Liquid $v_f$ | Evap. $v_{fg}$ | Sat. Vapor $v_g$ | Sat. Liquid $u_f$ | Evap. $u_{fg}$ | Sat. Vapor $u_g$ |
| 0.6113 | 0.01 | 0.001000 | 206.131 | 206.132 | 0 | 2375.3 | 2375.3 |
| 1 | 6.98 | 0.001000 | 129.20702 | 129.20802 | 29.29 | 2355.69 | 2384.98 |
| 1.5 | 13.03 | 0.001001 | 87.97913 | 87.98013 | 54.70 | 2338.63 | 2393.32 |
| 2 | 17.50 | 0.001001 | 67.00285 | 67.00385 | 73.47 | 2326.02 | 2399.48 |
| 2.5 | 21.08 | 0.001002 | 54.25285 | 54.25385 | 88.47 | 2315.93 | 2404.40 |
| 3 | 24.08 | 0.001003 | 45.66402 | 45.66502 | 101.03 | 2307.48 | 2408.51 |
| 4 | 28.96 | 0.001004 | 34.79915 | 34.80015 | 121.44 | 2293.73 | 2415.17 |
| 5 | 32.88 | 0.001005 | 28.19150 | 28.19251 | 137.79 | 2282.70 | 2420.49 |
| 7.5 | 40.29 | 0.001008 | 19.23674 | 19.23775 | 168.76 | 2261.74 | 2430.50 |
| 10 | 45.81 | 0.001010 | 14.67254 | 14.67355 | 191.79 | 2246.10 | 2437.89 |
| 15 | 53.97 | 0.001014 | 10.02117 | 10.02218 | 225.90 | 2222.83 | 2448.73 |
| 20 | 60.06 | 0.001017 | 7.64835 | 7.64937 | 251.35 | 2205.36 | 2456.71 |
| 25 | 64.97 | 0.001020 | 6.20322 | 6.20424 | 271.88 | 2191.21 | 2463.08 |
| 30 | 69.10 | 0.001022 | 5.22816 | 5.22918 | 289.18 | 2179.22 | 2468.40 |
| 40 | 75.87 | 0.001026 | 3.99243 | 3.99345 | 317.51 | 2159.49 | 2477.00 |
| 50 | 81.33 | 0.001030 | 3.23931 | 3.24034 | 340.42 | 2143.43 | 2483.85 |
| 75 | 91.77 | 0.001037 | 2.21607 | 2.21711 | 394.29 | 2112.39 | 2496.67 |
| 100 | 99.62 | 0.001043 | 1.69296 | 1.69400 | 417.33 | 2088.72 | 2506.06 |
| 125 | 105.99 | 0.001048 | 1.37385 | 1.37490 | 444.16 | 2069.32 | 2513.48 |
| 150 | 111.37 | 0.001053 | 1.15828 | 1.15933 | 466.92 | 2052.72 | 2519.64 |
| 175 | 116.06 | 0.001057 | 1.00257 | 1.00363 | 486.78 | 2038.12 | 2524.90 |
| 200 | 120.23 | 0.001061 | 0.88467 | 0.88573 | 504.47 | 2025.02 | 2529.49 |
| 225 | 124.00 | 0.001064 | 0.79219 | 0.79325 | 520.45 | 2013.10 | 2533.56 |
| 250 | 127.43 | 0.001067 | 0.71765 | 0.71871 | 535.08 | 2002.14 | 2537.21 |
| 275 | 130.60 | 0.001070 | 0.65624 | 0.65731 | 548.57 | 1991.95 | 2540.53 |
| 300 | 133.55 | 0.001073 | 0.60475 | 0.60582 | 561.13 | 1982.43 | 2543.55 |
| 325 | 136.30 | 0.001076 | 0.56093 | 0.56201 | 572.88 | 1973.46 | 2546.34 |
| 350 | 138.88 | 0.001079 | 0.52317 | 0.52425 | 583.93 | 1964.98 | 2548.92 |
| 375 | 141.32 | 0.001081 | 0.49029 | 0.49137 | 594.38 | 1956.93 | 2551.31 |
| 400 | 143.63 | 0.001084 | 0.46138 | 0.46246 | 604.29 | 1949.26 | 2553.55 |
| 450 | 147.93 | 0.001088 | 0.41289 | 0.41398 | 622.75 | 1934.87 | 2557.62 |
| 500 | 151.86 | 0.001093 | 0.37380 | 0.37489 | 639.66 | 1921.57 | 2561.23 |
| 550 | 155.48 | 0.001097 | 0.34159 | 0.34268 | 655.30 | 1909.17 | 2564.47 |
| 600 | 158.85 | 0.001101 | 0.31457 | 0.31567 | 669.88 | 1897.52 | 2567.40 |
| 650 | 162.01 | 0.001104 | 0.29158 | 0.29268 | 683.55 | 1886.51 | 2570.06 |
| 700 | 164.97 | 0.001108 | 0.27176 | 0.27286 | 696.43 | 1876.07 | 2572.49 |
| 750 | 167.77 | 0.001111 | 0.25449 | 0.25560 | 708.62 | 1866.11 | 2574.73 |
| 800 | 170.43 | 0.001115 | 0.23931 | 0.24043 | 720.20 | 1856.58 | 2576.79 |

TABLE B.1.2 (*Continued*)
**Saturated Water Pressure Entry**

| Press. (kPa) | Temp. (°C) | ENTHALPY, kJ/kg | | | ENTROPY, kJ/kg-K | | |
|---|---|---|---|---|---|---|---|
| | | Sat. Liquid $h_f$ | Evap. $h_{fg}$ | Sat. Vapor $h_g$ | Sat. Liquid $s_f$ | Evap. $s_{fg}$ | Sat. Vapor $s_g$ |
| 0.6113 | 0.01 | 0.00 | 2501.3 | 2501.3 | 0 | 9.1562 | 9.1562 |
| 1.0 | 6.98 | 29.29 | 2484.89 | 2514.18 | 0.1059 | 8.8697 | 8.9756 |
| 1.5 | 13.03 | 54.70 | 2470.59 | 2525.30 | 0.1956 | 8.6322 | 8.8278 |
| 2.0 | 17.50 | 73.47 | 2460.02 | 2533.49 | 0.2607 | 8.4629 | 8.7236 |
| 2.5 | 21.08 | 88.47 | 2451.56 | 2540.03 | 0.3120 | 8.3311 | 8.6431 |
| 3.0 | 24.08 | 101.03 | 2444.47 | 2545.50 | 0.3545 | 8.2231 | 8.5775 |
| 4.0 | 28.96 | 121.44 | 2432.93 | 2554.37 | 0.4226 | 8.0520 | 8.4746 |
| 5.0 | 32.88 | 137.79 | 2423.66 | 2561.45 | 0.4763 | 7.9187 | 8.3950 |
| 7.5 | 40.29 | 168.77 | 2406.02 | 2574.79 | 0.5763 | 7.6751 | 8.2514 |
| 10 | 45.81 | 191.81 | 2392.82 | 2584.63 | 0.6492 | 7.5010 | 8.1501 |
| 15 | 53.97 | 225.91 | 2373.14 | 2599.06 | 0.7548 | 7.2536 | 8.0084 |
| 20 | 60.06 | 251.38 | 2358.33 | 2609.70 | 0.8319 | 7.0766 | 7.9085 |
| 25 | 64.97 | 271.90 | 2346.29 | 2618.19 | 0.8930 | 6.9383 | 7.8313 |
| 30 | 69.10 | 289.21 | 2336.07 | 2625.28 | 0.9439 | 6.8247 | 7.7686 |
| 40 | 75.87 | 317.55 | 2319.19 | 2636.74 | 1.0258 | 6.6441 | 7.6700 |
| 50 | 81.33 | 340.47 | 2305.40 | 2645.87 | 1.0910 | 6.5029 | 7.5939 |
| 75 | 91.77 | 384.36 | 2278.59 | 2662.96 | 1.2129 | 6.2434 | 7.4563 |
| 100 | 99.62 | 417.44 | 2258.02 | 2675.46 | 1.3025 | 6.0568 | 7.3593 |
| 125 | 105.99 | 444.30 | 2241.05 | 2685.35 | 1.3739 | 5.9104 | 7.2843 |
| 150 | 111.37 | 467.08 | 2226.46 | 2693.54 | 1.4335 | 5.7897 | 7.2232 |
| 175 | 116.06 | 486.97 | 2213.57 | 2700.53 | 1.4848 | 5.6868 | 7.1717 |
| 200 | 120.23 | 504.68 | 2201.96 | 2706.63 | 1.5300 | 5.5970 | 7.1271 |
| 225 | 124.00 | 520.69 | 2191.35 | 2712.04 | 1.5705 | 5.5173 | 7.0878 |
| 250 | 127.43 | 535.34 | 2181.55 | 2716.89 | 1.6072 | 5.4455 | 7.0526 |
| 275 | 130.60 | 548.87 | 2172.42 | 2721.29 | 1.6407 | 5.3801 | 7.0208 |
| 300 | 133.55 | 561.45 | 2163.85 | 2725.30 | 1.6717 | 5.3201 | 6.9918 |
| 325 | 136.30 | 573.23 | 2155.76 | 2728.99 | 1.7005 | 5.2646 | 6.9651 |
| 350 | 138.88 | 584.31 | 2148.10 | 2732.40 | 1.7274 | 5.2130 | 6.9404 |
| 375 | 141.32 | 594.79 | 2140.79 | 2735.58 | 1.7527 | 5.1647 | 6.9174 |
| 400 | 143.63 | 604.73 | 2133.81 | 2738.53 | 1.7766 | 5.1193 | 6.8958 |
| 450 | 147.93 | 623.24 | 2120.67 | 2743.91 | 1.8206 | 5.0359 | 6.8565 |
| 500 | 151.86 | 640.21 | 2108.47 | 2748.67 | 1.8606 | 4.9606 | 6.8212 |
| 550 | 155.48 | 655.91 | 2097.04 | 2752.94 | 1.8972 | 4.8920 | 6.7892 |
| 600 | 158.85 | 670.54 | 2086.26 | 2756.80 | 1.9311 | 4.8289 | 6.7600 |
| 650 | 162.01 | 684.26 | 2076.04 | 2760.30 | 1.9627 | 4.7704 | 6.7330 |
| 700 | 164.97 | 697.20 | 2066.30 | 2763.50 | 1.9922 | 4.7158 | 6.7080 |
| 750 | 167.77 | 709.45 | 2056.98 | 2766.43 | 2.0199 | 4.6647 | 6.6846 |
| 800 | 170.43 | 721.10 | 2048.04 | 2769.13 | 2.0461 | 4.6166 | 6.6627 |

**TABLE B.1.2** (*continued*)
**Saturated Water Pressure Entry**

| Press. (kPa) | Temp. (°C) | SPECIFIC VOLUME, m³/kg | | | INTERNAL ENERGY, kJ/kg | | |
|---|---|---|---|---|---|---|---|
| | | Sat. Liquid $v_f$ | Evap. $v_{fg}$ | Sat. Vapor $v_g$ | Sat. Liquid $u_f$ | Evap. $u_{fg}$ | Sat. Vapor $u_g$ |
| 850 | 172.96 | 0.001118 | 0.22586 | 0.22698 | 731.25 | 1847.45 | 2578.69 |
| 900 | 175.38 | 0.001121 | 0.21385 | 0.21497 | 741.81 | 1838.65 | 2580.46 |
| 950 | 177.69 | 0.001124 | 0.20306 | 0.20419 | 751.94 | 1830.17 | 2582.11 |
| 1000 | 179.91 | 0.001127 | 0.19332 | 0.19444 | 761.67 | 1821.97 | 2583.64 |
| 1100 | 184.09 | 0.001133 | 0.17639 | 0.17753 | 780.08 | 1806.32 | 2586.40 |
| 1200 | 187.99 | 0.001139 | 0.16220 | 0.16333 | 797.27 | 1791.55 | 2588.82 |
| 1300 | 191.64 | 0.001144 | 0.15011 | 0.15125 | 813.42 | 1777.53 | 2590.95 |
| 1400 | 195.07 | 0.001149 | 0.13969 | 0.14084 | 828.68 | 1764.15 | 2592.83 |
| 1500 | 198.32 | 0.001154 | 0.13062 | 0.13177 | 843.14 | 1751.3 | 2594.5 |
| 1750 | 205.76 | 0.001166 | 0.11232 | 0.11349 | 876.44 | 1721.39 | 2597.83 |
| 2000 | 212.42 | 0.001177 | 0.09845 | 0.09963 | 906.42 | 1693.84 | 2600.26 |
| 2250 | 218.45 | 0.001187 | 0.08756 | 0.08875 | 933.81 | 1668.18 | 2601.98 |
| 2500 | 223.99 | 0.001197 | 0.07878 | 0.07998 | 959.09 | 1644.04 | 2603.13 |
| 2750 | 229.12 | 0.001207 | 0.07154 | 0.07275 | 982.65 | 1621.16 | 2603.81 |
| 3000 | 233.90 | 0.001216 | 0.06546 | 0.06668 | 1004.76 | 1599.34 | 2604.10 |
| 3250 | 238.38 | 0.001226 | 0.06029 | 0.06152 | 1025.62 | 1578.43 | 2604.04 |
| 3500 | 242.60 | 0.001235 | 0.05583 | 0.05707 | 1045.41 | 1558.29 | 2603.70 |
| 4000 | 250.40 | 0.001252 | 0.04853 | 0.04978 | 1082.28 | 1519.99 | 2602.27 |
| 5000 | 263.99 | 0.001286 | 0.03815 | 0.03944 | 1147.78 | 1449.34 | 2597.12 |
| 6000 | 275.64 | 0.001319 | 0.03112 | 0.03244 | 1205.41 | 1384.27 | 2589.69 |
| 7000 | 285.88 | 0.001351 | 0.02602 | 0.02737 | 1257.51 | 1322.97 | 2580.48 |
| 8000 | 295.06 | 0.001384 | 0.02213 | 0.02352 | 1305.54 | 1264.25 | 2569.79 |
| 9000 | 303.40 | 0.001418 | 0.01907 | 0.02048 | 1350.47 | 1207.28 | 2557.75 |
| 10000 | 311.06 | 0.001452 | 0.01657 | 0.01803 | 1393.00 | 1151.40 | 2544.41 |
| 11000 | 318.15 | 0.001489 | 0.01450 | 0.01599 | 1433.68 | 1096.06 | 2529.74 |
| 12000 | 324.75 | 0.001527 | 0.01274 | 0.01426 | 1472.92 | 1040.76 | 2513.67 |
| 13000 | 330.93 | 0.001567 | 0.01121 | 0.01278 | 1511.09 | 984.99 | 2496.08 |
| 14000 | 336.75 | 0.001611 | 0.00987 | 0.01149 | 1548.53 | 928.23 | 2476.76 |
| 15000 | 342.24 | 0.001658 | 0.00868 | 0.01034 | 1585.58 | 869.85 | 2455.43 |
| 16000 | 347.43 | 0.001711 | 0.00760 | 0.00931 | 1622.63 | 809.07 | 2431.70 |
| 17000 | 352.37 | 0.001770 | 0.00659 | 0.00836 | 1660.16 | 744.80 | 2404.96 |
| 18000 | 357.06 | 0.001840 | 0.00565 | 0.00749 | 1698.86 | 675.42 | 2374.28 |
| 19000 | 361.54 | 0.001924 | 0.00473 | 0.00666 | 1739.87 | 598.18 | 2338.05 |
| 20000 | 365.81 | 0.002035 | 0.00380 | 0.00583 | 1785.47 | 507.58 | 2293.05 |
| 21000 | 369.89 | 0.002206 | 0.00275 | 0.00495 | 1841.97 | 388.74 | 2230.71 |
| 22000 | 373.80 | 0.002808 | 0.00072 | 0.00353 | 1973.16 | 108.24 | 2081.39 |
| 22089 | 374.14 | 0.003155 | 0 | 0.00315 | 2029.58 | 0 | 2029.58 |

**TABLE B.1.2** (*Continued*)
*Saturated Water Pressure Entry*

| Press. (kPa) | Temp. (°C) | ENTHALPY, kJ/kg | | | ENTROPY, kJ/kg-K | | |
|---|---|---|---|---|---|---|---|
| | | Sat. Liquid $h_f$ | Evap. $h_{fg}$ | Sat. Vapor $h_g$ | Sat. Liquid $s_f$ | Evap. $s_{fg}$ | Sat. Vapor $s_g$ |
| 850 | 172.96 | 732.20 | 2039.43 | 2771.63 | 2.0709 | 4.5711 | 6.6421 |
| 900 | 175.38 | 742.82 | 2031.12 | 2773.94 | 2.0946 | 4.5280 | 6.6225 |
| 950 | 177.69 | 753.00 | 2023.08 | 2776.08 | 2.1171 | 4.4869 | 6.6040 |
| 1000 | 179.91 | 762.79 | 2015.29 | 2778.08 | 2.1386. | 4.4478 | 6.5864 |
| 1100 | 184.09 | 781.32 | 2000.36 | 2781.68 | 2.1791 | 4.3744 | 6.5535 |
| 1200 | 187.99 | 798.64 | 1986.19 | 2784.82 | 2.2165 | 4.3067 | 6.5233 |
| 1300 | 191.64 | 814.91 | 1972.67 | 2787.58 | 2.2514 | 4.2438 | 6.4953 |
| 1400 | 195.07 | 830.29 | 1959.72 | 2790.00 | 2.2842 | 4.1850 | 6.4692 |
| 1500 | 198.32 | 844.87 | 1947.28 | 2792.15 | 2.3150 | 4.1298 | 6.4448 |
| 1750 | 205.76 | 878.48 | 1917.95 | 2796.43 | 2.3851 | 4.0044 | 6.3895 |
| 2000 | 212.42 | 908.77 | 1890.74 | 2799.51 | 2.4473 | 3.8935 | 6.3408 |
| 2250 | 218.45 | 936.48 | 1865.19 | 2801.67 | 2.5034 | 3.7938 | 6.2971 |
| 2500 | 223.99 | 962.09 | 1840.98 | 2803.07 | 2.5546 | 3.7028 | 6.2574 |
| 2750 | 229.12 | 985.97 | 1817.89 | 2803.86 | 2.6018 | 3.6190 | 6.2208 |
| 3000 | 233.90 | 1008.41 | 1795.73 | 2804.14 | 2.6456 | 3.5412 | 6.1869 |
| 3250 | 238.38 | 1029.60 | 1774.37 | 2803.97 | 2.6866 | 3.4685 | 6.1551 |
| 3500 | 242.60 | 1049.73 | 1753.70 | 2803.43 | 2.7252 | 3.4000 | 6.1252 |
| 4000 | 250.40 | 1087.29 | 1714.09 | 2801.38 | 2.7963 | 3.2737 | 6.0700 |
| 5000 | 263.99 | 1154.21 | 1640.12 | 2794.33 | 2.9201 | 3.0532 | 5.9733 |
| 6000 | 275.64 | 1213.32 | 1571.00 | 2784.33 | 3.0266 | 2.8625 | 5.8891 |
| 7000 | 285.88 | 1266.97 | 1505.10 | 2772.07 | 3.1210 | 2.6922 | 5.8132 |
| 8000 | 295.06 | 1316.61 | 1441.33 | 2757.94 | 3.2067 | 2.5365 | 5.7431 |
| 9000 | 303.40 | 1363.23 | 1378.88 | 2742.11 | 3.2857 | 2.3915 | 5.6771 |
| 10000 | 311.06 | 1407.53 | 1317.14 | 2724.67 | 3.3595 | 2.2545 | 5.6140 |
| 11000 | 318.15 | 1450.05 | 1255.55 | 2705.60 | 3.4294 | 2.1233 | 5.5527 |
| 12000 | 324.75 | 1491.24 | 1193.59 | 2684.83 | 3.4961 | 1.9962 | 5.4923 |
| 13000 | 330.93 | 1531.46 | 1130.76 | 2662.22 | 3.5604 | 1.8718 | 5.4323 |
| 14000 | 336.75 | 1571.08 | 1066.47 | 2637.55 | 3.6231 | 1.7485 | 5.3716 |
| 15000 | 342.24 | 1610.45 | 1000.04 | 2610.49 | 3.6847 | 1.6250 | 5.3097 |
| 16000 | 347.43 | 1650.00 | 930.59 | 2580.59 | 3.7460 | 1.4995 | 5.2454 |
| 17000 | 352.37 | 1690.25 | 856.90 | 2547.15 | 3.8078 | 1.3698 | 5.1776 |
| 18000 | 357.06 | 1731.97 | 777.13 | 2509.09 | 3.8713 | 1.2330 | 5.1044 |
| 19000 | 361.54 | 1776.43 | 688.11 | 2464.54 | 3.9387 | 1.0841 | 5.0227 |
| 20000 | 365.81 | 1826.18 | 583.56 | 2409.74 | 4.0137 | 0.9132 | 4.9269 |
| 21000 | 369.89 | 1888.30 | 446.42 | 2334.72 | 4.1073 | 0.6942 | 4.8015 |
| 22000 | 373.80 | 2034.92 | 124.04 | 2158.97 | 4.3307 | 0.1917 | 4.5224 |
| 22089 | 374.14 | 2099.26 | 0 | 2099.26 | 4.4297 | 0 | 4.4297 |

**Table B.1.3**
*Superheated Vapor Water*

| Temp. (°C) | $v$ (m³/kg) | $u$ (kJ/kg) | $h$ (kJ/kg) | $s$ (kJ/kg-K) | $v$ (m³/kg) | $u$ (kJ/kg) | $h$ (kJ/kg) | $s$ (kJ/kg-K) |
|---|---|---|---|---|---|---|---|---|
| | $P = 10$ kPa (45.81) | | | | $P = 50$ kPa (81.33) | | | |
| Sat. | 14.67355 | 2437.89 | 2584.63 | 8.1501 | 3.24034 | 2483.85 | 2645.87 | 7.5939 |
| 50 | 14.86920 | 2443.87 | 2592.56 | 8.1749 | — | — | — | — |
| 100 | 17.19561 | 2515.50 | 2687.46 | 8.4479 | 3.41833 | 2511.61 | 2682.52 | 7.6947 |
| 150 | 19.51251 | 2587.86 | 2782.99 | 8.6881 | 3.88937 | 2585.61 | 2780.08 | 7.9400 |
| 200 | 21.82507 | 2661.27 | 2879.52 | 8.9037 | 4.35595 | 2659.85 | 2877.64 | 8.1579 |
| 250 | 24.13559 | 2735.95 | 2977.31 | 9.1002 | 4.82045 | 2734.97 | 2975.99 | 8.3555 |
| 300 | 26.44508 | 2812.06 | 3076.51 | 9.2812 | 5.28391 | 2811.33 | 3075.52 | 8.5372 |
| 400 | 31.06252 | 2968.89 | 3279.51 | 9.6076 | 6.20929 | 2968.43 | 3278.89 | 8.8641 |
| 500 | 35.67896 | 3132.26 | 3489.05 | 9.8977 | 7.13364 | 3131.94 | 3488.62 | 9.1545 |
| 600 | 40.29488 | 3302.45 | 3705.40 | 10.1608 | 8.05748 | 3302.22 | 3705.10 | 9.4177 |
| 700 | 44.91052 | 3479.63 | 3928.73 | 10.4028 | 8.98104 | 3479.45 | 3928.51 | 9.6599 |
| 800 | 49.52599 | 3663.84 | 4159.10 | 10.6281 | 9.90444 | 3663.70 | 4158.92 | 9.8852 |
| 900 | 54.14137 | 3855.03 | 4396.44 | 10.8395 | 10.82773 | 3854.91 | 4396.30 | 10.0967 |
| 1000 | 58.75669 | 4053.01 | 4640.58 | 11.0392 | 11.75097 | 4052.91 | 4640.46 | 10.2964 |
| 1100 | 63.37198 | 4257.47 | 4891.19 | 11.2287 | 12.67418 | 4257.37 | 4891.08 | 10.4858 |
| 1200 | 67.98724 | 4467.91 | 5147.78 | 11.4090 | 13.59737 | 4467.82 | 5147.69 | 10.6662 |
| 1300 | 72.60250 | 4683.68 | 5409.70 | 14.5810 | 14.52054 | 4683.58 | 5409.61 | 10.8382 |
| | 100 kPa (99.62) | | | | 200 kPa (120.23) | | | |
| Sat. | 1.69400 | 2506.06 | 2675.46 | 7.3593 | 0.88573 | 2529.49 | 2706.63 | 7.1271 |
| 150 | 1.93636 | 2582.75 | 2776.38 | 7.6133 | 0.95964 | 2576.87 | 2768.80 | 7.2795 |
| 200 | 2.17226 | 2658.05 | 2875.27 | 7.8342 | 1.08034 | 2654.39 | 2870.46 | 7.5066 |
| 250 | 2.40604 | 2733.73 | 2974.33 | 8.0332 | 1.19880 | 2731.22 | 2970.98 | 7.7085 |
| 300 | 2.63876 | 2810.41 | 3074.28 | 8.2157 | 1.31616 | 2808.55 | 3071.79 | 7.8926 |
| 400 | 3.10263 | 2967.85 | 3278.11 | 8.5434 | 1.54930 | 2966.69 | 3276.55 | 8.2217 |
| 500 | 3.56547 | 3131.54 | 3488.09 | 8.8341 | 1.78139 | 3130.75 | 3487.03 | 8.5132 |
| 600 | 4.02781 | 3301.94 | 3704.72 | 9.0975 | 2.01297 | 3301.36 | 3703.96 | 8.7769 |
| 700 | 4.48986 | 3479.24 | 3928.23 | 9.3398 | 2.24426 | 3478.81 | 3927.66 | 9.0194 |
| 800 | 4.95174 | 3663.53 | 4158.71 | 9.5652 | 2.47539 | 3663.19 | 4158.27 | 9.2450 |
| 900 | 5.41353 | 3854.77 | 4396.12 | 9.7767 | 2.70643 | 3854.49 | 4395.77 | 9.4565 |
| 1000 | 5.87526 | 4052.78 | 4640.31 | 9.9764 | 2.93740 | 4052.53 | 4640.01 | 9.6563 |
| 1100 | 6.33696 | 4257.25 | 4890.95 | 10.1658 | 3.16834 | 4257.01 | 4890.68 | 9.8458 |
| 1200 | 6.79863 | 4467.70 | 5147.56 | 10.3462 | 3.39927 | 4467.46 | 5147.32 | 10.0262 |
| 1300 | 7.26030 | 4683.47 | 5409.49 | 10.5182 | 3.63018 | 4683.23 | 5409.26 | 10.1982 |
| | 300 kPa (133.55) | | | | 400 kPa (143.63) | | | |
| Sat. | 0.60582 | 2543.55 | 2725.30 | 6.9918 | 0.46246 | 2553.55 | 2738.53 | 6.8958 |
| 150 | 0.63388 | 2570.79 | 2760.95 | 7.0778 | 0.47084 | 2564.48 | 2752.82 | 6.9299 |
| 200 | 0.71629 | 2650.65 | 2865.54 | 7.3115 | 0.53422 | 2646.83 | 2860.51 | 7.1706 |

TABLE B.1.3 (*continued*)
**Superheated Vapor Water**

| Temp. (°C) | v (m³/kg) | u (kJ/kg) | h (kJ/kg) | s (kJ/kg-K) | v (m³/kg) | u (kJ/kg) | h (kJ/kg) | s (kJ/kg-K) |
|---|---|---|---|---|---|---|---|---|
| | 300 kPa (133.55) | | | | 400 kPa (143.63) | | | |
| 250 | 0.79636 | 2728.69 | 2967.59 | 7.5165 | 0.59512 | 2726.11 | 2964.16 | 7.3788 |
| 300 | 0.87529 | 2806.69 | 3069.28 | 7.7022 | 0.65484 | 2804.81 | 3066.75 | 7.5661 |
| 400 | 1.03151 | 2965.53 | 3274.98 | 8.0329 | 0.77262 | 2964.36 | 3273.41 | 7.8984 |
| 500 | 1.18669 | 3129.95 | 3485.96 | 8.3250 | 0.88934 | 3129.15 | 3484.89 | 8.1912 |
| 600 | 1.34136 | 3300.79 | 3703.20 | 8.5892 | 1.00555 | 3300.22 | 3702.44 | 8.4557 |
| 700 | 1.49573 | 3478.38 | 3927.10 | 8.8319 | 1.12147 | 3477.95 | 3926.53 | 8.6987 |
| 800 | 1.64994 | 3662.85 | 4157.83 | 9.0575 | 1.23722 | 3662.51 | 4157.40 | 8.9244 |
| 900 | 1.80406 | 3854.20 | 4395.42 | 9.2691 | 1.35288 | 3853.91 | 4395.06 | 9.1361 |
| 1000 | 1.95812 | 4052.27 | 4639.71 | 9.4689 | 1.46847 | 4052.02 | 4639.41 | 9.3360 |
| 1100 | 2.11214 | 4256.77 | 4890.41 | 9.6585 | 1.58404 | 4256.53 | 4890.15 | 9.5255 |
| 1200 | 2.26614 | 4467.23 | 5147.07 | 9.8389 | 1.69958 | 4466.99 | 5146.83 | 9.7059 |
| 1300 | 2.42013 | 4682.99 | 5409.03 | 10.0109 | 1.81511 | 4682.75 | 5408.80 | 9.8780 |
| | 500 kPa (151.86) | | | | 600 kPa (158.85) | | | |
| Sat. | 0.37489 | 2561.23 | 2748.67 | 6.8212 | 0.31567 | 2567.40 | 2756.80 | 6.7600 |
| 200 | 0.42492 | 2642.91 | 2855.37 | 7.0592 | 0.35202 | 2638.91 | 2850.12 | 6.9665 |
| 250 | 0.47436 | 2723.50 | 2960.68 | 7.2708 | 0.39383 | 2720.86 | 2957.16 | 7.1816 |
| 300 | 0.52256 | 2802.91 | 3064.20 | 7.4598 | 0.43437 | 2801.00 | 3061.63 | 7.3723 |
| 350 | 0.57012 | 2882.59 | 3167.65 | 7.6328 | 0.47424 | 2881.12 | 3165.66 | 7.5463 |
| 400 | 0.61728 | 2963.19 | 3271.83 | 7.7937 | 0.51372 | 2962.02 | 3270.25 | 7.7078 |
| 500 | 0.71093 | 3128.35 | 3483.82 | 8.0872 | 0.59199 | 3127.55 | 3482.75 | 8.0020 |
| 600 | 0.80406 | 3299.64 | 3701.67 | 8.3521 | 0.66974 | 3299.07 | 3700.91 | 8.2673 |
| 700 | 0.89691 | 3477.52 | 3925.97 | 8.5952 | 0.74720 | 3477.08 | 3925.41 | 8.5107 |
| 800 | 0.98959 | 3662.17 | 4156.96 | 8.8211 | 0.82450 | 3661.83 | 4156.52 | 8.7367 |
| 900 | 1.08217 | 3853.63 | 4394.71 | 9.0329 | 0.90169 | 3853.34 | 4394.36 | 8.9485 |
| 1000 | 1.17469 | 4051.76 | 4639.11 | 9.2328 | 0.97883 | 4051.51 | 4638.81 | 9.1484 |
| 1100 | 1.26718 | 4256.29 | 4889.88 | 9.4224 | 1.05594 | 4256.05 | 4889.61 | 9.3381 |
| 1200 | 1.35964 | 4466.76 | 5146.58 | 9.6028 | 1.13302 | 4466.52 | 5146.34 | 9.5185 |
| 1300 | 1.45210 | 4682.52 | 5408.57 | 9.7749 | 1.21009 | 4682.28 | 5408.34 | 9.6906 |
| | 800 kPa (170.43) | | | | 1000 kPa (179.91) | | | |
| Sat. | 0.24043 | 2576.79 | 2769.13 | 6.6627 | 0.19444 | 2583.64 | 2778.08 | 6.5864 |
| 200 | 0.26080 | 2630.61 | 2839.25 | 6.8158 | 0.20596 | 2621.90 | 2827.86 | 6.6939 |
| 250 | 0.29314 | 2715.46 | 2949.97 | 7.0384 | 0.23268 | 2709.91 | 2942.59 | 6.9246 |
| 300 | 0.32411 | 2797.14 | 3056.43 | 7.2327 | 0.25794 | 2793.21 | 3051.15 | 7.1228 |
| 350 | 0.35439 | 2878.16 | 3161.68 | 7.4088 | 0.28247 | 2875.18 | 3157.65 | 7.3010 |
| 400 | 0.38426 | 2959.66 | 3267.07 | 7.5715 | 0.30659 | 2957.29 | 3263.88 | 7.4650 |
| 500 | 0.44331 | 3125.95 | 3480.60 | 7.8672 | 0.35411 | 3124.34 | 3478.44 | 7.7621 |
| 600 | 0.50184 | 3297.91 | 3699.38 | 8.1332 | 0.40109 | 3296.76 | 3697.85 | 8.0289 |

TABLE B.1.3 (*continued*)
*Superheated Vapor Water*

| Temp. (°C) | v (m³/kg) | u (kJ/kg) | h (kJ/kg) | s (kJ/kg-K) | v (m³/kg) | u (kJ/kg) | h (kJ/kg) | s (kJ/kg-K) |
|---|---|---|---|---|---|---|---|---|
| | | 800 kPa (170.43) | | | | 1000 kPa (179.91) | | |
| 700 | 0.56007 | 3476.22 | 3924.27 | 8.3770 | 0.44779 | 3475.35 | 3923.14 | 8.2731 |
| 800 | 0.61813 | 3661.14 | 4155.65 | 8.6033 | 0.49432 | 3660.46 | 4154.78 | 8.4996 |
| 900 | 0.67610 | 3852.77 | 4393.65 | 8.8153 | 0.54075 | 3852.19 | 4392.94 | 8.7118 |
| 1000 | 0.73401 | 4051.00 | 4638.20 | 9.0153 | 0.58712 | 4050.49 | 4637.60 | 8.9119 |
| 1100 | 0.79188 | 4255.57 | 4889.08 | 9.2049 | 0.63345 | 4255.09 | 4888.55 | 9.1016 |
| 1200 | 0.84974 | 4466.05 | 5145.85 | 9.3854 | 0.67977 | 4465.58 | 5145.36 | 9.2821 |
| 1300 | 0.90758 | 4681.81 | 5407.87 | 9.5575 | 0.72608 | 4681.33 | 5407.41 | 9.4542 |
| | | 1200 kPa (187.99) | | | | 1400 kPa (195.07) | | |
| Sat. | 0.16333 | 2588.82 | 2784.82 | 6.5233 | 0.14084 | 2592.83 | 2790.00 | 6.4692 |
| 200 | 0.16930 | 2612.74 | 2815.90 | 6.5898 | 0.14302 | 2603.09 | 2803.32 | 6.4975 |
| 250 | 0.19235 | 2704.20 | 2935.01 | 6.8293 | 0.16350 | 2698.32 | 2927.22 | 6.7467 |
| 300 | 0.21382 | 2789.22 | 3045.80 | 7.0316 | 0.18228 | 2785.16 | 3040.35 | 6.9533 |
| 350 | 0.23452 | 2872.16 | 3153.59 | 7.2120 | 0.20026 | 2869.12 | 3149.49 | 7.1359 |
| 400 | 0.25480 | 2954.90 | 3260.66 | 7.3773 | 0.21780 | 2952.50 | 3257.42 | 7.3025 |
| 500 | 0.29463 | 3122.72 | 3476.28 | 7.6758 | 0.25215 | 3121.10 | 3474.11 | 7.6026 |
| 600 | 0.33393 | 3295.60 | 3696.32 | 7.9434 | 0.28596 | 3294.44 | 3694.78 | 7.8710 |
| 700 | 0.37294 | 3474.48 | 3922.01 | 8.1881 | 0.31947 | 3473.61 | 3920.87 | 8.1160 |
| 800 | 0.41177 | 3659.77 | 4153.90 | 8.4149 | 0.35281 | 3659.09 | 4153.03 | 8.3431 |
| 900 | 0.45051 | 3851.62 | 4392.23 | 8.6272 | 0.38606 | 3851.05 | 4391.53 | 8.5555 |
| 1000 | 0.48919 | 4049.98 | 4637.00 | 8.8274 | 0.41924 | 4049.47 | 4636.41 | 8.7558 |
| 1100 | 0.52783 | 4254.61 | 4888.02 | 9.0171 | 0.45239 | 4254.14 | 4887.49 | 8.9456 |
| 1200 | 0.56646 | 4465.12 | 5144.87 | 9.1977 | 0.48552 | 4464.65 | 5144.38 | 9.1262 |
| 1300 | 0.60507 | 4680.86 | 5406.95 | 9.3698 | 0.51864 | 4680.39 | 5406.49 | 9.2983 |
| | | 1600 kPa (201.40) | | | | 1800 kPa (207.15) | | |
| Sat. | 0.12380 | 2595.95 | 2794.02 | 6.4217 | 0.11042 | 2598.38 | 2797.13 | 6.3793 |
| 250 | 0.14184 | 2692.26 | 2919.20 | 6.6732 | 0.12497 | 2686.02 | 2910.96 | 6.6066 |
| 300 | 0.15862 | 2781.03 | 3034.83 | 6.8844 | 0.14021 | 2776.83 | 3029.21 | 6.8226 |
| 350 | 0.17456 | 2866.05 | 3145.35 | 7.0693 | 0.15457 | 2862.95 | 3141.18 | 7.0099 |
| 400 | 0.19005 | 2950.09 | 3254.17 | 7.2373 | 0.16847 | 2947.66 | 3250.90 | 7.1793 |
| 500 | 0.22029 | 3119.47 | 3471.93 | 7.5389 | 0.19550 | 3117.84 | 3469.75 | 7.4824 |
| 600 | 0.24998 | 3293.27 | 3693.23 | 7.8080 | 0.22199 | 3292.10 | 3691.69 | 7.7523 |
| 700 | 0.27937 | 3472.74 | 3919.73 | 8.0535 | 0.24818 | 3471.87 | 3918.59 | 7.9983 |
| 800 | 0.30859 | 3658.40 | 4152.15 | 8.2808 | 0.27420 | 3657.71 | 4151.27 | 8.2258 |
| 900 | 0.33772 | 3850.47 | 4390.82 | 8.4934 | 0.30012 | 3849.90 | 4390.11 | 8.4386 |
| 1000 | 0.36678 | 4048.96 | 4635.81 | 8.6938 | 0.32598 | 4048.45 | 4635.21 | 8.6390 |
| 1100 | 0.39581 | 4253.66 | 4886.95 | 8.8837 | 0.35180 | 4253.18 | 4886.42 | 8.8290 |
| 1200 | 0.42482 | 4464.18 | 5143.89 | 9.0642 | 0.37761 | 4463.71 | 5143.40 | 9.0096 |
| 1300 | 0.45382 | 4679.92 | 5406.02 | 9.2364 | 0.40340 | 4679.44 | 5405.56 | 9.1817 |

TABLE B.1.3  (*continued*)
*Superheated Vapor Water*

| Temp. (°C) | $v$ (m³/kg) | $u$ (kJ/kg) | $h$ (kJ/kg) | $s$ (kJ/kg-K) | $v$ (m³/kg) | $u$ (kJ/kg) | $h$ (kJ/kg) | $s$ (kJ/kg-K) |
|---|---|---|---|---|---|---|---|---|
| | 2000 kPa (212.42) | | | | 2500 kPa (223.99) | | | |
| Sat. | 0.09963 | 2600.26 | 2799.51 | 6.3408 | 0.07998 | 2603.13 | 2803.07 | 6.2574 |
| 250 | 0.11144 | 2679.58 | 2902.46 | 6.5452 | 0.08700 | 2662.55 | 2880.06 | 6.4084 |
| 300 | 0.12547 | 2772.56 | 3023.50 | 6.7663 | 0.09890 | 2761.56 | 3008.81 | 6.6437 |
| 350 | 0.13857 | 2859.81 | 3136.96 | 6.9562 | 0.10976 | 2851.84 | 3126.24 | 6.8402 |
| 400 | 0.15120 | 2945.21 | 3247.60 | 7.1270 | 0.12010 | 2939.03 | 3239.28 | 7.0147 |
| 450 | 0.16353 | 3030.41 | 3357.48 | 7.2844 | 0.13014 | 3025.43 | 3350.77 | 7.1745 |
| 500 | 0.17568 | 3116.20 | 3467.55 | 7.4316 | 0.13998 | 3112.08 | 3462.04 | 7.3233 |
| 600 | 0.19960 | 3290.93 | 3690.14 | 7.7023 | 0.15930 | 3287.99 | 3686.25 | 7.5960 |
| 700 | 0.22323 | 3470.99 | 3917.45 | 7.9487 | 0.17832 | 3468.80 | 3914.59 | 7.8435 |
| 800 | 0.24668 | 3657.03 | 4150.40 | 8.1766 | 0.19716 | 3655.30 | 4148.20 | 8.0720 |
| 900 | 0.27004 | 3849.33 | 4389.40 | 8.3895 | 0.21590 | 3847.89 | 4387.64 | 8.2853 |
| 1000 | 0.29333 | 4047.94 | 4634.61 | 8.5900 | 0.23458 | 4046.67 | 4633.12 | 8.4860 |
| 1100 | 0.31659 | 4252.71 | 4885.89 | 8.7800 | 0.25322 | 4251.52 | 4884.57 | 8.6761 |
| 1200 | 0.33984 | 4463.25 | 5142.92 | 8.9606 | 0.27185 | 4462.08 | 5141.70 | 8.8569 |
| 1300 | 0.36306 | 4678.97 | 5405.10 | 9.1328 | 0.29046 | 4677.80 | 5403.95 | 9.0291 |
| | 3000 kPa (233.90) | | | | 4000 kPa (250.40) | | | |
| Sat. | 0.06668 | 2604.10 | 2804.14 | 6.1869 | 0.04978 | 2602.27 | 2801.38 | 6.0700 |
| 250 | 0.07058 | 2644.00 | 2855.75 | 6.2871 | — | — | — | — |
| 300 | 0.08114 | 2750.05 | 2993.48 | 6.5389 | 0.05884 | 2725.33 | 2960.68 | 6.3614 |
| 350 | 0.09053 | 2843.66 | 3115.25 | 6.7427 | 0.06645 | 2826.65 | 3092.43 | 6.5820 |
| 400 | 0.09936 | 2932.75 | 3230.82 | 6.9211 | 0.07341 | 2919.88 | 3213.51 | 6.7689 |
| 450 | 0.10787 | 3020.38 | 3344.00 | 7.0833 | 0.08003 | 3010.13 | 3330.23 | 6.9362 |
| 500 | 0.11619 | 3107.92 | 3456.48 | 7.2337 | 0.08643 | 3099.49 | 3445.21 | 7.0900 |
| 600 | 0.13243 | 3285.03 | 3682.34 | 7.5084 | 0.09885 | 3279.06 | 3674.44 | 7.3688 |
| 700 | 0.14838 | 3466.59 | 3911.72 | 7.7571 | 0.11095 | 3462.15 | 3905.94 | 7.6198 |
| 800 | 0.16414 | 3653.58 | 4146.00 | 7.9862 | 0.12287 | 3650.11 | 4141.59 | 7.8502 |
| 900 | 0.17980 | 3846.46 | 4385.87 | 8.1999 | 0.13469 | 3843.59 | 4382.34 | 8.0647 |
| 1000 | 0.19541 | 4045.40 | 4631.63 | 8.4009 | 0.14645 | 4042.87 | 4628.65 | 8.2661 |
| 1100 | 0.21098 | 4250.33 | 4883.26 | 8.5911 | 0.15817 | 4247.96 | 4880.63 | 8.4566 |
| 1200 | 0.22652 | 4460.92 | 5140.49 | 8.7719 | 0.16987 | 4458.60 | 5138.07 | 8.6376 |
| 1300 | 0.24206 | 4676.63 | 5402.81 | 8.9442 | 0.18156 | 4674.29 | 5400.52 | 8.8099 |

TABLE B.1.3 (*continued*)
*Superheated Vapor Water*

| Temp. (°C) | v (m³/kg) | u (kJ/kg) | h (kJ/kg) | s (kJ/kg-K) | v (m³/kg) | u (kJ/kg) | h (kJ/kg) | s (kJ/kg-K) |
|---|---|---|---|---|---|---|---|---|
| | 5000 kPa (263.99) | | | | 6000 kPa (275.64) | | | |
| Sat. | 0.03944 | 2597.12 | 2794.33 | 5.9733 | 0.03244 | 2589.69 | 2784.33 | 5.8891 |
| 300 | 0.04532 | 2697.94 | 2924.53 | 6.2083 | 0.03616 | 2667.22 | 2884.19 | 6.0673 |
| 350 | 0.05194 | 2808.67 | 3068.39 | 6.4492 | 0.04223 | 2789.61 | 3042.97 | 6.3334 |
| 400 | 0.05781 | 2906.58 | 3195.64 | 6.6458 | 0.04739 | 2892.81 | 3177.17 | 6.5407 |
| 450 | 0.06330 | 2999.64 | 3316.15 | 6.8185 | 0.05214 | 2988.90 | 3301.76 | 6.7192 |
| 500 | 0.06857 | 3090.92 | 3433.76 | 6.9758 | 0.05665 | 3082.20 | 3422.12 | 6.8802 |
| 550 | 0.07368 | 3181.82 | 3550.23 | 7.1217 | 0.06101 | 3174.57 | 3540.62 | 7.0287 |
| 600 | 0.07869 | 3273.01 | 3666.47 | 7.2588 | 0.06525 | 3266.89 | 3658.40 | 7.1676 |
| 700 | 0.08849 | 3457.67 | 3900.13 | 7.5122 | 0.07352 | 3453.15 | 3894.28 | 7.4234 |
| 800 | 0.09811 | 3646.62 | 4137.17 | 7.7440 | 0.08160 | 3643.12 | 4132.74 | 7.6566 |
| 900 | 0.10762 | 3840.71 | 4378.82 | 7.9593 | 0.08958 | 3837.84 | 4375.29 | 7.8727 |
| 1000 | 0.11707 | 4040.35 | 4625.69 | 8.1612 | 0.09749 | 4037.83 | 4622.74 | 8.0751 |
| 1100 | 0.12648 | 4245.61 | 4878.02 | 8.3519 | 0.10536 | 4243.26 | 4875.42 | 8.2661 |
| 1200 | 0.13587 | 4456.30 | 5135.67 | 8.5330 | 0.11321 | 4454.00 | 5133.28 | 8.4473 |
| 1300 | 0.14526 | 4671.96 | 5398.24 | 8.7055 | 0.12106 | 4669.64 | 5395.97 | 8.6199 |
| | 8000 kPa (295.06) | | | | 10000 kPa (311.06) | | | |
| Sat. | 0.02352 | 2569.79 | 2757.94 | 5.7431 | 0.01803 | 2544.41 | 2724.67 | 5.6140 |
| 300 | 0.02426 | 2590.93 | 2784.98 | 5.7905 | — | — | — | — |
| 350 | 0.02995 | 2747.67 | 2987.30 | 6.1300 | 0.02242 | 2699.16 | 2923.39 | 5.9442 |
| 400 | 0.03432 | 2863.75 | 3138.28 | 6.3633 | 0.02641 | 2832.38 | 3096.46 | 6.2119 |
| 450 | 0.03817 | 2966.66 | 3271.99 | 6.5550 | 0.02975 | 2943.32 | 3240.83 | 6.4189 |
| 500 | 0.04175 | 3064.30 | 3398.27 | 6.7239 | 0.03279 | 3045.77 | 3373.63 | 6.5965 |
| 550 | 0.04516 | 3159.76 | 3521.01 | 6.8778 | 0.03564 | 3144.54 | 3500.92 | 6.7561 |
| 600 | 0.04845 | 3254.43 | 3642.03 | 7.0205 | 0.03837 | 3241.68 | 3625.34 | 6.9028 |
| 700 | 0.05481 | 3444.00 | 3882.47 | 7.2812 | 0.04358 | 3434.72 | 3870.52 | 7.1687 |
| 800 | 0.06097 | 3636.08 | 4123.84 | 7.5173 | 0.04859 | 3628.97 | 4114.91 | 7.4077 |
| 900 | 0.06702 | 3832.08 | 4368.26 | 7.7350 | 0.05349 | 3826.32 | 4361.24 | 7.6272 |
| 1000 | 0.07301 | 4032.81 | 4616.87 | 7.9384 | 0.05832 | 4027.81 | 4611.04 | 7.8315 |
| 1100 | 0.07896 | 4238.60 | 4870.25 | 8.1299 | 0.06312 | 4233.97 | 4865.14 | 8.0236 |
| 1200 | 0.08489 | 4449.45 | 5128.54 | 8.3115 | 0.06789 | 4444.93 | 5123.84 | 8.2054 |
| 1300 | 0.09080 | 4665.02 | 5391.46 | 8.4842 | 0.07265 | 4660.44 | 5386.99 | 8.3783 |

TABLE B.1.3 (*continued*)
*Superheated Vapor Water*

| Temp. (°C) | v (m³/kg) | u (kJ/kg) | h (kJ/kg) | s (kJ/kg-K) | v (m³/kg) | u (kJ/kg) | h (kJ/kg) | s (kJ/kg-K) |
|---|---|---|---|---|---|---|---|---|
| | 15000 kPa (342.24) | | | | 20000 kPa (365.81) | | | |
| Sat. | 0.01034 | 2455.43 | 2610.49 | 5.3097 | 0.00583 | 2293.05 | 2409.74 | 4.9269 |
| 350 | 0.01147 | 2520.36 | 2692.41 | 5.4420 | — | — | — | — |
| 400 | 0.01565 | 2740.70 | 2975.44 | 5.8810 | 0.00994 | 2619.22 | 2818.07 | 5.5539 |
| 450 | 0.01845 | 2879.47 | 3156.15 | 6.1403 | 0.01270 | 2806.16 | 3060.06 | 5.9016 |
| 500 | 0.02080 | 2996.52 | 3308.53 | 6.3442 | 0.01477 | 2942.82 | 3238.18 | 6.1400 |
| 550 | 0.02293 | 3104.71 | 3448.61 | 6.5198 | 0.01656 | 3062.34 | 3393.45 | 6.3347 |
| 600 | 0.02491 | 3208.64 | 3582.30 | 6.6775 | 0.01818 | 3174.00 | 3537.57 | 6.5048 |
| 650 | 0.02680 | 3310.37 | 3712.32 | 6.8223 | 0.01969 | 3281.46 | 3675.32 | 6.6582 |
| 700 | 0.02861 | 3410.94 | 3840.12 | 6.9572 | 0.02113 | 3386.46 | 3809.09 | 6.7993 |
| 800 | 0.03210 | 3610.99 | 4092.43 | 7.2040 | 0.02385 | 3592.73 | 4069.80 | 7.0544 |
| 900 | 0.03546 | 3811.89 | 4343.75 | 7.4279 | 0.02645 | 3797.44 | 4326.37 | 7.2830 |
| 1000 | 0.03875 | 4015.41 | 4596.63 | 7.6347 | 0.02897 | 4003.12 | 4582.45 | 7.4925 |
| 1100 | 0.04200 | 4222.55 | 4852.56 | 7.8282 | 0.03145 | 4211.30 | 4840.24 | 7.6874 |
| 1200 | 0.04523 | 4433.78 | 5112.27 | 8.0108 | 0.03391 | 4422.81 | 5100.96 | 7.8706 |
| 1300 | 0.04845 | 4649.12 | 5375.94 | 8.1839 | 0.03636 | 4637.95 | 5365.10 | 8.0441 |
| | 30000 kPa | | | | 40000 kPa | | | |
| 375 | 0.001789 | 1737.75 | 1791.43 | 3.9303 | 0.001641 | 1677.09 | 1742.71 | 3.8289 |
| 400 | 0.002790 | 2067.34 | 2151.04 | 4.4728 | 0.001908 | 1854.52 | 1930.83 | 4.1134 |
| 425 | 0.005304 | 2455.06 | 2614.17 | 5.1503 | 0.002532 | 2096.83 | 2198.11 | 4.5028 |
| 450 | 0.006735 | 2619.30 | 2821.35 | 5.4423 | 0.003693 | 2365.07 | 2512.79 | 4.9459 |
| 500 | 0.008679 | 2820.67 | 3081.03 | 5.7904 | 0.005623 | 2678.36 | 2903.26 | 5.4699 |
| 550 | 0.010168 | 2970.31 | 3275.36 | 6.0342 | 0.006984 | 2869.69 | 3149.05 | 5.7784 |
| 600 | 0.011446 | 3100.53 | 3443.91 | 6.2330 | 0.008094 | 3022.61 | 3346.38 | 6.0113 |
| 650 | 0.012596 | 3221.04 | 3598.93 | 6.4057 | 0.009064 | 3158.04 | 3520.58 | 6.2054 |
| 700 | 0.013661 | 3335.84 | 3745.67 | 6.5606 | 0.009942 | 3283.63 | 3681.29 | 6.3750 |
| 800 | 0.015623 | 3555.60 | 4024.31 | 6.8332 | 0.011523 | 3517.89 | 3978.80 | 6.6662 |
| 900 | 0.017448 | 3768.48 | 4291.93 | 7.0717 | 0.012963 | 3739.42 | 4257.93 | 6.9150 |
| 1000 | 0.019196 | 3978.79 | 4554.68 | 7.2867 | 0.014324 | 3954.64 | 4527.59 | 7.1356 |
| 1100 | 0.020903 | 4189.18 | 4816.28 | 7.4845 | 0.015643 | 4167.38 | 4793.08 | 7.3364 |
| 1200 | 0.022589 | 4401.29 | 5078.97 | 7.6691 | 0.016940 | 4380.11 | 5057.72 | 7.5224 |
| 1300 | 0.024266 | 4615.96 | 5343.95 | 7.8432 | 0.018229 | 4594.28 | 5323.45 | 7.6969 |

**Table B.1.4**
*Compressed Liquid Water*

| Temp. (°C) | v (m³/kg) | u (kJ/kg) | h (kJ/kg) | s (kJ/kg-K) | v (m³/kg) | u (kJ/kg) | h (kJ/kg) | s (kJ/kg-K) |
|---|---|---|---|---|---|---|---|---|
| | 500 kPa (151.86) | | | | 2000 kPa (212.42) | | | |
| Sat. | 0.001093 | 639.66 | 640.21 | 1.8606 | 0.001177 | 906.42 | 908.77 | 2.4473 |
| 0.01 | 0.000999 | 0.01 | 0.51 | 0.0000 | 0.000999 | 0.03 | 2.03 | 0.0001 |
| 20 | 0.001002 | 83.91 | 84.41 | 0.2965 | 0.001001 | 83.82 | 85.82 | .2962 |
| 40 | 0.001008 | 167.47 | 167.98 | 0.5722 | 0.001007 | 167.29 | 169.30 | .5716 |
| 60 | 0.001017 | 251.00 | 251.51 | 0.8308 | 0.001016 | 250.73 | 252.77 | .8300 |
| 80 | 0.001029 | 334.73 | 335.24 | 1.0749 | 0.001028 | 334.38 | 336.44 | 1.0739 |
| 100 | 0.001043 | 418.80 | 419.32 | 1.3065 | 0.001043 | 418.36 | 420.45 | 1.3053 |
| 120 | 0.001060 | 503.37 | 503.90 | 1.5273 | 0.001059 | 502.84 | 504.96 | 1.5259 |
| 140 | 0.001080 | 588.66 | 589.20 | 1.7389 | 0.001079 | 588.02 | 590.18 | 1.7373 |
| 160 | — | — | — | — | 0.001101 | 674.14 | 676.34 | 1.9410 |
| 180 | — | — | — | — | 0.001127 | 761.46 | 763.71 | 2.1382 |
| 200 | — | — | — | — | 0.001156 | 850.30 | 852.61 | 2.3301 |
| | 5000 kPa (263.99) | | | | 10000 kPa (311.06) | | | |
| Sat | 0.001286 | 1147.78 | 1154.21 | 2.9201 | 0.001452 | 1393.00 | 1407.53 | 3.3595 |
| 0 | 0.000998 | 0.03 | 5.02 | 0.0001 | 0.000995 | 0.10 | 10.05 | 0.0003 |
| 20 | 0.001000 | 83.64 | 88.64 | 0.2955 | 0.000997 | 83.35 | 93.32 | 0.2945 |
| 40 | 0.001006 | 166.93 | 171.95 | 0.5705 | 0.001003 | 166.33 | 176.36 | 0.5685 |
| 60 | 0.001015 | 250.21 | 255.28 | 0.8284 | 0.001013 | 249.34 | 259.47 | 0.8258 |
| 80 | 0.001027 | 333.69 | 338.83 | 1.0719 | 0.001025 | 332.56 | 342.81 | 1.0687 |
| 100 | 0.001041 | 417.50 | 422.71 | 1.3030 | 0.001039 | 416.09 | 426.48 | 1.2992 |
| 120 | 0.001058 | 501.79 | 507.07 | 1.5232 | 0.001055 | 500.07 | 510.61 | 1.5188 |
| 140 | 0.001077 | 586.74 | 592.13 | 1.7342 | 0.001074 | 584.67 | 595.40 | 1.7291 |
| 160 | 0.001099 | 672.61 | 678.10 | 1.9374 | 0.001195 | 670.11 | 681.07 | 1.9316 |
| 180 | 0.001124 | 759.62 | 765.24 | 2.1341 | 0.001120 | 756.63 | 767.83 | 2.1274 |
| 200 | 0.001153 | 848.08 | 853.85 | 2.3254 | 0.001148 | 844.49 | 855.97 | 2.3178 |
| 220 | 0.001187 | 938.43 | 944.36 | 2.5128 | 0.001181 | 934.07 | 945.88 | 2.5038 |
| 240 | 0.001226 | 1031.34 | 1037.47 | 2.6978 | 0.001219 | 1025.94 | 1038.13 | 2.6872 |
| 260 | 0.001275 | 1127.92 | 1134.30 | 2.8829 | 0.001265 | 1121.03 | 1133.68 | 2.8698 |
| 280 | | | | | 0.001322 | 1220.90 | 1234.11 | 3.0547 |
| 300 | | | | | 0.001397 | 1328.34 | 1342.31 | 3.2468 |

| Temp. (°C) | $v$ (m³/kg) | $u$ (kJ/kg) | $h$ (kJ/kg) | $s$ (kJ/kg-K) | $v$ (m³/kg) | $u$ (kJ/kg) | $h$ (kJ/kg) | $s$ (kJ/kg-K) |
|---|---|---|---|---|---|---|---|---|
| | 15000 kPa (342.24) | | | | 20000 kPa (365.81) | | | |
| Sat. | 0.001658 | 1585.58 | 1610.45 | 3.6847 | 0.002035 | 1785.47 | 1826.18 | 4.0137 |
| 0 | 0.000993 | 0.15 | 15.04 | 0.0004 | 0.000990 | 0.20 | 20.00 | 0.0004 |
| 20 | 0.000995 | 83.05 | 97.97 | 0.2934 | 0.000993 | 82.75 | 102.61 | 0.2922 |
| 40 | 0.001001 | 165.73 | 180.75 | 0.5665 | 0.000999 | 165.15 | 185.14 | 0.5646 |
| 60 | 0.001011 | 248.49 | 263.65 | 0.8231 | 0.001008 | 247.66 | 267.82 | 0.8205 |
| 80 | 0.001022 | 331.46 | 346.79 | 1.0655 | 0.001020 | 330.38 | 350.78 | 1.0623 |
| 100 | 0.001036 | 414.72 | 430.26 | 1.2954 | 0.001034 | 413.37 | 434.04 | 1.2917 |
| 120 | 0.001052 | 498.39 | 514.17 | 1.5144 | 0.001050 | 496.75 | 517.74 | 1.5101 |
| 140 | 0.001071 | 582.64 | 598.70 | 1.7241 | 0.001068 | 580.67 | 602.03 | 1.7192 |
| 160 | 0.001092 | 667.69 | 684.07 | 1.9259 | 0.001089 | 665.34 | 687.11 | 1.9203 |
| 180 | 0.001116 | 753.74 | 770.48 | 2.1209 | 0.001112 | 750.94 | 773.18 | 2.1146 |
| 200 | 0.001143 | 841.04 | 858.18 | 2.3103 | 0.001139 | 837.70 | 860.47 | 2.3031 |
| 220 | 0.001175 | 929.89 | 947.52 | 2.4952 | 0.001169 | 925.89 | 949.27 | 2.4869 |
| 240 | 0.001211 | 1020.82 | 1038.99 | 2.6770 | 0.001205 | 1015.94 | 1040.04 | 2.6673 |
| 260 | 0.001255 | 1114.59 | 1133.41 | 2.8575 | 0.001246 | 1108.53 | 1133.45 | 2.8459 |
| 280 | 0.001308 | 1212.47 | 1232.09 | 3.0392 | 0.001297 | 1204.69 | 1230.62 | 3.0248 |
| 300 | 0.001377 | 1316.58 | 1337.23 | 3.2259 | 0.001360 | 1306.10 | 1333.29 | 3.2071 |
| 320 | 0.001472 | 1431.05 | 1453.13 | 3.4246 | 0.001444 | 1415.66 | 1444.53 | 3.3978 |
| 340 | 0.001631 | 1567.42 | 1591.88 | 3.6545 | 0.001568 | 1539.64 | 1571.01 | 3.6074 |
| 360 | | | | | 0.001823 | 1702.78 | 1739.23 | 3.8770 |
| | 30000 kPa | | | | 50000 kPa | | | |
| 0 | 0.000986 | 0.25 | 29.82 | 0.0001 | 0.000977 | 0.20 | 49.03 | −0.0014 |
| 20 | 0.000989 | 82.16 | 111.82 | 0.2898 | 0.000980 | 80.98 | 130.00 | 0.2847 |
| 40 | 0.000995 | 164.01 | 193.87 | 0.5606 | 0.000987 | 161.84 | 211.20 | 0.5526 |
| 60 | 0.001004 | 246.03 | 276.16 | 0.8153 | 0.000996 | 242.96 | 292.77 | 0.8051 |
| 80 | 0.001016 | 328.28 | 358.75 | 1.0561 | 0.001007 | 324.32 | 374.68 | 1.0439 |
| 100 | 0.001029 | 410.76 | 441.63 | 1.2844 | 0.001020 | 405.86 | 456.87 | 1.2703 |
| 120 | 0.001044 | 493.58 | 524.91 | 1.5017 | 0.001035 | 487.63 | 539.37 | 1.4857 |
| 140 | 0.001062 | 576.86 | 608.73 | 1.7097 | 0.001052 | 569.76 | 622.33 | 1.6915 |
| 160 | 0.001082 | 660.81 | 693.27 | 1.9095 | 0.001070 | 652.39 | 705.91 | 1.8890 |
| 180 | 0.001105 | 745.57 | 778.71 | 2.1024 | 0.001091 | 735.68 | 790.24 | 2.0793 |
| 200 | 0.001130 | 831.34 | 865.24 | 2.2892 | 0.001115 | 819.73 | 875.46 | 2.2634 |
| 220 | 0.001159 | 918.32 | 953.09 | 2.4710 | 0.001141 | 904.67 | 961.71 | 2.4419 |
| 240 | 0.001192 | 1006.84 | 1042.60 | 2.6489 | 0.001170 | 990.69 | 1049.20 | 2.6158 |
| 260 | 0.001230 | 1097.38 | 1134.29 | 2.8242 | 0.001203 | 1078.06 | 1138.23 | 2.7860 |
| 280 | 0.001275 | 1190.69 | 1228.96 | 2.9985 | 0.001242 | 1167.19 | 1229.26 | 2.9536 |
| 300 | 0.001330 | 1287.89 | 1327.80 | 3.1740 | 0.001286 | 1258.66 | 1322.95 | 3.1200 |
| 320 | 0.001400 | 1390.64 | 1432.63 | 3.3538 | 0.001339 | 1353.23 | 1420.17 | 3.2867 |
| 340 | 0.001492 | 1501.71 | 1546.47 | 3.5425 | 0.001403 | 1451.91 | 1522.07 | 3.4556 |
| 360 | 0.001627 | 1626.57 | 1675.36 | 3.7492 | 0.001484 | 1555.97 | 1630.16 | 3.6290 |
| 380 | 0.001869 | 1781.35 | 1837.43 | 4.0010 | 0.001588 | 1667.13 | 1746.54 | 3.8100 |

**Table B.1.5**

*Saturated Solid–Saturated Vapor, Water*

| Temp. (°C) | Press. (kPa) | Specific Volume, m³/kg | | | Internal Energy, kJ/kg | | |
|---|---|---|---|---|---|---|---|
| | | Sat. Solid $v_i$ | Evap. $v_{ig}$ | Sat. Vapor $v_g$ | Sat. Solid $u_i$ | Evap. $u_{ig}$ | Sat. Vapor $u_g$ |
| 0.01 | 0.6113 | 0.0010908 | 206.152 | 206.153 | −333.40 | 2708.7 | 2375.3 |
| 0 | 0.6108 | 0.0010908 | 206.314 | 206.315 | −333.42 | 2708.7 | 2375.3 |
| −2 | 0.5177 | 0.0010905 | 241.662 | 241.663 | −337.61 | 2710.2 | 2372.5 |
| −4 | 0.4376 | 0.0010901 | 283.798 | 283.799 | −341.78 | 2711.5 | 2369.8 |
| −6 | 0.3689 | 0.0010898 | 334.138 | 334.139 | −345.91 | 2712.9 | 2367.0 |
| −8 | 0.3102 | 0.0010894 | 394.413 | 394.414 | −350.02 | 2714.2 | 2364.2 |
| −10 | 0.2601 | 0.0010891 | 466.756 | 466.757 | −354.09 | 2715.5 | 2361.4 |
| −12 | 0.2176 | 0.0010888 | 553.802 | 553.803 | −358.14 | 2716.8 | 2358.7 |
| −14 | 0.1815 | 0.0010884 | 658.824 | 658.824 | −362.16 | 2718.0 | 2355.9 |
| −16 | 0.1510 | 0.0010881 | 785.906 | 785.907 | −366.14 | 2719.2 | 2353.1 |
| −18 | 0.1252 | 0.0010878 | 940.182 | 940.183 | −370.10 | 2720.4 | 2350.3 |
| −20 | 0.10355 | 0.0010874 | 1128.112 | 1128.113 | −374.03 | 2721.6 | 2347.5 |
| −22 | 0.08535 | 0.0010871 | 1357.863 | 1357.864 | −377.93 | 2722.7 | 2344.7 |
| −24 | 0.07012 | 0.0010868 | 1639.752 | 1639.753 | −381.80 | 2723.7 | 2342.0 |
| −26 | 0.05741 | 0.0010864 | 1986.775 | 1986.776 | −385.64 | 2724.8 | 2339.2 |
| −28 | 0.04684 | 0.0010861 | 2415.200 | 2415.201 | −389.45 | 2725.8 | 2336.4 |
| −30 | 0.03810 | 0.0010858 | 2945.227 | 2945.228 | −393.23 | 2726.8 | 2333.6 |
| −32 | 0.03090 | 0.0010854 | 3601.822 | 3601.823 | −396.98 | 2727.8 | 2330.8 |
| −34 | 0.02499 | 0.0010851 | 4416.252 | 4416.253 | −400.71 | 2728.7 | 2328.0 |
| −36 | 0.02016 | 0.0010848 | 5430.115 | 5430.116 | −404.40 | 2729.6 | 2325.2 |
| −38 | 0.01618 | 0.0010844 | 6707.021 | 6707.022 | −408.06 | 2730.5 | 2322.4 |
| −40 | 0.01286 | 0.0010841 | 8366.395 | 8366.396 | −411.70 | 2731.3 | 2319.6 |

**TABLE B.1.5** (*continued*)
*Saturated Solid–Saturated Vapor, Water*

| Temp. (°C) | Press. (kPa) | ENTHALPY, kJ/kg | | | ENTROPY, kJ/kg-K | | |
|---|---|---|---|---|---|---|---|
| | | Sat. Solid $h_i$ | Evap. $h_{ig}$ | Sat. Vapor $h_g$ | Sat. Solid $s_i$ | Evap. $s_{ig}$ | Sat. Vapor $s_g$ |
| 0.01 | 0.6113 | −333.40 | 2834.7 | 2501.3 | −1.2210 | 10.3772 | 9.1562 |
| 0 | 0.6108 | −333.42 | 2834.8 | 2501.3 | −1.2211 | 10.3776 | 9.1565 |
| −2 | 0.5177 | −337.61 | 2835.3 | 2497.6 | −1.2369 | 10.4562 | 9.2193 |
| −4 | 0.4376 | −341.78 | 2835.7 | 2494.0 | −1.2526 | 10.5358 | 9.2832 |
| −6 | 0.3689 | −345.91 | 2836.2 | 2490.3 | −1.2683 | 10.6165 | 9.3482 |
| −8 | 0.3102 | −350.02 | 2836.6 | 2486.6 | −1.2839 | 10.6982 | 9.4143 |
| −10 | 0.2601 | −354.09 | 2837.0 | 2482.9 | −1.2995 | 10.7809 | 9.4815 |
| −12 | 0.2176 | −358.14 | 2837.3 | 2479.2 | −1.3150 | 10.8648 | 9.5498 |
| −14 | 0.1815 | −362.16 | 2837.6 | 2475.5 | −1.3306 | 10.9498 | 9.6192 |
| −16 | 0.1510 | −366.14 | 2837.9 | 2471.8 | −1.3461 | 11.0359 | 9.6898 |
| −18 | 0.1252 | −370.10 | 2838.2 | 2468.1 | −1.3617 | 11.1233 | 9.7616 |
| −20 | 0.10355 | −374.03 | 2838.4 | 2464.3 | −1.3772 | 11.2120 | 9.8348 |
| −22 | 0.08535 | −377.93 | 2838.6 | 2460.6 | −1.3928 | 11.3020 | 9.9093 |
| −24 | 0.07012 | −381.80 | 2838.7 | 2456.9 | −1.4083 | 11.3935 | 9.9852 |
| −26 | 0.05741 | −385.64 | 2838.9 | 2453.2 | −1.4239 | 11.4864 | 10.0625 |
| −28 | 0.04684 | −389.45 | 2839.0 | 2449.5 | −1.4394 | 11.5808 | 10.1413 |
| −30 | 0.03810 | −393.23 | 2839.0 | 2445.8 | −1.4550 | 11.6765 | 10.2215 |
| −32 | 0.03090 | −396.98 | 2839.1 | 2442.1 | −1.4705 | 11.7733 | 10.3028 |
| −34 | 0.02499 | −400.71 | 2839.1 | 2438.4 | −1.4860 | 11.8713 | 10.3853 |
| −36 | 0.02016 | −404.40 | 2839.1 | 2434.7 | −1.5014 | 11.9704 | 10.4690 |
| −38 | 0.01618 | −408.06 | 2839.0 | 2431.0 | −1.5168 | 12.0714 | 10.5546 |
| −40 | 0.01286 | −411.70 | 2838.9 | 2427.2 | −1.5321 | 12.1768 | 10.6447 |

**TABLE B.2**
*Thermodynamic Properties of Ammonia*

**TABLE B.2.1**
*Saturated Ammonia*

| Temp. (°C) | Press. (kPa) | Specific Volume, m³/kg | | | Internal Energy, kJ/kg | | |
|---|---|---|---|---|---|---|---|
| | | Sat. Liquid $v_f$ | Evap. $v_{fg}$ | Sat. Vapor $v_g$ | Sat. Liquid $u_f$ | Evap. $u_{fg}$ | Sat. Vapor $u_g$ |
| −50 | 40.9 | 0.001424 | 2.62557 | 2.62700 | −43.82 | 1309.1 | 1265.2 |
| −45 | 54.5 | 0.001437 | 2.00489 | 2.00632 | −22.01 | 1293.5 | 1271.4 |
| −40 | 71.7 | 0.001450 | 1.55111 | 1.55256 | −0.10 | 1277.6 | 1277.4 |
| −35 | 93.2 | 0.001463 | 1.21466 | 1.21613 | 21.93 | 1261.3 | 1283.3 |
| −30 | 119.5 | 0.001476 | 0.96192 | 0.96339 | 44.08 | 1244.8 | 1288.9 |
| −25 | 151.6 | 0.001490 | 0.76970 | 0.77119 | 66.36 | 1227.9 | 1294.3 |
| −20 | 190.2 | 0.001504 | 0.62184 | 0.62334 | 88.76 | 1210.7 | 1299.5 |
| −15 | 236.3 | 0.001519 | 0.50686 | 0.50838 | 111.30 | 1193.2 | 1304.5 |
| −10 | 290.9 | 0.001534 | 0.41655 | 0.41808 | 133.96 | 1175.2 | 1309.2 |
| −5 | 354.9 | 0.001550 | 0.34493 | 0.34648 | 156.76 | 1157.0 | 1313.7 |
| 0 | 429.6 | 0.001566 | 0.28763 | 0.28920 | 179.69 | 1138.3 | 1318.0 |
| 5 | 515.9 | 0.001583 | 0.24140 | 0.24299 | 202.77 | 1119.2 | 1322.0 |
| 10 | 615.2 | 0.001600 | 0.20381 | 0.20541 | 225.99 | 1099.7 | 1325.7 |
| 15 | 728.6 | 0.001619 | 0.17300 | 0.17462 | 249.36 | 1079.7 | 1329.1 |
| 20 | 857.5 | 0.001638 | 0.14758 | 0.14922 | 272.89 | 1059.3 | 1332.2 |
| 25 | 1003.2 | 0.001658 | 0.12647 | 0.12813 | 296.59 | 1038.4 | 1335.0 |
| 30 | 1167.0 | 0.001680 | 0.10881 | 0.11049 | 320.46 | 1016.9 | 1337.4 |
| 35 | 1350.4 | 0.001702 | 0.09397 | 0.09567 | 344.50 | 994.9 | 1339.4 |
| 40 | 1554.9 | 0.001725 | 0.08141 | 0.08313 | 368.74 | 972.2 | 1341.0 |
| 45 | 1782.0 | 0.001750 | 0.07073 | 0.07248 | 393.19 | 948.9 | 1342.1 |
| 50 | 2033.1 | 0.001777 | 0.06159 | 0.06337 | 417.87 | 924.8 | 1342.7 |
| 55 | 2310.1 | 0.001804 | 0.05375 | 0.05555 | 442.79 | 899.9 | 1342.7 |
| 60 | 2614.4 | 0.001834 | 0.04697 | 0.04880 | 467.99 | 874.2 | 1342.1 |
| 65 | 2947.8 | 0.001866 | 0.04109 | 0.04296 | 493.51 | 847.4 | 1340.9 |
| 70 | 3312.0 | 0.001900 | 0.03597 | 0.03787 | 519.39 | 819.5 | 1338.9 |
| 75 | 3709.0 | 0.001937 | 0.03148 | 0.03341 | 545.70 | 790.4 | 1336.1 |
| 80 | 4140.5 | 0.001978 | 0.02753 | 0.02951 | 572.50 | 759.9 | 1332.4 |
| 85 | 4608.6 | 0.002022 | 0.02404 | 0.02606 | 599.90 | 727.8 | 1327.7 |
| 90 | 5115.3 | 0.002071 | 0.02093 | 0.02300 | 627.99 | 693.7 | 1321.7 |
| 95 | 5662.9 | 0.002126 | 0.01815 | 0.02028 | 656.95 | 657.4 | 1314.4 |
| 100 | 6253.7 | 0.002188 | 0.01565 | 0.01784 | 686.96 | 618.4 | 1305.3 |
| 105 | 6890.4 | 0.002261 | 0.01337 | 0.01564 | 718.30 | 575.9 | 1294.2 |
| 110 | 7575.7 | 0.002347 | 0.01128 | 0.01363 | 751.37 | 529.1 | 1280.5 |
| 115 | 8313.3 | 0.002452 | 0.00933 | 0.01178 | 786.82 | 476.2 | 1263.1 |
| 120 | 9107.2 | 0.002589 | 0.00744 | 0.01003 | 825.77 | 414.5 | 1240.3 |
| 125 | 9963.5 | 0.002783 | 0.00554 | 0.00833 | 870.69 | 337.7 | 1208.4 |
| 130 | 10891.6 | 0.003122 | 0.00337 | 0.00649 | 929.29 | 226.9 | 1156.2 |
| 132.3 | 11333.2 | 0.004255 | 0 | 0.00426 | 1037.62 | 0 | 1037.6 |

**TABLE B.2.1** (*continued*)
**Saturated Ammonia**

| Temp. (°C) | Press. (kPa) | ENTHALPY, kJ/kg | | | ENTROPY, kJ/kg-K | | |
|---|---|---|---|---|---|---|---|
| | | Sat. Liquid $h_f$ | Evap. $h_{fg}$ | Sat. Vapor $h_g$ | Sat. Liquid $s_f$ | Evap. $s_{fg}$ | Sat. Vapor $s_g$ |
| −50 | 40.9 | −43.76 | 1416.3 | 1372.6 | −0.1916 | 6.3470 | 6.1554 |
| −45 | 54.5 | −21.94 | 1402.8 | 1380.8 | −0.0950 | 6.1484 | 6.0534 |
| −40 | 71.7 | 0 | 1388.8 | 1388.8 | 0 | 5.9567 | 5.9567 |
| −35 | 93.2 | 22.06 | 1374.5 | 1396.5 | 0.0935 | 5.7715 | 5.8650 |
| −30 | 119.5 | 44.26 | 1359.8 | 1404.0 | 0.1856 | 5.5922 | 5.7778 |
| −25 | 151.6 | 66.58 | 1344.6 | 1411.2 | 0.2763 | 5.4185 | 5.6947 |
| −20 | 190.2 | 89.05 | 1329.0 | 1418.0 | 0.3657 | 5.2498 | 5.6155 |
| −15 | 236.3 | 111.66 | 1312.9 | 1424.6 | 0.4538 | 5.0859 | 5.5397 |
| −10 | 290.9 | 134.41 | 1296.4 | 1430.8 | 0.5408 | 4.9265 | 5.4673 |
| −5 | 354.9 | 157.31 | 1279.4 | 1436.7 | 0.6266 | 4.7711 | 5.3977 |
| 0 | 429.6 | 180.36 | 1261.8 | 1442.2 | 0.7114 | 4.6195 | 5.3309 |
| 5 | 515.9 | 203.58 | 1243.7 | 1447.3 | 0.7951 | 4.4715 | 5.2666 |
| 10 | 615.2 | 226.97 | 1225.1 | 1452.0 | 0.8779 | 4.3266 | 5.2045 |
| 15 | 728.6 | 250.54 | 1205.8 | 1456.3 | 0.9598 | 4.1846 | 5.1444 |
| 20 | 857.5 | 274.30 | 1185.9 | 1460.2 | 1.0408 | 4.0452 | 5.0860 |
| 25 | 1003.2 | 298.25 | 1165.2 | 1463.5 | 1.1210 | 3.9083 | 5.0293 |
| 30 | 1167.0 | 322.42 | 1143.9 | 1466.3 | 1.2005 | 3.7734 | 4.9738 |
| 35 | 1350.4 | 346.80 | 1121.8 | 1468.6 | 1.2792 | 3.6403 | 4.9196 |
| 40 | 1554.9 | 371.43 | 1098.8 | 1470.2 | 1.3574 | 3.5088 | 4.8662 |
| 45 | 1782.0 | 396.31 | 1074.9 | 1471.2 | 1.4350 | 3.3786 | 4.8136 |
| 50 | 2033.1 | 421.48 | 1050.0 | 1471.5 | 1.5121 | 3.2493 | 4.7614 |
| 55 | 2310.1 | 446.96 | 1024.1 | 1471.0 | 1.5888 | 3.1208 | 4.7095 |
| 60 | 2614.4 | 472.79 | 997.0 | 1469.7 | 1.6652 | 2.9925 | 4.6577 |
| 65 | 2947.8 | 499.01 | 968.5 | 1467.5 | 1.7415 | 2.8642 | 4.6057 |
| 70 | 3312.0 | 525.69 | 938.7 | 1464.4 | 1.8178 | 2.7354 | 4.3533 |
| 75 | 3709.0 | 552.88 | 907.2 | 1460.1 | 1.8943 | 2.6058 | 4.5001 |
| 80 | 4140.5 | 580.69 | 873.9 | 1454.6 | 1.9712 | 2.4746 | 4.4458 |
| 85 | 4608.6 | 609.21 | 838.6 | 1447.8 | 2.0488 | 2.3413 | 4.3901 |
| 90 | 5115.3 | 638.59 | 800.8 | 1439.4 | 2.1273 | 2.2051 | 4.3325 |
| 95 | 5662.9 | 668.99 | 760.2 | 1429.2 | 2.2073 | 2.0650 | 4.2723 |
| 100 | 6253.7 | 700.64 | 716.2 | 1416.9 | 2.2893 | 1.9195 | 4.2088 |
| 105 | 6890.4 | 733.87 | 668.1 | 1402.0 | 2.3740 | 1.7667 | 4.1407 |
| 110 | 7575.7 | 769.15 | 614.6 | 1383.7 | 2.4625 | 1.6040 | 4.0665 |
| 115 | 8313.3 | 807.21 | 553.8 | 1361.0 | 2.5566 | 1.4267 | 3.9833 |
| 120 | 9107.2 | 849.36 | 482.3 | 1331.7 | 2.6593 | 1.2268 | 3.8861 |
| 125 | 9963.5 | 898.42 | 393.0 | 1291.4 | 2.7775 | 0.9870 | 3.7645 |
| 130 | 10892 | 963.29 | 263.7 | 1227.0 | 2.9326 | 0.6540 | 3.5866 |
| 132.3 | 11333 | 1085.85 | 0 | 1085.9 | 3.2316 | 0 | 3.2316 |

TABLE B.2.2
*Superheated Ammonia*

| Temp. (°C) | v (m³/kg) | u (kJ/kg) | h (kJ/kg) | s (kJ/kg-K) | v (m³/kg) | u (kJ/kg) | h (kJ/kg) | s (kJ/kg-K) |
|---|---|---|---|---|---|---|---|---|
| | 50 kPa (−46.53) | | | | 100 kPa (−33.60) | | | |
| Sat. | 2.1752 | 1269.6 | 1378.3 | 6.0839 | 1.1381 | 1284.9 | 1398.7 | 5.8401 |
| −30 | 2.3448 | 1296.2 | 1413.4 | 6.2333 | 1.1573 | 1291.0 | 1406.7 | 5.8734 |
| −20 | 2.4463 | 1312.3 | 1434.6 | 6.3187 | 1.2101 | 1307.8 | 1428.8 | 5.9626 |
| −10 | 2.5471 | 1328.4 | 1455.7 | 6.4006 | 1.2621 | 1324.6 | 1450.8 | 6.0477 |
| 0 | 2.6474 | 1344.5 | 1476.9 | 6.4795 | 1.3136 | 1341.3 | 1472.6 | 6.1291 |
| 10 | 2.7472 | 1360.7 | 1498.1 | 6.5556 | 1.3647 | 1357.9 | 1494.4 | 6.2073 |
| 20 | 2.8466 | 1377.0 | 1519.3 | 6.6293 | 1.4153 | 1374.5 | 1516.1 | 6.2826 |
| 30 | 2.9458 | 1393.3 | 1540.6 | 6.7008 | 1.4657 | 1391.2 | 1537.7 | 6.3553 |
| 40 | 3.0447 | 1409.8 | 1562.0 | 6.7703 | 1.5158 | 1407.9 | 1559.5 | 6.4258 |
| 50 | 3.1435 | 1426.3 | 1583.5 | 6.8379 | 1.5658 | 1424.7 | 1581.2 | 6.4943 |
| 60 | 3.2421 | 1443.0 | 1605.1 | 6.9038 | 1.6156 | 1441.5 | 1603.1 | 6.5609 |
| 70 | 3.3406 | 1459.9 | 1626.9 | 6.9682 | 1.6653 | 1458.5 | 1625.1 | 6.6258 |
| 80 | 3.4390 | 1476.9 | 1648.8 | 7.0312 | 1.7148 | 1475.6 | 1647.1 | 6.6892 |
| 100 | 3.6355 | 1511.4 | 1693.2 | 7.1533 | 1.8137 | 1510.3 | 1691.7 | 6.8120 |
| 120 | 3.8318 | 1546.6 | 1738.2 | 7.2708 | 1.9124 | 1545.7 | 1736.9 | 6.9300 |
| 140 | 4.0280 | 1582.5 | 1783.9 | 7.3842 | 2.0109 | 1581.7 | 1782.8 | 7.0439 |
| 160 | 4.2240 | 1619.2 | 1830.4 | 7.4941 | 2.1093 | 1618.5 | 1829.4 | 7.1540 |
| 180 | 4.4199 | 1656.7 | 1877.7 | 7.6008 | 2.2075 | 1656.0 | 1876.8 | 7.2609 |
| 200 | 4.6157 | 1694.9 | 1925.7 | 7.7045 | 2.3057 | 1694.3 | 1924.9 | 7.3648 |
| | 150 kPa (−25.22) | | | | 200 kPa (−18.86) | | | |
| Sat. | 0.7787 | 1294.1 | 1410.9 | 5.6983 | 0.5946 | 1300.6 | 1419.6 | 5.5979 |
| −20 | 0.7977 | 1303.3 | 1422.9 | 5.7465 | — | — | — | — |
| −10 | 0.8336 | 1320.7 | 1445.7 | 5.8349 | 0.6193 | 1316.7 | 1440.6 | 5.6791 |
| 0 | 0.8689 | 1337.9 | 1468.3 | 5.9189 | 0.6465 | 1334.5 | 1463.8 | 5.7659 |
| 10 | 0.9037 | 1355.0 | 1490.6 | 5.9992 | 0.6732 | 1352.1 | 1486.8 | 5.8484 |
| 20 | 0.9382 | 1372.0 | 1512.8 | 6.0761 | 0.6995 | 1369.5 | 1509.4 | 5.9270 |
| 30 | 0.9723 | 1389.0 | 1534.9 | 6.1502 | 0.7255 | 1386.8 | 1531.9 | 6.0025 |
| 40 | 1.0062 | 1406.0 | 1556.9 | 6.2217 | 0.7513 | 1404.0 | 1554.3 | 6.0751 |
| 50 | 1.0398 | 1423.0 | 1578.9 | 6.2910 | 0.7769 | 1421.3 | 1576.6 | 6.1453 |
| 60 | 1.0734 | 1440.0 | 1601.0 | 6.3583 | 0.8023 | 1438.5 | 1598.9 | 6.2133 |
| 70 | 1.1068 | 1457.2 | 1623.2 | 6.4238 | 0.8275 | 1455.8 | 1621.3 | 6.2794 |
| 80 | 1.1401 | 1474.4 | 1645.4 | 6.4877 | 0.8527 | 1473.1 | 1643.7 | 6.3437 |
| 100 | 1.2065 | 1509.3 | 1690.2 | 6.6112 | 0.9028 | 1508.2 | 1688.8 | 6.4679 |
| 120 | 1.2726 | 1544.8 | 1735.6 | 6.7297 | 0.9527 | 1543.8 | 1734.4 | 6.5869 |
| 140 | 1.3386 | 1580.9 | 1781.7 | 6.8439 | 1.0024 | 1580.1 | 1780.6 | 6.7015 |
| 160 | 1.4044 | 1617.8 | 1828.4 | 6.9544 | 1.0519 | 1617.0 | 1827.4 | 6.8123 |
| 180 | 1.4701 | 1655.4 | 1875.9 | 7.0615 | 1.1014 | 1654.7 | 1875.0 | 6.9196 |
| 200 | 1.5357 | 1693.7 | 1924.1 | 7.1656 | 1.1507 | 1693.2 | 1923.3 | 7.0239 |
| 220 | 1.6013 | 1732.9 | 1973.1 | 7.2670 | 1.2000 | 1732.4 | 1972.4 | 7.1255 |

**TABLE B.2.2** (*continued*)
*Superheated Ammonia*

| Temp. (°C) | v (m³/kg) | u (kJ/kg) | h (kJ/kg) | s (kJ/kg-K) | v (m³/kg) | u (kJ/kg) | h (kJ/kg) | s (kJ/kg-K) |
|---|---|---|---|---|---|---|---|---|
| | 300 kPa (−9.24) | | | | 400 kPa (−1.89) | | | |
| Sat | 0.40607 | 1309.9 | 1431.7 | 5.4565 | 0.30942 | 1316.4 | 1440.2 | 5.3559 |
| 0 | 0.42382 | 1327.5 | 1454.7 | 5.5420 | 0.31227 | 1320.2 | 1445.1 | 5.3741 |
| 10 | 0.44251 | 1346.1 | 1478.9 | 5.6290 | 0.32701 | 1339.9 | 1470.7 | 5.4663 |
| 20 | 0.46077 | 1364.4 | 1502.6 | 5.7113 | 0.34129 | 1359.1 | 1495.6 | 5.5525 |
| 30 | 0.47870 | 1382.3 | 1526.0 | 5.7896 | 0.35520 | 1377.7 | 1519.8 | 5.6338 |
| 40 | 0.49636 | 1400.1 | 1549.0 | 5.8645 | 0.36884 | 1396.1 | 1543.6 | 5.7111 |
| 50 | 0.51382 | 1417.8 | 1571.9 | 5.9365 | 0.38226 | 1414.2 | 1567.1 | 5.7850 |
| 60 | 0.53111 | 1435.4 | 1594.7 | 6.0060 | 0.39550 | 1432.2 | 1590.4 | 5.8560 |
| 70 | 0.54827 | 1453.0 | 1617.5 | 6.0732 | 0.40860 | 1450.1 | 1613.6 | 5.9244 |
| 80 | 0.56532 | 1470.6 | 1640.2 | 6.1385 | 0.42160 | 1468.0 | 1636.7 | 5.9907 |
| 100 | 0.59916 | 1506.1 | 1685.8 | 6.2642 | 0.44732 | 1503.9 | 1682.8 | 6.1179 |
| 120 | 0.63276 | 1542.0 | 1731.8 | 6.3842 | 0.47279 | 1540.1 | 1729.2 | 6.2390 |
| 140 | 0.66618 | 1578.5 | 1778.3 | 6.4996 | 0.49808 | 1576.8 | 1776.0 | 6.3552 |
| 160 | 0.69946 | 1615.6 | 1825.4 | 6.6109 | 0.52323 | 1614.1 | 1823.4 | 6.4671 |
| 180 | 0.73263 | 1653.4 | 1873.2 | 6.7188 | 0.54827 | 1652.1 | 1871.4 | 6.5755 |
| 200 | 0.76572 | 1692.0 | 1921.7 | 6.8235 | 0.57321 | 1690.8 | 1920.1 | 6.6806 |
| 220 | 0.79872 | 1731.3 | 1970.9 | 6.9254 | 0.59809 | 1730.3 | 1969.5 | 6.7828 |
| 240 | 0.83167 | 1771.4 | 2020.9 | 7.0247 | 0.62289 | 1770.5 | 2019.6 | 6.8825 |
| 260 | 0.86455 | 1812.2 | 2071.6 | 7.1217 | 0.64764 | 1811.4 | 2070.5 | 6.9797 |
| | 500 kPa (4.13) | | | | 600 kPa (9.28) | | | |
| Sat. | 0.25035 | 1321.3 | 1446.5 | 5.2776 | 0.21038 | 1325.2 | 1451.4 | 5.2133 |
| 10 | 0.25757 | 1333.5 | 1462.3 | 5.3340 | 0.21115 | 1326.7 | 1453.4 | 5.2205 |
| 20 | 0.26949 | 1353.6 | 1488.3 | 5.4244 | 0.22154 | 1347.9 | 1480.8 | 5.3156 |
| 30 | 0.28103 | 1373.0 | 1513.5 | 5.5090 | 0.23152 | 1368.2 | 1507.1 | 5.4037 |
| 40 | 0.29227 | 1392.0 | 1538.1 | 5.5889 | 0.24118 | 1387.8 | 1532.5 | 5.4862 |
| 50 | 0.30328 | 1410.6 | 1562.2 | 5.6647 | 0.25059 | 1406.9 | 1557.3 | 5.5641 |
| 60 | 0.31410 | 1429.0 | 1586.1 | 5.7373 | 0.25981 | 1425.7 | 1581.6 | 5.6383 |
| 70 | 0.32478 | 1447.3 | 1609.6 | 5.8070 | 0.26888 | 1444.3 | 1605.7 | 5.7094 |
| 80 | 0.33535 | 1465.4 | 1633.1 | 5.8744 | 0.27783 | 1462.8 | 1629.5 | 5.7778 |
| 100 | 0.35621 | 1501.7 | 1679.8 | 6.0031 | 0.29545 | 1499.5 | 1676.8 | 5.9081 |
| 120 | 0.37681 | 1538.2 | 1726.6 | 6.1253 | 0.31281 | 1536.3 | 1724.0 | 6.0314 |
| 140 | 0.39722 | 1575.2 | 1773.8 | 6.2422 | 0.32997 | 1573.5 | 1771.5 | 6.1491 |
| 160 | 0.41748 | 1612.7 | 1821.4 | 6.3548 | 0.34699 | 1611.2 | 1819.4 | 6.2623 |
| 180 | 0.43764 | 1650.8 | 1869.6 | 6.4636 | 0.36389 | 1649.5 | 1867.8 | 6.3717 |
| 200 | 0.45771 | 1689.6 | 1918.5 | 6.5691 | 0.38071 | 1688.5 | 1916.9 | 6.4776 |
| 220 | 0.47770 | 1729.2 | 1968.1 | 6.6717 | 0.39745 | 1728.2 | 1966.6 | 6.5806 |
| 240 | 0.49763 | 1769.5 | 2018.3 | 6.7717 | 0.41412 | 1768.6 | 2017.0 | 6.6808 |
| 260 | 0.51749 | 1810.6 | 2069.3 | 6.8692 | 0.43073 | 1809.8 | 2068.2 | 6.7786 |

**TABLE B.2.2** (*continued*)
*Superheated Ammonia*

| Temp. (°C) | v (m³/kg) | u (kJ/kg) | h (kJ/kg) | s (kJ/kg-K) | v (m³/kg) | u (kJ/kg) | h (kJ/kg) | s (kJ/kg-K) |
|---|---|---|---|---|---|---|---|---|
| | 800 kPa (17.85) | | | | 1000 kPa (24.90) | | | |
| Sat. | 0.15958 | 1330.9 | 1458.6 | 5.1110 | 0.12852 | 1334.9 | 1463.4 | 5.0304 |
| 20 | 0.16138 | 1335.8 | 1464.9 | 5.1328 | — | — | — | — |
| 30 | 0.16947 | 1358.0 | 1493.5 | 5.2287 | 0.13206 | 1347.1 | 1479.1 | 5.0826 |
| 40 | 0.17720 | 1379.0 | 1520.8 | 5.3171 | 0.13868 | 1369.8 | 1508.5 | 5.1778 |
| 50 | 0.18465 | 1399.3 | 1547.0 | 5.3996 | 0.14499 | 1391.3 | 1536.3 | 5.2654 |
| 60 | 0.19189 | 1419.0 | 1572.5 | 5.4774 | 0.15106 | 1412.1 | 1563.1 | 5.3471 |
| 70 | 0.19896 | 1438.3 | 1597.5 | 5.5513 | 0.15695 | 1432.2 | 1589.1 | 5.4240 |
| 80 | 0.20590 | 1457.4 | 1622.1 | 5.6219 | 0.16270 | 1451.9 | 1614.6 | 5.4971 |
| 100 | 0.21949 | 1495.0 | 1670.6 | 5.7555 | 0.17389 | 1490.5 | 1664.3 | 5.6342 |
| 120 | 0.23280 | 1532.5 | 1718.7 | 5.8811 | 0.18477 | 1528.6 | 1713.4 | 5.7622 |
| 140 | 0.24590 | 1570.1 | 1766.9 | 6.0006 | 0.19545 | 1566.8 | 1762.2 | 5.8834 |
| 160 | 0.25886 | 1608.2 | 1815.3 | 6.1150 | 0.20597 | 1605.2 | 1811.2 | 5.9992 |
| 180 | 0.27170 | 1646.8 | 1864.2 | 6.2254 | 0.21638 | 1644.2 | 1860.5 | 6.1105 |
| 200 | 0.28445 | 1686.1 | 1913.6 | 6.3322 | 0.22669 | 1683.7 | 1910.4 | 6.2182 |
| 220 | 0.29712 | 1726.0 | 1963.7 | 6.4358 | 0.23693 | 1723.9 | 1960.8 | 6.3226 |
| 240 | 0.30973 | 1766.7 | 2014.5 | 6.5367 | 0.24710 | 1764.8 | 2011.9 | 6.4241 |
| 260 | 0.32228 | 1808.1 | 2065.9 | 6.6350 | 0.25720 | 1806.4 | 2063.6 | 6.5229 |
| 280 | 0.33477 | 1850.2 | 2118.0 | 6.7310 | 0.26726 | 1848.8 | 2116.0 | 6.6194 |
| 300 | 0.34722 | 1893.1 | 2170.9 | 6.8248 | 0.27726 | 1891.8 | 2169.1 | 6.7137 |
| | 1200 kPa (30.94) | | | | 1400 kPa (36.26) | | | |
| Sat. | 0.10751 | 1337.8 | 1466.8 | 4.9635 | 0.09231 | 1339.8 | 1469.0 | 4.9060 |
| 40 | 0.11287 | 1360.0 | 1495.4 | 5.0564 | 0.09432 | 1349.5 | 1481.6 | 4.9463 |
| 50 | 0.11846 | 1383.0 | 1525.1 | 5.1497 | 0.09942 | 1374.2 | 1513.4 | 5.0462 |
| 60 | 0.12378 | 1404.8 | 1553.3 | 5.2357 | 0.10423 | 1397.2 | 1543.1 | 5.1370 |
| 70 | 0.12890 | 1425.8 | 1580.5 | 5.3159 | 0.10882 | 1419.2 | 1571.5 | 5.2209 |
| 80 | 0.13387 | 1446.2 | 1606.8 | 5.3916 | 0.11324 | 1440.3 | 1598.8 | 5.2994 |
| 100 | 0.14347 | 1485.8 | 1658.0 | 5.5325 | 0.12172 | 1481.0 | 1651.4 | 5.4443 |
| 120 | 0.15275 | 1524.7 | 1708.0 | 5.6631 | 0.12986 | 1520.7 | 1702.5 | 5.5775 |
| 140 | 0.16181 | 1563.3 | 1757.5 | 5.7860 | 0.13777 | 1559.9 | 1752.8 | 5.7023 |
| 160 | 0.17071 | 1602.2 | 1807.1 | 5.9031 | 0.14552 | 1599.2 | 1802.9 | 5.8208 |
| 180 | 0.17950 | 1641.5 | 1856.9 | 6.0156 | 0.15315 | 1638.8 | 1853.2 | 5.9343 |
| 200 | 0.18819 | 1681.3 | 1907.1 | 6.1241 | 0.16068 | 1678.9 | 1903.8 | 6.0437 |
| 220 | 0.19680 | 1721.8 | 1957.9 | 6.2292 | 0.16813 | 1719.6 | 1955.0 | 6.1495 |
| 240 | 0.20534 | 1762.9 | 2009.3 | 6.3313 | 0.17551 | 1761.0 | 2006.7 | 6.2523 |
| 260 | 0.21382 | 1804.7 | 2061.3 | 6.4308 | 0.18283 | 1803.0 | 2059.0 | 6.3523 |
| 280 | 0.22225 | 1847.3 | 2114.0 | 6.5278 | 0.19010 | 1845.8 | 2111.9 | 6.4498 |
| 300 | 0.23063 | 1890.6 | 2167.3 | 6.6225 | 0.19732 | 1889.3 | 2165.5 | 6.5450 |
| 320 | 0.23897 | 1934.6 | 2221.3 | 6.7151 | 0.20450 | 1933.5 | 2219.8 | 6.6380 |

**TABLE B.2.2** (*continued*)
*Superheated Ammonia*

| Temp. (°C) | v (m³/kg) | u (kJ/kg) | h (kJ/kg) | s (kJ/kg-K) | v (m³/kg) | u (kJ/kg) | h (kJ/kg) | s (kJ/kg-K) |
|---|---|---|---|---|---|---|---|---|
| | 1600 kPa (41.03) | | | | 2000 kPa (49.37) | | | |
| Sat. | 0.08079 | 1341.2 | 1470.5 | 4.8553 | 0.06444 | 1342.6 | 1471.5 | 4.7680 |
| 50 | 0.08506 | 1364.9 | 1501.0 | 4.9510 | 0.06471 | 1344.5 | 1473.9 | 4.7754 |
| 60 | 0.08951 | 1389.3 | 1532.5 | 5.0472 | 0.06875 | 1372.3 | 1509.8 | 4.8848 |
| 70 | 0.09372 | 1412.3 | 1562.3 | 5.1351 | 0.07246 | 1397.8 | 1542.7 | 4.9821 |
| 80 | 0.09774 | 1434.3 | 1590.6 | 5.2167 | 0.07595 | 1421.6 | 1573.5 | 5.0707 |
| 100 | 0.10539 | 1476.2 | 1644.8 | 5.3659 | 0.08248 | 1466.1 | 1631.1 | 5.2294 |
| 120 | 0.11268 | 1516.6 | 1696.9 | 5.5018 | 0.08861 | 1508.3 | 1685.5 | 5.3714 |
| 140 | 0.11974 | 1556.4 | 1748.0 | 5.6286 | 0.09447 | 1549.3 | 1738.2 | 5.5022 |
| 160 | 0.12662 | 1596.1 | 1798.7 | 5.7485 | 0.10016 | 1589.9 | 1790.2 | 5.6251 |
| 180 | 0.13339 | 1636.1 | 1849.5 | 5.8631 | 0.10571 | 1630.6 | 1842.0 | 5.7420 |
| 200 | 0.14005 | 1676.5 | 1900.5 | 5.9734 | 0.11116 | 1671.6 | 1893.9 | 5.8540 |
| 220 | 0.14663 | 1717.4 | 1952.0 | 6.0800 | 0.11652 | 1713.1 | 1946.1 | 5.9621 |
| 240 | 0.15314 | 1759.0 | 2004.1 | 6.1834 | 0.12182 | 1755.2 | 1998.8 | 6.0668 |
| 260 | 0.15959 | 1801.3 | 2056.7 | 6.2839 | 0.12705 | 1797.9 | 2052.0 | 6.1685 |
| 280 | 0.16599 | 1844.3 | 2109.9 | 6.3819 | 0.13224 | 1841.3 | 2105.8 | 6.2675 |
| 300 | 0.17234 | 1888.0 | 2163.7 | 6.4775 | 0.13737 | 1885.4 | 2160.1 | 6.3641 |
| 320 | 0.17865 | 1932.4 | 2218.2 | 6.5710 | 0.14246 | 1930.2 | 2215.1 | 6.4583 |
| 340 | 0.18492 | 1977.5 | 2273.4 | 6.6624 | 0.14751 | 1975.6 | 2270.7 | 6.5505 |
| 360 | 0.19115 | 2023.3 | 2329.1 | 6.7519 | 0.15253 | 2021.8 | 2326.8 | 6.6406 |
| | 5000 kPa (88.90) | | | | 10000 kPa (125.20) | | | |
| Sat. | 0.02365 | 1323.2 | 1441.4 | 4.3454 | 0.00826 | 1206.8 | 1289.4 | 3.7587 |
| 100 | 0.02636 | 1369.7 | 1501.5 | 4.5091 | — | — | — | — |
| 120 | 0.03024 | 1435.1 | 1586.3 | 4.7306 | — | — | — | — |
| 140 | 0.03350 | 1489.8 | 1657.3 | 4.9068 | 0.01195 | 1341.8 | 1461.3 | 4.1839 |
| 160 | 0.03643 | 1539.5 | 1721.7 | 5.0591 | 0.01461 | 1432.2 | 1578.3 | 4.4610 |
| 180 | 0.03916 | 1586.9 | 1782.7 | 5.1968 | 0.01666 | 1500.6 | 1667.2 | 4.6617 |
| 200 | 0.04174 | 1633.1 | 1841.8 | 5.3245 | 0.01842 | 1560.3 | 1744.5 | 4.8287 |
| 220 | 0.04422 | 1678.9 | 1900.0 | 5.4450 | 0.02001 | 1615.8 | 1816.0 | 4.9767 |
| 240 | 0.04662 | 1724.8 | 1957.9 | 5.5600 | 0.02150 | 1669.2 | 1884.2 | 5.1123 |
| 260 | 0.04895 | 1770.9 | 2015.6 | 5.6704 | 0.02290 | 1721.6 | 1950.6 | 5.2392 |
| 280 | 0.05123 | 1817.4 | 2073.6 | 5.7771 | 0.02424 | 1773.6 | 2015.9 | 5.3596 |
| 300 | 0.05346 | 1864.5 | 2131.8 | 5.8805 | 0.02552 | 1825.5 | 2080.7 | 5.4746 |
| 320 | 0.05565 | 1912.1 | 2190.3 | 5.9809 | 0.02676 | 1877.6 | 2145.2 | 5.5852 |
| 340 | 0.05779 | 1960.3 | 2249.2 | 6.0786 | 0.02796 | 1930.0 | 2209.6 | 5.6921 |
| 360 | 0.05990 | 2009.1 | 2308.6 | 6.1738 | 0.02913 | 1982.8 | 2274.1 | 5.7955 |
| 380 | 0.06198 | 2058.5 | 2368.4 | 6.2668 | 0.03026 | 2036.1 | 2338.7 | 5.8960 |
| 400 | 0.06403 | 2108.4 | 2428.6 | 6.3576 | 0.03137 | 2089.8 | 2403.5 | 5.9937 |
| 420 | 0.06606 | 2159.0 | 2489.3 | 6.4464 | 0.03245 | 2143.9 | 2468.5 | 6.0888 |
| 440 | 0.06806 | 2210.1 | 2550.4 | 6.5334 | 0.03351 | 2198.5 | 2533.7 | 6.1815 |

*Thermodynamic Properties of R-12*

*Saturated R-12*

| Temp. (°C) | Press. (kPa) | SPECIFIC VOLUME, m³/kg | | | INTERNAL ENERGY, kJ/kg | | |
|---|---|---|---|---|---|---|---|
| | | Sat. Liquid $v_f$ | Evap. $v_{fg}$ | Sat. Vapor $v_g$ | Sat. Liquid $u_f$ | Evap. $u_{fg}$ | Sat. Vapor $u_g$ |
| −90 | 2.8 | 0.000608 | 4.41494 | 4.41555 | −43.29 | 177.20 | 133.91 |
| −80 | 6.2 | 0.000617 | 2.13773 | 2.13835 | −34.73 | 172.54 | 137.82 |
| −70 | 12.3 | 0.000627 | 1.12665 | 1.12728 | −26.14 | 167.94 | 141.81 |
| −60 | 22.6 | 0.000637 | 0.63727 | 0.63791 | −17.50 | 163.36 | 145.86 |
| −50 | 39.1 | 0.000648 | 0.38246 | 0.38310 | −8.80 | 158.76 | 149.95 |
| −45 | 50.4 | 0.000654 | 0.30203 | 0.30268 | −4.43 | 156.44 | 152.01 |
| −40 | 64.2 | 0.000659 | 0.24125 | 0.24191 | −0.04 | 154.11 | 154.07 |
| −35 | 80.7 | 0.000666 | 0.19473 | 0.19540 | 4.37 | 151.77 | 156.13 |
| −30 | 100.4 | 0.000672 | 0.15870 | 0.15937 | 8.79 | 149.40 | 158.19 |
| −29.8 | 101.3 | 0.000672 | 0.15736 | 0.15803 | 8.98 | 149.30 | 158.28 |
| −25 | 123.7 | 0.000679 | 0.13049 | 0.13117 | 13.24 | 147.01 | 160.25 |
| −20 | 150.9 | 0.000685 | 0.10816 | 0.10885 | 17.71 | 144.59 | 162.31 |
| −15 | 182.6 | 0.000693 | 0.09033 | 0.09102 | 22.20 | 142.15 | 164.35 |
| −10 | 219.1 | 0.000700 | 0.07595 | 0.07665 | 26.72 | 139.67 | 166.39 |
| −5 | 261.0 | 0.000708 | 0.06426 | 0.06496 | 31.26 | 137.16 | 168.42 |
| 0 | 308.6 | 0.000716 | 0.05467 | 0.05539 | 35.83 | 134.61 | 170.44 |
| 5 | 362.6 | 0.000724 | 0.04676 | 0.04749 | 40.43 | 132.01 | 172.44 |
| 10 | 423.3 | 0.000733 | 0.04018 | 0.04091 | 45.06 | 129.36 | 174.42 |
| 15 | 491.4 | 0.000743 | 0.03467 | 0.03541 | 49.73 | 126.65 | 176.38 |
| 20 | 567.3 | 0.000752 | 0.03003 | 0.03078 | 54.45 | 123.87 | 178.32 |
| 25 | 651.6 | 0.000763 | 0.02609 | 0.02685 | 59.21 | 121.03 | 180.23 |
| 30 | 744.9 | 0.000774 | 0.02273 | 0.02351 | 64.02 | 118.09 | 182.11 |
| 35 | 847.7 | 0.000786 | 0.01986 | 0.02064 | 68.88 | 115.06 | 183.95 |
| 40 | 960.7 | 0.000798 | 0.01737 | 0.01817 | 73.82 | 111.92 | 185.74 |
| 45 | 1084.3 | 0.000811 | 0.01522 | 0.01603 | 78.83 | 108.66 | 187.49 |
| 50 | 1219.3 | 0.000826 | 0.01334 | 0.01417 | 83.93 | 105.24 | 189.17 |
| 55 | 1366.3 | 0.000841 | 0.01170 | 0.01254 | 89.12 | 101.66 | 190.78 |
| 60 | 1525.9 | 0.000858 | 0.01025 | 0.01111 | 94.43 | 97.88 | 192.31 |
| 65 | 1698.8 | 0.000877 | 0.00897 | 0.00985 | 99.87 | 93.86 | 193.73 |
| 70 | 1885.8 | 0.000897 | 0.00783 | 0.00873 | 105.46 | 89.56 | 195.03 |
| 75 | 2087.5 | 0.000920 | 0.00680 | 0.00772 | 111.23 | 84.94 | 196.17 |
| 80 | 2304.6 | 0.000946 | 0.00588 | 0.00682 | 117.21 | 79.90 | 197.11 |
| 85 | 2538.0 | 0.000976 | 0.00503 | 0.00600 | 123.45 | 74.34 | 197.80 |
| 90 | 2788.5 | 0.001012 | 0.00425 | 0.00526 | 130.02 | 68.12 | 198.14 |
| 95 | 3056.9 | 0.001056 | 0.00351 | 0.00456 | 137.01 | 60.98 | 197.99 |
| 100 | 3344.1 | 0.001113 | 0.00279 | 0.00390 | 144.59 | 52.48 | 197.07 |
| 105 | 3650.9 | 0.001197 | 0.00205 | 0.00324 | 153.15 | 41.58 | 194.73 |
| 110 | 3978.5 | 0.001364 | 0.00110 | 0.00246 | 164.12 | 24.08 | 188.20 |
| 112.0 | 4116.8 | 0.001792 | 0 | 0.00179 | 176.06 | 0 | 176.06 |

**TABLE B.3.1** (*continued*)
*Saturated R-12*

| Temp. (°C) | Press. (kPa) | ENTHALPY, kJ/kg | | | ENTROPY, kJ/kg-K | | |
|---|---|---|---|---|---|---|---|
| | | Sat. Liquid $h_f$ | Evap. $h_{fg}$ | Sat. Vapor $h_g$ | Sat. Liquid $s_f$ | Evap. $s_{fg}$ | Sat. Vapor $s_g$ |
| −90 | 2.8 | −43.28 | 189.75 | 146.46 | −0.2086 | 1.0359 | 0.8273 |
| −80 | 6.2 | −34.72 | 185.74 | 151.02 | −0.1631 | 0.9616 | 0.7984 |
| −70 | 12.3 | −26.13 | 181.76 | 155.64 | −0.1198 | 0.8947 | 0.7749 |
| −60 | 22.6 | −17.49 | 177.77 | 160.29 | −0.0783 | 0.8340 | 0.7557 |
| −50 | 39.1 | −8.78 | 173.73 | 164.95 | −0.0384 | 0.7785 | 0.7401 |
| −45 | 50.4 | −4.40 | 171.68 | 167.28 | −0.0190 | 0.7524 | 0.7334 |
| −40 | 64.2 | 0 | 169.59 | 169.59 | 0 | 0.7274 | 0.7274 |
| −35 | 80.7 | 4.42 | 167.48 | 171.90 | 0.0187 | 0.7032 | 0.7219 |
| −30 | 100.4 | 8.86 | 165.34 | 174.20 | 0.0371 | 0.6799 | 0.7170 |
| −29.8 | 101.3 | 9.05 | 165.24 | 174.29 | 0.0379 | 0.6790 | 0.7168 |
| −25 | 123.7 | 13.33 | 163.15 | 176.48 | 0.0552 | 0.6574 | 0.7126 |
| −20 | 150.9 | 17.82 | 160.92 | 178.74 | 0.0731 | 0.6356 | 0.7087 |
| −15 | 182.6 | 22.33 | 158.64 | 180.97 | 0.0906 | 0.6145 | 0.7051 |
| −10 | 219.1 | 26.87 | 156.31 | 183.19 | 0.1080 | 0.5940 | 0.7019 |
| −5 | 261.0 | 31.45 | 153.93 | 185.37 | 0.1251 | 0.5740 | 0.6991 |
| 0 | 308.6 | 36.05 | 151.48 | 187.53 | 0.1420 | 0.5545 | 0.6965 |
| 5 | 362.6 | 40.69 | 148.96 | 189.65 | 0.1587 | 0.5355 | 0.6942 |
| 10 | 423.3 | 45.37 | 146.37 | 191.74 | 0.1752 | 0.5169 | 0.6921 |
| 15 | 491.4 | 50.10 | 143.68 | 193.78 | 0.1915 | 0.4986 | 0.6902 |
| 20 | 567.3 | 54.87 | 140.91 | 195.78 | 0.2078 | 0.4806 | 0.6884 |
| 25 | 651.6 | 59.70 | 138.03 | 197.73 | 0.2239 | 0.4629 | 0.6868 |
| 30 | 744.9 | 64.59 | 135.03 | 199.62 | 0.2399 | 0.4454 | 0.6853 |
| 35 | 847.7 | 69.55 | 131.90 | 201.45 | 0.2559 | 0.4280 | 0.6839 |
| 40 | 960.7 | 74.59 | 128.61 | 203.20 | 0.2718 | 0.4107 | 0.6825 |
| 45 | 1084.3 | 79.71 | 125.16 | 204.87 | 0.2877 | 0.3934 | 0.6811 |
| 50 | 1219.3 | 84.94 | 121.51 | 206.45 | 0.3037 | 0.3760 | 0.6797 |
| 55 | 1366.3 | 90.27 | 117.65 | 207.92 | 0.3197 | 0.3585 | 0.6782 |
| 60 | 1525.9 | 95.74 | 113.52 | 209.26 | 0.3358 | 0.3407 | 0.6765 |
| 65 | 1698.8 | 101.36 | 109.10 | 210.46 | 0.3521 | 0.3226 | 0.6747 |
| 70 | 1885.8 | 107.15 | 104.33 | 211.48 | 0.3686 | 0.3040 | 0.6726 |
| 75 | 2087.5 | 113.15 | 99.14 | 212.29 | 0.3854 | 0.2847 | 0.6702 |
| 80 | 2304.6 | 119.39 | 93.44 | 212.83 | 0.4027 | 0.2646 | 0.6672 |
| 85 | 2538.0 | 125.93 | 87.11 | 213.04 | 0.4204 | 0.2432 | 0.6636 |
| 90 | 2788.5 | 132.84 | 79.96 | 212.80 | 0.4389 | 0.2202 | 0.6590 |
| 95 | 3056.9 | 140.23 | 71.71 | 211.94 | 0.4583 | 0.1948 | 0.6531 |
| 100 | 3344.1 | 148.31 | 61.81 | 210.12 | 0.4793 | 0.1656 | 0.6449 |
| 105 | 3650.9 | 157.52 | 49.05 | 206.57 | 0.5028 | 0.1297 | 0.6325 |
| 110 | 3978.5 | 169.55 | 28.44 | 197.99 | 0.5333 | 0.0742 | 0.6076 |
| 112.0 | 4116.8 | 183.43 | 0 | 183.43 | 0.5689 | 0 | 0.5689 |

**TABLE B.3.2**
*Superheated R-12*

| Temp. (°C) | v (m³/kg) | u (kJ/kg) | h (kJ/kg) | s (kJ/kg-K) | v (m³/kg) | u (kJ/kg) | h (kJ/kg) | s (kJ/kg-K) |
|---|---|---|---|---|---|---|---|---|
| | 50 kPa (−45.18) | | | | 100 kPa (−33.10) | | | |
| Sat. | 0.30515 | 151.94 | 167.19 | 0.7336 | 0.15999 | 158.15 | 174.15 | 0.7171 |
| −30 | 0.32738 | 159.18 | 175.55 | 0.7691 | 0.16006 | 158.20 | 174.21 | 0.7174 |
| −20 | 0.34186 | 164.08 | 181.17 | 0.7917 | 0.16770 | 163.22 | 179.99 | 0.7406 |
| −10 | 0.35623 | 169.08 | 186.89 | 0.8139 | 0.17522 | 168.32 | 185.84 | 0.7633 |
| 0 | 0.37051 | 174.18 | 192.70 | 0.8356 | 0.18265 | 173.50 | 191.77 | 0.7854 |
| 10 | 0.38472 | 179.38 | 198.61 | 0.8568 | 0.18999 | 178.77 | 197.77 | 0.8070 |
| 20 | 0.39886 | 184.67 | 204.62 | 0.8776 | 0.19728 | 184.13 | 203.85 | 0.8281 |
| 30 | 0.41296 | 190.06 | 210.71 | 0.8981 | 0.20451 | 189.57 | 210.02 | 0.8488 |
| 40 | 0.42701 | 195.54 | 216.89 | 0.9181 | 0.21169 | 195.09 | 216.26 | 0.8691 |
| 50 | 0.44103 | 201.11 | 223.16 | 0.9378 | 0.21884 | 200.70 | 222.58 | 0.8889 |
| 60 | 0.45502 | 206.76 | 229.51 | 0.9572 | 0.22596 | 206.39 | 228.98 | 0.9084 |
| 70 | 0.46898 | 212.50 | 235.95 | 0.9762 | 0.23305 | 212.15 | 235.46 | 0.9276 |
| 80 | 0.48292 | 218.31 | 242.46 | 0.9949 | 0.24011 | 218.00 | 242.01 | 0.9464 |
| 90 | 0.49684 | 224.21 | 249.05 | 1.0133 | 0.24716 | 223.91 | 248.63 | 0.9649 |
| 100 | 0.51074 | 230.18 | 255.71 | 1.0314 | 0.25419 | 229.90 | 255.32 | 0.9831 |
| 110 | 0.52463 | 236.22 | 262.45 | 1.0493 | 0.26121 | 235.96 | 262.08 | 1.0009 |
| 120 | 0.53851 | 242.33 | 269.26 | 1.0668 | 0.26821 | 242.09 | 268.91 | 1.0185 |
| | 200 kPa (−12.53) | | | | 400 kPa (8.15) | | | |
| Sat. | 0.08354 | 165.36 | 182.07 | 0.7035 | 0.04321 | 173.69 | 190.97 | 0.6928 |
| 0 | 0.08861 | 172.08 | 189.80 | 0.7325 | — | — | — | — |
| 10 | 0.09255 | 177.51 | 196.02 | 0.7548 | 0.04363 | 174.76 | 192.21 | 0.6972 |
| 20 | 0.09642 | 183.00 | 202.28 | 0.7766 | 0.04584 | 180.57 | 198.91 | 0.7204 |
| 30 | 0.10023 | 188.55 | 208.60 | 0.7978 | 0.04797 | 186.39 | 205.58 | 0.7428 |
| 40 | 0.10399 | 194.17 | 214.97 | 0.8184 | 0.05005 | 192.23 | 212.25 | 0.7645 |
| 50 | 0.10771 | 199.86 | 221.41 | 0.8387 | 0.05207 | 198.11 | 218.94 | 0.7855 |
| 60 | 0.11140 | 205.62 | 227.90 | 0.8585 | 0.05406 | 204.03 | 225.65 | 0.8060 |
| 70 | 0.11506 | 211.45 | 234.46 | 0.8779 | 0.05601 | 209.99 | 232.40 | 0.8259 |
| 80 | 0.11869 | 217.35 | 241.09 | 0.8969 | 0.05794 | 216.01 | 239.19 | 0.8454 |
| 90 | 0.12230 | 223.31 | 247.77 | 0.9156 | 0.05985 | 222.08 | 246.02 | 0.8645 |
| 100 | 0.12590 | 229.34 | 254.53 | 0.9339 | 0.06173 | 228.20 | 252.89 | 0.8831 |
| 110 | 0.12948 | 235.44 | 261.34 | 0.9519 | 0.06360 | 234.37 | 259.81 | 0.9015 |
| 120 | 0.13305 | 241.60 | 268.21 | 0.9696 | 0.06546 | 240.60 | 266.79 | 0.9194 |
| 130 | 0.13661 | 247.82 | 275.15 | 0.9870 | 0.06730 | 246.89 | 273.81 | 0.9370 |
| 140 | 0.14016 | 254.11 | 282.14 | 1.0042 | 0.06913 | 253.22 | 280.88 | 0.9544 |
| 150 | 0.14370 | 260.45 | 289.19 | 1.0210 | 0.07095 | 259.61 | 287.99 | 0.9714 |

TABLE B.3.2  (*continued*)
Superheated R-12

| Temp. (°C) | $v$ (m³/kg) | $u$ (kJ/kg) | $h$ (kJ/kg) | $s$ (kJ/kg-K) | $v$ (m³/kg) | $u$ (kJ/kg) | $h$ (kJ/kg) | $s$ (kJ/kg-K) |
|---|---|---|---|---|---|---|---|---|
| | 500 kPa (15.60) | | | | 1000 kPa (41.64) | | | |
| Sat. | 0.03482 | 176.62 | 194.03 | 0.6899 | 0.01744 | 186.32 | 203.76 | 0.6820 |
| 30 | 0.03746 | 185.23 | 203.96 | 0.7235 | — | — | — | — |
| 40 | 0.03921 | 191.20 | 210.81 | 0.7457 | — | — | — | — |
| 50 | 0.04091 | 197.19 | 217.64 | 0.7672 | 0.01837 | 191.95 | 210.32 | 0.7026 |
| 60 | 0.04257 | 203.20 | 224.48 | 0.7881 | 0.01941 | 198.56 | 217.97 | 0.7259 |
| 70 | 0.04418 | 209.24 | 231.33 | 0.8083 | 0.02040 | 205.09 | 225.49 | 0.7481 |
| 80 | 0.04577 | 215.32 | 238.21 | 0.8281 | 0.02134 | 211.57 | 232.91 | 0.7695 |
| 90 | 0.04734 | 221.44 | 245.11 | 0.8473 | 0.02225 | 218.03 | 240.28 | 0.7900 |
| 100 | 0.04889 | 227.61 | 252.05 | 0.8662 | 0.02313 | 224.48 | 247.61 | 0.8100 |
| 110 | 0.05041 | 233.83 | 259.03 | 0.8847 | 0.02399 | 230.94 | 254.93 | 0.8293 |
| 120 | 0.05193 | 240.09 | 266.06 | 0.9028 | 0.02483 | 237.41 | 262.25 | 0.8482 |
| 130 | 0.05343 | 246.41 | 273.12 | 0.9205 | 0.02566 | 243.91 | 269.57 | 0.8665 |
| 140 | 0.05492 | 252.77 | 280.23 | 0.9379 | 0.02647 | 250.43 | 276.90 | 0.8845 |
| 150 | 0.05640 | 259.19 | 287.39 | 0.9550 | 0.02728 | 256.98 | 284.26 | 0.9021 |
| 160 | 0.05788 | 265.65 | 294.59 | 0.9718 | 0.02807 | 263.56 | 291.63 | 0.9193 |
| 170 | 0.05934 | 272.16 | 301.83 | 0.9884 | 0.02885 | 270.18 | 299.04 | 0.9362 |
| 180 | 0.06080 | 278.72 | 309.12 | 1.0046 | 0.02963 | 276.84 | 306.47 | 0.9528 |
| | 1500 kPa (59.22) | | | | 2000 kPa (72.88) | | | |
| Sat. | 0.01132 | 192.08 | 209.06 | 0.6768 | 0.00813 | 195.70 | 211.97 | 0.6713 |
| 70 | 0.01226 | 200.05 | 218.44 | 0.7046 | — | — | — | — |
| 80 | 0.01305 | 207.16 | 226.73 | 0.7284 | 0.00870 | 201.61 | 219.02 | 0.6914 |
| 90 | 0.01377 | 214.11 | 234.77 | 0.7508 | 0.00941 | 209.41 | 228.23 | 0.7171 |
| 100 | 0.01446 | 220.95 | 242.65 | 0.7722 | 0.01003 | 216.87 | 236.94 | 0.7408 |
| 110 | 0.01512 | 227.73 | 250.41 | 0.7928 | 0.01061 | 224.11 | 245.34 | 0.7630 |
| 120 | 0.01575 | 234.47 | 258.10 | 0.8126 | 0.01116 | 231.21 | 253.53 | 0.7841 |
| 130 | 0.01636 | 241.20 | 265.74 | 0.8318 | 0.01168 | 238.22 | 261.58 | 0.8043 |
| 140 | 0.01696 | 247.91 | 273.35 | 0.8504 | 0.01217 | 245.18 | 269.53 | 0.8238 |
| 150 | 0.01754 | 254.63 | 280.94 | 0.8686 | 0.01265 | 252.10 | 277.41 | 0.8426 |
| 160 | 0.01811 | 261.36 | 288.52 | 0.8863 | 0.01312 | 259.00 | 285.24 | 0.8609 |
| 170 | 0.01867 | 268.10 | 296.11 | 0.9036 | 0.01357 | 265.90 | 293.04 | 0.8787 |
| 180 | 0.01922 | 274.87 | 303.70 | 0.9205 | 0.01401 | 272.79 | 300.82 | 0.8961 |
| 190 | 0.01977 | 281.65 | 311.31 | 0.9371 | 0.01445 | 279.69 | 308.59 | 0.9131 |
| 200 | 0.02031 | 288.47 | 318.93 | 0.9534 | 0.01488 | 286.61 | 316.36 | 0.9297 |
| 210 | 0.02084 | 295.31 | 326.58 | 0.9694 | 0.01530 | 293.54 | 324.14 | 0.9459 |
| 220 | 0.02137 | 302.19 | 334.24 | 0.9851 | 0.01572 | 300.49 | 331.92 | 0.9619 |

TABLE B.4
*Thermodynamic Properties of R-22*

TABLE B.4.1
*Saturated R-22*

| Temp. (°C) | Press. (kPa) | Specific Volume, m³/kg | | | Internal Energy, kJ/kg | | |
|---|---|---|---|---|---|---|---|
| | | Sat. Liquid $v_f$ | Evap. $v_{fg}$ | Sat. Vapor $v_g$ | Sat. Liquid $u_f$ | Evap. $u_{fg}$ | Sat. Vapor $u_g$ |
| −70 | 20.5 | 0.000670 | 0.94027 | 0.94094 | −30.62 | 230.13 | 199.51 |
| −65 | 28.0 | 0.000676 | 0.70480 | 0.70547 | −25.68 | 227.21 | 201.54 |
| −60 | 37.5 | 0.000682 | 0.53647 | 0.53715 | −20.68 | 224.25 | 203.57 |
| −55 | 49.5 | 0.000689 | 0.41414 | 0.41483 | −15.62 | 221.21 | 205.59 |
| −50 | 64.4 | 0.000695 | 0.32386 | 0.32456 | −10.50 | 218.11 | 207.61 |
| −45 | 82.7 | 0.000702 | 0.25629 | 0.25699 | −5.32 | 214.94 | 209.62 |
| −40.8 | 101.3 | 0.000708 | 0.21191 | 0.21261 | −0.87 | 212.18 | 211.31 |
| −40 | 104.9 | 0.000709 | 0.20504 | 0.20575 | −0.07 | 211.68 | 211.60 |
| −35 | 131.7 | 0.000717 | 0.16568 | 0.16640 | 5.23 | 208.34 | 213.57 |
| −30 | 163.5 | 0.000725 | 0.13512 | 0.13584 | 10.61 | 204.91 | 215.52 |
| −25 | 201.0 | 0.000733 | 0.11113 | 0.11186 | 16.04 | 201.39 | 217.44 |
| −20 | 244.8 | 0.000741 | 0.09210 | 0.09284 | 21.55 | 197.78 | 219.32 |
| −15 | 295.7 | 0.000750 | 0.07688 | 0.07763 | 27.11 | 194.07 | 221.18 |
| −10 | 354.3 | 0.000759 | 0.06458 | 0.06534 | 32.74 | 190.25 | 222.99 |
| −5 | 421.3 | 0.000768 | 0.05457 | 0.05534 | 38.44 | 186.33 | 224.77 |
| 0 | 497.6 | 0.000778 | 0.04636 | 0.04714 | 44.20 | 182.30 | 226.50 |
| 5 | 583.8 | 0.000789 | 0.03957 | 0.04036 | 50.03 | 178.15 | 228.17 |
| 10 | 680.7 | 0.000800 | 0.03391 | 0.03471 | 55.92 | 173.87 | 229.79 |
| 15 | 789.1 | 0.000812 | 0.02918 | 0.02999 | 61.88 | 169.47 | 231.35 |
| 20 | 909.9 | 0.000824 | 0.02518 | 0.02600 | 67.92 | 164.92 | 232.85 |
| 25 | 1043.9 | 0.000838 | 0.02179 | 0.02262 | 74.04 | 160.22 | 234.26 |
| 30 | 1191.9 | 0.000852 | 0.01889 | 0.01974 | 80.23 | 155.35 | 235.59 |
| 35 | 1354.8 | 0.000867 | 0.01640 | 0.01727 | 86.53 | 150.30 | 236.82 |
| 40 | 1533.5 | 0.000884 | 0.01425 | 0.01514 | 92.92 | 145.02 | 237.94 |
| 45 | 1729.0 | 0.000902 | 0.01238 | 0.01328 | 99.42 | 139.50 | 238.93 |
| 50 | 1942.3 | 0.000922 | 0.01075 | 0.01167 | 106.06 | 133.70 | 239.76 |
| 55 | 2174.4 | 0.000944 | 0.00931 | 0.01025 | 112.85 | 127.56 | 240.41 |
| 60 | 2426.6 | 0.000969 | 0.00803 | 0.00900 | 119.83 | 121.01 | 240.84 |
| 65 | 2699.9 | 0.000997 | 0.00689 | 0.00789 | 127.04 | 113.94 | 240.98 |
| 70 | 2995.9 | 0.001030 | 0.00586 | 0.00689 | 134.54 | 106.22 | 240.76 |
| 75 | 3316.1 | 0.001069 | 0.00491 | 0.00598 | 142.44 | 97.61 | 240.05 |
| 80 | 3662.3 | 0.001118 | 0.00403 | 0.00515 | 150.92 | 87.71 | 238.63 |
| 85 | 4036.8 | 0.001183 | 0.00317 | 0.00436 | 160.32 | 75.78 | 236.10 |
| 90 | 4442.5 | 0.001282 | 0.00228 | 0.00356 | 171.51 | 59.90 | 231.41 |
| 95 | 4883.5 | 0.001521 | 0.00103 | 0.00255 | 188.93 | 29.89 | 218.83 |
| 96.0 | 4969.0 | 0.001906 | 0 | 0.00191 | 203.07 | 0 | 203.07 |

TABLE **B.4.1**  (*continued*)
*Saturated R-22*

| Temp. (°C) | Press. (kPa) | ENTHALPY, kJ/kg | | | ENTROPY, kJ/kg-K | | |
|---|---|---|---|---|---|---|---|
| | | Sat. Liquid $h_f$ | Evap. $h_{fg}$ | Sat. Vapor $h_g$ | Sat. Liquid $s_f$ | Evap. $s_{fg}$ | Sat. Vapor $s_g$ |
| −70 | 20.5 | −30.61 | 249.43 | 218.82 | −0.1401 | 1.2277 | 1.0876 |
| −65 | 28.0 | −25.66 | 246.93 | 221.27 | −0.1161 | 1.1862 | 1.0701 |
| −60 | 37.5 | −20.65 | 244.35 | 223.70 | −0.0924 | 1.1463 | 1.0540 |
| −55 | 49.5 | −15.59 | 241.70 | 226.12 | −0.0689 | 1.1079 | 1.0390 |
| −50 | 64.4 | −10.46 | 238.96 | 228.51 | −0.0457 | 1.0708 | 1.0251 |
| −45 | 82.7 | −5.26 | 236.13 | 230.87 | −0.0227 | 1.0349 | 1.0122 |
| −40.8 | 101.3 | −0.80 | 233.65 | 232.85 | −0.0034 | 1.0053 | 1.0019 |
| −40 | 104.9 | 0 | 233.20 | 233.20 | 0 | 1.0002 | 1.0002 |
| −35 | 131.7 | 5.33 | 230.16 | 235.48 | 0.0225 | 0.9664 | 0.9889 |
| −30 | 163.5 | 10.73 | 227.00 | 237.73 | 0.0449 | 0.9335 | 0.9784 |
| −25 | 201.0 | 16.19 | 223.73 | 239.92 | 0.0670 | 0.9015 | 0.9685 |
| −20 | 244.8 | 21.73 | 220.33 | 242.06 | 0.0890 | 0.8703 | 0.9593 |
| −15 | 295.7 | 27.33 | 216.80 | 244.13 | 0.1107 | 0.8398 | 0.9505 |
| −10 | 354.3 | 33.01 | 213.13 | 246.14 | 0.1324 | 0.8099 | 0.9422 |
| −5 | 421.3 | 38.76 | 209.32 | 248.09 | 0.1538 | 0.7806 | 0.9344 |
| 0 | 497.6 | 44.59 | 205.36 | 249.95 | 0.1751 | 0.7518 | 0.9269 |
| 5 | 583.8 | 50.49 | 201.25 | 251.73 | 0.1963 | 0.7235 | 0.9197 |
| 10 | 680.7 | 56.46 | 196.96 | 253.42 | 0.2173 | 0.6956 | 0.9129 |
| 15 | 789.1 | 62.52 | 192.49 | 255.02 | 0.2382 | 0.6680 | 0.9062 |
| 20 | 909.9 | 68.67 | 187.84 | 256.51 | 0.2590 | 0.6407 | 0.8997 |
| 25 | 1043.9 | 74.91 | 182.97 | 257.88 | 0.2797 | 0.6137 | 0.8934 |
| 30 | 1191.9 | 81.25 | 177.87 | 259.12 | 0.3004 | 0.5867 | 0.8871 |
| 35 | 1354.8 | 87.70 | 172.52 | 260.22 | 0.3210 | 0.5598 | 0.8809 |
| 40 | 1533.5 | 94.27 | 166.88 | 261.15 | 0.3417 | 0.5329 | 0.8746 |
| 45 | 1729.0 | 100.98 | 160.91 | 261.90 | 0.3624 | 0.5058 | 0.8682 |
| 50 | 1942.3 | 107.85 | 154.58 | 262.43 | 0.3832 | 0.4783 | 0.8615 |
| 55 | 2174.4 | 114.91 | 147.80 | 262.71 | 0.4042 | 0.4504 | 0.8546 |
| 60 | 2426.6 | 122.18 | 140.50 | 262.68 | 0.4255 | 0.4217 | 0.8472 |
| 65 | 2699.9 | 129.73 | 132.55 | 262.28 | 0.4472 | 0.3920 | 0.8391 |
| 70 | 2995.9 | 137.63 | 123.77 | 261.40 | 0.4695 | 0.3607 | 0.8302 |
| 75 | 3316.1 | 145.99 | 113.90 | 259.89 | 0.4927 | 0.3272 | 0.8198 |
| 80 | 3662.3 | 155.01 | 102.47 | 257.49 | 0.5173 | 0.2902 | 0.8075 |
| 85 | 4036.8 | 165.09 | 88.60 | 253.69 | 0.5445 | 0.2474 | 0.7918 |
| 90 | 4442.5 | 177.20 | 70.04 | 247.24 | 0.5767 | 0.1929 | 0.7695 |
| 95 | 4883.5 | 196.36 | 34.93 | 231.28 | 0.6273 | 0.0949 | 0.7222 |
| 96.0 | 4969.0 | 212.54 | 0 | 212.54 | 0.6708 | 0 | 0.6708 |

**TABLE B.4.2**
*Superheated R-22*

| Temp. (°C) | v (m³/kg) | u (kJ/kg) | h (kJ/kg) | s (kJ/kg-K) | v (m³/kg) | u (kJ/kg) | h (kJ/kg) | s (kJ/kg-K) |
|---|---|---|---|---|---|---|---|---|
| | 50 kPa (−54.80) | | | | 100 kPa (−41.03) | | | |
| Sat. | 0.41077 | 205.67 | 226.21 | 1.0384 | 0.21525 | 211.19 | 232.72 | 1.0026 |
| −40 | 0.44063 | 212.69 | 234.72 | 1.0762 | 0.21633 | 211.70 | 233.34 | 1.0052 |
| −30 | 0.46064 | 217.57 | 240.60 | 1.1008 | 0.22675 | 216.68 | 239.36 | 1.0305 |
| −20 | 0.48054 | 222.56 | 246.59 | 1.1250 | 0.23706 | 221.76 | 245.47 | 1.0551 |
| −10 | 0.50036 | 227.66 | 252.68 | 1.1485 | 0.24728 | 226.94 | 251.67 | 1.0791 |
| 0 | 0.52010 | 232.87 | 258.87 | 1.1717 | 0.25742 | 232.21 | 257.96 | 1.1026 |
| 10 | 0.53977 | 238.19 | 265.18 | 1.1943 | 0.26748 | 237.60 | 264.35 | 1.1256 |
| 20 | 0.55939 | 243.62 | 271.59 | 1.2166 | 0.27750 | 243.08 | 270.83 | 1.1481 |
| 30 | 0.57897 | 249.17 | 278.12 | 1.2385 | 0.28747 | 248.67 | 277.42 | 1.1702 |
| 40 | 0.59851 | 254.82 | 284.74 | 1.2600 | 0.29739 | 254.36 | 284.10 | 1.1919 |
| 50 | 0.61801 | 260.58 | 291.48 | 1.2811 | 0.30729 | 260.16 | 290.89 | 1.2132 |
| 60 | 0.63749 | 266.44 | 298.32 | 1.3020 | 0.31715 | 266.06 | 297.77 | 1.2342 |
| 70 | 0.65694 | 272.42 | 305.26 | 1.3225 | 0.32699 | 272.06 | 304.76 | 1.2548 |
| 80 | 0.67636 | 278.50 | 312.31 | 1.3428 | 0.33680 | 278.16 | 311.84 | 1.2752 |
| 90 | 0.69577 | 284.68 | 319.47 | 1.3627 | 0.34660 | 284.37 | 319.03 | 1.2952 |
| 100 | 0.71516 | 290.96 | 326.72 | 1.3824 | 0.35637 | 290.67 | 326.31 | 1.3150 |
| 110 | 0.73454 | 297.34 | 334.07 | 1.4019 | 0.36614 | 297.07 | 333.69 | 1.3345 |
| | 150 kPa (−32.02) | | | | 200 kPa (−25.12) | | | |
| Sat. | 0.14727 | 214.74 | 236.83 | 0.9826 | 0.11237 | 217.39 | 239.87 | 0.9688 |
| −20 | 0.15585 | 220.94 | 244.32 | 1.0129 | 0.11520 | 220.10 | 243.14 | 0.9818 |
| −10 | 0.16288 | 226.20 | 250.63 | 1.0373 | 0.12065 | 225.44 | 249.57 | 1.0068 |
| 0 | 0.16982 | 231.55 | 257.02 | 1.0612 | 0.12600 | 230.87 | 256.07 | 1.0310 |
| 10 | 0.17670 | 236.99 | 263.50 | 1.0844 | 0.13129 | 236.38 | 262.63 | 1.0546 |
| 20 | 0.18352 | 242.53 | 270.06 | 1.1072 | 0.13651 | 241.97 | 269.27 | 1.0776 |
| 30 | 0.19028 | 248.17 | 276.71 | 1.1295 | 0.14168 | 247.66 | 275.99 | 1.1002 |
| 40 | 0.19701 | 253.90 | 283.45 | 1.1514 | 0.14681 | 253.43 | 282.80 | 1.1222 |
| 50 | 0.20370 | 259.73 | 290.29 | 1.1729 | 0.15190 | 259.31 | 289.69 | 1.1439 |
| 60 | 0.21036 | 265.67 | 297.22 | 1.1940 | 0.15696 | 265.27 | 296.66 | 1.1652 |
| 70 | 0.21700 | 271.70 | 304.25 | 1.2148 | 0.16200 | 271.33 | 303.73 | 1.1861 |
| 80 | 0.22361 | 277.83 | 311.37 | 1.2353 | 0.16701 | 277.49 | 310.89 | 1.2066 |
| 90 | 0.23020 | 284.05 | 318.58 | 1.2554 | 0.17200 | 283.74 | 318.14 | 1.2269 |
| 100 | 0.23678 | 290.38 | 325.90 | 1.2753 | 0.17697 | 290.09 | 325.48 | 1.2468 |
| 110 | 0.24333 | 296.80 | 333.30 | 1.2948 | 0.18193 | 296.53 | 332.91 | 1.2665 |
| 120 | 0.24988 | 303.32 | 340.80 | 1.3142 | 0.18688 | 303.06 | 340.44 | 1.2858 |
| 130 | 0.25642 | 309.93 | 348.39 | 1.3332 | 0.19181 | 309.69 | 348.05 | 1.3050 |

TABLE B.4.2  (*continued*)
*Superheated R-22*

| Temp. (°C) | v (m³/kg) | u (kJ/kg) | h (kJ/kg) | s (kJ/kg-K) | v (m³/kg) | u (kJ/kg) | h (kJ/kg) | s (kJ/kg-K) |
|---|---|---|---|---|---|---|---|---|
| | 300 kPa (−14.61) | | | | 400 kPa (−6.52) | | | |
| Sat. | 0.07657 | 221.32 | 244.29 | 0.9499 | 0.05817 | 224.23 | 247.50 | 0.9367 |
| −10 | 0.07834 | 223.88 | 247.38 | 0.9617 | — | — | — | — |
| 0 | 0.08213 | 229.47 | 254.10 | 0.9868 | 0.06013 | 228.00 | 252.05 | 0.9536 |
| 10 | 0.08583 | 235.11 | 260.86 | 1.0111 | 0.06306 | 233.80 | 259.02 | 0.9787 |
| 20 | 0.08947 | 240.83 | 267.67 | 1.0347 | 0.06591 | 239.64 | 266.01 | 1.0029 |
| 30 | 0.09305 | 246.62 | 274.53 | 1.0577 | 0.06871 | 245.55 | 273.03 | 1.0265 |
| 40 | 0.09659 | 252.48 | 281.46 | 1.0802 | 0.07146 | 251.51 | 280.09 | 1.0494 |
| 50 | 0.10009 | 258.43 | 288.46 | 1.1022 | 0.07416 | 257.54 | 287.21 | 1.0717 |
| 60 | 0.10355 | 264.47 | 295.54 | 1.1238 | 0.07683 | 263.65 | 294.39 | 1.0936 |
| 70 | 0.10699 | 270.59 | 302.69 | 1.1449 | 0.07947 | 269.84 | 301.63 | 1.1150 |
| 80 | 0.11040 | 276.80 | 309.92 | 1.1657 | 0.08209 | 276.11 | 308.94 | 1.1361 |
| 90 | 0.11379 | 283.10 | 317.24 | 1.1861 | 0.08468 | 282.46 | 316.33 | 1.1567 |
| 100 | 0.11716 | 289.49 | 324.64 | 1.2062 | 0.08725 | 288.89 | 323.80 | 1.1770 |
| 110 | 0.12052 | 295.97 | 332.13 | 1.2260 | 0.08981 | 295.41 | 331.34 | 1.1969 |
| 120 | 0.12387 | 302.54 | 339.70 | 1.2455 | 0.09236 | 302.02 | 338.96 | 1.2165 |
| 130 | 0.12720 | 309.20 | 347.36 | 1.2648 | 0.09489 | 308.71 | 346.66 | 1.2359 |
| 140 | 0.13052 | 315.95 | 355.10 | 1.2837 | 0.09741 | 315.48 | 354.45 | 1.2550 |
| | 500 kPa (0.15) | | | | 600 kPa (5.88) | | | |
| Sat. | 0.04692 | 226.55 | 250.00 | 0.9267 | 0.03929 | 228.46 | 252.04 | 0.9185 |
| 10 | 0.04936 | 232.43 | 257.11 | 0.9522 | 0.04018 | 231.00 | 255.11 | 0.9295 |
| 20 | 0.05175 | 238.42 | 264.30 | 0.9772 | 0.04228 | 237.15 | 262.52 | 0.9552 |
| 30 | 0.05408 | 244.44 | 271.48 | 1.0013 | 0.04431 | 243.30 | 269.89 | 0.9799 |
| 40 | 0.05636 | 250.51 | 278.69 | 1.0247 | 0.04628 | 249.48 | 277.25 | 1.0038 |
| 50 | 0.05859 | 256.63 | 285.93 | 1.0474 | 0.04820 | 255.70 | 284.62 | 1.0270 |
| 60 | 0.06079 | 262.82 | 293.22 | 1.0696 | 0.05008 | 261.97 | 292.02 | 1.0495 |
| 70 | 0.06295 | 269.08 | 300.55 | 1.0913 | 0.05193 | 268.30 | 299.46 | 1.0715 |
| 80 | 0.06509 | 275.40 | 307.95 | 1.1126 | 0.05375 | 274.69 | 306.94 | 1.0930 |
| 90 | 0.06721 | 281.81 | 315.41 | 1.1334 | 0.05555 | 281.14 | 314.48 | 1.1140 |
| 100 | 0.06930 | 288.29 | 322.94 | 1.1539 | 0.05733 | 287.67 | 322.07 | 1.1347 |
| 110 | 0.07138 | 294.85 | 330.54 | 1.1740 | 0.05909 | 294.27 | 329.73 | 1.1549 |
| 120 | 0.07345 | 301.49 | 338.21 | 1.1937 | 0.06084 | 300.95 | 337.46 | 1.1748 |
| 130 | 0.07550 | 308.21 | 345.96 | 1.2132 | 0.06258 | 307.71 | 345.26 | 1.1944 |
| 140 | 0.07755 | 315.02 | 353.79 | 1.2324 | 0.06430 | 314.54 | 353.12 | 1.2137 |
| 150 | 0.07958 | 321.90 | 361.69 | 1.2513 | 0.06601 | 321.46 | 361.07 | 1.2327 |
| 160 | 0.08160 | 328.87 | 369.67 | 1.2699 | 0.06772 | 328.45 | 369.08 | 1.2514 |

**TABLE B.4.2** (*continued*)
***Superheated R-22***

| Temp. (°C) | v (m³/kg) | u (kJ/kg) | h (kJ/kg) | s (kJ/kg-K) | v (m³/kg) | u (kJ/kg) | h (kJ/kg) | s (kJ/kg-K) |
|---|---|---|---|---|---|---|---|---|
| | 800 kPa (15.47) | | | | 1000 kPa (23.42) | | | |
| Sat. | 0.02958 | 231.50 | 255.16 | 0.9056 | 0.02364 | 233.82 | 257.46 | 0.8954 |
| 20 | 0.03037 | 234.44 | 258.74 | 0.9179 | — | — | — | — |
| 30 | 0.03203 | 240.91 | 266.53 | 0.9440 | 0.02460 | 238.31 | 262.91 | 0.9136 |
| 40 | 0.03363 | 247.34 | 274.24 | 0.9690 | 0.02599 | 245.05 | 271.04 | 0.9400 |
| 50 | 0.03518 | 253.77 | 281.91 | 0.9931 | 0.02732 | 251.72 | 279.05 | 0.9651 |
| 60 | 0.03667 | 260.21 | 289.55 | 1.0164 | 0.02860 | 258.37 | 286.97 | 0.9893 |
| 70 | 0.03814 | 266.69 | 297.20 | 1.0391 | 0.02984 | 265.02 | 294.86 | 1.0126 |
| 80 | 0.03957 | 273.21 | 304.87 | 1.0611 | 0.03104 | 271.69 | 302.73 | 1.0352 |
| 90 | 0.04097 | 279.79 | 312.57 | 1.0826 | 0.03221 | 278.39 | 310.60 | 1.0572 |
| 100 | 0.04236 | 286.42 | 320.30 | 1.1036 | 0.03336 | 285.12 | 318.49 | 1.0786 |
| 110 | 0.04373 | 293.11 | 328.09 | 1.1242 | 0.03449 | 291.91 | 326.41 | 1.0996 |
| 120 | 0.04508 | 299.86 | 335.93 | 1.1444 | 0.03561 | 298.75 | 334.36 | 1.1200 |
| 130 | 0.04641 | 306.69 | 343.82 | 1.1642 | 0.03671 | 305.65 | 342.36 | 1.1401 |
| 140 | 0.04774 | 313.59 | 351.78 | 1.1837 | 0.03780 | 312.61 | 350.41 | 1.1599 |
| 150 | 0.04905 | 320.56 | 359.80 | 1.2029 | 0.03887 | 319.64 | 358.51 | 1.1792 |
| 160 | 0.05036 | 327.60 | 367.89 | 1.2218 | 0.03994 | 326.74 | 366.68 | 1.1983 |
| 170 | 0.05166 | 334.72 | 376.04 | 1.2404 | 0.04100 | 333.90 | 374.90 | 1.2171 |
| | 1200 kPa (30.26) | | | | 1400 kPa (36.31) | | | |
| Sat. | 0.01960 | 235.66 | 259.18 | 0.8868 | 0.01668 | 237.12 | 260.48 | 0.8792 |
| 40 | 0.02085 | 242.58 | 267.60 | 0.9141 | 0.01712 | 239.89 | 263.86 | 0.8901 |
| 50 | 0.02205 | 249.55 | 276.01 | 0.9405 | 0.01825 | 247.22 | 272.77 | 0.9181 |
| 60 | 0.02319 | 256.43 | 284.26 | 0.9657 | 0.01930 | 254.38 | 281.40 | 0.9444 |
| 70 | 0.02428 | 263.28 | 292.42 | 0.9898 | 0.02029 | 261.45 | 289.86 | 0.9694 |
| 80 | 0.02534 | 270.10 | 300.51 | 1.0131 | 0.02125 | 268.45 | 298.20 | 0.9934 |
| 90 | 0.02636 | 276.94 | 308.57 | 1.0356 | 0.02217 | 275.44 | 306.47 | 1.0165 |
| 100 | 0.02736 | 283.79 | 316.62 | 1.0574 | 0.02306 | 282.42 | 314.70 | 1.0388 |
| 110 | 0.02833 | 290.68 | 324.68 | 1.0788 | 0.02393 | 289.42 | 322.92 | 1.0606 |
| 120 | 0.02929 | 297.61 | 332.76 | 1.0996 | 0.02477 | 296.44 | 331.13 | 1.0817 |
| 130 | 0.03024 | 304.59 | 340.87 | 1.1199 | 0.02561 | 303.50 | 339.35 | 1.1024 |
| 140 | 0.03117 | 311.62 | 349.02 | 1.1399 | 0.02643 | 310.61 | 347.60 | 1.1226 |
| 150 | 0.03208 | 318.71 | 357.21 | 1.1595 | 0.02723 | 317.76 | 355.89 | 1.1424 |
| 160 | 0.03299 | 325.86 | 365.45 | 1.1787 | 0.02803 | 324.97 | 364.21 | 1.1618 |
| 170 | 0.03389 | 333.07 | 373.74 | 1.1977 | 0.02882 | 332.23 | 372.57 | 1.1809 |
| 180 | 0.03479 | 340.35 | 382.09 | 1.2163 | 0.02960 | 339.55 | 380.99 | 1.1997 |
| 190 | 0.03567 | 347.69 | 390.50 | 1.2346 | 0.03037 | 346.94 | 389.45 | 1.2182 |

TABLE B.4.2  (*continued*)
*Superheated R-22*

| Temp. (°C) | $v$ (m³/kg) | $u$ (kJ/kg) | $h$ (kJ/kg) | $s$ (kJ/kg-K) | $v$ (m³/kg) | $u$ (kJ/kg) | $h$ (kJ/kg) | $s$ (kJ/kg-K) |
|---|---|---|---|---|---|---|---|---|
| | 1600 kPa (41.75) | | | | 2000 kPa (51.28) | | | |
| Sat. | 0.01446 | 238.30 | 261.43 | 0.8724 | 0.01129 | 239.95 | 262.53 | 0.8598 |
| 50 | 0.01535 | 244.70 | 269.26 | 0.8969 | — | — | — | — |
| 60 | 0.01635 | 252.20 | 278.36 | 0.9246 | 0.01213 | 247.29 | 271.56 | 0.8873 |
| 70 | 0.01728 | 259.52 | 287.17 | 0.9507 | 0.01301 | 255.29 | 281.31 | 0.9161 |
| 80 | 0.01817 | 266.73 | 295.80 | 0.9755 | 0.01381 | 263.02 | 290.64 | 0.9429 |
| 90 | 0.01901 | 273.88 | 304.30 | 0.9992 | 0.01456 | 270.57 | 299.70 | 0.9682 |
| 100 | 0.01983 | 281.00 | 312.73 | 1.0221 | 0.01528 | 278.02 | 308.57 | 0.9923 |
| 110 | 0.02061 | 288.12 | 321.10 | 1.0442 | 0.01596 | 285.40 | 317.32 | 1.0155 |
| 120 | 0.02138 | 295.25 | 329.46 | 1.0658 | 0.01662 | 292.75 | 325.99 | 1.0378 |
| 130 | 0.02213 | 302.39 | 337.81 | 1.0867 | 0.01726 | 300.09 | 334.61 | 1.0594 |
| 140 | 0.02287 | 309.57 | 346.16 | 1.1072 | 0.01788 | 307.44 | 343.20 | 1.0805 |
| 150 | 0.02359 | 316.79 | 354.54 | 1.1272 | 0.01849 | 314.80 | 351.78 | 1.1010 |
| 160 | 0.02430 | 324.06 | 362.95 | 1.1469 | 0.01909 | 322.19 | 360.37 | 1.1211 |
| 170 | 0.02501 | 331.37 | 371.39 | 1.1661 | 0.01967 | 329.62 | 368.97 | 1.1407 |
| 180 | 0.02570 | 338.74 | 379.87 | 1.1851 | 0.02025 | 337.09 | 377.60 | 1.1600 |
| 190 | 0.02639 | 346.17 | 388.40 | 1.2037 | 0.02082 | 344.61 | 386.25 | 1.1788 |
| 200 | 0.02707 | 353.66 | 396.97 | 1.2220 | 0.02138 | 352.17 | 394.94 | 1.1974 |
| | 3000 kPa (70.09) | | | | 4000 kPa (84.53) | | | |
| Sat. | 0.00688 | 240.75 | 261.38 | 0.8300 | 0.00443 | 236.40 | 254.13 | 0.7935 |
| 80 | 0.00775 | 251.29 | 274.53 | 0.8678 | — | — | — | — |
| 90 | 0.00847 | 260.65 | 286.04 | 0.9000 | 0.00504 | 245.48 | 265.63 | 0.8254 |
| 100 | 0.00910 | 269.37 | 296.66 | 0.9288 | 0.00580 | 257.78 | 281.00 | 0.8672 |
| 110 | 0.00967 | 277.72 | 306.74 | 0.9555 | 0.00641 | 268.13 | 293.75 | 0.9009 |
| 120 | 0.01021 | 285.84 | 316.47 | 0.9805 | 0.00692 | 277.58 | 305.27 | 0.9306 |
| 130 | 0.01072 | 293.80 | 325.96 | 1.0044 | 0.00739 | 286.52 | 316.08 | 0.9578 |
| 140 | 0.01120 | 301.67 | 335.27 | 1.0272 | 0.00782 | 295.14 | 326.42 | 0.9831 |
| 150 | 0.01166 | 309.47 | 344.47 | 1.0492 | 0.00823 | 303.54 | 336.45 | 1.0071 |
| 160 | 0.01211 | 317.24 | 353.58 | 1.0705 | 0.00861 | 311.81 | 346.25 | 1.0300 |
| 170 | 0.01255 | 325.00 | 362.65 | 1.0912 | 0.00898 | 319.97 | 355.89 | 1.0520 |
| 180 | 0.01298 | 332.75 | 371.68 | 1.1113 | 0.00933 | 328.08 | 365.41 | 1.0732 |
| 190 | 0.01339 | 340.52 | 380.70 | 1.1310 | 0.00968 | 336.15 | 374.85 | 1.0939 |
| 200 | 0.01380 | 348.31 | 389.71 | 1.1502 | 0.01001 | 344.20 | 384.24 | 1.1139 |
| 210 | 0.01420 | 356.12 | 398.73 | 1.1691 | 0.01033 | 352.25 | 393.59 | 1.1335 |
| 220 | 0.01460 | 363.98 | 407.77 | 1.1876 | 0.01065 | 360.31 | 402.93 | 1.1526 |
| 230 | 0.01499 | 371.87 | 416.83 | 1.2058 | 0.01097 | 368.38 | 412.25 | 1.1713 |

**Table B.5**
*Thermodynamic Properties of R-134a*

**Table B.5.1**
*Saturated R-134a*

| Temp. (°C) | Press. (kPa) | Specific Volume, m³/kg | | | Internal Energy, kJ/kg | | |
|---|---|---|---|---|---|---|---|
| | | Sat. Liquid $v_f$ | Evap. $v_{fg}$ | Sat. Vapor $v_g$ | Sat. Liquid $u_f$ | Evap. $u_{fg}$ | Sat. Vapor $u_g$ |
| −70 | 8.3 | 0.000675 | 1.97207 | 1.97274 | 119.46 | 218.74 | 338.20 |
| −65 | 11.7 | 0.000679 | 1.42915 | 1.42983 | 123.18 | 217.76 | 340.94 |
| −60 | 16.3 | 0.000684 | 1.05199 | 1.05268 | 127.52 | 216.19 | 343.71 |
| −55 | 22.2 | 0.000689 | 0.78609 | 0.78678 | 132.36 | 214.14 | 346.50 |
| −50 | 29.9 | 0.000695 | 0.59587 | 0.59657 | 137.60 | 211.71 | 349.31 |
| −45 | 39.6 | 0.000701 | 0.45783 | 0.45853 | 143.15 | 208.99 | 352.15 |
| −40 | 51.8 | 0.000708 | 0.35625 | 0.35696 | 148.95 | 206.05 | 355.00 |
| −35 | 66.8 | 0.000715 | 0.28051 | 0.28122 | 154.93 | 202.93 | 357.86 |
| −30 | 85.1 | 0.000722 | 0.22330 | 0.22402 | 161.06 | 199.67 | 360.73 |
| −26.3 | 101.3 | 0.000728 | 0.18947 | 0.19020 | 165.73 | 197.16 | 362.89 |
| −25 | 107.2 | 0.000730 | 0.17957 | 0.18030 | 167.30 | 196.31 | 363.61 |
| −20 | 133.7 | 0.000738 | 0.14576 | 0.14649 | 173.65 | 192.85 | 366.50 |
| −15 | 165.0 | 0.000746 | 0.11932 | 0.12007 | 180.07 | 189.32 | 369.39 |
| −10 | 201.7 | 0.000755 | 0.09845 | 0.09921 | 186.57 | 185.70 | 372.27 |
| −5 | 244.5 | 0.000764 | 0.08181 | 0.08257 | 193.14 | 182.01 | 375.15 |
| 0 | 294.0 | 0.000773 | 0.06842 | 0.06919 | 199.77 | 178.24 | 378.01 |
| 5 | 350.9 | 0.000783 | 0.05755 | 0.05833 | 206.48 | 174.38 | 380.85 |
| 10 | 415.8 | 0.000794 | 0.04866 | 0.04945 | 213.25 | 170.42 | 383.67 |
| 15 | 489.5 | 0.000805 | 0.04133 | 0.04213 | 220.10 | 166.35 | 386.45 |
| 20 | 572.8 | 0.000817 | 0.03524 | 0.03606 | 227.03 | 162.16 | 389.19 |
| 25 | 666.3 | 0.000829 | 0.03015 | 0.03098 | 234.04 | 157.83 | 391.87 |
| 30 | 771.0 | 0.000843 | 0.02587 | 0.02671 | 241.14 | 153.34 | 394.48 |
| 35 | 887.6 | 0.000857 | 0.02224 | 0.02310 | 248.34 | 148.68 | 397.02 |
| 40 | 1017.0 | 0.000873 | 0.01915 | 0.02002 | 255.65 | 143.81 | 399.46 |
| 45 | 1160.2 | 0.000890 | 0.01650 | 0.01739 | 263.08 | 138.71 | 401.79 |
| 50 | 1318.1 | 0.000908 | 0.01422 | 0.01512 | 270.63 | 133.35 | 403.98 |
| 55 | 1491.6 | 0.000928 | 0.01224 | 0.01316 | 278.33 | 127.68 | 406.01 |
| 60 | 1681.8 | 0.000951 | 0.01051 | 0.01146 | 286.19 | 121.66 | 407.85 |
| 65 | 1889.9 | 0.000976 | 0.00899 | 0.00997 | 294.24 | 115.22 | 409.46 |
| 70 | 2117.0 | 0.001005 | 0.00765 | 0.00866 | 302.51 | 108.27 | 410.78 |
| 75 | 2364.4 | 0.001038 | 0.00645 | 0.00749 | 311.06 | 100.68 | 411.74 |
| 80 | 2633.6 | 0.001078 | 0.00537 | 0.00645 | 319.96 | 92.26 | 412.22 |
| 85 | 2926.2 | 0.001128 | 0.00437 | 0.00550 | 329.35 | 82.67 | 412.01 |
| 90 | 3244.5 | 0.001195 | 0.00341 | 0.00461 | 339.51 | 71.24 | 410.75 |
| 95 | 3591.5 | 0.001297 | 0.00243 | 0.00373 | 351.17 | 56.25 | 407.42 |
| 100 | 3973.2 | 0.001557 | 0.00108 | 0.00264 | 368.55 | 28.19 | 396.74 |
| 101.2 | 4064.0 | 0.001969 | 0 | 0.00197 | 382.97 | 0 | 382.97 |

**TABLE B.5.1** (*continued*)
*Saturated R-134a*

| Temp. (°C) | Press. (kPa) | ENTHALPY, kJ/kg | | | ENTROPY, kJ/kg-K | | |
|---|---|---|---|---|---|---|---|
| | | Sat. Liquid $h_f$ | Evap. $h_{fg}$ | Sat. Vapor $h_g$ | Sat. Liquid $s_f$ | Evap. $s_{fg}$ | Sat. Vapor $s_g$ |
| −70 | 8.3 | 119.47 | 235.15 | 354.62 | 0.6645 | 1.1575 | 1.8220 |
| −65 | 11.7 | 123.18 | 234.55 | 357.73 | 0.6825 | 1.1268 | 1.8094 |
| −60 | 16.3 | 127.53 | 233.33 | 360.86 | 0.7031 | 1.0947 | 1.7978 |
| −55 | 22.2 | 132.37 | 231.63 | 364.00 | 0.7256 | 1.0618 | 1.7874 |
| −50 | 29.9 | 137.62 | 229.54 | 367.16 | 0.7493 | 1.0286 | 1.7780 |
| −45 | 39.6 | 143.18 | 227.14 | 370.32 | 0.7740 | 0.9956 | 1.7695 |
| −40 | 51.8 | 148.98 | 224.50 | 373.48 | 0.7991 | 0.9629 | 1.7620 |
| −35 | 66.8 | 154.98 | 221.67 | 376.64 | 0.8245 | 0.9308 | 1.7553 |
| −30 | 85.1 | 161.12 | 218.68 | 379.80 | 0.8499 | 0.8994 | 1.7493 |
| −26.3 | 101.3 | 165.80 | 216.36 | 382.16 | 0.8690 | 0.8763 | 1.7453 |
| −25 | 107.2 | 167.38 | 215.57 | 382.95 | 0.8754 | 0.8687 | 1.7441 |
| −20 | 133.7 | 173.74 | 212.34 | 386.08 | 0.9007 | 0.8388 | 1.7395 |
| −15 | 165.0 | 180.19 | 209.00 | 389.20 | 0.9258 | 0.8096 | 1.7354 |
| −10 | 201.7 | 186.72 | 205.56 | 392.28 | 0.9507 | 0.7812 | 1.7319 |
| −5 | 244.5 | 193.32 | 202.02 | 395.34 | 0.9755 | 0.7534 | 1.7288 |
| 0 | 294.0 | 200.00 | 198.36 | 398.36 | 1.0000 | 0.7262 | 1.7262 |
| 5 | 350.9 | 206.75 | 194.57 | 401.32 | 1.0243 | 0.6995 | 1.7239 |
| 10 | 415.8 | 213.58 | 190.65 | 404.23 | 1.0485 | 0.6733 | 1.7218 |
| 15 | 489.5 | 220.49 | 186.58 | 407.07 | 1.0725 | 0.6475 | 1.7200 |
| 20 | 572.8 | 227.49 | 182.35 | 409.84 | 1.0963 | 0.6220 | 1.7183 |
| 25 | 666.3 | 234.59 | 177.92 | 412.51 | 1.1201 | 0.5967 | 1.7168 |
| 30 | 771.0 | 241.79 | 173.29 | 415.08 | 1.1437 | 0.5716 | 1.7153 |
| 35 | 887.6 | 249.10 | 168.42 | 417.52 | 1.1673 | 0.5465 | 1.7139 |
| 40 | 1017.0 | 256.54 | 163.28 | 419.82 | 1.1909 | 0.5214 | 1.7123 |
| 45 | 1160.2 | 264.11 | 157.85 | 421.96 | 1.2145 | 0.4962 | 1.7106 |
| 50 | 1318.1 | 271.83 | 152.08 | 423.91 | 1.2381 | 0.4706 | 1.7088 |
| 55 | 1491.6 | 279.72 | 145.93 | 425.65 | 1.2619 | 0.4447 | 1.7066 |
| 60 | 1681.8 | 287.79 | 139.33 | 427.13 | 1.2857 | 0.4182 | 1.7040 |
| 65 | 1889.9 | 296.09 | 132.21 | 428.30 | 1.3099 | 0.3910 | 1.7008 |
| 70 | 2117.0 | 304.64 | 124.47 | 429.11 | 1.3343 | 0.3627 | 1.6970 |
| 75 | 2364.4 | 313.51 | 115.94 | 429.45 | 1.3592 | 0.3330 | 1.6923 |
| 80 | 2633.6 | 322.79 | 106.40 | 429.19 | 1.3849 | 0.3013 | 1.6862 |
| 85 | 2926.2 | 332.65 | 95.45 | 428.10 | 1.4117 | 0.2665 | 1.6782 |
| 90 | 3244.5 | 343.38 | 82.31 | 425.70 | 1.4404 | 0.2267 | 1.6671 |
| 95 | 3591.5 | 355.83 | 64.98 | 420.81 | 1.4733 | 0.1765 | 1.6498 |
| 100 | 3973.2 | 374.74 | 32.47 | 407.21 | 1.5228 | 0.0870 | 1.6098 |
| 101.2 | 4064.0 | 390.98 | 0 | 390.98 | 1.5658 | 0 | 1.5658 |

TABLE B.5.2
*Superheated R-134a*

| Temp. (°C) | v (m³/kg) | u (kJ/kg) | h (kJ/kg) | s (kJ/kg-K) | v (m³/kg) | u (kJ/kg) | h (kJ/kg) | s (kJ/kg-K) |
|---|---|---|---|---|---|---|---|---|
| | 50 kPa (−40.67) | | | | 100 kPa (−26.54) | | | |
| Sat. | 0.36889 | 354.61 | 373.06 | 1.7629 | 0.19257 | 362.73 | 381.98 | 1.7456 |
| −20 | 0.40507 | 368.57 | 388.82 | 1.8279 | 0.19860 | 367.36 | 387.22 | 1.7665 |
| −10 | 0.42222 | 375.53 | 396.64 | 1.8582 | 0.20765 | 374.51 | 395.27 | 1.7978 |
| 0 | 0.43921 | 382.63 | 404.59 | 1.8878 | 0.21652 | 381.76 | 403.41 | 1.8281 |
| 10 | 0.45608 | 389.90 | 412.70 | 1.9170 | 0.22527 | 389.14 | 411.67 | 1.8578 |
| 20 | 0.47287 | 397.32 | 420.96 | 1.9456 | 0.23392 | 396.66 | 420.05 | 1.8869 |
| 30 | 0.48958 | 404.90 | 429.38 | 1.9739 | 0.24250 | 404.31 | 428.56 | 1.9155 |
| 40 | 0.50623 | 412.64 | 437.96 | 2.0017 | 0.25101 | 412.12 | 437.22 | 1.9436 |
| 50 | 0.52284 | 420.55 | 446.70 | 2.0292 | 0.25948 | 420.08 | 446.03 | 1.9712 |
| 60 | 0.53941 | 428.63 | 455.60 | 2.0563 | 0.26791 | 428.20 | 454.99 | 1.9985 |
| 70 | 0.55595 | 436.86 | 464.66 | 2.0831 | 0.27631 | 436.47 | 464.10 | 2.0255 |
| 80 | 0.57247 | 445.26 | 473.88 | 2.1096 | 0.28468 | 444.89 | 473.36 | 2.0521 |
| 90 | 0.58896 | 453.82 | 483.26 | 2.1358 | 0.29302 | 453.47 | 482.78 | 2.0784 |
| 100 | 0.60544 | 462.53 | 492.81 | 2.1617 | 0.30135 | 462.21 | 492.35 | 2.1044 |
| 110 | 0.62190 | 471.41 | 502.50 | 2.1874 | 0.30967 | 471.11 | 502.07 | 2.1301 |
| 120 | 0.63835 | 480.44 | 512.36 | 2.2128 | 0.31797 | 480.16 | 511.95 | 2.1555 |
| 130 | 0.65479 | 489.63 | 522.37 | 2.2379 | 0.32626 | 489.36 | 521.98 | 2.1807 |
| | 150 kPa (−17.29) | | | | 200 kPa (−10.22) | | | |
| Sat. | 0.13139 | 368.06 | 387.77 | 1.7372 | 0.10002 | 372.15 | 392.15 | 1.7320 |
| −10 | 0.13602 | 373.44 | 393.84 | 1.7606 | 0.10013 | 372.31 | 392.34 | 1.7328 |
| 0 | 0.14222 | 380.85 | 402.19 | 1.7917 | 0.10501 | 379.91 | 400.91 | 1.7647 |
| 10 | 0.14828 | 388.36 | 410.60 | 1.8220 | 0.10974 | 387.55 | 409.50 | 1.7956 |
| 20 | 0.15424 | 395.98 | 419.11 | 1.8515 | 0.11436 | 395.27 | 418.15 | 1.8256 |
| 30 | 0.16011 | 403.71 | 427.73 | 1.8804 | 0.11889 | 403.10 | 426.87 | 1.8549 |
| 40 | 0.16592 | 411.59 | 436.47 | 1.9088 | 0.12335 | 411.04 | 435.71 | 1.8836 |
| 50 | 0.17168 | 419.60 | 445.35 | 1.9367 | 0.12776 | 419.11 | 444.66 | 1.9117 |
| 60 | 0.17740 | 427.76 | 454.37 | 1.9642 | 0.13213 | 427.31 | 453.74 | 1.9394 |
| 70 | 0.18308 | 436.06 | 463.53 | 1.9913 | 0.13646 | 435.65 | 462.95 | 1.9666 |
| 80 | 0.18874 | 444.52 | 472.83 | 2.0180 | 0.14076 | 444.14 | 472.30 | 1.9935 |
| 90 | 0.19437 | 453.13 | 482.28 | 2.0444 | 0.14504 | 452.78 | 481.79 | 2.0200 |
| 100 | 0.19999 | 461.89 | 491.89 | 2.0705 | 0.14930 | 461.56 | 491.42 | 2.0461 |
| 110 | 0.20559 | 470.80 | 501.64 | 2.0963 | 0.15355 | 470.50 | 501.21 | 2.0720 |
| 120 | 0.21117 | 479.87 | 511.54 | 2.1218 | 0.15777 | 479.58 | 511.13 | 2.0976 |
| 130 | 0.21675 | 489.08 | 521.60 | 2.1470 | 0.16199 | 488.81 | 521.21 | 2.1229 |
| 140 | 0.22231 | 498.45 | 531.80 | 2.1720 | 0.16620 | 498.19 | 531.43 | 2.1479 |

**Table B.5.2** (*continued*)
*Superheated R-134a*

| Temp. (°C) | $v$ (m³/kg) | $u$ (kJ/kg) | $h$ (kJ/kg) | $s$ (kJ/kg-K) | $v$ (m³/kg) | $u$ (kJ/kg) | $h$ (kJ/kg) | $s$ (kJ/kg-K) |
|---|---|---|---|---|---|---|---|---|
| | | 300 kPa (0.56) | | | | 400 kPa (8.84) | | |
| Sat. | 0.06787 | 378.33 | 398.69 | 1.7259 | 0.05136 | 383.02 | 403.56 | 1.7223 |
| 10 | 0.07111 | 385.84 | 407.17 | 1.7564 | 0.05168 | 383.98 | 404.65 | 1.7261 |
| 20 | 0.07441 | 393.80 | 416.12 | 1.7874 | 0.05436 | 392.22 | 413.97 | 1.7584 |
| 30 | 0.07762 | 401.81 | 425.10 | 1.8175 | 0.05693 | 400.45 | 423.22 | 1.7895 |
| 40 | 0.08075 | 409.90 | 434.12 | 1.8468 | 0.05940 | 408.70 | 432.46 | 1.8195 |
| 50 | 0.08382 | 418.09 | 443.23 | 1.8755 | 0.06181 | 417.03 | 441.75 | 1.8487 |
| 60 | 0.08684 | 426.39 | 452.44 | 1.9035 | 0.06417 | 425.44 | 451.10 | 1.8772 |
| 70 | 0.08982 | 434.82 | 461.76 | 1.9311 | 0.06648 | 433.95 | 460.55 | 1.9051 |
| 80 | 0.09277 | 443.37 | 471.21 | 1.9582 | 0.06877 | 442.58 | 470.09 | 1.9325 |
| 90 | 0.09570 | 452.07 | 480.78 | 1.9850 | 0.07102 | 451.34 | 479.75 | 1.9595 |
| 100 | 0.09861 | 460.90 | 490.48 | 2.0113 | 0.07325 | 460.22 | 489.52 | 1.9860 |
| 110 | 0.10150 | 469.87 | 500.32 | 2.0373 | 0.07547 | 469.24 | 499.43 | 2.0122 |
| 120 | 0.10437 | 478.99 | 510.30 | 2.0631 | 0.07767 | 478.40 | 509.46 | 2.0381 |
| 130 | 0.10723 | 488.26 | 520.43 | 2.0885 | 0.07985 | 487.69 | 519.63 | 2.0636 |
| 140 | 0.11008 | 497.66 | 530.69 | 2.1136 | 0.08202 | 497.13 | 529.94 | 2.0889 |
| 150 | 0.11292 | 507.22 | 541.09 | 2.1385 | 0.08418 | 506.71 | 540.38 | 2.1139 |
| 160 | 0.11575 | 516.91 | 551.64 | 2.1631 | 0.08634 | 516.43 | 550.97 | 2.1386 |
| | | 500 kPa (15.66) | | | | 600 kPa (21.52) | | |
| Sat. | 0.04126 | 386.82 | 407.45 | 1.7198 | 0.03442 | 390.01 | 410.66 | 1.7179 |
| 20 | 0.04226 | 390.52 | 411.65 | 1.7342 | — | — | — | — |
| 30 | 0.04446 | 398.99 | 421.22 | 1.7663 | 0.03609 | 397.44 | 419.09 | 1.7461 |
| 40 | 0.04656 | 407.44 | 430.72 | 1.7971 | 0.03796 | 406.11 | 428.88 | 1.7779 |
| 50 | 0.04858 | 415.91 | 440.20 | 1.8270 | 0.03974 | 414.75 | 438.59 | 1.8084 |
| 60 | 0.05055 | 424.44 | 449.72 | 1.8560 | 0.04145 | 423.41 | 448.28 | 1.8379 |
| 70 | 0.05247 | 433.06 | 459.29 | 1.8843 | 0.04311 | 432.13 | 457.99 | 1.8666 |
| 80 | 0.05435 | 441.77 | 468.94 | 1.9120 | 0.04473 | 440.93 | 467.76 | 1.8947 |
| 90 | 0.05620 | 450.59 | 478.69 | 1.9392 | 0.04632 | 449.82 | 477.61 | 1.9222 |
| 100 | 0.05804 | 459.53 | 488.55 | 1.9660 | 0.04788 | 458.82 | 487.55 | 1.9492 |
| 110 | 0.05985 | 468.60 | 498.52 | 1.9924 | 0.04943 | 467.94 | 497.59 | 1.9758 |
| 120 | 0.06164 | 477.79 | 508.61 | 2.0184 | 0.05095 | 477.18 | 507.75 | 2.0019 |
| 130 | 0.06342 | 487.13 | 518.83 | 2.0440 | 0.05246 | 486.55 | 518.03 | 2.0277 |
| 140 | 0.06518 | 496.59 | 529.19 | 2.0694 | 0.05396 | 496.05 | 528.43 | 2.0532 |
| 150 | 0.06694 | 506.20 | 539.67 | 2.0945 | 0.05544 | 505.69 | 538.95 | 2.0784 |
| 160 | 0.06869 | 515.95 | 550.29 | 2.1193 | 0.05692 | 515.46 | 549.61 | 2.1033 |
| 170 | 0.07043 | 525.83 | 561.04 | 2.1438 | 0.05839 | 525.36 | 560.40 | 2.1279 |

TABLE **B.5.2** (*continued*)
*Superheated R-134a*

| Temp. (°C) | v (m³/kg) | u (kJ/kg) | h (kJ/kg) | s (kJ/kg-K) | v (m³/kg) | u (kJ/kg) | h (kJ/kg) | s (kJ/kg-K) |
|---|---|---|---|---|---|---|---|---|
| | | 800 kPa (31.30) | | | | 1000 kPa (39.37) | | |
| Sat. | 0.02571 | 395.15 | 415.72 | 1.7150 | 0.02038 | 399.16 | 419.54 | 1.7125 |
| 40 | 0.02711 | 403.17 | 424.86 | 1.7446 | 0.02047 | 399.78 | 420.25 | 1.7148 |
| 50 | 0.02861 | 412.23 | 435.11 | 1.7768 | 0.02185 | 409.39 | 431.24 | 1.7494 |
| 60 | 0.03002 | 421.20 | 445.22 | 1.8076 | 0.02311 | 418.78 | 441.89 | 1.7818 |
| 70 | 0.03137 | 430.17 | 455.27 | 1.8373 | 0.02429 | 428.05 | 452.34 | 1.8127 |
| 80 | 0.03268 | 439.17 | 465.31 | 1.8662 | 0.02542 | 437.29 | 462.70 | 1.8425 |
| 90 | 0.03394 | 448.22 | 475.38 | 1.8943 | 0.02650 | 446.53 | 473.03 | 1.8713 |
| 100 | 0.03518 | 457.35 | 485.50 | 1.9218 | 0.02754 | 455.82 | 483.36 | 1.8994 |
| 110 | 0.03639 | 466.58 | 495.70 | 1.9487 | 0.02856 | 465.18 | 493.74 | 1.9268 |
| 120 | 0.03758 | 475.92 | 505.99 | 1.9753 | 0.02956 | 474.62 | 504.17 | 1.9537 |
| 130 | 0.03876 | 485.37 | 516.38 | 2.0014 | 0.03053 | 484.16 | 514.69 | 1.9801 |
| 140 | 0.03992 | 494.94 | 526.88 | 2.0271 | 0.03150 | 493.81 | 525.30 | 2.0061 |
| 150 | 0.04107 | 504.64 | 537.50 | 2.0525 | 0.03244 | 503.57 | 536.02 | 2.0318 |
| 160 | 0.04221 | 514.46 | 548.23 | 2.0775 | 0.03338 | 513.46 | 546.84 | 2.0570 |
| 170 | 0.04334 | 524.42 | 559.09 | 2.1023 | 0.03431 | 523.46 | 557.77 | 2.0820 |
| 180 | 0.04446 | 534.51 | 570.08 | 2.1268 | 0.03523 | 533.60 | 568.83 | 2.1067 |
| | | 1200 kPa (46.31) | | | | 1400 kPa (52.42) | | |
| Sat. | 0.01676 | 402.37 | 422.49 | 1.7102 | 0.01414 | 404.98 | 424.78 | 1.7077 |
| 50 | 0.01724 | 406.15 | 426.84 | 1.7237 | — | — | — | — |
| 60 | 0.01844 | 416.08 | 438.21 | 1.7584 | 0.01503 | 413.03 | 434.08 | 1.7360 |
| 70 | 0.01953 | 425.74 | 449.18 | 1.7908 | 0.01608 | 423.20 | 445.72 | 1.7704 |
| 80 | 0.02055 | 435.27 | 459.92 | 1.8217 | 0.01704 | 433.09 | 456.94 | 1.8026 |
| 90 | 0.02151 | 444.74 | 470.55 | 1.8514 | 0.01793 | 442.83 | 467.93 | 1.8333 |
| 100 | 0.02244 | 454.20 | 481.13 | 1.8801 | 0.01878 | 452.50 | 478.79 | 1.8628 |
| 110 | 0.02333 | 463.71 | 491.70 | 1.9081 | 0.01958 | 462.17 | 489.59 | 1.8914 |
| 120 | 0.02420 | 473.27 | 502.31 | 1.9354 | 0.02036 | 471.87 | 500.38 | 1.9192 |
| 130 | 0.02504 | 482.91 | 512.97 | 1.9621 | 0.02112 | 481.63 | 511.19 | 1.9463 |
| 140 | 0.02587 | 492.65 | 523.70 | 1.9884 | 0.02186 | 491.46 | 522.05 | 1.9730 |
| 150 | 0.02669 | 502.48 | 534.51 | 2.0143 | 0.02258 | 501.37 | 532.98 | 1.9991 |
| 160 | 0.02750 | 512.43 | 545.43 | 2.0398 | 0.02329 | 511.39 | 543.99 | 2.0248 |
| 170 | 0.02829 | 522.50 | 556.44 | 2.0649 | 0.02399 | 521.51 | 555.10 | 2.0502 |
| 180 | 0.02907 | 532.68 | 567.57 | 2.0898 | 0.02468 | 531.75 | 566.30 | 2.0752 |

TABLE B.5.2  (*continued*)
**Superheated R-134a**

| Temp. (°C) | $v$ (m³/kg) | $u$ (kJ/kg) | $h$ (kJ/kg) | $s$ (kJ/kg-K) | $v$ (m³/kg) | $u$ (kJ/kg) | $h$ (kJ/kg) | $s$ (kJ/kg-K) |
|---|---|---|---|---|---|---|---|---|
| | 1600 kPa (57.90) | | | | 2000 kPa (67.48) | | | |
| Sat. | 0.01215 | 407.11 | 426.54 | 1.7051 | 0.00930 | 410.15 | 428.75 | 1.6991 |
| 60 | 0.01239 | 409.49 | 429.32 | 1.7135 | — | — | — | — |
| 70 | 0.01345 | 420.37 | 441.89 | 1.7507 | 0.00958 | 413.37 | 432.53 | 1.7101 |
| 80 | 0.01438 | 430.72 | 453.72 | 1.7847 | 0.01055 | 425.20 | 446.30 | 1.7497 |
| 90 | 0.01522 | 440.79 | 465.15 | 1.8166 | 0.01137 | 436.20 | 458.95 | 1.7850 |
| 100 | 0.01601 | 450.71 | 476.33 | 1.8469 | 0.01211 | 446.78 | 471.00 | 1.8177 |
| 110 | 0.01676 | 460.57 | 487.39 | 1.8762 | 0.01279 | 457.12 | 482.69 | 1.8487 |
| 120 | 0.01748 | 470.42 | 498.39 | 1.9045 | 0.01342 | 467.34 | 494.19 | 1.8783 |
| 130 | 0.01817 | 480.30 | 509.37 | 1.9321 | 0.01403 | 477.51 | 505.57 | 1.9069 |
| 140 | 0.01884 | 490.23 | 520.38 | 1.9591 | 0.01461 | 487.68 | 516.90 | 1.9346 |
| 150 | 0.01949 | 500.24 | 531.43 | 1.9855 | 0.01517 | 497.89 | 528.22 | 1.9617 |
| 160 | 0.02013 | 510.33 | 542.54 | 2.0115 | 0.01571 | 508.15 | 539.57 | 1.9882 |
| 170 | 0.02076 | 520.52 | 553.73 | 2.0370 | 0.01624 | 518.48 | 550.96 | 2.0142 |
| 180 | 0.02138 | 530.81 | 565.02 | 2.0622 | 0.01676 | 528.89 | 562.42 | 2.0398 |
| | 3000 kPa (86.20) | | | | 4000 kPa (100.33) | | | |
| Sat. | 0.00528 | 411.83 | 427.67 | 1.6759 | 0.00252 | 394.86 | 404.94 | 1.6036 |
| 90 | 0.00575 | 418.93 | 436.19 | 1.6995 | — | — | — | — |
| 100 | 0.00665 | 433.77 | 453.73 | 1.7472 | — | — | — | — |
| 110 | 0.00734 | 446.48 | 468.50 | 1.7862 | 0.00428 | 429.74 | 446.84 | 1.7148 |
| 120 | 0.00792 | 458.27 | 482.04 | 1.8211 | 0.00500 | 445.97 | 465.99 | 1.7642 |
| 130 | 0.00845 | 469.58 | 494.91 | 1.8535 | 0.00556 | 459.63 | 481.87 | 1.8040 |
| 140 | 0.00893 | 480.61 | 507.39 | 1.8840 | 0.00603 | 472.19 | 496.29 | 1.8394 |
| 150 | 0.00937 | 491.49 | 519.62 | 1.9133 | 0.00644 | 484.15 | 509.92 | 1.8720 |
| 160 | 0.00980 | 502.30 | 531.70 | 1.9415 | 0.00683 | 495.77 | 523.07 | 1.9027 |
| 170 | 0.01021 | 513.09 | 543.71 | 1.9689 | 0.00718 | 507.19 | 535.92 | 1.9320 |
| 180 | 0.01060 | 523.89 | 555.69 | 1.9956 | 0.00752 | 518.51 | 548.57 | 1.9603 |
| | 6000 kPa | | | | 10000 kPa | | | |
| 90 | 0.001059 | 328.34 | 334.70 | 1.4081 | 0.000991 | 320.72 | 330.62 | 1.3856 |
| 100 | 0.001150 | 346.71 | 353.61 | 1.4595 | 0.001040 | 336.45 | 346.85 | 1.4297 |
| 110 | 0.001307 | 368.06 | 375.90 | 1.5184 | 0.001100 | 352.74 | 363.73 | 1.4744 |
| 120 | 0.001698 | 396.59 | 406.78 | 1.5979 | 0.001175 | 369.69 | 381.44 | 1.5200 |
| 130 | 0.002396 | 426.81 | 441.18 | 1.6843 | 0.001272 | 387.44 | 400.16 | 1.5670 |
| 140 | 0.002985 | 448.34 | 466.25 | 1.7458 | 0.001400 | 405.97 | 419.98 | 1.6155 |
| 150 | 0.003439 | 465.19 | 485.82 | 1.7926 | 0.001564 | 424.99 | 440.63 | 1.6649 |
| 160 | 0.003814 | 479.89 | 502.77 | 1.8322 | 0.001758 | 443.77 | 461.34 | 1.7133 |
| 170 | 0.004141 | 493.45 | 518.30 | 1.8676 | 0.001965 | 461.65 | 481.30 | 1.7589 |
| 180 | 0.004435 | 506.35 | 532.96 | 1.9004 | 0.002172 | 478.40 | 500.12 | 1.8009 |

TABLE B.6
*Thermodynamic Properties of Nitrogen*

TABLE B.6.1
*Saturated Nitrogen*

| Temp. (K) | Press. (kPa) | SPECIFIC VOLUME, m³/kg | | | INTERNAL ENERGY, kJ/kg | | |
|---|---|---|---|---|---|---|---|
| | | Sat. Liquid $v_f$ | Evap. $v_{fg}$ | Sat. Vapor $v_g$ | Sat. Liquid $u_f$ | Evap. $u_{fg}$ | Sat. Vapor $u_g$ |
| 63.1 | 12.5 | 0.001150 | 1.48074 | 1.48189 | −150.92 | 196.86 | 45.94 |
| 65 | 17.4 | 0.001160 | 1.09231 | 1.09347 | −147.19 | 194.37 | 47.17 |
| 70 | 38.6 | 0.001191 | 0.52513 | 0.52632 | −137.13 | 187.54 | 50.40 |
| 75 | 76.1 | 0.001223 | 0.28052 | 0.28174 | −127.04 | 180.47 | 53.43 |
| 77.3 | 101.3 | 0.001240 | 0.21515 | 0.21639 | −122.27 | 177.04 | 54.76 |
| 80 | 137.0 | 0.001259 | 0.16249 | 0.16375 | −116.86 | 173.06 | 56.20 |
| 85 | 229.1 | 0.001299 | 0.10018 | 0.10148 | −106.55 | 165.20 | 58.65 |
| 90 | 360.8 | 0.001343 | 0.06477 | 0.06611 | −96.06 | 156.76 | 60.70 |
| 95 | 541.1 | 0.001393 | 0.04337 | 0.04476 | −85.35 | 147.60 | 62.25 |
| 100 | 779.2 | 0.001452 | 0.02975 | 0.03120 | −74.33 | 137.50 | 63.17 |
| 105 | 1084.6 | 0.001522 | 0.02066 | 0.02218 | −62.89 | 126.18 | 63.29 |
| 110 | 1467.6 | 0.001610 | 0.01434 | 0.01595 | −50.81 | 113.11 | 62.31 |
| 115 | 1939.3 | 0.001729 | 0.00971 | 0.01144 | −37.66 | 97.36 | 59.70 |
| 120 | 2513.0 | 0.001915 | 0.00608 | 0.00799 | −22.42 | 76.63 | 54.21 |
| 125 | 3208.0 | 0.002355 | 0.00254 | 0.00490 | −0.83 | 40.73 | 39.90 |
| 126.2 | 3397.8 | 0.003194 | 0 | 0.00319 | 18.94 | 0 | 18.94 |

| Temp. (K) | Press. (kPa) | ENTHALPY, kJ/kg | | | ENTROPY, kJ/kg-K | | |
|---|---|---|---|---|---|---|---|
| | | Sat. Liquid $h_f$ | Evap. $h_{fg}$ | Sat. Vapor $h_g$ | Sat. Liquid $s_f$ | Evap. $s_{fg}$ | Sat. Vapor $s_g$ |
| 63.1 | 12.5 | −150.91 | 215.39 | 64.48 | 2.4234 | 3.4109 | 5.8343 |
| 65 | 17.4 | −147.17 | 213.38 | 66.21 | 2.4816 | 3.2828 | 5.7645 |
| 70 | 38.6 | −137.09 | 207.79 | 70.70 | 2.6307 | 2.9684 | 5.5991 |
| 75 | 76.1 | −126.95 | 201.82 | 74.87 | 2.7700 | 2.6909 | 5.4609 |
| 77.3 | 101.3 | −122.15 | 198.84 | 76.69 | 2.8326 | 2.5707 | 5.4033 |
| 80 | 137.0 | −116.69 | 195.32 | 78.63 | 2.9014 | 2.4415 | 5.3429 |
| 85 | 229.1 | −106.25 | 188.15 | 81.90 | 3.0266 | 2.2135 | 5.2401 |
| 90 | 360.8 | −95.58 | 180.13 | 84.55 | 3.1466 | 2.0015 | 5.1480 |
| 95 | 541.1 | −84.59 | 171.07 | 86.47 | 3.2627 | 1.8007 | 5.0634 |
| 100 | 779.2 | −73.20 | 160.68 | 87.48 | 3.3761 | 1.6068 | 4.9829 |
| 105 | 1084.6 | −61.24 | 148.59 | 87.35 | 3.4883 | 1.4151 | 4.9034 |
| 110 | 1467.6 | −48.45 | 134.15 | 85.71 | 3.6017 | 1.2196 | 4.8213 |
| 115 | 1939.3 | −34.31 | 116.19 | 81.88 | 3.7204 | 1.0104 | 4.7307 |
| 120 | 2513.0 | −17.61 | 91.91 | 74.30 | 3.8536 | 0.7659 | 4.6195 |
| 125 | 3208.0 | 6.73 | 48.88 | 55.60 | 4.0399 | 0.3910 | 4.4309 |
| 126.2 | 3397.8 | 29.79 | 0 | 29.79 | 4.2193 | 0 | 4.2193 |

TABLE **B.6.2**

*Superheated Nitrogen*

| Temp. (K) | v (m³/kg) | u (kJ/kg) | h (kJ/kg) | s (kJ/kg-K) | v (m³/kg) | u (kJ/kg) | h (kJ/kg) | s (kJ/kg-K) |
|---|---|---|---|---|---|---|---|---|
| | 100 kPa (77.24 K) | | | | 200 kPa (83.62 K) | | | |
| Sat. | 0.21903 | 54.70 | 76.61 | 5.4059 | 0.11520 | 58.01 | 81.05 | 5.2673 |
| 100 | 0.29103 | 72.84 | 101.94 | 5.6944 | 0.14252 | 71.73 | 100.24 | 5.4775 |
| 120 | 0.35208 | 87.94 | 123.15 | 5.8878 | 0.17397 | 87.14 | 121.93 | 5.6753 |
| 140 | 0.41253 | 102.95 | 144.20 | 6.0501 | 0.20476 | 102.33 | 143.28 | 5.8399 |
| 160 | 0.47263 | 117.91 | 165.17 | 6.1901 | 0.23519 | 117.40 | 164.44 | 5.9812 |
| 180 | 0.53254 | 132.83 | 186.09 | 6.3132 | 0.26542 | 132.41 | 185.49 | 6.1052 |
| 200 | 0.59231 | 147.74 | 206.97 | 6.4232 | 0.29551 | 147.37 | 206.48 | 6.2157 |
| 220 | 0.65199 | 162.63 | 227.83 | 6.5227 | 0.32552 | 162.31 | 227.41 | 6.3155 |
| 240 | 0.71161 | 177.51 | 248.67 | 6.6133 | 0.35546 | 177.23 | 248.32 | 6.4064 |
| 260 | 0.77118 | 192.39 | 269.51 | 6.6967 | 0.38535 | 192.14 | 269.21 | 6.4900 |
| 280 | 0.83072 | 207.26 | 290.33 | 6.7739 | 0.41520 | 207.04 | 290.08 | 6.5674 |
| 300 | 0.89023 | 222.14 | 311.16 | 6.8457 | 0.44503 | 221.93 | 310.94 | 6.6393 |
| 350 | 1.03891 | 259.35 | 363.24 | 7.0063 | 0.51952 | 259.18 | 363.09 | 6.8001 |
| 400 | 1.18752 | 296.66 | 415.41 | 7.1456 | 0.59392 | 296.52 | 415.31 | 6.9396 |
| 450 | 1.33607 | 334.16 | 467.77 | 7.2690 | 0.66827 | 334.04 | 467.70 | 7.0630 |
| 500 | 1.48458 | 371.95 | 520.41 | 7.3799 | 0.74258 | 371.85 | 520.37 | 7.1740 |
| 600 | 1.78154 | 448.79 | 626.94 | 7.5741 | 0.89114 | 448.71 | 626.94 | 7.3682 |
| 700 | 2.07845 | 527.74 | 735.58 | 7.7415 | 1.03965 | 527.68 | 735.61 | 7.5357 |
| 800 | 2.37532 | 609.07 | 846.60 | 7.8897 | 1.18812 | 609.02 | 846.64 | 7.6839 |
| 900 | 2.67217 | 692.79 | 960.01 | 8.0232 | 1.33657 | 692.75 | 960.07 | 7.8175 |
| 1000 | 2.96900 | 778.78 | 1075.68 | 8.1451 | 1.48501 | 778.74 | 1075.75 | 7.9393 |

| Temp. (K) | $v$ (m³/kg) | $u$ (kJ/kg) | $h$ (kJ/kg) | $s$ (kJ/kg-K) | $v$ (m³/kg) | $u$ (kJ/kg) | $h$ (kJ/kg) | $s$ (kJ/kg-K) |
|---|---|---|---|---|---|---|---|---|
| | 400 kPa (91.22 K) | | | | 600 kPa (96.37 K) | | | |
| Sat | 0.05992 | 61.13 | 85.10 | 5.1268 | 0.04046 | 62.57 | 86.85 | 5.0411 |
| 100 | 0.06806 | 69.30 | 96.52 | 5.2466 | 0.04299 | 66.41 | 92.20 | 5.0957 |
| 120 | 0.08486 | 85.48 | 119.42 | 5.4556 | 0.05510 | 83.73 | 116.79 | 5.3204 |
| 140 | 0.10085 | 101.06 | 141.40 | 5.6250 | 0.06620 | 99.75 | 139.47 | 5.4953 |
| 160 | 0.11647 | 116.38 | 162.96 | 5.7690 | 0.07689 | 115.34 | 161.47 | 5.6422 |
| 180 | 0.13186 | 131.55 | 184.30 | 5.8947 | 0.08734 | 130.69 | 183.10 | 5.7696 |
| 200 | 0.14712 | 146.64 | 205.49 | 6.0063 | 0.09766 | 145.91 | 204.50 | 5.8823 |
| 220 | 0.16228 | 161.68 | 226.59 | 6.1069 | 0.10788 | 161.04 | 225.76 | 5.9837 |
| 240 | 0.17738 | 176.67 | 247.62 | 6.1984 | 0.11803 | 176.11 | 246.92 | 6.0757 |
| 260 | 0.19243 | 191.64 | 268.61 | 6.2824 | 0.12813 | 191.13 | 268.01 | 6.1601 |
| 280 | 0.20745 | 206.58 | 289.56 | 6.3600 | 0.13820 | 206.13 | 289.05 | 6.2381 |
| 300 | 0.22244 | 221.52 | 310.50 | 6.4322 | 0.14824 | 221.11 | 310.06 | 6.3105 |
| 350 | 0.25982 | 258.85 | 362.78 | 6.5934 | 0.17326 | 258.52 | 362.48 | 6.4722 |
| 400 | 0.29712 | 296.25 | 415.10 | 6.7331 | 0.19819 | 295.97 | 414.89 | 6.6121 |
| 450 | 0.33437 | 333.81 | 467.56 | 6.8567 | 0.22308 | 333.57 | 467.42 | 6.7359 |
| 500 | 0.37159 | 371.65 | 520.28 | 6.9678 | 0.24792 | 371.45 | 520.20 | 6.8471 |
| 600 | 0.44595 | 448.55 | 626.93 | 7.1622 | 0.29755 | 448.40 | 626.93 | 7.0416 |
| 700 | 0.52025 | 527.55 | 735.65 | 7.3298 | 0.34712 | 527.43 | 735.70 | 7.2093 |
| 800 | 0.59453 | 608.92 | 846.73 | 7.4781 | 0.39666 | 608.82 | 846.82 | 7.3576 |
| 900 | 0.66878 | 692.67 | 960.19 | 7.6117 | 0.44618 | 692.59 | 960.30 | 7.4912 |
| 1000 | 0.74302 | 778.68 | 1075.89 | 7.7335 | 0.49568 | 778.61 | 1076.02 | 7.6131 |
| | 800 kPa (100.38 K) | | | | 1000 kPa (103.73 K) | | | |
| Sat. | 0.03038 | 63.21 | 87.52 | 4.9768 | 0.02416 | 63.35 | 87.51 | 4.9237 |
| 120 | 0.04017 | 81.88 | 114.02 | 5.2191 | 0.03117 | 79.91 | 111.08 | 5.1357 |
| 140 | 0.04886 | 98.41 | 137.50 | 5.4002 | 0.03845 | 97.02 | 135.47 | 5.3239 |
| 160 | 0.05710 | 114.28 | 159.95 | 5.5501 | 0.04522 | 113.20 | 158.42 | 5.4772 |
| 180 | 0.06509 | 129.82 | 181.89 | 5.6793 | 0.05173 | 128.94 | 180.67 | 5.6082 |
| 200 | 0.07293 | 145.17 | 203.51 | 5.7933 | 0.05809 | 144.43 | 202.52 | 5.7234 |
| 220 | 0.08067 | 160.40 | 224.94 | 5.8954 | 0.06436 | 159.76 | 224.11 | 5.8263 |
| 240 | 0.08835 | 175.54 | 246.23 | 5.9880 | 0.07055 | 174.98 | 245.53 | 5.9194 |
| 260 | 0.09599 | 190.63 | 267.42 | 6.0728 | 0.07670 | 190.13 | 266.83 | 6.0047 |
| 280 | 0.10358 | 205.68 | 288.54 | 6.1511 | 0.08281 | 205.23 | 288.04 | 6.0833 |
| 300 | 0.11115 | 220.70 | 309.62 | 6.2238 | 0.08889 | 220.29 | 309.18 | 6.1562 |
| 350 | 0.12998 | 258.19 | 362.17 | 6.3858 | 0.10401 | 257.86 | 361.87 | 6.3187 |
| 400 | 0.14873 | 295.69 | 414.68 | 6.5260 | 0.11905 | 295.42 | 414.47 | 6.4591 |
| 500 | 0.18609 | 371.25 | 520.12 | 6.7613 | 0.14899 | 371.04 | 520.04 | 6.6947 |
| 600 | 0.22335 | 448.24 | 626.93 | 6.9560 | 0.17883 | 448.09 | 626.92 | 6.8895 |
| 700 | 0.26056 | 527.31 | 735.76 | 7.1237 | 0.20862 | 527.19 | 735.81 | 7.0573 |
| 800 | 0.29773 | 608.73 | 846.91 | 7.2721 | 0.23837 | 608.63 | 847.00 | 7.2057 |
| 900 | 0.33488 | 692.52 | 960.42 | 7.4058 | 0.26810 | 692.44 | 960.54 | 7.3394 |
| 1000 | 0.37202 | 778.55 | 1076.16 | 7.5277 | 0.29782 | 778.49 | 1076.30 | 7.4614 |

**Table B.6.2** (*continued*)
**Superheated Nitrogen**

| Temp. (K) | v (m³/kg) | u (kJ/kg) | h (kJ/kg) | s (kJ/kg-K) | v (m³/kg) | u (kJ/kg) | h (kJ/kg) | s (kJ/kg-K) |
|---|---|---|---|---|---|---|---|---|
| | 1500 kPa (110.38 K) | | | | 2000 kPa (115.58 K) | | | |
| Sat. | 0.01555 | 62.17 | 85.51 | 4.8148 | 0.01100 | 59.25 | 81.25 | 4.7193 |
| 120 | 0.01899 | 74.26 | 102.75 | 4.9650 | 0.01260 | 66.90 | 92.10 | 4.8116 |
| 140 | 0.02452 | 93.36 | 130.15 | 5.1767 | 0.01752 | 89.37 | 124.40 | 5.0618 |
| 160 | 0.02937 | 110.44 | 154.50 | 5.3394 | 0.02144 | 107.55 | 150.43 | 5.2358 |
| 180 | 0.03393 | 126.71 | 177.60 | 5.4755 | 0.02503 | 124.42 | 174.48 | 5.3775 |
| 200 | 0.03832 | 142.56 | 200.03 | 5.5937 | 0.02844 | 140.66 | 197.53 | 5.4989 |
| 220 | 0.04260 | 158.14 | 222.05 | 5.6987 | 0.03174 | 156.52 | 219.99 | 5.6060 |
| 240 | 0.04682 | 173.57 | 243.80 | 5.7933 | 0.03496 | 172.15 | 242.08 | 5.7021 |
| 260 | 0.05099 | 188.87 | 265.36 | 5.8796 | 0.03814 | 187.62 | 263.90 | 5.7894 |
| 280 | 0.05512 | 204.10 | 286.78 | 5.9590 | 0.04128 | 202.97 | 285.53 | 5.8696 |
| 300 | 0.05922 | 219.27 | 308.10 | 6.0325 | 0.04440 | 218.24 | 307.03 | 5.9438 |
| 350 | 0.06940 | 257.03 | 361.13 | 6.1960 | 0.05209 | 256.21 | 360.39 | 6.1083 |
| 400 | 0.07949 | 294.73 | 413.96 | 6.3371 | 0.05971 | 294.05 | 413.47 | 6.2500 |
| 450 | 0.08953 | 332.53 | 466.82 | 6.4616 | 0.06727 | 331.95 | 466.49 | 6.3750 |
| 500 | 0.09953 | 370.54 | 519.84 | 6.5733 | 0.07480 | 370.05 | 519.65 | 6.4870 |
| 600 | 0.11948 | 447.71 | 626.92 | 6.7685 | 0.08980 | 447.33 | 626.93 | 6.6825 |
| 700 | 0.13937 | 526.89 | 735.94 | 6.9365 | 0.10474 | 526.59 | 736.07 | 6.8507 |
| 800 | 0.15923 | 608.39 | 847.22 | 7.0851 | 0.11965 | 608.14 | 847.45 | 6.9994 |
| 900 | 0.17906 | 692.24 | 960.83 | 7.2189 | 0.13454 | 692.04 | 961.13 | 7.1333 |
| 1000 | 0.19889 | 778.32 | 1076.65 | 7.3409 | 0.14942 | 778.16 | 1077.01 | 7.2553 |
| | 3000 kPa (123.61 K) | | | | 10000 kPa | | | |
| Sat. | 0.00582 | 46.03 | 63.47 | 4.5032 | — | — | — | — |
| 140 | 0.01038 | 79.98 | 111.13 | 4.8706 | 0.00200 | 0.84 | 20.87 | 4.0373 |
| 160 | 0.01350 | 101.35 | 141.85 | 5.0763 | 0.00291 | 47.44 | 76.52 | 4.4088 |
| 180 | 0.01614 | 119.68 | 168.09 | 5.2310 | 0.00402 | 82.44 | 122.65 | 4.6813 |
| 200 | 0.01857 | 136.78 | 192.49 | 5.3596 | 0.00501 | 108.21 | 158.35 | 4.8697 |
| 220 | 0.02088 | 153.24 | 215.88 | 5.4711 | 0.00590 | 129.86 | 188.88 | 5.0153 |
| 240 | 0.02312 | 169.30 | 238.66 | 5.5702 | 0.00672 | 149.42 | 216.64 | 5.1362 |
| 260 | 0.02531 | 185.10 | 261.02 | 5.6597 | 0.00749 | 167.77 | 242.72 | 5.2406 |
| 280 | 0.02746 | 200.72 | 283.09 | 5.7414 | 0.00824 | 185.34 | 267.69 | 5.3331 |
| 300 | 0.02958 | 216.21 | 304.94 | 5.8168 | 0.00895 | 202.38 | 291.90 | 5.4167 |
| 350 | 0.03480 | 254.57 | 358.96 | 5.9834 | 0.01067 | 243.57 | 350.26 | 5.5967 |
| 400 | 0.03993 | 292.70 | 412.50 | 6.1264 | 0.01232 | 283.59 | 406.79 | 5.7477 |
| 500 | 0.05008 | 369.06 | 519.29 | 6.3647 | 0.01551 | 362.42 | 517.48 | 5.9948 |
| 600 | 0.06013 | 446.57 | 626.95 | 6.5609 | 0.01861 | 441.47 | 627.58 | 6.1955 |
| 700 | 0.07012 | 525.99 | 736.35 | 6.7295 | 0.02167 | 521.96 | 738.65 | 6.3667 |
| 800 | 0.08008 | 607.67 | 847.92 | 6.8785 | 0.02470 | 604.42 | 851.43 | 6.5172 |
| 900 | 0.09003 | 691.65 | 961.73 | 7.0125 | 0.02771 | 689.02 | 966.15 | 6.6523 |
| 1000 | 0.09996 | 777.85 | 1077.72 | 7.1347 | 0.03072 | 775.68 | 1082.84 | 6.7753 |

# Ideal-Gas Specific Heat

Three types of energy storage or possession were identified in Section 2.6, of which two, translation and intramolecular energy, are associated with the individual molecules. These comprise the ideal-gas model, with the third type, the system intermolecular potential energy, then accounting for the behavior of real (nonideal-gas) substances. This appendix deals with the ideal-gas contributions. Since these contribute to the energy, and therefore also the enthalpy, they also contribute to the specific heat of each gas. The different possibilities can be grouped according to the intramolecular energy contributions as follows:

## MONATOMIC GASES (INERT GASES AR, HE, NE, XE, KR, ALSO N, O, H, CL, F, . . .)

$$h = h_{\text{translation}} + h_{\text{electronic}} = h_t + h_e$$

$$\frac{dh}{dT} = \frac{dh_t}{dT} + \frac{dh_e}{dT}, \qquad C_{P0} = C_{P0t} + C_{P0e} = \frac{5}{2}R + f_e(T)$$

where the electronic contribution, $f_e(T)$, is usually small, except at very high $T$ (common exceptions are O, Cl, F).

## DIATOMIC AND LINEAR POLYATOMIC GASES ($N_2$, $O_2$, CO, OH, . . . , $CO_2$, $N_2O$, . . .)

In addition to translational and electronic contributions to specific heat, these also have molecular rotation (about the center of mass of the molecule) and also $(3a - 5)$ independent modes of molecular vibration of the $a$ atoms in the molecule relative to one another, such that

$$C_{P0} = C_{P0t} + C_{P0r} + C_{P0v} + C_{P0e} = \frac{5}{2}R + R + f_v(T) + f_e(T)$$

where the vibrational contribution is

$$f_v(T) = R \sum_{i=1}^{3a-5} [x_i^2 e^{x_i}/(e^{x_i} - 1)^2], \qquad x_i = \frac{\theta_i}{T}$$

and the electronic contribution, $f_e(T)$, is usually small, except at very high $T$ (common exceptions are $O_2$, NO, OH).

---

**EXAMPLE C.1**

$N_2$, $3a - 5 = 1$ vibrational mode, with $\theta_i = 3392$ K.

At $T = 300$ K, $C_{P0} = 0.742 + 0.2968 + 0.0005 + \approx 0 = 1.0393$ kJ/kg K.

At $T = 1000$ K, $C_{P0} = 0.742 + 0.2968 + 0.123 + \approx 0 = 1.1618$ kJ/kg K.

(an increase of 11.8% from 300 K).

---

**EXAMPLE C.2**

$CO_2$, $3a - 5 = 4$ vibrational modes, with $\theta_i = 960$ K, 960 K, 1993 K, 3380 K

At $T = 300$ K, $C_{P0} = 0.4723 + 0.1889 + 0.1826 + \approx 0 = 0.8438$ kJ/kg K.

At $T = 1000$ K, $C_{P0} = 0.4723 + 0.1889 + 0.5659 + \approx 0 = 1.2271$ kJ/kg K.

(an increase of 45.4% from 300 K).

---

## NONLINEAR POLYATOMIC MOLECULES ($H_2O$, $NH_3$, $CH_4$, $C_2H_6$, . . .)

Contributions to specific heat are similar to those for linear molecules, except that the rotational contribution is larger and there are $(3a - 6)$ independent vibrational modes, such that

$$C_{P0} = C_{P0t} + C_{P0r} + C_{P0v} + C_{P0e} = \frac{5}{2} R + \frac{3}{2} R + f_v(T) + f_e(T)$$

where the vibrational contribution is

$$f_v(T) = R \sum_{i=1}^{3a-6} [x_i^2 e^{x_i}/(e^{x_i} - 1)^2], \qquad x_i = \frac{\theta_i}{T}$$

and $f_e(T)$ is usually small, except at very high temperatures.

---

**EXAMPLE C.3**

$CH_4$, $3a - 6 = 9$ vibrational modes, with $\theta_i = 4196$ K, 2207 K (two modes), 1879 K (three), 4343 K (three)

At $T = 300$ K, $C_{P0} = 1.2958 + 0.7774 + 0.1527 + \approx 0 = 2.2259$ kJ/kg K.

At $T = 1000$ K, $C_{P0} = 1.2958 + 0.7774 + 2.4022 + \approx 0 = 4.4754$ kJ/kg K.

(an increase of 101.1% from 300 K).

# Appendix D

# Equations of State

Some of the most used pressure-explicit equations of state can be shown in a form with two parameters. This form is known as a cubic equation of state and contains as a special case the ideal-gas law:

$$P = \frac{RT}{v - b} - \frac{a}{v^2 + cbv + db^2}$$

where $(a, b)$ are parameters and $(c, d)$ define the model as shown in the following table with the acentric factor $(\omega)$ and

$$b = b_0 RT_c/P_c \qquad \text{and} \qquad a = a_0 R^2 T_c^2/P_c$$

The acentric factor is defined by the saturation pressure at a reduced temperature $T_r = 0.7$

$$\omega = -\frac{\ln P_r^{\text{sat}} \text{ at } T_r = 0.7}{\ln 10} - 1$$

TABLE D.1
**Equations of State**

| Model | $c$ | $d$ | $b_0$ | $a_0$ |
|---|---|---|---|---|
| Ideal gas | 0 | 0 | 0 | 0 |
| van der Waals | 0 | 0 | 1/8 | 27/64 |
| Redlich–Kwong | 1 | 0 | 0.08664 | $0.42748\, T_r^{-1/2}$ |
| Soave | 1 | 0 | 0.08664 | $0.42748\,[1 + f(1 - T_r^{1/2})]^2$ |
| Peng–Robinson | 2 | −1 | 0.0778 | $0.45724\,[1 + f(1 - T_r^{1/2})]^2$ |

$$f = 0.48 + 1.574\omega - 0.176\omega^2 \qquad \text{for Soave}$$

$$f = 0.37464 + 1.54226\omega - 0.26992\omega^2 \qquad \text{for Peng–Robinson}$$

**TABLE D.2**
*Empirical Constants for Benedict–Webb–Rubin Equation**

| Gas | Formula | $A_0$ | $B_0$ | $C_0 \times 10^{-6}$ | $a$ | $b$ | $c \times 10^{-6}$ | $\alpha \times 10^3$ | $\gamma \times 10^2$ |
|---|---|---|---|---|---|---|---|---|---|
| Methane | $CH_4$ | 1.85500 | 0.042600 | 0.022570 | 0.49400 | 0.00338004 | 0.002545 | 0.124359 | 0.600 |
| Ethylene | $C_2H_4$ | 3.33958 | 0.0556833 | 0.131140 | 0.25900 | 0.008600 | 0.021120 | 0.17800 | 0.923 |
| Ethane | $C_2H_6$ | 4.15556 | 0.0627724 | 0.179592 | 0.34516 | 0.011122 | 0.032767 | 0.243389 | 1.180 |
| Propylene | $C_3H_6$ | 6.11220 | 0.0850647 | 0.439182 | 0.774056 | 0.0187059 | 0.102611 | 0.455696 | 1.829 |
| Propane | $C_3H_8$ | 6.872 | 0.097313 | 0.508256 | 0.94770 | 0.022500 | 0.12900 | 0.607175 | 2.200 |
| $n$-Butane | $C_4H_{10}$ | 10.0847 | 0.124361 | 0.992830 | 1.88231 | 0.0399983 | 0.316400 | 1.10132 | 3.400 |
| $n$-Pentane | $C_5H_{12}$ | 12.1794 | 0.156751 | 2.12121 | 4.07480 | 0.066812 | 0.82417 | 1.81000 | 4.750 |
| $n$-Hexane | $C_6H_{14}$ | 14.4373 | 0.177813 | 3.31935 | 7.11671 | 0.109131 | 1.51276 | 2.81086 | 6.66849 |
| $n$-Heptane | $C_7H_{16}$ | 17.5206 | 0.199005 | 4.74574 | 10.36475 | 0.151954 | 2.47000 | 4.35611 | 9.000 |
| Nitrogen | $N_2$ | 1.19250 | 0.04580 | 0.0058891 | 0.01490 | 0.00198154 | 0.000548064 | 0.291545 | 0.750 |
| Oxygen | $O_2$ | 1.49880 | 0.046524 | 0.0038617 | -0.040507 | -0.000027963 | -0.00020376 | 0.008641 | 0.359 |
| Ammonia | $NH_3$ | 3.78928 | 0.0516461 | 0.178567 | 0.10354 | 0.000719561 | 0.000157536 | 0.00465189 | 1.980 |
| Carbon dioxide | $CO_2$ | 2.67340 | 0.045628 | 0.11333 | 0.051689 | 0.0030819 | 0.0070672 | 0.11271 | 0.494 |

*Units: atmospheres, liters, moles, K, gas constant $R = 0.08206$.

**TABLE D.3**
**The Lee–Kesler Equation of State**

The Lee–Kesler generalized equation of state is

$$Z = \frac{P_r v_r'}{T_r} = 1 + \frac{B}{v_r'} + \frac{C}{v_r'^2} + \frac{D}{v_r'^5} + \frac{c_4}{T_r^3 v_r'^2} + \left(\beta + \frac{\gamma}{v_r'^2}\right)\exp\left(-\frac{\gamma}{v_r'^2}\right)$$

$$B = b_1 - \frac{b_2}{T_r} - \frac{b_3}{T_r^2} - \frac{b_4}{T_r^3}$$

$$C = c_1 - \frac{c_2}{T_r} + \frac{c_3}{T_r^3}$$

$$D = d_1 + \frac{d_2}{T_r}$$

in which

$$T_r = \frac{T}{T_c}, \quad P_r = \frac{P}{P_c}, \quad v_r' = \frac{v}{RT_c/P_c}$$

The set of constants is as follows:

| Constant | Simple Fluids | Constant | Simple Fluids |
|----------|---------------|----------|---------------|
| $b_1$ | 0.118 119 3 | $c_3$ | 0.0 |
| $b_2$ | 0.265 728 | $c_4$ | 0.042 724 |
| $b_3$ | 0.154 790 | $d_1 \times 10^4$ | 0.155 488 |
| $b_4$ | 0.030 323 | $d_2 \times 10^4$ | 0.623 689 |
| $c_1$ | 0.023 674 4 | $\beta$ | 0.653 92 |
| $c_2$ | 0.018 698 4 | $\gamma$ | 0.060 167 |

**TABLE D.4**
**Saturated Liquid–Vapor Compressibilities, Lee–Kesler Simple Fluid**

| $T_r$ | 0.40 | 0.50 | 0.60 | 0.70 | 0.80 | 0.85 | 0.90 | 0.95 | 1 |
|-------|------|------|------|------|------|------|------|------|---|
| $P_r$ sat | 2.7E-4 | 4.6E-3 | 0.028 | 0.099 | 0.252 | 0.373 | 0.532 | 0.737 | 1 |
| $Z_f$ | 6.5E-5 | 9.5E-4 | 0.0052 | 0.017 | 0.042 | 0.062 | 0.090 | 0.132 | 0.29 |
| $Z_g$ | 0.999 | 0.988 | 0.957 | 0.897 | 0.807 | 0.747 | 0.673 | 0.569 | 0.29 |

**TABLE D.5**
**Acentric Factor for Some Substances**

| Substance | | $\omega$ | Substance | | $\omega$ |
|-----------|--|----------|-----------|--|----------|
| Ammonia | $NH_3$ | 0.25 | Water | $H_2O$ | 0.344 |
| Argon | Ar | 0.001 | n-Butane | $C_4H_{10}$ | 0.199 |
| Bromine | $Br_2$ | 0.108 | Ethane | $C_2H_6$ | 0.099 |
| Helium | He | −0.365 | Methane | $CH_4$ | 0.011 |
| Hydrogen | $H_2$ | −0.218 | R-32 | | 0.277 |
| Nitrogen | $N_2$ | 0.039 | R-125 | | 0.305 |

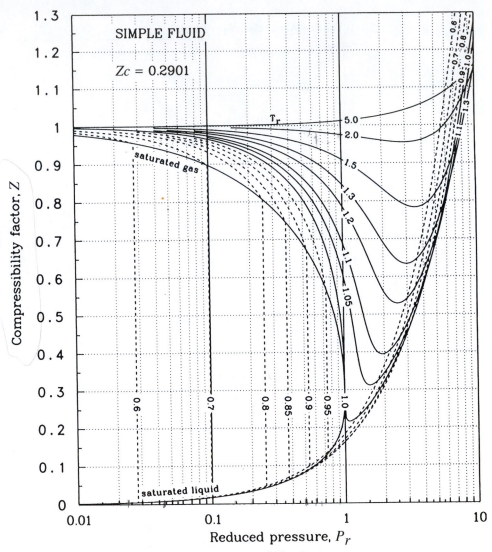

**FIGURE D.1** Lee–Kesler Simple Fluid Compressibility Factor.

# Appendix E

# Figures

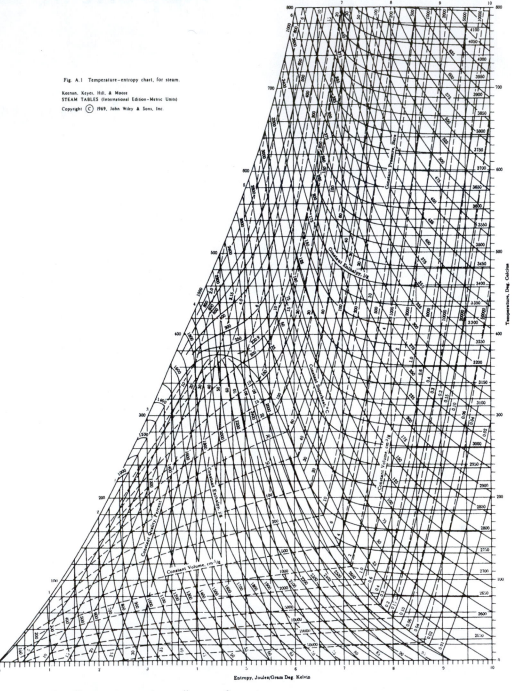

FIGURE E.1 Temperature–entropy diagram for water.

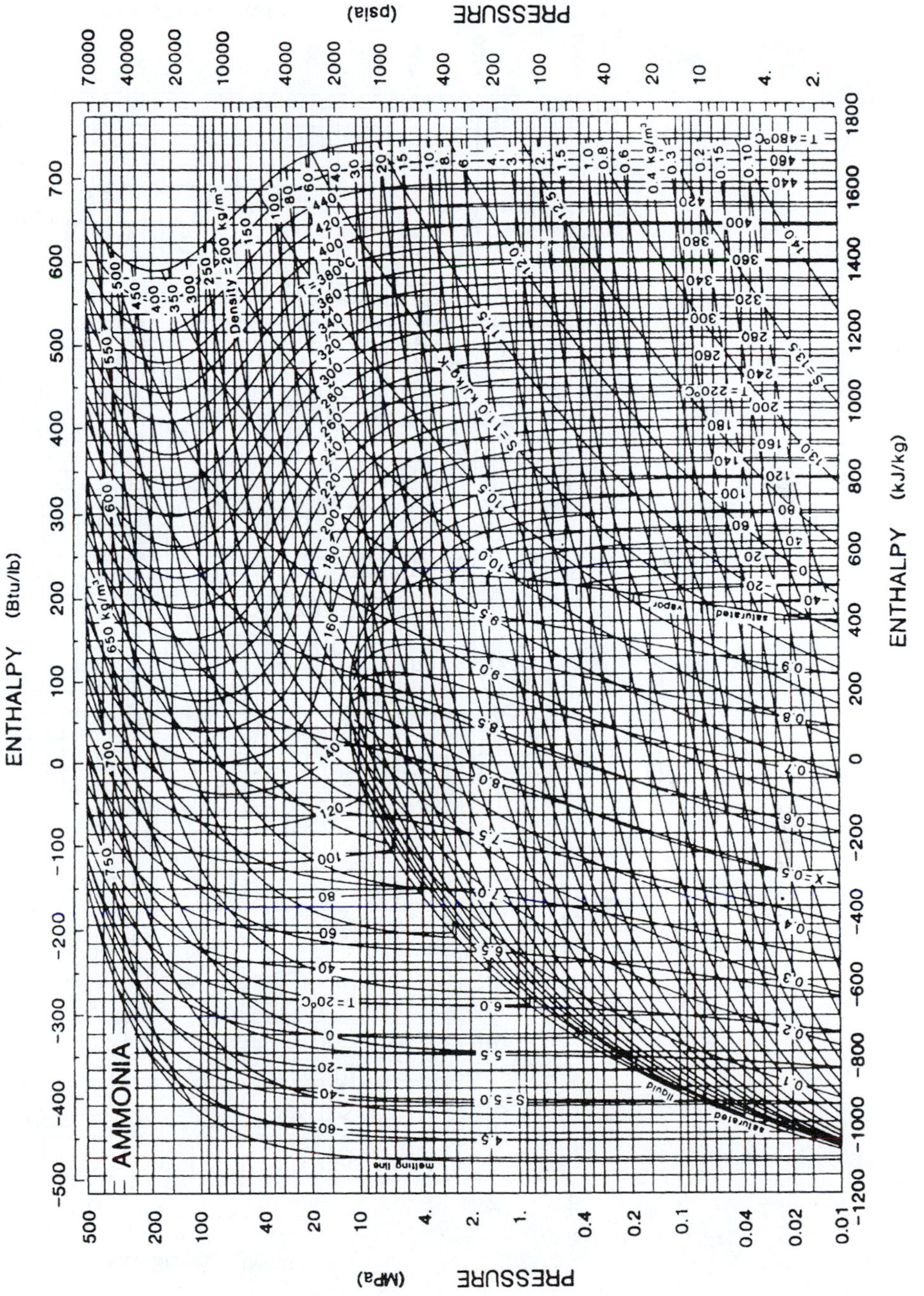

**FIGURE E.2** Pressure–enthalpy diagram for ammonia.

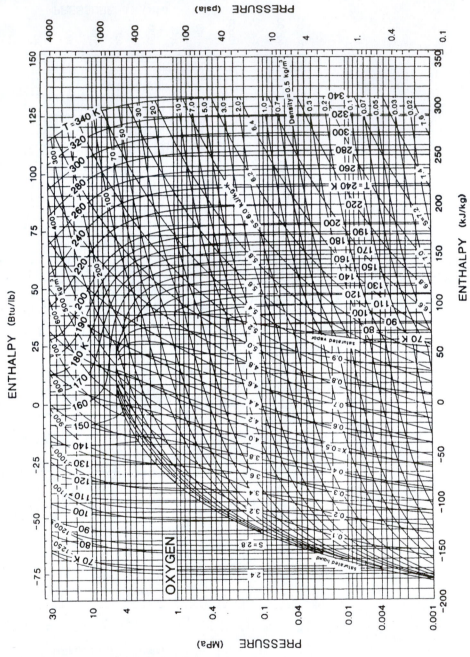

**FIGURE E.3** Pressure–enthalpy diagram for oxygen.

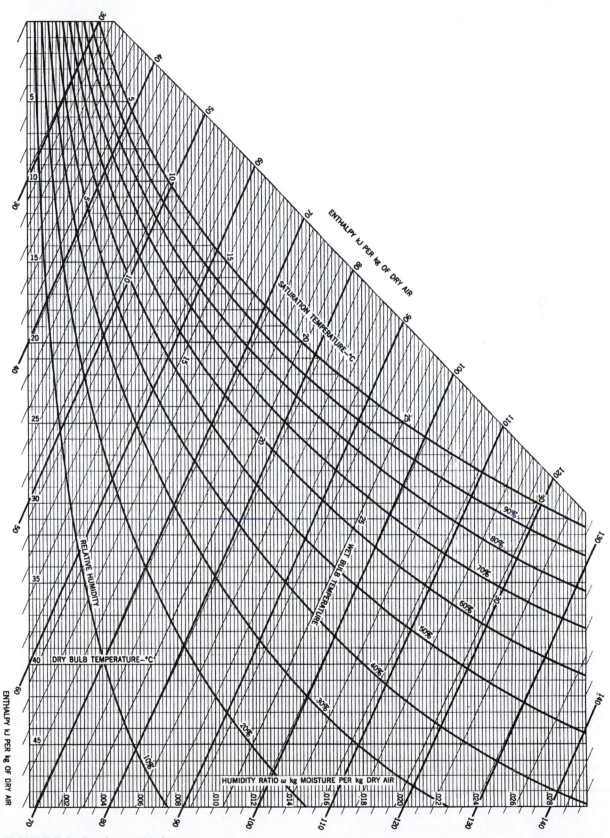

**FIGURE E.4** Psychrometric chart.

565

# Appendix **F**

# English Unit Supplement

The system of units in current use in the United States is termed the English Engineering System. In this system, the unit of time is the second, discussed in Section 2.5. The basic unit of length is the foot (ft), which is defined at present in terms of the meter as

$$1 \text{ ft} = 0.3048 \text{ m}$$

The inch (in.) is defined in terms of the foot:

$$12 \text{ in.} = 1 \text{ ft}$$

These and other conversions are listed in Table A.1. The unit of mass in this system is the pound mass (lbm), originally the mass of a certain platinum cylinder kept in the Tower of London, but now defined in terms of the kilogram as

$$1 \text{ lbm} = 0.453592 \text{ kg}$$

A related unit is the pound mole (lb mol), which is an amount of substance in pounds mass numerically equal to the molecular weight of that substance. It is important to distinguish between a pound mole and a mole (gram mole).

In the English Engineering System of Units, the unit of force is the pound force (lbf), defined as the force with which the standard pound mass is attracted to the earth under conditions of standard acceleration of gravity, which is that at 45° latitude and sea level elevation, 9.806 65 $m/s^2$ or 32.1740 $ft/s^2$. Thus, it follows from Newton's second law, that

$$1 \text{ lbf} = 32.174 \text{ lbm ft/s}^2$$

which is a necessary factor for the purpose of units conversion and consistency. Note that we must be careful to distinguish between a lbm and a lbf, and we do not use the term pound alone. For gravitational accelerations that are not too different from the standard value (32.174 $ft/s^2$ or 9.80665 $n/s^2$), the lbf is approximately numerically equal to the lbm. This is the reason that people often drop the designation, which can lead to confusion and inconsistencies. To illustrate, let us calculate the force due to gravity on a one pound mass at a location where the acceleration due to gravity is 32.14 ft/s² (about 10,000 ft above sea level):

$$F = ma = \frac{1 \text{ lbm} \times 32.14 \text{ ft/s}^2}{32.174 \text{ lbm ft/lbf s}^2}$$

$$= 0.999 \text{ lbf}$$

The temperature scale in the English Engineering System is the Fahrenheit scale, on which the normal freezing point of water (0°C) is 32°F, and the normal boiling point (100°C) is 212°F. Thus a 100° change on the Celsius scale corresponds to a 180° change on the Fahrenheit scale. The absolute temperature scale related to the Fahrenheit scale is named the Rankine scale, which is designated R. The relation between the two is

$$R = F + 459.67$$

Following the basic units defined above, the unit for pressure (force per unit area) is the pound force per square inch (lbf/in.² = psi). Also used is the standard atmosphere, where

$$1 \text{ atm} = 14.6959 \text{ lbf/in.}^2 \ (=101.325 \text{ kPa})$$

Since the volume unit is the ft³, the units for specific volume are ft³/lbm or ft³/lb mol, and the Universal Gas Constant $\bar{R}$ is

$$\bar{R} = 1545 \frac{(\text{lbf/ft}^2) \times (\text{ft}^3/\text{lb mol})}{R}$$

$$= 1545 \frac{\text{ft lbf}}{\text{lb mol} \cdot R} = 1.98589 \frac{\text{Btu}}{\text{lb mol} \cdot R}$$

For any particular substance on a unit lbm basis,

$$R = \frac{\bar{R}}{M} \left( \frac{\text{ft lbf}}{\text{lbm} \cdot R} \right)$$

Values of $R$ for various gases are listed in Table G4.

The basic energy unit (force × length) is the foot pound force (ft lbf), but in most thermodynamic applications, a more convenient (because of magnitude) unit is the British Thermal Unit (Btu). This unit was originally defined as the amount of energy required to raise 1 lbm of water from 59.5 F to 60.5 F, but it is now defined more precisely by the relation

$$1 \text{ Btu} = 778.17 \text{ ft lbf}$$

Two particular types of energy that we deal with are kinetic energy and potential energy; each of these is associated with an integral involving Newton's Second Law and, therefore, involves units conversion. Proceeding as in Section 5.2 for kinetic energy, we get

$$KE = \int F \, dx = \int m \frac{d\mathbf{V}}{dx} \frac{dx}{dt} dx = \int m\mathbf{V} \, d\mathbf{V} = \frac{m\mathbf{V}^2}{2}$$

To illustrate, for a mass of 5 lbm and velocity of 200 ft/s, the kinetic energy

$$KE = \frac{5 \text{ lbm} \times (200 \text{ ft/s})^2}{2 \times 32.174 \dfrac{\text{lbm ft}}{\text{lbf s}^2} \times 778 \dfrac{\text{ft lbf}}{\text{Btu}}} = 4.0 \text{ Btu}$$

For potential energy, we have

$$PE = \int F\, d\mathbf{Z} = \int mg\, d\mathbf{Z} = mg\mathbf{Z}$$

To illustrate, for a mass of 5 lbm, elevation of 200 ft, and local gravitational acceleration of 32.16 ft/s², the potential energy is

$$PE = \frac{5 \text{ lbm} \times 32.16 \text{ ft/s}^2 \times 200 \text{ ft}}{32.174 \dfrac{\text{lbm ft}}{\text{lbf s}^2} \times 778 \dfrac{\text{ft lbf}}{\text{Btu}}} = 1.28 \text{ Btu}$$

The rate of energy change, or power, unit in the English system is the horsepower (hp), defined as

$$1\text{hp} = 550 \frac{\text{ft lbf}}{\text{s}} = 2544 \frac{\text{Btu}}{\text{h}} = 745.7 \text{ W}$$

Finally, the entropy (energy/temperature) unit in this system is Btu/R, or per unit mass is Btu/lbm · R.

## English Unit Problems

A set of English unit problems has been selected from the corresponding SI unit chapter problems to cover the use of English units. These problems are chosen for the use of the units and their conversions together with the use of all the English unit tables. Although the problem set does not represent a complete coverage of all the text material we have selected problems for all chapters so that the use of English units can be done in nearly all of the subjects included in the book.

# ENGLISH UNIT PROBLEMS

**2.1** A 2500-lbm car moving at 15 mi/h is accelerated at a constant rate of 15 ft/s² up to a speed of 50 mi/h. What are the force and total time required?

**2.2** A car of mass 4000 lbm travels with a velocity of 60 mi/h. Find the kinetic energy. How high in the standard gravitational field should it be lifted to have the same potential energy?

**2.3** One pound mass of diatomic oxygen ($O_2$ molecular weight 32) is contained in a 100-gal tank. Find the specific volume on both a mass and mole basis ($v$ and $\bar{v}$).

**2.4** A 7-ft-m tall steel cylinder has a cross-sectional area of 15 ft². At the bottom, with a height of 2 ft, is liquid water on top of which is a 4-ft-high layer of gasoline. The gasoline surface is exposed to atmospheric air at 14.7 psia. What is the highest pressure in the water?

**2.5** A piston/cylinder with cross-sectional area of 0.1 ft² has a piston mass of 200 lbm resting on the stops, as shown in Fig. P2.45. With an outside atmospheric pressure of 1 atm, what should the water pressure be to lift the piston?

**2.6** The pressure gage on an air tank shows 10 psi when the diver is 30 ft down in the ocean. At what depth will the gage pressure be zero? What does that mean?

**2.7** Blue manometer fluid of density 60 lbm/ft³ shows a column height difference of 3 inches vacuum with one end attached to a pipe and the other end open to the atmosphere at 14.5 psia. What is the absolute pressure in the pipe?

**3.1** A substance is at 300 psia, 65 F in a rigid tank. Using only critical properties, can the phase of the mass be determined if the substance is nitrogen, water, or propane?

**3.2** Determine the missing property of ($P$, $T$, $v$, and $x$ if applicable) for water at:
a. 600 psia, 500 F
a. 15 psia, 30 F

**3.3** Give the phase and the missing property of $P$, $T$, $v$, and $x$ for R-134a at:
a. $T = -10\ F, P = 18$ psia
b. $P = 40$ psia, $v = 1.3$ ft³/lbm

**3.4** Determine the missing property of ($P$, $T$, $v$, and $x$ if applicable) for water at:

a. 680 psia, 0.03 ft³/lbm

b. 150 psia, 320 F

c. 400 F, 3 ft³/lbm

**3.5** A sealed rigid vessel has volume of 35 ft³ and contains 2 lbm of water at 200 F. The vessel is now heated. If a safety pressure valve is installed, at what pressure should the valve be set to have a maximum temperature of 400 F?

**3.6** Two tanks are connected together as shown in Fig. P3.49, both containing water. Tank $A$ is at 30 lbf/in², $v = 8$ ft³/lbm, $V = 40$ ft³, and tank $B$ contains 8 lbm at 80 lbf/in², 750 F. The valve is now opened and the two come to a uniform state. Find the final specific volume.

**3.7** Air in an internal combustion engine is at 440 F, 150 psia with a volume of 10 in³. Now combustion heats it to 2700 R in a constant volume process. What is the mass of air and how high does the pressure become?

**3.8** A 35 ft³ rigid tank has propane at 15 psia, 540 R and connected by a valve to another tank of 20 ft³ with propane at 40 psia, 720 R. The valve is opened and the two tanks come to a uniform state at 600 R. What is the final pressure?

**3.9** Determine the mass of an ethane gas stored in a 25-ft³ tank at 250 F, 440 lbf/in² using the compressibility chart. Estimate the error (%) if the ideal gas model is used.

**4.1** Work as $F \, \Delta x$ has units of lbf-ft, what is that in Btu?

**4.2** A work of 2.5 Btu must be delivered on a rod from a pneumatic piston/cylinder where the air pressure is limited to 75 psia. What diameter cylinder should I have to restrict the rod motion to maximum 2 ft?

**4.3** A force of 300 lbf moves a truck with 40 mi/h up a hill. What is the power?

**4.4** A cylinder fitted with a frictionless piston contains 0.1 lbm of superheated refrigerant R-134a vapor at 100 psi, 300 F. The setup is cooled at constant pressure until the R-134a reaches a quality of 25%. Calculate the work done in the process.

**4.5** A piston cylinder contains 1 lbm of air at 100 psia, 900 R. The air expands in a process where $P$ is linearly related to volume to a final state of 15 psia, 500 R. Find the work in the process.

**4.6** A piston/cylinder has 15 ft of liquid 70 F water on top of the piston ($m = 0$) with cross-sectional area of 1 ft² (see Fig. P2.39). Air is let in under the piston that rises and pushes the water out over the top edge. Find the

necessary work to push all the water out and plot the process in a $P$–$V$ diagram.

**4.7** Helium gas expands from 20 psia, 630 R and 1 ft³ to 15 psia in a polytropic process with $n = 1.667$. How much work is given out?

**4.8** A water heater is covered up with insulation boards over a total surface area of 30 ft². The inside board surface is at 175 F, the outside surface is at 70 F, and the board material has a conductivity of 0.05 Btu/h ft F. How thick should the board be to limit the heat transfer loss to 720 Btu/h?

**5.1** What is 1 cal in English units; what is 1 ft-lbf in Btu?

**5.2** Airplane takeoff from an aircraft carrier is assisted by a steam driven piston/cylinder with an average pressure of 200 psia. A 38 500 lbm airplane should be accelerated from zero to a speed of 100 ft/s with 30% of the energy coming from the steam piston. Find the needed piston displacement volume.

**5.3** Give the phase and missing properties ($P$, $T$, $v$, $u$, $h$, and $x$) for water at:

a. 270 F, $u = 1000$ Btu/lbm

b. 1000 psia, $u = 300$ Btu/lbm

c. 60 psia, 340 F

**5.4** Give the phase and missing properties ($P$, $T$, $v$, $u$, $h$ and $x$) for R-134a at:

a. 30 F, 75 psia

b. 140 F, $h = 175$ Btu/lbm

c. 400 psia, $u = 200$ F

**5.5** A cylinder fitted with a frictionless piston contains 4 lbm of superheated refrigerant R-134a vapor at 400 lbf/in², 200 F. The cylinder is now cooled so that the R-134a remains at constant pressure until it reaches a quality of 75%. Calculate the heat transfer in the process.

**5.6** A constant-pressure piston/cylinder has 2 lbm of water at 1100 F and 2.26 ft³. It is now cooled to occupy 1/10 of the original volume. Find the heat transfer in the process.

**5.7** A water-filled reactor with volume of 50 ft³ is at 2000 lbf/in², 560 F and placed inside a containment room, as shown in Fig. P5.40. The room is well insulated and initially evacuated. Due to a failure, the reactor ruptures and the water fills the containment room. Find the minimum room volume so the final pressure does not exceed 30 lbf/in².

**5.8** An engine, shown in Fig. P5.68, consists of a 200 lbm cast iron block with a 40 lbm aluminum head, 40 lbm steel parts, 10 lbm engine oil, and 12 lbm glycerine (antifreeze). Everything begins at 40 F and as the engine starts, it absorbs a net of 7000 Btu before it reaches a steady uniform temperature. We want to know how hot it becomes.

**5.9** A piston cylinder of carbon dioxide gas at constant 1 atm and 2000 R is heated to 2200 R. Solve for the heat transfer using the heat capacity from Table F.4 and repeat using Table F.6

**5.10** Do the previous problem for nitrogen gas.

**5.11** A closed cylinder is divided into two rooms by a frictionless piston held in place by a pin, as shown in Fig. P5.81. Room $A$ has 0.3 ft$^3$ air at 14.7 lbf/in$^2$, 90 F, and room $B$ has 10 ft$^3$ saturated water vapor at 90 F. The pin is pulled, releasing the piston, and both rooms come to equilibrium at 90 F. Considering a control mass of the air and water, determine the work done by the system and the heat transfer to the cylinder.

**5.12** Helium gas expands from 20 psia, 600 R and 9 ft$^3$ to 15 psia in a polytropic process with $n = 1.667$. How much heat transfer is involved?

**5.13** Oxygen at 50 lbf/in$^2$, 200 F is in a piston/cylinder arrangement with a volume of 4 ft3. It is now compressed in a polytropic process with exponent, $n = 1.2$, to a final temperature of 400 F. Calculate the heat transfer for the process.

**5.14** Water is in a piston/cylinder maintaining constant $P$ at 330 F, quality 90%, with a volume of 4 ft$^3$. A heater is turned on heating the water with 10 000 Btu/h. What is the elapsed time to vaporize all the liquid?

**5.15** A computer in a closed room of volume 5000 ft$^3$ dissipates energy at a rate of 7.5 kW. The room has 100 lbm of wood, 50 lbm of steel and air, with all material at 540 R, 1 atm. Assuming all the mass heats up uniformly how long time will it take to increase the temperature 20 F?

**5.16** A crane uses 7000 Btu/h to raise a 200-lbm box 60 ft. How much time does it take?

**6.1** Liquid water at 60 F flows out of a nozzle straight up 40 ft. What is nozzle $\mathbf{V}_{exit}$?

**6.2** R-134a at 90 F, 125 psia is throttled to come out cold at 10 F. What is the exit $P$?

**6.3** In a boiler you vaporize some liquid water at 103 psia flowing at 3 ft/s. What is the velocity of the saturated vapor at 103 psia if the pipe size is the same? Can the flow then be constant $P$?

**6.4** Air at 60 ft/s, 480 R, 11 psia with 10 lbm/s flows into a jet engine and it flows out at 1500 ft/s, 1440 R, 11 psia. What is the change (power) in flow of kinetic energy?

**6.5** Nitrogen gas flows into a convergent nozzle at 30 lbf/in$^2$, 600 R and very low velocity. It flows out of the nozzle at 15 lbf/in$^2$, 500 R. If the nozzle is Insulated, find the exit velocity.

**6.6** A compressor in a new refrigerator receives R-134a at $-10$ F and $x = 1$. The exit is at 150 psia, 140 F. Neglect kinetic energies and find the specific work.

**6.7** A small water pump is used in an irrigation system. The pump takes water in from a river at 50 F, 1 atm at a rate of 10 lbm/s. The exit line enters a pipe that goes up to an elevation 60 ft above the pump and river, where the water runs into an open channel. Assume that the process is adiabatic and that the water stays at 50 F. Find the required pump work.

**6.8** An automotive radiator has glycerine at 200 F enter and return at 130 F as shown in Fig. P6.72. Air flows in at 68 F and leaves at 77 F. If the radiator should transfer 33 hp, what is the mass flow rate of the glycerine and what is the volume flow rate of air in at 15 psia?

**6.9** The following data are for a simple steam power plant as shown in Fig. P6.82.

| State | 1 | 2 | 3 | 4 | 5 | 6 | 7 |
|---|---|---|---|---|---|---|---|
| $P$ lbf/in$^2$ | 900 | 890 | 860 | 830 | 800 | 1.5 | 1.4 |
| $T$ F | | 115 | 350 | 920 | 900 | | 110 |
| $h$, Btu/lbm | | 85.3 | 323 | 1468 | 1456 | 1029 | 78 |

State 6 has $x_6 = 0.92$ and velocity of 600 ft/s. The rate of steam flow is 200 000 lbm/h, with 400 hp input to the pump. Piping diameters are 8 in. from the steam generator to the turbine and 3 in. from the condenser to the steam generator. Determine the power output of the turbine and the heat transfer rate in the condenser.

**6.10** An initially empty cylinder is filled with air from 70 F, 15 psia until it is full. Assuming no heat transfer, is the final temperature larger, equal to, or smaller than 70 F? Does the final $T$ depends on the size of the cylinder?

**6.11** An empty canister of volume 0.05 ft$^3$ is filled with R-134a from a line flowing saturated liquid R-134a at 40 F. The filling is done quickly, so it is adiabatic. How much mass of R-134a is in the canister? It is now placed on a storage shelf where it slowly heats up to room temperature 70 F. What is the final pressure?

**7.1** A gasoline engine produces 20 hp using 35 Btu/s of heat transfer from burning fuel. What is its thermal efficiency and how much power is rejected to the ambient?

**7.2** A refrigerator removes 1.5 Btu from the cold space using 1 Btu work input. How much energy goes into the kitchen and what is its coefficient of performance?

**7.3** An air conditioner provides 1 lbm/s of air at 60 F cooled from the outside air at 95 F. Estimate the amount of power needed to operate the unit assuming it has a coefficient of performance that is half of what a Carnot unit would have.

**7.4** A farmer runs a heat pump with a 2 kW motor. It should keep a chicken hatchery at 90 F, which loses energy at a rate of 10 Btu/s to the colder ambient $T_{amb}$. What is the minimum coefficient of performance that will be acceptable for the heat pump?

**7.5** A large heat pump should upgrade 5 000 Btu/s of heat at 185 F to be delivered as heat at 300 F. Suppose the actual heat pump has a COP of 2.5. How much power is needed to drive the unit? For the same COP, how high a high temperature would a Carnot heat pump have if I assume the same low $T$?

**8.1** Find, the missing properties of ($P$, $T$, $u$, $h$, $s$, and $x$) for water at:

a. 150 F, $s = 1.75$ Btu/lbm R

b. 1500 psia, $u = 1350$ Btu/lbm

c. 280 F, $s = 2$ Btu/lbm R

**8.2** Find the missing properties of ($P$, $T$, $u$, $h$, $s$, and $x$) for R-134a at:

a. 80 F, 80 psia

b. 10 F, $x = 0.45$

c. 100 psia, $h = 180$ Btu/lbm

**8.3** Water at 30 lbf/in.$^2$, $x = 1.0$ is compressed in a piston/cylinder to 140 lbf/in.$^2$, 600 F in a reversible process. Find the sign for the work and the sign for the heat transfer.

**8.4** One pound-mass of water at 600 F expands against a piston in a cylinder until it reaches ambient pressure, 14.7 lbf/in.$^2$, at which point the water has a quality of 90%. It may be assumed that the expansion is reversible and adiabatic.

a. What was the initial pressure in the cylinder?

b. How much work is done by the water?

**8.5** A cylinder containing R-134a at 60 F, 30 psia has an initial volume of 1 ft$^3$. A piston compresses the R-134a in a reversible isothermal process to a state of saturated vapor. Calculate the required work and heat transfer to accomplish this process.

**8.6** A foundry form box with 50 lbm of 400 F hot sand is dumped into a bucket with 2 ft$^3$ water at 60 F. Assuming no heat transfer with the surroundings and no boiling away of liquid water, calculate the net entropy change for the process.

**8.7** Nitrogen gas in a piston/cylinder at 500 R and 1 atm with a volume of 1 ft$^3$ is compressed in a reversible adiabatic process to a final temperature of 1000 R. Find the final pressure and volume using constant heat capacity from Table F.4.

**8.8** Nitrogen gas in a piston/cylinder at 500 R and 1 atm with a volume of 1 ft$^3$ is compressed in a reversible adiabatic process to a final temperature of 1000 R. Find the final pressure and volume using Table F.6.

**8.9** Helium in a piston/cylinder at 70 F, 15 lbf/in.$^2$, is brought to 720 R in a reversible polytropic process with exponent $n = 1.25$. You may assume helium is an ideal gas with constant specific heat. Find the final pressure and both the specific heat transfer and specific work.

**8.10** One kg of air at 540 R is mixed with one kg air at 720 R in a process at a constant 15 psia and $Q = 0$. Find the final $T$ and the entropy generation in the process.

**8.11** One lbm of air at 15 psia is mixed with one lbm air at 30 psia, both at 540 R, in a rigid insulated tank. Find the final state ($P$, $T$) and the entropy generation in the process.

**8.12** A farmer runs a heat pump using 2.5 hp of power input. It keeps a chicken hatchery at a constant 86 F while the room loses 20 Btu/s to the colder outside ambient at 50 F. What is the rate of entropy generated in the heat pump? What is the rate of entropy generated in the heat loss process?

**9.1** Steam enters a turbine at 400 psia, 900 F and expands in a reversible adiabatic process with an exhaust at 130 F. The turbine produces a power output of 800 Btu/s and kinetic/potential energies can be neglected. Find the mass flow rate.

**9.2** The exit nozzle in a jet engine receives air at 2100 R, 20 psia, with negligible kinetic energy. The exit pressure is 10 psia, and the process is reversible and adiabatic. Use constant heat capacity at 77 F to find the exit velocity.

**9.3** An expander receives 1 lbm/s air at 300 psia, 540 R, with an exit state of 60 psia, 540 R. Assume the process is reversible and isothermal. Find the rates of heat transfer and work, neglecting kinetic and potential energy changes.

**9.4** Two flowstreams of water, one at 100 lbf/in.$^2$, saturated vapor, and the other at 100 lbf/in.$^2$, 1000 F, mix adiabatically in a steady flow process to produce a single flow out at 100 lbf/in.$^2$, 600 F. Find the total entropy generation for this process.

**9.5** A condenser in a power plant receives 10 lbm/s steam at 130 F, quality 90%, and rejects the heat to cooling water with an average temperature of 62 F. Find the power given to the cooling water in this constant pressure process and the total rate of entropy generation when condenser exit is saturated liquid.

**9.6** A small pump takes in water at 70 F, 14.7 lbf/in.$^2$, and pumps it to 250 lbf/in.$^2$, at a flow rate of 200 lbm/min. Find the required pump power input.

**9.7** An initially empty 5 ft$^3$ tank is filled with air from 70 F, 15 psia until it is full. Assume no heat transfer and find the final mass and the entropy generation.

**9.8** An empty canister of volume 0.05 ft$^3$ is filled with R-134a from a line flowing saturated liquid R-134a at 40 F. The filling is done quickly, so it is adiabatic. How much mass of R-134a is in the canister. How much entropy was generated?

**9.9** A steam turbine inlet is at 200 psia, 900 F. The exit is at 40 psia. What is the lowest possible exit temperature? Which efficiency does that correspond to?

**9.10** A steam turbine inlet is at 200 psia, 900 F. The exit is at 40 psia. What is the highest possible exit temperature? Which efficiency does that correspond to?

**9.11** A steam turbine inlet is at 200 psia, 900 F. The exit is at 40 psia, 600 F. What is the isentropic efficiency?

**9.12** A flow of air at 150 psia, 540 R is throttled to 75 psia. What is the irreversibility? What is the drop in flow availability?

**9.13** A steam turbine inlet is at 200 psia, 900 F. The actual exit is at 40 psia with an actual work of 143.61 Btu/lbm. What is its second law efficiency?

**9.14** A heat exchanger increases the availability of 6 lbm/s water by 800 Btu/lbm using 20 lbm/s air coming in at 2500 R and leaving with 250 Btu/lbm less availability. What is the irreversibility and the second law efficiency?

**10.1** A gas mixture at 250 F, 18 lbf/in.$^2$ is 50% $N_2$, 30% $H_2O$, and 20% $O_2$ on a mole basis. Find the mass fractions, the mixture gas constant, and the volume for 10 lbm of mixture.

**10.2** A new refrigerant R-410a is a mixture of R-32 and R-125 in a 1:1 mass ratio. What is the overall molecular weight, the gas constant, and the ratio of specific heats for such a mixture?

**10.3** A pipe flows 1.5 lbm/s of a mixture with mass fractions of 40% $CO_2$ and 60% $N_2$ at 60 lbf/in.$^2$, 540 R. Heating tape is wrapped around a section of pipe with insulation added, and 2 Btu/s electrical power is heating the pipe flow. Find the mixture exit temperature.

**10.4** A mixture of 4 lbm oxygen and 4 lbm of argon is in an insulated piston cylinder arrangement at 14.7 lbt/in.$^2$, 540 R. The piston now compresses the mixture to half its initial volume. Find the final pressure, temperature, and the piston work.

**10.5** Carbon dioxide gas flow, 2 lbm/s, at 1200 R is mixed with a nitrogen gas flow, 1 lbm/s, at 600 R at a constant pressure of 1 atm. Find the exit temperature and the entropy generation rate using constant heat capacity from Table F.4.

**10.6** Solve the previous problem using values from table F.6.

**10.7** Consider a volume of 2000 ft$^3$ that contains an air–water vapor mixture at 14.7 lbf/in.$^2$, 60 F, and 40% relative humidity. Find the mass of water and the humidity ratio. What is the dew point of the mixture?

**10.8** If I have air at 14.7 psia and (a) 15 F, (b) 115 F, and (c) 230 F, what is the maximum absolute humidity I can have?

**10.9** Consider at 35 ft$^3$/s flow of atmospheric air at 14.7 psia, 80 F, and 80% relative humidity. Assume this flows into a basement room where it cools to 60 F at 14.7 psia. How much liquid will condense out?

**10.10** A flow of moist air from a domestic furnace, state 1 in Fig. P10.82. is at 120 F, 10% relative humidity with a flow rate of 0.1bm/s dry air. A small electric heater adds steam at 212 F, 14.7 psia, generated from tap water at 60 F. Up in the living room the flow comes out at state 4: 90 F, 60% relative humidity. Find the power needed for the electric heater and the heat transfer to the flow from state 1 to state 4.

**11.1** A steam power plant, as shown in Fig. 11.3, operating in a Rankine cycle has saturated vapor at 600 lbf/in.$^2$ leaving the boiler. The turbine exhausts to the condenser operating at 2.225 lbf/in.$^2$. Find the specific work and heat transfer in each of the ideal components and the cycle efficiency.

**11.2** A Rankine cycle uses R-134a as the working substance and powered by solar energy. It heats the R-134a to 240 F at 750 psia in the boiler/superheater. The condenser is water cooled and the exit kept at 70 F. Find $T$, $P$, and $x$ if applicable for all four states in the cycle.

**11.3** Find the thermal efficiency for the Rankine cycle in Problem 11.2.

**11.4** The power plant in Problem 11.167 is modified to have a superheater section following the boiler so the steam leaves the superheater at 600 lbf/in.$^2$, 700 F. Find the specific work and heat transfer in each of the ideal components and the cycle efficiency.

**11.5** A closed feedwater heater in a regenerative steam power cycle heats 40 lbm/s of water from 200 F, 2000 lbf/in.², to 450 F, 2000 lbf/in.². The extraction steam from the turbine enters the heater at 500 lbf/in.², 550 F and leaves as saturated liquid. What is the required mass flow rate of the extraction steam?

**11.6** A Brayton cycle produces 14 000 Btu/s with an inlet state of 60 F, 14.5 psia and a compression ratio of 16:1. The heat added in the combustion is 400 Btu/lbm. Find the highest temperature and the mass flow rate of air assuming cold air properties.

**11.7** Do the previous problem using properties from Table F.5.

**11.8** An ideal regenerator is used in the Brayton cycle of Problem 11.6. Find the modified cycle efficiency.

**11.9** The turbine in a jet engine receives air at 2200 R, 220 lbf/in.². It exhausts to a nozzle at 35 lbf/in.², which in turn exhausts to the atmosphere at 14.7 lbf/in.². The isentropic efficiency of the turbine is 85%, and the nozzle efficiency is 95%. Find the nozzle inlet temperature and the nozzle exit velocity. Assume negligible kinetic energy out of the turbine.

**11.10** A gasoline engine has a volumetric compression ratio of 10 and before compression has air at 520 R, 12.2 psia, in the cylinder. The combustion peak pressure is 900 psia. Assume cold air properties. What is the highest temperature in the cycle? Find the temperature at the beginning of the exhaust (heat rejection) and the overall cycle efficiency.

**11.11** A 6 cylinder 4 stroke minivan engine of total displacement of 220 in³ has a compression ratio of 10:1 and is running at 2000 RPM. The state before compression is 7 psia, 500 R and the combustion adds 430 Btu/lbm to the charge. Use cold air properties and find the cycle maximum temperature, the thermal efficiency.

**11.12** Find the mean effective pressure and the total power output from the engine in the previous problem.

**11.13** A diesel engine has a bore of 4 in., a stroke of 4.3 in., and a compression ratio of 19:1 running at 2000 RPM (revolutions per minute). Each cycle takes two revolutions and has a mean effective pressure of 200 lbf/in.². With a total of six cylinders find the engine power in Btu/s and horsepower, hp.

**11.14** A diesel engine has a compression ratio of 20:1 with an inlet state of 520 R, 11 psia. The combustion adds 500 Btu/lbm to the charge. Find the highest $P$ and $T$ in the cycle and the mean effective pressure.

**11.15** A car air conditioner (refrigerator) in 70 F ambient uses R-134a and I want to have cold air at 20 F produced.

What is the minimum high $P$ and the maximum low $P$ it can use?

**11.16** Consider an ideal refrigeration cycle that has a condenser pressure of 150 psia and an evaporator temperature of 10 F with R-134a as the working substance. Find the cycle COP.

**11.17** Consider the refrigerator in the previous problem having an actual compressor exit state of 150 psia, 160 F. Find the non-ideal cycle COP and the specific entropy generation in all processes inside the R-134a.

**12.1** What is the enthalpy of formation for oxygen as $O_2$? If O? for $CO_2$?

**12.2** Pentane is burned with 120% theoretical air in a constant-pressure process at 14.7 lbf/in.². The products are cooled to ambient temperature, 70 F. How much mass of water is condensed per pound-mass of fuel? Repeat the problem, assuming that the air used in the combustion has a relative humidity of 90%.

**12.3** A rigid vessel initially contains 2-pound moles of carbon and 2-pound moles of oxygen at 77 F, 30 lbf/in.². Combustion occurs, and the resulting products consist of a 1-pound mole of carbon dioxide, 1-pound mole of carbon monoxide, and excess oxygen at a temperature of 1800 R. Determine the final pressure in the vessel and the heat transfer from the vessel during the process.

**12.4** One alternative to using petroleum or natural gas as fuels is ethanol ($C_2H_5OH$), which is commonly produced from grain by fermentation. Consider a combustion process in which liquid ethanol is burned with 120% theoretical air in a steady-flow process. The reactants enter the combustion chamber at 77 F, and the products exit at 140 F, 14.7 lbf/in.². Calculate the heat transfer per pound mole of ethanol, using the enthalpy of formation of ethanol gas plus the generalized tables or charts.

**12.5** Pentene, $C_5H_{10}$ is burned with pure oxygen in a steady-state process. The products at one point are brought to 1300 R and used in a heat exchanger, where they are cooled to 77 F. Find the specific heat transfer in the heat exchanger.

**12.6** Methane, $CH_4$, is burned in a steady-state process with two different oxidizers: case **A**—pure oxygen, $O_2$ and case **B**—a mixture of $O_2 + x$Ar. The reactants are supplied at $T_0$, $P_0$, and the products are at 3200 R in both cases. Find the required equivalence ratio in case **A** and the amount of argon, $x$, for a stoichiometric ratio in case **B**.

**12.7** A burner receives a mixture of two fuels with mass fraction 40% n-butane and 60% methanol, both vapor. The fuel is burned with stoichiometric air. Find the product composition and the lower heating value of this fuel mixture (Btu/lbm fuel mix).

**12.8** Hydrogen gas is burned with pure oxygen in a steady-flow burner where both reactants are supplied in a stoichiometric ratio at the reference pressure and temperature. What is the adiabatic flame temperature?

**12.9** Acetylene gas at 77 F, 14.7 lbf/in.$^2$ is fed to the head of a cutting torch. Calculate the adiabatic flame temperature if the acetylene is burned with 100% theoretical air at 77 F.

**12.10** Solve the previous problem using 100% theoretical oxygen at 77 F.

**12.11** Ethene, $C_2H_4$, burns with 150% theoretical air in a steady-state constant-pressure process, with reactants entering at $P_0$, $T_0$. Find the adiabatic flame temperature.

**12.12** Methane is burned with air, both of which are supplied at the reference conditions. There is enough excess air to give a flame temperature of 3200 R. What are the percent theoretical air and the irreversibility in the process?

**12.13** Graphite, C, at $P_0$, $T_0$ is burned with air coming in at $P_0$, 900 R, in a ratio so the products exit at $P_0$, 2200 R. Find the equivalence ratio, the percent theoretical air, and the total irreversibly.

**12.14** In Example 12.13, a basic hydrogen–oxygen fuel cell reaction was analyzed at 25°C, 100 kPa. Repeat this calculation, assuming that the fuel cell operates on air at 77 F, 14.7 lbf/in.$^2$, instead of on pure oxygen at this state.

**12.15** Calculate the equilibrium constant for the reaction $O_2 \rightleftharpoons 2O$ at temperatures of 537 R and 10 800 R.

**12.16** Pure oxygen is heated from 77 F to 5300 F in a steady-state process at a constant pressure of 30 lbf/in.$^2$. Find the exit composition and the heat transfer.

**12.17** The equilibrium reaction with methane as $CH_4 \rightleftharpoons C + 2H_2$ has ln $K = -0.3362$ at 1440 R and ln $K = -4.607$ at 1080 R. By noting the relation of $K$ to temperature, show how you would interpolate ln $K$ in $(1/T)$ to find $K$ at 1260 R and compare that to a linear interpolation.

**12.18** A gas mixture of 1 pound mol carbon monoxide, 1 pound mol nitrogen, and 1 pound mol oxygen at 77 F, 20 lbf/in.$^2$, is heated in a constant-pressure flow process. The exit mixture can be assumed to be in chemical equilibrium with $CO_2$, $CO$, $O_2$, and $N_2$ present. The mole fraction of $CO_2$ at this point is 0.176. Calculate the heat transfer for the process.

**12.19** An important step in the manufacture of chemical fertilizer is the production of ammonia, according to the reaction $N_2 + 3H_2 \rightleftharpoons 2NH_3$

a. Calculate the equilibrium constant for this reaction at 300 F.

b. For an initial composition of 25% nitrogen, 75% hydrogen, on a mole basis, calculate the equilibrium composition at 300 F, 750 lbf/in.$^2$.

**12.20** Consider the reaction $2 CO_2 \Leftrightarrow 2 CO + O_2$ obtained after heating 1 lbmol $CO_2$ to 5400 R. Find the equilibrium constant from the shift in Gibbs function and verify its value with the entry in Table A.11.

**12.21** Consider the previous problem and find the equilibrium mole fraction of CO at 5400 R and 1 atm.

# Appendix G

# English Unit Tables

## Table G.1
### Critical Constants (English Units)

| Substance | Formula | Molec. Weight | Temp. (R) | Pressure (lbf/in.$^2$) | Volume (ft$^3$/lbm) |
|---|---|---|---|---|---|
| Ammonia | $NH_3$ | 17.031 | 729.9 | 1646 | 0.0682 |
| Argon | $Ar$ | 39.948 | 271.4 | 706 | 0.0300 |
| Bromine | $Br_2$ | 159.808 | 1058.4 | 1494 | 0.0127 |
| Carbon dioxide | $CO_2$ | 44.010 | 547.4 | 1070 | 0.0342 |
| Carbon monoxide | $CO$ | 28.010 | 239.2 | 508 | 0.0533 |
| Chlorine | $Cl_2$ | 70.906 | 750.4 | 1157 | 0.0280 |
| Fluorine | $F_2$ | 37.997 | 259.7 | 757 | 0.0279 |
| Helium | $He$ | 4.003 | 9.34 | 32.9 | 0.2300 |
| Hydrogen (normal) | $H_2$ | 2.016 | 59.76 | 188.6 | 0.5170 |
| Krypton | $Kr$ | 83.800 | 376.9 | 798 | 0.0174 |
| Neon | $Ne$ | 20.183 | 79.92 | 400 | 0.0330 |
| Nitric oxide | $NO$ | 30.006 | 324.0 | 940 | 0.0308 |
| Nitrogen | $N_2$ | 28.013 | 227.2 | 492 | 0.0514 |
| Nitrogen dioxide | $NO_2$ | 46.006 | 775.8 | 1465 | 0.0584 |
| Nitrous oxide | $N_2O$ | 44.013 | 557.3 | 1050 | 0.0354 |
| Oxygen | $O_2$ | 31.999 | 278.3 | 731 | 0.0367 |
| Sulfur dioxide | $SO_2$ | 64.063 | 775.4 | 1143 | 0.0306 |
| Water | $H_2O$ | 18.015 | 1165.1 | 3208 | 0.0508 |
| Xenon | $Xe$ | 131.300 | 521.5 | 847 | 0.0144 |
| Acetylene | $C_2H_2$ | 26.038 | 554.9 | 891 | 0.0693 |
| Benzene | $C_6H_6$ | 78.114 | 1012.0 | 709 | 0.0531 |
| *n*-Butane | $C_4H_{10}$ | 58.124 | 765.4 | 551 | 0.0703 |
| Chlorodifluoroethane (142b) | $CH_3CClF_2$ | 100.495 | 738.5 | 616 | 0.0368 |
| Chlorodifluoromethane (22) | $CHClF_2$ | 86.469 | 664.7 | 721 | 0.0307 |
| Dichlorodifluoroethane (141) | $CH_3CCl_2F$ | 116.950 | 866.7 | 658 | 0.0345 |
| Dichlorotrifluoroethane (123) | $CHCl_2CF_3$ | 152.930 | 822.4 | 532 | 0.0291 |
| Difluoroethane (152a) | $CHF_2CH_3$ | 66.050 | 695.5 | 656 | 0.0435 |
| Difluoromethane (32) | $CH_2F_2$ | 52.024 | 632.3 | 838 | 0.0378 |
| Ethane | $C_2H_6$ | 30.070 | 549.7 | 708 | 0.0790 |
| Ethyl alcohol | $C_2H_5OH$ | 46.069 | 925.0 | 891 | 0.0581 |
| Ethylene | $C_2H_4$ | 28.054 | 508.3 | 731 | 0.0744 |
| *n*-Heptane | $C_7H_{16}$ | 100.205 | 972.5 | 397 | 0.0691 |
| *n*-Hexane | $C_6H_{14}$ | 86.178 | 913.5 | 437 | 0.0688 |
| Methane | $CH_4$ | 16.043 | 342.7 | 667 | 0.0990 |
| Methyl alcohol | $CH_3OH$ | 32.042 | 922.7 | 1173 | 0.0590 |
| *n*-Octane | $C_8H_{18}$ | 114.232 | 1023.8 | 361 | 0.0690 |
| Pentafluoroethane (125) | $CHF_2CF_3$ | 120.022 | 610.6 | 525 | 0.0282 |
| *n*-Pentane | $C_5H_{12}$ | 72.151 | 845.5 | 489 | 0.0675 |
| Propane | $C_3H_8$ | 44.094 | 665.6 | 616 | 0.0964 |
| Propene | $C_3H_6$ | 42.081 | 656.8 | 667 | 0.0689 |
| Tetrafluoroethane (134a) | $CF_3CH_2F$ | 102.030 | 673.6 | 589 | 0.0311 |

TABLE **G.2**
*Properties of Selected Solids at 77 F*

| Substance | $\rho$ (lbm/ft$^3$) | $C_p$ (Btu/lbm R) |
|---|---|---|
| Asphalt | 132.3 | 0.225 |
| Brick, common | 112.4 | 0.20 |
| Carbon, diamond | 202.9 | 0.122 |
| Carbon, graphite | 125–156 | 0.146 |
| Coal | 75–95 | 0.305 |
| Concrete | 137 | 0.21 |
| Glass, plate | 156 | 0.191 |
| Glass, wool | 1.25 | 0.158 |
| Granite | 172 | 0.212 |
| Ice (32 F) | 57.2 | 0.487 |
| Paper | 43.7 | 0.287 |
| Plexiglas | 73.7 | 0.344 |
| Polystyrene | 57.4 | 0.549 |
| Polyvinyl chloride | 86.1 | 0.229 |
| Rubber, soft | 68.7 | 0.399 |
| Sand, dry | 93.6 | 0.191 |
| Salt, rock | 130–156 | 0.2196 |
| Silicon | 145.5 | 0.167 |
| Snow, firm | 35 | 0.501 |
| Wood, hard (oak) | 44.9 | 0.301 |
| Wood, soft (pine) | 31.8 | 0.33 |
| Wool | 6.24 | 0.411 |
| **Metals** | | |
| Aluminum, duralumin | 170 | 0.215 |
| Brass, 60-40 | 524 | 0.0898 |
| Copper, commercial | 518 | 0.100 |
| Gold | 1205 | 0.03082 |
| Iron, cast | 454 | 0.100 |
| Iron, 304 St Steel | 488 | 0.110 |
| Lead | 708 | 0.031 |
| Magnesium, 2% Mn | 111 | 0.239 |
| Nickel, 10% Cr | 541 | 0.1066 |
| Silver, 99.9% Ag | 657 | 0.0564 |
| Sodium | 60.6 | 0.288 |
| Tin | 456 | 0.0525 |
| Tungsten | 1205 | 0.032 |
| Zinc | 446 | 0.0927 |

TABLE **G.3**
*Properties of Some Liquids at 77 F*

| Substance | $\rho$ (lbm/ft$^3$) | $C_p$ (Btu/lbm R) |
|---|---|---|
| Ammonia | 37.7 | 1.151 |
| Benzene | 54.9 | 0.41 |
| Butane | 34.7 | 0.60 |
| $CCl_4$ | 98.9 | 0.20 |
| $CO_2$ | 42.5 | 0.69 |
| Ethanol | 48.9 | 0.59 |
| Gasoline | 46.8 | 0.50 |
| Glycerine | 78.7 | 0.58 |
| Kerosene | 50.9 | 0.48 |
| Methanol | 49.1 | 0.61 |
| *n*-octane | 43.2 | 0.53 |
| Oil, engine | 55.2 | 0.46 |
| Oil, light | 57 | 0.43 |
| Propane | 31.8 | 0.61 |
| R-12 | 81.8 | 0.232 |
| R-22 | 74.3 | 0.30 |
| R-32 | 60 | 0.463 |
| R-125 | 74.4 | 0.337 |
| R-134a | 75.3 | 0.34 |
| Water | 62.2 | 1.00 |
| **Liquid Metals** | | |
| Bismuth, Bi | 627 | 0.033 |
| Lead, Pb | 665 | 0.038 |
| Mercury, Hg | 848 | 0.033 |
| NaK (56/44) | 55.4 | 0.27 |
| Potassium, K | 51.7 | 0.193 |
| Sodium, Na | 58 | 0.33 |
| Tin, Sn | 434 | 0.057 |
| Zinc, Zn | 410 | 0.12 |

## Table G.4
*Properties of Various Ideal Gases at 77 F, 1 atm\* (English Units)*

| Gas | Chemical Formula | Molecular Mass | $R$ (ft-lbf/lbm R) | $\rho \times 10^3$ (lbm/ft$^3$) | $C_{p0}$ (Btu/lbm R) | $C_{v0}$ (Btu/lbm R) | $k$ $C_{p0}/C_{v0}$ |
|---|---|---|---|---|---|---|---|
| Steam | $H_2O$ | 18.015 | 85.76 | 1.442 | 0.447 | 0.337 | 1.327 |
| Acetylene | $C_2H_2$ | 26.038 | 59.34 | 65.55 | 0.406 | 0.330 | 1.231 |
| Air | — | 28.97 | 53.34 | 72.98 | 0.240 | 0.171 | 1.400 |
| Ammonia | $NH_3$ | 17.031 | 90.72 | 43.325 | 0.509 | 0.392 | 1.297 |
| Argon | Ar | 39.948 | 38.68 | 100.7 | 0.124 | 0.0745 | 1.667 |
| Butane | $C_4H_{10}$ | 58.124 | 26.58 | 150.3 | 0.410 | 0.376 | 1.091 |
| Carbon dioxide | $CO_2$ | 44.01 | 35.10 | 110.8 | 0.201 | 0.156 | 1.289 |
| Carbon monoxide | CO | 28.01 | 55.16 | 70.5 | 0.249 | 0.178 | 1.399 |
| Ethane | $C_2H_6$ | 30.07 | 51.38 | 76.29 | 0.422 | 0.356 | 1.186 |
| Ethanol | $C_2H_5OH$ | 46.069 | 33.54 | 117.6 | 0.341 | 0.298 | 1.145 |
| Ethylene | $C_2H_4$ | 28.054 | 55.07 | 71.04 | 0.370 | 0.299 | 1.237 |
| Helium | He | 4.003 | 386.0 | 10.08 | 1.240 | 0.744 | 1.667 |
| Hydrogen | $H_2$ | 2.016 | 766.5 | 5.075 | 3.394 | 2.409 | 1.409 |
| Methane | $CH_4$ | 16.043 | 96.35 | 40.52 | 0.538 | 0.415 | 1.299 |
| Methanol | $CH_3OH$ | 32.042 | 48.22 | 81.78 | 0.336 | 0.274 | 1.227 |
| Neon | Ne | 20.183 | 76.55 | 50.81 | 0.246 | 0.148 | 1.667 |
| Nitric oxide | NO | 30.006 | 51.50 | 75.54 | 0.237 | 0.171 | 1.387 |
| Nitrogen | $N_2$ | 28.013 | 55.15 | 70.61 | 0.249 | 0.178 | 1.400 |
| Nitrous oxide | $N_2O$ | 44.013 | 35.10 | 110.8 | 0.210 | 0.165 | 1.274 |
| *n*-octane | $C_8H_{18}$ | 114.23 | 13.53 | 5.74 | 0.409 | 0.391 | 1.044 |
| Oxygen | $O_2$ | 31.999 | 48.28 | 80.66 | 0.220 | 0.158 | 1.393 |
| Propane | $C_3H_8$ | 44.094 | 35.04 | 112.9 | 0.401 | 0.356 | 1.126 |
| R-12 | $CCl_2F_2$ | 120.914 | 12.78 | 310.9 | 0.147 | 0.131 | 1.126 |
| R-22 | $CHClF_2$ | 86.469 | 17.87 | 221.0 | 0.157 | 0.134 | 1.171 |
| R-32 | $CF_2H_2$ | 52.024 | 29.70 | 132.6 | 0.196 | 0.158 | 1.242 |
| R-125 | $CHF_2CF_3$ | 120.022 | 12.87 | 307.0 | 0.189 | 0.172 | 1.097 |
| R-134a | $CF_3CH_2F$ | 102.03 | 15.15 | 262.2 | 0.203 | 0.184 | 1.106 |
| Sulfur dioxide | $SO_2$ | 64.059 | 24.12 | 163.4 | 0.149 | 0.118 | 1.263 |
| Sulfur trioxide | $SO_3$ | 80.053 | 19.30 | 204.3 | 0.152 | 0.127 | 1.196 |

\*Or saturation pressure if it is less than 1 atm.

**TABLE G.5**

*Ideal-Gas Properties of Air (English Units), Standard Entropy at 1 atm = 101.325 kPa = 14.696 lbf/in.²*

| T (R) | u (Btu/lbm) | h (Btu/lbm) | $s_T^0$ (Btu/lbm R) | T (R) | u (Btu/lbm) | h (Btu/lbm) | $s_T^0$ (Btu/lbm R) |
|---|---|---|---|---|---|---|---|
| 400 | 68.212 | 95.634 | 1.56788 | 1950 | 357.243 | 490.928 | 1.96404 |
| 440 | 75.047 | 105.212 | 1.59071 | 2000 | 367.642 | 504.755 | 1.97104 |
| 480 | 81.887 | 114.794 | 1.61155 | 2050 | 378.096 | 518.636 | 1.97790 |
| 520 | 88.733 | 124.383 | 1.63074 | 2100 | 388.602 | 532.570 | 1.98461 |
| 536.67 | 91.589 | 128.381 | 1.63831 | 2150 | 399.158 | 546.554 | 1.99119 |
| 540 | 92.160 | 129.180 | 1.63979 | 2200 | 409.764 | 560.588 | 1.99765 |
| 560 | 95.589 | 133.980 | 1.64852 | 2300 | 431.114 | 588.793 | 2.01018 |
| 600 | 102.457 | 143.590 | 1.66510 | 2400 | 452.640 | 617.175 | 2.02226 |
| 640 | 109.340 | 153.216 | 1.68063 | 2500 | 474.330 | 645.721 | 2.03391 |
| 680 | 116.242 | 162.860 | 1.69524 | 2600 | 496.175 | 674.421 | 2.04517 |
| 720 | 123.167 | 172.528 | 1.70906 | 2700 | 518.165 | 703.267 | 2.05606 |
| 760 | 130.118 | 182.221 | 1.72216 | 2800 | 540.286 | 732.244 | 2.06659 |
| 800 | 137.099 | 191.944 | 1.73463 | 2900 | 562.532 | 761.345 | 2.07681 |
| 840 | 144.114 | 201.701 | 1.74653 | 3000 | 584.895 | 790.564 | 2.08671 |
| 880 | 151.165 | 211.494 | 1.75791 | 3100 | 607.369 | 819.894 | 2.09633 |
| 920 | 158.255 | 221.327 | 1.76884 | 3200 | 629.948 | 849.328 | 2.10567 |
| 960 | 165.388 | 231.202 | 1.77935 | 3300 | 652.625 | 878.861 | 2.11476 |
| 1000 | 172.564 | 241.121 | 1.78947 | 3400 | 675.396 | 908.488 | 2.12361 |
| 1040 | 179.787 | 251.086 | 1.79924 | 3500 | 698.257 | 938.204 | 2.13222 |
| 1080 | 187.058 | 261.099 | 1.80868 | 3600 | 721.203 | 968.005 | 2.14062 |
| 1120 | 194.378 | 271.161 | 1.81783 | 3700 | 744.230 | 997.888 | 2.14880 |
| 1160 | 201.748 | 281.273 | 1.82670 | 3800 | 767.334 | 1027.848 | 2.15679 |
| 1200 | 209.168 | 291.436 | 1.83532 | 3900 | 790.513 | 1057.882 | 2.16459 |
| 1240 | 216.640 | 301.650 | 1.84369 | 4000 | 813.763 | 1087.988 | 2.17221 |
| 1280 | 224.163 | 311.915 | 1.85184 | 4100 | 837.081 | 1118.162 | 2.17967 |
| 1320 | 231.737 | 322.231 | 1.85977 | 4200 | 860.466 | 1148.402 | 2.18695 |
| 1360 | 239.362 | 332.598 | 1.86751 | 4300 | 883.913 | 1178.705 | 2.19408 |
| 1400 | 247.037 | 343.016 | 1.87506 | 4400 | 907.422 | 1209.069 | 2.20106 |
| 1440 | 254.762 | 353.483 | 1.88243 | 4500 | 930.989 | 1239.492 | 2.20790 |
| 1480 | 262.537 | 364.000 | 1.88964 | 4600 | 954.613 | 1269.972 | 2.21460 |
| 1520 | 270.359 | 374.565 | 1.89668 | 4700 | 978.292 | 1300.506 | 2.22117 |
| 1560 | 278.230 | 385.177 | 1.90357 | 4800 | 1002.023 | 1331.093 | 2.22761 |
| 1600 | 286.146 | 395.837 | 1.91032 | 4900 | 1025.806 | 1361.732 | 2.23392 |
| 1650 | 296.106 | 409.224 | 1.91856 | 5000 | 1049.638 | 1392.419 | 2.24012 |
| 1700 | 306.136 | 422.681 | 1.92659 | 5100 | 1073.518 | 1423.155 | 2.24621 |
| 1750 | 316.232 | 436.205 | 1.93444 | 5200 | 1097.444 | 1453.936 | 2.25219 |
| 1800 | 326.393 | 449.794 | 1.94209 | 5300 | 1121.414 | 1484.762 | 2.25806 |
| 1850 | 336.616 | 463.445 | 1.94957 | 5400 | 1145.428 | 1515.632 | 2.26383 |
| 1900 | 346.901 | 477.158 | 1.95689 | | | | |

**TABLE G.6**
*Ideal-Gas Properties of Various Substances (English Units), Entropies at 1 atm Pressure*

| | NITROGEN, DIATOMIC ($N_2$) $\overline{h}^0_{f,537} = 0$ Btu/lb mol $M = 28.013$ | | NITROGEN, MONATOMIC (N) $\overline{h}^0_{f,537} = 203\ 216$ Btu/lb mol $M = 14.007$ | |
|---|---|---|---|---|
| $T$ R | $\overline{h} - \overline{h}^0_{537}$ Btu/lb mol | $\overline{s}^0_T$ Btu/lbmol R | $\overline{h} - \overline{h}^0_{537}$ Btu/lb mol | $\overline{s}^0_T$ Btu/lbmol R |
| 0 | −3727 | 0 | −2664 | 0 |
| 200 | −2341 | 38.877 | −1671 | 31.689 |
| 400 | −950 | 43.695 | −679 | 35.130 |
| 537 | 0 | 45.739 | 0 | 36.589 |
| 600 | 441 | 46.515 | 314 | 37.143 |
| 800 | 1837 | 48.524 | 1307 | 38.571 |
| 1000 | 3251 | 50.100 | 2300 | 39.679 |
| 1200 | 4693 | 51.414 | 3293 | 40.584 |
| 1400 | 6169 | 52.552 | 4286 | 41.349 |
| 1600 | 7681 | 53.561 | 5279 | 42.012 |
| 1800 | 9227 | 54.472 | 6272 | 42.597 |
| 2000 | 10804 | 55.302 | 7265 | 43.120 |
| 2200 | 12407 | 56.066 | 8258 | 43.593 |
| 2400 | 14034 | 56.774 | 9251 | 44.025 |
| 2600 | 15681 | 57.433 | 10244 | 44.423 |
| 2800 | 17345 | 58.049 | 11237 | 44.791 |
| 3000 | 19025 | 58.629 | 12230 | 45.133 |
| 3200 | 20717 | 59.175 | 13223 | 45.454 |
| 3400 | 22421 | 59.691 | 14216 | 45.755 |
| 3600 | 24135 | 60.181 | 15209 | 46.038 |
| 3800 | 25857 | 60.647 | 16202 | 46.307 |
| 4000 | 27587 | 61.090 | 17195 | 46.562 |
| 4200 | 29324 | 61.514 | 18189 | 46.804 |
| 4400 | 31068 | 61.920 | 19183 | 47.035 |
| 4600 | 32817 | 62.308 | 20178 | 47.256 |
| 4800 | 34571 | 62.682 | 21174 | 47.468 |
| 5000 | 36330 | 63.041 | 22171 | 47.672 |
| 5500 | 40745 | 63.882 | 24670 | 48.148 |
| 6000 | 45182 | 64.654 | 27186 | 48.586 |
| 6500 | 49638 | 65.368 | 29724 | 48.992 |
| 7000 | 54109 | 66.030 | 32294 | 49.373 |
| 7500 | 58595 | 66.649 | 34903 | 49.733 |
| 8000 | 63093 | 67.230 | 37559 | 50.076 |
| 8500 | 67603 | 67.777 | 40270 | 50.405 |
| 9000 | 72125 | 68.294 | 43040 | 50.721 |
| 9500 | 96658 | 68.784 | 45875 | 51.028 |
| 10000 | 81203 | 69.250 | 48777 | 51.325 |

**Table G.6** (*continued*)
*Ideal-Gas Properties of Various Substances (English Units), Entropies at 1 atm Pressure*

| | OXYGEN, DIATOMIC ($O_2$) $\overline{h}^0_{f,537} = 0$ Btu/lb mol $M = 31.999$ | | OXYGEN, MONATOMIC (O) $\overline{h}^0_{f,537} = 107\ 124$ Btu/lb mol $M = 16.00$ | |
|---|---|---|---|---|
| $T$ R | $\overline{h} - \overline{h}^0_{537}$ Btu/lb mol | $\overline{s}^0_T$ Btu/lbmol R | $\overline{h} - \overline{h}^0_{537}$ Btu/lb mol | $\overline{s}^0_T$ Btu/lbmol R |
| 0 | −3733 | 0 | −2891 | 0 |
| 200 | −2345 | 42.100 | −1829 | 33.041 |
| 400 | −955 | 46.920 | −724 | 36.884 |
| 537 | 0 | 48.973 | 0 | 38.442 |
| 600 | 446 | 49.758 | 330 | 39.023 |
| 800 | 1881 | 51.819 | 1358 | 40.503 |
| 1000 | 3366 | 53.475 | 2374 | 41.636 |
| 1200 | 4903 | 54.876 | 3383 | 42.556 |
| 1400 | 6487 | 56.096 | 4387 | 43.330 |
| 1600 | 8108 | 57.179 | 5389 | 43.999 |
| 1800 | 9761 | 58.152 | 6389 | 44.588 |
| 2000 | 11438 | 59.035 | 7387 | 45.114 |
| 2200 | 13136 | 59.844 | 8385 | 45.589 |
| 2400 | 14852 | 60.591 | 9381 | 46.023 |
| 2600 | 16584 | 61.284 | 10378 | 46.422 |
| 2800 | 18329 | 61.930 | 11373 | 46.791 |
| 3000 | 20088 | 62.537 | 12369 | 47.134 |
| 3200 | 21860 | 63.109 | 13364 | 47.455 |
| 3400 | 23644 | 63.650 | 14359 | 47.757 |
| 3600 | 25441 | 64.163 | 15354 | 48.041 |
| 3800 | 27250 | 64.652 | 16349 | 48.310 |
| 4000 | 29071 | 65.119 | 17344 | 48.565 |
| 4200 | 30904 | 65.566 | 18339 | 48.808 |
| 4400 | 32748 | 65.995 | 19334 | 49.039 |
| 4600 | 34605 | 66.408 | 20330 | 49.261 |
| 4800 | 36472 | 66.805 | 21327 | 49.473 |
| 5000 | 38350 | 67.189 | 22325 | 49.677 |
| 5500 | 43091 | 68.092 | 24823 | 50.153 |
| 6000 | 47894 | 68.928 | 27329 | 50.589 |
| 6500 | 52751 | 69.705 | 29847 | 50.992 |
| 7000 | 57657 | 70.433 | 32378 | 51.367 |
| 7500 | 62608 | 71.116 | 34924 | 51.718 |
| 8000 | 67600 | 71.760 | 37485 | 52.049 |
| 8500 | 72633 | 72.370 | 40063 | 52.362 |
| 9000 | 77708 | 72.950 | 42658 | 52.658 |
| 9500 | 82828 | 73.504 | 45270 | 52.941 |
| 10000 | 87997 | 74.034 | 47897 | 53.210 |

TABLE G.6 (*continued*)
*Ideal-Gas Properties of Various Substances (English Units), Entropies at 1 atm Pressure*

| | CARBON DIOXIDE ($CO_2$) $\bar{h}^0_{f,537} = -169\ 184$ Btu/lb mol $M = 44.01$ | | CARBON MONOXIDE (CO) $\bar{h}^0_{f,537} = -47\ 518$ Btu/lb mol $M = 28.01$ | |
|---|---|---|---|---|
| $T$ R | $\bar{h} - \bar{h}^0_{537}$ Btu/lb mol | $\bar{s}^0_T$ Btu/lbmol R | $\bar{h} - \bar{h}^0_{537}$ Btu/lb mol | $\bar{s}^0_T$ Btu/lbmol R |
| 0 | −4026 | 0 | −3728 | 0 |
| 200 | −2636 | 43.466 | −2343 | 40.319 |
| 400 | −1153 | 48.565 | −951 | 45.137 |
| 537 | 0 | 51.038 | 0 | 47.182 |
| 600 | 573 | 52.047 | 441 | 47.959 |
| 800 | 2525 | 54.848 | 1842 | 49.974 |
| 1000 | 4655 | 57.222 | 3266 | 51.562 |
| 1200 | 6927 | 59.291 | 4723 | 52.891 |
| 1400 | 9315 | 61.131 | 6220 | 54.044 |
| 1600 | 11798 | 62.788 | 7754 | 55.068 |
| 1800 | 14358 | 64.295 | 9323 | 55.992 |
| 2000 | 16982 | 65.677 | 10923 | 56.835 |
| 2200 | 19659 | 66.952 | 12549 | 57.609 |
| 2400 | 22380 | 68.136 | 14197 | 58.326 |
| 2600 | 25138 | 69.239 | 15864 | 58.993 |
| 2800 | 27926 | 70.273 | 17547 | 59.616 |
| 3000 | 30741 | 71.244 | 19243 | 60.201 |
| 3200 | 33579 | 72.160 | 20951 | 60.752 |
| 3400 | 36437 | 73.026 | 22669 | 61.273 |
| 3600 | 39312 | 73.847 | 24395 | 61.767 |
| 3800 | 42202 | 74.629 | 26128 | 62.236 |
| 4000 | 45105 | 75.373 | 27869 | 62.683 |
| 4200 | 48021 | 76.084 | 29614 | 63.108 |
| 4400 | 50948 | 76.765 | 31366 | 63.515 |
| 4600 | 53885 | 77.418 | 33122 | 63.905 |
| 4800 | 56830 | 78.045 | 34883 | 64.280 |
| 5000 | 59784 | 78.648 | 36650 | 64.641 |
| 5500 | 67202 | 80.062 | 41089 | 65.487 |
| 6000 | 74660 | 81.360 | 45548 | 66.263 |
| 6500 | 82155 | 82.560 | 50023 | 66.979 |
| 7000 | 89682 | 83.675 | 54514 | 67.645 |
| 7500 | 97239 | 84.718 | 59020 | 68.267 |
| 8000 | 104823 | 85.697 | 63539 | 68.850 |
| 8500 | 112434 | 86.620 | 68069 | 69.399 |
| 9000 | 120071 | 87.493 | 72610 | 69.918 |
| 9500 | 127734 | 88.321 | 77161 | 70.410 |
| 10000 | 135426 | 89.110 | 81721 | 70.878 |

**Table G.6** (*continued*)
*Ideal-Gas Properties of Various Substances (English Units), Entropies at 1 atm Pressure*

| | WATER ($H_2O$) $\overline{h}_{f,537}^0 = -103\ 966$ Btu/lb mol $M = 18.015$ | | HYDROXYL (OH) $\overline{h}_{f,537}^0 = 16\ 761$ Btu/lb mol $M = 17.007$ | |
|---|---|---|---|---|
| $T$ R | $\overline{h} - \overline{h}_{537}^0$ Btu/lb mol | $\overline{s}_T^0$ Btu/lbmol R | $\overline{h} - \overline{h}_{537}^0$ Btu/lb mol | $\overline{s}_T^0$ Btu/lbmol R |
| 0 | −4528 | 0 | −3943 | 0 |
| 200 | −2686 | 37.209 | −2484 | 36.521 |
| 400 | −1092 | 42.728 | −986 | 41.729 |
| 537 | 0 | 45.076 | 0 | 43.852 |
| 600 | 509 | 45.973 | 452 | 44.649 |
| 800 | 2142 | 48.320 | 1870 | 46.689 |
| 1000 | 3824 | 50.197 | 3280 | 48.263 |
| 1200 | 5566 | 51.784 | 4692 | 49.549 |
| 1400 | 7371 | 53.174 | 6112 | 50.643 |
| 1600 | 9241 | 54.422 | 7547 | 51.601 |
| 1800 | 11178 | 55.563 | 9001 | 52.457 |
| 2000 | 13183 | 56.619 | 10477 | 53.235 |
| 2200 | 15254 | 57.605 | 11978 | 53.950 |
| 2400 | 17388 | 58.533 | 13504 | 54.614 |
| 2600 | 19582 | 59.411 | 15054 | 55.235 |
| 2800 | 21832 | 60.245 | 16627 | 55.817 |
| 3000 | 24132 | 61.038 | 18220 | 56.367 |
| 3200 | 26479 | 61.796 | 19834 | 56.887 |
| 3400 | 28867 | 62.520 | 21466 | 57.382 |
| 3600 | 31293 | 63.213 | 23114 | 57.853 |
| 3800 | 33756 | 63.878 | 24777 | 58.303 |
| 4000 | 36251 | 64.518 | 26455 | 58.733 |
| 4200 | 38774 | 65.134 | 28145 | 59.145 |
| 4400 | 41325 | 65.727 | 29849 | 59.542 |
| 4600 | 43899 | 66.299 | 31563 | 59.922 |
| 4800 | 46496 | 66.852 | 33287 | 60.289 |
| 5000 | 49114 | 67.386 | 35021 | 60.643 |
| 5500 | 55739 | 68.649 | 39393 | 61.477 |
| 6000 | 62463 | 69.819 | 43812 | 62.246 |
| 6500 | 69270 | 70.908 | 48272 | 62.959 |
| 7000 | 76146 | 71.927 | 52767 | 63.626 |
| 7500 | 83081 | 72.884 | 57294 | 64.250 |
| 8000 | 90069 | 73.786 | 61851 | 64.838 |
| 8500 | 97101 | 74.639 | 66434 | 65.394 |
| 9000 | 104176 | 75.448 | 71043 | 65.921 |
| 9500 | 111289 | 76.217 | 75677 | 66.422 |
| 10000 | 118440 | 76.950 | 80335 | 66.900 |

Table G.6 (*continued*)
*Ideal-Gas Properties of Various Substances (English Units), Entropies at 1 atm Pressure*

| | HYDROGEN ($H_2$) $\overline{h}^0_{f,537} = 0$ Btu/lb mol $M = 2.016$ | | HYDROGEN, MONATOMIC (H) $\overline{h}^0_{f,537} = 93\ 723$ Btu/lb mol $M = 1.008$ | |
|---|---|---|---|---|
| $T$ R | $\overline{h} - \overline{h}^0_{537}$ Btu/lb mol | $\overline{s}^0_T$ Btu/lbmol R | $\overline{h} - \overline{h}^0_{537}$ Btu/lb mol | $\overline{s}^0_T$ Btu/lbmol R |
| 0 | −3640 | 0 | −2664 | 0 |
| 200 | −2224 | 24.703 | −1672 | 22.473 |
| 400 | −927 | 29.193 | −679 | 25.914 |
| 537 | 0 | 31.186 | 0 | 27.373 |
| 600 | 438 | 31.957 | 314 | 27.927 |
| 800 | 1831 | 33.960 | 1307 | 29.355 |
| 1000 | 3225 | 35.519 | 2300 | 30.463 |
| 1200 | 4622 | 36.797 | 3293 | 31.368 |
| 1400 | 6029 | 37.883 | 4286 | 32.134 |
| 1600 | 7448 | 38.831 | 5279 | 32.797 |
| 1800 | 8884 | 39.676 | 6272 | 33.381 |
| 2000 | 10337 | 40.441 | 7265 | 33.905 |
| 2200 | 11812 | 41.143 | 8258 | 34.378 |
| 2400 | 13309 | 41.794 | 9251 | 34.810 |
| 2600 | 14829 | 42.401 | 10244 | 35.207 |
| 2800 | 16372 | 42.973 | 11237 | 35.575 |
| 3000 | 17938 | 43.512 | 12230 | 35.917 |
| 3200 | 19525 | 44.024 | 13223 | 36.238 |
| 3400 | 21133 | 44.512 | 14215 | 36.539 |
| 3600 | 22761 | 44.977 | 15208 | 36.823 |
| 3800 | 24407 | 45.422 | 16201 | 37.091 |
| 4000 | 26071 | 45.849 | 17194 | 37.346 |
| 4200 | 27752 | 46.260 | 18187 | 37.588 |
| 4400 | 29449 | 46.655 | 19180 | 37.819 |
| 4600 | 31161 | 47.035 | 20173 | 38.040 |
| 4800 | 32887 | 47.403 | 21166 | 38.251 |
| 5000 | 34627 | 47.758 | 22159 | 38.454 |
| 5500 | 39032 | 48.598 | 24641 | 38.927 |
| 6000 | 43513 | 49.378 | 27124 | 39.359 |
| 6500 | 48062 | 50.105 | 29606 | 39.756 |
| 7000 | 52678 | 50.789 | 32088 | 40.124 |
| 7500 | 57356 | 51.434 | 34571 | 40.467 |
| 8000 | 62094 | 52.045 | 37053 | 40.787 |
| 8500 | 66889 | 52.627 | 39535 | 41.088 |
| 9000 | 71738 | 53.182 | 42018 | 41.372 |
| 9500 | 76638 | 53.712 | 44500 | 41.640 |
| 10000 | 81581 | 54.220 | 46982 | 41.895 |

**Table G.6** (*continued*)

*Ideal-Gas Properties of Various Substances (English Units), Entropies at 1 atm Pressure*

| | Nitric Oxide (NO) $\bar{h}^0_{f,537} = 38\ 818$ Btu/lb mol $M = 30.006$ | | Nitrogen Dioxide (NO$_2$) $\bar{h}^0_{f,537} = 14\ 230$ Btu/lb mol $M = 46.005$ | |
|---|---|---|---|---|
| **T** **R** | $\bar{h} - \bar{h}^0_{537}$ **Btu/lb mol** | $\bar{s}^0_T$ **Btu/lbmol R** | $\bar{h} - \bar{h}^0_{537}$ **Btu/lb mol** | $\bar{s}^0_T$ **Btu/lbmol R** |
| 0 | −3952 | 0 | −4379 | 0 |
| 200 | −2457 | 43.066 | −2791 | 49.193 |
| 400 | −979 | 48.207 | −1172 | 54.789 |
| 537 | 0 | 50.313 | 0 | 57.305 |
| 600 | 451 | 51.107 | 567 | 58.304 |
| 800 | 1881 | 53.163 | 2469 | 61.034 |
| 1000 | 3338 | 54.788 | 4532 | 63.333 |
| 1200 | 4834 | 56.152 | 6733 | 65.337 |
| 1400 | 6372 | 57.337 | 9044 | 67.118 |
| 1600 | 7948 | 58.389 | 11442 | 68.718 |
| 1800 | 9557 | 59.336 | 13905 | 70.168 |
| 2000 | 11193 | 60.198 | 16421 | 71.493 |
| 2200 | 12853 | 60.989 | 18978 | 72.712 |
| 2400 | 14532 | 61.719 | 21567 | 73.838 |
| 2600 | 16228 | 62.397 | 24182 | 74.885 |
| 2800 | 17937 | 63.031 | 26819 | 75.861 |
| 3000 | 19657 | 63.624 | 29473 | 76.777 |
| 3200 | 21388 | 64.183 | 32142 | 77.638 |
| 3400 | 23128 | 64.710 | 34823 | 78.451 |
| 3600 | 24875 | 65.209 | 37515 | 79.220 |
| 3800 | 26629 | 65.684 | 40215 | 79.950 |
| 4000 | 28389 | 66.135 | 42923 | 80.645 |
| 4200 | 30154 | 66.565 | 45637 | 81.307 |
| 4400 | 31924 | 66.977 | 48358 | 81.940 |
| 4600 | 33698 | 67.371 | 51083 | 82.545 |
| 4800 | 35476 | 67.750 | 53813 | 83.126 |
| 5000 | 37258 | 68.113 | 56546 | 83.684 |
| 5500 | 41726 | 68.965 | 63395 | 84.990 |
| 6000 | 46212 | 69.746 | 70260 | 86.184 |
| 6500 | 50714 | 70.467 | 77138 | 87.285 |
| 7000 | 55229 | 71.136 | 84026 | 88.306 |
| 7500 | 59756 | 71.760 | 90923 | 89.258 |
| 8000 | 64294 | 72.346 | 97826 | 90.149 |
| 8500 | 68842 | 72.898 | 104735 | 90.986 |
| 9000 | 73401 | 73.419 | 111648 | 91.777 |
| 9500 | 77968 | 73.913 | 118565 | 92.525 |
| 10000 | 82544 | 74.382 | 125485 | 93.235 |

**Table G.7**
*Thermodynamic Properties of Water*
**Table G.7.1**
*Saturated Water*

| Temp. (F) | Press. (psia) | Specific Volume, ft³/lbm | | | Internal Energy, Btu/lbm | | |
|---|---|---|---|---|---|---|---|
| | | Sat. Liquid $v_f$ | Evap. $v_{fg}$ | Sat. Vapor $v_g$ | Sat. Liquid $u_f$ | Evap. $u_{fg}$ | Sat. Vapor $u_g$ |
| 32 | 0.0887 | 0.01602 | 3301.6545 | 3301.6705 | 0 | 1021.21 | 1021.21 |
| 35 | 0.100 | 0.01602 | 2947.5021 | 2947.5181 | 2.99 | 1019.20 | 1022.19 |
| 40 | 0.122 | 0.01602 | 2445.0713 | 2445.0873 | 8.01 | 1015.84 | 1023.85 |
| 45 | 0.147 | 0.01602 | 2036.9527 | 2036.9687 | 13.03 | 1012.47 | 1025.50 |
| 50 | 0.178 | 0.01602 | 1703.9867 | 1704.0027 | 18.05 | 1009.10 | 1027.15 |
| 60 | 0.256 | 0.01603 | 1206.7283 | 1206.7443 | 28.08 | 1002.36 | 1030.44 |
| 70 | 0.363 | 0.01605 | 867.5791 | 867.5952 | 38.09 | 995.64 | 1033.72 |
| 80 | 0.507 | 0.01607 | 632.6739 | 632.6900 | 48.08 | 988.91 | 1036.99 |
| 90 | 0.699 | 0.01610 | 467.5865 | 467.6026 | 58.06 | 982.18 | 1040.24 |
| 100 | 0.950 | 0.01613 | 349.9602 | 349.9764 | 68.04 | 975.43 | 1043.47 |
| 110 | 1.276 | 0.01617 | 265.0548 | 265.0709 | 78.01 | 968.67 | 1046.68 |
| 120 | 1.695 | 0.01620 | 203.0105 | 203.0267 | 87.99 | 961.88 | 1049.87 |
| 130 | 2.225 | 0.01625 | 157.1419 | 157.1582 | 97.96 | 955.07 | 1053.03 |
| 140 | 2.892 | 0.01629 | 122.8567 | 122.8730 | 107.95 | 948.21 | 1056.16 |
| 150 | 3.722 | 0.01634 | 96.9611 | 96.9774 | 117.94 | 941.32 | 1059.26 |
| 160 | 4.745 | 0.01639 | 77.2079 | 77.2243 | 127.94 | 934.39 | 1062.32 |
| 170 | 5.997 | 0.01645 | 61.9983 | 62.0148 | 137.94 | 927.41 | 1065.35 |
| 180 | 7.515 | 0.01651 | 50.1826 | 50.1991 | 147.96 | 920.38 | 1068.34 |
| 190 | 9.344 | 0.01657 | 40.9255 | 40.9421 | 157.99 | 913.29 | 1071.29 |
| 200 | 11.530 | 0.01663 | 33.6146 | 33.6312 | 168.03 | 906.15 | 1074.18 |
| 210 | 14.126 | 0.01670 | 27.7964 | 27.8131 | 178.09 | 898.95 | 1077.04 |
| 212.0 | 14.696 | 0.01672 | 26.7864 | 26.8032 | 180.09 | 897.51 | 1077.60 |
| 220 | 17.189 | 0.01677 | 23.1325 | 23.1492 | 188.16 | 891.68 | 1079.84 |
| 230 | 20.781 | 0.01685 | 19.3677 | 19.3846 | 198.25 | 884.33 | 1082.58 |
| 240 | 24.968 | 0.01692 | 16.3088 | 16.3257 | 208.36 | 876.91 | 1085.27 |
| 250 | 29.823 | 0.01700 | 13.8077 | 13.8247 | 218.48 | 869.41 | 1087.90 |
| 260 | 35.422 | 0.01708 | 11.7503 | 11.7674 | 228.64 | 861.82 | 1090.46 |
| 270 | 41.848 | 0.01717 | 10.0483 | 10.0655 | 238.81 | 854.14 | 1092.95 |
| 280 | 49.189 | 0.01726 | 8.6325 | 8.6498 | 249.02 | 846.35 | 1095.37 |
| 290 | 57.535 | 0.01735 | 7.4486 | 7.4660 | 259.25 | 838.46 | 1097.71 |
| 300 | 66.985 | 0.01745 | 6.4537 | 6.4712 | 269.51 | 830.45 | 1099.96 |
| 310 | 77.641 | 0.01755 | 5.6136 | 5.6312 | 279.80 | 822.32 | 1102.13 |
| 320 | 89.609 | 0.01765 | 4.9010 | 4.9186 | 290.13 | 814.07 | 1104.20 |
| 330 | 103.00 | 0.01776 | 4.2938 | 4.3115 | 300.50 | 805.68 | 1106.17 |
| 340 | 117.94 | 0.01787 | 3.7742 | 3.7921 | 310.90 | 797.14 | 1108.04 |
| 350 | 134.54 | 0.01799 | 3.3279 | 3.3459 | 321.35 | 788.45 | 1109.80 |

**TABLE G.7.1**  (*continued*)
*Saturated Water*

| Temp. (F) | Press. (psia) | ENTHALPY, Btu/lbm | | | ENTROPY, Btu/lbm R | | |
|---|---|---|---|---|---|---|---|
| | | Sat. Liquid $h_f$ | Evap. $h_{fg}$ | Sat. Vapor $h_g$ | Sat. Liquid $s_f$ | Evap. $s_{fg}$ | Sat. Vapor $s_g$ |
| 32 | 0.0887 | 0 | 1075.38 | 1075.39 | 0 | 2.1869 | 2.1869 |
| 35 | 0.100 | 2.99 | 1073.71 | 1076.70 | 0.0061 | 2.1703 | 2.1764 |
| 40 | 0.122 | 8.01 | 1070.89 | 1078.90 | 0.0162 | 2.1430 | 2.1591 |
| 45 | 0.147 | 13.03 | 1068.06 | 1081.10 | 0.0262 | 2.1161 | 2.1423 |
| 50 | 0.178 | 18.05 | 1065.24 | 1083.29 | 0.0361 | 2.0898 | 2.1259 |
| 60 | 0.256 | 28.08 | 1059.59 | 1087.67 | 0.0555 | 2.0388 | 2.0943 |
| 70 | 0.363 | 38.09 | 1053.95 | 1092.04 | 0.0746 | 1.9896 | 2.0642 |
| 80 | 0.507 | 48.08 | 1048.31 | 1096.39 | 0.0933 | 1.9423 | 2.0356 |
| 90 | 0.699 | 58.06 | 1042.65 | 1100.72 | 0.1116 | 1.8966 | 2.0083 |
| 100 | 0.950 | 68.04 | 1036.98 | 1105.02 | 0.1296 | 1.8526 | 1.9822 |
| 110 | 1.276 | 78.01 | 1031.28 | 1109.29 | 0.1473 | 1.8101 | 1.9574 |
| 120 | 1.695 | 87.99 | 1025.55 | 1113.54 | 0.1646 | 1.7690 | 1.9336 |
| 130 | 2.225 | 97.97 | 1019.78 | 1117.75 | 0.1817 | 1.7292 | 1.9109 |
| 140 | 2.892 | 107.96 | 1013.96 | 1121.92 | 0.1985 | 1.6907 | 1.8892 |
| 150 | 3.722 | 117.95 | 1008.10 | 1126.05 | 0.2150 | 1.6533 | 1.8683 |
| 160 | 4.745 | 127.95 | 1002.18 | 1130.14 | 0.2313 | 1.6171 | 1.8484 |
| 170 | 5.997 | 137.96 | 996.21 | 1134.17 | 0.2473 | 1.5819 | 1.8292 |
| 180 | 7.515 | 147.98 | 990.17 | 1138.15 | 0.2631 | 1.5478 | 1.8109 |
| 190 | 9.344 | 158.02 | 984.06 | 1142.08 | 0.2786 | 1.5146 | 1.7932 |
| 200 | 11.530 | 168.07 | 977.87 | 1145.94 | 0.2940 | 1.4822 | 1.7762 |
| 210 | 14.126 | 178.13 | 971.61 | 1149.74 | 0.3091 | 1.4507 | 1.7599 |
| 212.0 | 14.696 | 180.13 | 970.35 | 1150.49 | 0.3121 | 1.4446 | 1.7567 |
| 220 | 17.189 | 188.21 | 965.26 | 1153.47 | 0.3240 | 1.4201 | 1.7441 |
| 230 | 20.781 | 198.31 | 958.81 | 1157.12 | 0.3388 | 1.3901 | 1.7289 |
| 240 | 24.968 | 208.43 | 952.27 | 1160.70 | 0.3533 | 1.3609 | 1.7142 |
| 250 | 29.823 | 218.58 | 945.61 | 1164.19 | 0.3677 | 1.3324 | 1.7001 |
| 260 | 35.422 | 228.75 | 938.84 | 1167.59 | 0.3819 | 1.3044 | 1.6864 |
| 270 | 41.848 | 238.95 | 931.95 | 1170.90 | 0.3960 | 1.2771 | 1.6731 |
| 280 | 49.189 | 249.17 | 924.93 | 1174.10 | 0.4098 | 1.2504 | 1.6602 |
| 290 | 57.535 | 259.43 | 917.76 | 1177.19 | 0.4236 | 1.2241 | 1.6477 |
| 300 | 66.985 | 269.73 | 910.45 | 1180.18 | 0.4372 | 1.1984 | 1.6356 |
| 310 | 77.641 | 280.06 | 902.98 | 1183.03 | 0.4507 | 1.1731 | 1.6238 |
| 320 | 89.609 | 290.43 | 895.34 | 1185.76 | 0.4640 | 1.1483 | 1.6122 |
| 330 | 103.00 | 300.84 | 887.52 | 1188.36 | 0.4772 | 1.1238 | 1.6010 |
| 340 | 117.94 | 311.29 | 879.51 | 1190.80 | 0.4903 | 1.0997 | 1.5900 |
| 350 | 134.54 | 321.80 | 871.30 | 1193.10 | 0.5033 | 1.0760 | 1.5793 |

**TABLE G.7.1** (*continued*)
*Saturated Water*

| Temp. (F) | Press. (psia) | SPECIFIC VOLUME, ft³/lbm | | | INTERNAL ENERGY, Btu/lbm | | |
|---|---|---|---|---|---|---|---|
| | | Sat. Liquid $v_f$ | Evap. $v_{fg}$ | Sat. Vapor $v_g$ | Sat. Liquid $u_f$ | Evap. $u_{fg}$ | Sat. Vapor $u_g$ |
| 360 | 152.93 | 0.01811 | 2.9430 | 2.9611 | 331.83 | 779.60 | 1111.43 |
| 370 | 173.24 | 0.01823 | 2.6098 | 2.6280 | 342.37 | 770.57 | 1112.94 |
| 380 | 195.61 | 0.01836 | 2.3203 | 2.3387 | 352.95 | 761.37 | 1114.31 |
| 390 | 220.17 | 0.01850 | 2.0680 | 2.0865 | 363.58 | 751.97 | 1115.55 |
| 400 | 347.08 | 0.01864 | 1.8474 | 1.8660 | 374.26 | 742.37 | 1116.63 |
| 410 | 276.48 | 0.01878 | 1.6537 | 1.6725 | 385.00 | 732.56 | 1117.56 |
| 420 | 308.52 | 0.01894 | 1.4833 | 1.5023 | 395.80 | 722.52 | 1118.32 |
| 430 | 343.37 | 0.01909 | 1.3329 | 1.3520 | 406.67 | 712.24 | 1118.91 |
| 440 | 381.18 | 0.01926 | 1.1998 | 1.2191 | 417.61 | 701.71 | 1119.32 |
| 450 | 422.13 | 0.01943 | 1.0816 | 1.1011 | 428.63 | 690.90 | 1119.53 |
| 460 | 466.38 | 0.01961 | 0.9764 | 0.9961 | 439.73 | 679.82 | 1119.55 |
| 470 | 514.11 | 0.01980 | 0.8826 | 0.9024 | 450.92 | 668.43 | 1119.35 |
| 480 | 565.50 | 0.02000 | 0.7986 | 0.8186 | 462.21 | 656.72 | 1118.93 |
| 490 | 620.74 | 0.02021 | 0.7233 | 0.7435 | 473.60 | 644.67 | 1118.28 |
| 500 | 680.02 | 0.02043 | 0.6556 | 0.6761 | 485.11 | 632.26 | 1117.37 |
| 510 | 743.53 | 0.02066 | 0.5946 | 0.6153 | 496.75 | 619.46 | 1116.21 |
| 520 | 811.48 | 0.02091 | 0.5395 | 0.5604 | 508.53 | 606.23 | 1114.76 |
| 530 | 884.07 | 0.02117 | 0.4896 | 0.5108 | 520.46 | 592.56 | 1113.02 |
| 540 | 961.51 | 0.02145 | 0.4443 | 0.4658 | 532.56 | 578.39 | 1110.95 |
| 550 | 1044.02 | 0.02175 | 0.4031 | 0.4249 | 544.85 | 563.69 | 1108.54 |
| 560 | 1131.85 | 0.02207 | 0.3656 | 0.3876 | 557.35 | 548.42 | 1105.76 |
| 570 | 1225.21 | 0.02241 | 0.3312 | 0.3536 | 570.07 | 532.50 | 1102.56 |
| 580 | 1324.37 | 0.02278 | 0.2997 | 0.3225 | 583.05 | 515.87 | 1098.91 |
| 590 | 1429.58 | 0.02318 | 0.2707 | 0.2939 | 596.31 | 498.44 | 1094.76 |
| 600 | 1541.13 | 0.02362 | 0.2440 | 0.2676 | 609.91 | 480.11 | 1090.02 |
| 610 | 1659.32 | 0.02411 | 0.2193 | 0.2434 | 623.87 | 460.76 | 1084.63 |
| 620 | 1784.48 | 0.02465 | 0.1963 | 0.2209 | 638.26 | 440.20 | 1078.46 |
| 630 | 1916.96 | 0.02525 | 0.1747 | 0.2000 | 653.17 | 418.22 | 1071.38 |
| 640 | 2057.17 | 0.02593 | 0.1545 | 0.1804 | 668.68 | 394.52 | 1063.20 |
| 650 | 2205.54 | 0.02673 | 0.1353 | 0.1620 | 684.96 | 368.66 | 1053.63 |
| 660 | 2362.59 | 0.02766 | 0.1169 | 0.1446 | 702.24 | 340.02 | 1042.26 |
| 670 | 2528.88 | 0.02882 | 0.0990 | 0.1278 | 720.91 | 307.52 | 1028.43 |
| 680 | 2705.09 | 0.03031 | 0.0809 | 0.1112 | 741.70 | 269.26 | 1010.95 |
| 690 | 2891.99 | 0.03248 | 0.0618 | 0.0943 | 766.34 | 220.82 | 987.16 |
| 700 | 3090.47 | 0.03665 | 0.0377 | 0.0743 | 801.66 | 145.92 | 947.57 |
| 705.4 | 3203.79 | 0.05053 | 0 | 0.0505 | 872.56 | 0 | 872.56 |

TABLE G.7.1  (*continued*)
*Saturated Water*

| Temp. (F) | Press. (psia) | ENTHALPY, Btu/lbm | | | ENTROPY, Btu/lbm R | | |
|---|---|---|---|---|---|---|---|
| | | Sat. Liquid $h_f$ | Evap. $h_{fg}$ | Sat. Vapor $h_g$ | Sat. Liquid $s_f$ | Evap. $s_{fg}$ | Sat. Vapor $s_g$ |
| 360 | 152.93 | 332.35 | 862.88 | 1195.23 | 0.5162 | 1.0526 | 1.5688 |
| 370 | 173.24 | 342.95 | 854.24 | 1197.19 | 0.5289 | 1.0295 | 1.5584 |
| 380 | 195.61 | 353.61 | 845.36 | 1198.97 | 0.5416 | 1.0067 | 1.5483 |
| 390 | 220.17 | 364.33 | 836.23 | 1200.56 | 0.5542 | 0.9841 | 1.5383 |
| 400 | 247.08 | 375.11 | 826.84 | 1201.95 | 0.5667 | 0.9617 | 1.5284 |
| 410 | 276.48 | 385.96 | 817.17 | 1203.13 | 0.5791 | 0.9395 | 1.5187 |
| 420 | 308.52 | 396.89 | 807.20 | 1204.09 | 0.5915 | 0.9175 | 1.5090 |
| 430 | 343.37 | 407.89 | 796.93 | 1204.82 | 0.6038 | 0.8957 | 1.4995 |
| 440 | 381.18 | 418.97 | 786.34 | 1205.31 | 0.6160 | 0.8740 | 1.4900 |
| 450 | 422.13 | 430.15 | 775.40 | 1205.54 | 0.6282 | 0.8523 | 1.4805 |
| 460 | 466.38 | 441.42 | 764.09 | 1205.51 | 0.6404 | 0.8308 | 1.4711 |
| 470 | 514.11 | 452.80 | 752.40 | 1205.20 | 0.6525 | 0.8093 | 1.4618 |
| 480 | 565.50 | 464.30 | 740.30 | 1204.60 | 0.6646 | 0.7878 | 1.4524 |
| 490 | 620.74 | 475.92 | 727.76 | 1203.68 | 0.6767 | 0.7663 | 1.4430 |
| 500 | 680.02 | 487.68 | 714.76 | 1202.44 | 0.6888 | 0.7447 | 1.4335 |
| 510 | 743.53 | 499.59 | 701.27 | 1200.86 | 0.7009 | 0.7232 | 1.4240 |
| 520 | 811.48 | 511.67 | 687.25 | 1198.92 | 0.7130 | 0.7015 | 1.4144 |
| 530 | 884.07 | 523.93 | 672.66 | 1196.58 | 0.7251 | 0.6796 | 1.4048 |
| 540 | 961.51 | 536.38 | 657.45 | 1193.83 | 0.7374 | 0.6576 | 1.3950 |
| 550 | 1044.02 | 549.05 | 641.58 | 1190.63 | 0.7496 | 0.6354 | 1.3850 |
| 560 | 1131.85 | 561.97 | 624.98 | 1186.95 | 0.7620 | 0.6129 | 1.3749 |
| 570 | 1225.21 | 575.15 | 607.59 | 1182.74 | 0.7745 | 0.5901 | 1.3646 |
| 580 | 1324.37 | 588.63 | 589.32 | 1177.95 | 0.7871 | 0.5668 | 1.3539 |
| 590 | 1429.58 | 602.45 | 570.06 | 1172.51 | 0.7999 | 0.5431 | 1.3430 |
| 600 | 1541.13 | 616.64 | 549.71 | 1166.35 | 0.8129 | 0.5187 | 1.3317 |
| 610 | 1659.32 | 631.27 | 528.08 | 1159.36 | 0.8262 | 0.4937 | 1.3199 |
| 620 | 1784.48 | 646.40 | 505.00 | 1151.41 | 0.8397 | 0.4677 | 1.3075 |
| 630 | 1916.96 | 662.12 | 480.21 | 1142.33 | 0.8537 | 0.4407 | 1.2943 |
| 640 | 2057.17 | 678.55 | 453.33 | 1131.89 | 0.8681 | 0.4122 | 1.2803 |
| 650 | 2205.54 | 695.87 | 423.89 | 1119.76 | 0.8831 | 0.3820 | 1.2651 |
| 660 | 2362.59 | 714.34 | 391.13 | 1105.47 | 0.8990 | 0.3493 | 1.2483 |
| 670 | 2528.88 | 734.39 | 353.83 | 1088.23 | 0.9160 | 0.3132 | 1.2292 |
| 680 | 2705.09 | 756.87 | 309.77 | 1066.64 | 0.9350 | 0.2718 | 1.2068 |
| 690 | 2891.99 | 783.72 | 253.88 | 1037.60 | 0.9575 | 0.2208 | 1.1783 |
| 700 | 3090.47 | 822.61 | 167.47 | 990.09 | 0.9901 | 0.1444 | 1.1345 |
| 705.4 | 3203.79 | 902.52 | 0 | 902.52 | 1.0580 | 0 | 1.0580 |

## Table G.7.2
### Superheated Vapor Water

| Temp. (F) | v (ft³/lbm) | u (Btu/lbm) | h (Btu/lbm) | s (Btu/lbm R) | v (ft³/lbm) | u (Btu/lbm) | h (Btu/lbm) | s (Btu/lbm R) |
|---|---|---|---|---|---|---|---|---|
| | 1 psia (101.70) | | | | 5 psia (162.20) | | | |
| Sat. | 333.58 | 1044.02 | 1105.75 | 1.9779 | 73.531 | 1062.99 | 1131.03 | 1.8441 |
| 200 | 392.51 | 1077.49 | 1150.12 | 2.0507 | 78.147 | 1076.25 | 1148.55 | 1.8715 |
| 240 | 416.42 | 1091.22 | 1168.28 | 2.0775 | 83.001 | 1090.25 | 1167.05 | 1.8987 |
| 280 | 440.32 | 1105.02 | 1186.50 | 2.1028 | 87.831 | 1104.27 | 1185.53 | 1.9244 |
| 320 | 464.19 | 1118.92 | 1204.82 | 2.1269 | 92.645 | 1118.32 | 1204.04 | 1.9487 |
| 360 | 488.05 | 1132.92 | 1223.23 | 2.1499 | 97.447 | 1132.42 | 1222.59 | 1.9719 |
| 400 | 511.91 | 1147.02 | 1241.75 | 2.1720 | 102.24 | 1146.61 | 1241.21 | 1.9941 |
| 440 | 535.76 | 1161.23 | 1260.37 | 2.1932 | 107.03 | 1160.89 | 1259.92 | 2.0154 |
| 500 | 571.53 | 1182.77 | 1288.53 | 2.2235 | 114.21 | 1182.50 | 1288.17 | 2.0458 |
| 600 | 631.13 | 1219.30 | 1336.09 | 2.2706 | 126.15 | 1219.10 | 1335.82 | 2.0930 |
| 700 | 690.72 | 1256.65 | 1384.47 | 2.3142 | 138.08 | 1256.50 | 1384.26 | 2.1367 |
| 800 | 750.30 | 1294.86 | 1433.70 | 2.3549 | 150.01 | 1294.73 | 1433.53 | 2.1774 |
| 900 | 809.88 | 1333.94 | 1483.81 | 2.3932 | 161.94 | 1333.84 | 1483.68 | 2.2157 |
| 1000 | 869.45 | 1373.93 | 1534.82 | 2.4294 | 173.86 | 1373.85 | 1534.71 | 2.2520 |
| 1100 | 929.03 | 1414.83 | 1586.75 | 2.4638 | 185.78 | 1414.77 | 1586.66 | 2.2864 |
| 1200 | 988.60 | 1456.67 | 1639.61 | 2.4967 | 197.70 | 1456.61 | 1639.53 | 2.3192 |
| 1300 | 1048.17 | 1499.43 | 1693.40 | 2.5281 | 209.62 | 1499.38 | 1693.33 | 2.3507 |
| 1400 | 1107.74 | 1543.13 | 1748.12 | 2.5584 | 221.53 | 1543.09 | 1748.06 | 2.3809 |
| | 10 psia (193.19) | | | | 14.696 psia (211.99) | | | |
| Sat. | 38.424 | 1072.21 | 1143.32 | 1.7877 | 26.803 | 1077.60 | 1150.49 | 1.7567 |
| 200 | 38.848 | 1074.67 | 1146.56 | 1.7927 | — | — | — | — |
| 240 | 41.320 | 1089.03 | 1165.50 | 1.8205 | 27.999 | 1087.87 | 1164.02 | 1.7764 |
| 280 | 43.768 | 1103.31 | 1184.31 | 1.8467 | 29.687 | 1102.40 | 1183.14 | 1.8030 |
| 320 | 46.200 | 1117.56 | 1203.05 | 1.8713 | 31.359 | 1116.83 | 1202.11 | 1.8280 |
| 360 | 48.620 | 1131.81 | 1221.78 | 1.8948 | 33.018 | 1131.22 | 1221.01 | 1.8516 |
| 400 | 51.032 | 1146.10 | 1240.53 | 1.9171 | 34.668 | 1145.62 | 1239.90 | 1.8741 |
| 440 | 53.438 | 1160.46 | 1259.34 | 1.9385 | 36.313 | 1160.05 | 1258.80 | 1.8956 |
| 500 | 57.039 | 1182.16 | 1287.71 | 1.9690 | 38.772 | 1181.83 | 1287.27 | 1.9262 |
| 600 | 63.027 | 1218.85 | 1335.48 | 2.0164 | 42.857 | 1218.61 | 1335.16 | 1.9737 |
| 700 | 69.006 | 1256.30 | 1384.00 | 2.0601 | 46.932 | 1256.12 | 1383.75 | 2.0175 |
| 800 | 74.978 | 1294.58 | 1433.32 | 2.1009 | 51.001 | 1294.43 | 1433.13 | 2.0584 |
| 900 | 80.946 | 1333.72 | 1483.51 | 2.1392 | 55.066 | 1333.60 | 1483.35 | 2.0967 |
| 1000 | 86.912 | 1373.74 | 1534.57 | 2.1755 | 59.128 | 1373.65 | 1534.44 | 2.1330 |
| 1100 | 92.875 | 1414.68 | 1586.54 | 2.2099 | 63.188 | 1414.60 | 1586.44 | 2.1674 |
| 1200 | 98.837 | 1456.53 | 1639.43 | 2.2428 | 67.247 | 1456.47 | 1639.34 | 2.2003 |
| 1300 | 104.798 | 1499.32 | 1693.25 | 2.2743 | 71.304 | 1499.26 | 1693.17 | 2.2318 |
| 1400 | 110.759 | 1543.03 | 1747.99 | 2.3045 | 75.361 | 1542.98 | 1747.92 | 2.2620 |
| 1500 | 116.718 | 1587.67 | 1803.66 | 2.3337 | 79.417 | 1587.63 | 1803.60 | 2.2912 |
| 1600 | 122.678 | 1633.24 | 1860.25 | 2.3618 | 83.473 | 1633.20 | 1860.20 | 2.3194 |

**TABLE G.7.2** (*continued*)
*Superheated Vapor Water*

| Temp. (F) | v (ft³/lbm) | u (Btu/lbm) | h (Btu/lbm) | s (Btu/lbm R) | v (ft³/lbm) | u (Btu/lbm) | h (Btu/lbm) | s (Btu/lbm R) |
|---|---|---|---|---|---|---|---|---|
| | 20 psia (227.96) | | | | 40 psia (267.26) | | | |
| Sat. | 20.091 | 1082.02 | 1156.38 | 1.7320 | 10.501 | 1092.27 | 1170.00 | 1.6767 |
| 240 | 20.475 | 1086.54 | 1162.32 | 1.7405 | — | — | — | — |
| 280 | 21.734 | 1101.36 | 1181.80 | 1.7676 | 10.711 | 1097.31 | 1176.59 | 1.6857 |
| 320 | 22.976 | 1116.01 | 1201.04 | 1.7929 | 11.360 | 1112.81 | 1196.90 | 1.7124 |
| 360 | 24.206 | 1130.55 | 1220.14 | 1.8168 | 11.996 | 1127.98 | 1216.77 | 1.7373 |
| 400 | 25.427 | 1145.06 | 1239.17 | 1.8395 | 12.623 | 1142.95 | 1236.38 | 1.7606 |
| 440 | 26.642 | 1159.59 | 1258.19 | 1.8611 | 13.243 | 1157.82 | 1255.84 | 1.7827 |
| 500 | 28.456 | 1181.46 | 1286.78 | 1.8919 | 14.164 | 1180.06 | 1284.91 | 1.8140 |
| 600 | 31.466 | 1218.35 | 1334.80 | 1.9395 | 15.685 | 1217.33 | 1333.43 | 1.8621 |
| 700 | 34.466 | 1255.91 | 1383.47 | 1.9834 | 17.196 | 1255.14 | 1382.42 | 1.9063 |
| 800 | 37.460 | 1294.27 | 1432.91 | 2.0243 | 18.701 | 1293.65 | 1432.08 | 1.9474 |
| 900 | 40.450 | 1333.47 | 1483.17 | 2.0626 | 20.202 | 1332.96 | 1482.50 | 1.9859 |
| 1000 | 43.437 | 1373.54 | 1534.30 | 2.0989 | 21.700 | 1373.12 | 1533.74 | 2.0222 |
| 1100 | 46.422 | 1414.51 | 1586.32 | 2.1334 | 23.196 | 1414.16 | 1585.86 | 2.0568 |
| 1200 | 49.406 | 1456.39 | 1639.24 | 2.1663 | 24.690 | 1456.09 | 1638.85 | 2.0897 |
| 1300 | 52.389 | 1499.19 | 1693.08 | 2.1978 | 26.184 | 1498.94 | 1692.75 | 2.1212 |
| 1400 | 55.371 | 1542.92 | 1747.85 | 2.2280 | 27.677 | 1542.70 | 1747.56 | 2.1515 |
| 1500 | 58.352 | 1587.58 | 1803.54 | 2.2572 | 29.169 | 1587.38 | 1803.29 | 2.1807 |
| 1600 | 61.333 | 1633.15 | 1860.14 | 2.2854 | 30.660 | 1632.97 | 1859.92 | 2.2089 |
| | 60 psia (292.73) | | | | 80 psia (312.06) | | | |
| Sat. | 7.177 | 1098.33 | 1178.02 | 1.6444 | 5.474 | 1102.56 | 1183.61 | 1.6214 |
| 320 | 7.485 | 1109.46 | 1192.56 | 1.6633 | 5.544 | 1105.95 | 1188.02 | 1.6270 |
| 360 | 7.924 | 1125.31 | 1213.29 | 1.6893 | 5.886 | 1122.53 | 1209.67 | 1.6541 |
| 400 | 8.353 | 1140.77 | 1233.52 | 1.7134 | 6.217 | 1138.53 | 1230.56 | 1.6790 |
| 440 | 8.775 | 1156.01 | 1253.44 | 1.7360 | 6.541 | 1154.15 | 1250.98 | 1.7022 |
| 500 | 9.399 | 1178.64 | 1283.00 | 1.7678 | 7.017 | 1177.19 | 1281.07 | 1.7346 |
| 600 | 10.425 | 1216.31 | 1332.06 | 1.8165 | 7.794 | 1215.28 | 1330.66 | 1.7838 |
| 700 | 11.440 | 1254.35 | 1381.37 | 1.8609 | 8.561 | 1253.57 | 1380.31 | 1.8285 |
| 800 | 12.448 | 1293.03 | 1431.24 | 1.9022 | 9.322 | 1292.41 | 1430.40 | 1.8700 |
| 900 | 13.452 | 1332.46 | 1481.82 | 1.9408 | 10.078 | 1331.95 | 1481.14 | 1.9087 |
| 1000 | 14.454 | 1372.71 | 1533.19 | 1.9773 | 10.831 | 1372.29 | 1532.63 | 1.9453 |
| 1100 | 15.454 | 1413.81 | 1585.39 | 2.0119 | 11.583 | 1413.46 | 1584.93 | 1.9799 |
| 1200 | 16.452 | 1455.80 | 1638.46 | 2.0448 | 12.333 | 1455.51 | 1638.08 | 2.0129 |
| 1300 | 17.449 | 1498.69 | 1692.42 | 2.0764 | 13.082 | 1498.43 | 1692.09 | 2.0445 |
| 1400 | 18.445 | 1542.48 | 1747.28 | 2.1067 | 13.830 | 1542.26 | 1746.99 | 2.0749 |
| 1500 | 19.441 | 1587.18 | 1803.04 | 2.1359 | 14.577 | 1586.99 | 1802.79 | 2.1041 |
| 1600 | 20.436 | 1632.79 | 1859.70 | 2.1641 | 15.324 | 1632.62 | 1859.48 | 2.1323 |
| 1800 | 22.426 | 1726.69 | 1975.69 | 2.2178 | 16.818 | 1726.54 | 1975.50 | 2.1861 |
| 2000 | 24.415 | 1824.02 | 2095.10 | 2.2685 | 18.310 | 1823.88 | 2094.94 | 2.2367 |

**Table G.7.2** (*continued*)
*Superheated Vapor Water*

| Temp. (F) | v (ft³/lbm) | u (Btu/lbm) | h (Btu/lbm) | s (Btu/lbm R) | v (ft³/lbm) | u (Btu/lbm) | h (Btu/lbm) | s (Btu/lbm R) |
|---|---|---|---|---|---|---|---|---|
| | \multicolumn{4}{} 100 psia (327.85) | | | | 150 psia (358.47) | | | |
| Sat. | 4.4340 | 1105.76 | 1187.81 | 1.6034 | 3.0163 | 1111.19 | 1194.91 | 1.5704 |
| 350 | 4.5917 | 1115.39 | 1200.36 | 1.6191 | — | — | — | — |
| 400 | 4.9344 | 1136.21 | 1227.53 | 1.6517 | 3.2212 | 1130.10 | 1219.51 | 1.5997 |
| 450 | 5.2646 | 1156.20 | 1253.62 | 1.6812 | 3.4547 | 1151.47 | 1247.36 | 1.6312 |
| 500 | 5.5866 | 1175.72 | 1279.10 | 1.7085 | 3.6789 | 1171.93 | 1274.04 | 1.6598 |
| 550 | 5.9032 | 1195.02 | 1304.25 | 1.7340 | 3.8970 | 1191.88 | 1300.05 | 1.6862 |
| 600 | 6.2160 | 1214.23 | 1329.26 | 1.7582 | 4.1110 | 1211.58 | 1325.69 | 1.7110 |
| 700 | 6.8340 | 1252.78 | 1379.24 | 1.8033 | 4.5309 | 1250.78 | 1376.55 | 1.7568 |
| 800 | 7.4455 | 1291.78 | 1429.56 | 1.8449 | 4.9441 | 1290.21 | 1427.44 | 1.7989 |
| 900 | 8.0528 | 1331.45 | 1480.47 | 1.8838 | 5.3529 | 1330.18 | 1478.76 | 1.8381 |
| 1000 | 8.6574 | 1371.87 | 1532.08 | 1.9204 | 5.7590 | 1370.83 | 1530.68 | 1.8750 |
| 1100 | 9.2599 | 1413.12 | 1584.47 | 1.9551 | 6.1630 | 1412.24 | 1583.31 | 1.9098 |
| 1200 | 9.8610 | 1455.21 | 1637.69 | 1.9882 | 6.5655 | 1454.47 | 1636.71 | 1.9430 |
| 1300 | 10.4610 | 1498.18 | 1691.76 | 2.0198 | 6.9670 | 1497.55 | 1690.93 | 1.9747 |
| 1400 | 11.0602 | 1542.04 | 1746.71 | 2.0502 | 7.3677 | 1541.49 | 1745.99 | 2.0052 |
| 1500 | 11.6588 | 1586.79 | 1802.54 | 2.0794 | 7.7677 | 1586.30 | 1801.91 | 2.0345 |
| 1600 | 12.2570 | 1632.44 | 1859.25 | 2.1076 | 8.1673 | 1632.00 | 1858.70 | 2.0627 |
| 1800 | 13.4525 | 1726.38 | 1975.32 | 2.1614 | 8.9657 | 1726.00 | 1974.86 | 2.1165 |
| 2000 | 14.6472 | 1823.74 | 2094.78 | 2.2120 | 9.7633 | 1823.38 | 2094.38 | 2.1672 |
| | \multicolumn{4}{} 200 psia (381.86) | | | | 300 psia (417.42) | | | |
| Sat. | 2.2892 | 1114.55 | 1199.28 | 1.5464 | 1.5441 | 1118.14 | 1203.86 | 1.5115 |
| 400 | 2.3609 | 1123.45 | 1210.83 | 1.5600 | — | — | — | — |
| 450 | 2.5477 | 1146.44 | 1240.73 | 1.5938 | 1.6361 | 1135.37 | 1226.20 | 1.5365 |
| 500 | 2.7238 | 1167.96 | 1268.77 | 1.6238 | 1.7662 | 1159.47 | 1257.52 | 1.5701 |
| 550 | 2.8932 | 1188.65 | 1295.72 | 1.6512 | 1.8878 | 1181.85 | 1286.65 | 1.5997 |
| 600 | 3.0580 | 1208.87 | 1322.05 | 1.6767 | 2.0041 | 1203.24 | 1314.50 | 1.6266 |
| 700 | 3.3792 | 1248.76 | 1373.82 | 1.7234 | 2.2269 | 1244.63 | 1368.26 | 1.6751 |
| 800 | 3.6932 | 1288.62 | 1425.31 | 1.7659 | 2.4421 | 1285.41 | 1420.99 | 1.7187 |
| 900 | 4.0029 | 1328.90 | 1477.04 | 1.8055 | 2.6528 | 1326.31 | 1473.58 | 1.7589 |
| 1000 | 4.3097 | 1369.77 | 1529.28 | 1.8425 | 2.8604 | 1367.65 | 1526.45 | 1.7964 |
| 1100 | 4.6145 | 1411.36 | 1582.15 | 1.8776 | 3.0660 | 1409.60 | 1579.80 | 1.8317 |
| 1200 | 4.9178 | 1453.73 | 1635.74 | 1.9109 | 3.2700 | 1452.24 | 1633.77 | 1.8653 |
| 1300 | 5.2200 | 1496.91 | 1690.10 | 1.9427 | 3.4730 | 1495.63 | 1688.43 | 1.8972 |
| 1400 | 5.5214 | 1540.93 | 1745.28 | 1.9732 | 3.6751 | 1539.82 | 1743.84 | 1.9279 |
| 1500 | 5.8222 | 1585.81 | 1801.29 | 2.0025 | 3.8767 | 1584.82 | 1800.03 | 1.9573 |
| 1600 | 6.1225 | 1631.55 | 1858.15 | 2.0308 | 4.0777 | 1630.66 | 1857.04 | 1.9857 |
| 1800 | 6.7223 | 1725.62 | 1974.41 | 2.0847 | 4.4790 | 1724.85 | 1973.50 | 2.0396 |
| 2000 | 7.3214 | 1823.02 | 2093.99 | 2.1354 | 4.8794 | 1822.32 | 2093.20 | 2.0904 |

Table G.7.2  (*continued*)
*Superheated Vapor Water*

| Temp. (F) | v (ft³/lbm) | u (Btu/lbm) | h (Btu/lbm) | s (Btu/lbm R) | v (ft³/lbm) | u (Btu/lbm) | h (Btu/lbm) | s (Btu/lbm R) |
|---|---|---|---|---|---|---|---|---|
| | 400 psia (444.69) | | | | 600 psia (486.33) | | | |
| Sat. | 1.1619 | 1119.44 | 1205.45 | 1.4856 | 0.7702 | 1118.54 | 1204.06 | 1.4464 |
| 450 | 1.1745 | 1122.63 | 1209.57 | 1.4901 | — | — | — | — |
| 500 | 1.2843 | 1150.11 | 1245.17 | 1.5282 | 0.7947 | 1127.97 | 1216.21 | 1.4592 |
| 550 | 1.3834 | 1174.56 | 1276.95 | 1.5605 | 0.8749 | 1158.23 | 1255.36 | 1.4990 |
| 600 | 1.4760 | 1197.33 | 1306.58 | 1.5892 | 0.9456 | 1184.50 | 1289.49 | 1.5320 |
| 700 | 1.6503 | 1240.38 | 1362.54 | 1.6396 | 1.0728 | 1231.51 | 1350.62 | 1.5871 |
| 800 | 1.8163 | 1282.14 | 1416.59 | 1.6844 | 1.1900 | 1275.42 | 1407.55 | 1.6343 |
| 900 | 1.9776 | 1323.69 | 1470.07 | 1.7252 | 1.3021 | 1318.36 | 1462.92 | 1.6766 |
| 1000 | 2.1357 | 1365.51 | 1523.59 | 1.7632 | 1.4108 | 1361.15 | 1517.79 | 1.7155 |
| 1100 | 2.2917 | 1407.81 | 1577.44 | 1.7989 | 1.5173 | 1404.20 | 1572.66 | 1.7519 |
| 1200 | 2.4462 | 1450.73 | 1631.79 | 1.8327 | 1.6222 | 1447.68 | 1627.80 | 1.7861 |
| 1300 | 2.5995 | 1494.34 | 1686.76 | 1.8648 | 1.7260 | 1491.74 | 1683.38 | 1.8186 |
| 1400 | 2.7520 | 1538.70 | 1742.40 | 1.8956 | 1.8289 | 1536.44 | 1739.51 | 1.8497 |
| 1500 | 2.9039 | 1583.83 | 1798.78 | 1.9251 | 1.9312 | 1581.84 | 1796.26 | 1.8794 |
| 1600 | 3.0553 | 1629.77 | 1855.93 | 1.9535 | 2.0330 | 1627.98 | 1853.71 | 1.9080 |
| 1700 | 3.2064 | 1676.52 | 1913.86 | 1.9810 | 2.1345 | 1674.88 | 1911.87 | 1.9355 |
| 1800 | 3.3573 | 1724.08 | 1972.59 | 2.0076 | 2.2357 | 1722.55 | 1970.78 | 1.9622 |
| 2000 | 3.6585 | 1821.61 | 2092.41 | 2.0584 | 2.4375 | 1820.20 | 2090.84 | 2.0131 |
| | 800 psia (518.36) | | | | 1000 psia (544.74) | | | |
| Sat. | 0.5691 | 1115.02 | 1199.26 | 1.4160 | 0.4459 | 1109.86 | 1192.37 | 1.3903 |
| 550 | 0.6154 | 1138.83 | 1229.93 | 1.4469 | 0.4534 | 1114.77 | 1198.67 | 1.3965 |
| 600 | 0.6776 | 1170.10 | 1270.41 | 1.4861 | 0.5140 | 1153.66 | 1248.76 | 1.4450 |
| 650 | 0.7324 | 1197.22 | 1305.64 | 1.5186 | 0.5637 | 1184.74 | 1289.06 | 1.4822 |
| 700 | 0.7829 | 1222.08 | 1337.98 | 1.5471 | 0.6080 | 1212.03 | 1324.54 | 1.5135 |
| 800 | 0.8764 | 1268.45 | 1398.19 | 1.5969 | 0.6878 | 1261.21 | 1388.49 | 1.5664 |
| 900 | 0.9640 | 1312.88 | 1455.60 | 1.6408 | 0.7610 | 1307.26 | 1448.08 | 1.6120 |
| 1000 | 1.0482 | 1356.71 | 1511.88 | 1.6807 | 0.8305 | 1352.17 | 1505.86 | 1.6530 |
| 1100 | 1.1300 | 1400.52 | 1567.81 | 1.7178 | 0.8976 | 1396.77 | 1562.88 | 1.6908 |
| 1200 | 1.2102 | 1444.60 | 1623.76 | 1.7525 | 0.9630 | 1441.46 | 1619.67 | 1.7260 |
| 1300 | 1.2892 | 1489.11 | 1679.97 | 1.7854 | 1.0272 | 1486.45 | 1676.53 | 1.7593 |
| 1400 | 1.3674 | 1534.17 | 1736.59 | 1.8167 | 1.0905 | 1531.88 | 1733.67 | 1.7909 |
| 1500 | 1.4448 | 1579.85 | 1793.74 | 1.8467 | 1.1531 | 1577.84 | 1791.21 | 1.8210 |
| 1600 | 1.5218 | 1626.19 | 1851.49 | 1.8754 | 1.2152 | 1624.40 | 1849.27 | 1.8499 |
| 1700 | 1.5985 | 1673.25 | 1909.89 | 1.9031 | 1.2769 | 1671.61 | 1907.91 | 1.8777 |
| 1800 | 1.6749 | 1721.03 | 1968.98 | 1.9298 | 1.3384 | 1719.51 | 1967.18 | 1.9046 |
| 2000 | 1.8271 | 1818.80 | 2089.28 | 1.9808 | 1.4608 | 1817.41 | 2087.74 | 1.9557 |

**TABLE G.7.2** (*continued*)
*Superheated Vapor Water*

| Temp. (F) | v (ft³/lbm) | u (Btu/lbm) | h (Btu/lbm) | s (Btu/lbm R) | v (ft³/lbm) | u (Btu/lbm) | h (Btu/lbm) | s (Btu/lbm R) |
|---|---|---|---|---|---|---|---|---|
| | 1500 psia (596.38) | | | | 2000 psia (635.99) | | | |
| Sat. | 0.2769 | 1091.81 | 1168.67 | 1.3358 | 0.1881 | 1066.63 | 1136.25 | 1.2861 |
| 650 | 0.3329 | 1146.95 | 1239.34 | 1.4012 | 0.2057 | 1091.06 | 1167.18 | 1.3141 |
| 700 | 0.3716 | 1183.44 | 1286.60 | 1.4429 | 0.2487 | 1147.74 | 1239.79 | 1.3782 |
| 750 | 0.4049 | 1214.13 | 1326.52 | 1.4766 | 0.2803 | 1187.32 | 1291.07 | 1.4216 |
| 800 | 0.4350 | 1241.79 | 1362.53 | 1.5058 | 0.3071 | 1220.13 | 1333.80 | 1.4562 |
| 850 | 0.4631 | 1267.69 | 1396.23 | 1.5321 | 0.3312 | 1249.46 | 1372.03 | 1.4860 |
| 900 | 0.4897 | 1292.53 | 1428.46 | 1.5562 | 0.3534 | 1276.78 | 1407.58 | 1.5126 |
| 1000 | 0.5400 | 1340.43 | 1490.32 | 1.6001 | 0.3945 | 1328.10 | 1474.09 | 1.5598 |
| 1100 | 0.5876 | 1387.16 | 1550.26 | 1.6398 | 0.4325 | 1377.17 | 1537.23 | 1.6017 |
| 1200 | 0.6334 | 1433.45 | 1609.25 | 1.6765 | 0.4685 | 1425.19 | 1598.58 | 1.6398 |
| 1300 | 0.6778 | 1479.68 | 1667.82 | 1.7108 | 0.5031 | 1472.74 | 1658.95 | 1.6751 |
| 1400 | 0.7213 | 1526.06 | 1726.28 | 1.7431 | 0.5368 | 1520.15 | 1718.81 | 1.7082 |
| 1500 | 0.7641 | 1572.77 | 1784.86 | 1.7738 | 0.5697 | 1567.64 | 1778.48 | 1.7395 |
| 1600 | 0.8064 | 1619.90 | 1843.72 | 1.8301 | 0.6020 | 1615.37 | 1838.18 | 1.7692 |
| 1700 | 0.8482 | 1667.53 | 1902.98 | 1.8312 | 0.6340 | 1663.45 | 1898.08 | 1.7976 |
| 1800 | 0.8899 | 1715.73 | 1962.73 | 1.8582 | 0.6656 | 1711.97 | 1958.32 | 1.8248 |
| 1900 | 0.9313 | 1764.53 | 2023.03 | 1.8843 | 0.6971 | 1760.99 | 2018.99 | 1.8511 |
| 2000 | 0.9725 | 1813.97 | 2083.91 | 1.9096 | 0.7284 | 1810.56 | 2080.15 | 1.8765 |
| | 4000 psia | | | | 8000 psia | | | |
| 650 | 0.02447 | 657.71 | 675.82 | 0.8574 | 0.02239 | 627.01 | 660.16 | 0.8278 |
| 700 | 0.02867 | 742.13 | 763.35 | 0.9345 | 0.02418 | 688.59 | 724.39 | 0.8844 |
| 750 | 0.06332 | 960.69 | 1007.56 | 1.1395 | 0.02671 | 755.67 | 795.21 | 0.9441 |
| 800 | 0.10523 | 1095.04 | 1172.93 | 1.2740 | 0.03061 | 830.67 | 875.99 | 1.0095 |
| 850 | 0.12833 | 1156.47 | 1251.46 | 1.3352 | 0.03706 | 915.81 | 970.67 | 1.0832 |
| 900 | 0.14623 | 1201.47 | 1309.71 | 1.3789 | 0.04657 | 1003.68 | 1072.63 | 1.1596 |
| 950 | 0.16152 | 1239.20 | 1358.75 | 1.4143 | 0.05721 | 1079.59 | 1164.28 | 1.2259 |
| 1000 | 0.17520 | 1272.94 | 1402.62 | 1.4449 | 0.06722 | 1141.04 | 1240.55 | 1.2791 |
| 1100 | 0.19954 | 1333.90 | 1481.60 | 1.4973 | 0.08445 | 1236.84 | 1361.85 | 1.3595 |
| 1200 | 0.22129 | 1390.11 | 1553.91 | 1.5423 | 0.09892 | 1314.18 | 1460.62 | 1.4210 |
| 1300 | 0.24137 | 1443.72 | 1622.38 | 1.5823 | 0.11161 | 1382.27 | 1547.50 | 1.4718 |
| 1400 | 0.26029 | 1495.73 | 1688.39 | 1.6188 | 0.12309 | 1444.85 | 1627.08 | 1.5158 |
| 1500 | 0.27837 | 1546.73 | 1752.78 | 1.6525 | 0.13372 | 1503.78 | 1701.74 | 1.5549 |
| 1600 | 0.29586 | 1597.12 | 1816.11 | 1.6841 | 0.14373 | 1560.12 | 1772.89 | 1.5904 |
| 1700 | 0.31291 | 1647.17 | 1878.79 | 1.7138 | 0.15328 | 1614.58 | 1841.49 | 1.6229 |
| 1800 | 0.32964 | 1697.11 | 1941.11 | 1.7420 | 0.16251 | 1667.69 | 1908.27 | 1.6531 |
| 1900 | 0.34616 | 1747.10 | 2003.32 | 1.7689 | 0.17151 | 1719.85 | 1973.75 | 1.6815 |
| 2000 | 0.36251 | 1797.27 | 2065.60 | 1.7948 | 0.18034 | 1771.38 | 2038.36 | 1.7083 |

**TABLE G.7.3**
*Compressed Liquid Water*

| Temp. (F) | v (ft³/lbm) | u (Btu/lbm) | h (Btu/lbm) | s (Btu/lbm R) | v (ft³/lbm) | u (Btu/lbm) | h (Btu/lbm) | s (Btu/lbm R) |
|---|---|---|---|---|---|---|---|---|
| | 500 psia (467.12) | | | | 1000 psia (544.74) | | | |
| Sat. | 0.01975 | 447.69 | 449.51 | 0.6490 | 0.02159 | 538.37 | 542.36 | 0.74318 |
| 32 | 0.01599 | 0.00 | 1.48 | 0.0000 | 0.01597 | 0.02 | 2.98 | 0.0000 |
| 50 | 0.01599 | 18.02 | 19.50 | 0.0360 | 0.01599 | 17.98 | 20.94 | 0.0359 |
| 75 | 0.0160 | 42.98 | 44.46 | 0.0838 | 0.0160 | 42.87 | 45.83 | 0.0836 |
| 100 | 0.0161 | 67.87 | 69.36 | 0.1293 | 0.0161 | 67.70 | 70.67 | 0.1290 |
| 125 | 0.0162 | 92.75 | 94.24 | 0.1728 | 0.0162 | 92.52 | 95.51 | 0.1724 |
| 150 | 0.0163 | 117.66 | 119.17 | 0.2146 | 0.0163 | 117.37 | 120.39 | 0.2141 |
| 175 | 0.0165 | 142.62 | 144.14 | 0.2547 | 0.0164 | 142.28 | 145.32 | 0.2542 |
| 200 | 0.0166 | 167.64 | 168.18 | 0.2934 | 0.0166 | 167.25 | 170.32 | 0.2928 |
| 225 | 0.0168 | 192.76 | 194.31 | 0.3308 | 0.0168 | 192.30 | 195.40 | 0.3301 |
| 250 | 0.0170 | 217.99 | 219.56 | 0.3670 | 0.0169 | 217.46 | 220.60 | 0.3663 |
| 275 | 0.0172 | 243.36 | 244.95 | 0.4022 | 0.0171 | 242.77 | 245.94 | 0.4014 |
| 300 | 0.0174 | 268.91 | 270.52 | 0.4364 | 0.0174 | 268.24 | 271.45 | 0.4355 |
| 325 | 0.0177 | 294.68 | 296.32 | 0.4698 | 0.0176 | 293.91 | 297.17 | 0.4688 |
| 350 | 0.0180 | 320.70 | 322.36 | 0.5025 | 0.0179 | 319.83 | 323.14 | 0.5014 |
| 375 | 0.0183 | 347.01 | 348.70 | 0.5345 | 0.0182 | 346.02 | 349.39 | 0.5333 |
| 400 | 0.0186 | 373.68 | 375.40 | 0.5660 | 0.0185 | 372.55 | 375.98 | 0.5647 |
| 425 | 0.0190 | 400.77 | 402.52 | 0.5971 | 0.0189 | 399.47 | 402.97 | 0.5957 |
| 450 | 0.0194 | 428.39 | 430.19 | 0.6280 | 0.0193 | 426.89 | 430.47 | 0.6263 |
| | 2000 psia (635.99) | | | | 8000 psia | | | |
| Sat. | 0.02565 | 662.38 | 671.87 | 0.8622 | — | — | — | — |
| 50 | 0.01592 | 17.91 | 23.80 | 0.0357 | 0.01563 | 17.38 | 40.52 | 0.0342 |
| 75 | 0.0160 | 42.66 | 48.57 | 0.0832 | 0.0157 | 41.42 | 64.65 | 0.0804 |
| 100 | 0.0160 | 67.36 | 73.30 | 0.1284 | 0.01577 | 65.49 | 88.83 | 0.1246 |
| 125 | 0.0161 | 92.07 | 98.04 | 0.1716 | 0.01586 | 89.62 | 113.10 | 0.1670 |
| 150 | 0.0162 | 116.82 | 122.84 | 0.2132 | 0.01597 | 113.81 | 137.45 | 0.2078 |
| 175 | 0.0164 | 141.62 | 147.68 | 0.2531 | 0.01610 | 138.04 | 161.87 | 0.2471 |
| 200 | 0.0165 | 166.48 | 172.60 | 0.2916 | 0.01623 | 162.31 | 186.34 | 0.2849 |
| 225 | 0.0167 | 191.42 | 197.59 | 0.3288 | 0.01639 | 186.61 | 210.87 | 0.3214 |
| 250 | 0.0169 | 216.45 | 222.69 | 0.3648 | 0.01655 | 210.97 | 235.47 | 0.3567 |
| 275 | 0.0171 | 241.61 | 247.93 | 0.3998 | 0.01675 | 235.39 | 260.16 | 0.3909 |
| 300 | 0.0173 | 266.92 | 273.33 | 0.4337 | 0.01693 | 259.91 | 284.97 | 0.4241 |
| 325 | 0.0176 | 292.42 | 298.92 | 0.4669 | 0.01714 | 284.53 | 309.91 | 0.4564 |
| 350 | 0.0178 | 318.14 | 324.74 | 0.4993 | 0.01737 | 309.29 | 335.01 | 0.4878 |
| 400 | 0.0184 | 370.38 | 377.20 | 0.5621 | 0.01788 | 359.26 | 385.73 | 0.5486 |
| 450 | 0.0192 | 424.03 | 431.13 | 0.6231 | 0.01848 | 409.94 | 437.30 | 0.6069 |
| 500 | 0.0201 | 479.84 | 487.29 | 0.6832 | 0.01918 | 461.56 | 489.95 | 0.6633 |
| 600 | 0.0233 | 605.37 | 613.99 | 0.8086 | 0.02106 | 569.36 | 600.53 | 0.7728 |

TABLE G.7.4
*Saturated Solid–Saturated Vapor, Water (English Units)*

| Temp. (F) | Press. (lbf/in.$^2$) | SPECIFIC VOLUME, ft$^3$/lbm | | INTERNAL ENERGY, Btu/lbm | | |
|---|---|---|---|---|---|---|
| | | Sat. Solid $v_i$ | Sat. Vapor $v_g \times 10^{-3}$ | Sat. Solid $u_i$ | Evap. $u_{ig}$ | Sat. Vapor $u_g$ |
| 32.02 | 0.08866 | 0.017473 | 3.302 | −143.34 | 1164.5 | 1021.2 |
| 32 | 0.08859 | 0.01747 | 3.305 | −143.35 | 1164.5 | 1021.2 |
| 30 | 0.08083 | 0.01747 | 3.607 | −144.35 | 1164.9 | 1020.5 |
| 25 | 0.06406 | 0.01746 | 4.505 | −146.84 | 1165.7 | 1018.9 |
| 20 | 0.05051 | 0.01745 | 5.655 | −149.31 | 1166.5 | 1017.2 |
| 15 | 0.03963 | 0.01745 | 7.133 | −151.75 | 1167.3 | 1015.6 |
| 10 | 0.03093 | 0.01744 | 9.043 | −154.16 | 1168.1 | 1013.9 |
| 5 | 0.02402 | 0.01743 | 11.522 | −156.56 | 1168.8 | 1012.2 |
| 0 | 0.01855 | 0.01742 | 14.761 | −158.93 | 1169.5 | 1010.6 |
| −5 | 0.01424 | 0.01742 | 19.019 | −161.27 | 1170.2 | 1008.9 |
| −10 | 0.01086 | 0.01741 | 24.657 | −163.59 | 1170.8 | 1007.3 |
| −15 | 0.00823 | 0.01740 | 32.169 | −165.89 | 1171.5 | 1005.6 |
| −20 | 0.00620 | 0.01740 | 42.238 | −168.16 | 1172.1 | 1003.9 |
| −25 | 0.00464 | 0.01739 | 55.782 | −170.40 | 1172.7 | 1002.3 |
| −30 | 0.00346 | 0.01738 | 74.046 | −172.63 | 1173.2 | 1000.6 |
| −35 | 0.00256 | 0.01737 | 98.890 | −174.82 | 1173.8 | 998.9 |
| −40 | 0.00187 | 0.01737 | 134.017 | −177.00 | 1174.3 | 997.3 |

TABLE G.7.4  (*continued*)
*Saturated Solid–Saturated Vapor, Water (English Units)*

| Temp. (F) | Press. (lbf/in.²) | ENTHALPY, Btu/lbm | | | ENTROPY, Btu/lbm R | | |
|---|---|---|---|---|---|---|---|
| | | Sat. Solid $h_i$ | Evap. $h_{ig}$ | Sat. Vapor $h_g$ | Sat. Solid $s_i$ | Evap. $s_{ig}$ | Sat. Vapor $s_g$ |
| 32.02 | 0.08866 | −143.34 | 1218.7 | 1075.4 | −0.2916 | 2.4786 | 2.1869 |
| 32 | 0.08859 | −143.35 | 1218.7 | 1075.4 | −0.2917 | 2.4787 | 2.1870 |
| 30 | 0.08083 | −144.35 | 1218.8 | 1074.5 | −0.2938 | 2.4891 | 2.1953 |
| 25 | 0.06406 | −146.84 | 1219.1 | 1072.3 | −0.2990 | 2.5154 | 2.2164 |
| 20 | 0.05051 | −149.31 | 1219.4 | 1070.1 | −0.3042 | 2.5422 | 2.2380 |
| 15 | 0.03963 | −151.75 | 1219.6 | 1067.9 | −0.3093 | 2.5695 | 2.2601 |
| 10 | 0.03093 | −154.16 | 1219.8 | 1065.7 | −0.3145 | 2.5973 | 2.2827 |
| 5 | 0.02402 | −156.56 | 1220.0 | 1063.5 | −0.3197 | 2.6256 | 2.3059 |
| 0 | 0.01855 | −158.93 | 1220.2 | 1061.2 | −0.3248 | 2.6544 | 2.3296 |
| −5 | 0.01424 | −161.27 | 1220.3 | 1059.0 | −0.3300 | 2.6839 | 2.3539 |
| −10 | 0.01086 | −163.59 | 1220.4 | 1056.8 | −0.3351 | 2.7140 | 2.3788 |
| −15 | 0.00823 | −165.89 | 1220.5 | 1054.6 | −0.3403 | 2.7447 | 2.4044 |
| −20 | 0.00620 | −168.16 | 1220.5 | 1052.4 | −0.3455 | 2.7761 | 2.4307 |
| −25 | 0.00464 | −170.40 | 1220.6 | 1050.2 | −0.3506 | 2.8081 | 2.4575 |
| −30 | 0.00346 | −172.63 | 1220.6 | 1048.0 | −0.3557 | 2.8406 | 2.4849 |
| −35 | 0.00256 | −174.82 | 1220.6 | 1045.7 | −0.3608 | 2.8737 | 2.5129 |
| −40 | 0.00187 | −177.00 | 1220.5 | 1043.5 | −0.3659 | 2.9084 | 2.5425 |

| Temp. F | Press. (psia) | SPECIFIC VOLUME, ft³/lbm | | | INTERNAL ENERGY, Btu/lbm | | |
|---|---|---|---|---|---|---|---|
| | | Sat. Liquid $v_f$ | Evap. $v_{fg}$ | Sat. Vapor $v_g$ | Sat. Liquid $u_f$ | Evap. $u_{fg}$ | Sat. Vapor $u_g$ |
| −100 | 0.951 | 0.01077 | 39.5032 | 39.5139 | 50.47 | 94.15 | 144.62 |
| −90 | 1.410 | 0.01083 | 27.3236 | 27.3345 | 52.03 | 93.89 | 145.92 |
| −80 | 2.047 | 0.01091 | 19.2731 | 19.2840 | 53.96 | 93.27 | 147.24 |
| −70 | 2.913 | 0.01101 | 13.8538 | 13.8648 | 56.19 | 92.38 | 148.57 |
| −60 | 4.067 | 0.01111 | 10.1389 | 10.1501 | 58.64 | 91.26 | 149.91 |
| −50 | 5.575 | 0.01122 | 7.5468 | 7.5580 | 61.27 | 89.99 | 151.26 |
| −40 | 7.511 | 0.01134 | 5.7066 | 5.7179 | 64.04 | 88.58 | 152.62 |
| −30 | 9.959 | 0.01146 | 4.3785 | 4.3900 | 66.90 | 87.09 | 153.99 |
| −20 | 13.009 | 0.01159 | 3.4049 | 3.4165 | 69.83 | 85.53 | 155.36 |
| −15.3 | 14.696 | 0.01166 | 3.0350 | 3.0466 | 71.25 | 84.76 | 156.02 |
| −10 | 16.760 | 0.01173 | 2.6805 | 2.6922 | 72.83 | 83.91 | 156.74 |
| 0 | 21.315 | 0.01187 | 2.1340 | 2.1458 | 75.88 | 82.24 | 158.12 |
| 10 | 26.787 | 0.01202 | 1.7162 | 1.7282 | 78.96 | 80.53 | 159.50 |
| 20 | 33.294 | 0.01218 | 1.3928 | 1.4050 | 82.09 | 78.78 | 160.87 |
| 30 | 40.962 | 0.01235 | 1.1398 | 1.1521 | 85.25 | 76.99 | 162.24 |
| 40 | 49.922 | 0.01253 | 0.9395 | 0.9520 | 88.45 | 75.16 | 163.60 |
| 50 | 60.311 | 0.01271 | 0.7794 | 0.7921 | 91.68 | 73.27 | 164.95 |
| 60 | 72.271 | 0.01291 | 0.6503 | 0.6632 | 94.95 | 71.32 | 166.28 |
| 70 | 85.954 | 0.01313 | 0.5451 | 0.5582 | 98.27 | 69.31 | 167.58 |
| 80 | 101.515 | 0.01335 | 0.4588 | 0.4721 | 101.63 | 67.22 | 168.85 |
| 90 | 119.115 | 0.01360 | 0.3873 | 0.4009 | 105.04 | 65.04 | 170.09 |
| 100 | 138.926 | 0.01387 | 0.3278 | 0.3416 | 108.51 | 62.77 | 171.28 |
| 110 | 161.122 | 0.01416 | 0.2777 | 0.2919 | 112.03 | 60.38 | 172.41 |
| 120 | 185.890 | 0.01448 | 0.2354 | 0.2499 | 115.62 | 57.85 | 173.48 |
| 130 | 213.425 | 0.01483 | 0.1993 | 0.2142 | 119.29 | 55.17 | 174.46 |
| 140 | 243.932 | 0.01523 | 0.1684 | 0.1836 | 123.04 | 52.30 | 175.34 |
| 150 | 277.630 | 0.01568 | 0.1415 | 0.1572 | 126.89 | 49.21 | 176.11 |
| 160 | 314.758 | 0.01620 | 0.1181 | 0.1343 | 130.86 | 45.85 | 176.71 |
| 170 | 355.578 | 0.01683 | 0.0974 | 0.1142 | 134.99 | 42.12 | 177.11 |
| 180 | 400.392 | 0.01760 | 0.0787 | 0.0963 | 139.32 | 37.91 | 177.23 |
| 190 | 449.572 | 0.01862 | 0.0614 | 0.0801 | 143.97 | 32.94 | 176.90 |
| 200 | 503.624 | 0.02013 | 0.0444 | 0.0645 | 149.19 | 26.59 | 175.79 |
| 210 | 563.438 | 0.02334 | 0.0238 | 0.0471 | 156.18 | 16.17 | 172.34 |
| 214.1 | 589.953 | 0.03153 | 0 | 0.0315 | 164.65 | 0 | 164.65 |

**TABLE G.8.1** (*continued*)
*Saturated R-134a*

| Temp. (F) | Press. (psia) | ENTHALPY, Btu/lbm | | | ENTROPY, Btu/lbm R | | |
|---|---|---|---|---|---|---|---|
| | | Sat. Liquid $h_f$ | Evap. $h_{fg}$ | Sat. Vapor $h_g$ | Sat. Liquid $s_f$ | Evap. $s_{fg}$ | Sat. Vapor $s_g$ |
| −100 | 0.951 | 50.47 | 101.10 | 151.57 | 0.1563 | 0.2811 | 0.4373 |
| −90 | 1.410 | 52.04 | 101.02 | 153.05 | 0.1605 | 0.2733 | 0.4338 |
| −80 | 2.047 | 53.97 | 100.58 | 154.54 | 0.1657 | 0.2649 | 0.4306 |
| −70 | 2.913 | 56.19 | 99.85 | 156.04 | 0.1715 | 0.2562 | 0.4277 |
| −60 | 4.067 | 58.65 | 98.90 | 157.55 | 0.1777 | 0.2474 | 0.4251 |
| −50 | 5.575 | 61.29 | 97.77 | 159.06 | 0.1842 | 0.2387 | 0.4229 |
| −40 | 7.511 | 64.05 | 96.52 | 160.57 | 0.1909 | 0.2300 | 0.4208 |
| −30 | 9.959 | 66.92 | 95.16 | 162.08 | 0.1976 | 0.2215 | 0.4191 |
| −20 | 13.009 | 69.86 | 93.72 | 163.59 | 0.2044 | 0.2132 | 0.4175 |
| −15.3 | 14.696 | 71.28 | 93.02 | 164.30 | 0.2076 | 0.2093 | 0.4169 |
| −10 | 16.760 | 72.87 | 92.22 | 165.09 | 0.2111 | 0.2051 | 0.4162 |
| 0 | 21.315 | 75.92 | 90.66 | 166.58 | 0.2178 | 0.1972 | 0.4150 |
| 10 | 26.787 | 79.02 | 89.04 | 168.06 | 0.2244 | 0.1896 | 0.4140 |
| 20 | 33.294 | 82.16 | 87.36 | 169.53 | 0.2310 | 0.1821 | 0.4132 |
| 30 | 40.962 | 85.34 | 85.63 | 170.98 | 0.2375 | 0.1749 | 0.4124 |
| 40 | 49.922 | 88.56 | 83.83 | 172.40 | 0.2440 | 0.1678 | 0.4118 |
| 50 | 60.311 | 91.82 | 81.97 | 173.79 | 0.2504 | 0.1608 | 0.4112 |
| 60 | 72.271 | 95.13 | 80.02 | 175.14 | 0.2568 | 0.1540 | 0.4108 |
| 70 | 85.954 | 98.48 | 77.98 | 176.46 | 0.2631 | 0.1472 | 0.4103 |
| 80 | 101.515 | 101.88 | 75.84 | 177.72 | 0.2694 | 0.1405 | 0.4099 |
| 90 | 119.115 | 105.34 | 73.58 | 178.92 | 0.2757 | 0.1339 | 0.4095 |
| 100 | 138.926 | 108.86 | 71.19 | 180.06 | 0.2819 | 0.1272 | 0.4091 |
| 110 | 161.122 | 112.46 | 68.66 | 181.11 | 0.2882 | 0.1205 | 0.4087 |
| 120 | 185.890 | 116.12 | 65.95 | 182.07 | 0.2945 | 0.1138 | 0.4082 |
| 130 | 213.425 | 119.88 | 63.04 | 182.92 | 0.3008 | 0.1069 | 0.4077 |
| 140 | 243.932 | 123.73 | 59.90 | 183.63 | 0.3071 | 0.0999 | 0.4070 |
| 150 | 277.630 | 127.70 | 56.49 | 184.18 | 0.3135 | 0.0926 | 0.4061 |
| 160 | 314.758 | 131.81 | 52.73 | 184.53 | 0.3200 | 0.0851 | 0.4051 |
| 170 | 355.578 | 136.09 | 48.53 | 184.63 | 0.3267 | 0.0771 | 0.4037 |
| 180 | 400.392 | 140.62 | 43.74 | 184.36 | 0.3336 | 0.0684 | 0.4020 |
| 190 | 449.572 | 145.52 | 38.05 | 183.56 | 0.3409 | 0.0586 | 0.3995 |
| 200 | 503.624 | 151.07 | 30.73 | 181.80 | 0.3491 | 0.0466 | 0.3957 |
| 210 | 563.438 | 158.61 | 18.65 | 177.26 | 0.3601 | 0.0278 | 0.3879 |
| 214.1 | 589.953 | 168.09 | 0 | 168.09 | 0.3740 | 0 | 0.3740 |

**Table G.8.2**
*Superheated R-134a*

| Temp. (F) | v (ft³/lbm) | u (Btu/lbm) | h (Btu/lbm) | s (Btu/lbm R) | v (ft³/lbm) | u (Btu/lbm) | h (Btu/lbm) | s (Btu/lbm R) |
|---|---|---|---|---|---|---|---|---|
| | | 5 psia (−53.51) | | | | 15 psia (−14.44) | | |
| Sat. | 8.3676 | 150.78 | 158.53 | 0.4236 | 2.9885 | 156.13 | 164.42 | 0.4168 |
| −20 | 9.1149 | 156.03 | 164.47 | 0.4377 | — | — | — | — |
| 0 | 9.5533 | 159.27 | 168.11 | 0.4458 | 3.1033 | 158.58 | 167.19 | 0.4229 |
| 20 | 9.9881 | 162.58 | 171.83 | 0.4537 | 3.2586 | 162.01 | 171.06 | 0.4311 |
| 40 | 10.4202 | 165.99 | 175.63 | 0.4615 | 3.4109 | 165.51 | 174.97 | 0.4391 |
| 60 | 10.8502 | 169.48 | 179.52 | 0.4691 | 3.5610 | 169.07 | 178.95 | 0.4469 |
| 80 | 11.2786 | 173.06 | 183.50 | 0.4766 | 3.7093 | 172.70 | 183.00 | 0.4545 |
| 100 | 11.7059 | 176.73 | 187.56 | 0.4840 | 3.8563 | 176.41 | 187.12 | 0.4620 |
| 120 | 12.1322 | 180.49 | 191.71 | 0.4913 | 4.0024 | 180.20 | 191.31 | 0.4694 |
| 140 | 12.5578 | 184.33 | 195.95 | 0.4985 | 4.1476 | 184.08 | 195.59 | 0.4767 |
| 160 | 12.9828 | 188.27 | 200.28 | 0.5056 | 4.2922 | 188.03 | 199.95 | 0.4838 |
| 180 | 13.4073 | 192.29 | 204.69 | 0.5126 | 4.4364 | 192.07 | 204.39 | 0.4909 |
| 200 | 13.8314 | 196.39 | 209.19 | 0.5195 | 4.5801 | 196.19 | 208.91 | 0.4978 |
| 220 | 14.2551 | 200.58 | 213.77 | 0.5263 | 4.7234 | 200.40 | 213.51 | 0.5047 |
| 240 | 14.6786 | 204.86 | 218.44 | 0.5331 | 4.8665 | 204.68 | 218.19 | 0.5115 |
| 260 | 15.1019 | 209.21 | 223.19 | 0.5398 | 5.0093 | 209.05 | 222.96 | 0.5182 |
| 280 | 15.5250 | 213.65 | 228.02 | 0.5464 | 5.1519 | 213.50 | 227.80 | 0.5248 |
| 300 | 15.9478 | 218.17 | 232.93 | 0.5530 | 5.2943 | 218.03 | 232.72 | 0.5314 |
| 320 | 16.3706 | 222.78 | 237.92 | 0.5595 | 5.4365 | 222.64 | 237.73 | 0.5379 |
| | | 30 psia (15.15) | | | | 40 psia (28.83) | | |
| Sat. | 1.5517 | 160.21 | 168.82 | 0.4136 | 1.1787 | 162.08 | 170.81 | 0.4125 |
| 20 | 1.5725 | 161.09 | 169.82 | 0.4157 | — | — | — | — |
| 40 | 1.6559 | 164.73 | 173.93 | 0.4240 | 1.2157 | 164.18 | 173.18 | 0.4173 |
| 60 | 1.7367 | 168.41 | 178.05 | 0.4321 | 1.2796 | 167.95 | 177.42 | 0.4256 |
| 80 | 1.8155 | 172.14 | 182.21 | 0.4400 | 1.3413 | 171.74 | 181.67 | 0.4336 |
| 100 | 1.8929 | 175.92 | 186.43 | 0.4477 | 1.4015 | 175.57 | 185.95 | 0.4414 |
| 120 | 1.9691 | 179.77 | 190.70 | 0.4552 | 1.4604 | 179.46 | 190.27 | 0.4490 |
| 140 | 2.0445 | 183.68 | 195.03 | 0.4625 | 1.5184 | 183.41 | 194.65 | 0.4565 |
| 160 | 2.1192 | 187.68 | 199.44 | 0.4697 | 1.5757 | 187.43 | 199.09 | 0.4637 |
| 180 | 2.1933 | 191.74 | 203.92 | 0.4769 | 1.6324 | 191.52 | 203.60 | 0.4709 |
| 200 | 2.2670 | 195.89 | 208.48 | 0.4839 | 1.6886 | 195.69 | 208.18 | 0.4780 |
| 220 | 2.3403 | 200.12 | 213.11 | 0.4908 | 1.7444 | 199.93 | 212.84 | 0.4849 |
| 240 | 2.4133 | 204.42 | 217.82 | 0.4976 | 1.7999 | 204.24 | 217.57 | 0.4918 |
| 260 | 2.4860 | 208.80 | 222.61 | 0.5044 | 1.8552 | 208.64 | 222.37 | 0.4985 |
| 280 | 2.5585 | 213.27 | 227.47 | 0.5110 | 1.9102 | 213.11 | 227.25 | 0.5052 |
| 300 | 2.6309 | 217.81 | 232.41 | 0.5176 | 1.9650 | 217.66 | 232.20 | 0.5118 |
| 320 | 2.7030 | 222.42 | 237.43 | 0.5241 | 2.0196 | 222.28 | 237.23 | 0.5184 |
| 340 | 2.7750 | 227.12 | 242.53 | 0.5306 | 2.0741 | 226.99 | 242.34 | 0.5248 |
| 360 | 2.8469 | 231.89 | 247.70 | 0.5370 | 2.1285 | 231.76 | 247.52 | 0.5312 |

TABLE **G.8.2**  (*continued*)
*Superheated R-134a*

| Temp. (F) | v (ft³/lbm) | u (Btu/lbm) | h (Btu/lbm) | s (Btu/lbm R) | v (ft³/lbm) | u (Btu/lbm) | h (Btu/lbm) | s (Btu/lbm R) |
|---|---|---|---|---|---|---|---|---|
| | 60 psia (49.72) | | | | 80 psia (65.81) | | | |
| Sat. | 0.7961 | 164.91 | 173.75 | 0.4113 | 0.5996 | 167.04 | 175.91 | 0.4105 |
| 60 | 0.8204 | 166.95 | 176.06 | 0.4157 | — | — | — | — |
| 80 | 0.8657 | 170.89 | 180.51 | 0.4241 | 0.6262 | 169.97 | 179.24 | 0.4168 |
| 100 | 0.9091 | 174.85 | 184.94 | 0.4322 | 0.6617 | 174.06 | 183.86 | 0.4252 |
| 120 | 0.9510 | 178.82 | 189.38 | 0.4400 | 0.6954 | 178.15 | 188.44 | 0.4332 |
| 140 | 0.9918 | 182.85 | 193.86 | 0.4476 | 0.7279 | 182.25 | 193.03 | 0.4410 |
| 160 | 1.0318 | 186.92 | 198.38 | 0.4550 | 0.7595 | 186.39 | 197.64 | 0.4485 |
| 180 | 1.0712 | 191.06 | 202.95 | 0.4623 | 0.7903 | 190.58 | 202.28 | 0.4559 |
| 200 | 1.1100 | 195.26 | 207.59 | 0.4694 | 0.8205 | 194.83 | 206.98 | 0.4632 |
| 220 | 1.1484 | 199.54 | 212.29 | 0.4764 | 0.8503 | 199.14 | 211.72 | 0.4702 |
| 240 | 1.1865 | 203.88 | 217.05 | 0.4833 | 0.8796 | 203.51 | 216.53 | 0.4772 |
| 260 | 1.2243 | 208.30 | 221.89 | 2.4902 | 0.9087 | 207.95 | 221.41 | 0.4841 |
| 280 | 1.2618 | 212.79 | 226.80 | 0.4969 | 0.9375 | 212.47 | 226.34 | 0.4909 |
| 300 | 1.2991 | 217.36 | 231.78 | 0.5035 | 0.9661 | 217.05 | 231.35 | 0.4975 |
| 320 | 1.3362 | 222.00 | 236.83 | 0.1501 | 0.9945 | 221.71 | 236.43 | 0.5041 |
| 340 | 1.3732 | 226.71 | 241.96 | 0.5166 | 1.0227 | 226.44 | 241.58 | 0.5107 |
| 360 | 1.4100 | 231.51 | 247.16 | 0.5230 | 1.0508 | 231.24 | 246.80 | 0.5171 |
| 380 | 1.4468 | 236.37 | 252.43 | 0.5294 | 1.0788 | 236.12 | 252.09 | 0.5235 |
| 400 | 1.4834 | 241.31 | 257.78 | 0.5357 | 1.1066 | 241.07 | 257.46 | 0.5298 |
| | 100 psia (79.08) | | | | 125 psia (93.09) | | | |
| Sat. | 0.4794 | 168.74 | 177.61 | 0.4100 | 0.3814 | 170.46 | 179.28 | 0.4094 |
| 80 | 0.4809 | 168.93 | 177.83 | 0.4104 | — | — | — | — |
| 100 | 0.5122 | 173.20 | 182.68 | 0.4192 | 0.3910 | 172.01 | 181.06 | 0.4126 |
| 120 | 0.5414 | 177.42 | 187.44 | 0.4276 | 0.4171 | 176.43 | 186.08 | 0.4214 |
| 140 | 0.5691 | 181.62 | 192.15 | 0.4356 | 0.4413 | 180.77 | 190.98 | 0.4297 |
| 160 | 0.5957 | 185.84 | 196.86 | 0.4433 | 0.4642 | 185.10 | 195.84 | 0.4377 |
| 180 | 0.6215 | 190.08 | 201.58 | 0.4508 | 0.4861 | 189.43 | 200.68 | 0.4454 |
| 200 | 0.6466 | 194.38 | 206.34 | 0.4581 | 0.5073 | 193.79 | 205.52 | 0.4529 |
| 220 | 0.6712 | 198.72 | 211.15 | 0.4653 | 0.5278 | 198.19 | 210.40 | 0.4601 |
| 240 | 0.6954 | 203.13 | 216.00 | 0.4723 | 0.5480 | 202.64 | 215.32 | 0.4673 |
| 260 | 0.7193 | 207.60 | 220.91 | 0.4792 | 0.5677 | 207.15 | 220.28 | 0.4743 |
| 280 | 0.7429 | 212.14 | 225.88 | 0.4861 | 0.5872 | 211.72 | 225.30 | 0.4811 |
| 300 | 0.7663 | 216.74 | 230.92 | 0.4928 | 0.6064 | 216.35 | 230.38 | 0.4879 |
| 320 | 0.7895 | 221.42 | 236.03 | 0.4994 | 0.6254 | 221.05 | 235.51 | 0.4946 |
| 340 | 0.8125 | 226.16 | 241.20 | 0.5060 | 0.6442 | 225.81 | 240.71 | 0.5012 |
| 360 | 0.8353 | 230.98 | 246.44 | 0.5124 | 0.6629 | 230.65 | 245.98 | 0.5077 |
| 380 | 0.8580 | 235.87 | 251.75 | 0.5188 | 0.6814 | 235.56 | 251.32 | 0.5141 |
| 400 | 0.8806 | 240.83 | 257.13 | 0.5252 | 0.6998 | 240.53 | 256.72 | 0.5205 |

**Table G.8.2** (*continued*)
*Superheated R-134a*

| Temp. (F) | v (ft³/lbm) | u (Btu/lbm) | h (Btu/lbm) | s (Btu/lbm R) | v (ft³/lbm) | u (Btu/lbm) | h (Btu/lbm) | s (Btu/lbm R) |
|---|---|---|---|---|---|---|---|---|
| | 150 psia (105.13) | | | | 200 psia (125.25) | | | |
| Sat. | 0.3150 | 171.87 | 180.61 | 0.4089 | 0.2304 | 174.00 | 182.53 | 0.4080 |
| 120 | 0.3332 | 175.33 | 184.57 | 0.4159 | — | — | — | — |
| 140 | 0.3554 | 179.85 | 189.72 | 0.4246 | 0.2459 | 177.72 | 186.82 | 0.4152 |
| 160 | 0.3761 | 184.31 | 194.75 | 0.4328 | 0.2645 | 182.54 | 192.33 | 0.4242 |
| 180 | 0.3955 | 188.74 | 199.72 | 0.4407 | 0.2814 | 187.23 | 197.64 | 0.4327 |
| 200 | 0.4141 | 193.18 | 204.67 | 0.4484 | 0.2971 | 191.86 | 202.85 | 0.4407 |
| 220 | 0.4321 | 197.64 | 209.63 | 0.4558 | 0.3120 | 196.46 | 208.01 | 0.4484 |
| 240 | 0.4496 | 202.14 | 214.62 | 0.4630 | 0.3262 | 201.08 | 213.15 | 0.4559 |
| 260 | 0.4666 | 206.69 | 219.64 | 0.4701 | 0.3400 | 205.72 | 218.31 | 0.4631 |
| 280 | 0.4833 | 211.29 | 224.70 | 0.4770 | 0.3534 | 210.40 | 223.48 | 0.4702 |
| 300 | 0.4998 | 215.95 | 229.82 | 0.4838 | 0.3664 | 215.13 | 228.69 | 0.4772 |
| 320 | 0.5160 | 220.67 | 235.00 | 0.4906 | 0.3792 | 219.91 | 233.94 | 0.4840 |
| 340 | 0.5320 | 225.46 | 240.23 | 0.4972 | 0.3918 | 224.74 | 239.24 | 0.4907 |
| 360 | 0.5479 | 230.32 | 245.52 | 0.5037 | 0.4042 | 229.64 | 244.60 | 0.4973 |
| 380 | 0.5636 | 235.24 | 250.88 | 0.5102 | 0.4165 | 234.60 | 250.01 | 0.5038 |
| 400 | 0.5792 | 240.23 | 256.31 | 0.5166 | 0.4286 | 239.62 | 255.48 | 0.5103 |
| | 250 psia (141.87) | | | | 300 psia (156.14) | | | |
| Sat. | 0.1783 | 175.50 | 183.75 | 0.4068 | 0.1428 | 176.50 | 184.43 | 0.4055 |
| 160 | 0.1955 | 180.42 | 189.46 | 0.4162 | 0.1467 | 177.70 | 185.84 | 0.4078 |
| 180 | 0.2117 | 185.49 | 195.28 | 0.4255 | 0.1637 | 183.44 | 192.53 | 0.4184 |
| 200 | 0.2261 | 190.38 | 200.84 | 0.4340 | 0.1779 | 188.71 | 198.59 | 0.4278 |
| 220 | 0.2394 | 195.18 | 206.26 | 0.4421 | 0.1905 | 193.77 | 204.35 | 0.4364 |
| 240 | 0.2519 | 199.94 | 211.60 | 0.4498 | 0.2020 | 198.72 | 209.93 | 0.4445 |
| 260 | 0.2638 | 204.70 | 216.90 | 0.4573 | 0.2128 | 203.62 | 215.43 | 0.4522 |
| 280 | 0.2752 | 209.47 | 222.21 | 0.4646 | 0.2230 | 208.50 | 220.88 | 0.4597 |
| 300 | 0.2863 | 214.27 | 227.52 | 0.4717 | 0.2328 | 213.39 | 226.31 | 0.4669 |
| 320 | 0.2971 | 219.12 | 232.86 | 0.4786 | 0.2423 | 218.30 | 231.75 | 0.4740 |
| 340 | 0.3076 | 224.01 | 238.24 | 0.4854 | 0.2515 | 223.25 | 237.21 | 0.4809 |
| 360 | 0.3180 | 228.95 | 243.66 | 0.4921 | 0.2605 | 228.24 | 242.70 | 0.4877 |
| 380 | 0.3282 | 233.95 | 249.13 | 0.4987 | 0.2693 | 233.29 | 248.24 | 0.4944 |
| 400 | 0.3382 | 239.01 | 254.65 | 0.5052 | 0.2779 | 238.38 | 253.81 | 0.5009 |

TABLE G.8.2 (*continued*)
*Superheated R-134a*

| Temp. (F) | $v$ (ft³/lbm) | $u$ (Btu/lbm) | $h$ (Btu/lbm) | $s$ (Btu/lbm R) | $v$ (ft³/lbm) | $u$ (Btu/lbm) | $h$ (Btu/lbm) | $s$ (Btu/lbm R) |
|---|---|---|---|---|---|---|---|---|
| | 400 psia (179.92) | | | | 500 psia (199.36) | | | |
| Sat. | 0.0965 | 177.23 | 184.37 | 0.4020 | 0.0655 | 175.90 | 181.96 | 0.3960 |
| 180 | 0.0966 | 177.26 | 184.41 | 0.4020 | — | — | — | — |
| 200 | 0.1146 | 184.44 | 192.92 | 0.4152 | 0.0666 | 176.38 | 182.54 | 0.3969 |
| 220 | 0.1277 | 190.41 | 199.86 | 0.4255 | 0.0867 | 185.78 | 193.80 | 0.4137 |
| 240 | 0.1386 | 195.92 | 206.19 | 0.4347 | 0.0990 | 192.46 | 201.62 | 0.4251 |
| 260 | 0.1484 | 201.21 | 212.20 | 0.4432 | 0.1089 | 198.40 | 208.47 | 0.4347 |
| 280 | 0.1573 | 206.38 | 218.03 | 0.4512 | 0.1174 | 204.00 | 214.86 | 0.4435 |
| 300 | 0.1657 | 211.49 | 223.76 | 0.4588 | 0.1252 | 209.41 | 220.99 | 0.4517 |
| 320 | 0.1737 | 216.58 | 229.44 | 0.4662 | 0.1323 | 214.74 | 226.98 | 0.4594 |
| 340 | 0.1813 | 221.68 | 235.09 | 0.4733 | 0.1390 | 220.01 | 232.87 | 0.4669 |
| 360 | 0.1886 | 226.79 | 240.75 | 0.4803 | 0.1454 | 225.27 | 238.73 | 0.4741 |
| 380 | 0.1957 | 231.93 | 246.42 | 0.4872 | 0.1516 | 230.53 | 244.56 | 0.4812 |
| 400 | 0.2027 | 237.12 | 252.12 | 0.4939 | 0.1575 | 235.82 | 250.39 | 0.4880 |
| | 750 psia | | | | 1000 psia | | | |
| 180 | 0.01640 | 136.22 | 138.49 | 0.3285 | 0.01593 | 134.77 | 137.71 | 0.3262 |
| 200 | 0.01786 | 144.85 | 147.32 | 0.3421 | 0.01700 | 142.70 | 145.84 | 0.3387 |
| 220 | 0.02069 | 155.27 | 158.14 | 0.3583 | 0.01851 | 151.26 | 154.69 | 0.3519 |
| 240 | 0.03426 | 173.83 | 178.58 | 0.3879 | 0.02102 | 160.95 | 164.84 | 0.3666 |
| 260 | 0.05166 | 187.78 | 194.95 | 0.4110 | 0.02603 | 172.59 | 177.40 | 0.3843 |
| 280 | 0.06206 | 196.16 | 204.77 | 0.4244 | 0.03411 | 184.70 | 191.01 | 0.4029 |
| 300 | 0.06997 | 203.08 | 212.79 | 0.4351 | 0.04208 | 194.58 | 202.37 | 0.4181 |
| 320 | 0.07662 | 209.37 | 220.00 | 0.4445 | 0.04875 | 202.67 | 211.69 | 0.4302 |
| 340 | 0.08250 | 215.33 | 226.78 | 0.4531 | 0.05441 | 209.79 | 219.86 | 0.4406 |
| 360 | 0.08786 | 221.11 | 233.30 | 0.4611 | 0.05938 | 216.36 | 227.35 | 0.4498 |
| 380 | 0.09284 | 226.78 | 239.66 | 0.4688 | 0.06385 | 222.61 | 234.43 | 0.4583 |
| 400 | 0.09753 | 232.39 | 245.92 | 0.4762 | 0.06797 | 228.67 | 241.25 | 0.4664 |

## Table G.9
*Enthalpy of Formation and Absolute Entropy of Various Substances at 77 F, 1 atm Pressure*

| Substance | Formula | M | State | $\bar{h}_f^0$ Btu/lbmol | $\bar{s}_f^0$ Btu/lbmol R |
|---|---|---|---|---|---|
| Water | $H_2O$ | 18.015 | gas | −103 966 | 45.076 |
| Water | $H_2O$ | 18.015 | liq | −122 885 | 16.707 |
| Hydrogen peroxide | $H_2O_2$ | 34.015 | gas | −58 515 | 55.623 |
| Ozone | $O_3$ | 47.998 | gas | +61 339 | 57.042 |
| Carbon (graphite) | C | 12.011 | solid | 0 | 1.371 |
| Carbon monoxide | CO | 28.011 | gas | −47 518 | 47.182 |
| Carbon dioxide | $CO_2$ | 44.010 | gas | −169 184 | 51.038 |
| Methane | $CH_4$ | 16.043 | gas | −32 190 | 44.459 |
| Acetylene | $C_2H_2$ | 26.038 | gas | +97 477 | 47.972 |
| Ethene | $C_2H_4$ | 28.054 | gas | +22.557 | 52.360 |
| Ethane | $C_2H_6$ | 30.070 | gas | −36 432 | 54.812 |
| Propene | $C_3H_6$ | 42.081 | gas | +8 783 | 63.761 |
| Propane | $C_3H_8$ | 44.094 | gas | −44 669 | 64.442 |
| Butane | $C_4H_{10}$ | 58.124 | gas | −54 256 | 73.215 |
| Pentane | $C_5H_{12}$ | 72.151 | gas | −62 984 | 83.318 |
| Benzene | $C_6H_6$ | 78.114 | gas | +35 675 | 64.358 |
| Hexane | $C_6H_{14}$ | 86.178 | gas | −71 926 | 92.641 |
| Heptane | $C_7H_{16}$ | 100.205 | gas | −80 782 | 102.153 |
| *n*-Octane | $C_8H_{18}$ | 114.232 | gas | −89 682 | 111.399 |
| *n*-Octane | $C_8H_{18}$ | 114.232 | liq | −107 526 | 86.122 |
| Methanol | $CH_3OH$ | 32.042 | gas | −86 543 | 57.227 |
| Ethanol | $C_2H_5OH$ | 46.069 | gas | −101 032 | 67.434 |
| Ammonia | $NH_3$ | 17.031 | gas | −19 656 | 45.969 |
| T-T-Diesel | $C_{14.4}H_{24.9}$ | 198.06 | liq | −74 807 | 125.609 |
| Sulfur | S | 32.06 | solid | 0 | 7.656 |
| Sulfur dioxide | $SO_2$ | 64.059 | gas | −127 619 | 59.258 |
| Sulfur trioxide | $SO_3$ | 80.058 | gas | −170 148 | 61.302 |
| Nitrogen oxide | $N_2O$ | 44.013 | gas | +35 275 | 52.510 |
| Nitromethane | $CH_3NO_2$ | 61.04 | liq | −48 624 | 41.034 |

# Answers to Selected Problems

| | |
|---|---|
| 2.18 | 0.406 kmol |
| 2.21 | 3583 N |
| 2.24 | 21.7 m/s$^2$ |
| 2.27 | 1530 N |
| 2.30 | 198 kPa |
| 2.33 | 700 N |
| 2.36 | 332 kPa |
| 2.42 | 0.12 kPa |
| 2.45 | 100.73 kPa |
| 2.48 | 1.77 kPa, 13.2 mm |
| 2.51 | 1346 kPa |
| 2.54 | 8.72 m |
| 2.57 | 370 kPa |
| 2.60 | 6.9 mm |
| 3.12 | 1234 k, 1356 k |
| 3.15 | 190 k |
| 3.18 | vapor, solid |
| 3.24 | a. vapor |
| | b. liquid |
| | c. mixture |
| | d. vapor |
| 3.27 | 1200 kPa, undefined, 2033 kPa, 0.0326 m$^3$/kg |
| 3.30 | mixture, 2513 kPa, 0.672 gas, 10000 kPa, undefined |
| 3.33 | 431 kPa |
| 3.36 | 35.7 kg |
| 3.39 | 771 kPa, 1318 kPa, 0.468 |
| 3.42 | 212°C, more vapor |
| 3.45 | rise, fall |
| 3.48 | 100%, −50% |
| 3.51 | 1.19 kg, 0.83 kg, 1.81 kg |
| 3.54 | 96.4 kPa |
| 3.57 | 0.697 kg, 3000 kPa |
| 3.60 | 2.75%, 9.4%, 25.3% |
| 3.63 | 3.2%, 9.9%, 23.4% |
| 3.66 | No |
| 3.69 | 4.5%, 1.4% |
| 3.72 | 0.47 m$^3$ |
| 3.75 | 925 kPa, 0.007 m$^3$/kg |
| 3.78 | under |
| 3.84 | 101.98 K, 102.15 K |

| | |
|---|---|
| 4.6 | $F_2 = 4 F_1$ |
| 4.15 | 9807 J |
| 4.18 | 0.083 m, 0.028 m |
| 4.21 | 303 m |
| 4.24 | 3.17 kPa, 0.0361, 0 kJ |
| 4.27 | 1160 K |
| 4.30 | 0.71 m$^3$, −291 kJ |
| 4.33 | 211.7°C, 0.1 m$^3$, 0 kJ |
| 4.36 | 1.72 m$^3$, 264 kJ |
| 4.39 | 877 K, 357.5 kJ/kg |
| 4.42 | −74.5 kJ |
| 4.45 | −7.165 kJ |
| 4.48 | 1.969, −51.8 kJ/kg |
| 4.51 | −49.4 kJ |
| 4.54 | 1.29 m$^3$, 215 kJ |
| 4.57 | 1.55 MPa, 0.5 m$^3$, 80 kJ |
| 4.60 | 829°C, 25.4 m$^3$, 3390 kJ |
| 4.63 | 778 kJ |
| 4.66 | 12.6 J |
| 4.72 | 20 kW |
| 4.75 | 186 W |
| 4.78 | 74 W, 1063 kJ |
| 4.84 | 5400 W |
| 4.87 | 203 kJ |
| 4.90 | 16 kW/m$^2$ |
| 5.9 | 30.89 kJ, 0.0386 m$^3$ |
| 5.12 | 4.29 m$^3$ |
| 5.15 | 9.9 m/s, −5.49 m |
| 5.18 | a. 3973 kPa, 1664 kJ/kg |
| | b. undefined, 114 kJ/kg |
| | c. −337.6 kJ/kg, 0.00109 m$^3$/kg |
| | d. 32.5°C, 405 kJ/kg |
| 5.21 | a. −0.03, 0.47 kJ/kg |
| | b. −0.12, 1.88 kJ/kg |
| | c. −1.19, 18.67 kJ/kg |
| 5.24 | a. mixture L+V |
| | b. compressed liquid |
| | c. superheated vapor |
| | d. superheated vapor |
| | e. mixture L+V |

| | |
|---|---|
| 5.27 | 133.6°C, 0 kJ, −1148 kJ |
| 5.30 | 721 kJ |
| 5.33 | 165 kJ |
| 5.36 | 878 kJ, 0 kJ |
| 5.39 | 114 kJ |
| 5.42 | 180°C, 1684 kJ |
| 5.45 | 3387 kJ, 285 kJ |
| 5.48 | 25.51 MJ |
| 5.51 | −7421 kJ |
| 5.54 | 0.274, 759 kJ |
| 5.57 | 2611 kJ |
| 5.60 | 65°C |
| 5.63 | 41.8 MJ |
| 5.66 | 5.03 kJ |
| 5.69 | 1 kJ/kg K, 14%, 21% |
| 5.72 | 53.7°C, 43.7°C |
| 5.75 | 36 kJ/kg, 45 kJ/kg |
| 5.78 | 261 kJ, 444 kJ |
| 5.81 | 2.3 kg, 3.48 kg, 736 K, 613 kPa |
| 5.84 | 472°C, 476°C, 501°C |
| 5.87 | 0.03 m$^3$, 1913 kPa, −83.5 kJ, −320 kJ |
| 5.90 | −2.67 kJ |
| 5.93 | 73.7 kJ/kg |
| 5.96 | 845 kPa, 459 K |
| 5.99 | 1491 kPa, 41.5 kJ, 1025 kJ |
| 5.102 | 0.23 m/s |
| 5.105 | 249 s |
| 5.108 | 15 h |
| 5.111 | 2.45 kW |
| 5.114 | 5.9 kW, 267 N |
| 6.9 | 1624 m/s, no |
| 6.12 | 6.4 m/s |
| 6.15 | 17.2 m/s |
| 6.18 | 582 m/s |
| 6.21 | 124 kPa, 320 K |
| 6.24 | 345 K |
| 6.27 | 7.6°C |
| 6.30 | −20°C, 3680% |
| 6.33 | 624 kW |
| 6.36 | −9.9 kW |
| 6.39 | 98.6 kW |
| 6.42 | 12 kg/s |
| 6.45 | 317 kJ/kg, −307 kJ/kg |
| 6.48 | 1400 kJ/kg |
| 6.51 | 0.84 kW, 1.01 kW |
| 6.54 | 0.224 kg/s |
| 6.57 | 20.2°C, 44 m/s |
| 6.60 | −11 kW |
| 6.63 | 22 MW |
| 6.66 | 18 MW |

| | |
|---|---|
| 6.69 | 0.27 kg/s |
| 6.72 | 0.258 kg/s, 4.19 m$^3$/s |
| 6.75 | 0.076 kg/s |
| 6.78 | 2.07 kg/s |
| 6.81 | 1389 K |
| 6.84 | 13.75 MW, 67 MW |
| 6.87 | −0.55 kW, −8.665 kW, 5.42 kW |
| 6.90 | 16.96 kg, −469 kJ |
| 6.93 | 520°C, 0.342 m$^3$ |
| 6.96 | 11.61 kg, 573 kPa |
| 6.99 | 41 MJ |
| 6.102 | 270.4 MJ |
| 7.12 | 0.43, 20 kW |
| 7.15 | 2.91 |
| 7.18 | 1.3 kW, 750 W |
| 7.24 | 2.17 |
| 7.27 | 10.9 kW |
| 7.33 | Irreversible |
| 7.39 | 0.15 |
| 7.42 | 1.39 kW |
| 7.45 | impossible |
| 7.48 | 15.2 kJ, 20 s |
| 7.51 | refrigerator claim OK deep freezer not OK |
| 7.54 | 100 MJ |
| 7.57 | 0.91 MW |
| 7.60 | 1.9°C |
| 7.63 | 11, 7.8, 6 |
| 7.66 | $\dot{W} = K(T_H - T_L)^2/T_H$ |
| 7.69 | 5.4°C |
| 7.75 | 0.27 m$^3$/kg, 320 kPa, 125 kJ/kg |
| 8.3 | all constant |
| 8.6 | $v \sim C$, rest increases |
| 8.15 | a. impossible |
| | b. OK |
| | c. OK |
| 8.18 | 4.05, 6.545, −1.237 all kJ/kg K |
| 8.21 | $\Delta u$'s 23.2, 26, 28.3 kJ/kg |
| | $\Delta s$'s 0.77, 1.1, 1.85 kJ/kg K |
| 8.24 | a. 0.966, 350.9 kPa |
| | b. 232 kPa, $x$ undefined |
| | c. 42.6°C, 1.1975 kJ/kg K |
| 8.27 | 1940 kJ/kg, 0.46 kJ |
| 8.30 | both positive |
| 8.33 | 16.94 kJ, 225.7 kJ, 225.7 kJ, yes |
| 8.36 | 0.383 m$^3$/kg, 237.3°C |
| 8.39 | 26.4°C |
| 8.42 | 171.95°C, −132 kJ/kg |
| 8.45 | −0.675 kJ |
| 8.48 | −463 kJ, −4360 kJ |
| 8.51 | 0.385 m$^3$ |

8.54 334.6 kJ/kg, 1.0 kJ/kg K, 334.4 kJ/kg, 1.0 kJ/kg K

8.57 65.2°C, 0.023 kJ/K

8.60 1353 kJ out, 0.979 kJ/K

8.63 316 K, 0.0728 kJ/K

8.66 2.909, 2.7249, 2.3352 all kJ/kg K

8.69 2015 kPa, 0.0116 m$^3$

8.72 660.8 kJ, 0.661 kJ/K

8.75 a. 697.3 K, −253.3 kJ/kg
    b. 789.7 K, −254.5 kJ/kg

8.78 143 K, −623.6 kJ/kg

8.81 48.6 cm, 0.294 kJ

8.84 both negative

8.87 negative

8.90 1.81 kJ, −0.963 kJ

8.93 147 kJ, −91 kJ

8.96 6.2 J/K, 0.5 J/K

8.99 875 kJ, 6955 kJ, 5.3 kJ/K

8.102 yes

8.105 −112 kJ, −1485 kJ, 0.29 kJ/K

8.108 180 kPa, 1500 K, 0.369 kJ/K

8.111 161 kJ, 852 kJ, 0.6 kJ/K

8.114 0.365 kJ/K

8.117 0.2 kJ/K

8.120 0.1 kW/K, 0.1 kW/K

8.123 3.1 W/K, 0.27 W/K

9.15 0.728 kg/s

9.18 $q = 1891$ kJ/kg, $w \simeq 0$

9.21 2849 kJ/kg

9.24 633 m/s

9.27 27 MW

9.30 Isentropic processes, 357 K, 359 m/s

9.33 $w_T = 1569$ kJ/kg, $w_{p,\,in} = 20.1, 0.428$

9.36 14 m/s, 10.1 m

9.39 6.08 MPa, 25.3°C

9.42 2.36 kW

9.45 a. 16.7 kW, 45°C
    b. 0.3 kW, −9.6°C

9.48 17.6°C, 100.67 kPa

9.51 18442 kPa, −849 kJ/kg, −104 kJ/kg

9.54 Yes

9.57 0.0686 kJ/kg K

9.60 0.0154 kW/K

9.63 326.7 K, 0.0358 kW/K

9.66 120.2°C, 1.54 kW/K

9.69 443 K, 0.0228 kW/K

9.72 0.506 kJ/kg K

9.75 125.5 kg, 0.0887 m$^3$, 0.0113 m$^3$, irreversible

9.78 14 m/s, 49 m$^2$/s$^2$, 13.1 J/K

9.81 −0.39 kJ, 0.0185 kJ/K

9.84 0.361 kg, 333°C, 0.186 kJ/K

9.87 85%, 0.149 kJ/kg K

9.90 87%

9.93 89%

9.96 70%

9.99 71.7%

9.102 1230 kJ/kg, 0.5 kJ/kg K

9.105 28.5 kJ/kg, 6°C

9.108 313 K, 129 kPa

9.111 No

9.114 −38.9 kJ/kg

9.117 No

9.120 94.4 kW, 63.9 kW

9.123 20.45 kJ/kg, 20.45 kJ/kg

9.126 274 kW

9.129 6.24 kJ/kg

9.132 −755 kW, 305 kW

9.135 924 K

9.138 0.92

9.141 0.885, 0.906

9.144 −47.8 kJ/kg, 0.0444 kJ/kg K, 13.24 kJ/kg

9.147 3003 kPa

10.9 0.668, 0.234, 0.098, 0.454 kg

10.12 2.325 mole $O_2$/mole $CH_4$

10.15 0.2513 kJ/kg K, 1.0 m$^3$

10.18 2.57 kg/m$^3$, 0.2682 kJ/kg K, 6.43 kg/s

10.21 1.675 m$^3$, 373 kJ

10.24 335 K, 306 kPa

10.27 901 kJ/kg, 1.354 kJ/kg K

10.30 −8.8 kJ, −8.8 kJ

10.33 1056 K, 1.31 kJ/kg K

10.36 353 K, 134 kJ/kg

10.39 316 K, 242 kPa

10.42 13236 kPa, 0.933, −1596 kJ, −1988 kJ, −6.766 kJ/K

10.45 5.76 kJ/kmol K

10.48 1283 kW, 1.01 kW/K

10.51 1070 K

10.54 1.9 kW/K

10.57 302 K, 1.87 kW/K

10.60 0.00081, 0.0313, 1.572

10.63 0.445 kg

10.66 0.0125, −195 kW

10.69 0.00773, 0.0155 kg/s, 29°C

10.72 96°C

10.75 0.068 kg, 84.95 kPa, −740.8 kJ

10.78 35 kJ/kg air, 0.0099 kg/kg air

10.81 1.7°C, 0.0288 kg/kg air, −1.05 kW

10.84 33.3 kg/s, 0.323 kg/s

10.87 18.4 kg/h, −4.21 kW

10.90 a. 0.016, 22.3°C,
     b. 0.0106, 15°C
     c. 57%, 14.4°C
     d. 0.017, 86%

10.93 beach A

10.96 94%

10.99 3.77, 6.43 kJ/kg air

10.102 0.006, 18%, 18°C, 8.4 kJ/kg air, 37.2 kJ/kg air,
      28.5 kJ/kg air

11.3 3.0, 2609, 846, 1766 all kJ/kg, 32%

11.6 35%, 56%

11.9 758°C, 4.82 kg/s

11.12 10.3%

11.15 32.6 MW, 4358 kg/s

11.18 36%, $x = 0.92$

11.21 36%, 896 kJ/kg

11.24 0.0437, 198 kJ/kg

11.27 34.5%, 866 kJ/kg

11.30 21.57 kg/s, 44.8 MW, 30.7%, 52.6%

11.33 34.7%

11.36 3.0 kJ/kg K

11.39 40.3°C, 29.2 MW, 11.6 MW

11.42 22 MW, 62.5°C

11.45 22.3

11.48 3.04 MW, 7.32 MW, 48.4%

11.51 Yes

11.54 166 MW, 0.4, 58%

11.57 253 kJ/kg, 271 kJ/kg

11.60 1012 m/s

11.63 824 K, 602 m/s

11.66 623 m/s, 1012 m/s

11.69 232 kPa, 462 kJ/kg, 58.5%

11.72 2047 K, 815 K, 60%

11.75 23.8, 518 kJ/kg in, 15129 kPa

11.78 19.3, 61.9%

11.81 6298 kPa, 551 kJ/kg, 65.3%

11.84 20.9, 895 kPa

11.87 3.2, 3.17

11.90 44.9 kJ/kg, 3.77

11.93 388.5 kW, 4300 kW

11.96 64%, 4.44

11.99 0.09 W/K, 0.04 W/K, valve, inside

11.102 0.57

11.105 11.4, 52.9%

11.111 4358 kg/s, 5703 kW water, 286 kW ocean, 5%

11.114 26.66 kg/s, 18303 kW source, 2 kW intake,
      4302 kW exhaust

12.12 11 $H_2O$ + 10 $CO_2$ + 87.42 $N_2$ + 7.75 $O_2$

12.15 2.39 $H_2O$ + 1.47 $CO_2$ + 12.1 $N_2$ + 0.53 $O_2$ for
      1 mole gas

12.18 2.324 $H_2O$ + 1 $CO_2$ + 11.28 $N_2$ + 1 $O_2$, 53.8°C

12.21 1.25 kg, 1.49 kg

12.24 1197 MJ/kmol

12.27 −463 272 kJ/kmol

12.30 0.1475, 9.575

12.33 −3842 MJ/kmol

12.36 −1215 MJ/kmol

12.39 1.8 $CO_2$ + 2.8 $H_2O$ + 10.69 $N_2$, 30941 kJ/kg

12.42 −585 MJ/kmol

12.45 1843 K

12.48 169%

12.51 1480 K

12.54 2909 K

12.57 4991 K, 24.6 MJ/kmol

12.60 1838 kJ/kmol K

12.63 4991 K, 79.7 kJ/kmol K

12.66 1.06 V

12.69 nearly no dissociation

12.72 no

12.75 2980 K

12.78 49.7% $H_2$, 50.3% H by mole

12.81 1444 K

12.84 36.2% $H_2O$, 21.3% $O_2$ and 42.5% $H_2$

12.87 176 811 kJ

13.3 109 W/m$^2$K

13.6 187.5 W, 97 W

13.9 2.22 W/mK, 0.03 K/W

13.12 2 m$^2$

13.15 1.7 m$^2$K/W, 0.82 m$^2$ k/W, 21 kW

13.18 58.4°C

13.21 240 W

13.24 45.6°C

13.27 28.3°C, 31 W/m

13.30 387 W

13.33 397 W

13.36 41.8 W

13.39 102 K/W, 40.7 K/W, 502°C

13.42 63.2°C

13.45 60 W

13.48 75%, 19

13.51 20.5, 28

13.54 48°C

13.57 39.8°C

13.60 50.7°C, 55.7°C

13.63 yes, 1143 sec

13.66 1 h 28 min

13.69 5392 W/m$^2$K

13.72 42.6°C, 26.4°C

13.75 0.74 m

13.78 5 kW/m$^2$, 2.5 kW/m$^2$

13.81  3.28 kW
13.84  37.9 kW/m
13.87  0.0265 mK/W, 470 m
13.90  73.5°C
13.93  127.5°C
13.96  80°C
13.99  2033 kPa
13.102  0.05, 0.05, 0.1, 2.5, 3.125 all K/W, 9.44 W
13.105  59°C, 163°C
13.108  28% lower

# APPENDIX F, ENGLISH UNIT PROBLEMS

2.3  0.1337 ft³/lbm, 4.3 ft³/mol
2.6  53 ft
3.3  a. compr. liquid, 0.01173 ft³/lbm
     b. sup. vapor, 67 F
3.6  10.5 psia
3.9  48 lbm, 10% error
4.3  22.6 Btu/s
4.6  49.8 Btu
4.8  2.6 in.
5.2  62 ft³
5.5  −78 Btu
5.8  197 F
5.11  0, −12.6 Btu
5.14  32.9 s
6.3  728 ft/s, no

6.6  24.6 Btu/lbm
6.9  33 000 hp, −1.92 × 10⁸ Btu/h
6.11  3.6 lbm, 86 psia
7.2  2.5 Btu, 1.5
7.5  3333 Btu/s, 1074 R
8.3  negative, positive
8.6  1.29 Btu/R
8.9  69.4 psia, −236 Btu/lbm, −377 Btu/lbm
8.12  0.00088 Btu/s R, 0.0026 Btu/s R
9.3  60 Btu/s, 60 Btu/s
9.6  2.3 Btu/s  = 3.3 hp
9.9  484 F, 100%
9.12  25.5 Btu/lbm, 25.5 Btu/lbm
9.14  200 Btu/s, 96%
10.3  546 R
10.7  0.66 lbm, 0.00436, 35.5 F
10.10  1.235 Btu/s, −0.78 Btu/s
11.1  1.8, 1104, 360, 746 all in Btu/lbm, 32.5%
11.4  1.8, 1253, 424, 829 all in Btu/lbm, 33.7%
11.7  2600 R, 67.2 lbm/s
11.10  3836 R, 1527 R, 60.2%
11.13  116 Btu/s  = 164 hp
11.16  3.72
12.3  126 psia, 195 945 Btu (out)
12.6  0.145, 9.689
12.9  5236 R
12.12  151%, 114 570 Btu/lbmol
12.15  ln K = −186; ln K = 5.13
12.18  75 360 Btu
12.21  0.36

# Index